VOLUME SIX HUNDRED AND FORTY NINE

METHODS IN ENZYMOLOGY

Pore-Forming Toxins

METHODS IN ENZYMOLOGY

Editors-in-Chief

ANNA MARIE PYLE
Departments of Molecular, Cellular and Developmental Biology and Department of Chemistry
Investigator, Howard Hughes Medical Institute
Yale University

DAVID W. CHRISTIANSON
Roy and Diana Vagelos Laboratories
Department of Chemistry
University of Pennsylvania
Philadelphia, PA

Founding Editors

SIDNEY P. COLOWICK and NATHAN O. KAPLAN

VOLUME SIX HUNDRED AND FORTY NINE

METHODS IN ENZYMOLOGY

Pore-Forming Toxins

Edited by

ALEJANDRO P. HEUCK
Department of Biochemistry and Molecular Biology, University of Massachusetts, Amherst, MA, United States

ELSEVIER

ACADEMIC PRESS
An imprint of Elsevier

Academic Press is an imprint of Elsevier
50 Hampshire Street, 5th Floor, Cambridge, MA 02139, United States
525 B Street, Suite 1650, San Diego, CA 92101, United States
The Boulevard, Langford Lane, Kidlington, Oxford OX5 1GB, United Kingdom
125 London Wall, London, EC2Y 5AS, United Kingdom

First edition 2021

ISBN: 978-0-12-823858-5
ISSN: 0076-6879

For information on all Academic Press publications
visit our website at https://www.elsevier.com/books-and-journals

Publisher: Zoe Kruze
Editorial Project Manager:
Federico Paulo S. Mendoza
Production Project Manager: James Selvam
Cover Designer: Alan Studholme

Typeset by SPi Global, India

Contents

Contributors

Saša Aden
Department for Molecular Biology and Nanobiotechnology, National Institute of Chemistry, Ljubljana, Slovenia

Mazdak A. Bakshloo
CY Cergy Paris Université, CNRS, LAMBE, Cergy; Université Paris-Saclay, Univ Evry, CNRS, LAMBE, Evry-Courcouronnes, France

Gregor Anderluh
Department for Molecular Biology and Nanobiotechnology, National Institute of Chemistry, Ljubljana, Slovenia

K. Ganapathy Ayappa
Department of Chemical Engineering; Centre for Biosystems Science and Engineering, Indian Institute of Science, Bengaluru, India

Gerhard Baaken
Ionera Technologies GmbH, Freiburg, Germany

Amit Behera
Department of Chemical Engineering, Indian Institute of Science, Bengaluru, India

Jan C. Behrends
Laboratory for Membrane Physiology and Technology, Department of Physiology, Faculty of Medicine, University of Freiburg; Freiburg Centre for Interactive Materials and Bioinspired Technologies (FIT); Freiburg Centre for Materials Research, Freiburg, Germany

Jose M.M. Caaveiro
Department of Global Healthcare, Graduate School of Pharmaceutical Sciences, Kyushu University, Fukuoka-shi, Fukuoka, Japan

Michelle P. Christie
Department of Biochemistry and Molecular Biology, Bio21 Molecular Science and Biotechnology Institute, The University of Melbourne, Parkville, VIC, Australia

Antonio De la Torre-Cabrera
Departamento de Bioquímica y Biología Molecular, Instituto Biofisika, UPV/EHU-CSIC, Leioa, Spain

Rajat Desikan
Department of Chemical Engineering, Indian Institute of Science, Bengaluru, India

Pascal Didier
UMR 7021 CNRS, Université de Strasbourg, Illkirch, France

Michelle A. Dunstone
Department of Biochemistry and Molecular Biology, Biomedicine Discovery Institute, Monash University, Clayton, VIC, Australia

Tobias Enßlen
Laboratory for Membrane Physiology and Technology, Department of Physiology, Faculty of Medicine, University of Freiburg, Freiburg, Germany

Jordan C. Evans
Department of Microbiology and Immunology, The University of Oklahoma Health Sciences Center, Oklahoma City, OK, United States

Nicole S. Galenkamp
Groningen Biomolecular Sciences & Biotechnology Institute, University of Groningen, Groningen, The Netherlands

Sara García-Linares
Department of Biochemistry and Molecular Biology, Complutense University, Madrid, Spain

José G. Gavilanes
Department of Biochemistry and Molecular Biology, Complutense University, Madrid, Spain

Robert J.C. Gilbert
Division of Structural Biology, Wellcome Centre for Human Genetics, University of Oxford, Oxford, United Kingdom

Hanling Guo
Program in Molecular and Cellular Biology, University of Massachusetts, Amherst, MA, United States

Katharine Hammond
National Physical Laboratory, Teddington; London Centre for Nanotechnology; Department of Physics & Astronomy, University College London, London, United Kingdom

Alejandro P. Heuck
Department of Biochemistry and Molecular Biology, University of Massachusetts, Amherst, MA, United States

Adrian W. Hodel
Killer Cell Biology Laboratory, Peter MacCallum Cancer Centre, Melbourne, VIC, Australia

Bart W. Hoogenboom
London Centre for Nanotechnology; Department of Physics & Astronomy, University College London, London, United Kingdom

Bo Huang
Institute of Biophysics, Chinese Academy of Sciences, Beijing, China

Fang Jiao
Department of Anesthesiology; Department of Physiology and Biophysics, Weill Cornell Medicine, New York, NY, United States

Kristen A. Johnson
Department of Molecular Genetics, University of Texas Southwestern Medical Center, Dallas, TX, United States

Bronte A. Johnstone
Department of Biochemistry and Molecular Biology, Bio21 Molecular Science and Biotechnology Institute, The University of Melbourne, Parkville, VIC, Australia

Shinako Kakuda
Department of Biochemistry and Cell Biology, Stony Brook University, Stony Brook, NY, United States

Toshihide Kobayashi
UMR 7021 CNRS, Université de Strasbourg, Illkirch, France

Alexander Kyrychenko
Institute of Chemistry and School of Chemistry, V. N. Karazin Kharkiv National University, Kharkiv, Ukraine

Alexey S. Ladokhin
Department of Biochemistry and Molecular Biology, Kansas University Medical Center, Kansas City, KS, United States

Bingchen Li
Department of Biochemistry and Cell Biology, Stony Brook University, Stony Brook, NY, United States

Jialing Lin
Department of Biochemistry and Molecular Biology; Stephenson Cancer Center, University of Oklahoma Health Sciences Center, Oklahoma City, OK, United States

Erwin London
Department of Biochemistry and Cell Biology, Stony Brook University, Stony Brook, NY, United States

Florian L.R. Lucas
Groningen Biomolecular Sciences & Biotechnology Institute, University of Groningen, Groningen, The Netherlands

Giovanni Maglia
Groningen Biomolecular Sciences & Biotechnology Institute, University of Groningen, Groningen, The Netherlands

Prabal K. Maiti
Centre for Condensed Matter Theory, Department of Physics, Indian Institute of Science, Bengaluru, India

Álvaro Martínez-del-Pozo
Department of Biochemistry and Molecular Biology, Complutense University, Madrid, Spain

Yves Mély
UMR 7021 CNRS, Université de Strasbourg, Illkirch, France

Craig J. Morton
Department of Biochemistry and Molecular Biology, Bio21 Molecular Science and Biotechnology Institute, The University of Melbourne, Parkville, VIC, Australia

Motohide Murate
UMR 7021 CNRS, Université de Strasbourg, Illkirch, France

Abdelghani Oukhaled
CY Cergy Paris Université, CNRS, LAMBE, Cergy; Université Paris-Saclay, Univ Evry, CNRS, LAMBE, Evry-Courcouronnes, France

Hadjer Ouldali
CY Cergy Paris Université, CNRS, LAMBE, Cergy; Université Paris-Saclay, Univ Evry, CNRS, LAMBE, Evry-Courcouronnes, France

Juan Palacios-Ortega
Department of Biochemistry and Molecular Biology, Complutense University, Madrid, Spain; Biochemistry, Faculty of Science and Engineering, Åbo Akademi University, Turku, Finland

Michael W. Parker
Department of Biochemistry and Molecular Biology, Bio21 Molecular Science and Biotechnology Institute, The University of Melbourne, Parkville; St. Vincent's Institute of Medical Research, Fitzroy, VIC, Australia

Christopher Parperis
Department of Chemistry, Britannia House, King's College London, London, United Kingdom

Manuela Pastoriza-Gallego
CY Cergy Paris Université, CNRS, LAMBE, Cergy; Université Paris-Saclay, Univ Evry, CNRS, LAMBE, Evry-Courcouronnes, France

Fabien Piguet
CY Cergy Paris Université, CNRS, LAMBE, Cergy; Université Paris-Saclay, Univ Evry, CNRS, LAMBE, Evry-Courcouronnes, France

Arun Radhakrishnan
Department of Molecular Genetics, University of Texas Southwestern Medical Center, Dallas, TX, United States

Ludovic Richert
UMR 7021 CNRS, Université de Strasbourg, Illkirch, France

Esperanza Rivera-de-Torre
Department of Biochemistry and Molecular Biology, Complutense University, Madrid, Spain; Department of Biotechnology and Biomedicine, Technical University of Denmark, Kongens Lyngby, Denmark

Mykola V. Rodnin
Department of Biochemistry and Molecular Biology, Kansas University Medical Center, Kansas City, KS, United States

David Rodriguez-Larrea
Departamento de Bioquímica y Biología Molecular, Instituto Biofisika, UPV/EHU-CSIC, Leioa, Spain

Yi Ruan
Zhejiang Provincial Key Laboratory & Collaborative Innovation Center for Quantum Precision Measurement, College of Science, Zhejiang University of Technology, Hangzhou, China

Simon Scheuring
Department of Anesthesiology; Department of Physiology and Biophysics, Weill Cornell Medicine, New York, NY, United States

J. Peter Slotte
Biochemistry, Faculty of Science and Engineering, Åbo Akademi University, Turku, Finland

Tina Snoj
Department for Molecular Biology and Nanobiotechnology, National Institute of Chemistry, Ljubljana, Slovenia

Bradley A. Spicer
Department of Biochemistry and Molecular Biology, Biomedicine Discovery Institute, Monash University, Clayton, VIC, Australia

Monasadat Talarimoghari
Laboratory for Membrane Physiology and Technology, Department of Physiology, Faculty of Medicine, University of Freiburg, Freiburg, Germany

Yuzhou Tang
Program in Molecular and Cellular Biology, University of Massachusetts, Amherst, MA, United States

Nario Tomishige
UMR 7021 CNRS, Université de Strasbourg, Illkirch, France

Hideaki Tsuge
Faculty of Life Sciences; Institute for Protein Dynamics; Center for Molecular Research in Infectious Diseases, Kyoto Sangyo University, Kyoto, Japan

Kouhei Tsumoto
Department of Bioengineering, Graduate School of Engineering; Department of Chemistry and Biotechnology, School of Engineering, The University of Tokyo, Bunkyo-ku; Institute of Medical Science, The University of Tokyo, Minato-ku, Tokyo, Japan

Rodney K. Tweten
Department of Microbiology and Immunology, The University of Oklahoma Health Sciences Center, Oklahoma City, OK, United States

Nieck J. van der Heide
Groningen Biomolecular Sciences & Biotechnology Institute, University of Groningen, Groningen, The Netherlands

Victor Vasquez-Montes
Department of Biochemistry and Molecular Biology, Kansas University Medical Center, Kansas City, KS, United States

Arjan J. Vermeulen
Department of Biochemistry and Molecular Biology, University of Massachusetts, Amherst, MA, United States

Mark I. Wallace
Department of Chemistry, Britannia House, King's College London, London, United Kingdom

Carsten Wloka
Groningen Biomolecular Sciences & Biotechnology Institute, University of Groningen, Groningen, The Netherlands

Tomohito Yamada
Faculty of Life Sciences, Kyoto Sangyo University, Kyoto, Japan

Ekaterina Zaitseva
Ionera Technologies GmbH, Freiburg, Germany

Xuejun C. Zhang
Institute of Biophysics, Chinese Academy of Sciences, Beijing, China

Zhi Zhang
Department of Biochemistry and Molecular Biology, University of Oklahoma Health Sciences Center, Oklahoma City, OK, United States

Preface

The present volume on Pore-Forming Toxins (PFTs) is intended for both, the expert and the novice in the field of membrane proteins. The expert will find latest advances on the techniques used to study of PFTs in vivo and in vitro. Those who are starting in this field will be exposed to a historical perspective on how different techniques cooperated and evolved to unveil the intricate molecular mechanism of these fascinating proteins. The detailed protocols provided in each section of this volume will help researchers to select the technique(s) that better fit their needs based on the complexity and properties of the system under investigation.

PFTs were discovered more than a century ago based on their ability to lyse erythrocytes and other mammalian cells, and it was in the 1960s when the first large, round pores were observed using electron microscopy (EM). The development of protein engineering tools in the 70s and 80s catapulted the structural analysis of PFT; X-ray structures dominated the field in the late 80s and 90s, and the advances on cryo-EM techniques started to reveal the structure of the large membrane–protein complexes by the end of the century. Meanwhile, site-directed labeling of PFTs with fluorophores and spin labels, as well as conductance measurements on planar lipid bilayer were employed to unveil the detailed mechanistic aspects of pore formation. More recently, single-molecule fluorescence and atomic force microscopy (AFM) techniques uncovered the dynamic aspects of the pore formation process with unprecedented detail. Still, many gaps exist in our understanding of how these PFTs work, and the new advances in computational approaches and molecular dynamics (MD) simulations are being deployed to fill these gaps.

In Chapter 1, M.W. Parker and colleagues provide a fascinating description of the challenges overcame to obtain the first crystal structure of a colicin domain during the late 80s, and on how X-ray crystallography has contributed over the years to our understanding of the PFTs mechanism.

J.C. Evans and R.K. Tweten describe in Chapter 2 how protein engineering tools developed in the 70s and 80s boosted the studies on the mechanism of pore formation. They illustrate the power of site-directed mutagenesis with their elegant studies on Perfringolysin O, a prototype of the cholesterol-dependent cytolysins.

In Chapter 3, R.J.C. Gilbert highlights the role that EM has played since the 80s in the discovery of PFTs, and how today cryo-EM plays a prominent role on the analysis of the three-dimensional structures adopted by PFTs in lipid systems. This chapter introduces the procedures used to study PFTs using cryo-EM, from sample preparation and image recording all through the interpretation of the 3D maps, including troubleshooting of typical problems that could be encountered with this procedures.

The next two chapters provide prominent examples of the structures of large oligomeric complexes of PFTs recently solved by cryo-EM structures. In Chapter 4, B.A. Spicer and M.A. Dunstone reveal how they overcame the challenges of working with a metastable, monomeric PFTs—the human complement component 9 (C9)—that contains multiple disulfide bonds and glycosylations. In Chapter 5, T. Yamada and H. Tsuge describe a method to prepare iota B pore complex that does not required model membranes. The authors used ethanol to induce the formation of the iota B pore that is subsequently solubilized with a detergent to avoid aggregation. Purified oligomers are incubated with the iota A component to reconstitute the whole AB-toxin complex and studied using cryo-EM.

The following two chapters provide examples of how AFM was employed to directly observe—via high-resolution images—how PFTs are incorporated into growing oligomers, and how they insert into membranes. In Chapter 6, B.W. Hoogenboom and colleagues describe the preparation of supported lipid bilayers, the setup of the AFM instrument, the imaging, and the analysis of the data proving examples of potential artifacts encountered while using this technique to study PFTs. In Chapter 7, S. Scheuring and colleagues uses high-speed AFM to study the dynamic aspects of oligomer assembly and pore formation. The ability to image pore formation in the subsecond time scale promises to unveil the detailed kinetics of oligomer assembly and prepore to pore transitions.

Liposomes have played a fundamental role in the biophysical characterization of the PFTs mechanisms of action. However, liposomes lack the lipid asymmetry and electrochemical gradients present in live cell membranes, and they are too small to be seen using light microscopy. The size problem was overcame by the use of Giant Unilamellar Vesicles in PFTs studies, described in Chapter 8 by G. Anderluh and colleagues. In Chapter 9, E. London and colleagues describe a procedure to prepare asymmetric lipid vesicles, revealing the important role that lipid asymmetry plays in pore formation.

The interaction of PFTs with lipids plays a fundamental role during the conformational rearrangement of these proteins. In Chapter 10, J.M.M. Caaveiro and K. Tsumoto illustrate how the studies on actinoporin–lipid interactions developed into biotechnological and biomedical applications. In Chapter 11, S. García-Linares and colleagues describe biophysical approaches they have used to unveil details of the actinoporin–lipid interactions, and how these approaches can be used to study other PFTs.

Some pore-forming proteins have resisted structural characterization. These proteins are rich in α-helical content and involved in the translocation of other proteins across membranes. In Chapter 12, A.S. Ladokhin and colleagues describe the elegant biophysical approaches they used to gain insights into the intricate mechanism of diphtheria toxin translocation domain using model systems. The next two chapters provide examples on structural studies of pore-forming proteins in vivo. In Chapter 13, J. Lin and colleagues speak of their efforts to characterize the structural arrangement of the Bcl-2 family of proteins in mitochondrial membranes using disulfide crosslinking, whereas in Chapter 14, my colleagues and I discuss the efforts carried out to properly study the topological arrangement of the type 3 secretion translocators in membranes of human cells.

In Chapter 15, C. Parperis and M.I. Wallace describe the incredible power of single-molecule imaging to study PFTs, particularly when the technique is combined with lipid bilayers formed at the interface of lipid droplets.

Some pore-forming proteins are metastable or produce heterogeneous intermediates, and therefore are not suitable for high-resolution structural analysis. MD simulation is rapidly evolving and helping researchers to understand the mechanistic details of these proteins. In Chapter 16, G. Ayappa and colleagues provide examples on how MD has provided molecular insights into different aspect of the pore formation mechanism, including the dynamics of membrane lipids during toxin binding and formation of the pore.

The last five chapters of the volume illustrate how PFTs emerged as powerful analytical tools to study lipids in cell membranes. Examples of PFTs developed as tools to image lipids and lipid domains in membranes are described in Chapter 17 by T. Kobayashi and colleagues. Furthermore, in Chapter 18, K.A. Johnson and A. Radhakrishnan describe the use of PFTs as probes to specifically study sphingomyelin and cholesterol in cells, and in Chapter 19, G. Maglia and colleagues show how ClyA constitutes a highly specific nanopore sensor to detect protein–substrate interactions. In Chapter 20, J.C. Behrends, A. Oukhaled, and colleagues describe the versatility of toxin nanopores to size and sequence biomolecules. This volume ends with the contribution of A. De la Torre-Cabrera and D. Rodriguez-Larrea (Chapter 21) who feature the use of PFTs pores in the study of cotranslational protein folding.

I would like to thank all authors for their hard work and insightful contributions preparing and peer reviewing the chapters included in this volume.

Alejandro P. Heuck
Department of Biochemistry and Molecular Biology
University of Massachusetts
Amherst, MA, United States

CHAPTER ONE

X-ray crystallography shines a light on pore-forming toxins

Bronte A. Johnstone[a], Michelle P. Christie[a,†], Craig J. Morton[a,†], and Michael W. Parker[a,b,*,†]

[a]Department of Biochemistry and Molecular Biology, Bio21 Molecular Science and Biotechnology Institute, The University of Melbourne, Parkville, VIC, Australia
[b]St. Vincent's Institute of Medical Research, Fitzroy, VIC, Australia
*Corresponding author: e-mail address: mwp@unimelb.edu.au

Contents

Abstract

A common form of cellular attack by pathogenic bacteria is to secrete pore-forming toxins (PFTs). Capable of forming transmembrane pores in various biological membranes, PFTs have also been identified in a diverse range of other organisms such as sea anemones, earthworms and even mushrooms and trees. The mechanism of pore formation by PFTs is associated with substantial conformational changes in going from the water-soluble to transmembrane states of the protein. The determination of the crystal structures for numerous PFTs has shed much light on our understanding of these proteins. Other than elucidating the atomic structural details of PFTs and the conformational changes that must occur for pore formation, crystal structures have revealed structural homology that has led to the discovery of new PFTs and new PFT families.

†Contributed equally.

Methods in Enzymology, Volume 649
ISSN 0076-6879
https://doi.org/10.1016/bs.mie.2021.01.001

Here we review some key crystallographic results together with complimentary approaches for studying PFTs. We discuss how these studies have impacted our understanding of PFT function and guided research into biotechnical applications.

1. Introduction

For many decades, X-ray crystallography has been the predominant method of high-resolution structure determination of proteins, providing almost 150,000 available protein structures in the Protein Data Bank (PDB) (http:www.rcsb.org) (Bhat et al., 2001). The water-soluble state of pore-forming toxins (PFTs) has proved to be greatly amenable for structure determination by X-ray crystallography. This has resulted in hundreds of crystal structures that have contributed significantly to our understanding of PFT atomic structures while providing a basis for hypotheses and subsequent experiments regarding the mechanisms of pore formation. These studies have led to a classification of PFTs being either an α- or β-PFT depending on whether the transmembrane pore is formed by α-helices or a β-barrel, respectively. Furthermore, these structures have unexpectedly revealed many new PFT families, where members from different families have been found to be related by structure despite little sequence identity. For some PFTs, crystal structures of other states, such as the pore state, intermediate "prepore" complexes and early stage oligomers, have been successfully visualized. As a result, for some PFT families we have been able to visualize the conformational changes that occur during the process of pore formation in atomic detail. However, for other PFTs crystallization of the pore has remained elusive due to the technical challenges involved, with the requirement for membranes or lipids to initiate and stabilize pore formation and heterogeneity of pore sizes being two of the major problems that commonly arise.

Here, we highlight some of the key crystallographic studies that have contributed to our understanding of PFTs, concentrating on some of the major protein families of bacterial PFTs. In particular, we focus on the crystallographic methods employed to obtain the crystal structures, including circumstances where difficulties have presented and ongoing challenges when it comes to studying PFTs by X-ray crystallography. Furthermore, we discuss how some of the crystal structures have guided the engineering of PFTs for biotechnical applications.

2. α-PFTs

2.1 Pore-forming colicins

Colicins are bacterial secreted antimicrobial proteins (bacteriocins) that are produced by *Escherichia coli* and kill related bacterial species. The plasmid-encoded proteins are secreted into the extracellular medium, bind to receptors on the outer membrane of target cells and subsequently translocate across the outer membrane into the periplasm. There they exhibit lethal activity, which may involve pore formation or further translocation into the cytoplasm where they exhibit DNase, RNase or phosphatase activity. Rather than classification based on activity, colicins are classified on the basis of the system used for translocation across the outer membrane, where group A colicins utilize the Tol system and group B colicins exploit the Ton system (Braun, Patzer, & Hantke, 2002; Lazzaroni, Dubuisson, & Vianney, 2002). Here, we restrict our discussion to pore-forming colicins.

The very first PFT structure to be solved was for the pore-forming domain of colicin A (PDB: 1COL) (Parker, Pattus, Tucker, & Tsernoglou, 1989; Parker, Postma, Pattus, Tucker, & Tsernoglou, 1992). The structure was originally published using a model derived from diffraction data collected in-house (Parker et al., 1989), with a refined structure subsequently achieved using data collected using synchrotron radiation (Parker et al., 1992). Achieving the first PFT structure encompassed many challenges. Firstly, the crystal structure of the pore-forming domain was published long before that of a full-length colicin. Colicins contain a characteristic intrinsically disordered N-terminal region; a feature that is well known for preventing successful crystallization. By exposing purified colicin A to proteolysis with thermolysin, a well-ordered C-terminal domain was produced, previously determined to be the pore-forming domain (Martinez, Lazdunski, & Pattus, 1983; Ohno-Iwashita & Imahori, 1982). The fragment was shown to possess pore-forming activity comparable to the wild-type protein (Tucker, Pattus, & Tsernoglou, 1986).

Following crystallization, attempts to generate isomorphous derivatives with heavy atoms for experimental phasing proved unsuccessful, despite over 30 attempts (Parker et al., 1989). Given that the C-terminal domain of colicin A lacked cysteines, site-directed mutagenesis was undertaken to introduce cysteine residues to which mercury derivatives could bind (Tucker et al., 1989). This approach had not been previously reported for an unknown protein structure. Five mutants were trialed, directed at an α-helix that does not contribute to pore-forming activity (Baty et al., 1987). One particular mutant, with a serine-to-cysteine mutation and an additional repeat decapeptide prior to the mutation, yielded diffracting crystals that successfully generated phases to 3 Å resolution using isomorphous and anomalous scattering contributions. Refinement of the atomic model was achieved using a distance-restrained, least-squares procedure in combination with molecular dynamics (MD), a technique that had only recently been applied to crystallographic refinement (Brünger, Kuriyan, & Karplus, 1987). Application of MD dramatically reduced the timeframe for refinement and produced a well-refined model; however, the model generated was not without problems, which had to be corrected for with rebuilding and further refinement (Postma, Parker, & Tsernoglou, 1989).

The crystal structure of the colicin A pore-forming domain (Fig. 1A) revealed a molecule composed of ten α-helices that arrange into three layers, forming a globin-like topology (Holm & Sander, 1993). Furthermore, colicin adopted an inside-out membrane protein fold with a hydrophobic interior and amphipathic exterior. The observation of a hydrophobic hairpin within the hydrophobic interior gave rise to a model of membrane insertion termed the "umbrella" model (Parker et al., 1992). The proposed mechanism involved the hydrophobic loop of the hairpin initially inserting into the membrane, which necessitated opening of the structure and spontaneous insertion of the entire hairpin into the lipid bilayer, with the remaining helices remaining on the membrane surface. While a substantial amount of

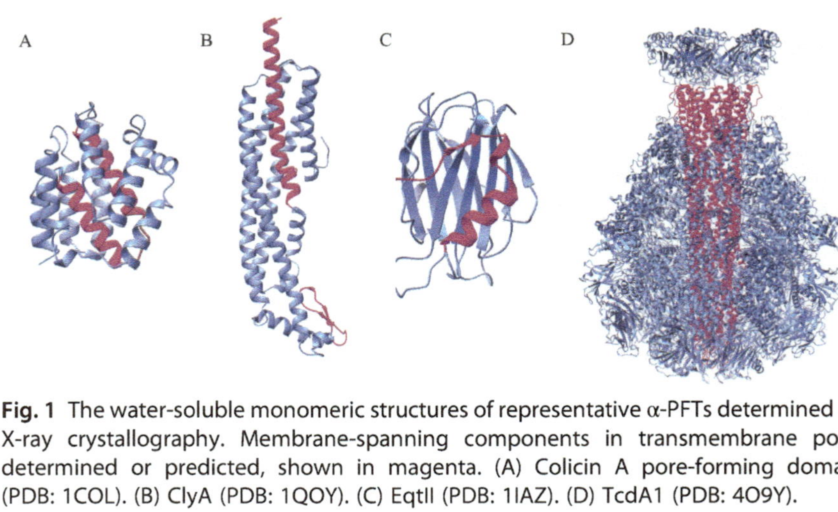

Fig. 1 The water-soluble monomeric structures of representative α-PFTs determined by X-ray crystallography. Membrane-spanning components in transmembrane pore, determined or predicted, shown in magenta. (A) Colicin A pore-forming domain (PDB: 1COL). (B) ClyA (PDB: 1QOY). (C) EqtII (PDB: 1IAZ). (D) TcdA1 (PDB: 4O9Y).

experimental data supports the umbrella model (Brunden, Uratani, & Cramer, 1984; Duché, Baty, Chartier, & Letellier, 1994; Kim, Valentine, Opella, Schendel, & Cramer, 1998; Lakey et al., 1991; Luo, Yao, & Hong, 2005; Pulagam & Steinhoff, 2013; Rath, Bousché, Merrill, Cramer, & Rothschild, 1991), other data failed to support the model. This resulted in the proposal of the "penknife" model, which differs from the umbrella model primarily in the orientation of the helical hairpin, being parallel to the membrane as opposed to perpendicular (Duché, Parker, González-Mañas, Pattus, & Baty, 1994; Lakey, Duché, González-Mañas, Baty, & Pattus, 1993). However, subsequent studies questioned the use of a non-physiological lipid in this study and suggested the penknife model was an artifact. The lack of a high-resolution structure for the channel form has meant a mechanism is yet to be confirmed, although it is generally accepted, based on available evidence, that the umbrella model is likely an intermediate in the physiological mechanism.

The first crystal structure for a complete colicin molecule was for colicin Ia (PDB: 1CII) (Wiener, Freymann, Ghosh, & Stroud, 1997). The structure was solved by multiple isomorphous replacement (MIR), including four heavy-atom derivative crystals of the wild-type protein and additional mercury derivative crystals of a serine to cysteine mutant (Ghosh, Mel, & Stroud, 1994). Colicin Ia is a considerably elongated (210 Å) "Y"-shaped molecule with an extended coiled-coil structure composed almost entirely of α-helices. Three discrete structural domains are observed; the N-terminal receptor-binding (R) domain, the translocation (T) domain, and the C-terminal pore-forming (C) domain, which displays a similar structure to the pore-forming domain of colicin A.

A number of other pore-forming colicins have also been crystallized and their structure solved, either of the entire molecule or of sub-domains. These include colicin E1 (Elkins, Bunker, Cramer, & Stauffacher, 1997), colicin N (PDB: 1A87) (Vetter et al., 1998) and colicin B (PDB: 1RH1) (Hilsenbeck et al., 2004). Common to all colicins, even those with functions other than pore-formation, is the three-domain structure; however, assignment of the domains by structure alone has proved ambiguous, sometimes requiring proteolysis or differential scanning calorimetry experiments to determine domain boundaries (Baty et al., 1988; Griko, Zakharov, & Cramer, 2000; Ohno-Iwashita & Imahori, 1982). The structure of the pore domain is similar across pore-forming colicins (root-mean-square deviation (RMSD) on alpha carbon atoms <3 Å), with all possessing the buried hydrophobic α-helical hairpin, suggesting a common mechanism for pore formation.

For colicins crystallized with the receptor-binding domain, a lack of electron density for the N-terminal region is a common observation. As examples, no density was observed for the N-terminal domain in colicin N (Vetter et al., 1998) and the first 22 residues of colicin Ia are not included in the final model (Wiener et al., 1997). However, the N-terminal domain has been observed to be stabilized upon binding the corresponding receptor, an observation that has been utilized for crystallographic studies. A crystal structure for the N-terminal receptor-binding domain of colicin Ia bound to its receptor in the outer membrane, colicin I receptor (Cir), was determined in 2007 (PDB: 2HDI) (Buchanan et al., 2007). The complex was achieved by refolding the insoluble colicin Ia R-domain from inclusion bodies in the presence of Cir. The structure of Cir contains an N-terminal plug domain within the transmembrane large β-barrel (PDB 2HDF) (Buchanan et al., 2007). Similar to previous structures of nuclease colicins with their receptors (Kurisu et al., 2003; Sharma et al., 2007), binding of colicin Ia to Cir was not observed to displace the plug domain, suggesting that these receptors are not utilized by the colicins for outer membrane translocation. It was later shown that a second copy of Cir is recruited to serve as the outer membrane translocator (Jakes & Finkelstein, 2010), a mechanism that differs from other group A colicins.

The fold first identified in the pore-forming domain of colicins has been subsequently observed in a number of other proteins including diphtheria toxin (Choe et al., 1992), produced by pathogenic strains of *Corynebacterium diphtheriae*, the mammalian BAX and BCL-2 homologous antagonist/killer (BAK) family (Brouwer et al., 2014; Czabotar et al., 2013; Moldoveanu et al., 2006; Muchmore et al., 1996) and a large family of insecticidal crystal proteins (Cry toxins) from *Bacillus thuringiensis* (Li, Carroll, & Ellar, 1991; Xu, Wang, Yu, & Sun, 2014). A very similar multi-domain, elongated structure to colicins has also been identified in a group of bacteriocins produced by *Pseudomonas aeruginosa* known as pyocins (Barreteau et al., 2012; McCaughey et al., 2014), with pore-forming pyocins sharing the canonical fold of the pore-forming domain of colicins (Behrens et al., 2020).

2.2 ClyA family

Another *E. coli*-derived α-PFT is cytolysin A (ClyA), also referred to as hemolysin E (HlyE) or less commonly silent hemolysin (SheA) (del Castillo, Leal, Moreno, & del Castillo, 1997; Green & Baldwin, 1997). ClyA is produced by pathogenic strains of *E. coli*, where it provides hemolytic activity (del Castillo et al., 1997; Ludwig, Bauer, Benz, Bergmann, & Goebel, 1999; Oscarsson, Mizunoe, Uhlin, & Haydon, 1996), with homologs also existing in *Salmonella typhi* (del Castillo et al., 1997) and *Shigella flexneri* (Oscarsson et al., 2002). ClyA is the prototype of a wider family of structurally homologous cation-selective pores (Ludwig et al., 1999; Mueller, Grauschopf, Maier, Glockshuber, & Ban, 2009). While ClyA is a one-component PFT, the family also encompasses two- and three-component toxins (Bräuning et al., 2018; Ganash et al., 2013; Madegowda, Eswaramoorthy, Burley, & Swaminathan, 2008; Schubert, Vetter, Prumbaum, Penczek, & Raunser, 2018; Vigneux et al., 2007; Wagner, Lin, Borst, & Miller, 2013; Wilson et al., 2019).

The soluble, monomeric structure of ClyA (PDB: 1QOY) revealed a novel toxin structure that was the first of the ClyA PFT family of structures (Wallace et al., 2000). An initial model was derived using MIR from diffraction of five different derivative crystals collected at room temperature. Data from native crystals, diffracting to 2.0 Å resolution, were subsequently collected under cryogenic conditions and phases deduced using the initial model. Despite variation in unit cell dimensions, and therefore large isomorphous differences, the final ClyA model derived was identical to the initial model (with a RMSD of 0.56 Å).

The crystal structure of ClyA revealed an elongated molecule that is divided into two domains, referred to as the head and tail domains (Figs. 1B and 2A). While the tail domain adopts a five α-helical bundle motif, the head domain contains a small β-hairpin flanked by two short α-helices. Overall, the structure is almost entirely α-helical, with the β-hairpin, referred to as the β-tongue motif, being the exception to this. This motif is highly hydrophobic and is buried within the interior of the head domain. The β-tongue was predicted at the time to be the most likely site of membrane interaction (del Castillo et al., 1997; Oscarsson et al., 1999; Wallace et al., 2000). It is now established that pore formation by ClyA involves detachment of the β-tongue motif from the protein core, conversion to an α-helix followed by membrane insertion. However, it does not fully penetrate the membrane or line the pore itself, instead penetrating only into a single leaflet of the membrane (Mueller et al., 2009; Peng, de Souza Santos, Li, Tomchick, & Orth, 2019).

There are numerous crystal structures of other members of the ClyA PFT family. These include the one-component members Cry6Aa (PDB: 5KUC, 5KUD, 5GHE) (Dementiev et al., 2016), AB21 (PDB: 6FPD) (Komárek et al., 2018) and MakA (PDB: 6EZV) (Dongre et al., 2018) from

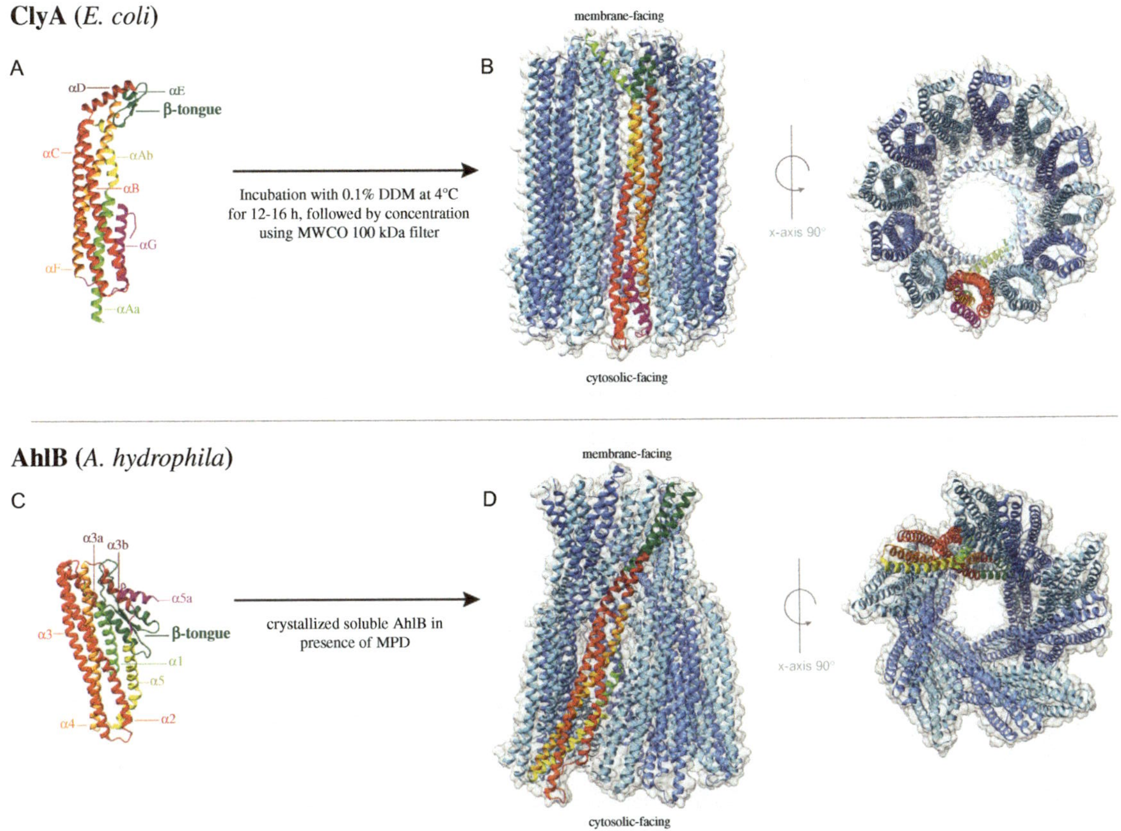

Fig. 2 Crystal structures of the soluble and transmembrane pore of ClyA family members. (A) The soluble monomer of ClyA (PDB: 1QOY) contains a tail domain composed of a five-helical bundle and a head domain consisting of a small β-hairpin (green) flanked by two short α-helices. (B) Formation of the dodecameric ClyA (PDB: 2WCD) involves substantial rearrangements. This includes a β-strand-to-α-helix transition of the β-tongue, which inserts into a single leaflet of the membrane, while αA1 inserts into the membrane to form the membrane-spanning pore. (C) The soluble monomer of AhlB (PDB: 6GRK), a component from a

B. thuringiensis, Agaricus bisporus and *Vibrio cholerae*, respectively, as well as a crystal structures for at least one component of multi-component family members. For the two-component members, both A and B subunits have been solved for XaxAB from *Xenorhabdus nematophila* (PDB: 6GY8, 6GY7) (Schubert et al., 2018) and YaxAB from *Yersinia enterocolitica* (PDB: 6EK7, 6EK8) (Bräuning et al., 2018), with a structure also solved for PaxB from *Photorhabdus luminescens* (PDB 6EK4) (Bräuning et al., 2018). Lastly, there are now also crystal structures of HblB (PDB: 2NRJ) (Madegowda et al., 2008) and NheA (PDB 4K1P) (Ganash et al., 2013) from two different *Bacillus cereus* tripartite toxins in addition to structures for the AhlB (PDB: 6GRK) (Fig. 2C) and AhlC (PDB: 6H2E) components from a tripartite toxin from *Aeromonas hydrophila* (Wilson et al., 2019). Because of the low sequence identity (15–45%) between family members (Bräuning & Groll, 2018), the determination of a crystal structure was the critical step in revealing a protein belonged to the ClyA PFT family. Not surprisingly, again given the low sequence identity, almost all ClyA-related proteins were solved by experimental phasing methods, in particular selenomethionine (SeMet)-SAD methods (Bräuning et al., 2018; Dementiev et al., 2016; Dongre et al., 2018; Ganash et al., 2013; Madegowda et al., 2008; Wilson et al., 2019), with the exception of YaxB which was solved by molecular replacement (MR) with PaxB as the search model (40% sequence identity) (Bräuning et al., 2018).

Despite the low sequence identity within the ClyA family, all structures display the canonical five-helical bundle fold within the tail domain. Small differences, such as a varying length of the N-terminal helix (αA in ClyA), can be observed across the structures. Some family members, including XaxA and XaxB, have an additional "neck" domain that results in a more elongated molecule (Schubert et al., 2018). Intriguingly, structures of YaxB, XaxB and PaxB lack a β-tongue motif (Bräuning et al., 2018; Schubert et al., 2018). Instead, analysis of the YaxB and XaxB protomers in cryo-EM pore structures reveal that the protein rearranges to expose a single amphipathic helix that forms the head domain in the protomer that is capable of spanning the membrane bilayer (Bräuning et al., 2018; Schubert et al., 2018).

The crystal structure of the ClyA transmembrane pore (PDB: 2WCD) (Mueller et al., 2009) was the first structure of an α-PFT pore. Previously, the addition of the detergent *n*-dodecyl β-D-maltoside (DDM) to ClyA was observed to spontaneously induce the formation of pore complexes (Eifler et al., 2006), which could be used for crystallization. ClyA was incubated with 0.1% DDM at 4 °C for 12–16 h, prior to concentration with a

MWCO of 100 kDa, also ensuring the absence of any existing monomeric species. Crystals of the transmembrane ClyA pore were grown using standard vapor diffusion methods. While native crystals displayed severe anisotropic diffraction, efforts to derivatize crystals for phasing revealed that the addition of sodium ethylmercurithiosalicylate decreased this anisotropy significantly. This also allowed a preliminary model to be solved using single-wavelength anomalous dispersion (SAD). The preliminary model, achieved using monoclinic crystals, was then utilized for MR with a triclinic crystal form to achieve the final model. Owing to the anisotropy of the native crystals, only derivatized crystals were used to determine the structure. However, superimposition of the crystal structure with a recently solved cryo-EM structure to 2.8 Å resolution (Peng et al., 2019) suggests distortion of an α-helix within the structure due to binding of two mercury atoms. While the overall fold was not affected, this may highlight a need to verify structural models achieved by derivatized crystals with models determined using diffraction data from native crystals.

The crystal structure of ClyA transmembrane pore revealed a dodecameric complex. EM studies however, revealed tridecameric and tetradecameric complexes in addition to the dodecamer (Eifler et al., 2006; Peng et al., 2019; Tzokov et al., 2006). This suggests only the dodecamer was observed in the crystal structure as a result of being self-selected for by the crystallization conditions. Within the dodecameric complex, ClyA protomers are arranged in a parallel, head-to-head arrangement where all helices, except for αG, contribute to the protomer-protomer interface (Fig. 2B). Interestingly, this interface surface is not observed in the soluble form of ClyA, suggesting that ClyA converts to an intermediate state resembling the protomer structure prior to oligomerization. Key characteristics of the pore channel include constriction at the top of the channel, narrowing the opening to 35 Å, and the lumen of the pore containing an excess of negatively-charged residues, supporting previous studies suggesting cation selectivity (Ludwig et al., 1999). The crystal structure of the pore revealed the substantial conformational changes that occur during the transition from the soluble monomer to the transmembrane pore, where 55% of residues move position and rearrangements of almost 140 Å are observed (Mueller et al., 2009). Furthermore, 16% of residues change secondary structure and become α-helical, increasing the helical content from 79% to 83% in agreement with circular dichroism data (Eifler et al., 2006). As a result of the large rearrangements, the originally hidden β-tongue becomes exposed, undergoes a β-strand-to-α-helix transition and merges with helices αC and αD that insert into a single leaflet of the membrane. The membrane-spanning component of the pore is provided by the N-terminal helix (αA1), which is only able to span the membrane following structural rearrangement.

Not only did the ClyA transmembrane pore crystal structure provide a major insight into understanding pore formation by ClyA, but it also inspired exploitation of the structural knowledge for biotechnological applications (Soskine et al., 2012). ClyA nanopores have been engineered to recognize a wide range of molecular targets, including ubiquitin, avidin and various enzymes, DNA and protein-DNA complexes and even supramolecular complexes (Biesemans, Soskine, & Maglia, 2015; Franceschini, Brouns, Willems, Carlon, & Maglia, 2016; Li, Lee, Shorkey, Chen, & Chen, 2020; Lu et al., 2018; Nomidis, Hooyberghs, Maglia, & Carlon, 2018; Soskine et al., 2012; Soskine, Biesemans, De Maeyer, & Maglia, 2013; Soskine, Biesemans, & Maglia, 2015; Van Meervelt, Soskine, & Maglia, 2014; Willems et al., 2019; Willems et al., 2020; Wloka et al., 2017).

More recently, a second ClyA-family pore crystal structure has been obtained, which was the decameric transmembrane pore formed by the AhlB component from the tripartite AhlABC toxin (PDB: 6GRJ) (Wilson et al., 2019) (Fig. 2D). Unlike the ClyA pore structure, the crystal structure was achieved by growing crystals in the presence of 2-methyl-2,4-pentanediol (MPD). Despite only 8% sequence identity between ClyA and AhlB, the soluble states of each protein resemble each other (RMSD of 3.9 Å) with both consisting of the typical ClyA helical fold (Fig. 2A and C). Comparison of the AhlB pore with the soluble monomer reveal similar major rearrangements to that observed for ClyA, where the β-tongue and helices α3a and α4a transition to α-helices to form the head of the pore channel (Fig. 2D). However, as a tripartite toxin, the β-tongue of AhlB is not proposed to be involved in initial membrane leaflet insertion, as in ClyA. Instead, the AhlC component is proposed to carry out this function. While a model for the assembly of the AhlABC toxin has been proposed, a structure of the fully assembled AhlABC toxin is likely required to answer some of the numerous outstanding questions regarding this recently identified ClyA family member.

2.3 Actinoporins

The eukaryotic actinoporins are a multi-gene family of hemolytic PFTs that are produced by species of sea anemone. Actinoporin family members are approximately 20 kDa, basic and generally cysteine-less proteins that share a high degree of sequence identity (60–90%). Formation of the cation-selective actinoporin pores has a requirement for sphingomyelin-containing membranes, to which actinoporins bind with high affinity (Bakrac et al., 2008; Bernheimer & Avigad, 1976).

The crystal structure of equinatoxin II (Eqt II) from the species *Actinia equina* (PDB: 1IAZ) (Athanasiadis, Anderluh, Macek, & Turk, 2001) was the first structure of a pore-forming protein not only from the actinoporin family, but also the first from a eukaryote. Solved by MIR to 1.9 Å resolution, a single-domain structure was revealed, consisting of a β-sandwich formed from two six-stranded β-sheets (Fig. 1C). This fold is flanked by two short α-helices, one per β-sheet, which are orientated perpendicularly with respect to one another. The N-terminal α-helix, which is amphipathic, mobile and flexible, was later shown to detach from the β-sandwich core and insert into the membrane to form the α-helical pore during pore formation (Kristan, Viero, Dalla Serra, Macek, & Anderluh, 2009). Furthermore, the small region of residues N-terminal to the helix that are non-helical in the monomer becomes helical and merges with the existing helix, so that the inserted helix is 25 residues in length (Tanaka, Caaveiro, Morante, González-Mañas, & Tsumoto, 2015). The structure of EqtII also revealed an exposed tryptophan-rich cluster of aromatic residues, which were predicted to play a role in membrane binding (Athanasiadis et al., 2001). These residues have since been shown to be important for the initial step of membrane binding, where in the two-step mechanism of pore formation binding is succeeded by insertion of the N-terminal helix (Hong et al., 2002).

A number of subsequent crystal structures of water-soluble actinoporin family members have been obtained, including sticholysin II (Stn II) (PDB: 1GWY, 1O72, 1O71) (Mancheño, Martín-Benito, Martínez-Ripoll, Gavilanes, & Hermoso, 2003), fragaceatoxin C (FraC) (PDB: 3LIM, 3VWI, 3W9P 3ZWG, 3ZWJ, 4TSL, 4TSN, 4TSO, 4TSQ, 4TSP, 5GWF) (Mechaly et al., 2011; Tanaka et al., 2015; Tanaka, Caaveiro, Morante, & Tsumoto, 2017) and fragaceatoxin E (FraE) (PDB: 6K2G) (Morante et al., 2019). Similar to EqtII, all crystallized actinoporins have been purified from sea anemones (Athanasiadis et al., 2001; Bellomio et al., 2009; Macek & Lebez, 1988;Mechaly et al., 2011 ; Morante et al., 2019), with the exception of FraC, which was expressed as a recombinant protein in *E. coli* (Tanaka et al., 2015). Given the high sequence identity between family members, it is not surprising that all were solved by MR; however, the structure of Stn II almost required experimental phasing approaches when the structure originally could not be solved by MR with the structure of EqtII (Mancheño, Martínez-Ripoll, Gavilanes, & Hermoso, 2002). The structure was finally solved by switching from the MR program package AMoRe to the emerging program at the time MOLREP, which incorporated full automation with new approaches and improvements compared to AMoRe (Vagin & Teplyakov, 1997).

The various monomeric actinoporin crystal structures are strikingly similar (RMSD between 0.2 and 0.9 Å) (Mechaly et al., 2011; Morante et al., 2019). The region of most variability between the structures is the N-terminal helix, which is also the least conserved region in terms of sequence. Intriguingly, actinoporins display differences in both binding affinity and hemolytic activity despite very high sequence and structural similarity. This variability has been shown to be attributable to slight disparity in both the sequence and structure, particularly the hydrophobic and electrostatic properties, of the N-terminal helix (Carretero et al., 2018; Cilli et al., 2007; Monastyrnaya et al., 2010; Ros et al., 2015).

The structural details regarding the affinity of actinoporins for sphingomyelin were first identified in a co-crystal structure of Stn II with phosphocholine (POC) (PDB: 1O72) (Mancheño et al., 2003), which is the head-group moiety for sphingomyelin. Soaking native crystals with 20 mM POC for 20 s led to a crystal structure which revealed binding within an amphipathic cavity where a large cluster of residues interact with both the choline and phosphate moieties of POC. X-ray crystallography also contributed to the understanding of actinoporins with the crystal structures of FraC in its water-soluble monomeric, lipid-bound monomeric, intermediate dimeric and octameric states (Tanaka et al., 2015) (Fig. 3). The lipid-bound monomer was prepared by co-crystallizing FraC with 1,2-dihexanoyl-sn-glycero-3-phosphocholine (DHPC), while the dimeric form was co-crystallized with POC. The co-crystallization with POC, rather than soaking in POC as for Stn II, likely accounts for the observance of the dimeric form. To generate the octomeric pore, FraC was incubated with lipid vesicles from which the pore structures were extracted using the detergent Triton X-100. The pore structure was sufficiently stable to allow further isolation of the complex by ion-exchange and size-exclusion chromatography in the presence of the detergent *N*-*N*-Dimethyldodecylamine *N*-oxide (LDAO), after which crystallization was conducted using the lipidic mesophase method with DDM in place of LDAO.

The study revealed novel details regarding the conformation of the pore, including the role of lipids in completing and stabilizing the structure, which were bound with such affinity that surprisingly they were not displaced during any of the detergent extraction, purification or crystallization procedures. In addition, the conformational changes that occur between monomer and protomer were deduced. These changes were remarkably minor

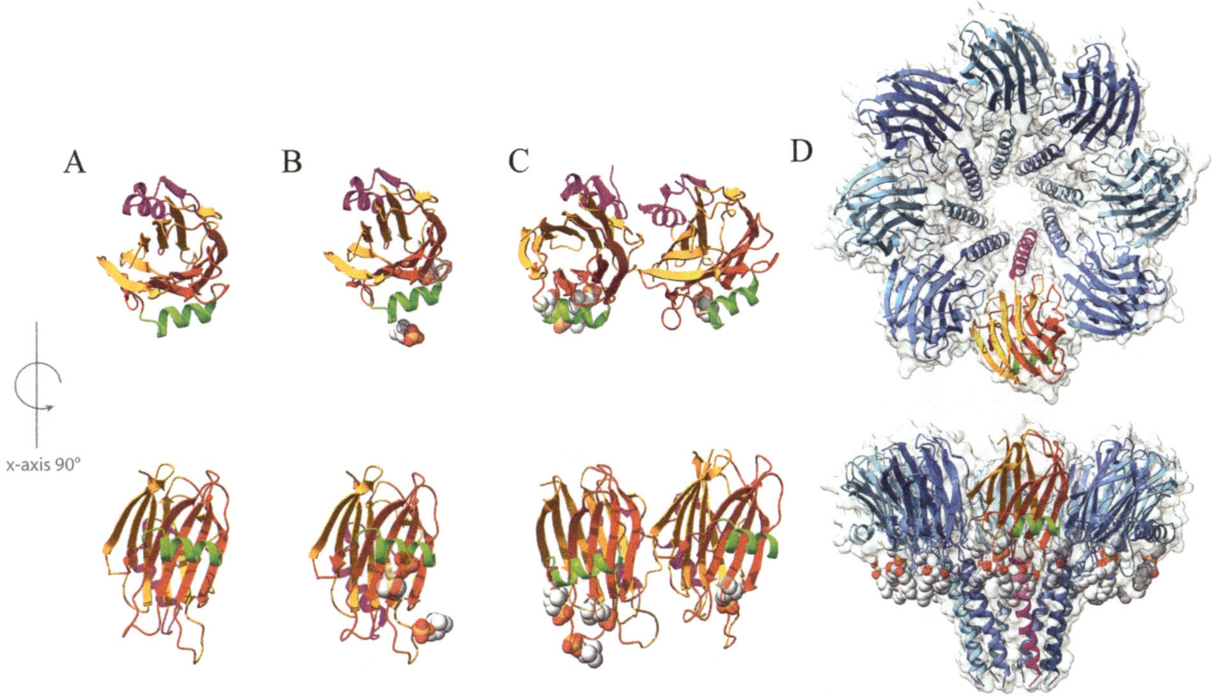

Fig. 3 The mechanism of FraC pore formation, illustrated by crystallography. (A) The crystal structure of the water-soluble FraC monomer (PDB: 3VWI) consists of a β-sandwich fold formed by two β-sheets (orange, red), flanked by two helices (green, magenta (N-terminal helix)). (B) The co-crystal structure of FraC with DHPC (cpk coloring) (PDB: 4TSQ) reveals binding of DHPC on the membrane-facing plane of the molecule. (C) Co-crystallization with POC (PDB: 4TSN) reveals an intermediate assembly dimeric structure with a conformation between that observed for FraC monomer or protomers. (D) The transmembrane pore of FraC (PDB: 4TSY) is an octameric complex, completed and stabilized by lipid molecules (shown in cpk).

compared to other PFTs; the β-core remained unchanged and the major conformational change was the >50 Å movement of the N-terminal helix to form the pore lining, as suggested by previous data (Hong, Gutierrez-Aguirre, et al., 2002; Kristan et al., 2004; Kristan, Viero, Macek, Dalla Serra, & Anderluh, 2007). Furthermore, the dimer structure likely represents an intermediate state wherein the N-terminal helix is not detached from the protein core, but other smaller conformational changes are observed in the protomer. This suggests that actinoporins assemble in a sequential manner with no prepore intermediate.

The pore form of FraC has properties, such as the narrow constriction at the base of the channel, that make it viable for nanopore applications (Huang, Willems, Soskine, Wloka, & Maglia, 2017; Mutter, Volarić, Szymanski, Feringa, & Maglia, 2019; Restrepo-Pérez, Wong, Maglia, Dekker, & Joo, 2019; Wloka, Mutter, Soskine, & Maglia, 2016; Zhao et al., 2019). One particularly exciting application is the engineering of FraC nanopores with the aim to develop a high-throughput single-molecule method for protein sequencing (Huang, Voet, & Maglia, 2019). By tuning the size of the pore, engineered FraC nanopores currently have the ability to differentially identify peptides that differ by only 44 Da.

2.4 Tc toxins

Toxin complex (Tc) toxins are a family of tripartite (ABC) toxins derived from various insect and human pathogenic bacteria. First identified in *P. luminescens* (Bowen et al., 1998; ffrench-Constant & Bowen, 2000), analogs have also been identified in *Pseudomonas, Xenorhabdus, Serratia* and *Yersinia* species (Boemare, Akhurst, & Mourant, 1993; Dodd, Hurst, Glare, O'Callaghan, & Ronson, 2006; Morgan, Sergeant, Ellis, Ousley, & Jarrett, 2001). The three subunits that contribute to the heterotrimeric complex structure of Tc toxins are referred to as TcA, TcB and TcC, with all three subunits required for toxic activity. Despite this, the actual toxin component of the complex is the C-terminal product derived from TcC following autoproteolysis (Busby, Panjikar, Landsberg, Hurst, & Lott, 2013; Meusch et al., 2014), which is shielded by a cocoon formed by a heterodimer of TcB and TcC subunits (Busby et al., 2013; Meusch et al., 2014). TcA is the membrane penetrating component, forming a channel through which the toxic enzyme is translocated, in addition to mediating cell adhesion via receptor-binding domains (Gatsogiannis et al., 2013). Unlike many of the other PFTs discussed in this chapter, Tc toxins are not secreted as soluble monomers. Instead, the soluble form of the pore component TcA is a pre-formed pentameric complex, referred to as the prepore. Structural information regarding the fully assembled Tc toxins has been primarily derived from cryo-EM studies. In contrast, high resolution crystal structures of the TcA subunit or the TcB-TcC heterodimer have revealed the atomic details of the prepore state of Tc toxins.

A major achievement in understanding Tc toxins was the first, and still only, crystal structure of the TcA subunit (TcdA1) from *P. luminescens* (PDB ID: 4O9Y) (Meusch et al., 2014). Previously, only low resolution electron microscopy structures of TcA subunits from *P. luminescens* (Gatsogiannis et al., 2013) or the Tc toxins XptA1 (Lee et al., 2007) and YenTcA (Landsberg et al., 2011) had been achieved; however, the low resolution structure of TcdA1 (Gatsogiannis et al., 2013) played a critical role in obtaining the TcdA1 crystal structure. A hybrid approach was used, bringing together X-ray crystallography and cryo-EM, where the structure was solved by MR with the search model being a poly-alanine model constructed from the TcdA1 cryo-EM structure. This circumvented the need for experimental phasing approaches for crystals that already possessed other challenges. This included the difficulty of collecting high-resolution data due to the very long unit cell axes (303 Å × 321 Å × 667 Å) and highly anisotropic diffraction. Furthermore, collection of an adequate dataset required the use of a highly focused beam and collecting a redundant low-resolution dataset that was merged with four datasets collected at higher resolution over a very small wedge (40°). Overall, this yielded a crystal structure at a moderate resolution of 4.0 Å.

TcdA1 crystallized as a homopentameric complex (Fig. 1D), in contrast to past suggestions of a tetrameric complex from low-resolution negative stain EM of XptA1 (Lee et al., 2007), although this has since been shown to also be pentameric (Leidreiter et al., 2019). Each protomer of the bell-shaped complex consists of eight domains, where six form the completely α-helical shell that contains a neuraminidase-like domain and four receptor binding domains. The other two domains of the protomer form the pore-forming component of the complex. This includes the TcB-binding domain and a 200 Å helical channel, the luminal surface of which is negatively charged, supporting previous evidence revealing a cation selective pore channel (Lang, Konukiewitz, Aktories, & Benz, 2013). The shell and channel components of TcdA1 are connected via a linker, which after the opening of the TcB-TcC cocoon acts as the spring in the syringe-like injection mechanism of Tc toxins (Gatsogiannis et al., 2016; Meusch et al., 2014).

A number of TcB-TcC heterodimer crystal structures have been published, corresponding to Tc toxins from *Y. entomophaga* and *P. luminescens* (Busby et al., 2013; Gatsogiannis et al., 2018; Meusch et al., 2014). The first of these to be solved was YenB–YenC2 from *Y. entomophaga* in 2013 (PDB: 4IGL), with the crystal structure solved by SAD using a combination of datasets from SeMet crystals and dodecabromohexatantalum-soaked crystals (Busby et al., 2013). All structures show an identical fold of stacked β-barrels that converge into a large cocoon that is enclosed by a β-propellor at one end and the RHS repeat-associated core domain at the other end. This RHS core domain is an aspartate autoprotease, where the catalytic dyad of the protease cleaves the hypervariable region (HVR) of the TcC subunit encapsulated within the outer cocoon to form the toxic enzyme (Busby et al., 2013).

In all crystal structures of the heterodimer published to date, the HVR is not resolved, thus the structure remains elusive, although some have suggested that it exists in either a partially or fully unfolded state (Roderer, Hofnagel, Benz, & Raunser, 2019). The lack of sequence and functional conservation in the HVR between Tc family members, has further hindered our understanding regarding the requirements for translocation of the enzyme. However, analysis of various engineered TcB-TcC complexes where the HVR was replaced with different cargo proteins has helped determine some of these requirements. X-ray crystallography aided in these studies by revealing the molecular details for scenarios where an observed lack of translocation for particular cargo proteins could not be understood otherwise (Roderer, Schubert, Sitsel, & Raunser, 2019). Furthermore, this showed that Tc toxins can be engineered to deliver cargo proteins of interest, a concept that could prove to have potential in biotechnical applications in the future.

3. β-PFTs

3.1 Aerolysin-like family

The aerolysin-like family of toxins are a diverse family of β-PFTs found in both eukaryotes and prokaryotes, including many pathogenic bacteria such as *A. hydrophila, Clostridium perfringens and B. thuringiensis*. These toxins are usually secreted as inactive protoxins that are activated by proteolytic cleavage of a propeptide to oligomerize into pores typically consisting of seven-, eight- or nine- protomers (Szczesny et al., 2011).

The first crystal structure obtained for this class of toxins was for the protoxin aerolysin, produced by the bacterium *A. hydrophila* (PDB:

1PRE) (Parker et al., 1994). The structure of proaerolysin was solved by MIR, with native crystals derivatized by soaking with gold, mercury, lead or uranyl derivatives (Parker et al., 1994; Tucker, Parker, Tsernoglou, & Buckley, 1990). Anomalous scattering data were collected from all four derivatives. The initial experimental map was vastly improved by solvent flattening and enabled the building of a partial model which was further improved by density modification and MD refinement. Iterative rounds of model building and refinement yielded a final structure to 2.5 Å resolution (Parker et al., 1994).

The structure of proaerolysin revealed an elongated molecule made up of four domains (Figs. 4A and 5A). Domain 1 forms a separate lobe that protrudes out from the rest of the molecule and is implicated in receptor binding (Rossjohn et al., 1998). The larger lobe is made up primarily of β-strands, arranged into eight antiparallel β-sheets, containing Domains 2–4. Domain 2 has been shown to bind glycans in the glycosylphophatidylinositol (GPI) anchored proteins (Hong et al., 2002). Initially, based on modeling of the crystal structure into electron micrographs of the assembled pore, Domain 4 was proposed as the membrane spanning region (Parker et al., 1994). Since then it has been shown that an amphipathic β-hairpin loop, referred to as the pre-stem loop, is the membrane spanning region (Iacovache et al., 2006). The core antiparallel β-sheets along with the membrane inserting loop were

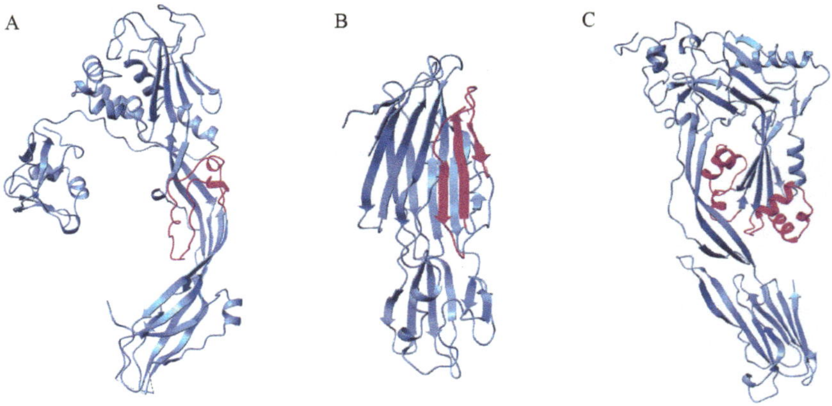

Fig. 4 The water-soluble structures of representative β-PFTs determined by crystallography. Only monomeric subunits shown. Membrane spanning-components in transmembrane pore shown in magenta. (A) Proaerolysin (PDB: 1PRE). (B) *S. aureus* α-hemolysin (PDB: 1IDJ). (C) Perfringolysin O (PDB: 1PFO).

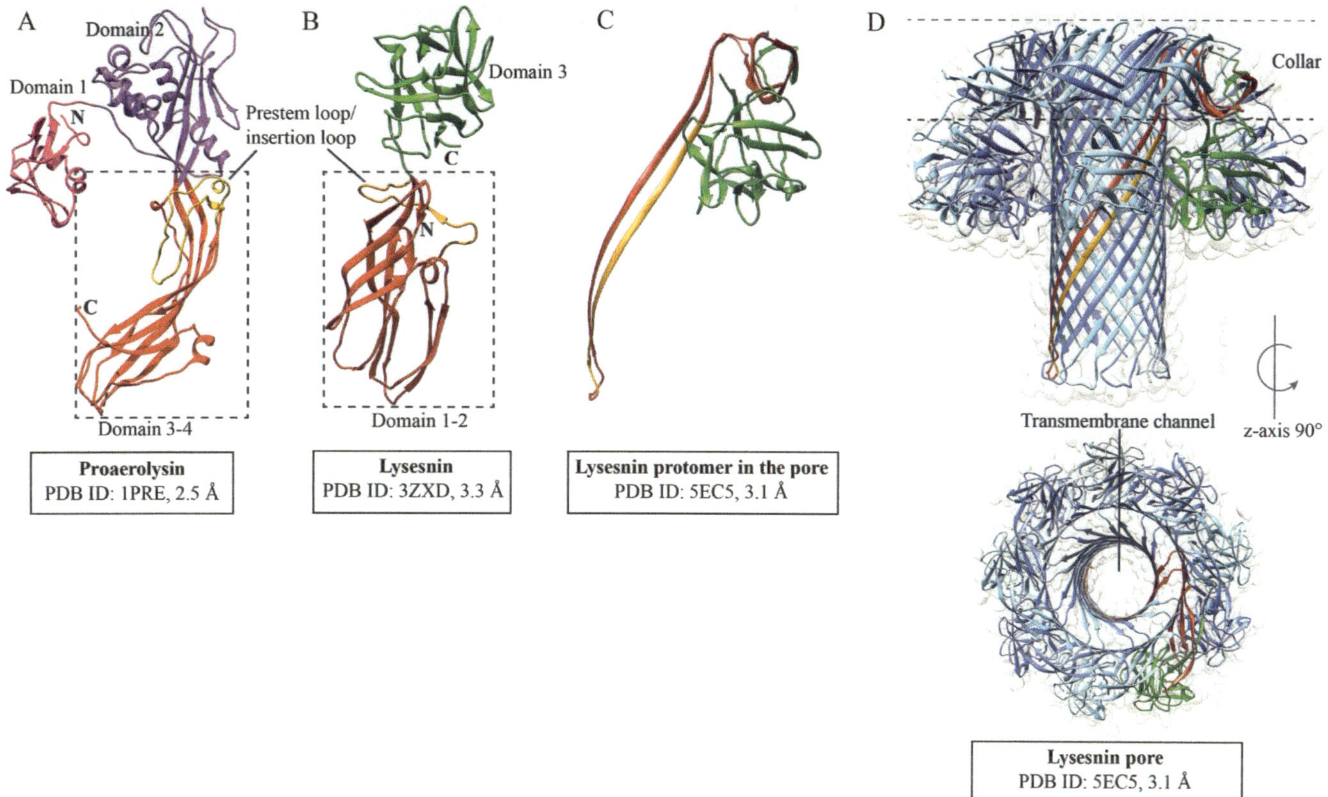

Fig. 5 Structures of aerolysin family members. (A) Proaerolysin (PDB: 1PRE) and (B) the soluble form of Lysenin (PDB: 3ZXD), highlighting the aerolysin fold (red) and the prestem or insertion loop (yellow). Aerolysin has two additional domains: Domains 1 (pink) and 2 (purple) which are involved in receptor binding, while lysenin has a single receptor binding domain (green). (C) Lysenin protomer in its membrane inserted conformation and (D) the lysenin pore (PDB: 5EC5) showing the structural rearrangements associated with Domains 1 and 2 and the insertion loop upon pore formation.

found to be conserved among many other proteins. Termed the aerolysin fold, it is a defining feature of members of this family (Szczesny et al., 2011).

The structures of many other PFTs bearing the aerolysin-like fold have since been solved by X-ray crystallography. These include the *C. perfringens* epsilon toxin (PDB: 1UYJ, 3ZJX) (Bokori-Brown et al., 2013; Cole et al., 2004) and enterotoxin (PDB: 3AM2, 2XH6, 2YHJ, 3ZIX, 3ZIW, 4P5H) (Briggs et al., 2011; Yelland et al., 2014), parasporin-1 (PDB: 2ZTB) (Akiba et al., 2009) and Cry51Aa1 (PDB: 4PKM, 5HD2) (Gowda et al., 2016; Xu et al., 2015) from *B. thuringiensis*, monalysin from *Pseudomonas entemophila* (PDB: 4MJT, 4MKO, 4MKQ) (Blemont et al., 2013; Leone et al., 2015) and BinB (PDB: 3WA1) (Srisucharitpanit et al., 2014) from *Lysinibacillus sphaericus*. Structures have also been solved for the non-bacterial aerolysin-like PFTs: hemolytic lectin (LSL) (PDB: 1W3A, 1W3G) (Mancheño, Tateno, Goldstein, Martínez-Ripoll, & Hermoso, 2005), lysenin (PDB: 3ZXD, 3ZX7, 3ZXG, 5GAQ) (De Colibus et al., 2012; Podobnik et al., 2016) (Fig. 5B), Dln1 (5DI0, 4ZNQ, 4ZNR, 4ZNO) (Jia et al., 2015) from *Laetiporous sulphureous*, *Eisenia fetida* and *Zebrafish*, respectively. In addition, the crystal structure of BmALP3 from *Bombina maxima,* an aerolysin-like protein lacking membrane pore-forming activity has been solved (PDB: 6LH8, 6LHZ) (Wang et al., 2020).

Many of these toxins were classed aerolysin-like only after crystal structure determination due to low sequence similarity (generally <25%) between family members. Crystal structures also revealed the presence and location of receptor binding domains, where observed differences in these domains account for different host cell specificities. An exception to this being monalysin (Leone et al., 2015) which does not contain a receptor binding domain. Instead, the protoxin exists as a dimer of nonamers arranged such that the membrane inserting regions are buried. This observation was confirmed by cryo-EM as well as size-exclusion chromatography multi-angle laser light scattering. These studies suggested that proteolytic activation of the toxin released the two assembled nonamers so that they could be recruited onto host cell membranes directly, excluding the need for receptor binding as a means of concentrating proteins on host membranes (Leone et al., 2015). Variations based on the secreted form were also observed. Some toxins, such as monalysin, are secreted as inactive protoxins that require proteolytic cleavage of a propeptide for activation (Leone et al., 2015) while others such as Dln1 (Jia et al., 2015) are produced in their active form.

Given the low sequence similarity between family members, structure determination required experimental phasing. The most common strategy involved soaking heavy atoms such as uranium, mercury and platinum into native crystals. Alternatively, Se-Met labeled recombinant proteins were used with the crystal structures solved either by SAD (De Colibus et al., 2012; Kitadokoro et al., 2011), MAD (Cole et al., 2004) or MIR with anomalous scattering (De Colibus et al., 2012). Structure resolution of the enterotoxin from *C. perfringens* combined SAD data from a samarium derivatized crystal with MR. The search model used for MR was the C-terminal region (K197-F319) of the toxin which had been solved previously (PDB ID; 2QUO) (Van Itallie, Betts, Smedley, McClane, & Anderson, 2008). This strategy yielded high quality maps that enabled the unequivocal tracing of the length of the chain from residues 36–319 (Kitadokoro et al., 2011). In contrast, Briggs et al. solved the structure of enterotoxin by MR alone having built a chimeric model by substituting the C-terminal region of enterotoxin into domain I of Ha3b, a subcomponent of the botulinum type C progenitor toxin (PDB ID: 2ZS6) (Nakamura et al., 2009). A solution was obtained by mutating all non-conserved residues to alanine. This, together with NCS averaging and density modification, provided an electron density map of sufficient quality for model building (Briggs et al., 2011).

Several structures of aerolysin family members in complex with their receptors also exist and provide the structural basis for toxin receptor binding. Toxin-receptor complexes were obtained either by soaking ligands into native crystals (De Colibus et al., 2012; Degiacomi et al., 2013; Jia et al., 2015) or by co-crystallization: (De Colibus et al., 2012; Jia et al., 2015; Yelland et al., 2014). Apart from revealing the receptor binding site, these crystal structures also explained the molecular basis for ligand specificity as was the case for Dln1 where Dln1:sugar structures (PDB IDs: 4ZNQ, 4ZNR, 4ZNO) (Jia et al., 2015) revealed the toxin's preference for mannose. Interestingly, locating the sugar binding site in proaerolysin (PDB ID: 3C0O) (Degiacomi et al., 2013), together with molecular modeling, helped place aerolysin protomers on membranes revealing conformational changes associated with oligomerization (Degiacomi et al., 2013).

Crystal structures of mutant toxins, solved by MR, have provided insight into the function and mechanism of pore formation. Crystal structures of aerolysin Y221G, a prepore-trapped mutant, together with MD simulations and electron microscopy provided insight into the conformational changes associated with pore formation (Degiacomi et al., 2013). Structures of the mutant H132D, which is unable to oligomerize, (PDB ID: 3G4N) (Iacovache et al., 2011) showed that the propeptide remains attached to the activated toxin even after cleavage and helped identify key interactions between Domain 4 and the propeptide. The H149A Epsilon toxin mutant was crystallized to verify conformation of a less toxic variant prior to undertaking biochemical studies (PDB ID: 3ZJX) (Bokori-Brown et al., 2013).

A mutant lysenin (E84Q/E85K/E92Q/E97S/D126G) initially generated for optimal translocation of DNA, was also able to form more stable oligomers. Therefore this mutant was used for crystallographic studies of the nonameric lysenin pore (PDB ID:5EC5) (Podobnik et al., 2016). Liposomes were used as a platform for pore assembly and pores were detergent extracted from liposomes using 10% DDM. Prior to crystallization the solubilized pores were buffer exchanged by size exclusion chromatography in a stepwise manner to a buffer solution containing 0.01% DDM. Crystallization was by vapor diffusion and crystals were derivatized by soaking in a mercury derivative. Structural resolution was by SAD (Podobnik et al., 2016). The crystal structure revealed that the pore shared the same mushroom shaped-structure as other aerolysin-like toxins (Iacovache et al., 2016) (Fig. 5D). Comparison of soluble lysenin with a lysensin protomer within the pore structure highlights the dramatic domain rearrangements associated with pore formation (Fig. 5B and C). To form the pore, three of the seven β-strands from the N-terminal aerolysin domain unfold before rearranging into a twisted β-hairpin that lines the lysenin pore. Although the remaining β-strands are structurally intact, they tilt by 45° with respect to the C-terminal receptor binding Domain 3 to form the collar of the membrane inserted pore. The lysenin pore structure also showed that Domain 3 is involved in protomer-protomer interactions revealing an additional role for this domain in oligomerization.

Crystal structures have enabled the engineering of the aerolysin family of toxins for biotechnological applications. Due to the highly charged nature of the pore, narrow pore diameter, pore stability and pH insensitivity, aerolysin has been studied extensively for its application as a biosensor (Cao, Liao, Yu, Tian, & Long, 2017). Although not yet commercially available, reports have shown that aerolysin can be engineered to detect a variety of molecules including oligonucleotides (Cao et al., 2017), methylated oligonucleotides (Yu, Cao, & Long, 2017), peptides and proteins (Li, Cao, Yang, & Long, 2019) and cofactors (Li, Wang, Ying, & Long, 2019). The aerolysin-like toxin Cry51Aa2 produced by *B. thuringiensis* has been engineered for improved insecticidal properties in the production of transgenic cotton plants (Gowda et al., 2016).

3.2 Hemolysin/Leukocidin family

While not a normal part of the human microbiota, *Staphylococcus aureus* is often found asymptomatically colonizing the skin or upper respiratory tract of healthy individuals. As well as being a commensal bacterium it is also an opportunistic pathogen, causing a range of diseases including skin and soft tissue infection and sepsis. Its ability to efficiently infect host tissues can largely be attributed to the diverse set of β-barrel PFTs that it secretes as virulence factors (Seilie & Bubeck Wardenburg, 2017). In *S. aureus* this family of proteins includes several bi-component leukocidins and one of the most well characterized PFTs, α-hemolysin. The protein family extends to PFTs from other human pathogens such as the NetB and Delta toxin proteins from *C. perfringens* and, more distantly, the *V. cholerae* cytolysin (VCC).

From a structural viewpoint the hemolysin/leukocidin toxins are unique as a class of β-barrel PFTs, in that the first crystal structure determined was of an intact toxin pore, rather than a soluble monomer which was determined many years later. Studies of the interaction of purified α-hemolysin from *S. aureus* with artificial membrane models demonstrated that the toxin readily formed stable heptamers in the presence of simple detergents like deoxycholate (Bhakdi, Fussle, & Tranum-Jensen, 1981). Sparse matrix crystallization screens that incorporated detergents, specifically intended for membrane proteins (Song & Gouaux, 1997), resulted in several different crystal forms for the heptamer and the final structure was solved to 1.8 Å resolution (PDB: 7AHL) by MIR using uranium, osmium and iridium derivatives (Song et al., 1996). The structure of the soluble hemolysin monomer wasn't determined for another 17 years, firstly as an antibody complex (PDB: 4IDJ) (Foletti et al., 2013) then as a free protein carrying a point mutation (H35A, PDB: 4YHD) (Sugawara et al., 2015) (Fig. 4B).

Features of the original α-hemolysin pore structure (Song et al., 1996) were named by the authors (Fig. 6) as the β-sandwich and Rim domains, making up the main body of the protein, with the Stem domain connected to the β-sandwich via a so-called Triangle region. At the N-terminus of the protein a short stretch of protein hooks across to interact with the adjacent monomer, forming what they termed the Amino latch. The β-sandwich and Rim domains create the Cap structure in the pore, with the Rim domain touching and partly embedded in the surface of the cell membrane. The Stem domain forms the transmembrane pore, a sheet consisting of 14 antiparallel β-strands with a pore size of around 26 Å. In the monomer

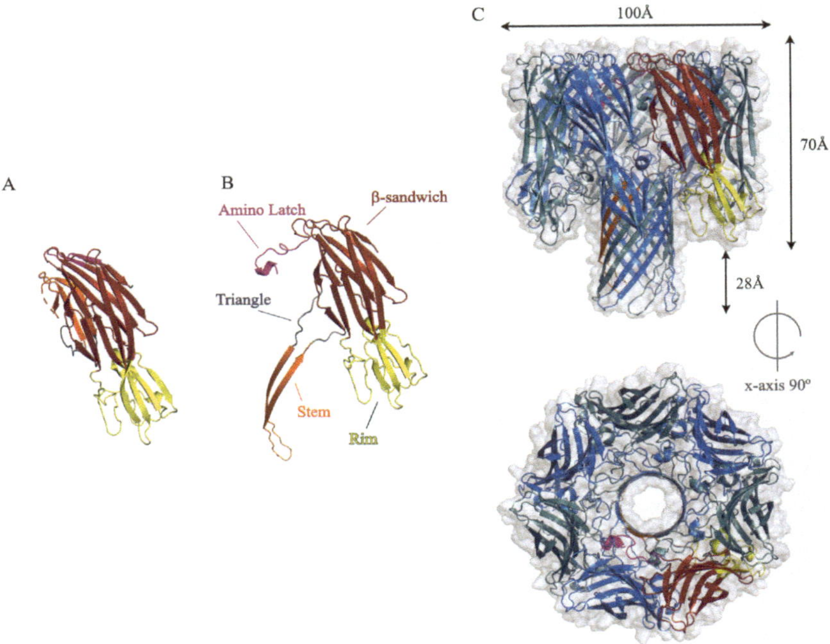

Fig. 6 Structure of (A) the *S. aureus* α-hemolysin soluble monomer (PDB: 4YHD) and (B) a protomer taken from the pore structure (PDB: 7AHL), colored by region and showing the restructuring of the Amino latch and Stem regions. The compact Stem in the soluble monomer unfolds to form an extended two-strand β-hairpin that, in the (C) heptameric pore forms the transmembrane β-barrel.

structures, the Amino latch forms a strand on the outside of the β-sandwich domain, with the Stem loop folded back as a small additional sheet in the same domain.

Structures for several other single component toxins from this family have been determined including the soluble monomer of Delta toxin (PDB: 2YGT) (Huyet et al., 2013) and both the soluble monomer and heptameric pore forms of NetB (PDB: 4I0N, 4H56) (Savva et al., 2013; Yan et al., 2013) and VCC (PDB: 1XEZ, 3O44) (De & Olson, 2011; Olson & Gouaux, 2005). These structures were primarily solved by MR using the previously published structures of some of the bi-component toxins (see below) except for VCC, where the sequence identity of ~16% to proteins of known structure was too low. Hence both VCC structures were solved by MAD analysis of SeMet-labeled VCC samples. Despite there being only four methionine residues in the 716 residue monomer, the map quality was sufficient to allow automatic tracing of 75% of the protein, with the rest of the monomer structure completed manually (Olson & Gouaux, 2005). The heptamer structure (PDB 3O44) used both the MAD-derived phase information and the existing monomeric structure for solving (De & Olson, 2011).

For the bi-component toxins, the two components are usually expressed as part of a single open reading frame, and classed as F or S subunits based on their sequence similarity to the subunits of the Panton-Valentine leukocidin (PVL), which originally received their designations from being in the "fast" or "slow" fractions during chromatography (Woodin, 1959). In the case of γ-hemolysin there are three subunits; two S-subunits HlgA, HlgC and a single F-subunit HlgB (Prevost et al., 1995; Tomita & Kamio, 1997). This allows assembly of two different active toxins, the classic γ-hemolysin HlgAB and a less hemolytic but more leukotoxic HlgCB (Tomita & Kamio, 1997). In each of the bi-component toxins, both subunits are required for pore formation, with the S subunit making the initial interaction with a target receptor, where it then recruits F subunits to form a heterodimer. Oligomerization of these heterodimers leads to octameric prepores which then insert to form the β-barrel pore (Seilie & Bubeck Wardenburg, 2017).

Structures have been determined for a number of these bi-component toxins, both for soluble F and S subunits and for some of the hetero-oligomeric pore and prepore forms. The first (PDB: 1LKF) was for HlgC, also known as LukF, solved by MAD using SeMet-labeled protein (Olson, Nariya, Yokota, Kamio, & Gouaux, 1999). This was followed by another F subunit structure, that of LukF-PV from PVL (PDB: 1PVL), solved to 2.0 Å resolution by MAD using a crystal soaked with $(NH_4)_2IrCl_6$ (Pedelacq et al., 1999). The first of the S-component structures, for LukS-PV, then followed (PDB: 1T5R), which was solved by MR using the LukF-PV structure (Guillet et al., 2004). Interestingly the LukS-PV structure contains eight protomers in the asymmetric unit, though as packed monomers rather than an octomeric pore. A disulphide cross-linked HlgA-HlgB dimer (PDB: 2QK7) was the next structure solved, using MR (Roblin et al., 2008) followed by the first complete bi-component pore structure (PDB: 3B07) for the same HlgA-HlgB subunit pairing (Yamashita et al., 2011). The most recent bi-component toxin structure was the HlgC-HlgB octomeric prepore (PDB: 4P1X) making use of a double HlgC mutation W177A/R198A that stabilizes the prepore form of the complex (Yamashita et al., 2014).

The existence of structures for members of the hemolysin/leukocidin family across many of the stages of pore formation provides deep insight into the conformational changes that occur and the details of the interactions that activate the process (Kaneko & Kamio, 2004; Yamashita et al., 2011; Yamashita et al., 2014). Assembly of the prepore occurs through oligomerization of monomers bound to the cellular surface, with monomer-monomer interactions causing the release of the Amino latch. The latch both stabilizes the monomer-monomer interface and destabilizes the Stem domain from its soluble conformation packed against the main β-sandwich. Further oligomerization assembles the appropriate number of subunits into the prepore, in which the stem loop has fully dissociated from its original position, but has only formed an initial membrane-anchored stub of the full stem with the rest of the stem loop disordered (Yamashita et al., 2014). The full pore forms with the stem completely inserting into the membrane, probably in a concerted single action. The transmembrane pore is consistent in size between the heptameric toxins and the octameric bi-component toxins, although the 14 antiparallel β-strands of the heptameric pores such as α-hemolysin and NetB is replaced with 16 strands for octameric pores created by the bi-component toxins.

The regular size, stability and ease of manipulation of the α-hemolysin pore revealed by crystallographic studies has provided the basis for technological applications, the most prominent of which has been the development of nanopore molecular sequencers for analyzing nucleotides (Howorka, Cheley, & Bayley, 2001; Maglia, Restrepo, Mikhailova, & Bayley, 2008). This technology, commercialized by Oxford Nanopore Technologies Ltd. in 2005, has revolutionized DNA and RNA sequencing emphasizing the important contribution of crystallographic analysis of PFTs to diverse areas of scientific endeavor.

3.3 CDC/MACPF/STNX superfamily

Cholesterol-dependent cytolysins (CDCs) are a family of PFTs that form very large β-barrel pores that are typically greater than 250 Å in diameter and consist of 35-40 monomers. Derived predominantly from a diverse range of Gram-positive bacterial species belonging to the *Bacillus, Clostridium, Streptococcus, Listeria, Arcanobacterium, Gardnerella* and *Lactobacillus* genera (Hotze & Tweten, 2012), CDCs have also recently been identified in some Gram-negative bacterial species (Hotze et al., 2013). Initial membrane binding is dependent on cholesterol-rich membranes, with the exception of CDCs that bind the human cell-surface glycoprotein CD59 (Gelber, Aguilar, Lewis, & Ratner, 2008; Giddings, Zhao, Sims, & Tweten, 2004; Wickham et al., 2011). After binding to cholesterol, or CD59, CDC monomers oligomerize into a stable, prepore intermediate, which is yet to fully insert into the membrane. Transition to the fully inserted pore requires numerous conformational changes; most significantly, unfurling of two helical bundles from each monomer into the β-strands that form the final β-barrel pore (Shatursky et al., 1999; Shepard et al., 1998). Currently, no CDC pore has been successfully crystallized, likely resulting from the heterogeneity of the large oligomeric pores due to small variations in the number of monomers involved. As a result, achieving the structures for CDC pore complexes will likely rely on cryo-EM methods, as were used to solve the structure of the pneumolysin (PLY) pore (van Pee et al., 2017). In contrast, there are a numerous monomeric CDC structures, which have revealed the key structural characteristics of this PFT family and helped drive what is now a deep understanding of the mechanism, albeit incomplete, of pore formation by CDCs.

The monomeric structure of perfringolysin O (PFO), the archetypical toxin of the family, was the first to be solved by crystallography (PDB: 1PFO) (Rossjohn, Feil, McKinstry, Tweten, & Parker, 1997). Crystals of PFO were originally very small and highly twinned; however, the addition of dioxane, a "poison" that partially inhibits nucleation (McPherson, Nguyen, Cudney, & Larson, 2011), enabled large crystals with no twinning and diffraction to 2.7 Å resolution (Feil, Rossjohn, Rohde, Tweten, & Parker, 1996). The structure was solved by MIR with four heavy metal derivatives, allowing an initial map to be achieved, the phases of which were substantially improved by solvent flattening and flipping, likely owing to the relatively high solvent content of the crystals (66%) (Rossjohn et al., 1997).

PFO is an elongated molecule, with its longest length 115 Å, that is rich in β-sheet (Figs. 4C and 7A). The structure consists of four domains (D1–D4), with D1, D2 and D3 being discontinuous and D4 being the only discrete, continuous domain. Analysis of the initial structure resulted in a region of highly conserved tryptophan-rich sequence in D4, also known as undecapeptide (UDP) sequence, being proposed to be the cholesterol-binding site of the molecule, in addition to potentially acting as a membrane-penetrating dagger (Rossjohn et al., 1997). It is now known that the UDP does not bind cholesterol or penetrate deeply into the membrane during pore formation. Instead, it only inserts slightly into the membrane during the initial membrane binding steps, providing membrane anchorage, and has also been shown to have an allosteric role in the mechanism (Dowd, Farrand, & Tweten, 2012; Heuck, Hotze, Tweten, & Johnson, 2000; Polekhina, Giddings, Tweten, & Parker, 2005). The cholesterol-recognition motif (CRM) has been identified as a pair

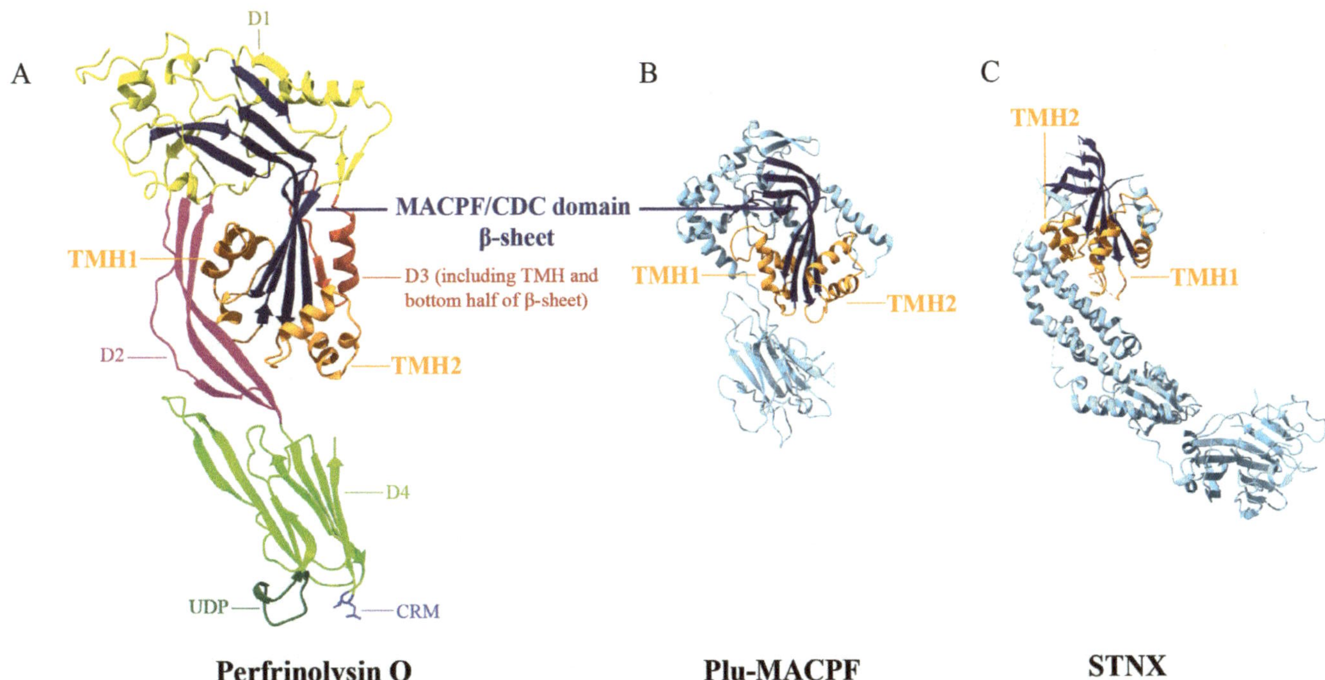

Fig. 7 Crystal structures of MACPF/CDC proteins reveal a structurally related superfamily of pore-forming proteins. (A) The structure of perfringolysin O (PFO) (PDB: 1PFO) revealed a four-domain structure (D1–D4), with a twisted β-sheet (navy) flanked by two transmembrane helices (TMH, orange), which unfurl to generate β-strands that form the β-barrel pore. (B) The structure of the bacterial MACPF protein Plu-MACPF (PDB: 2QP2). (C) The structure of stonefish toxin STNX (PDB: 4WVM).

of threonine-leucine residues in D4 (Farrand, LaChapelle, Hotze, Johnson, & Tweten, 2010). The structure of PFO also revealed two α-helical bundles located on either side of a β-sheet in D3, which were subsequently shown to undergo a α-helix-to-β-strand transition to generate amphipathic β-hairpins that insert into the membrane and form the final β-barrel pore (Shatursky et al., 1999; Shepard et al., 1998).

In addition to PFO, crystal structures for the water-soluble monomer of intermedilysin (ILY) (PDB: 1S3R, 4BIK, 5IMW, 5IMT) (Johnson, Brooks, Smith, Lea, & Bubeck, 2013; Lawrence et al., 2016; Polekhina et al., 2005), anthrolysin O (ALO) (PDB: 3CQF) (Bourdeau et al., 2009), suilysin (SLY) (PDB: 3HVN) (Xu et al., 2010), streptolysin O (SLO) (PDB: 4HSC) (Feil, Ascher, Kuiper, Tweten, & Parker, 2014), listerolysin O (LLO) (PDB: 4CDB) (Köster et al., 2014), PLY (4QQA, 4ZGH, 5CR6, 5AOD, 4QQQ, 5AOE, 5AOF) (Lawrence et al., 2015; Marshall et al., 2015; Park, Park, Bong, & Lee, 2016; van Pee et al., 2017; van Pee, Mulvihill, Müller, & Yildiz, 2016), vaginolysin (VLY) (PDB: 5IMY) (Lawrence et al., 2016) and desulfolysin (DLY) (PDB: 6NAL) (Wade et al., 2019) have been determined. All structures show high structural similarity with PFO, which is not surprising given CDCs share 40–80% sequence similarity. Differences between CDC structures to highlight include the lack of structural conservation within the UDP and differences in the degree to which the molecule is bent, due to variability in the angle between D1–D3 and D4. Despite the high structural similarity within CDCs, not all CDC crystal structures determined subsequent to PFO could be solved simply by MR with the PFO structure, presumably due in part to the variations in domain-domain orientations. For example, the first crystal structure achieved for ILY, and the second CDC structure overall, required a combination of SIR and MAD after MR attempts proved unsuccessful, despite 40% sequence identity between PFO and ILY (Polekhina et al., 2005; Polekhina, Giddings, Tweten, & Parker, 2004). Similarly, the structures of SLY and DLY required SeMet-based MAD phasing (Wade et al., 2019; Xu, Huang, et al., 2010).

An interesting feature observed in the crystal structures of PLY (Lawrence et al., 2015; Marshall et al., 2015), ILY (CD59-bound) (Johnson et al., 2013; Lawrence et al., 2016) and LLO (Köster et al., 2014) is the linear packing of monomers within the crystal lattice, essentially forming elongated linear CDC oligomers. These oligomers have been suggested to resemble an early state within the process of pore formation (Lawrence et al., 2015; Marshall et al., 2015), with the three aforementioned crystal structures forming monomer-monomer interactions through equivalent regions of the CDC molecule (Marshall et al., 2015). Furthermore, the number of contacts between monomers increase from ILY to LLO and finally PLY, suggesting these structures may even represent different stages within a hypothetical model for CDC oligomerization (Lawrence et al., 2015). SDS-AGE analysis of PLY crystals revealed the presence of stable, SDS-resistant oligomers (Lawrence et al., 2015), similar to the SDS-resistant prepore structures observed for CDCs in general.

While many CDCs have been crystallized, others have failed crystallization attempts. For example, exhaustive crystallization trials for non-CD59-bound VLY (Lawrence et al., 2016; Pleckaityte, 2020) and full-length lectinolysin (LLY) (Lawrence and Parker, unpublished) have not been able to produce diffraction quality crystals. However, LLY is a particularly interesting CDC, being the only family member to possess an additional domain (termed D0 or LLY^{lec}), which has been shown to bind fucosylated ligands (Farrand et al., 2008). To determine the position of this domain in reference to the four characteristic CDC domains, crystallographic studies of LLY^{lec} were conducted in conjunction with small angle x-ray scattering (SAXS) of full-length LLY. Fitting of monomeric CDC crystallographic models together with the LLY^{lec} crystal structure into the ab initio envelope generated from SAXS analysis revealed the additional domain to be located at the top of the molecule, abutting against the N-terminus of LLY within D1 (Feil et al., 2012).

One very surprising discovery from crystallographic studies, has been the identification of a common fold between CDCs and mammalian membrane attack complex/perforin family (MACPF) proteins. This was revealed in the first structures of MACPF proteins solved in 2007, the bacterial Plu-MACPF from *P. luminescens* (PDB: 2QP2) (Rosado et al., 2007) and the mammalian complement component C8α (PDB: 2QQH) (Hadders, Beringer, & Gros, 2007). This fold, commonly referred to as the MACPF/CDC domain, incorporates the highly twisted β-sheet spanning D3 and D1 in CDCs, and associated helical bundles (Fig. 7). Despite this structural similarity, no sequence similarity exists except for a conserved diglycine motif. A number of subsequent MACPF crystal structures have since been solved, with the family incorporating members from mammalian eukaryotes, non-mammalian eukaryotes, apicomplexan parasites and bacteria (Hadders et al., 2012; Law et al., 2010; Lovelace, Cooper, Sodetz, & Lebioda, 2011; Lukoyanova et al., 2015; Ni et al., 2018; Ni et al., 2020; Ni, Harlos, & Gilbert, 2016; Spicer et al., 2018; Xu et al., 2010; Zaitseva et al., 2019). The MACPF/CDC superfamily was extended further in 2015 when the crystal structure of stonustoxin (SNTX), a lethal toxin in stonefish venom, was determined (PDB: 4WVM) (Ellisdon et al., 2015). The N-terminal domain of SNTX is homologous with the MACPF/CDC domain, with phylogenetic analysis suggesting that SNTX may be representative of a third branch within the wider MACPF/CDC/STNX superfamily (Ellisdon et al., 2015).

4. Conclusion

Since the crystal structure of the first PFT in 1989, X-ray crystallography has exponentially increased our understanding of PFT structure and function. Crystal structures have continuously revealed structural similarity between PFTs where there is no sequence identity, and our current collection of PFTs may prove to only be a small fraction of what exists. While we have discussed much of the input of crystallography to PFT understanding here, we note that some PFT families have been omitted due to a lack of space, in particular the clostridial and *Bacillus* binary toxins (such as anthrax). The structures of these toxins have been previously reviewed by Parker and Feil (2005) to which we refer readers for more information. Some of the remaining questions about toxin function will likely continue to be answered in part by crystallography; however, other gaps in our structural knowledge may have to be answered using other complimentary techniques such as cryo-EM where protein crystals are difficult to grow. Overall, crystallography is very likely to remain a cornerstone in our investigation of PFTs for years to come, where methods and lessons learnt from previous crystallographic studies of PFTs will help accelerate future endeavors, including the development of PFTs for biotechnological applications.

Acknowledgments

We thank Professor Rod Tweten for a wonderful collaboration on pore-forming toxins for more than 20 years. This work was supported by grants from the Australian Research Council to M.W·P and C.J.M (DP160101874, DP200102871). Funding from the Victorian Government Operational Infrastructure Support Scheme to St Vincent's Institute is acknowledged. M.W.P. is a National Health and Medical Research Council of Australia Research Fellow (APP1117183). B.A.J. is supported by an Australian Government Research Training Program (RTP) Scholarship.

References

Akiba, T., Abe, Y., Kitada, S., Kusaka, Y., Ito, A., Ichimatsu, T., et al. (2009). Crystal structure of the parasporin-2 Bacillus thuringiensis toxin that recognizes cancer cells. *Journal of Molecular Biology*, *386*(1), 121–133. https://doi.org/10.1016/j.jmb.2008.12.002.

Athanasiadis, A., Anderluh, G., Macek, P., & Turk, D. (2001). Crystal structure of the soluble form of equinatoxin II, a pore-forming toxin from the sea anemone Actinia equina. *Structure*, *9*(4), 341–346. https://doi.org/10.1016/s0969-2126(01)00592-5.
Bakrac, B., Gutiérrez-Aguirre, I., Podlesek, Z., Sonnen, A. F., Gilbert, R. J., Macek, P., et al. (2008). Molecular determinants of sphingomyelin specificity of a eukaryotic pore-forming toxin. *The Journal of Biological Chemistry*, *283*(27), 18665–18677. https://doi.org/10.1074/jbc.M708747200.
Barreteau, H., Tiouajni, M., Graille, M., Josseaume, N., Bouhss, A., Patin, D., et al. (2012). Functional and structural characterization of PaeM, a colicin M-like bacteriocin produced by Pseudomonas aeruginosa. *The Journal of Biological Chemistry*, *287*(44), 37395–37405. https://doi.org/10.1074/jbc.M112.406439.
Baty, D., Frenette, M., Lloubès, R., Geli, V., Howard, S. P., Pattus, F., et al. (1988). Functional domains of colicin A. *Molecular Microbiology*, *2*(6), 807–811. https://doi.org/10.1111/j.1365-2958.1988.tb00092.x.
Baty, D., Knibiehler, M., Verheij, H., Pattus, F., Shire, D., Bernadac, A., et al. (1987). Site-directed mutagenesis of the COOH-terminal region of colicin A: Effect on secretion and voltage-dependent channel activity. *Proceedings of the National Academy of Sciences of the United States of America*, *84*(5), 1152–1156. https://doi.org/10.1073/pnas.84.5.1152.
Behrens, H. M., Lowe, E. D., Gault, J., Housden, N. G., Kaminska, R., Weber, T. M., et al. (2020). Pyocin S5 import into Pseudomonas aeruginosa reveals a generic mode of bacteriocin transport. *MBio*, *11*(2). https://doi.org/10.1128/mBio.03230-19.
Bellomio, A., Morante, K., Barlic, A., Gutiérrez-Aguirre, I., Viguera, A. R., & González-Mañas, J. M. (2009). Purification, cloning and characterization of fragaceatoxin C, a novel actinoporin from the sea anemone Actinia fragacea. *Toxicon*, *54*(6), 869–880. https://doi.org/10.1016/j.toxicon.2009.06.022.
Bernheimer, A. W., & Avigad, L. S. (1976). Properties of a toxin from the sea anemone Stoichacis helianthus, including specific binding to sphingomyelin. *Proceedings of the National Academy of Sciences of the United States of America*, *73*(2), 467–471. https://doi.org/10.1073/pnas.73.2.467.
Bhakdi, S., Fussle, R., & Tranum-Jensen, J. (1981). Staphylococcal alpha-toxin: oligomerization of hydrophilic monomers to form amphiphilic hexamers induced through contact with deoxycholate detergent micelles. *Proceedings of the National Academy of Sciences of the United States of America*, *78*(9), 5475–5479. https://doi.org/10.1073/pnas.78.9.5475.
Bhat, T. N., Bourne, P., Feng, Z., Gilliland, G., Jain, S., Ravichandran, V., et al. (2001). The PDB data uniformity project. *Nucleic Acids Research*, *29*(1), 214–218. https://doi.org/10.1093/nar/29.1.214.
Biesemans, A., Soskine, M., & Maglia, G. (2015). A protein rotaxane controls the translocation of proteins across a ClyA nanopore. *Nano Letters*, *15*(9), 6076–6081. https://doi.org/10.1021/acs.nanolett.5b02309.
Blemont, M., Vincentelli, R., Kellenberger, C., Opota, O., Lemaitre, B., Roussel, A., et al. (2013). Crystallization and preliminary X-ray analysis of monalysin, a novel β-pore-forming toxin from the entomopathogen Pseudomonas entomophila. *Acta Crystallographica. Section F, Structural Biology and Crystallization Communications*, *69*(Pt. 8), 930–933. https://doi.org/10.1107/s174430911301885x.
Boemare, N. E., Akhurst, R. J., & Mourant, R. G. (1993). DNA relatedness between Xenorhabdus spp. (Enterobacteriaceae), symbiotic bacteria of entomopathogenic nematodes, and a proposal to transfer xenorhabdus luminescens to a new genus, photorhabdus gen. nov. *International Journal of Systematic and Evolutionary Microbiology*, *43*(2), 249–255. https://doi.org/10.1099/00207713-43-2-249.
Bokori-Brown, M., Kokkinidou, M. C., Savva, C. G., Costa, S. F.d., Naylor, C. E., Cole, A. R., et al. (2013). Clostridium perfringens epsilon toxin H149A mutant as a

platform for receptor binding studies. *Protein Science*, *22*(5), 650–659. https://doi.org/10.1002/pro.2250.

Bourdeau, R. W., Malito, E., Chenal, A., Bishop, B. L., Musch, M. W., Villereal, M. L., et al. (2009). Cellular functions and X-ray structure of anthrolysin O, a cholesterol-dependent cytolysin secreted by Bacillus anthracis. *The Journal of Biological Chemistry*, *284*(21), 14645–14656. https://doi.org/10.1074/jbc.M807631200.

Bowen, D., Rocheleau, T. A., Blackburn, M., Andreev, O., Golubeva, E., Bhartia, R., et al. (1998). Insecticidal toxins from the bacterium Photorhabdus luminescens. *Science*, *280*(5372), 2129–2132. https://doi.org/10.1126/science.280.5372.2129.

Braun, V., Patzer, S. I., & Hantke, K. (2002). Ton-dependent colicins and microcins: modular design and evolution. *Biochimie*, *84*(5-6), 365–380. https://doi.org/10.1016/s0300-9084(02)01427-x.

Bräuning, B., Bertosin, E., Praetorius, F., Ihling, C., Schatt, A., Adler, A., et al. (2018). Structure and mechanism of the two-component α-helical pore-forming toxin YaxAB. *Nature Communications*, *9*(1), 1806. https://doi.org/10.1038/s41467-018-04139-2.

Bräuning, B., & Groll, M. (2018). Structural and mechanistic features of ClyA-like α-pore-forming toxins. *Toxins (Basel)*, *10*(9), 343. https://doi.org/10.3390/toxins10090343.

Briggs, D. C., Naylor, C. E., Smedley, J. G., Lukoyanova, N., Robertson, S., Moss, D. S., et al. (2011). Structure of the food-poisoning clostridium perfringens enterotoxin reveals similarity to the aerolysin-like pore-forming toxins. *Journal of Molecular Biology*, *413*(1), 138–149.

Brouwer, J. M., Westphal, D., Dewson, G., Robin, A. Y., Uren, R. T., Bartolo, R., et al. (2014). Bak core and latch domains separate during activation, and freed core domains form symmetric homodimers. *Molecular Cell*, *55*(6), 938–946. https://doi.org/10.1016/j.molcel.2014.07.016.

Brunden, K. R., Uratani, Y., & Cramer, W. A. (1984). Dependence of the conformation of a colicin E1 channel-forming peptide on acidic pH and solvent polarity. *The Journal of Biological Chemistry*, *259*(12), 7682–7687.

Brünger, A. T., Kuriyan, J., & Karplus, M. (1987). Crystallographic R factor refinement by molecular dynamics. *Science*, *235*(4787), 458–460. https://doi.org/10.1126/science.235.4787.458.

Buchanan, S. K., Lukacik, P., Grizot, S., Ghirlando, R., Ali, M. M., Barnard, T. J., et al. (2007). Structure of colicin I receptor bound to the R-domain of colicin Ia: Implications for protein import. *The EMBO Journal*, *26*(10), 2594–2604. https://doi.org/10.1038/sj.emboj.7601693.

Busby, J. N., Panjikar, S., Landsberg, M. J., Hurst, M. R., & Lott, J. S. (2013). The BC component of ABC toxins is an RHS-repeat-containing protein encapsulation device. *Nature*, *501*(7468), 547–550. https://doi.org/10.1038/nature12465.

Cao, C., Liao, D. F., Yu, J., Tian, H., & Long, Y. T. (2017). Construction of an aerolysin nanopore in a lipid bilayer for single-oligonucleotide analysis. *Nature Protocols*, *12*(9), 1901–1911. https://doi.org/10.1038/nprot.2017.077.

Carretero, G. P. B., Vicente, E. F., Cilli, E. M., Alvarez, C. M., Jenssen, H., & Schreier, S. (2018). Dissecting the mechanism of action of actinoporins. Role of the N-terminal amphipathic α-helix in membrane binding and pore activity of sticholysins I and II. *PLoS One*, *13*(8), e0202981. https://doi.org/10.1371/journal.pone.0202981.

Choe, S., Bennett, M. J., Fujii, G., Curmi, P. M., Kantardjieff, K. A., Collier, R. J., et al. (1992). The crystal structure of diphtheria toxin. *Nature*, *357*(6375), 216–222. https://doi.org/10.1038/357216a0.

Cilli, E. M., Pigossi, F. T., Crusca, E., Jr., Ros, U., Martinez, D., Lanio, M. E., et al. (2007). Correlations between differences in amino-terminal sequences and different hemolytic activity of sticholysins. *Toxicon*, *50*(8), 1201–1204. https://doi.org/10.1016/j.toxicon.2007.07.013.

Cole, A. R., Gibert, M., Popoff, M., Moss, D. S., Titball, R. W., & Basak, A. K. (2004). Clostridium perfringens ε-toxin shows structural similarity to the pore-forming toxin aerolysin. *Nature Structural & Molecular Biology*, *11*(8), 797–798. https://doi.org/10.1038/nsmb804.

Czabotar, P. E., Westphal, D., Dewson, G., Ma, S., Hockings, C., Fairlie, W. D., et al. (2013). Bax crystal structures reveal how BH3 domains activate Bax and nucleate its oligomerization to induce apoptosis. *Cell*, *152*(3), 519–531. https://doi.org/10.1016/j.cell.2012.12.031.

De Colibus, L., Sonnen, A. F. P., Morris, K. J., Siebert, C. A., Abrusci, P., Plitzko, J., et al. (2012). Structures of lysenin reveal a shared evolutionary origin for pore-forming proteins and its mode of sphingomyelin recognition. *Structure*, *20*(9), 1498–1507. https://doi.org/10.1016/j.str.2012.06.011.

De, S., & Olson, R. (2011). Crystal structure of the Vibrio cholerae cytolysin heptamer reveals common features among disparate pore-forming toxins. *Proceedings of the National Academy of Sciences of the United States of America*, *108*(18), 7385–7390. https://doi.org/10.1073/pnas.1017442108.

Degiacomi, M. T., Iacovache, I., Pernot, L., Chami, M., Kudryashev, M., Stahlberg, H., et al. (2013). Molecular assembly of the aerolysin pore reveals a swirling membrane-insertion mechanism. *Nature Chemical Biology*, *9*(10), 623–629. https://doi.org/10.1038/nchembio.

del Castillo, F. J., Leal, S. C., Moreno, F., & del Castillo, I. (1997). The Escherichia coli K-12 sheA gene encodes a 34-kDa secreted haemolysin. *Molecular Microbiology*, *25*(1), 107–115. https://doi.org/10.1046/j.1365-2958.1997.4391813.x.

Dementiev, A., Board, J., Sitaram, A., Hey, T., Kelker, M. S., Xu, X., et al. (2016). The pesticidal Cry6Aa toxin from Bacillus thuringiensis is structurally similar to HlyE-family alpha pore-forming toxins. *BMC Biology*, *14*(1), 71. https://doi.org/10.1186/s12915-016-0295-9.

Dodd, S. J., Hurst, M. R., Glare, T. R., O'Callaghan, M., & Ronson, C. W. (2006). Occurrence of sep insecticidal toxin complex genes in Serratia spp. and Yersinia frederiksenii. *Applied and Environmental Microbiology*, *72*(10), 6584–6592. https://doi.org/10.1128/aem.00954-06.

Dongre, M., Singh, B., Aung, K. M., Larsson, P., Miftakhova, R., Persson, K., et al. (2018). Flagella-mediated secretion of a novel Vibrio cholerae cytotoxin affecting both vertebrate and invertebrate hosts. *Communications Biology*, *1*, 59. https://doi.org/10.1038/s42003-018-0065-z.

Dowd, K. J., Farrand, A. J., & Tweten, R. K. (2012). The cholesterol-dependent cytolysin signature motif: a critical element in the allosteric pathway that couples membrane binding to pore assembly. *PLoS Pathogens*, *8*(7), e1002787. https://doi.org/10.1371/journal.ppat.1002787.

Duché, D., Baty, D., Chartier, M., & Letellier, L. (1994). Unfolding of colicin A during its translocation through the Escherichia coli envelope as demonstrated by disulfide bond engineering. *The Journal of Biological Chemistry*, *269*(40), 24820–24825.

Duché, D., Parker, M. W., González-Mañas, J. M., Pattus, F., & Baty, D. (1994). Uncoupled steps of the colicin A pore formation demonstrated by disulfide bond engineering. *The Journal of Biological Chemistry*, *269*(9), 6332–6339.

Eifler, N., Vetsch, M., Gregorini, M., Ringler, P., Chami, M., Philippsen, A., et al. (2006). Cytotoxin ClyA from Escherichia coli assembles to a 13-meric pore independent of its redox-state. *The EMBO Journal*, *25*(11), 2652–2661. https://doi.org/10.1038/sj.emboj.7601130.

Elkins, P., Bunker, A., Cramer, W. A., & Stauffacher, C. V. (1997). A mechanism for toxin insertion into membranes is suggested by the crystal structure of the channel-forming domain of colicin E1. *Structure*, *5*(3), 443–458. https://doi.org/10.1016/s0969-2126(97)00200-1.

Ellisdon, A. M., Reboul, C. F., Panjikar, S., Huynh, K., Oellig, C. A., Winter, K. L., et al. (2015). Stonefish toxin defines an ancient branch of the perforin-like superfamily. *Proceedings of the National Academy of Sciences of the United States of America, 112*(50), 15360–15365. https://doi.org/10.1073/pnas.1507622112.

Farrand, A. J., LaChapelle, S., Hotze, E. M., Johnson, A. E., & Tweten, R. K. (2010). Only two amino acids are essential for cytolytic toxin recognition of cholesterol at the membrane surface. *Proceedings of the National Academy of Sciences of the United States of America, 107*(9), 4341–4346. https://doi.org/10.1073/pnas.0911581107.

Farrand, S., Hotze, E., Friese, P., Hollingshead, S. K., Smith, D. F., Cummings, R. D., et al. (2008). Characterization of a streptococcal cholesterol-dependent cytolysin with a lewis y and b specific lectin domain. *Biochemistry, 47*(27), 7097–7107. https://doi.org/10.1021/bi8005835.

Feil, S. C., Ascher, D. B., Kuiper, M. J., Tweten, R. K., & Parker, M. W. (2014). Structural studies of Streptococcus pyogenes streptolysin O provide insights into the early steps of membrane penetration. *Journal of Molecular Biology, 426*(4), 785–792. https://doi.org/10.1016/j.jmb.2013.11.020.

Feil, S. C., Lawrence, S., Mulhern, T. D., Holien, J. K., Hotze, E. M., Farrand, S., et al. (2012). Structure of the lectin regulatory domain of the cholesterol-dependent cytolysin lectinolysin reveals the basis for its lewis antigen specificity. *Structure, 20*(2), 248–258. https://doi.org/10.1016/j.str.2011.11.017.

Feil, S. C., Rossjohn, J., Rohde, K., Tweten, R. K., & Parker, M. W. (1996). Crystallization and preliminary X-ray analysis of a thiol-activated cytolysin. *FEBS Letters, 397*(2-3), 290–292. https://doi.org/10.1016/s0014-5793(96)01200-8.

ffrench-Constant, R. H., & Bowen, D. J. (2000). Novel insecticidal toxins from nematode-symbiotic bacteria. *Cellular and Molecular Life Sciences, 57*(5), 828–833. https://doi.org/10.1007/s000180050044.

Foletti, D., Strop, P., Shaughnessy, L., Hasa-Moreno, A., Casas, M. G., Russell, M., et al. (2013). Mechanism of action and in vivo efficacy of a human-derived antibody against Staphylococcus aureus alpha-hemolysin. *Journal of Molecular Biology, 425*(10), 1641–1654. https://doi.org/10.1016/j.jmb.2013.02.008.

Franceschini, L., Brouns, T., Willems, K., Carlon, E., & Maglia, G. (2016). DNA translocation through nanopores at physiological ionic strengths requires precise nanoscale engineering. *ACS Nano, 10*(9), 8394–8402. https://doi.org/10.1021/acsnano.6b03159.

Ganash, M., Phung, D., Sedelnikova, S. E., Lindbäck, T., Granum, P. E., & Artymiuk, P. J. (2013). Structure of the NheA component of the Nhe toxin from Bacillus cereus: implications for function. *PLoS One, 8*(9), e74748. https://doi.org/10.1371/journal.pone.0074748.

Gatsogiannis, C., Lang, A. E., Meusch, D., Pfaumann, V., Hofnagel, O., Benz, R., et al. (2013). A syringe-like injection mechanism in Photorhabdus luminescens toxins. *Nature, 495*(7442), 520–523. https://doi.org/10.1038/nature11987.

Gatsogiannis, C., Merino, F., Prumbaum, D., Roderer, D., Leidreiter, F., Meusch, D., et al. (2016). Membrane insertion of a Tc toxin in near-atomic detail. *Nature Structural & Molecular Biology, 23*(10), 884–890. https://doi.org/10.1038/nsmb.3281.

Gatsogiannis, C., Merino, F., Roderer, D., Balchin, D., Schubert, E., Kuhlee, A., et al. (2018). Tc toxin activation requires unfolding and refolding of a β-propeller. *Nature, 563*(7730), 209–213. https://doi.org/10.1038/s41586-018-0556-6.

Gelber, S. E., Aguilar, J. L., Lewis, K. L., & Ratner, A. J. (2008). Functional and phylogenetic characterization of Vaginolysin, the human-specific cytolysin from Gardnerella vaginalis. *Journal of Bacteriology, 190*(11), 3896–3903. https://doi.org/10.1128/jb.01965-07.

Ghosh, P., Mel, S. F., & Stroud, R. M. (1994). The domain structure of the ion channel-forming protein colicin Ia. *Nature Structural Biology, 1*(9), 597–604. https://doi.org/10.1038/nsb0994-597.

Giddings, K. S., Zhao, J., Sims, P. J., & Tweten, R. K. (2004). Human CD59 is a receptor for the cholesterol-dependent cytolysin intermedilysin. *Nature Structural & Molecular Biology*, *11*(12), 1173–1178. https://doi.org/10.1038/nsmb862.

Gowda, A., Rydel, T. J., Wollacott, A. M., Brown, R. S., Akbar, W., Clark, T. L., et al. (2016). A transgenic approach for controlling Lygus in cotton. *Nature Communications*, 7(1), 12213. https://doi.org/10.1038/ncomms12213.

Green, J., & Baldwin, M. L. (1997). The molecular basis for the differential regulation of the hlyE-encoded haemolysin of Escherichia coli by FNR and HlyX lies in the improved activating region 1 contact of HlyX. *Microbiology*, *143*(Pt. 12), 3785–3793. https://doi.org/10.1099/00221287-143-12-3785.

Griko, Y. V., Zakharov, S. D., & Cramer, W. A. (2000). Structural stability and domain organization of colicin E1. *Journal of Molecular Biology*, *302*(4), 941–953. https://doi.org/10.1006/jmbi.2000.4504.

Guillet, V., Roblin, P., Werner, S., Coraiola, M., Menestrina, G., Monteil, H., et al. (2004). Crystal structure of leucotoxin S component: New insight into the Staphylococcal beta-barrel pore-forming toxins. *The Journal of Biological Chemistry*, *279*(39), 41028–41037. https://doi.org/10.1074/jbc.M406904200.

Hadders, M. A., Beringer, D. X., & Gros, P. (2007). Structure of C8alpha-MACPF reveals mechanism of membrane attack in complement immune defense. *Science*, *317*(5844), 1552–1554. https://doi.org/10.1126/science.1147103.

Hadders, M. A., Bubeck, D., Roversi, P., Hakobyan, S., Forneris, F., Morgan, B. P., et al. (2012). Assembly and regulation of the membrane attack complex based on structures of C5b6 and sC5b9. *Cell Reports*, *1*(3), 200–207. https://doi.org/10.1016/j.celrep.2012.02.003.

Heuck, A. P., Hotze, E. M., Tweten, R. K., & Johnson, A. E. (2000). Mechanism of membrane insertion of a multimeric beta-barrel protein: Perfringolysin O creates a pore using ordered and coupled conformational changes. *Molecular Cell*, *6*(5), 1233–1242. https://doi.org/10.1016/s1097-2765(00)00119-2.

Hilsenbeck, J. L., Park, H., Chen, G., Youn, B., Postle, K., & Kang, C. (2004). Crystal structure of the cytotoxic bacterial protein colicin B at 2.5 A resolution. *Molecular Microbiology*, *51*(3), 711–720. https://doi.org/10.1111/j.1365-2958.2003.03884.x.

Holm, L., & Sander, C. (1993). Structural alignment of globins, phycocyanins and colicin A. *FEBS Letters*, *315*(3), 301–306. https://doi.org/10.1016/0014-5793(93)81183-z.

Hong, Q., Gutierrez-Aguirre, I., Barlic, A., Malovrh, P., Kristan, K., Podlesek, Z., et al. (2002). Two-step membrane binding by Equinatoxin II, a pore-forming toxin from the sea anemone, involves an exposed aromatic cluster and a flexible helix. *The Journal of Biological Chemistry*, *277*(44), 41916–41924. https://doi.org/10.1074/jbc.M204625200.

Hong, Y., Ohishi, K., Inoue, N., Kang, J. Y., Shime, H., Horiguchi, Y., et al. (2002). Requirement of N-glycan on GPI-anchored proteins for efficient binding of aerolysin but not Clostridium septicum alpha-toxin. *The EMBO Journal*, *21*(19), 5047–5056. https://doi.org/10.1093/emboj/cdf508.

Hotze, E. M., Le, H. M., Sieber, J. R., Bruxvoort, C., McInerney, M. J., & Tweten, R. K. (2013). Identification and characterization of the first cholesterol-dependent cytolysins from Gram-negative bacteria. *Infection and Immunity*, *81*(1), 216–225. https://doi.org/10.1128/iai.00927-12.

Hotze, E. M., & Tweten, R. K. (2012). Membrane assembly of the cholesterol-dependent cytolysin pore complex. *Biochimica et Biophysica Acta*, *1818*(4), 1028–1038. https://doi.org/10.1016/j.bbamem.2011.07.036.

Howorka, S., Cheley, S., & Bayley, H. (2001). Sequence-specific detection of individual DNA strands using engineered nanopores. *Nature Biotechnology*, *19*(7), 636–639. https://doi.org/10.1038/90236.

Huang, G., Voet, A., & Maglia, G. (2019). FraC nanopores with adjustable diameter identify the mass of opposite-charge peptides with 44 dalton resolution. *Nature Communications, 10*(1), 835. https://doi.org/10.1038/s41467-019-08761-6.

Huang, G., Willems, K., Soskine, M., Wloka, C., & Maglia, G. (2017). Electro-osmotic capture and ionic discrimination of peptide and protein biomarkers with FraC nanopores. *Nature Communications, 8*(1), 935. https://doi.org/10.1038/s41467-017-01006-4.

Huyet, J., Naylor, C. E., Savva, C. G., Gibert, M., Popoff, M. R., & Basak, A. K. (2013). Structural insights into Clostridium perfringens delta toxin pore formation. *PLoS One, 8*(6), e66673. https://doi.org/10.1371/journal.pone.0066673.

Iacovache, I., Carlo, S. D., Cirauqui, N., Peraro, M. D., van der Goot, F. G., & Zuber, B. (2016). Cryo-EM structure of aerolysin variants reveals a novel protein fold and the pore-formation process. *Nature communications*, 7(1), 12062. https://doi.org/10.1038/ncomms12062. PMID - 27405240.

Iacovache, I., Degiacomi, M. T., Pernot, L., Ho, S., Schiltz, M., Peraro, M. D., et al. (2011). Dual chaperone role of the C-terminal propeptide in folding and oligomerization of the pore-forming toxin aerolysin. *PLoS Pathogens*, 7(7), e1002135. https://doi.org/10.1371/journal.ppat.1002135.

Iacovache, I., Paumard, P., Scheib, H., Lesieur, C., Sakai, N., Matile, S., et al. (2006). A rivet model for channel formation by aerolysin-like pore-forming toxins. *The EMBO Journal, 25*(3), 457–466. https://doi.org/10.1038/sj.emboj.7600959.

Jakes, K. S., & Finkelstein, A. (2010). The colicin Ia receptor, Cir, is also the translocator for colicin Ia. *Molecular Microbiology, 75*(3), 567–578. https://doi.org/10.1111/j.1365-2958.2009.06966.x.

Jia, N., Liu, N., Cheng, W., Jiang, Y. L., Sun, H., Chen, L. L., et al. (2015). Structural basis for receptor recognition and pore formation of a zebrafish aerolysin-like protein. *EMBO Reports, 17*(2), 235–248. https://doi.org/10.15252/embr.201540851.

Johnson, S., Brooks, N. J., Smith, R. A., Lea, S. M., & Bubeck, D. (2013). Structural basis for recognition of the pore-forming toxin intermedilysin by human complement receptor CD59. *Cell Reports, 3*(5), 1369–1377. https://doi.org/10.1016/j.celrep.2013.04.029.

Kaneko, J., & Kamio, Y. (2004). Bacterial two-component and hetero-heptameric pore-forming cytolytic toxins: structures, pore-forming mechanism, and organization of the genes. *Bioscience, Biotechnology, and Biochemistry, 68*(5), 981–1003. https://doi.org/10.1271/bbb.68.981.

Kim, Y., Valentine, K., Opella, S. J., Schendel, S. L., & Cramer, W. A. (1998). Solid-state NMR studies of the membrane-bound closed state of the colicin E1 channel domain in lipid bilayers. *Protein Science*, 7(2), 342–348. https://doi.org/10.1002/pro.5560070214.

Kitadokoro, K., Nishimura, K., Kamitani, S., Fukui-Miyazaki, A., Toshima, H., Abe, H., et al. (2011). Crystal structure of Clostridium perfringens enterotoxin displays features of beta-pore-forming toxins. *The Journal of Biological Chemistry, 286*(22), 19549–19555. https://doi.org/10.1074/jbc.m111.228478.

Komárek, J., Ivanov Kavková, E., Houser, J., Horáčková, A., Ždánská, J., Demo, G., et al. (2018). Structure and properties of AB21, a novel Agaricus bisporus protein with structural relation to bacterial pore-forming toxins. *Proteins, 86*(9), 897–911. https://doi.org/10.1002/prot.25522.

Köster, S., van Pee, K., Hudel, M., Leustik, M., Rhinow, D., Kühlbrandt, W., et al. (2014). Crystal structure of listeriolysin O reveals molecular details of oligomerization and pore formation. *Nature Communications, 5*, 3690. https://doi.org/10.1038/ncomms4690.

Kristan, K., Podlesek, Z., Hojnik, V., Gutiérrez-Aguirre, I., Guncar, G., Turk, D., et al. (2004). Pore formation by equinatoxin, a eukaryotic pore-forming toxin, requires a flexible N-terminal region and a stable beta-sandwich. *The Journal of Biological Chemistry, 279*(45), 46509–46517. https://doi.org/10.1074/jbc.M406193200.

Kristan, K., Viero, G., Macek, P., Dalla Serra, M., & Anderluh, G. (2007). The equinatoxin N-terminus is transferred across planar lipid membranes and helps to stabilize the transmembrane pore. *The FEBS Journal, 274*(2), 539–550. https://doi.org/10.1111/j.1742-4658.2006.05608.x.

Kristan, K. C., Viero, G., Dalla Serra, M., Macek, P., & Anderluh, G. (2009). Molecular mechanism of pore formation by actinoporins. *Toxicon, 54*(8), 1125–1134. https://doi.org/10.1016/j.toxicon.2009.02.026.

Kurisu, G., Zakharov, S. D., Zhalnina, M. V., Bano, S., Eroukova, V. Y., Rokitskaya, T. I., et al. (2003). The structure of BtuB with bound colicin E3 R-domain implies a translocon. *Nature Structural Biology, 10*(11), 948–954. https://doi.org/10.1038/nsb997.

Lakey, J. H., Duché, D., González-Mañas, J. M., Baty, D., & Pattus, F. (1993). Fluorescence energy transfer distance measurements. The hydrophobic helical hairpin of colicin A in the membrane bound state. *Journal of Molecular Biology, 230*(3), 1055–1067. https://doi.org/10.1006/jmbi.1993.1218.

Lakey, J. H., Massotte, D., Heitz, F., Dasseux, J. L., Faucon, J. F., Parker, M. W., et al. (1991). Membrane insertion of the pore-forming domain of colicin A. A spectroscopic study. *European Journal of Biochemistry, 196*(3), 599–607. https://doi.org/10.1111/j.1432-1033.1991.tb15855.x.

Landsberg, M. J., Jones, S. A., Rothnagel, R., Busby, J. N., Marshall, S. D., Simpson, R. M., et al. (2011). 3D structure of the Yersinia entomophaga toxin complex and implications for insecticidal activity. *Proceedings of the National Academy of Sciences of the United States of America, 108*(51), 20544–20549. https://doi.org/10.1073/pnas.1111155108.

Lang, A. E., Konukiewitz, J., Aktories, K., & Benz, R. (2013). TcdA1 of Photorhabdus luminescens: Electrophysiological analysis of pore formation and effector binding. *Biophysical Journal, 105*(2), 376–384. https://doi.org/10.1016/j.bpj.2013.06.003.

Law, R. H., Lukoyanova, N., Voskoboinik, I., Caradoc-Davies, T. T., Baran, K., Dunstone, M. A., et al. (2010). The structural basis for membrane binding and pore formation by lymphocyte perforin. *Nature, 468*(7322), 447–451. https://doi.org/10.1038/nature09518.

Lawrence, S. L., Feil, S. C., Morton, C. J., Farrand, A. J., Mulhern, T. D., Gorman, M. A., et al. (2015). Crystal structure of Streptococcus pneumoniae pneumolysin provides key insights into early steps of pore formation. *Scientific Reports, 5*, 14352. https://doi.org/10.1038/srep14352.

Lawrence, S. L., Gorman, M. A., Feil, S. C., Mulhern, T. D., Kuiper, M. J., Ratner, A. J., et al. (2016). Structural basis for receptor recognition by the human CD59-responsive cholesterol-dependent cytolysins. *Structure, 24*(9), 1488–1498. https://doi.org/10.1016/j.str.2016.06.017.

Lazzaroni, J. C., Dubuisson, J. F., & Vianney, A. (2002). The Tol proteins of Escherichia coli and their involvement in the translocation of group A colicins. *Biochimie, 84*(5-6), 391–397. https://doi.org/10.1016/s0300-9084(02)01419-0.

Lee, S. C., Stoilova-McPhie, S., Baxter, L., Fülöp, V., Henderson, J., Rodger, A., et al. (2007). Structural characterisation of the insecticidal toxin XptA1, reveals a 1.15 MDa tetramer with a cage-like structure. *Journal of Molecular Biology, 366*(5), 1558–1568. https://doi.org/10.1016/j.jmb.2006.12.057.

Leidreiter, F., Roderer, D., Meusch, D., Gatsogiannis, C., Benz, R., & Raunser, S. (2019). Common architecture of Tc toxins from human and insect pathogenic bacteria. *Science Advances, 5*(10), eaax6497. https://doi.org/10.1126/sciadv.aax6497.

Leone, P., Bebeacua, C., Opota, O., Kellenberger, C., Klaholz, B., Orlov, I., et al. (2015). X-ray and cryo-electron microscopy structures of monalysin pore-forming toxin reveal multimerization of the pro-form. *The Journal of Biological Chemistry, 290*(21), 13191–13201. https://doi.org/10.1074/jbc.m115.646109.

Li, J. D., Carroll, J., & Ellar, D. J. (1991). Crystal structure of insecticidal delta-endotoxin from Bacillus thuringiensis at 2.5 A resolution. *Nature, 353*(6347), 815–821. https://doi.org/10.1038/353815a0.

Li, M.-Y., Wang, Y.-Q., Ying, Y.-L., & Long, Y.-T. (2019). Revealing the transient conformations of a single flavin adenine dinucleotide using an aerolysin nanopore. *Chemical Science, 10*(44), 10400–10404. https://doi.org/10.1039/C9SC03163D.

Li, S., Cao, C., Yang, J., & Long, Y.-T. (2019). Detection of peptides with different charges and lengths by using the aerolysin nanopore. *ChemElectroChem, 6*(1), 126–129. https://doi.org/10.1002/celc.201800288.

Li, X., Lee, K. H., Shorkey, S., Chen, J., & Chen, M. (2020). Different anomeric sugar bound states of maltose binding protein resolved by a cytolysin A nanopore tweezer. *ACS Nano, 14*(2), 1727–1737. https://doi.org/10.1021/acsnano.9b07385.

Lovelace, L. L., Cooper, C. L., Sodetz, J. M., & Lebioda, L. (2011). Structure of human C8 protein provides mechanistic insight into membrane pore formation by complement. *The Journal of Biological Chemistry, 286*(20), 17585–17592. https://doi.org/10.1074/jbc.M111.219766.

Lu, B., Stokes, C., Fahie, M., Chen, M., Golovchenko, J. A., & Hau, L. V. (2018). Protein motion and configurations in a form-fitting nanopore: Avidin in ClyA. *Biophysical Journal, 115*(5), 801–808. https://doi.org/10.1016/j.bpj.2018.07.024.

Ludwig, A., Bauer, S., Benz, R., Bergmann, B., & Goebel, W. (1999). Analysis of the SlyA-controlled expression, subcellular localization and pore-forming activity of a 34 kDa haemolysin (ClyA) from Escherichia coli K-12. *Molecular Microbiology, 31*(2), 557–567. https://doi.org/10.1046/j.1365-2958.1999.01196.x.

Lukoyanova, N., Kondos, S. C., Farabella, I., Law, R. H., Reboul, C. F., Caradoc-Davies, T. T., et al. (2015). Conformational changes during pore formation by the perforin-related protein pleurotolysin. *PLoS Biology, 13*(2), e1002049. https://doi.org/10.1371/journal.pbio.1002049.

Luo, W., Yao, X., & Hong, M. (2005). Large structure rearrangement of colicin ia channel domain after membrane binding from 2D 13C spin diffusion NMR. *Journal of the American Chemical Society, 127*(17), 6402–6408. https://doi.org/10.1021/ja0433121.

Macek, P., & Lebez, D. (1988). Isolation and characterization of three lethal and hemolytic toxins from the sea anemone Actinia equina L. *Toxicon, 26*(5), 441–451. https://doi.org/10.1016/0041-0101(88)90183-3.

Madegowda, M., Eswaramoorthy, S., Burley, S. K., & Swaminathan, S. (2008). X-ray crystal structure of the B component of Hemolysin BL from Bacillus cereus. *Proteins, 71*(2), 534–540. https://doi.org/10.1002/prot.21888.

Maglia, G., Restrepo, M. R., Mikhailova, E., & Bayley, H. (2008). Enhanced translocation of single DNA molecules through alpha-hemolysin nanopores by manipulation of internal charge. *Proceedings of the National Academy of Sciences of the United States of America, 105*(50), 19720–19725. https://doi.org/10.1073/pnas.0808296105.

Mancheño, J. M., Martín-Benito, J., Martínez-Ripoll, M., Gavilanes, J. G., & Hermoso, J. A. (2003). Crystal and electron microscopy structures of sticholysin II actinoporin reveal insights into the mechanism of membrane pore formation. *Structure, 11*(11), 1319–1328. https://doi.org/10.1016/j.str.2003.09.019.

Mancheño, J. M., Martínez-Ripoll, M., Gavilanes, J. G., & Hermoso, J. A. (2002). Crystallization and preliminary X-ray diffraction studies of the water-soluble state of the pore-forming toxin sticholysin II from the sea anemone Stichodactyla helianthus. *Acta Crystallographica. Section D, Biological Crystallography, 58*(Pt. 7), 1229–1231. https://doi.org/10.1107/s090744490200762x.

Mancheño, J. M., Tateno, H., Goldstein, I. J., Martínez-Ripoll, M., & Hermoso, J. A. (2005). Structural analysis of the laetiporus sulphureus hemolytic pore-forming lectin

in complex with sugars. *Journal of Biological Chemistry, 280*(17), 17251–17259. https://doi.org/10.1074/jbc.m413933200.
Marshall, J. E., Faraj, B. H., Gingras, A. R., Lonnen, R., Sheikh, M. A., El-Mezgueldi, M., et al. (2015). The crystal structure of pneumolysin at 2.0 Å resolution reveals the molecular packing of the pre-pore complex. *Scientific Reports, 5*, 13293. https://doi.org/10.1038/srep13293.
Martinez, M. C., Lazdunski, C., & Pattus, F. (1983). Isolation, molecular and functional properties of the C-terminal domain of colicin A. *The EMBO Journal, 2*(9), 1501–1507.
McCaughey, L. C., Grinter, R., Josts, I., Roszak, A. W., Waløen, K. I., Cogdell, R. J., et al. (2014). Lectin-like bacteriocins from Pseudomonas spp. utilise D-rhamnose containing lipopolysaccharide as a cellular receptor. *PLoS Pathogens, 10*(2), e1003898. https://doi.org/10.1371/journal.ppat.1003898.
McPherson, A., Nguyen, C., Cudney, R., & Larson, S. B. (2011). The role of small molecule additives and chemical modification in protein crystallization. *Crystal Growth & Design, 11*(5), 1469–1474. https://doi.org/10.1021/cg101308r.
Mechaly, A. E., Bellomio, A., Gil-Cartón, D., Morante, K., Valle, M., González-Mañas, J. M., et al. (2011). Structural insights into the oligomerization and architecture of eukaryotic membrane pore-forming toxins. *Structure, 19*(2), 181–191. https://doi.org/10.1016/j.str.2010.11.013.
Meusch, D., Gatsogiannis, C., Efremov, R. G., Lang, A. E., Hofnagel, O., Vetter, I. R., et al. (2014). Mechanism of Tc toxin action revealed in molecular detail. *Nature, 508*(7494), 61–65. https://doi.org/10.1038/nature13015.
Moldoveanu, T., Liu, Q., Tocilj, A., Watson, M., Shore, G., & Gehring, K. (2006). The X-ray structure of a BAK homodimer reveals an inhibitory zinc binding site. *Molecular Cell, 24*(5), 677–688. https://doi.org/10.1016/j.molcel.2006.10.014.
Monastyrnaya, M., Leychenko, E., Isaeva, M., Likhatskaya, G., Zelepuga, E., Kostina, E., et al. (2010). Actinoporins from the sea anemones, tropical Radianthus macrodactylus and northern Oulactis orientalis: Comparative analysis of structure-function relationships. *Toxicon, 56*(8), 1299–1314. https://doi.org/10.1016/j.toxicon.2010.07.011.
Morante, K., Bellomio, A., Viguera, A. R., González-Mañas, J. M., Tsumoto, K., & Caaveiro, J. M. M. (2019). The isolation of new pore-forming toxins from the sea anemone Actinia fragacea provides insights into the mechanisms of actinoporin evolution. *Toxins (Basel), 11*(7), 401. https://doi.org/10.3390/toxins11070401.
Morgan, J. A., Sergeant, M., Ellis, D., Ousley, M., & Jarrett, P. (2001). Sequence analysis of insecticidal genes from Xenorhabdus nematophilus PMFI296. *Applied and Environmental Microbiology, 67*(5), 2062–2069. https://doi.org/10.1128/aem.67.5.2062-2069.2001.
Muchmore, S. W., Sattler, M., Liang, H., Meadows, R. P., Harlan, J. E., Yoon, H. S., et al. (1996). X-ray and NMR structure of human Bcl-xL, an inhibitor of programmed cell death. *Nature, 381*(6580), 335–341. https://doi.org/10.1038/381335a0.
Mueller, M., Grauschopf, U., Maier, T., Glockshuber, R., & Ban, N. (2009). The structure of a cytolytic alpha-helical toxin pore reveals its assembly mechanism. *Nature, 459*(7247), 726–730. https://doi:org/10.1038/nature08026.
Mutter, N. L., Volarić, J., Szymanski, W., Feringa, B. L., & Maglia, G. (2019). Reversible photocontrolled nanopore assembly. *Journal of the American Chemical Society, 141*(36), 14356–14363. https://doi.org/10.1021/jacs.9b06998.
Nakamura, T., Kotani, M., Tonozuka, T., Ide, A., Oguma, K., & Nishikawa, A. (2009). Crystal structure of the HA3 subcomponent of Clostridium botulinum type C progenitor toxin. *Journal of Molecular Biology, 385*(4), 1193–1206. https://doi.org/10.1016/j.jmb.2008.11.039.
Ni, T., Harlos, K., & Gilbert, R. (2016). Structure of astrotactin-2: A conserved vertebrate-specific and perforin-like membrane protein involved in neuronal development. *Open Biology, 6*(5), 160053. https://doi.org/10.1098/rsob.160053.

Ni, T., Jiao, F., Yu, X., Aden, S., Ginger, L., Williams, S. I., et al. (2020). Structure and mechanism of bactericidal mammalian perforin-2, an ancient agent of innate immunity. *Science Advances*, *6*(5), eaax8286. https://doi.org/10.1126/sciadv.aax8286.

Ni, T., Williams, S. I., Rezelj, S., Anderluh, G., Harlos, K., Stansfeld, P. J., et al. (2018). Structures of monomeric and oligomeric forms of the Toxoplasma gondii perforin-like protein 1. *Science Advances*, *4*(3), eaaq0762. https://doi.org/10.1126/sciadv.aaq0762.

Nomidis, S. K., Hooyberghs, J., Maglia, G., & Carlon, E. (2018). DNA capture into the ClyA nanopore: diffusion-limited versus reaction-limited processes. *Journal of Physics. Condensed Matter*, *30*(30), 304001. https://doi.org/10.1088/1361-648X/aacc01.

Ohno-Iwashita, Y., & Imahori, K. (1982). Assignment of the functional loci in the colicin E1 molecule by characterization of its proteolytic fragments. *The Journal of Biological Chemistry*, *257*(11), 6446–6451.

Olson, R., & Gouaux, E. (2005). Crystal structure of the Vibrio cholerae cytolysin (VCC) pro-toxin and its assembly into a heptameric transmembrane pore. *Journal of Molecular Biology*, *350*(5), 997–1016. https://doi.org/10.1016/j.jmb.2005.05.045.

Olson, R., Nariya, H., Yokota, K., Kamio, Y., & Gouaux, E. (1999). Crystal structure of staphylococcal LukF delineates conformational changes accompanying formation of a transmembrane channel. *Nature Structural Biology*, *6*(2), 134–140. https://doi.org/10.1038/5821.

Oscarsson, J., Mizunoe, Y., Li, L., Lai, X. H., Wieslander, A., & Uhlin, B. E. (1999). Molecular analysis of the cytolytic protein ClyA (SheA) from Escherichia coli. *Molecular Microbiology*, *32*(6), 1226–1238. https://doi.org/10.1046/j.1365-2958.1999.01435.x.

Oscarsson, J., Mizunoe, Y., Uhlin, B. E., & Haydon, D. J. (1996). Induction of haemolytic activity in Escherichia coli by the slyA gene product. *Molecular Microbiology*, *20*(1), 191–199. https://doi.org/10.1111/j.1365-2958.1996.tb02500.x.

Oscarsson, J., Westermark, M., Löfdahl, S., Olsen, B., Palmgren, H., Mizunoe, Y., et al. (2002). Characterization of a pore-forming cytotoxin expressed by Salmonella enterica serovars typhi and paratyphi A. *Infection and Immunity*, *70*(10), 5759–5769. https://doi.org/10.1128/iai.70.10.5759-5769.2002.

Park, S. A., Park, Y. S., Bong, S. M., & Lee, K. S. (2016). Structure-based functional studies for the cellular recognition and cytolytic mechanism of pneumolysin from Streptococcus pneumoniae. *Journal of Structural Biology*, *193*(2), 132–140. https://doi.org/10.1016/j.jsb.2015.12.002.

Parker, M. W., Buckley, J. T., Postma, J. P. M., Tucker, A. D., Leonard, K., Pattus, F., et al. (1994). Structure of the Aeromonas toxin proaerolysin in its water-soluble and membrane-channel states. *Nature*, *367*(6460), 292–295. https://doi.org/10.1038/367292a0.

Parker, M. W., & Feil, S. C. (2005). Pore-forming protein toxins: from structure to function. *Progress in Biophysics and Molecular Biology*, *88*(1), 91–142. https://doi.org/10.1016/j.pbiomolbio.2004.01.009.

Parker, M. W., Pattus, F., Tucker, A. D., & Tsernoglou, D. (1989). Structure of the membrane-pore-forming fragment of colicin A. *Nature*, *337*(6202), 93–96. https://doi.org/10.1038/337093a0.

Parker, M. W., Postma, J. P., Pattus, F., Tucker, A. D., & Tsernoglou, D. (1992). Refined structure of the pore-forming domain of colicin A at 2.4 A resolution. *Journal of Molecular Biology*, *224*(3), 639–657. https://doi.org/10.1016/0022-2836(92)90550-4.

Pedelacq, J. D., Maveyraud, L., Prevost, G., Baba-Moussa, L., Gonzalez, A., Courcelle, E., et al. (1999). The structure of a Staphylococcus aureus leucocidin component (LukF-PV) reveals the fold of the water-soluble species of a family of transmembrane pore-forming toxins. *Structure*, 7(3), 277–287. https://doi.org/10.1016/s0969-2126(99)80038-0.

Peng, W., de Souza Santos, M., Li, Y., Tomchick, D. R., & Orth, K. (2019). High-resolution cryo-EM structures of the E. coli hemolysin ClyA oligomers. *PLoS One*, *14*(5), e0213423. https://doi.org/10.1371/journal.pone.0213423.

Pleckaityte, M. (2020). Cholesterol-dependent cytolysins produced by vaginal bacteria: Certainties and controversies. *Frontiers in Cellular and Infection Microbiology*, *9*, 452. https://doi.org/10.3389/fcimb.2019.00452.

Podobnik, M., Savory, P., Rojko, N., Kisovec, M., Wood, N., Hambley, R., et al. (2016). Crystal structure of an invertebrate cytolysin pore reveals unique properties and mechanism of assembly. *Nature Communications*, 7(1), 11598. https://doi.org/10.1038/ncomms11598.

Polekhina, G., Giddings, K. S., Tweten, R. K., & Parker, M. W. (2004). Crystallization and preliminary X-ray analysis of the human-specific toxin intermedilysin. *Acta Crystallographica. Section D, Biological Crystallography*, *60*(Pt. 2), 347–349. https://doi.org/10.1107/s0907444903027240.

Polekhina, G., Giddings, K. S., Tweten, R. K., & Parker, M. W. (2005). Insights into the action of the superfamily of cholesterol-dependent cytolysins from studies of intermedilysin. *Proceedings of the National Academy of Sciences of the United States of America*, *102*(3), 600–605. https://doi.org/10.1073/pnas.0403229101.

Postma, J. P., Parker, M. W., & Tsernoglou, D. (1989). Application of molecular dynamics in the crystallographic refinement of colicin A. *Acta Crystallographica. Section A*, *45*(Pt. 7), 471–477. https://doi.org/10.1107/s0108767389002552.

Prevost, G., Cribier, B., Couppie, P., Petiau, P., Supersac, G., Finck-Barbancon, V., et al. (1995). Panton-valentine leucocidin and gamma-hemolysin from Staphylococcus aureus ATCC 49775 are encoded by distinct genetic loci and have different biological activities. *Infection and Immunity*, *63*(10), 4121–4129. https://doi.org/10.1128/IAI.63.10.4121-4129.1995.

Pulagam, L. P., & Steinhoff, H. J. (2013). Acidic pH-induced membrane insertion of colicin A into E. coli natural lipids probed by site-directed spin labeling. *Journal of Molecular Biology*, *425*(10), 1782–1794. https://doi.org/10.1016/j.jmb.2013.01.037.

Rath, P., Bousché, O., Merrill, A. R., Cramer, W. A., & Rothschild, K. J. (1991). Fourier transform infrared evidence for a predominantly alpha-helical structure of the membrane bound channel forming COOH-terminal peptide of colicin E1. *Biophysical Journal*, *59*(3), 516–522. https://doi.org/10.1016/s0006-3495(91)82268-8.

Restrepo-Pérez, L., Wong, C. H., Maglia, G., Dekker, C., & Joo, C. (2019). Label-free detection of post-translational modifications with a nanopore. *Nano Letters*, *19*(11), 7957–7964. https://doi.org/10.1021/acs.nanolett.9b03134.

Roblin, P., Guillet, V., Joubert, O., Keller, D., Erard, M., Maveyraud, L., et al. (2008). A covalent S-F heterodimer of leucotoxin reveals molecular plasticity of beta-barrel pore-forming toxins. *Proteins*, *71*(1), 485–496. https://doi.org/10.1002/prot.21900.

Roderer, D., Hofnagel, O., Benz, R., & Raunser, S. (2019). Structure of a Tc holotoxin pore provides insights into the translocation mechanism. *Proceedings of the National Academy of Sciences*, *116*(46), 23083. https://doi.org/10.1073/pnas.1909821116.

Roderer, D., Schubert, E., Sitsel, O., & Raunser, S. (2019). Towards the application of Tc toxins as a universal protein translocation system. *Nature Communications*, *10*(1), 5263. https://doi.org/10.1038/s41467-019-13253-8.

Ros, U., Rodríguez-Vera, W., Pedrera, L., Valiente, P. A., Cabezas, S., Lanio, M. E., et al. (2015). Differences in activity of actinoporins are related with the hydrophobicity of their N-terminus. *Biochimie*, *116*, 70–78. https://doi.org/10.1016/j.biochi.2015.06.024.

Rosado, C. J., Buckle, A. M., Law, R. H., Butcher, R. E., Kan, W. T., Bird, C. H., et al. (2007). A common fold mediates vertebrate defense and bacterial attack. *Science*, *317*(5844), 1548–1551. https://doi.org/10.1126/science.1144706.

Rossjohn, J., Feil, S. C., McKinstry, W. J., Tsernoglou, D., Goot, G. V. D., Buckley, J. T., et al. (1998). Aerolysin—A paradigm for membrane insertion of beta-sheet protein toxins? *Journal of Structural Biology*, *121*(2), 92–100. https://doi.org/10.1006/jsbi.1997.3947.

Rossjohn, J., Feil, S. C., McKinstry, W. J., Tweten, R. K., & Parker, M. W. (1997). Structure of a cholesterol-binding, thiol-activated cytolysin and a model of its membrane form. *Cell*, *89*(5), 685–692. https://doi.org/10.1016/s0092-8674(00)80251-2.

Savva, C. G., Fernandes da Costa, S. P., Bokori-Brown, M., Naylor, C. E., Cole, A. R., Moss, D. S., et al. (2013). Molecular architecture and functional analysis of NetB, a pore-forming toxin from Clostridium perfringens. *The Journal of Biological Chemistry*, *288*(5), 3512–3522. https://doi.org/10.1074/jbc.M112.430223.

Schubert, E., Vetter, I. R., Prumbaum, D., Penczek, P. A., & Raunser, S. (2018). Membrane insertion of α-xenorhabdolysin in near-atomic detail. *eLife*, *7*, e38017. https://doi.org/10.7554/eLife.38017.

Seilie, E. S., & Bubeck Wardenburg, J. (2017). Staphylococcus aureus pore-forming toxins: The interface of pathogen and host complexity. *Seminars in Cell & Developmental Biology*, *72*, 101–116. https://doi.org/10.1016/j.semcdb.2017.04.003.

Sharma, O., Yamashita, E., Zhalnina, M. V., Zakharov, S. D., Datsenko, K. A., Wanner, B. L., et al. (2007). Structure of the complex of the colicin E2 R-domain and its BtuB receptor. The outer membrane colicin translocon. *The Journal of Biological Chemistry*, *282*(32), 23163–23170. https://doi.org/10.1074/jbc.M703004200.

Shatursky, O., Heuck, A. P., Shepard, L. A., Rossjohn, J., Parker, M. W., Johnson, A. E., et al. (1999). The mechanism of membrane insertion for a cholesterol-dependent cytolysin: A novel paradigm for pore-forming toxins. *Cell*, *99*(3), 293–299. https://doi.org/10.1016/s0092-8674(00)81660-8.

Shepard, L. A., Heuck, A. P., Hamman, B. D., Rossjohn, J., Parker, M. W., Ryan, K. R., et al. (1998). Identification of a membrane-spanning domain of the thiol-activated pore-forming toxin Clostridium perfringens perfringolysin O: An alpha-helical to beta-sheet transition identified by fluorescence spectroscopy. *Biochemistry*, *37*(41), 14563–14574. https://doi.org/10.1021/bi981452f.

Song, L., & Gouaux, J. E. (1997). [4] Membrane protein crystallization: Application of sparse matrices to the alpha-hemolysin heptamer. *Methods in Enzymology*, *276*, 60–74. https://doi.org/10.1016/S0076-6879(97)76050-6.

Song, L., Hobaugh, M. R., Shustak, C., Cheley, S., Bayley, H., & Gouaux, J. E. (1996). Structure of staphylococcal alpha-hemolysin, a heptameric transmembrane pore. *Science*, *274*(5294), 1859–1866. https://www.ncbi.nlm.nih.gov/pubmed/8943190.

Soskine, M., Biesemans, A., De Maeyer, M., & Maglia, G. (2013). Tuning the size and properties of ClyA nanopores assisted by directed evolution. *Journal of the American Chemical Society*, *135*(36), 13456–13463. https://doi.org/10.1021/ja4053398.

Soskine, M., Biesemans, A., & Maglia, G. (2015). Single-molecule analyte recognition with ClyA nanopores equipped with internal protein adaptors. *Journal of the American Chemical Society*, *137*(17), 5793–5797. https://doi.org/10.1021/jacs.5b01520.

Soskine, M., Biesemans, A., Moeyaert, B., Cheley, S., Bayley, H., & Maglia, G. (2012). An engineered ClyA nanopore detects folded target proteins by selective external association and pore entry. *Nano Letters*, *12*(9), 4895–4900. https://doi.org/10.1021/nl3024438.

Spicer, B. A., Law, R. H. P., Caradoc-Davies, T. T., Ekkel, S. M., Bayly-Jones, C., Pang, S.-S., et al. (2018). The first transmembrane region of complement component-9 acts as a brake on its self-assembly. *Nature Communications*, *9*(1), 3266. https://doi.org/10 1038/s41467-018-05717-0.

Srisucharitpanit, K., Yao, M., Promdonkoy, B., Chimnaronk, S., Tanaka, I., & Boonserm, P. (2014). Crystal structure of BinB: A receptor binding component of the binary toxin from Lysinibacillus sphaericus. *Proteins*, *82*(10), 2703–2712. https://doi.org/10.1002/prot.24636.

Sugawara, T., Yamashita, D., Kato, K., Peng, Z., Ueda, J., Kaneko, J., et al. (2015). Structural basis for pore-forming mechanism of staphylococcal alpha-hemolysin. *Toxicon*, *108*, 226–231. https://doi.org/10.1016/j.toxicon.2015.09.033.

Szczesny, P., Iacovache, I., Muszewska, A., Ginalski, K., Goot, F. G.v.d., & Grynberg, M. (2011). Extending the aerolysin family: From bacteria to vertebrates. *PLoS One*, *6*(6), e20349. https://doi.org/10.1371/journal.pone.0020349.

Tanaka, K., Caaveiro, J. M., Morante, K., González-Mañas, J. M., & Tsumoto, K. (2015). Structural basis for self-assembly of a cytolytic pore lined by protein and lipid. *Nature Communications*, *6*, 6337. https://doi.org/10.1038/ncomms7337.

Tanaka, K., Caaveiro, J. M. M., Morante, K., & Tsumoto, K. (2017). Haemolytic actinoporins interact with carbohydrates using their lipid-binding module. *Philosophical Transactions of the Royal Society of London. Series B, Biological Sciences*, *372*(1726). https://doi.org/10.1098/rstb.2016.0216.

Tomita, T., & Kamio, Y. (1997). Molecular biology of the pore-forming cytolysins from Staphylococcus aureus, alpha- and gamma-hemolysins and leukocidin. *Bioscience, Biotechnology, and Biochemistry*, *61*(4), 565–572. https://doi.org/10.1271/bbb.61.565.

Tucker, A. D., Baty, D., Parker, M. W., Pattus, F., Lazdunski, C., & Tsernoglou, D. (1989). Crystallographic phases through genetic engineering: Experiences with colicin A. *Protein Engineering*, *2*(6), 399–405. https://doi.org/10.1093/protein/2.6.399.

Tucker, A. D., Parker, M. W., Tsernoglou, D., & Buckley, J. T. (1990). Crystallization of a proform of aerolysin, a hole-forming toxin from Aeromonas hydrophila. *Journal of Molecular Biology*, *212*(4), 561–562. https://doi.org/10.1016/0022-2836(90)90222-8.

Tucker, A. D., Pattus, F., & Tsernoglou, D. (1986). Crystallization of the C-terminal domain of colicin A carrying the voltage-dependent pore activity of the protein. *Journal of Molecular Biology*, *190*(1), 133–134. https://doi.org/10.1016/0022-2836(86)90084-7.

Tzokov, S. B., Wyborn, N. R., Stillman, T. J., Jamieson, S., Czudnochowski, N., Artymiuk, P. J., et al. (2006). Structure of the hemolysin E (HlyE, ClyA, and SheA) channel in its membrane-bound form. *The Journal of Biological Chemistry*, *281*(32), 23042–23049. https://doi.org/10.1074/jbc.M602421200.

Vagin, A., & Teplyakov, A. (1997). MOLREP: An automated program for molecular replacement. *Journal of Applied Crystallography*, *30*(6), 1022–1025. https://doi.org/10.1107/S0021889897006766.

Van Itallie, C. M., Betts, L., Smedley, J. G., McClane, B. A., & Anderson, J. M. (2008). Structure of the claudin-binding domain of clostridium perfringens enterotoxin. *Journal of Biological Chemistry*, *283*(1), 268–274. https://doi.org/10.1074/jbc.M708066200.

Van Meervelt, V., Soskine, M., & Maglia, G. (2014). Detection of two isomeric binding configurations in a protein-aptamer complex with a biological nanopore. *ACS Nano*, *8*(12), 12826–12835. https://doi.org/10.1021/nn506077e.

van Pee, K., Mulvihill, E., Müller, D. J., & Yildiz, Ö. (2016). Unraveling the pore-forming steps of pneumolysin from Streptococcus pneumoniae. *Nano Letters*, *16*(12), 7915–7924. https://doi.org/10.1021/acs.nanolett.6b04219.

van Pee, K., Neuhaus, A., D'Imprima, E., Mills, D. J., Kühlbrandt, W., & Yildiz, Ö. (2017). CryoEM structures of membrane pore and prepore complex reveal cytolytic mechanism of pneumolysin. *eLife*, *6*, e23644. https://doi.org/10.7554/eLife.23644.

Vetter, I. R., Parker, M. W., Tucker, A. D., Lakey, J. H., Pattus, F., & Tsernoglou, D. (1998). Crystal structure of a colicin N fragment suggests a model for toxicity. *Structure*, *6*(7), 863–874. https://doi.org/10.1016/s0969-2126(98)00088-4.

Vigneux, F., Zumbihl, R., Jubelin, G., Ribeiro, C., Poncet, J., Baghdiguian, S., et al. (2007). The xaxAB genes encoding a new apoptotic toxin from the insect pathogen Xenorhabdus nematophila are present in plant and human pathogens. *The Journal of Biological Chemistry*, *282*(13), 9571–9580. https://doi.org/10.1074/jbc.M604301200.

Wade, K. R., Lawrence, S. L., Farrand, A. J., Hotze, E. M., Kuiper, M. J., Gorman, M. A., et al. (2019). The structural basis for a transition state that regulates pore formation in a bacterial toxin. *MBio*, *10*(2). https://doi.org/10.1128/mBio.00538-19.

Wagner, N. J., Lin, C. P., Borst, L. B., & Miller, V. L. (2013). YaxAB, a Yersinia enterocolitica pore-forming toxin regulated by RovA. *Infection and Immunity*, *81*(11), 4208–4219. https://doi.org/10.1128/iai.00781-13.

Wallace, A. J., Stillman, T. J., Atkins, A., Jamieson, S. J., Bullough, P. A., Green, J., et al. (2000). E. coli hemolysin E (HlyE, ClyA, SheA): X-ray crystal structure of the toxin and observation of membrane pores by electron microscopy. *Cell*, *100*(2), 265–276. https://doi.org/10.1016/s0092-8674(00)81564-0.

Wang, Q., Bian, X., Zeng, L., Pan, F., Liu, L., Liang, J., et al. (2020). A cellular endolysosome-modulating pore-forming protein from a toad is negatively regulated by its paralog under oxidizing conditions. *The Journal of Biological Chemistry*, *295*(30), 10293–10306. https://doi.org/10.1074/jbc.RA120.013556.

Wickham, S. E., Hotze, E. M., Farrand, A. J., Polekhina, G., Nero, T. L., Tomlinson, S., et al. (2011). Mapping the intermedilysin-human CD59 receptor interface reveals a deep correspondence with the binding site on CD59 for complement binding proteins C8alpha and C9. *The Journal of Biological Chemistry*, *286*(23), 20952–20962. https://doi.org/10.1074/jbc.M111.237446.

Wiener, M., Freymann, D., Ghosh, P., & Stroud, R. M. (1997). Crystal structure of colicin Ia. *Nature*, *385*(6615), 461–464. https://doi.org/10.1038/385461a0.

Willems, K., Ruić, D., Biesemans, A., Galenkamp, N. S., Van Dorpe, P., & Maglia, G. (2019). Engineering and modeling the electrophoretic trapping of a single protein inside a nanopore. *ACS Nano*, *13*(9), 9980–9992. https://doi.org/10.1021/acsnano.8b09137.

Willems, K., Ruić, D., F, L. R. L., Barman, U., Verellen, N., Hofkens, J., et al. (2020). Accurate modeling of a biological nanopore with an extended continuum framework. *Nanoscale*, *12*(32), 16775–16795. https://doi.org/10.1039/d0nr03114c.

Wilson, J. S., Churchill-Angus, A. M., Davies, S. P., Sedelnikova, S. E., Tzokov, S. B., Rafferty, J. B., et al. (2019). Identification and structural analysis of the tripartite α-pore forming toxin of Aeromonas hydrophila. *Nature Communications*, *10*(1), 2900. https://doi.org/10.1038/s41467-019-10777-x.

Wloka, C., Mutter, N. L., Soskine, M., & Maglia, G. (2016). Alpha-helical fragaceatoxin C nanopore engineered for double-stranded and single-stranded nucleic acid analysis. *Angewandte Chemie International Edition*, *55*(40), 12494–12498. https://doi.org/10.1002/anie.201606742.

Wloka, C., Van Meervelt, V., van Gelder, D., Danda, N., Jager, N., Williams, C. P., et al. (2017). Label-free and real-time detection of protein ubiquitination with a biological nanopore. *ACS Nano*, *11*(5), 4387–4394. https://doi.org/10.1021/acsnano.6b07760.

Woodin, A. M. (1959). Fractionation of a leucocidin from Staphylococcus aureus. *The Biochemical Journal*, *73*, 225–237. https://doi.org/10.1042/bj0730225.

Xu, C., Chinte, U., Chen, L., Yao, Q., Meng, Y., Zhou, D., et al. (2015). Crystal structure of Cry51Aa1: A potential novel insecticidal aerolysin-type β-pore-forming toxin from Bacillus thuringiensis. *Biochemical and Biophysical Research Communications*, *462*(3), 184–189. https://doi.org/10.1016/j.bbrc.2015.04.068.

Xu, C., Wang, B.-C., Yu, Z., & Sun, M. (2014). Structural insights into Bacillus thuringiensis Cry, Cyt and parasporin toxins. *Toxins*, *6*(9), 2732–2770. https://doi.org/10.3390/toxins6092732.

Xu, L., Huang, B., Du, H., Zhang, X. C., Xu, J., Li, X., et al. (2010). Crystal structure of cytotoxin protein suilysin from Streptococcus suis. *Protein & Cell*, *1*(1), 96–105. https://doi.org/10.1007/s13238-010-0012-3.

Xu, Q., Abdubek, P., Astakhova, T., Axelrod, H. L., Bakolitsa, C., Cai, X., et al. (2010). Structure of a membrane-attack complex/perforin (MACPF) family protein from the human gut symbiont Bacteroides thetaiotaomicron. *Acta Crystallographica. Section F, Structural Biology and Crystallization Communications*, *66*(Pt. 10), 1297–1305. https://doi.org/10.1107/s1744309110023055.

Yamashita, D., Sugawara, T., Takeshita, M., Kaneko, J., Kamio, Y., Tanaka, I., et al. (2014). Molecular basis of transmembrane beta-barrel formation of staphylococcal pore-forming toxins. *Nature Communications*, *5*, 4897. https://doi.org/10.1038/ncomms5897.

Yamashita, K., Kawai, Y., Tanaka, Y., Hirano, N., Kaneko, J., Tomita, N., et al. (2011). Crystal structure of the octameric pore of staphylococcal gamma-hemolysin reveals the beta-barrel pore formation mechanism by two components. *Proceedings of the National Academy of Sciences of the United States of America*, *108*(42), 17314–17319. https://doi.org/10.1073/pnas.1110402108.

Yan, X. X., Porter, C. J., Hardy, S. P., Steer, D., Smith, A. I., Quinsey, N. S., et al. (2013). Structural and functional analysis of the pore-forming toxin NetB from Clostridium perfringens. *MBio*, *4*(1), e00019-00013. https://doi.org/10.1128/mBio.00019-13.

Yelland, T. S., Naylor, C. E., Bagoban, T., Savva, C. G., Moss, D. S., McClane, B. A., et al. (2014). Structure of a C. perfringens enterotoxin mutant in complex with a modified Claudin-2 extracellular loop 2. *Journal of Molecular Biology*, *426*(18), 3134–3147. https://doi.org/10.1016/j.jmb.2014.07.001.

Yu, J., Cao, C., & Long, Y. T. (2017). Selective and sensitive detection of methylcytosine by aerolysin nanopore under serum condition. *Analytical Chemistry*, *89*(21), 11685–11689. https://doi.org/10.1021/acs.analchem.7b03133.

Zaitseva, J., Vaknin, D., Krebs, C., Doroghazi, J., Milam, S. L., Balasubramanian, D., et al. (2019). Structure-function characterization of an insecticidal protein GNIP1Aa, a member of an MACPF and β-tripod families. *Proceedings of the National Academy of Sciences of the United States of America*, *116*(8), 2897–2906. https://doi.org/10.1073/pnas.1815547116.

Zhao, S., Restrepo-Pérez, L., Soskine, M., Maglia, G., Joo, C., Dekker, C., et al. (2019). Electro-mechanical conductance modulation of a nanopore using a removable gate. *ACS Nano*, *13*(2), 2398–2409. https://doi.org/10.1021/acsnano.8b09266.

CHAPTER TWO

How protein engineering has revealed the molecular mechanisms of pore-forming toxins

Jordan C. Evans and Rodney K. Tweten*
Department of Microbiology and Immunology, The University of Oklahoma Health Sciences Center, Oklahoma City, OK, United States
*Corresponding author: e-mail address: rod-tweten@ouhsc.edu

Contents

Methods in Enzymology, Volume 649
ISSN 0076-6879
https://doi.org/10.1016/bs.mie.2021.01.012

Abstract

Pore-forming proteins are found in prokaryotes, vertebrates, and invertebrates, and when involved in pathogenic processes they are classified as pore-forming toxins (PFTs). The use of gene engineering methods in combination with the information provided by the high-resolution crystal structures of the PFTs have allowed investigators to gain a deep understanding of their pore-forming mechanisms. In this chapter, we discuss how protein engineering has helped us and others to reveal the molecular mechanisms of pore formation by prokaryotic PFTs with an emphasis on our experiences with the cholesterol-dependent cytolysins (CDCs).

1. Introduction

The mechanisms by which bacterial pore-forming toxins (PFTs) bind to and create a pore in a membrane has been a subject of interest for well over a century, but only in the late 20th century have the tools been developed that have provided the research community the means to reveal these mechanisms at the molecular level. Tools developed in the late 1970s and early 1980s were essential to the field of protein engineering. Two of these essential technologies were the ability to clone and express proteins in *Escherichia coli* (Cohen, Chang, Boyer, & Helling, 1973), and the ability to synthesize DNA oligonucleotides (Matteucci & Caruthers, 1992). The subsequent development of polymerase chain reaction (Mullis et al., 1986)—based methods for protein engineering (Howorka & Bayley, 1998; Jones & Howard, 1991; Li & Mullins, 2002) were essential to the ability to rapidly and efficiently site-specifically mutate genes encoding the protein of choice. Since those discoveries the ability to introduce mutations and to synthesize entire genes has become relatively routine. This has greatly enhanced our ability to understand the mechanisms of pore-forming proteins, and proteins in general, at the molecular level.

PFTs are currently the largest known class of bacterial protein toxins, which may reflect the fact that forming a pore in a competitor's membrane was a simple and effective mechanism to kill off a competitor in the early evolutionary history of microbes. Bacterial and eukaryotic PFTs can form pores in target membranes by a number of methods, but the most numerous are those that assemble and insert a β-barrel pore, which are termed β-barrel PFT or β-PFTs. One of the initial conundrums of the β-PFTs was the assembly of the membrane-embedded pore complex, as most β-PFTs are initially secreted as soluble monomeric proteins. The pores of β-PFTs are typically assembled by membrane binding followed by oligomerization of several of the soluble toxin monomers (reviewed in Dal Peraro & van der Goot, 2016).

The first structure of a β-PFT pore, which unambiguously demonstrated how β-PFTs form a pore, was that of *Staphylococcus aureus* α-hemolysin (Song et al., 1996). This remarkably beautiful structure revealed that the pore assumed mushroom-shaped structure composed of 7 α-hemolysin monomers wherein the mushroom stem was the β-barrel structure and the cap remained above the surface of the membrane bilayer. However, at the time the structure of the pore was solved, the structure of the soluble α-hemolysin monomer was unknown, leaving the mechanism for how the monomer structure changed upon pore formation unknown. Only upon solving the structure for the soluble monomer of the closely related leukocidin F PFT from *S. aureus* were these changes revealed (Olson, Nariya, Yokota, Kamio, & Gouaux, 1999). Although these elegant studies revealed the changes in the monomer structure that occurred to create the pore, which were surprisingly few for α-hemolysin, they did not reveal the processes that triggered this transformation or any possible intermediate states the monomers might undergo to form the pore. Nonetheless, the α-hemolysin structure established the fundamental architecture of the β-barrel pore much like the initial discovery of diphtheria toxin's mechanism (reviewed in Collier, 1975) formed the basis of our understanding of how A-B toxins function. However, like the A-B toxins, which exhibit a wide number of diverse mechanisms by which the B fragment delivers the A fragment into the cytosol, it was clear that the α-hemolysin mechanism was just one example of how the β-PFTs assembled a pore in the membrane.

Although the crystal and cryo-electron microscopy (EM) structures of PFTs are extremely useful for understanding static states of the pore-forming process, thus far, there have been no high-resolution structures of any intermediate states of pore formation. This may change with single particle approaches that can be used to obtain high-resolution structures of the many published intermediate states of various PFTs that have been generated by mutation of one or more residues of their structures. It is also currently not possible using atomic resolution imaging to observe dynamic processes in real time within the protein structure that occur during the process of pore formation. Structures also cannot derive information on the intramolecular and intermolecular interactions that drive the process of pore formation or various thermodynamic features of the pore forming process. However, the combination of structural information with protein engineering methods have provided insights into the process of pore formation that have yielded a depth of understanding previously unattainable.

This review will discuss protein engineering methods that have been developed and used to study the mechanism of pore formation by the β-PFTs.

This is not a review of pore-forming proteins themselves or the many groups that have studied these proteins, but instead will discuss what protein engineering has revealed about their mechanism. This review will focus mostly on protein engineering methods we have employed over the last 30 years to understand the pore-forming process of a number of pore-forming bacterial toxins with an emphasis on the large family of bacterial pore-forming proteins that were originally termed the thiol-activated toxins (or cytolysins) but were renamed by us to the current designation of cholesterol-dependent cytolysins (CDCs) (Shatursky et al., 1999; Tweten, Parker, & Johnson, 2001).

Our studies have largely focused on the use of site-specific modification of the β-PFT perfringolysin (PFO) from *Clostridium perfringens*, although we will bring in other examples when appropriate. It should be noted herein that there is a single cysteine residue in the undecapeptide of PFO and most other CDCs that can be changed to an alanine without noticeably affecting it pore-forming activity: therefore, all further references to PFO should be assumed to be based on this cysteine-less derivative (Shepard et al., 1998). We have used PFO as the archetype for the CDCs for several reasons, many of which were the result of luck. In retrospect, it has been a relatively well-behaved, stable protein which can be purified in large quantities (i.e., tens of milligrams). An example of the "luck" in choosing PFO to study is exemplified when this author (Tweten) first purified the recombinant form of PFO in about 1994: I set up a couple of crystallization trials, the first I had ever attempted, and had crystals of PFO within 18 h. Upon a collaborative arrangement with Mike Parker (which continues to this day), crystals suitable for the solution of the structure were generated and the structure solved within about 6 months, which was published in 1998 (Feil, Rossjohn, Rohde, Tweten, & Parker, 1996; Rossjohn, Feil, McKinstry, Tweten, & Parker, 1997). The next successful crystal structure for a protein of this class did not occur until 2005 (Polekhina, Giddings, Tweten, & Parker, 2004, 2005), which was followed by several other structures over the next decade (Bourdeau et al., 2009; Feil, Ascher, Kuiper, Tweten, & Parker, 2014; Lawrence et al., 2015; van Pee, Mulvihill, Muller, & Yildiz, 2016; Xu et al., 2010).

For the purposes of this chapter, we have presented the pore-forming mechanism of PFO, as we understand it today, in Fig. 1. This is not a comprehensive review of all of the instances where protein engineering has revealed insights into all pore-forming proteins, rather we wish to provide examples of how we have used protein engineering to reveal aspects of

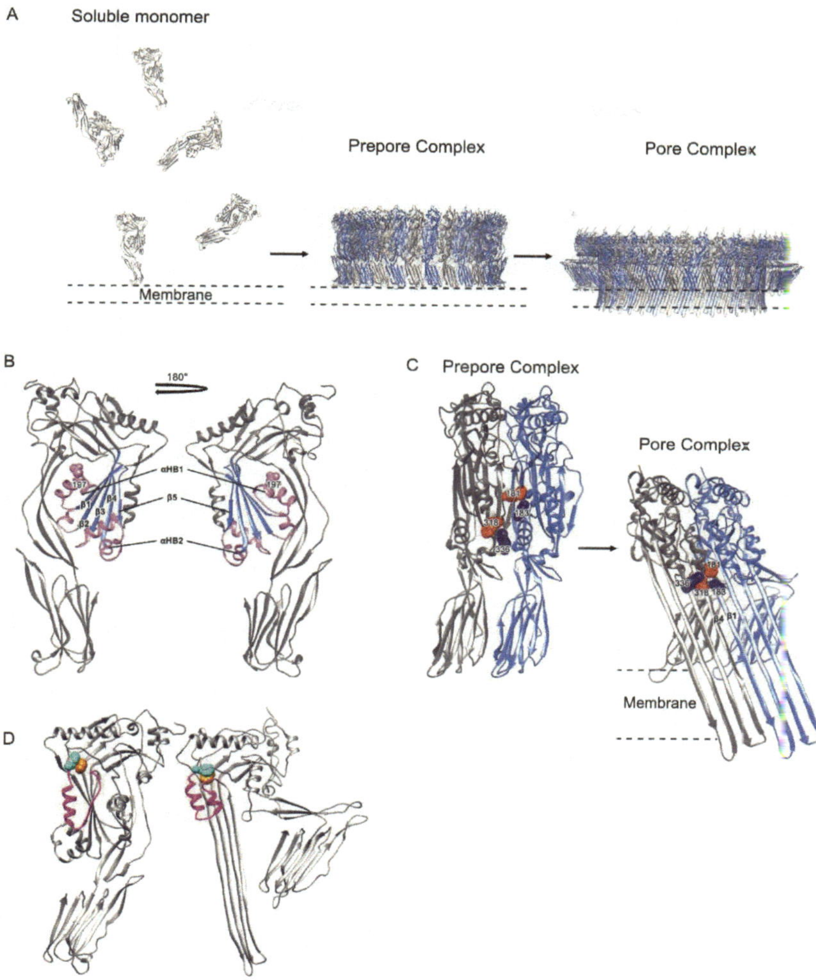

Fig. 1 Pore Formation of Perfringolysin O. (A) Secreted soluble monomers of PFO bind to membrane using cholesterol as the receptor. Membrane bound monomers will oligomerize into circular prepore complexes, composed of as many as 40 monomers, on the surface of the membrane prior to undergoing large structural rearrangements to insert a β-barrel through the membrane forming an aqueous pore. (B) The crystal structure of soluble monomeric PFO (PDB 1PFO) detailing the membrane spanning portion of the protein. A 5-stranded β-sheet (light blue ribbon) is flanked by αHBs (pink ribbon) on either side, and αHB1 forms stabilizing contacts with the core of the protein mediated by residue N197 (pink spheres). (C) Intermolecular interactions between monomers within the prepore complex are necessary in order to undergo the large structural transitions require for the prepore to convert to the pore. In the prepore, residue F318 on β4

(Continued)

the PFO pore-forming mechanism, as well as the limitations and problems we have encountered. It is hoped that this information will be useful to future investigators studying pore-forming proteins.

2. Alanine or cysteine scanning of bacterial PFTs

The advent of automated methods to rapidly produce oligonucleotides, which could be used to rapidly mutate the genes of PFTs, allowed the generation of a library of point mutants at every position in a protein. The first use of large-scale scanning mutagenesis for a PFT was accomplished by Walker and Bayley who generated 166 cysteine mutants in *S. aureus* α-hemolysin (Walker & Bayley, 1995), which was shown later to form a heptameric β-barrel pore (Song et al., 1996). As discussed below, the use of cysteine instead of alanine to scan a protein structure offers the advantage that the cysteine sulfhydryl can be specifically modified with any number of sulfhydryl-specific probes. In their study, they identified a number of sites where the introduction of cysteine did not affect activity until they were modified with a probe. Using this approach, they were able to identify a number of functional regions of α-hemolysin that were involved in binding, oligomerization and pore-formation. As a note, the Bayley lab over the years has been on the cutting edge of using and developing techniques to engineer pore-forming proteins. One of the more intriguing uses was to engineer them to be used in nanopore-based DNA sequencing (Howorka, Cheley, & Bayley, 2001).

Fig. 1—Cont'd of one monomer forms a π-stacking interaction with Y181 on β1 of a neighboring monomer (red spheres). This interaction promotes the disruption of αHB2 which is then refolded into a β-hairpin. Likewise, K336 at the tip of β5 forms an electrostatic interaction with E183 on β1 of the neighboring monomer (dark blue spheres). This interaction likely occurs temporarily and applies force on the central β-sheet and disrupts the stabilizing contacts formed by αHB1with the core of the protein. This instigates αHB1 to unfurl into a β-hairpin and the lowering of the entire complex to insert the newly β-barrel into the membrane. PFO was modeled onto the 4.5 Å cryo-EM structure of the pneumolysin pore complex (PDB 5LY6) (van Pee et al., 2017), and as such the residue placement is intended only as a visual aid and is not necessarily accurate. (D) Shown is the αβ loop on the soluble monomer (left panel) and its transition to the helix-turn-helix (HTH) in the monomer form the pore complex. Also shown are aromatic residues F318 (cyan) and F230 (orange), which form a flexible anchor at one end of the αβ loop (Evans et al., 2020). All structural representations were generated using UCSF Chimera (Pettersen et al., 2004).

We initially used the same approach on the structure of the pore-forming anthrax protective antigen (PA without its propeptide for a total of 568 amino acids), which, like the study of Walker et al. (Walker & Bayley, 1995) also provided a plethora of data on those residues that affected various aspects of the mechanism of anthrax toxin (Mourez et al., 2003). The generation of this mutant library was accomplished in approximately 6–8 months since each substitution required sequencing of the clone to confirm the integrity of the mutation and to ensure that no secondary mutations had been introduced. With more modern techniques this timeframe would likely be significantly reduced for similarly sized proteins. Subsequently, we also cysteine scanned the pore-forming α-toxin from *Clostridium septicum* (Melton-Witt, Bentsen, & Tweten, 2006) as well as PFO (unpublished data). These scans, which were carried out nearly 15 years ago, continue to be useful, as we have delved deeper into the pore-forming mechanisms of these toxins and these libraries provide ready access to mutants that may be used in studies of their mechanisms.

2.1 Advantages of using cysteine to scan proteins

Cysteine, rather than alanine, was used in these scans since it introduces a site-specific mutation that was amenable to labeling these proteins at these sites with the available sulfhydryl-specific probes with various functionalities. These mutants were primarily used to identify functional domains wherein one or more mutation caused a decrease or loss of pore-forming activity by disrupting a specific process such as receptor binding, oligomerization, or the insertion of the β-barrel pore (Melton-Witt et al., 2006; Mourez et al., 2003). However, not all mutants result in loss of activity, some significantly increase pore-forming activity (Burns, Morton, Parker, & Tweten, 2019; Wade et al., 2019). Hence, this type of large-scale approach to identifying residues that affect the function of these PFTs, as well as other proteins that are amenable to this approach, can provide a significant amount of information that can serve as a basis for the deeper investigation of a protein's function.

3. The introduction of probes using the cysteine sulfhydryl and their uses

Most secreted bacterial β-PFTs contain no cysteines, or a single cysteine that can be changed to another residue without significantly impacting its activity: the best example of the latter are the CDCs, which typically contain a single conserved cysteine that can be changed to alanine without a significant impact on their pore-forming activity (Shepard et al., 1998), even though its chemical modification dramatically decreases activity (Iwamoto, Ohno-Iwashita, & Ando, 1987): the basis by which its chemical modification alters its activity has yet to be satisfactorily explained. Since the cysteine sulfhydryl presents a unique target for chemical modification it can be used to introduce a number of useful probes into the protein structure that can interrogate the local environment of that residue. Furthermore, for those probes that respond to changes in their environment, or that can be used to measure changes in the distance between the probes, can be followed in real time, which can reveal the dynamics of specific conformational changes during pore formation. Using cysteine-substitution and modification of the sulfhydryl with various probes (e.g., fluorescent and electron spin probes) has led to numerous revelations about the structural transitions that pore-forming proteins, as they proceed from a soluble monomer to a membrane-embedded pore (e.g., Burns et al., 2019; Dowd, Farrand, & Tweten, 2012; Harris, Sims, & Tweten, 1991; Heuck, Hotze, Tweten, & Johnson, 2000; Heuck, Tweten, & Johnson, 2003; Hotze et al., 2002, 2012; Ramachandran, Tweten, & Johnson, 2004), which would have not been otherwise possible.

3.1 Limitations and problems when introducing probes onto the cysteine sulfhydryl

The use of cysteine engineering and its modification with probes, however, comes with some limitations. The first of which is that the protein cannot have any cysteines unless they can be changed to another residue without significantly impacting the pore-forming mechanism, or are not accessible to modification (i.e., buried in the protein structure or are in a disulfide). The presence of disulfides in a protein do not present a problem to the placement of a single cysteine into the protein structure and chemically modifying the sulfhydryl with a probe, unless its introduction causes a problem with the formation of the native disulfide(s) during protein folding. As discussed below, we have introduced engineered disulfides into PFO to lock it into a specific stage of pore formation and then introduced a third cysteine which could be derivatized with a probe to follow specific conformational changes upon reduction of the disulfide.

Although this review is not about the use of probes and their properties, it is important to note some of the published and unpublished problems and solutions that we have encountered in their use to modify engineered cysteines within the structure of the PFO. One of the obvious problems is that although the substitution of a cysteine may not significantly alter the function of a PFT its modification with a probe could have a much larger positive or negative impact on function. A change in the function of the pore-forming protein after modification of the cysteine sulfhydryl with a probe can also be informative. As an example, we found that the chemical modification of the sulfhydryl of cysteine substituted for an asparagine (N197, Fig. 1), or its substitution with tryptophan in PFO dramatically changed its pore-forming properties and largely eliminated its temperature dependence (Wade et al., 2019). The structure of the tryptophan mutant showed that it disrupted a critical water network that was necessarily broken in the wildtype protein to initiate the prepore to pore transition (Wade et al., 2019). Most of the time, however, we have experienced only small changes in activity after a residue has been modified (Shepard et al., 1998), despite a few cases in which we have seen significant increases or losses in pore-forming activity after modification (Heuck et al., 2000). As stated above, these changes can provide important insights into the mechanism of pore-formation. It should be noted that we often observe that PFO derivatives with an engineered cysteine substitution, which has been derivatized with a fluorescent probe, tend to aggregate to a small extent over time in solution. These aggregates can often be removed by centrifugation and do not appear to affect the remaining soluble toxin.

Modifying the cysteine sulfhydryl is typically accomplished using a variety of chemistries: some of the more commonly used sulfhydryl-specific chemical groups are probes that are coupled to maleimides, haloacetyls, haloacetamides and methanethiosulfonates. These form covalent bonds with the cysteine sulfhydryl and often have optimal pH ranges for maximizing labeling efficiency of the sulfhydryl and minimizing off-target effects such as the labeling of free amines. There are other sulfhydryl-specific probes that form a disulfide bond with the cysteine sulfhydryl such as pyridyldithiol, although its use is largely restricted to crosslinkers or biotin probes. A particularly useful book for sulfhydryl-specific fluorescent probes and other probes and crosslinkers can be obtained free from Thermo-Fisher (https://www.thermofisher.com/in/en/home/references/molecular-probes-the-handbook.html) whereas Toronto Research (https://www.trc-canada.com) has a large selection of spin and charged probes: The former are useful for probing a protein's structure by electron paramagnetic resonance (Hubbell, Lopez, Altenbach, & Yang, 2013). Although the labeling of the free sulfhydryls of engineered cysteines seems straightforward, there can be problems with the efficient labeling the cysteine residue with a probe. We have encountered some residues that are less accessible or inaccessible to the probe due to steric hindrance or the local environment of the cysteine sulfhydryl. One of the approaches we originally used with PFO to overcome this problem was to make the labeling buffer 3 M in guanidinium chloride to open up the structure sufficiently to allow the probe access to the sulfhydryl group (Shepard et al., 1998). In our case, the use of the guanidinium chloride did not have any apparent effect on its pore-forming activity, although this may not be the case for all pore-forming proteins. It is also possible that lower concentrations of guanidinium chloride or other chaotropic agents might work in this regard.

3.2 Mapping the membrane-spanning domains of PFTs using probe-labeled cysteines

One of the earliest uses of probe-labeled engineered cysteines in a PFT was done by Valeva et al. (Valeva et al., 1996) who mapped a putative membrane-embedded region of *S. aureus* α-hemolysin by individually mutating residues within its structure and modifying them with an environmentally sensitive fluorescent probe. They employed the environmentally sensitive fluorescent dye acrylodan, which is hydrophobic and tends to partition into the membrane even though a cysteine-substituted residue side-chain faces away from the membrane. Therefore, the structure of this membrane spanning region could not be determined, although the subsequent crystal structure revealed it to form a β-barrel pore (Song et al., 1996). Using a similar approach Shepard et al. (Shepard et al., 1998) mapped the residues of the membrane spanning regions of the β-PFT PFO. One important difference between these two studies, however, was the choice of the environmentally sensitive fluorescent probe: Shepard et al. used NBD [*N*,*N'*-dimethyl-*N*-(iodoacetyl)-*N'*-(7- nitrobenz-2-oxa-1,3-diazolyl)ethylenediamine)], which has an amphipathic nature so that does not spontaneously force the insertion of the modified side-chain of cysteine into the membrane if the side-chain resides near the membrane surface. Shepard et al. (Shepard et al., 1998) and subsequently Shatursky et al. (Shatursky et al., 1999) were able to show unambiguously that the membrane-spanning domain of PFO adopted a β-barrel structure by substituting individual cysteine residues along the length of two α-helical bundles (αHBs) (Fig. 1B) and modifying them with NBD. The αHBs were found to refold into two amphipathic β-strands to form a β-barrel pore. The NBD-modified cysteine-substituted residues that entered the membrane vs those that faced the aqueous pore lumen exhibited dramatic differences in their emission intensities, fluorescence lifetimes, and accessibility to collisional quenchers, which were restricted to either the membrane or the aqueous milieu. The β-barrel structure of the PFO pore was confirmed nearly 20 years later by the cryo-EM structure of the closely related *Streptococcus pneumoniae* pneumolysin pore (van Pee et al., 2017). Hence, it is important to select probes that are suited to what is being measured and to understand the caveats of the probe and its use.

3.3 The identification of other intra- and intermolecular interactions in PFO pore formation by labeled cysteines

One of the first significant intermolecular interactions between the monomers of the PFO oligomeric pore complex began with identification of an aromatic residue in β-strand 1 of PFO (Y181) by Hotze et al. (Hotze et al., 2002). When this residue was substituted with alanine or other non-aromatic residues it prevented the prepore to pore transition. Subsequently, Ramachandran et al. (Ramachandran et al., 2004) noted the presence another aromatic residue, F318, in β-strand 4 (Fig. 1C). Both residues were mutated to cysteine and labeled with the fluorescent probe pyrene, which increases emission near 450 nm when they stack and form an excimer. Once the labeled monomers oligomerized, pyrene excimer emission was detected, indicating that the two aromatic residues formed an intermolecular π-stacking interaction between β-strands 1 and 4 of neighboring monomers upon oligomerization. Together, with later studies by Burns et al. (Burns et al., 2019), this interaction has been shown to be critical in the initiation of the refolding of the αHB2 into one of the two membrane-spanning β-hairpins, which form the β-barrel pore and is described below in more detail.

The use of chemical labels with a charge has also been useful to elucidate electrostatic interactions critical for PFO's pore-forming ability. Substituting either of the charged residues K336 or E183 with alanine of PFO halted the progress of pore formation, like the aromatic pair described above, at the prepore stage. When the K336C mutant was allowed to form a prepore and then subsequently labeled with the positively charged molecule MTS-ET ([2-(Trimethylammonium)ethyl]methanethiosulfonate, bromide), pore conversion was observed. The reverse of this, mutant E183C labeled with negatively charged MTS-ES ((2-sulfonatoethyl)methanethiosulfonate), produced the same result. This studied revealed what we now know to be a temporary, yet critical electrostatic interaction between these two residues, which along with the two aromatic residues described above provides the energy necessary to initiate the transition of the prepore to the pore complex (Wade et al., 2015) (Fig. 1B and C).

4. Engineering disulfide bridges into PFTs

4.1 Trapping PFT in intermediate states of pore-formation

The ability to introduce disulfide bridges into the structure of PFTs has been useful in defining various stages of their mechanism, which also allows the possibility of imaging these intermediate states by various molecular imaging methods. One of the early uses of engineered disulfide bridges in a PFT was that of van der Goot et al. (van der Goot, Hardie, Parker, & Buckley, 1994) who used a disulfide bridge to covalently crosslink the aerolysin C-terminal propeptide to the main body of the toxin, which showed it inhibited pore-forming activity when oxidized but activity could be restored by its reduction that allowed it to dissociate from the toxin. Hotze et al. (Hotze et al., 2001) used an engineered disulfide to trap PFO in an oligomeric prepore state, the formation of which for the CDCs was controversial at that time, but has been accepted as a necessary intermediate in the pore-forming mechanism of the CDCs. They locked the membrane spanning domain of PFO with a disulfide to the backbone of the toxin, which allowed the membrane-bound monomers to form their large oligomeric prepore complex (35–40 monomers) but which was unable to assemble and insert the β-barrel pore. It is important to note that this particular prepore state was sensitive to dissociation by SDS whereas the pore complex is not. This will also be important in the next section where a disulfide trapped mutant was trapped in an SDS-stable state, which suggested it was trapped at a later stage of the prepore assembly. When released by the reduction of the disulfide the β-barrel formed much more rapidly than wildtype toxin demonstrating that the prepore was primed for the assembly and insertion of the β-barrel pore. This same disulfide locked mutant was also instrumental in demonstrating that the prepore complex of PFO underwent a collapse toward the surface of the bilayer to insert the β-barrel pore (Czajkowsky, Hotze, Shao, & Tweten, 2004; Ramachandran, Tweten, & Johnson, 2005): the disulfide trapped prepore was determined to collapse 35–40 Å toward the membrane after the reduction of the disulfide. This collapse was necessary to allow the β-barrel to span the bilayer since the membrane-spanning β-hairpins were only long enough to reach the membrane surface without this collapse. Subsequently, there have been many uses for engineered disulfides to prevent specific conformational stages of PFTs from proceeding (e.g., Iacovache et al., 2006; Jacquez et al., 2015; Kawate & Gouaux, 2003; Kristan et al., 2004; Lukoyanova et al., 2015; Melton, Parker, Rossjohn, Buckley, & Tweten, 2004; Ramachandran et al., 2004).

Most of our disulfide-trapped mutants of PFO (like that described above), if allowed to first form the prepore before reducing the disulfide, could be triggered to form the pore when reduced. However, one disulfide-trapped intermediate revealed that a specific structural transition had to occur at a specific time during the prepore to pore conversion, or it remained trapped in a stable prepore state after the reduction of the disulfide (Evans et al., 2020). In this specific case, two interacting aromatics form a flexible interaction that anchors one end of the αβ loop in PFO (Fig. 1D). This loop necessarily swings out of the way to allow the formation of the necessary intermolecular interactions between membrane-bound monomers to form the pore and allow its transition to the pore (Ramachandran et al., 2004). When these two aromatics (Fig. 1D) were substituted with cysteines they formed an intramolecular disulfide, which trapped PFO in an SDS-stable prepore state, in contrast to the SDS-sensitive prepore found for the Y181A mutant that eliminated the intermolecular π-stacking interaction (Hotze et al., 2002). The SDS-stable prepore suggested that it was trapped in a later stage of prepore assembly, showing that specific conformational changes and/or interactions must occur at specific times of the prepore assembly. If the disulfide in this mutant (Fig. 1D) was reduced in the soluble monomer, the mutant made the transition to the pore similar to native PFO. In contrast to the above studies wherein the disulfide-trapped prepore trapped complex rapidly converted to the pore when reduced, this mutant prevented conversion of the prepore to the pore when reduced. This result revealed that the αβ loop (Fig. 1) must be displaced at a specific time during the assembly of the prepore complex. If this occurred too late the αβ loop (specifically the short β-strand 5 of the αβ loop) remained trapped at the monomer-monomer interface and the remaining conformational changes necessary for the assembly of the β-barrel pore could not proceed. This likely prevented the β-strand to α-helical transition to occur in β-strand 5, which moves the αβ loop away from the monomer-monomer interface and results in the formation of the helix-turn-helix motif in the pore complex (Fig. 1D). Hence, this disulfide revealed the timing of this transition must occur at a specific time during the assembly of the prepore in order for the subsequent interactions and conformational transitions necessary for the formation of the β-barrel pore.

4.2 Resolving the structure of a transmembrane β-barrel using disulfide crosslinking

Disulfide formation has also been useful in delineating the structure of the β-barrel pore in PFO. Sato et al. (Sato, Tweten, & Johnson, 2013) used disulfide scanning to determine how the residues of the β-hairpins of the monomers within the pore complex were aligned in the β-barrel pore. This procedure involved the substitution of each residue along the two β-strands (β-strands 1 and 4 in Fig. 2) of the two monomers, which formed the intermolecular backbone hydrogen bonds upon the formation of the β-barrel pore (Fig. 2). Equimolar amounts of one of the cysteine mutants in β-strand 1 was individually mixed with each cysteine mutant in β-strand 4. Because of the distance limitations for disulfide bond formation the formation of disulfides would only occur if two cysteines were substituted for residues that formed backbone hydrogen bonds in the β-barrel pore (Fig. 2). Both mutants would then be added to liposomes and allowed to form pores. The addition of tetrathionate was used to promote the formation of disulfides. The assembly of the pore results in a Gaussian distribution of monomer pairings but sufficient numbers of one monomer with a cysteine in β-strand 1 pairing with a monomer containing a cysteine in β-strand 4 would form a disulfide (assuming that they juxtaposed in the β-barrel). The formation of the disulfides was detected by the presence of stable disulfide-linked dimers of PFO by SDS-polyacrylamide gel electrophoresis. This approach also showed that the observed pattern of disulfide formation required that the β-barrel pore β-hairpins adopt a 20° right handed twist with respect to the axis of the β-barrel: an observation that was confirmed 4 years later by cryo-EM structure of the related pneumolysin pore (van Pee et al., 2017).

4.3 Predicting disulfide formation from structural information

In general, disulfide formation between two engineered cysteines can be used to identify any pair of residues that are in close proximity, providing that the conformational restrictions of the cysteine side-chain of the two cysteines allow the formation of the disulfide. This may be less of a problem if the regions containing the cysteines exhibit some flexibility. In nearly all of our studies (e.g., Hotze et al., 2001; Ramachandran et al., 2004; Sato et al., 2013) disulfides formed when they were not expected to form by predictive programs for engineered disulfides between specific residues (Craig & Dombkowski, 2013) of a protein. Hence, the lesson here is that the failure of predictive models to suggest the formation of a disulfide

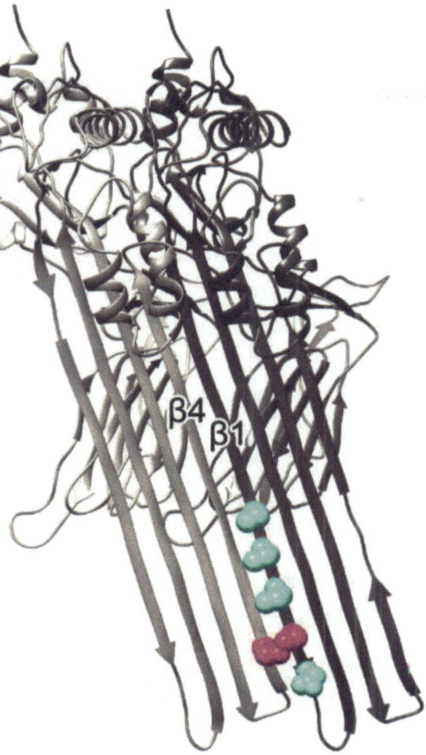

Fig. 2 Disulfide formation reveals the structure of the β-barrel pore. Shown is an example of the use of disulfide formation to reveal how the membrane-spanning β-strands of the β-barrel pore aligned with each other in the β-barrel pore. A pair of dimers are shown from the PFO pore complex modeled on the PLY cryoEM pore structure (Burns et al., 2019; van Pee et al., 2017). PFO mutants carrying 1 cysteine mutant in β-strand 4 (β4) would be individually mixed with each mutant carrying a single cysteine mutant in β-strand 1 (β1). These mixtures would then be subjected to SDS polyacrylamide gel electrophoresis. Those cysteines that were juxtaposed formed stable disulfide (magenta atoms) crosslinked dimers that could only be separated by reducing agent. These pairings could only form if the β-hairpins were tilted about 20° (Sato et al., 2013), which was shown several years later by the pore structure of the closely related CDC pneumolysin (van Pee et al., 2017).

between two adjacent cysteine-substituted residues should not discourage one from attempting it experimentally. In our experience, it appears that these types of prediction programs only provide the relative probability that a pair of residues, when substituted with cysteines, will form a disulfide. If the pair of residues are predicted to form a disulfide then the odds are probably better that they will. It is also important to determine the efficiency of disulfide formation, which can be quantified by using any number of sulfhydryl-specific probes to quantify free sulfhydryl groups (e.g., Ellman's reagent Ellman, 1959), especially if a third cysteine is to be introduced and modified with a probe.

These studies show that the use of engineered disulfide bridges to lock PFTs at various stages of their pore-forming mechanism have revealed major structural transitions that would not have otherwise been possible, unless one would be lucky enough to identify a point mutation that resulted in the same phenotype. Furthermore, the difference between trapping a protein at some intermediate conformational state with a point mutation vs an engineered disulfide is that the latter can be reduced to allow the conformation change to proceed. It is important to emphasize that this type of approach would not be likely to succeed without a crystal structure since a structure allows one to make predictions of how specific substitutions may affect conformational changes that impact the pore-forming mechanism.

5. Insights into the pore formatting mechanism by altering the amino acid side-chain properties

Perhaps one of the most useful approaches of protein engineering is to reveal the contribution of side-chain mediated interactions of amino acids within a pore-forming protein (or any protein), which is achieved by substituting residues with side-chains of different chemical properties. This versatile approach can be used in many ways, including to gather useful information about whether two or more residues form an interaction that is critical to a specific stage of the pore-forming mechanism.

5.1 Identifying intermolecular interactions of the oligomeric pore complex that drive the assembly of the β-barrel pore

Before demonstrating how single amino acid substitutions revealed critical aspects of the PFO prepore to pore transition it is important to point out a specific structural feature of the soluble monomer: the domain 3 core β-sheet exhibits a severe twist (Fig. 1). This twist must be relieved to allow the formation of the β-barrel pore: the core β-sheet must be "flattened" to form the upper wall of the β-barrel pore. This twist is largely maintained by interactions between αHB1 and domains 1 and 2 (Wade et al., 2019), and by interactions between αHB2 and the core β-sheet itself (Burns et al., 2019).

Understanding how this twist is relieved and its importance to the prepore to pore transition began with studies in 2002 and 2004 (Hotze et al., 2002; Ramachandran et al., 2004) that revealed an essential intermolecular interaction, which was critical to this process. Hotze et al. and Ramachandran et al. (Hotze et al., 2002; Ramachandran et al., 2004) showed that pore-forming activity was lost when Y181 (Fig. 1C), in β-strand 1 of the core β-sheet of PFO, was substituted with alanine. The loss of this interaction trapped PFO in a prepore state (Hotze et al., 2002). Ramachandran et al. (Ramachandran et al., 2004) further showed that Y181 formed an intermolecular π-stacking interaction with F318 in β-strand 4 (Fig. 1C) of an adjacent monomer in the pore complex. This stacking interaction was originally proposed to align the β-hairpins of the juxtaposed monomers within the prepore complex (Ramachandran et al., 2004) and its loss trapped PFO in a prepore state. However, over a decade later, Burns et al. (Burns et al., 2019) identified a mutation in αHB2 that restored wildtype pore-forming activity to mutants that could not form the stacking interaction. They revealed that substituting specific residues, which are important to the stability of the interface between αHB2 and the domain 3 core β-sheet (Fig. 1B and C), restored pore-forming activity to mutants in which the π-stacking interaction was eliminated. These PFO derivatives weakened the interaction between αHB2 with the core β-sheet (determined by changes in the melting temperature of PFO, which is largely dictated by the unfolding of domain 3 Wade et al., 2015). These results strongly suggested that the energy supplied by the formation of the π-stacking interaction was necessary to break the interaction of αHB2 with the core-β sheet thereby allowing it to refold into one of the two membrane-spanning β-hairpins (Shatursky et al., 1999). The other interaction that worked in tandem with the π-stacking interaction was an intermolecular electrostatic interaction between K336 and E183 (described in detail below) (Wade et al., 2015).

Wade et al. (Wade et al., 2015) showed that an intermolecular electrostatic interaction between E183 in β-strand 1 and K336 on the αβ loop (Fig. 1) was necessary for the disruption of the interface between αHB1 and domains 1 and 2. Similar to the loss of the π-stacking interaction, the loss of the intermolecular electrostatic interaction between E183 and K336 also trapped PFO in a prepore state. Over 20 years ago the crystal structure of the soluble PFO monomer revealed that this interface is largely composed of a network of water molecules that bridge residues in αHB1 to residues in domains 1 and 2 (Rossjohn et al., 1997). It further showed that the nonpolar residues at this interface are not in complementary interactions (i.e., hydrophobic-hydrophobic) suggesting an inherent instability of this interface. This interface, like that between αHB2 and the core β-sheet, is also necessarily broken to allow αHB1 to refold, along with αHB2, into one of the two membranes-spanning hairpins that form the β-barrel pore

(Shepard et al., 1998). Therefore, the interactions at these two interfaces are sufficient to maintain the structure of domain 3 and the twist in the core β-sheet in the soluble monomer but must be disrupted to allow αHB1 and 2 to refold into the β-hairpins of the β-barrel pore. Early work by Heuck et al. (Heuck et al., 2000) showed that altering N197 to cysteine increased its rate of pore-formation, which increased dramatically when labeled with the fluorescent probe NBD. Interestingly, N197 is located at the core of the αHB1-domains 1,2 interface and the bridging water network. Nearly 20 years later Wade et al. (Wade et al., 2019) showed that in the crystal structure of the PFO mutant, wherein N197 was substituted with tryptophan (N197W), that the bridging waters were nearly completely lost from this interface, although the local structure of the protein backbone and amino acid side-chains was largely unperturbed. Similar to the mutants in αHB2 that restored activity to mutants lacking the π-stacking interaction the N197W mutation restored greater than wildtype pore-forming activity to the PFO mutant that lacked the ability to form the intermolecular electrostatic interaction (Wade et al., 2015).

These revelations were only possible because we had generated amino acid substitutions at these interfaces that eliminated some but not all of the stabilizing the interactions, which allowed pore formation to proceed in the absence of either the π-stacking or electrostatic interactions. Destabilizing these interfaces eliminated the need for the energy supplied by the formation of either the π-stacking or the electrostatic interactions to disrupt those interfaces by flattening the core β-sheet. It is important to note that amino acid substitutions at the interface between αHB1 and domains 1 and 2, which restored activity to mutants that lacked the ability to form the intermolecular electrostatic interaction, did not restore activity to mutant proteins that lacked the ability to form the π-stacking interaction. Likewise, amino acid substitutions in αHB2 that restored activity to mutant proteins that lacked the intermolecular π-stacking interaction did not restore activity to mutants that lacked the intermolecular electrostatic interaction. Hence, these studies revealed that the π-stacking and electrostatic interactions worked in tandem to bring together β-strands 1 and 4 of the adjacent monomers in the oligomeric complex, which relieved the severe twist in the domain 3 and flattened the core β-sheet. This effectively disrupted the interface interactions of αHB1 and 2 that maintained their structure and allowed then to refold into the membrane-spanning β-hairpins of the β-barrel pore. These studies also led to a greater understanding of the basis for the temperature dependence of the pore-forming process, as described in the next section.

5.2 Mutations that revealed the structural basis for the temperature-dependence of pore formation

Early studies by Shepard et al. (Shepard, Shatursky, Johnson, & Tweten, 2000) showed that PFO could be trapped in its prepore state at low temperature (4 °C), which was the first demonstration that the PFO pore-forming mechanism was sensitive to low temperature and that it formed a large prepore complex prior to inserting the β-barrel pore. Wade et al. (Wade et al., 2019) subsequently showed that the introduction of the N197W mutation into PFO, which, as described above, results in the loss of the bridging water network at the interface between αHB1 and domains 1 and 2, significantly increased its pore-forming activity and, unexpectedly, largely eliminated the temperature-dependence of PFO's pore-forming activity. Native PFO quickly loses the ability to form pores below 30 °C (Wade et al., 2019) and remains trapped in a prepore state at low temperatures (Shepard et al., 2000). It is virtually inactive at temperatures below 15 °C whereas the N197W mutant did not lose significant activity down to 9 °C. These studies suggest that as temperature decreases, the stability provided by the of hydrogen bonds of the interface bridging water network, prevents the disengagement of αHB1 from its interface with domain 2 thereby trapping PFO in prepore state.

Although these substitutions experimentally revealed fundamental features regulating or restricting the pore-forming activity of PFO, they also provided a possible explanation for how the related CDC from *Desulfobulbus propionicus* (DLY), an inhabitant of colder climates, retains significant activity down to 6 °C (Wade et al., 2019). The DLY interface between its αHB1 and domains 1 and 2 is significantly different from that in PFO: compared to PFO there are half as many bridging waters at this interface in the crystal structure of DLY. This difference is partly due to increased number of bulky nonpolar residues: about half of the interface residues in DLY are nonpolar compared to less than 40% in PFO. Importantly, in DLY the hydrophobic interactions generally form complementary (hydrophobic–hydrophobic) interactions whereas the nonpolar residues in PFO are not in complementary interactions (Rossjohn et al., 1997). The bridging waters at this interface are also reduced from 12 in PFO to 7 in DLY. These observations may explain the difference in low temperature sensitivity of pore formation between the native PFO and DLY. More energy is required to break the stronger hydrogen-bond network present in PFO, explaining the requirement of a higher temperature to trigger the disengagement of αHB1 from domains 1 and 2, compared to DLY. On the other hand, the hydrophobic nature of the DLY interface may allow it to function at lower temperatures since it exhibits fewer bridging waters and a substantial increase in complementary nonpolar interactions. Although the nature of the hydrophobic effect remains unclear, it is thought to be important to the process of cold denaturation of proteins, as it is known that the stability of the hydrophobic effect decreases with temperature (Privalov, 1990; Sanfelice, Morandi, Pastore, Niccolai, & Temussi, 2015). Hence, the nonpolar nature of the DLY αHB1-domains 1,2 interface may explain its ability to function at much lower temperatures than PFO, as the effect of cold denaturation may destabilize this interface making it possible for DLY to remain active across a much broader range of temperatures. This scenario was supported by the observation that the substitution of a single hydrophobic residue in the αHB1 of DLY (L162), which is positioned within a core of 4 hydrophobic residues in domain 2 that bridge it αHB1 to domain 2, significantly reduced its activity at low temperatures but it retained wildtype DLY activity at temperatures above 20 °C (Wade et al., 2019). Hence, these studies revealed insight into how the CDCs functioned and how they can evolve to alter their specific activity of pore formation and how the CDC structure can adapt to function optimally at different temperatures.

6. Concluding remarks

The investigation of pore forming proteins by means of protein engineering has enabled researchers to understand in fine detail many of the mechanistic steps taken by these secreted soluble proteins, as they transform into membrane embedded, often large, protein complexes. Structural techniques such as X-ray crystallography and cryo-EM remain important tools for the field and have been critical for hypothesis generation and the confirmation of prior biochemical studies as discussed above. Until structural technologies improve to the extent of real-time visualization of protein dynamics and interaction at atomic resolution, the tools of protein engineering combined with molecular biology, biochemistry, structural biology and biophysics will remain indispensable for the study of PFTs and related proteins. We hope our experiences, as detailed in this review, will serve as a resource for investigators in this field.

Acknowledgments

I would like to acknowledge two incredibly productive, long-term collaborations with Professors Michael Parker and Art Johnson (the latter now retired) and the individuals in their lab groups, which began nearly a quarter of a century ago and continues to this day with Professor Parker. Their keen insights and enthusiasm made the deep insights into the molecular mechanisms of these pore-forming proteins possible. These studies have been and continue to be supported by the NIH/NIAID (4R37AI037657-25).

References

Bourdeau, R. W., Malito, E., Chenal, A., Bishop, B. L., Musch, M. W., Villereal, M. L., et al. (2009). Cellular functions and X-ray structure of anthrolysin O, a cholesterol-dependent cytolysin secreted by bacillus anthracis. *The Journal of Biological Chemistry, 284*(21), 14645–14656. https://doi.org/M807631200 (pii) https://doi.org/10.1074/jbc.M807631200.

Burns, J. R., Morton, C. J., Parker, M. W., & Tweten, R. K. (2019). An intermolecular pi-stacking interaction drives conformational changes necessary to beta-barrel formation in a pore-forming toxin. *mBio, 10*(4). https://doi.org/10.1128/mBio.01017-19.

Cohen, S. N., Chang, A. C., Boyer, H. W., & Helling, R. B. (1973). Construction of biologically functional bacterial plasmids in vitro. *Proceedings. National Academy of Sciences. United States of America, 70*(11), 3240–3244. https://doi.org/10.1073/pnas.70.11.3240.

Collier, R. J. (1975). Diphtheria toxin: Mode of action and structure. *Bacteriological Reviews, 39*(1), 54–85.

Craig, D. B., & Dombkowski, A. A. (2013). Disulfide by Design 2.0: A web-based tool for disulfide engineering in proteins. *BMC Bioinformatics, 14*, 346. https://doi.org/10.1186/1471-2105-14-346.

Czajkowsky, D. M., Hotze, E. M., Shao, Z., & Tweten, R. K. (2004). Vertical collapse of a cytolysin prepore moves its transmembrane β-hairpins to the membrane. *The EMBO Journal, 23*(16), 3206–3215.

Dal Peraro, M., & van der Goot, F. G. (2016). Pore-forming toxins: Ancient, but never really out of fashion. *Nature Reviews. Microbiology, 14*(2), 77–92. https://doi.org/10.1038/nrmicro.2015.3.

Dowd, K. J., Farrand, A. J., & Tweten, R. K. (2012). The cholesterol-dependent cytolysin signature motif: A critical element in the allosteric pathway that couples membrane binding to pore assembly. *PLoS Pathogens, 8*(7), e1002787. https://doi.org/10.1371/journal.ppat.1002787.

Ellman, G. L. (1959). Tissue sulfhydryl groups. *Archives of Biochemistry and Biophysics, 82*(1), 70–77. https://doi.org/10.1016/0003-9861(59)90090-6.

Evans, J. C., Johnstone, B. A., Lawrence, S. L., Morton, C. J., Christie, M. P., Parker, M. W., et al. (2020). A key motif in the cholesterol-dependent cytolysins reveals a large family of related proteins. *mBio, 11*(5). https://doi.org/10.1128/mBio.02351-20.

Feil, S. C., Ascher, D. B., Kuiper, M. J., Tweten, R. K., & Parker, M. W. (2014). Structural studies of *Streptococcus pyogenes* streptolysin O provide insights into the early steps of membrane penetration. *Journal of Molecular Biology, 426*(4), 785–792. https://doi.org/10.1016/j.jmb.2013.11.020.

Feil, S. C., Rossjohn, J., Rohde, K., Tweten, R. K., & Parker, M. W. (1996). Crystallization and preliminary X-ray analysis of a thiol-activated cytolysin. *FEBS Letters, 397*(2–3), 290–292.

Harris, R. W., Sims, P. J., & Tweten, R. K. (1991). Kinetic aspects of the aggregation of *Clostridium perfringens* theta toxin on erythrocyte membranes: A fluorescence energy transfer study. *The Journal of Biological Chemistry, 266*(11), 6936–6941.

Heuck, A. P., Hotze, E., Tweten, R. K., & Johnson, A. E. (2000). Mechanism of membrane insertion of a multimeric b-barrel protein: Perfringolysin O creates a pore using ordered and coupled conformational changes. *Molecular Cell, 6*, 1233–1242.

Heuck, A. P., Tweten, R. K., & Johnson, A. E. (2003). Assembly and topography of the prepore complex in cholesterol-dependent cytolysins. *The Journal of Biological Chemistry, 278*(33), 31218–31225. https://doi.org/10.1074/jbc.M303151200.

Hotze, E. M., Heuck, A. P., Czajkowsky, D. M., Shao, Z., Johnson, A. E., & Tweten, R. K. (2002). Monomer-monomer interactions drive the prepore to pore conversion of a beta-barrel-forming cholesterol-dependent cytolysin. *The Journal of Biological Chemistry*, *277*(13), 11597–11605. https://doi.org/10.1074/jbc.M111039200.

Hotze, E. M., Wilson-Kubalek, E., Farrand, A. J., Bentsen, L., Parker, M. W., Johnson, A. E., et al. (2012). Monomer-monomer interactions propagate structural transitions necessary for pore formation by the cholesterol-dependent cytolysins. *The Journal of Biological Chemistry*, *287*(29), 24534–24543. https://doi.org/10.1074/jbc.M112.380139.

Hotze, E. M., Wilson-Kubalek, E. M., Rossjohn, J., Parker, M. W., Johnson, A. E., & Tweten, R. K. (2001). Arresting pore formation of a cholesterol-dependent cytolysin by disulfide trapping synchronizes the insertion of the transmembrane beta-sheet from a prepore intermediate. *The Journal of Biological Chemistry*, *276*(11), 8261–8268.

Howorka, S., & Bayley, H. (1998). Improved protocol for high-throughput cysteine scanning mutagenesis. *BioTechniques*, *25*(5), 764–766. 768, 770 passim https://doi.org/10.2144/98255bm03.

Howorka, S., Cheley, S., & Bayley, H. (2001). Sequence-specific detection of individual DNA strands using engineered nanopores. *Nature Biotechnology*, *19*(7), 636–639. https://doi.org/10.1038/90236.

Hubbell, W. L., Lopez, C. J., Altenbach, C., & Yang, Z. (2013). Technological advances in site-directed spin labeling of proteins. *Current Opinion in Structural Biology*, *23*(5), 725–733. https://doi.org/10.1016/j.sbi.2013.06.008.

Iacovache, I., Paumard, P., Scheib, H., Lesieur, C., Sakai, N., Matile, S., et al. (2006). A rivet model for channel formation by aerolysin-like pore-forming toxins. *The EMBO Journal*, *25*(3), 457–466.

Iwamoto, M., Ohno-Iwashita, Y., & Ando, S. (1987). Role of the essential thiol group in the thiol-activated cytolysin from *Clostridium perfringens*. *European Journal of Biochemistry*, *167*(3), 425–430.

Jacquez, P., Avila, G., Boone, K., Altiyev, A., Puschhof, J., Sauter, R., et al. (2015). The disulfide bond Cys255-Cys279 in the immunoglobulin-like domain of Anthrax toxin receptor 2 is required for membrane insertion of Anthrax protective antigen pore. *PLoS One*, *10*(6), e0130832. https://doi.org/10.1371/journal.pone.0130832.

Jones, D. H., & Howard, B. H. (1991). A rapid method for recombination and site-specific mutagenesis by placing homologous ends on DNA using polymerase chain reaction. *BioTechniques*, *10*(1), 62–66.

Kawate, T., & Gouaux, E. (2003). Arresting and releasing staphylococcal alpha-hemolysin at intermediate stages of pore formation by engineered disulfide bonds. *Protein Science*, *12*(5), 997–1006. https://doi.org/10.1110/ps.0231203.

Kristan, K., Podlesek, Z., Hojnik, V., Gutierrez-Aguirre, I., Guncar, G., Turk, D., et al. (2004). Pore formation by equinatoxin, a eukaryotic pore-forming toxin, requires a flexible N-terminal region and a stable beta-sandwich. *The Journal of Biological Chemistry*, *279*(45), 46509–46517. https://doi.org/10.1074/jbc.M406193200.

Lawrence, S. L., Feil, S. C., Morton, C. J., Farrand, A. J., Mulhern, T. D., Gorman, M. A., et al. (2015). Crystal structure of *Streptococcus pneumoniae* pneumolysin provides key insights into early steps of pore formation. *Scientific Reports*, *5*, 14352. https://doi.org/10.1038/srep14352.

Li, F., & Mullins, J. I. (2002). Site-directed mutagenesis facilitated by DpnI selection on hemimethylated DNA. *Methods in Molecular Biology*, *182*, 19–27. https://doi.org/10.1385/1-59259-194-9:019.

Lukoyanova, N., Kondos, S. C., Farabella, I., Law, R. H., Reboul, C. F., Caradoc-Davies, T. T., et al. (2015). Conformational changes during pore formation by the perforin-related protein pleurotolysin. *PLoS Biology*, *13*(2), e1002049. https://doi.org/10.1371/journal.pbio.1002049.

Matteucci, M. D., & Caruthers, M. H. (1992). Synthesis of deoxyoligonucleotides on a polymer support. 1981. *Biotechnology, 24*, 92–98.
Melton, J. A., Parker, M. W., Rossjohn, J., Buckley, J. T., & Tweten, R. K. (2004). The identification and structure of the membrane-spanning domain of the *Clostridium septicum* alpha toxin. *The Journal of Biological Chemistry, 279*(14), 14315–14322. https://doi.org/10.1074/jbc.M313758200.
Melton-Witt, J. A., Bentsen, L. M., & Tweten, R. K. (2006). Identification of functional domains of *Clostridium septicum* alpha toxin. *Biochemistry, 45*(48), 14347–14354. https://doi.org/10.1021/bi061334p.
Mourez, M., Yan, M., Lacy, D. B., Dillon, L., Bentsen, L., Marpoe, A., et al. (2003). Mapping dominant-negative mutations of anthrax protective antigen by scanning mutagenesis. *Proceedings. National Academy of Sciences. United States of America, 100*(24), 13803–13808.
Mullis, K., Faloona, F., Scharf, S., Saiki, R., Horn, G., & Erlich, H. (1986). Specific enzymatic amplification of DNA in vitro: The polymerase chain reaction. *Cold Spring Harbor Symposia on Quantitative Biology, 51*(Pt 1), 263–273. https://doi.org/10.1101/sqb.1986.051.01.032.
Olson, R., Nariya, H., Yokota, K., Kamio, Y., & Gouaux, E. (1999). Crystal structure of staphylococcal LukF delineates conformational changes accompanying formation of a transmembrane channel. *Nature Structural Biology, 6*(2), 134–140. https://doi.org/10.1038/5821.
Pettersen, E. F., Goddard, T. D., Huang, C. C., Couch, G. S., Greenblatt, D. M., Meng, E. C., et al. (2004). UCSF chimera- -a visualization system for exploratory research and analysis. *Journal of Computational Chemistry, 25*(13), 1605–1612. https://doi.org/10.1002/jcc.20084.
Polekhina, G., Giddings, K. S., Tweten, R. K., & Parker, M. W. (2004). Crystallization and preliminary X-ray analysis of the human-specific toxin intermedilysin. *Acta Crystallographica. Section D, Biological Crystallography, 60*(Pt 2), 347–352.
Polekhina, G., Giddings, K. S., Tweten, R. K., & Parker, M. W. (2005). Insights into the action of the superfamily of cholesterol-dependent cytolysins from studies of intermedilysin. *Proceedings. National Academy of Sciences. United States of America, 102*(3), 600–605. https://doi.org/Doi https://doi.org/10.1073/Pnas.0403229101.
Privalov, P. L. (1990). Cold denaturation of proteins. *Critical Reviews in Biochemistry and Molecular Biology, 25*(4), 281–305. https://doi.org/10.3109/10409239009090612.
Ramachandran, R., Tweten, R. K., & Johnson, A. E. (2004). Membrane-dependent conformational changes initiate cholesterol-dependent cytolysin oligomerization and intersubunit beta-strand alignment. *Nature Structural & Molecular Biology, 11*(8), 697–705. https://doi.org/10.1038/nsmb793.
Ramachandran, R., Tweten, R. K., & Johnson, A. E. (2005). The domains of a cholesterol-dependent cytolysin undergo a major FRET-detected rearrangement during pore formation. *Proceedings. National Academy of Sciences. United States of America, 102*(20), 7139–7144. https://doi.org/10.1073/pnas.0500556102.
Rossjohn, J., Feil, S. C., McKinstry, W. J., Tweten, R. K., & Parker, M. W. (1997). Structure of a cholesterol-binding thiol-activated cytolysin and a model of its membrane form. *Cell, 89*, 685–692.
Sanfelice, D., Morandi, E., Pastore, A., Niccolai, N., & Temussi, P. A. (2015). Cold denaturation unveiled: Molecular mechanism of the asymmetric unfolding of yeast Frataxin. *ChemPhysChem, 16*(17), 3599–3602. https://doi.org/10.1002/cphc.201500765.
Sato, T. K., Tweten, R. K., & Johnson, A. E. (2013). Disulfide-bond scanning reveals assembly state and beta-strand tilt angle of the PFO beta-barrel. *Nature Chemical Biology, 9*(6), 383–389. https://doi.org/10.1038/nchembio.1228.
Shatursky, O., Heuck, A. P., Shepard, L. A., Rossjohn, J., Parker, M. W., Johnson, A. E., et al. (1999). The mechanism of membrane insertion for a cholesterol-dependent cytolysin: A novel paradigm for pore-forming toxins. *Cell, 99*(3), 293–299. https://doi.org/10.1016/s0092-8674(00)81660-8.

Shepard, L. A., Heuck, A. P., Hamman, B. D., Rossjohn, J., Parker, M. W., Ryan, K. R., et al. (1998). Identification of a membrane-spanning domain of the thiol-activated pore-forming toxin *Clostridium perfringens* perfringolysin O: An α-helical to β-sheet transition identified by fluorescence spectroscopy. *Biochemistry*, *37*(41), 14563–14574.

Shepard, L. A., Shatursky, O., Johnson, A. E., & Tweten, R. K. (2000). The mechanism of assembly and insertion of the membrane complex of the cholesterol-dependent cytolysin perfringolysin O: Formation of a large prepore complex. *Biochemistry*, *39*(33), 10284–10293.

Song, L., Hobaugh, M. R., Shustak, C., Cheley, S., Bayley, H., & Gouaux, J. E. (1996). Structure of staphylococcal alpha-hemolysin, a heptameric transmembrane pore. *Science*, *274*(5294), 1859–1866.

Tweten, R. K., Parker, M. W., & Johnson, A. E. (2001). The cholesterol-dependent Cytolysins. In G. van der Goot (Ed.), *Pore-forming toxins* (pp. 15–33). Heidelberg: Springer-Verlag.

Valeva, A., Weisser, A., Walker, B., Kehoe, M., Bayley, H., Bhakdi, S., et al. (1996). Molecular architecture of a toxin pore: A 15-residue sequence lines the transmembrane channel of staphylococcal alpha-toxin. *The EMBO Journal*, *15*(8), 1857–1864.

Van der Goot, F. G., Hardie, K. R., Parker, M. W., & Buckley, J. T. (1994). The C-terminal peptide produced upon proteolytic activation of the cytolytic toxin aerolysin is not involved in channel formation. *The Journal of Biological Chemistry*, *269*(48), 30496–30501.

van Pee, K., Mulvihill, E., Muller, D. J., & Yildiz, O. (2016). Unraveling the pore-forming steps of Pneumolysin from Streptococcus pneumoniae. *Nano Letters*, *16*(12), 7915–7924. https://doi.org/10.1021/acs.nanolett.6b04219.

van Pee, K., Neuhaus, A., D'Imprima, E., Mills, D. J., Kuhlbrandt, W., & Yildiz, O. (2017). CryoEM structures of membrane pore and prepore complex reveal cytolytic mechanism of Pneumolysin. *eLife*, *6*. https://doi.org/10.7554/eLife.23644.

Wade, K. R., Hotze, E. M., Kuiper, M. J., Morton, C. J., Parker, M. W., & Tweten, R. K. (2015). An intermolecular electrostatic interaction controls the prepore-to-pore transition in a cholesterol-dependent cytolysin. *Proceedings. National Academy of Sciences. United States of America*, *112*(7), 2204–2209. https://doi.org/10.1073/pnas.1423754112.

Wade, K. R., Lawrence, S. L., Farrand, A. J., Hotze, E. M., Kuiper, M. J., Gorman, M. A., et al. (2019). The structural basis for a transition state that regulates pore formation in a bacterial toxin. *mBio*, *10*(2). https://doi.org/10.1128/mBio.00538-19.

Walker, B., & Bayley, H. (1995). Key residues for membrane binding, oligomerization, and pore forming activity of staphylococcal alpha-hemolysin identified by cysteine scanning mutagenesis and targeted chemical modification. *The Journal of Biological Chemistry*, *270*(39), 23065–23071.

Xu, L., Huang, B., Du, H., Zhang, X. C., Xu, J., Li, X., et al. (2010). Crystal structure of cytotoxin protein suilysin from Streptococcus suis. *Protein & Cell*, *1*(1), 96–105. https://doi.org/10.1007/s13238-010-0012-3.

CHAPTER THREE

Electron microscopy as a critical tool in the determination of pore forming mechanisms in proteins

Robert J.C. Gilbert*
Division of Structural Biology, Wellcome Centre for Human Genetics, University of Oxford, Oxford, United Kingdom
*Corresponding author: e-mail address: gilbert@strubi.ox.ac.uk

Contents

Abstract

Electron microscopy has consistently played an important role in the description of pore-forming protein systems. The discovery of pore-forming proteins has depended on visualization of the structural pores formed by their oligomeric protein complexes, and as electron microscopy has advanced technologically so has the degree of insight it has been able to give. This review considers a large number of published studies of pore-forming complexes in prepore and pore states determined using single-particle cryo-electron microscopy. Sample isolation and preparation, imaging and image analysis, structure determination and optimization of results are all discussed alongside

Methods in Enzymology, Volume 649
ISSN 0076-6879
https://doi.org/10.1016/bs.mie.2021.01.034

challenges which pore-forming proteins particularly present. The review also considers the use made of cryo-electron tomography to study pores within their membrane environment and which will prove an increasingly important approach for the future.

1. Introduction

This review, like others in this volume of *Methods in Enzymology*, focuses on pore-forming proteins or toxins defined as those proteins which are initially produced in the form of species free in solution, but which then transition to a membrane-inserted form via a self-association that triggers refolding needed to move from a hydrophilic to a lipophilic state. What this review does not consider are native membrane proteins which may be said to form pores (e.g. porins, ion channels) which have of course themselves been the subject of numerous successful applications of cryo-electron microscopy (cryo-EM) (e.g. (Laverty et al., 2019; Zhao, Chen, Swensen, Qian, & Gouaux, 2019)).

Pore-forming systems defined in this way present particular challenges for cryo-EM study, but also particular opportunities and there are aspects of their structure and mechanism that cryo-EM is peculiarly well-suited to address. Electron microscopy historically played an important role in the discovery of pore-forming systems of various kinds, via simple two-dimensional (2D) and negatively-stained imaging of membrane-bound and membrane-inserted assemblies (Amiguet, Brunner, & Tschopp, 1985; Bhakdi, Tranum-Jensen, & Sziegoleit, 1985; Borsos, Dourmashkin, & Humphrey, 1964; Buckley, 1992; Podack & Dennert, 1983; Podack & Tschopp, 1982; Tschopp, 1984; Wilmsen, Leonard, Tichelaar, Buckley, & Pattus, 1992). Negative stain and cryo-EM 2D imaging still have important roles to play in pore-forming protein research (Drucker et al., 2019; Kuwana, 2019; Kuwana, Olson, Kiosses, Peters, & Newmeyer, 2016; Leung et al., 2017; Pang et al., 2019; Praper, Sonnen, Kladnik, et al., 2011; Praper, Sonnen, Viero, et al., 2011), and cryo-EM is a beautiful way simply to observe what is happening in model lipid systems when pore-forming proteins bind to them, because of the absence of stain or drying, but with the increasing sophistication of sample preparation, electron microscope and detector technology there has been a systematic expansion of the kinds of questions electron microscopy is equipped to address in three and four dimensions (Bai, McMullan, & Scheres, 2015; Boyd & Bubeck, 2018; Orlova & Saibil, 2011). Nowadays, three-dimensional EM (3DEM) provides the opportunity for direct observation of subunit binding and association on membranes, the assembly of precursor complexes or prepores, the membrane insertion process, the changes in membrane structure that result, and the final pore complex, all at up to an effectively atomic resolution and in ways which allow for the resolution of sample heterogeneity and of sub-stoichiometric assemblies.

I draw on a broad selection of studies from the last 25 years (see Table I in Supplementary Material in the online version at https://doi.org/10.1016/bs.mie.2021.01.034). I apologize in advance for any studies that have been omitted—this is due to a need to give a balanced account of the specific range of systems which cryo-EM has been used to study. It also reflects a need to provide the reader with a useful and up-to-date survey of approaches they might wish to take in their own work. I wish to acknowledge how much I have benefited from working alongside a number of colleagues including Helen Saibil and Shaoxia Chen, who first introduced me to cryo-electron microscopy, and also Tao Ni, Xiulian Yu and Andreas Sonnen.

In addition to a recent review on the set up and running of a modern cryo-EM facility (Mills, 2021), the reader is directed to the reviews by Mancini and Fuller (2000) and by Orlova and Saibil (2011) for background on cryo-EM and to two other recent reviews (Boyd & Bubeck, 2018; Scott, Huang, Bann, & Taylor, 2018) which have in a more focused way considered the impact of cryo-EM on the study of pore-forming proteins.

2. Materials and equipment

The preparation of samples for cryo-EM imaging relies on the generation of a thin vitrified layer of water on a metallic microscope grid covered with a holey carbon film, by plunging into (most usually) liquid ethane cooled by a bath of liquid nitrogen (Fig. 1A). Historically this was achieved using homemade plunge devices but currently commercial systems are more-or-less universally used and examples include the ThermoFisher series of Vitrobots™ and the Gatan series of CryoPlunge™ devices. Critical parameters for blotting include whether it is one- or two-sided, the blotting time, any delay time and the level of humidity in the blotting chamber. These are parameters which need to be optimized for different samples, along with the sample concentration which will typically be low compared to other biophysical and structural methods ($<1\,\mathrm{mg\,mL^{-1}}$). The critical goal is a thin vitrified water film up to ~300 nm thick, depending on the sample type.

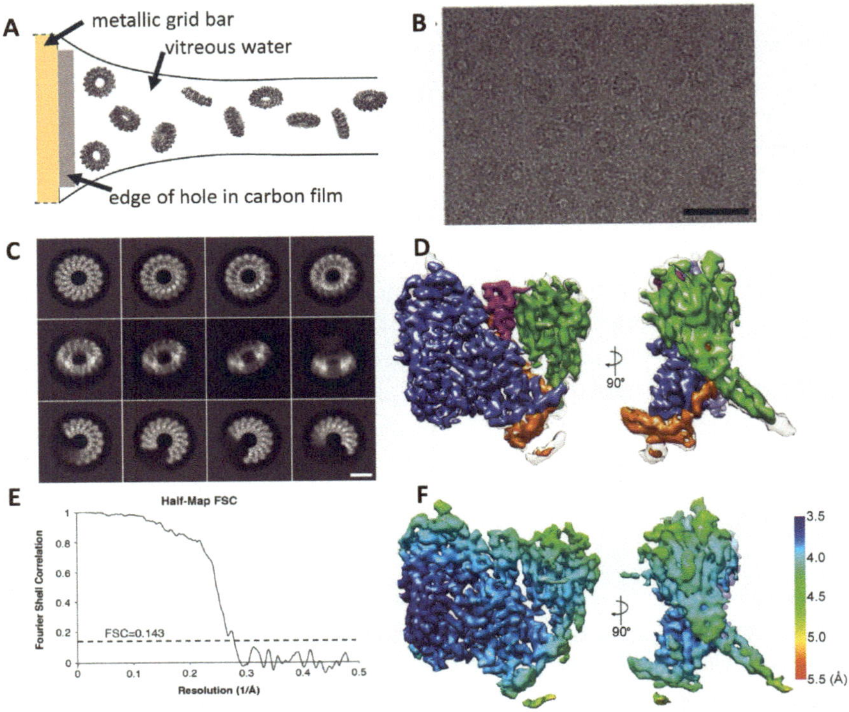

Fig. 1 (A) Schematic representation of single particles lying in random orientations within vitreous ice on a cryo-EM grid. (B) Experimental image of perforin-2 ectodomain prepore complexes in vitreous ice. Bar = 10 nm. (C) 2D class averages computed for images such as those seen in panel (B). (D) Isosurface representation of a single subunit from the perforin-2 prepore reconstruction. Blue: MACPF domain, purple: EGF repeat, green: P2 domain, orange: C-terminal tail. The map is contoured at two levels, solid/colored and transparent, to demonstrate the variable quality of the density. The resolution of the P2 domain was insufficient to build an atomic model for its whole length de novo, but the complementary solution of its crystal structure facilitated the construction of a complete perforin-2 ectodomain atomic model. (E) Cross- or gold-standard FSC to estimate overall resolution of this map. (F) 3D local resolution representation. For more details on this study, see Ni et al. (2020).

For example, imaging of pores in liposome membranes may require thicker films than imaging of single particles in lipid nanodiscs or detergent—but too thick a film and the contrast of the specimen against the background will be too low (Orlova & Saibil, 2011).

The vast majority of recent studies of pore-forming proteins have used Quantifoil® grids (Quantifoil Micro Tools GmbH), which have the advantage that the holes in the carbon film are of equal size and regularly spaced.

In a number of studies of pore-forming systems a layer of graphene oxide has been added to the Quantifoil grid, either by the manufacturer, or in the lab during sample preparation as with lysenin and intermedilysin (Bokori-Brown et al., 2016; Shah et al., 2020). Graphene oxide provides an, essentially transparent, supportive film that can be particularly valuable with more challenging cryo-EM samples (Palovcak et al., 2018) which include the mixed-phase preparations containing protein and detergent or lipid nanodiscs typical of studies of many pore-forming systems. It therefore is an improvement on other methods of providing a carbon backing to support a sample such as the addition of a thicker carbon layer which risks reducing specimen contrast (Jiang, Pentelute, Collier, & Zhou, 2015). An alternative commercial grid type that has been used for cryo-EM of pore-forming systems are C-flat™ grids (Jena BioScience GmbH) which, again, provide regularly spaced holes in the carbon film; these were used in the study of the YaxAB pore (Brauning et al., 2018).

The critical aspects of a microscope which will determine the quality of the data it provides and the resolution to which a structure may be determined include the electron source, the accelerating voltage, the stability of the sample holder or stage, and—most critically—the way in which images are recorded. The application of Field Emission Gun (FEG) technology in biomolecular imaging was transformative around 20 years ago, and is reflected in progress made in studies of pore-forming systems over the time since (Supplementary Material in the online version at https://doi.org/10.1016/bs.mie.2021.01.034 Table I). The benefits of a FEG include its smaller source size compared to earlier filaments such as LaB_6, and greater coherence and brightness with a very small spread of energies, thus limiting the effects of chromatic aberration on data quality (Orlova & Saibil, 2011). Recent further advances include the development of extreme or X-FEG and cold or CFEG technologies, with increasingly narrow energy spreads and the value of which has been demonstrated in determination of biological structures at an actually atomic resolution (Nakane et al., 2020; Yip, Fischer, Paknia, Chari, & Stark, 2020). X-FEG technology has already been applied to the study of three pore-forming systems, though not at improved resolution compared to what might have been achieved using a standard FEG system (Antoni et al., 2020; Roderer, Hofnagel, Benz, & Raunser, 2019; Schubert, Vetter, Prumbaum, Penczek, & Raunser, 2018; see also Gatsogiannis et al., 2016; Roderer et al., 2020). With all cryo-EM imaging, sample size (image number), and sample quality including homogeneity and stability are fundamental determinants of the resolution achievable, before any question of imaging technology arise.

Most recent structures determined by cryo-EM of pore-forming systems, reflecting the wider 3DEM field, have ultimately made use of a Titan Krios microscope (ThermoFisher) operating at 300 kV (Supplementary Material in the online version at https://doi.org/10.1016/bs.mie.2021.01.034 Table I). The high voltage is associated with a reduction in sample damage, due to faster transmission through the sample. *En route*, other microscopes are usually used—they may be more local or cheaper to access while a sample is optimized. In addition to the electron source, both phase plate and energy filter technology can be helpful in enhancing image quality. Phase plates assist by providing a phase shift in the unscattered beam that enhances defocus-derived phase contrast of the image (Orlova & Saibil, 2011; Wang & Fan, 2019), and their application is discussed further below. Energy filters also improve contrast, filtering out inelastically scattered electrons to enhance the imaging coherence (Orlova & Saibil, 2011), are more widely applied than phase plates and are particularly important in cryo-electron tomography, as in studies of pneumolysin pores in situ within liposome membranes (Sonnen, Plitzko, & Gilbert, 2014; van Pee et al., 2017).

The most important advance in EM technology that has enabled it to become a method for atomic structure determination is the development of direct electron detector cameras for image recording (Bai et al., 2015; McMullan, Faruqi, Clare, & Henderson, 2014). The detective quantum efficiency (DQE) describes the degree to which signal in an image is transferred to signal measured in a detector, and is defined as SNR_{out}^2/SNR_{in}^2. For the best image capture method previously available, SO163 film, the DQE was 0.3; whereas for the Falcon II direct electron detector it is 0.6, and for the K2 Summit 0.8 (McMullan et al., 2014). For newer cameras, Falcon-3 and Falcon-4, DQE is up to 0.9 and 1.0 respectively, at least at lower frequencies (Nakane et al., 2020). In addition to their DQE gain, direct electron detectors capture data as movies, with multiple frames per second allowing correction for beam-induced sample movement, and the weighting of frames depending on the dose received/radiation damage done (Bai et al., 2015; Wang et al., 2014). Direct electron detector technology and dose weighting have been critical to the results reported in recent studies of pore-forming proteins, and the advantages such detectors have brought over older methods of image capture such as SO163 and (sometimes) CCD use are clear (Supplementary Material in the online version at https://doi.org/10.1016/bs.mie.2021.01.034 Table I).

An increasing diversity of software are available for image analysis from cryo-EM data, including Relion (Scheres, 2012), CryoSparc (Punjani, Rubinstein, Fleet, & Brubaker, 2017), SPHIRE (Moriya et al., 2017) and EMAN2 (Bell, Chen, Baldwin, & Ludtke, 2016). These provide for motion correction, dose weighting, 2D classification, *ab initio* reconstruction, 3D classification and structure refinement, and also for post-processing of maps (for example to correct for smearing effects of defocus and other factors, see below under Quantification and Statistical Analysis). Exact workflows used will depend on local practice and preference and also to some extent on the challenges the sample poses, but each software package seeks to enable stepwise determination of 3D structures. Some standalone algorithms or approaches may still be needed to meet specific challenges, such as the suppression of noise while recovering features from maps in post-processing (Antoni et al., 2020; Ramlaul, Palmer, & Aylett, 2019).

To widen accessibility of cryo-EM and image processing, national centers of excellence have been established in many countries, such as eBIC in the United Kingdom (Clare et al., 2017), and integrated networks for trans-national access have developed, including Instruct-ERIC (https://instruct-eric.eu/).

3. Strategies for the application of cryo-EM to pore-forming proteins

3.1 Electron microscopy in the discovery of pore-forming proteins and as a two-dimensional imaging technique

The discovery and description of pore-forming protein systems such as the cholesterol-dependent cytolysins (Bhakdi et al., 1985), the complement membrane attack complex (Amiguet et al., 1985; Borsos et al., 1964; Podack & Tschopp, 1982; Tschopp, 1984) and, more recently, perforin-1 (Podack & Dennert, 1983) and aerolysin (Buckley, 1992; Wilmsen et al., 1992) benefited importantly from the use of negatively-stained electron microscopy to visualize pore-forming protein oligomers in two dimensions. Two-dimensional (2D) imaging by cryo-EM has its own particular value—for example where too few pore structures can be isolated for a reconstruction to be performed (Pang et al., 2019; Praper, Sonnen, Viero, et al., 2011) or in identifying changes to membrane structure occurring on interaction with a pore-forming protein (Drucker et al., 2019; Praper, Sonnen, Kladnik, et al., 2011). A particularly good example of the application of 2D cryo-EM to pore formation is in the study of pro-apoptotic Bax and Bak. It has become clear, based on crystallographic studies, that activation of Bax and Bak involves an unfurling of monomeric precursors to form a strand-swapped antiparallel dimer that grips the inner edge of a transmembrane proteolipid pore (Ugarte-Uribe & Garcia-Saez, 2017; Uren, Iyer, & Kluck, 2017). Self-association then occurs between the dimers but it does not seem to be into well-defined stoichiometric pore-forming assemblies, but to enable the creation of pleomorphic pores of various sizes in which lipids play a critical role (Bleicken et al., 2014; Cosentino & Garcia-Saez, 2017; Cowan et al., 2020; Uren et al., 2017). 2D cryo-EM has been important in allowing direct imaging of Bax/Bak pores formed in model lipid systems, and the counting of pore-forming dimers around pore edges (Kuwana, 2019; Kuwana et al., 2016). The particular value of cryo-EM for such a study is the way it allows structural changes in lipid bilayers to be imaged directly without artifacts that may occur from staining or dehydration as the rapid cooling of the samples ($\geq 10^4$ degrees/s) during sample preparation effectively fixes them in an instant (Orlova & Saibil, 2011). Because the cryo-EM images are projections, the whole structure is seen as viewed from one position and the signal in the images relates directly to the electron density (Coulomb potential) of the structures in view. Frequently for such studies cryo-EM will be combined with other imaging approaches such as fluorescence microscopy with Giant Unilamellar Vesicles, as has been done with Bax/Bak (Bleicken, Hantusch, Das, Frickey, & Garcia-Saez, 2017; Bleicken, Hofhaus, Ugarte-Uribe, Schroder, & Garcia-Saez, 2016) and perforin-1 (Praper, Sonnen, Kladnik, et al., 2011; Praper, Sonnen, Viero, et al., 2011).

3.2 Sample preparation: Starting material, oligomer assembly and isolation

Sample preparation will depend on the pore-forming protein in question but a number of general points can be made.

Most but not all pore-forming proteins exist in three states—monomer or another pre-complex (e.g. dimer) state, oligomeric prepore and pore. Intermediates between prepore and pore can also be identified and have been successfully visualized using cryo-EM (Leung et al., 2014; Lukoyanova et al., 2015).

Some pore-forming proteins exist as monomers in solution and await receptor recognition to enable a pore complex to form—for example pneumolysin (Gilbert et al., 1999; van Pee et al., 2017), intermedilysin (Boyd et al., 2016; Shah et al., 2020) and lysenin (Bokori-Brown et al., 2016). Others require a specific activation step, such as proteolysis (anthrax protective antigen, PA, and homologs (Anderson, Sheedlo, Jensen, & Lacy, 2020;

Jiang et al., 2015; Yamada et al., 2020), aerolysin (Iacovache et al., 2016)) or addition of an ion, such as Ca^{2+} (perforin-1 (Law et al., 2010)).

Some (pre)pore complexes will assemble spontaneously: PA forms a prepore assembly in solution once proteolysed, and can be induced to form a pore structure on transfer to a low pH buffer in the presence of detergent (Jiang et al., 2015); but can also be induced to form a pore without low pH/detergent treatment when presented with lipid nanodiscs to insert within (Hardenbrook et al., 2020). By contrast the homologous Iota toxin was found—via a screening process—to be induced to form a pore structure in solution by treatment with ethanol in the presence of detergent, after proteolytic activation (Yamada et al., 2020), while detergent alone enabled aerolysin to form a pore structure from its prepore state (Iacovache et al., 2016). The ectodomain of perforin-2 could be induced to form prepore complexes at pH 5.5 and pore complexes in the presence of detergent at pH 4.0 and below (Ni et al., 2020) but in another study prepore formation was spontaneous at pH 7.2, though in the presence of 10% glycerol, which may have had an effect (Pang et al., 2019). The complement membrane attack complex (MAC) pore forms spontaneously from active components (Menny et al., 2018).

Other pore assemblies need to be assembled in a membrane and then released, as with lysenin (Bokori-Brown et al., 2016) and Gasdermin A3 (Ruan, Xia, Liu, Lieberman, & Wu, 2018), but several recently have been formed directly in MSP protein-based lipid nanodiscs (Anderson et al., 2020; Efremov, Gatsogiannis, & Raunser, 2017; Hardenbrook et al., 2020; Roderer et al., 2019) before imaging *in situ*. In a more difficult case, the *C. perfringens* Epsilon toxin had to be assembled on target cells because a receptor or method to otherwise induce pore complexation remains unknown, and then released with detergent for single particle study (Savva et al., 2019).

If being used, the choice of detergent may prove critical to a study's success and quite often screening of different detergents/different detergent concentrations will be needed. Detergent can cause problems for the vitrification process, leading to non-vitreous ice states which are electron dense and mean the sample cannot be imaged: the choice of detergent also had a major influence on the size and homogeneity of Xax pores (Schubert et al., 2018). In our own work my group has found detergent screening essential and to be facilitated by use of fluorescence-based size-exclusion chromatography (Ni et al., 2020). Different concentrations of a single detergent can make complexes more or less stable (Yamada et al., 2020) and detergents may also need to be replaced with amphipols instead (Brauning et al., 2018; Rajesh, Knowles, & Overduin, 2011; Schubert et al., 2018; van Pee et al., 2017) or amphipols may be used directly as part of a pore complex's assembly in solution (Spicer et al., 2018). Extracted pore complexes may also need re-purification by, for example, size-exclusion or affinity chromatography, in order to provide a sample of appropriate homogeneity for structure determination. Prepore and pore complexes have also been imaged on or in liposome membranes, such as in studies of various members of the membrane attack complex-perforin/cholesterol-dependent cytolysin (MACPF/CDC) family (Gilbert et al., 1999; Pang et al., 2019; Sonnen et al., 2014; Tilley, Orlova, Gilbert, Andrew, & Saibil, 2005; van Pee et al., 2017), *Y. entomophaga* Tc pores and of Bax family protein pores (Kuwana, 2019; Kuwana et al., 2016). Salt-based buffers have universally been used, but care does need to be taken to minimize solutes which may contribute to high background in the sample images, such as high salt concentrations and additives like glycerol and sucrose.

Mutant forms of pore-forming protein may be of particular value for two reasons, firstly to provide a more stable sample for higher resolution imaging (Pang et al., 2019; Roderer et al., 2019), secondly to interrupt processes of pore formation at different stages between the prepore and pore states (Degiacomi et al., 2013; Iacovache et al., 2016; Leung et al., 2014, 2017; Lukoyanova et al., 2015). This allows stepwise dissection of the dramatic rearrangements needed for a protein to refold and partition into a membrane from an aqueous environment. Locking mutants can also help to prove mechanisms of pore formation inferred from prepore and pore states alone (Ni et al., 2020).

One specific artifact quite commonly observed in pore-forming protein complexes requires specific comment. In several cases prepore and pore complexes (and sometimes mixtures of the two) are observed stacked head-to-head or packed together in some other way which relates to their macroscopic geometry. In the case of aerolysin, identical prepore heptamers stack together but two other intermediates (post-prepore and quasi-pore) stack with one another—indeed the observation of the post-prepore state depended on the steric effects of the other, quasi-pore oligomer preventing both copies converting to it (Degiacomi et al., 2013; Iacovache et al., 2016). Two Gasdermin A3 oligomers also packed head-to-head, one in a pore state and the other apparently in some prepore state or another (Ruan et al., 2018). Perforin-2 prepores packed head-to-head were subjected to close analysis to show they have a variety of different non-specific relative positions, "rocking against" one another (Pang et al., 2019). Images of the MACPF/CDC pneumolysin also show head-to-head packing like this, including between pores inserted into liposomes (R.J.C. Gilbert, personal observation) and Drücker et al. have made a very interesting biophysical study of the effect of such interactions on the structures of multilamellar liposomes that used cryo-2DEM (Drucker et al., 2019). Finally, a single dataset for the *Clostridium difficile* toxin CDTb heptamer revealed two different kinds of foot-to-foot association (i.e. between the membrane-facing parts of the oligomer): one combining the prepore and the pore, the other two partially-inserted pores that serendipitously revealed an intermediate in membrane insertion (Anderson et al., 2020). The same assemblies were observed in a separate study, with the authors there showing these packing arrangements were present in solution using small-angle X-ray scattering, analytical ultracentrifugation and multi-angle laser light scattering (Xu et al., 2020). It is not very surprising that di-heptamers of CDTb are present in solution when tested biophysically, because cryo-EM is itself a (supercooled) solution method, but it is very unlikely that such packing arrangements have any biological significance as Xu et al. suggest (Xu et al., 2020). It is much more likely that these are effects associated with the presence of unnaturally high concentrations of oligomers possessing matching geometries and dimensions.

3.3 Recording images

As with any cryo-EM data collection process the sample should ideally consist of a monodisperse system—one complex, undergoing little or no aggregation and with individual objects sufficiently separated from one another that they can be analyzed separately (Fig. 1B). Data collection will follow established workflows in the imaging lab but critical factors will be to ensure a range of defocus values are collected, and to aim for large datasets of images wherever possible. Pore-forming protein complexes are frequently found as mixtures of states (see Sample Preparation, above and Supplementary Material in the online version at https://doi.org/10.1016/bs.mie.2021.01.034 Table I) and so large datasets will facilitate sorting into all the distinct states that can be identified among them. It is regularly the case that a single dataset can be used to determine a single reconstruction at a good resolution, but that, on closer inspection, a greater number of structures is found to be present within the image dataset than first thought.

The defocusing of images is necessary because of the weak contrast given by macromolecular complexes when viewed unstained against a background formed from vitreous water. Defocusing results in phase contrast due to interference between the scattered electrons contributing to image formation and the electrons passing through the sample without being scattered (Orlova & Saibil, 2011). The resulting Contrast Transfer Function modulates the signal such that alternate regions of Fourier space are present at opposite contrasts separated by nodes at which information is not preserved (Mancini & Fuller, 2000). Two steps are required to correct for this effect—assignment of the correct contrast to the full data range and combination of multiple different defocuses to compensate for the data missing at any one defocus (Mancini & Fuller, 2000; Orlova & Saibil, 2011). As discussed below, the defocus applied to enhance image contrast also leads to a loss of detail from 3DEM maps at a given resolution, but this can be corrected for. Images should be taken as close to focus as possible while achieving the contrast needed for their analysis.

3.4 Processing images

The key stages in image processing include correction for movement during the imaging process and dose weighting (Zheng et al., 2017), defocus/CTF determination, particle picking, two-dimensional (2D) classification, reconstruction determination, three-dimensional (3D) classification and potentially sub-volume/sub-region density analysis and reconstruction.

Automated particle picking can result in a high number of false positives, for example from mistaking the edge of grid holes or particles of ice for the pore complex. This can be ameliorated using an initial manual picking step to identify a subset of particles which can then be used to generate a set of templates to search for a much larger number of prepore/pore images, automatically (Brauning et al., 2018) but in other cases an automated process is sufficient from scratch. In any case, 2D classification has become a critical tool in the cleaning up of image datasets, particularly now that those datasets tend to have many times more numbers of particles than in the past. 2D classification allows the identification of false particles by visual inspection of the resulting 2D class averages which represent characteristic views (Fig. 1C), and in every case this would be an expected step in image processing of pore images. 2D classification may also allow already the identification of mixed populations of states in the data, as for example with *C. difficile* CDTb images which corresponded to "short" and "long" diheptamer stack barrel states

(Anderson et al., 2020). It may also be useful to exclude certain subsets of views—for example in the reconstruction of an anthrax PA_7LF_3 prepore assembly top views (looking along the barrel of the prepore assembly) were excluded (Antoni et al., 2020). By constrast, of course, in some cases only one kind of view with limited tilt out of plane may be available, as was the case in some earlier studies of pneumolysin, perforin-1, suilysin and pleurotolysin (Gilbert et al., 1999; Leung et al., 2014; Lukoyanova et al., 2015; Tilley et al., 2005). Such studies are termed pseudo-single particle because they rely on high object symmetry for 3D reconstruction to proceed and base map calculation on one view, or a small number of views from roughly the same perspective.

Reconstruction relies on identification of a set of relative angular coordinates describing the relationship of 2D projections to one another in 3D space. Practical options here will depend on the workflow being used and the reader is referred to individual publications and the relevant software for more information: including Relion (Scheres, 2012), CryoSparc (Punjani et al., 2017), SPHIRE (Moriya et al., 2017), XMIPP (de la Rosa-Trevin et al., 2013) and EMAN2 (Bell et al., 2016).

As has been emphasized several times, with pore-forming complexes as with any number of other 3DEM samples the likelihood of heterogeneity in the sample dataset means that 3D classification should be performed, and with sufficient degrees of freedom to enable the identification of all the different kinds of assemblies present. Thus, more possible 3D classes will commonly be obtained than in fact exist in the data, but the equivalent classes can then be combined, in the process identifying the balance of the distribution of states. Particularly useful examples among pore-forming proteins include the studies on perforin-2 (Ni et al., 2020; Pang et al., 2019), Gasdermin A3 (Ruan et al., 2018), the membrane attack complex (Menny et al., 2018), ClyA (Peng, de Souza Santos, Li, Tomchick, & Orth, 2019) and the PA_7LF_3 prepore (Antoni et al., 2020).

Symmetry applied to reconstructions should be tested, for example by determining structures at lower symmetry as well as higher (e.g. C2 as well as C16) in case the structure is better described by a lower symmetry operation, and C1 structures were determined alongside C7 structures for the CDTb and Iota toxin pores (Anderson et al., 2020; Yamada et al., 2020) as proper tests of the structures' determination. Some pores do not have rotational symmetry: both the MAC as a whole (Menny et al., 2018) and the Tc holotoxin from *P. luminescens* (Roderer et al., 2019), for example. Symmetry has also been usefully applied to sub-regions of maps, as in the intermedilysin-CD59 prepore structure (Shah et al., 2020). The application of the higher symmetry possessed by one part of the pore structure—thus smearing out another part which does not share that symmetry—can be a useful stepping stone on the way to a full unsymmetrized structure determination, as was the case in the structure of the anthrax PA_7EF pore (Hardenbrook et al., 2020). And reciprocally the determination of the structure of a region *not* following the symmetry of the rest of the assembly (Tc holotoxin (Roderer et al., 2019)) or of the asymmetric subunit (perforin-2 (Pang et al., 2019)) can be achieved too by focusing on that region alone. In the case of perforin-2 the 16 subunits forming the prepore assembly enabled 4,276,192 examples of those subunits to be identified, from which an optimized reconstruction could be determined for a selected 527,402 images (Pang et al., 2019).

Electron tomography, permitting the imaging of pores within membranes as 3D volumes without any assumption of symmetry, will prove increasingly important in the study of pore-forming proteins in coming years (Dunstone & de Marco, 2017). This is partly because of the way in which tomography has been revolutionized by the availability of direct electron detectors, which make resolutions achievable which were actually considered impossible in the past due to the electron dose required for tomographic data collection, which it was judged would damage the sample such that only ~20 Å maps might be obtainable (Henderson, 2004). Now, atomic models have been determined by tomographic studies of systems such as HIV (Schur et al., 2016) and there is no reason in principle why, in time, with large enough datasets, similar achievements may not be possible for pore-forming systems. For now, tomographic studies have been limited to determining structures of the pneumolysin pore within membranes (Sonnen et al., 2014; van Pee et al., 2017). These studies have played an important part in the evolution of our understanding of MACPF/CDC protein mechanism (Gilbert, Dalla Serra, Froelich, Wallace, & Anderluh, 2014; Vogele et al., 2019), precisely because they have not made assumptions about symmetry in determining 3D structures. Sub-tomogram averaging of extracted sub-volumes will be critical in any application of cryo-electron tomography to pore-forming systems that is to yield high resolution information (Ilca et al., 2015; Jiang & Tang, 2017; Schur et al., 2016).

3.5 Complementary techniques and integrative approaches

The focus of this article is cryo-EM but negative stain EM has very important roles to play in mapping out the mechanisms of pore-forming proteins, in sample optimization, and in testing hypotheses arising from cryo-EM data (Ni et al., 2020; Schubert et al., 2018). It also still has important discovery contributions to make, such as in its use to identify early assembly intermediates of the perforin-1 prepore, in combination with atomic force microscopy (AFM) (Leung et al., 2017).

AFM has been of particular value in complementing insights from cryo-EM, and the use of high-speed approaches has allowed real-time imaging of prepore and pore assembly and the post-assembly dynamics of functioning pores. Particular examples include imaging of suilysin in a combined single particle 3DEM/AFM study (Leung et al., 2014) complementary to a study by cryo-electron tomography of pneumolysin (Sonnen et al., 2014), and studies with listeriolysin (Podobnik et al., 2015; Ruan, Rezelj, Bedina Zavec, Anderluh, & Scheuring, 2016), perforin-1 (Leung et al., 2017) and perforin-2 (Ni et al., 2020; Pang et al., 2019).

As a structural technique cryo-EM is now one of a set of three methods, the others being X-ray crystallography and Nuclear Magnetic Resonance (NMR), which can provide 3D atomic resolution structures of macromolecular systems. The particular distinction of cryo-EM as applied to pore-forming proteins is that it can do so in a more native like context than crystallography, but for much larger systems than NMR. NMR is useful for the study of pore-forming peptides (Oestringer et al., 2018; OuYang et al., 2013; Schnell & Chou, 2008), for example, but the larger protein subunits of most pore-forming proteins and toxins are beyond the reach of solution NMR for structure determination. The first 3D atomic structure of a pore-forming toxin pore was determined for α-hemolysin using crystallography and other examples include both ClyA (Mueller, Grauschopf, Maier, Glockshuber, & Ban, 2009) and lysenin (Podobnik et al., 2016) which also have cryo-EM structures discussed in this review (Bokori-Brown et al., 2016; Peng et al., 2019). But as this review demonstrates 3DEM is probably now the preferred method for *de novo* pore structure determination.

The determination of structures of pore-forming proteins in their protomeric, pre-assembly state remains an important question for crystallography to address, and in a number of cases crystallographic structures of all (Schubert et al., 2018; Spicer et al., 2018) or part (Ni et al., 2020) of the protein have complemented the determination of its oligomeric state(s). The question of what the structure of a protein is before it oligomerises is an important one—it may not be the same as in the prepore state, for example perforin-2 oligomerises into a prepore with strand swapping between subunits (Ni et al., 2020; Pang et al., 2019)—but what is the structure of the strand involved in the isolated monomer? And the perforin-like protein from *Toxoplasma gondii* has an intrinsically disordered region linking its membrane-binding and pore-forming domains, so that its structure has not been determinable by crystallography, but has by small-angle X-ray scattering (revealing two different conformations in solution) (Ni et al., 2018).

Native mass spectroscopy has become a very powerful technique for the study of heterogeneous macromolecular complexes (Liko, Allison, Hopper, & Robinson, 2016) and including the study of membrane systems for which it has provided important insights (Chorev & Robinson, 2020). It, too, is a useful complement to cryo-EM for example in identifying the detergent-associated regions of a membrane pore (Podobnik et al., 2016) and detailed dissections of protein-lipid assemblies (Zhou et al., 2011).

4. Quantification and statistical analysis

4.1 Assessment of resolution

Quantitative assessment of map resolution in 3DEM structure determination makes use of the consistency with which the density has been reconstructed (Fig. 1D–F). For single-particle analyses this has almost exclusively consisted of a Fourier Shell Correlation approach (FSC), in which the correlation of two representative maps to one another is computed in segments of Fourier space, and then plotted against spatial frequency. Historically, and for many years, this analysis was performed using two halves of the same dataset which had been aligned to the same reference model. However, since in (cryo-)EM imaging the signal and noise in data are indistinguishable from one another, and since the alignment model being used inevitably has had strongly-represented features, both in principle and in practice this resulted in the fundamental problem of overfitting of data—referred to due to well-known test cases in the field as the "Einstein from noise" problem (Henderson, 2013). It is now universally recognized that a self-FSC calculation of this kind is unacceptable and does not provide an appropriate assessment of map quality, which must instead be determined using two properly independent maps deriving from two properly independent processes of alignment and reconstruction (Grigorieff, 2000; Scheres & Chen, 2012). Of the structural studies of pore-forming proteins reported in Table I (Supplementary Material in the online version at https://doi.org/10.1016/bs.mie.2021.01.034), all those published since 2015 have used such a cross-FSC or "gold standard" approach. Modern approaches to cryo-EM reconstruction, based as they are on an initial cleaning up of the dataset through 2D and sometimes 3D classification procedures, have then tended to rely on a featureless model to start the reconstruction process which, combined with the high level of signal found in modern cryo-EM datasets, has resulted in rapid determination of structures essentially shaped only by the information present in the particle images. One unusual case which is an exception to this regards a structure of the fully-loaded anthrax lethal factor prepore state, in which a single 3D model based on a combination of crystal structures was filtered to 30 Å to serve as a reference for 3D refinement. The fact that the *initial* structure then determined from a dataset cleaned by iterative 2D classification had a resolution of 3.5 Å (and that this was in effect a molecular replacement strategy with a 100% homologous model) highlights that *in this case* such an approach was unproblematic (Antoni et al., 2020), but it should only be taken with care as other examples, in which a detailed model has been used with apparent confidence not only to align particles from micrographs but to identify them in the first place, have produced spurious results (Henderson, 2013).

The question of what resolution value to assign a structure has also been a matter of significant debate, and given that higher resolution structures have tended to result in greater plaudits for a result and the arising publication there has historically been some implicit pressure to maximize resolution estimation. I confine myself here to the main two criteria which have been used, FSC = 0.5 and FSC = 0.134.

The choice of FSC = 0.5 to quantify the resolution of a structure was the most widespread when self-FSC assessments were considered acceptable, and equates to a signal-to-noise ratio substantially above zero at that resolution (Malhotra et al., 1998; Penczek, 2010; Rosenthal & Henderson, 2003). However this is to underestimate the resolution of an actual map (among other things because half of the data are being used to calculate the resolving power of the whole data set because of the absence of an external reference) (Rosenthal & Henderson, 2003). Rosenthal and Henderson showed that if a map computed from a full dataset were to be compared to a perfect reference (which does not exist) then the choice of correlation coefficient $C_{ref} = 0.5$ would represent a single objective threshold criterion and be equivalent to the crystallographic figure of merit (Rosenthal & Henderson, 2003). They furthermore showed empirically that for a real cryo-EM reconstruction (of the *Bacillus stearothermophilus* E2 icosahedral core from pyruvate dehydrogenase) when $C_{ref} = 0.5$, the correlation to a map computed from the E2 core crystal structure (which some might consider a near-perfect model) $C_{xray} = 0.5$, and the FSC between two properly independent E2 half maps was 0.143 (Rosenthal & Henderson, 2003). It is on this basis that the widely-used FSC = 0.143 criterion is being applied (Fig. 1E), and cryo-EM structures for pore-forming proteins (and of anything else) determined in recent years have used FSC = 0.143 to quantify their resolution by a single value (Penczek, 2020). In some cases the resolution at a "gold standard" cross-FSC = 0.5 is also being quoted, as a more conservative measure of map quality, and as in the case of the TcABC complex from *Photorhabdus luminescens* (Roderer et al., 2019) and the Xax (Schubert et al., 2018) pore-forming complexes. Of course, where a map is of sufficient quality to build an atomic model then the FSC between the final 3DEM map and the model built into it can also be computed (e.g. Ni et al., 2020; Schubert et al., 2018; Shah et al., 2020), is a further excellent validation of the quality of the map and model together and would look as with Rosenthal and Henderson's work for a threshold value at $C_{xray} = 0.5$ (Rosenthal & Henderson, 2003). A particularly good example is that of the *Clostridium perfringens* Iota toxin, where the resolutions determined between half maps at FSC = 0.143 and against the model at $C_{xray} = 0.5$ were the same, lending particular confidence in the results reported (Yamada et al., 2020).

But one of the key strengths of 3DEM is that a single resolution value is not the only estimate of map quality that can be given, and that a map of local resolution across an entire structure can be computed (Aiyer, Zhang, Baldwin, & Lyumkis, 2021; Cardone, Heymann, & Steven, 2013) (Fig. 1D and F). In all cases this should be reported where a medium-to-high resolution reconstruction has been determined, and provides for a helpful understanding of the local quality of a map (Fig. 1F), in combination with simple visualization (Can sidechains be seen? Is the handedness of the α--helices clear?) as well as the single-value measure of resolution (Fig. 1E). Thus, heatmaps are drawn over the surface of 3DEM reconstructions to represent a gradient in resolution to a given threshold (0.143) and, as is only to be expected, the cores of pore-forming structures—especially given their usual symmetry—display significantly better resolution than their peripheries (see Supplementary Material in the online version at https://doi.org/10.1016/bs.mie.2021.01.034 Table I and Fig. 1F). Of course, this can mean that while some areas of a structure are of high enough resolution for an atomic model to be built, others are not; a variety of approaches could be taken in such cases, including; (i) separate determination of the structure of a relatively disordered or flexible domain, for example by X-ray crystallography, that can then be fit into the cryo-EM map with high precision (Ni et al., 2020); (ii) cryo-EM sample improvement, including by using site-directed mutagenesis (Pang et al., 2019); (iii) the determination of a reconstruction for the relatively disordered domain on its own via sub-volume approaches (Roderer et al., 2019) (see Optimization and Troubleshooting below).

4.2 Analysis and interpretation of three-dimensional maps

Historically, the limited resolution achieved in 3DEM reconstructions meant that map interpretation benefitted greatly from the availability of high-resolution structures of components of the assembly. The distinctive shapes of many pore-forming protein subunits assisted this approach—they are frequently elongated and consist of combinations of characteristic domains. Nevertheless, there was a basic knowledge gap which made it hard at times to be certain that a fit was correct (Dudkina et al., 2016; Gilbert, Mikelj, Dalla Serra, Froelich, & Anderluh, 2013; Law et al., 2010) and before detailed characterization of a particular family of pore-forming proteins the mechanism of prepore-to-pore transition could not be guessed at (Iacovache et al., 2016; Parker et al., 1994). Software such as Chimera (Pettersen et al., 2004) is invaluable in easily fitting an atomic structure to a lower resolution 3DEM density map, but the correlation coefficients reported from it are just a guide to the relative quality of alternative fits and need to be considered with care (Gilbert et al., 2013; Law et al., 2010). Especially in the case of prepore assemblies where substantial refolding of the pore building blocks has not yet taken place, visual inspection of fit quality will be at least as important as any measure of it, with the distinctive shapes of subunits plainly visible within the oligomeric assembly (Tilley et al., 2005); and when the shared mechanism of a family of proteins is understood this can be extended from the prepore to the pore, with appropriate computational modeling (Lukoyanova et al., 2015; Serna, Giles, Morgan, & Bubeck, 2016). The precision with which an atomic model may be fit into a 3DEM density map is significantly greater than the map's own resolution, where the constraints of the density lead to definite orientation of the subunit; for example, it was estimated that a 23 Å resolution reconstruction of the Bluetongue virus enabled an atomic model to be assembled using the crystal structure of its VP7 trimer with a mean error of ~4 Å (Grimes et al., 1997), such that the atomic model supplied phase information to a far higher resolution than that of the EM map in subsequent crystallographic determination of the capsid structure (Grimes et al., 1998).

Rigid body fitting of pre-determined structures or homology models to 3DEM maps continues to be an important way to assess progress in structure determination, and to begin detailed interpretation of even high resolution 3DEM maps such as those into which full atomic models can be built. More sophisticated approaches include the use of normal mode or flexible fitting as provided by IMODFIT (Lopez-Blanco & Chacon, 2013), Namdinator (Kidmose et al., 2019) and FlexEM (Joseph et al., 2016; Topf et al., 2008) which can be especially useful in seeking to remodel a prepore subunit conformation into a pore one (Lukoyanova et al., 2015; Ni et al., 2020; Schubert et al., 2018). Such software is particularly useful for structure determination where the resolution of a map precludes *de novo* model building or manual rebuilding, for example in Coot (Emsley & Cowtan, 2004). Rosetta software has been particularly developed to enable modeling into mesoscopic resolutions of 3-5 Å (Wang et al., 2016) with the benefit of the vast quantity of structural knowledge now encapsulated in the Protein Data Bank (www.rcsb.org) and FlexEM has also proved applicable in such cases (Joseph et al., 2016) while ISOLDE makes use of computational molecular dynamics approaches to the same end (Croll, 2018) and was applied to the *Yersinia entomophaga* TcA pore structure (Piper et al., 2019). MODELER provides for homology modeling and more complex forms of protein structure prediction, and for flexible fitting into 3DEM maps including the *de novo* construction of loop regions between fitted domain structures (Webb & Sali, 2016, 2021) as in the determination of the MAC pore structure (Menny et al., 2018), and with the suilysin (Leung et al., 2014) and pleurotolysin (Lukoyanova et al., 2015) pore structures. Nevertheless, as the map resolutions achieved by 3DEM have approached those achieved by X-ray crystallography there has been a widespread tendency to turn to crystallographic programs for structure building and refinement and principal among these are Phenix, which has itself been adapted to enable real-space refinement appropriate for 3DEM structure determination at atomic resolution (Adams et al., 2010; Afonine et al., 2012).

5. Optimization and troubleshooting

5.1 Map optimization

A well-known feature of (cryo-)EM maps is that high-resolution features are down-weighted compared to lower resolution ones, as reflected in an envelope function modifying the amplitude profile in their Fourier transform. The origins of this effect are various but include inelastic scattering of electrons, the thickness of the water layer, the level of defocus used, and the methods associated with alignment, averaging and reconstruction (Mancini & Fuller, 2000; Orlova & Saibil, 2011; Rosenthal & Henderson, 2003; Savva et al., 2019). Such effects can be offset by the application of a negative B factor, as in the groundbreaking reconstruction of the Hepatitis B Virus capsid by Böttcher et al. which for the first time allowed the fold of a protein to be determined by single-particle cryo-EM (Bottcher, Wynne, & Crowther, 1997). At the time, use of a -500 Å^2 B factor seemed controversial to some, and conferred greater interpretability on the map reported by Böttcher et al. than that determined in a simultaneously-reported study from another lab which had a lower headline resolution (9 Å v. 7.4 Å (Bottcher et al., 1997; Conway et al., 1997)). Since then negative or inverse B factor correction has become a mainstream method for the improvement of map interpretability, and an experimental calculation of the appropriate measure to apply can be made from the gradient of a plot of *ln*N (N = number of particle images in a reconstruction) against 1/(resolution, Å)^2 (see (Rosenthal & Henderson, 2003; Zivanov et al., 2018)). A particularly useful example in context is the determination of the structure of *Clostridium perfringens* Epsilon toxin (Savva et al., 2019). Data were collected both without and with a Volta Phase Plate (Orlova & Saibil, 2011; Wang & Fan, 2019). The enhanced signal in the phase plate data, as expected, reduced signal decay with increasing resolution and resulted in a lower overall dampening of the density map features, requiring a smaller B factor correction (Savva et al., 2019). As reported in Table I (Supplementary Material in the online version at https://doi.org/10.1016/bs.mie.2021.01.034) most recent cryo-EM structures of pore-forming protein complexes have made use of B-factor sharpening to improve the appearance and interpretability of maps, and while some make general reference to the use of automated procedures (Hardenbrook et al., 2020) the explicit way in which other papers address the use of B-factor sharpening is to be welcomed (Savva et al., 2019), and it is impressive to see careful analysis of the smoothing effects on multiple different maps from the same analysis, in some cases (Pang et al., 2019). In another study the use of a Titan Krios microscope equipped with a spherical aberration (Cs) corrector (Barton et al., 2011; Fan et al., 2017; Fischer et al., 2016) enabled reconstructions to be performed with much lower B factor effects than in many studies because of the greatly improved contrast and smaller degree of defocusing needed (Yamada et al., 2020). Sharpening by B factor correction has also been combined with systematic downweighting of noise in maps within the Feature Enhanced Map protocol (Afonine et al., 2015), as applied to aerolysin (Iacovache et al., 2016).

5.2 Troubleshooting

One of the key limitations for cryo-EM studies of macromolecular systems has been the size of object being imaged. As recently as 2011 rather pessimistic lower limits could be identified for the accurate determination of cryo-EM structures, with even objects of 500kDa and more showing significant levels of error in orientation determination (Henderson et al., 2011). Of course, direct electron detectors have now greatly improved the situation such that objects of 100kDa and lower can have their molecular structures determined (Khoshouei, Radjainia, Baumeister, & Danev, 2017; Merk et al., 2016). The question of the lower limit for 3DEM structure determination is a practical one and will depend on the sample homogeneity, stability, the buffer conditions it is imaged in, the microscope it is imaged with, and the dataset size that can be collected, among other factors. If a target is proving hard to visualize then it may be possible to increase its size artificially in order to increase the ease with which its structure might be determined—such as with a targeted nanobody or megabody (Garcia-Nafria, Lee, Bai, Carpenter, & Tate, 2018; Laverty et al., 2019; Uchanski et al., 2021).

Another problem with cryo-EM which will apply commonly to pore-forming protein complexes is the existence of preferential orientations on the microscope grid. Of course, ironically, and as discussed already, the original determinations of MACPF/CDC pores made optimal use of such preferential orientations and by applying high symmetry to escape its effects on structural anisotropy (Gilbert et al., 1999; Leung et al., 2014; Lukoyanova et al., 2015; Tilley et al., 2005). If orientational bias arises then one approach to improve the evenness with which particles sample different views is to use polylysine treatment of the grids (Ortega, Singh, Ishikawa, Maurizi, & Steven, 2000) as applied to the anthrax pore structure (Jiang et al., 2015). Another is to tilt the microscope stage to allow a greater number of different views to be taken (Shah et al., 2020). In any case, modern single-particle 3DEM relies on as far as possible a complete set of possible views of an object, to enable a reconstruction to be made (Orlova & Saibil, 2011).

Finally, the hand of a reconstruction determined by 3DEM is arbitrary. This is due to the fact that projections of macromolecular complexes will look identical when lying in positions 180° to one another around any given axis of rotation. Rosenthal and Henderson described a helpful approach to the determination of the absolute hand of a map, at a time when single-particle cryo-EM could not hope to achieve atomic resolution in the vast majority of cases (Rosenthal & Henderson, 2003), a method which not only provided a route to accurate hand determination for novel 3DEM structures (Gilbert et al., 2005) but also to an estimation of the accuracy of angular assignments made to individual particles (Henderson et al., 2011). Nowadays, if the resolution of a map is too low to allow the hand to be identified based on prior knowledge (such as the right-hand twist of α-helices), then it should still be determined experimentally and if not its ambiguity should be clearly stated. Even relatively recently, two oppositely-handed structures of the same assembly were presented in successive papers in the journal *Science*, without comment (Arranz et al., 2012; Moeller, Kirchdoerfer, Potter, Carragher, & Wilson, 2012) though a subsequent atomic structure determined by cryo-EM has resolved the issue now (Fan et al., 2019; Moeller et al., 2012). During reconstruction processes the hand of maps may flip unexpectedly, and should be regularly checked on. In the case of the Yax pore the incorrect hand of the map could be corrected by reference to the expected left-handedness of a coiled coil (Brauning et al., 2018).

6. Conclusion

Three-dimensional cryo-EM of membrane pore-forming proteins now represents what is most likely the preferred method for the determination of their atomic structures. Table I (Supplementary Material in the online version at https://doi.org/10.1016/bs.mie.2021.01.034) summarizes the work reviewed in this article, provides a series of details for consideration as new targets are tackled and gives a snapshot of the current state of the field. Going forward, cryo-electron tomography applied to pores in increasingly native-like states will be a major frontier and lead to an increasing appreciation of the structural diversity and conformational dynamics of pore-forming complexes in the effects they have on the structures of biological membranes themselves.

References

Adams, P. D., Afonine, P. V., Bunkoczi, G., Chen, V. B., Davis, I. W., Echols, N., et al. (2010). PHENIX: A comprehensive python-based system for macromolecular structure solution. *Acta Crystallographica. Section D, Biological Crystallography*, *66*(Pt 2), 213–221. https://doi.org/10.1107/S0907444909052925.

Afonine, P. V., Grosse-Kunstleve, R. W., Echols, N., Headd, J. J., Moriarty, N. W., Mustyakimov, M., et al. (2012). Towards automated crystallographic structure refinement with phenix.refine. *Acta Crystallographica. Section D, Biological Crystallography*, *68*(Pt 4), 352–367. https://doi.org/10.1107/S0907444912001308.

Afonine, P. V., Moriarty, N. W., Mustyakimov, M., Sobolev, O. V., Terwilliger, T. C., Turk, D., et al. (2015). FEM: Feature-enhanced map. *Acta Crystallographica. Section D, Biological Crystallography*, *71*(Pt. 3), 646–666. https://doi.org/10.1107/S1399004714028132.

Aiyer, S., Zhang, C., Baldwin, P. R., & Lyumkis, D. (2021). Evaluating local and directional resolution of Cryo-EM density maps. *Methods in Molecular Biology*, *2215*, 161–187. https://doi.org/10.1007/978-1-0716-0966-8_8.

Amiguet, P., Brunner, J., & Tschopp, J. (1985). The membrane attack complex of complement: Lipid insertion of tubular and nontubular polymerized C9. *Biochemistry*, *24*(25), 7328–7334.

Anderson, D. M., Sheedlo, M. J., Jensen, J. L., & Lacy, D. B. (2020). Structural insights into the transition of Clostridioides difficile binary toxin from prepore to pore. *Nature Microbiology*, *5*(1), 102–107. https://doi.org/10.1038/s41564-019-0601-8.

Antoni, C., Quentin, D., Lang, A. E., Aktories, K., Gatsogiannis, C., & Raunser, S. (2020). Cryo-EM structure of the fully-loaded asymmetric anthrax lethal toxin in its heptameric pre-pore state. *PLoS Pathogens*, *16*(8), e1008530. https://doi.org/10.1371/journal.ppat.1008530.

Arranz, R., Coloma, R., Chichon, F. J., Conesa, J. J., Carrascosa, J. L., Valpuesta, J. M., et al. (2012). The structure of native influenza virion ribonucleoproteins. *Science*, *338*(6114), 1634–1637. https://doi.org/10.1126/science.1228172.

Bai, X. C., McMullan, G., & Scheres, S. H. (2015). How cryo-EM is revolutionizing structural biology. *Trends in Biochemical Sciences*, *40*(1), 49–57. https://doi.org/10.1016/j.tibs.2014.10.005.

Barton, B., Rhinow, D., Walter, A., Schroder, R., Benner, G., Majorovits, E., et al. (2011). In-focus electron microscopy of frozen-hydrated biological samples with a Boersch phase plate. *Ultramicroscopy*, *111*(12), 1696–1705. https://doi.org/10.1016/j.ultramic.2011.09.007.

Bell, J. M., Chen, M., Baldwin, P. R., & Ludtke, S. J. (2016). High resolution single particle refinement in EMAN2.1. *Methods*, *100*, 25–34. https://doi.org/10.1016/j.ymeth.2016.02.018.

Bhakdi, S., Tranum-Jensen, J., & Sziegoleit, A. (1985). Mechanism of membrane damage by streptolysin-O. *Infection and Immunity*, *47*(1), 52–60.

Bleicken, S., Hantusch, A., Das, K. K., Frickey, T., & Garcia-Saez, A. J. (2017). Quantitative interactome of a membrane Bcl-2 network identifies a hierarchy of complexes for apoptosis regulation. *Nature Communications*, *8*(1), 73. https://doi.org/10.1038/s41467-017-00086-6.

Bleicken, S., Hofhaus, G., Ugarte-Uribe, B., Schroder, R., & Garcia-Saez, A. J. (2016). cBid, Bax and Bcl-xL exhibit opposite membrane remodeling activities. *Cell Death & Disease*, *7*, e2121. https://doi.org/10.1038/cddis.2016.34.

Bleicken, S., Jeschke, G., Stegmueller, C., Salvador-Gallego, R., Garcia-Saez, A. J., & Bordignon, E. (2014). Structural model of active Bax at the membrane. *Molecular Cell*, *56*(4), 496–505. https://doi.org/10.1016/j.molcel.2014.09.022.

Bokori-Brown, M., Martin, T. G., Naylor, C. E., Basak, A. K., Titball, R. W., & Savva, C. G. (2016). Cryo-EM structure of lysenin pore elucidates membrane insertion by an aerolysin family protein. *Nature Communications*, *7*, 11293. https://doi.org/10.1038/ncomms11293.

Borsos, T., Dourmashkin, R. R., & Humphrey, J. H. (1964). Lesions in erythrocyte membranes caused by immune haemolysis. *Nature*, *202*, 251–252.

Bottcher, B., Wynne, S. A., & Crowther, R. A. (1997). Determination of the fold of the core protein of hepatitis B virus by electron cryomicroscopy. *Nature*, *386*(6620), 88–91. https://doi.org/10.1038/386088a0.

Boyd, C. M., & Bubeck, D. (2018). Advances in cryoEM and its impact on beta-pore forming proteins. *Current Opinion in Structural Biology*, *52*, 41–49. https://doi.org/10.1016/j.sbi.2018.07.010.

Boyd, C. M., Parsons, E. S., Smith, R. A., Seddon, J. M., Ces, O., & Bubeck, D. (2016). Disentangling the roles of cholesterol and CD59 in intermedilysin pore formation. *Scientific Reports*, *6*, 38446. https://doi.org/10.1038/srep38446

Brauning, B., Bertosin, E., Praetorius, F., Ihling, C., Schatt, A., Adler, A., et al. (2018). Structure and mechanism of the two-component alpha-helical pore-forming toxin YaxAB. *Nature Communications*, *9*(1), 1806. https://doi.org/10.1038/s41467-018-04139-2.

Buckley, J. T. (1992). The channel-forming toxin aerolysin. *FEMS Microbiology Immunology*, *5*(1–3), 13–17. https://doi.org/10.1111/j.1574-6968.1992.tb05881.x.

Cardone, G., Heymann, J. B., & Steven, A. C. (2013). One number does not fit all: Mapping local variations in resolution in cryo-EM reconstructions. *Journal of Structural Biology*, *184*(2), 226–236. https://doi.org/10.1016/j.jsb.2013.08.002.

Chorev, D. S., & Robinson, C. V. (2020). The importance of the membrane for biophysical measurements. *Nature Chemical Biology*, *16*(12), 1285–1292. https://doi.org/10.1038/s41589-020-0574-1.

Clare, D. K., Siebert, C. A., Hecksel, C., Hagen, C., Mordhorst, V., Grange, M., et al. (2017). Electron bio-imaging Centre (eBIC): The UK national research facility for biological electron microscopy. *Acta crystallographica. Section D, Structural biology*, *73*(Pt 6), 488–495. https://doi.org/10.1107/S2059798317007756.

Conway, J. F., Cheng, N., Zlotnick, A., Wingfield, P. T., Stahl, S. J., & Steven, A. C. (1997). Visualization of a 4-helix bundle in the hepatitis B virus capsid by cryo-electron microscopy. *Nature*, *386*(6620), 91–94. https://doi.org/10.1038/386091a0.

Cosentino, K., & Garcia-Saez, A. J. (2017). Bax and Bak pores: Are we closing the circle? *Trends in Cell Biology*, *27*(4), 266–275. https://doi.org/10.1016/j.tcb.2016.11.004.

Cowan, A. D., Smith, N. A., Sandow, J. J., Kapp, E. A., Rustam, Y. H., Murphy, J. M., et al. (2020). BAK core dimers bind lipids and can be bridged by them. *Nature Structural & Molecular Biology*, *27*(11), 1024–1031. https://doi.org/10.1038/s41594-020-0494-5.

Croll, T. I. (2018). ISOLDE: A physically realistic environment for model building into low-resolution electron-density maps. *Acta Crystallographica. Section D, Structural Biology*, *74*(Pt. 6), 519–530. https://doi.org/10.1107/S2059798318002425.

Degiacomi, M. T., Iacovache, I., Pernot, L., Chami, M., Kudryashev, M., Stahlberg, H., et al. (2013). Molecular assembly of the aerolysin pore reveals a swirling membrane-insertion mechanism. *Nature Chemical Biology*, *9*(10), 623–629. https://doi.org/10.1038/nchembio.1312.

de la Rosa-Trevin, J. M., Oton, J., Marabini, R., Zaldivar, A., Vargas, J., Carazo, J. M., et al. (2013). Xmipp 3.0: An improved software suite for image processing in electron microscopy. *Journal of Structural Biology*, *184*(2), 321–328. https://doi.org/10.1016/j.jsb.2013.09.015.

Drucker, P., Iacovache, I., Bachler, S., Zuber, B., Babiychuk, E. B., Dittrich, P. S., et al. (2019). Membrane deformation and layer-by-layer peeling of giant vesicles induced by the pore-forming toxin pneumolysin. *Biomaterials Science*, *7*(9), 3693–3705. https://doi.org/10.1039/c9bm00134d.

Dudkina, N. V., Spicer, B. A., Reboul, C. F., Conroy, P. J., Lukoyanova, N., Elmlund, H., et al. (2016). Structure of the poly-C9 component of the complement membrane attack complex. *Nature Communications*, *7*, 10588. https://doi.org/10.1038/ncomms10588.

Dunstone, M. A., & de Marco, A. (2017). Cryo-electron tomography: An ideal method to study membrane-associated proteins. *Philosophical Transactions of the Royal Society of London. Series B, Biological Sciences*, *372*(1726), 20160210. https://doi.org/10.1098/rstb.2016.0210.

Efremov, R. G., Gatsogiannis, C., & Raunser, S. (2017). Lipid nanodiscs as a tool for high-resolution structure determination of membrane proteins by single-particle Cryo-EM. *Methods in Enzymology*, *594*, 1–30. https://doi.org/10.1016/bs.mie.2017.05.007.

Emsley, P., & Cowtan, K. (2004). Coot: Model-building tools for molecular graphics. *Acta Crystallographica. Section D, Biological Crystallography*, *60*(Pt. 12 Pt. 1), 2126–2132.
Fan, H., Walker, A. P., Carrique, L., Keown, J. R., Serna Martin, I., Karia, D., et al. (2019). Structures of influenza A virus RNA polymerase offer insight into viral genome replication. *Nature*, *573*(7773), 287–290. https://doi.org/10.1038/s41586-019-1530-7.
Fan, X., Zhao, L., Liu, C., Zhang, J. C., Fan, K., Yan, X., et al. (2017). Near-atomic resolution structure determination in over-focus with volta phase plate by Cs-corrected cryo-EM. *Structure*, *25*(10), 1623–1630.e1623. https://doi.org/10.1016/j.str.2017.08.008.
Fischer, N., Neumann, P., Bock, L. V., Maracci, C., Wang, Z., Paleskava, A., et al. (2016). The pathway to GTPase activation of elongation factor SelB on the ribosome. *Nature*, *540*(7631), 80–85. https://doi.org/10.1038/nature20560.
Garcia-Nafria, J., Lee, Y., Bai, X., Carpenter, B., & Tate, C. G. (2018). Cryo-EM structure of the adenosine A2A receptor coupled to an engineered heterotrimeric G protein. *eLife*, 7, 35946. https://doi.org/10.7554/eLife.35946.
Gatsogiannis, C., Merino, F., Prumbaum, D., Roderer, D., Leidreiter, F., Meusch, D., et al. (2016). Membrane insertion of a Tc toxin in near-atomic detail. *Nature Structural & Molecular Biology*, *23*(10), 884–890. https://doi.org/10.1038/nsmb.3281.
Gilbert, R. J., Beales, L., Blond, D., Simon, M. N., Lin, B. Y., Chisari, F. V., et al. (2005). Hepatitis B small surface antigen particles are octahedral. *Proceedings of the National Academy of Sciences of the United States of America*, *102*(41), 14783–14788. https://doi.org/10.1073/pnas.0505062102.
Gilbert, R. J., Dalla Serra, M., Froelich, C. J., Wallace, M. I., & Anderluh, G. (2014). Membrane pore formation at protein-lipid interfaces. *Trends in Biochemical Sciences*, *39*, 510–516.
Gilbert, R. J., Jimenez, J. L., Chen, S., Tickle, I. J., Rossjohn, J., Parker, M., et al. (1999). Two structural transitions in membrane pore formation by pneumolysin, the pore-forming toxin of Streptococcus pneumoniae. *Cell*, *97*(5), 647–655.
Gilbert, R. J., Mikelj, M., Dalla Serra, M., Froelich, C. J., & Anderluh, G. (2013). Effects of MACPF/CDC proteins on lipid membranes. *Cellular and Molecular Life Sciences*, *70*(12), 2083–2098. https://doi.org/10.1007/s00018-012-1153-8.
Grigorieff, N. (2000). Resolution measurement in structures derived from single particles. *Acta Crystallographica. Section D, Biological Crystallography*, *56*(Pt. 10), 1270–1277.
Grimes, J. M., Burroughs, J. N., Gouet, P., Diprose, J. M., Malby, R., Zientara, S., et al. (1998). The atomic structure of the bluetongue virus core. *Nature*, *395*(6701), 470–478.
Grimes, J. M., Jakana, J., Ghosh, M., Basak, A. K., Roy, P., Chiu, W., et al. (1997). An atomic model of the outer layer of the bluetongue virus core derived from X-ray crystallography and electron cryomicroscopy. *Structure*, *5*(7), 885–893. https://doi.org/10.1016/s0969-2126(97)00243-8.
Hardenbrook, N. J., Liu, S., Zhou, K., Ghosal, K., Hong Zhou, Z., & Krantz, B. A. (2020). Atomic structures of anthrax toxin protective antigen channels bound to partially unfolded lethal and edema factors. *Nature Communications*, *11*(1), 840. https://doi.org/10.1038/s41467-020-14658-6.
Henderson, R. (2004). Realizing the potential of electron cryo-microscopy. *Quarterly Reviews of Biophysics*, *37*(1), 3–13. https://doi.org/10.1017/s0033583504003920.
Henderson, R. (2013). Avoiding the pitfalls of single particle cryo-electron microscopy: Einstein from noise. *Proceedings of the National Academy of Sciences of the United States of America*, *110*(45), 18037–18041. https://doi.org/10.1073/pnas.1314449110.
Henderson, R., Chen, S., Chen, J. Z., Grigorieff, N., Passmore, L. A., Ciccarelli, L., et al. (2011). Tilt-pair analysis of images from a range of different specimens in single-particle electron cryomicroscopy. *Journal of Molecular Biology*, *413*(5), 1028–1046. https://doi.org/10.1016/j.jmb.2011.09.008.

Iacovache, I., De Carlo, S., Cirauqui, N., Dal Peraro, M., van der Goot, F. G., & Zuber, B. (2016). Cryo-EM structure of aerolysin variants reveals a novel protein fold and the pore-formation process. *Nature Communications*, *7*, 12062. https://doi.org/10.1038/ncomms12062.

Ilca, S. L., Kotecha, A., Sun, X., Poranen, M. M., Stuart, D. I., & Huiskonen, J. T. (2015). Localized reconstruction of subunits from electron cryomicroscopy images of macromolecular complexes. *Nature Communications*, *6*, 8843. https://doi.org/10.1038/ncomms9843.

Jiang, J., Pentelute, B. L., Collier, R. J., & Zhou, Z. H. (2015). Atomic structure of anthrax protective antigen pore elucidates toxin translocation. *Nature*, *521*(7553), 545–549. https://doi.org/10.1038/nature14247.

Jiang, W., & Tang, L. (2017). Atomic cryo-EM structures of viruses. *Current Opinion in Structural Biology*, *46*, 122–129. https://doi.org/10.1016/j.sbi.2017.07.002.

Joseph, A. P., Malhotra, S., Burnley, T., Wood, C., Clare, D. K., Winn, M., et al. (2016). Refinement of atomic models in high resolution EM reconstructions using flex-EM and local assessment. *Methods*, *100*, 42–49. https://doi.org/10.1016/j.ymeth.2016.03.007.

Khoshouei, M., Radjainia, M., Baumeister, W., & Danev, R. (2017). Cryo-EM structure of haemoglobin at 3.2 A determined with the volta phase plate. *Nature Communications*, *8*, 16099. https://doi.org/10.1038/ncomms16099.

Kidmose, R. T., Juhl, J., Nissen, P., Boesen, T., Karlsen, J. K., & Pedersen, B. P. (2019). Namdinator—Automatic Molecular Dynamics flexible fitting of structural models into cryo-EM and crystallography experimental maps. *International Union of Crystallography Journal*, *6*(4), 526–531. https://doi.org/10.1107/s2052252519007619.

Kuwana, T. (2019). Cryo-electron microscopy to study Bax pores and MOMP. *Methods in Molecular Biology*, *1877*, 247–256. https://doi.org/10.1007/978-1-4939-8861-7_17.

Kuwana, T., Olson, N. H., Kiosses, W. B., Peters, B., & Newmeyer, D. D. (2016). Pro-apoptotic Bax molecules densely populate the edges of membrane pores. *Scientific Reports*, *6*, 27299. https://doi.org/10.1038/srep27299.

Laverty, D., Desai, R., Uchanski, T., Masiulis, S., Stec, W. J., Malinauskas, T., et al. (2019). Cryo-EM structure of the human alpha1beta3gamma2 GABAA receptor in a lipid bilayer. *Nature*, *565*(7740), 516–520. https://doi.org/10.1038/s41586-018-0833-4.

Law, R. H., Lukoyanova, N., Voskoboinik, I., Caradoc-Davies, T. T., Baran, K., Dunstone, M. A., et al. (2010). The structural basis for membrane binding and pore formation by lymphocyte perforin. *Nature*, *468*(7322), 447–451.

Leung, C., Dudkina, N. V., Lukoyanova, N., Hodel, A. W., Farabella, I., Pandurangan, A. P., et al. (2014). Stepwise visualization of membrane pore formation by suilysin, a bacterial cholesterol-dependent cytolysin. *eLife*, *3*, e04247. https://doi.org/10.7554/eLife.04247.

Leung, C., Hodel, A. W., Brennan, A. J., Lukoyanova, N., Tran, S., House, C. M., et al. (2017). Real-time visualization of perforin nanopore assembly. *Nature Nanotechnology*, *12*(5), 467–473. https://doi.org/10.1038/nnano.2016.303.

Liko, I., Allison, T. M., Hopper, J. T., & Robinson, C. V. (2016). Mass spectrometry guided structural biology. *Current Opinion in Structural Biology*, *40*, 136–144. https://doi.org/10.1016/j.sbi.2016.09.008.

Lopez-Blanco, J. R., & Chacon, P. (2013). iMODFIT: Efficient and robust flexible fitting based on vibrational analysis in internal coordinates. *Journal of Structural Biology*, *184*(2), 261–270. https://doi.org/10.1016/j.jsb.2013.08.010.

Lukoyanova, N., Kondos, S. C., Farabella, I., Law, R. H., Reboul, C. F., Caradoc-Davies, T. T., et al. (2015). Conformational changes during pore formation by the perforin-related protein pleurotolysin. *PLoS Biology*, *13*(2), e1002049. https://doi.org/10.1371/journal.pbio.1002049.

Malhotra, A., Penczek, P., Agrawal, R. K., Gabashvili, I. S., Grassucci, R. A., Junemann, R., et al. (1998). Escherichia coli 70 S ribosome at 15 A resolution by cryo-electron microscopy: Localization of fMet-tRNAfMet and fitting of L1 protein. *Journal of Molecular Biology*, *280*(1), 103–116. https://doi.org/10.1006/jmbi.1998.1859.

Mancini, E. J., & Fuller, S. D. (2000). Supplanting crystallography or supplementing microscopy? A combined approach to the study of an enveloped virus. *Acta Crystallographica. Section D, Biological Crystallography*, *56*(Pt 10), 1278–1287. https://doi.org/10.1107/s0907444900010817.

McMullan, G., Faruqi, A. R., Clare, D., & Henderson, R. (2014). Comparison of optimal performance at 300 keV of three direct electron detectors for use in low dose electron microscopy. *Ultramicroscopy*, *147*, 156–163. https://doi.org/10.1016/j.ultramic.2014.08.002.

Menny, A., Serna, M., Boyd, C. M., Gardner, S., Joseph, A. P., Morgan, B. P., et al. (2018). CryoEM reveals how the complement membrane attack complex ruptures lipid bilayers. *Nature Communications*, *9*(1), 5316. https://doi.org/10.1038/s41467-018-07653-5.

Merk, A., Bartesaghi, A., Banerjee, S., Falconieri, V., Rao, P., Davis, M. I., et al. (2016). Breaking cryo-EM resolution barriers to facilitate drug discovery. *Cell*, *165*(7), 1698–1707. https://doi.org/10.1016/j.cell.2016.05.040.

Mills, D. J. (2021). Setting up and operating a cryo-EM laboratory. *Quarterly Reviews of Biophysics*, *54*, e2. https://doi.org/10.1017/S003358352000013X.

Moeller, A., Kirchdoerfer, R. N., Potter, C. S., Carragher, B., & Wilson, I. A. (2012). Organization of the influenza virus replication machinery. *Science*, *338*(6114), 1631–1634. https://doi.org/10.1126/science.1227270.

Moriya, T., Saur, M., Stabrin, M., Merino, F., Voicu, H., Huang, Z., et al. (2017). High-resolution single particle analysis from electron cryo-microscopy images using SPHIRE. *Journal of Visualized Experiments*, (123), 55448. https://doi.org/10.3791/55448.

Mueller, M., Grauschopf, U., Maier, T., Glockshuber, R., & Ban, N. (2009). The structure of a cytolytic alpha-helical toxin pore reveals its assembly mechanism. *Nature*, *459*(7247), 726–730. https://doi.org/10.1038/nature08026.

Nakane, T., Kotecha, A., Sente, A., McMullan, G., Masiulis, S., Brown, P., et al. (2020). Single-particle cryo-EM at atomic resolution. *Nature*, *587*(7832), 152–156. https://doi.org/10.1038/s41586-020-2829-0.

Ni, T., Jiao, F., Yu, X., Aden, S., Ginger, L., Williams, S. I., et al. (2020). Structure and mechanism of bactericidal mammalian perforin-2, an ancient agent of innate immunity. *Science Advances*, *6*(5), eaax8286. https://doi.org/10.1126/sciadv.aax8286.

Ni, T., Williams, S. I., Rezelj, S., Anderluh, G., Harlos, K., Stansfeld, P. J., et al. (2018). Structures of monomeric and oligomeric forms of the toxoplasma gondii perforin-like protein 1. *Science Advances*, *4*(3), eaaq0762. https://doi.org/10.1126/sciadv.aaq0762.

Oestringer, B. P., Bolivar, J. H., Hensen, M., Claridge, J. K., Chipot, C., Dehez, F., et al. (2018). Re-evaluating the p7 viroporin structure. *Nature*, *562*(7727), E8–E18. https://doi.org/10.1038/s41586-018-0561-9.

Orlova, E. V., & Saibil, H. R. (2011). Structural analysis of macromolecular assemblies by electron microscopy. *Chemical Reviews*, *111*(12), 7710–7748. https://doi.org/10.1021/cr100353t.

Ortega, J., Singh, S. K., Ishikawa, T., Maurizi, M. R., & Steven, A. C. (2000). Visualization of substrate binding and translocation by the ATP-dependent protease, ClpXP. *Molecular Cell*, *6*(6), 1515–1521. https://doi.org/10.1016/s1097-2765(00)00148-9.

OuYang, B., Xie, S., Berardi, M. J., Zhao, X., Dev, J., Yu, W., et al. (2013). Unusual architecture of the p7 channel from hepatitis C virus. *Nature*, *498*(7455), 521–525. https://doi.org/10.1038/nature12283.

Palovcak, E., Wang, F., Zheng, S. Q., Yu, Z., Li, S., Betegon, M., et al. (2018). A simple and robust procedure for preparing graphene-oxide cryo-EM grids. *Journal of Structural Biology*, *204*(1), 80–84. https://doi.org/10.1016/j.jsb.2018.07.007.

Pang, S. S., Bayly-Jones, C., Radjainia, M., Spicer, B. A., Law, R. H. P., Hodel, A. W., et al. (2019). The cryo-EM structure of the acid activatable pore-forming immune effector macrophage-expressed gene 1. *Nature Communications*, *10*(1), 4288. https://doi.org/10.1038/s41467-019-12279-2.

Parker, M. W., Buckley, J. T., Postma, J. P., Tucker, A. D., Leonard, K., Pattus, F., et al. (1994). Structure of the Aeromonas toxin proaerolysin in its water-soluble and membrane-channel states. *Nature, 367*(6460), 292–295. https://doi.org/10.1038/367292a0.

Penczek, P. A. (2010). Resolution measures in molecular electron microscopy. *Methods in Enzymology, 482*, 73–100. https://doi.org/10.1016/S0076-6879(10)82003-8.

Penczek, P. A. (2020). Reliable cryo-EM resolution estimation with modified Fourier shell correlation. *IUCrJ*, 7(Pt. 6), 995–1008. https://doi.org/10.1107/S2052252520011574.

Peng, W., de Souza Santos, M., Li, Y., Tomchick, D. R., & Orth, K. (2019). High-resolution cryo-EM structures of the E. coli hemolysin ClyA oligomers. *PLoS One, 14*(5), e0213423. https://doi.org/10.1371/journal.pone.0213423.

Pettersen, E. F., Goddard, T. D., Huang, C. C., Couch, G. S., Greenblatt, D. M., Meng, E. C., et al. (2004). UCSF chimera—A visualization system for exploratory research and analysis. *Journal of Computational Chemistry, 25*(13), 1605–1612. https://doi.org/10.1002/jcc.20084.

Piper, S. J., Brillault, L., Rothnagel, R., Croll, T. I., Box, J. K., Chassagnon, I., et al. (2019). Cryo-EM structures of the pore-forming A subunit from the Yersinia entomophaga ABC toxin. *Nature Communications, 10*(1), 1952. https://doi.org/10.1038/s41467-019-09890-8.

Podack, E. R., & Dennert, G. (1983). Assembly of two types of tubules with putative cytolytic function by cloned natural killer cells. *Nature, 302*(5907), 442–445.

Podack, E. R., & Tschopp, J. (1982). Polymerization of the ninth component of complement (C9): Formation of poly(C9) with a tubular ultrastructure resembling the membrane attack complex of complement. *Proceedings of the National Academy of Sciences of the United States of America, 79*(2), 574–578.

Podobnik, M., Marchioretto, M., Zanetti, M., Bavdek, A., Kisovec, M., Cajnko, M. M., et al. (2015). Plasticity of listeriolysin O pores and its regulation by pH and a unique histidine. *Scientific Reports, 5*, 9623.

Podobnik, M., Savory, P., Rojko, N., Kisovec, M., Wood, N., Hambley, R., et al. (2016). Crystal structure of an invertebrate cytolysin pore reveals unique properties and mechanism of assembly. *Nature Communications*, 7, 11598. https://doi.org/10.1038/ncomms11598.

Praper, T., Sonnen, A. F., Kladnik, A., Andrighetti, A. O., Viero, G., Morris, K. J., et al. (2011). Perforin activity at membranes leads to invaginations and vesicle formation. *Proceedings of the National Academy of Sciences of the United States of America, 108*(52), 21016–21021.

Praper, T., Sonnen, A. F., Viero, G., Kladnik, A., Froelich, C. J., Anderluh, G., et al. (2011). Human perforin employs different avenues to damage membranes. *The Journal of Biological Chemistry, 286*, 2946–2955.

Punjani, A., Rubinstein, J. L., Fleet, D. J., & Brubaker, M. A. (2017). cryoSPARC: Algorithms for rapid unsupervised cryo-EM structure determination. *Nature Methods, 14*(3), 290–296. https://doi.org/10.1038/nmeth.4169.

Rajesh, S., Knowles, T., & Overduin, M. (2011). Production of membrane proteins without cells or detergents. *New Biotechnology, 28*(3), 250–254. https://doi.org/10.1016/j.nbt.2010.07.011.

Ramlaul, K., Palmer, C. M., & Aylett, C. H. S. (2019). A local agreement filtering algorithm for transmission EM reconstructions. *Journal of Structural Biology, 205*(1), 30–40. https://doi.org/10.1016/j.jsb.2018.11.011.

Roderer, D., Brocker, F., Sitsel, O., Kaplonek, P., Leidreiter, F., Seeberger, P. H., et al. (2020). Glycan-dependent cell adhesion mechanism of Tc toxins. *Nature Communications, 11*(1), 2694. https://doi.org/10.1038/s41467-020-16536-7.

Roderer, D., Hofnagel, O., Benz, R., & Raunser, S. (2019). Structure of a Tc holotoxin pore provides insights into the translocation mechanism. *Proceedings of the National Academy of Sciences of the United States of America, 116*(46), 23083–23090. https://doi.org/10.1073/pnas.1909821116.

Rosenthal, P. B., & Henderson, R. (2003). Optimal determination of particle orientation, absolute hand, and contrast loss in single-particle electron cryomicroscopy. *Journal of Molecular Biology*, *333*(4), 721–745.

Ruan, Y., Rezelj, S., Bedina Zavec, A., Anderluh, G., & Scheuring, S. (2016). Listeriolysin O membrane damaging activity involves arc formation and Lineaction—Implication for listeria monocytogenes escape from phagocytic vacuole. *PLoS Pathogens*, *12*(4), e1005597. https://doi.org/10.1371/journal.ppat.1005597.

Ruan, J., Xia, S., Liu, X., Lieberman, J., & Wu, H. (2018). Cryo-EM structure of the gasdermin A3 membrane pore. *Nature*, *557*(7703), 62–67. https://doi.org/10.1038/s41586-018-0058-6.

Savva, C. G., Clark, A. R., Naylor, C. E., Popoff, M. R., Moss, D. S., Basak, A. K., et al. (2019). The pore structure of Clostridium perfringens epsilon toxin. *Nature Communications*, *10*(1), 2641. https://doi.org/10.1038/s41467-019-10645-8.

Scheres, S. H. (2012). RELION: Implementation of a Bayesian approach to cryo-EM structure determination. *Journal of Structural Biology*, *180*(3), 519–530. https://doi.org/10.1016/j.jsb.2012.09.006.

Scheres, S. H., & Chen, S. (2012). Prevention of overfitting in cryo-EM structure determination. *Nature Methods*, *9*(9), 853–854. https://doi.org/10.1038/nmeth.2115.

Schnell, J. R., & Chou, J. J. (2008). Structure and mechanism of the M2 proton channel of influenza A virus. *Nature*, *451*(7178), 591–595. https://doi.org/10.1038/nature06531.

Schubert, E., Vetter, I. R., Prumbaum, D., Penczek, P. A., & Raunser, S. (2018). Membrane insertion of alpha-xenorhabdolysin in near-atomic detail. *eLife*, 7, 38017. https://doi.org/10.7554/eLife.38017.

Schur, F. K., Obr, M., Hagen, W. J., Wan, W., Jakobi, A. J., Kirkpatrick, J. M., et al. (2016). An atomic model of HIV-1 capsid-SP1 reveals structures regulating assembly and maturation. *Science*, *353*(6298), 506–508. https://doi.org/10.1126/science.aaf9620.

Scott, H., Huang, W., Bann, J. G., & Taylor, D. J. (2018). Advances in structure determination by cryo-EM to unravel membrane-spanning pore formation. *Protein Science*, *27*(9), 1544–1556. https://doi.org/10.1002/pro.3454.

Serna, M., Giles, J. L., Morgan, B. P., & Bubeck, D. (2016). Structural basis of complement membrane attack complex formation. *Nature Communications*, 7, 10587. https://doi.org/10.1038/ncomms10587.

Shah, N. R., Voisin, T. B., Parsons, E. S., Boyd, C. M., Hoogenboom, B. W., & Bubeck, D. (2020). Structural basis for tuning activity and membrane specificity of bacterial cytolysins. *Nature Communications*, *11*(1), 5818. https://doi.org/10.1038/s41467-020-19482-6.

Sonnen, A. F., Plitzko, J., & Gilbert, R. J. C. (2014). Incomplete pneumolysin oligomers form membrane pores. *Open Biology*, *4*, 140044.

Spicer, B. A., Law, R. H. P., Caradoc-Davies, T. T., Ekkel, S. M., Bayly-Jones, C., Pang, S. S., et al. (2018). The first transmembrane region of complement component-9 acts as a brake on its self-assembly. *Nature Communications*, *9*(1), 3266. https://doi.org/10.1038/s41467-018-05717-0.

Tilley, S. J., Orlova, E. V., Gilbert, R. J., Andrew, P. W., & Saibil, H. R. (2005). Structural basis of pore formation by the bacterial toxin pneumolysin. *Cell*, *121*(2), 247–256.

Topf, M., Lasker, K., Webb, B., Wolfson, H., Chiu, W., & Sali, A. (2008). Protein structure fitting and refinement guided by cryo-EM density. *Structure*, *16*(2), 295–307. https://doi.org/10.1016/j.str.2007.11.016.

Tschopp, J. (1984). Ultrastructure of the membrane attack complex of complement. Heterogeneity of the complex caused by different degree of C9 polymerization. *The Journal of Biological Chemistry*, *259*(12), 7857–7863.

Uchanski, T., Masiulis, S., Fischer, B., Kalichuk, V., Lopez-Sanchez, U., Zarkadas, E., et al. (2021). Megabodies expand the nanobody toolkit for protein structure determination by single-particle cryo-EM. *Nature Methods, 18*(1), 60–68. https://doi.org/10.1038/s41592-020-01001-6.

Ugarte-Uribe, B., & Garcia-Saez, A. J. (2017). Apoptotic foci at mitochondria: In and around Bax pores. *Philosophical Transactions of the Royal Society of London. Series B, Biological Sciences, 372*(1726), 20160217. https://doi.org/10.1098/rstb.2016.0217.

Uren, R. T., Iyer, S., & Kluck, R. M. (2017). Pore formation by dimeric Bak and Bax: An unusual pore? *Philosophical Transactions of the Royal Society of London. Series B, Biological Sciences, 372*(1726), 20160218. https://doi.org/10.1098/rstb.2016.0218.

van Pee, K., Neuhaus, A., D'Imprima, E., Mills, D. J., Kuhlbrandt, W., & Yildiz, O. (2017). CryoEM structures of membrane pore and prepore complex reveal cytolytic mechanism of Pneumolysin. *eLife, 6*, 23644. https://doi.org/10.7554/eLife.23644.

Vogele, M., Bhaskara, R. M., Mulvihill, E., van Pee, K., Yildiz, O., Kuhlbrandt, W., et al. (2019). Membrane perforation by the pore-forming toxin pneumolysin. *Proceedings of the National Academy of Sciences of the United States of America, 116*(27), 13352–13357. https://doi.org/10.1073/pnas.1904304116.

Wang, H. W., & Fan, X. (2019). Challenges and opportunities in cryo-EM with phase plate. *Current Opinion in Structural Biology, 58*, 175–182. https://doi.org/10.1016/j.sbi.2019.06.013.

Wang, Z., Hryc, C. F., Bammes, B., Afonine, P. V., Jakana, J., Chen, D. H., et al. (2014). An atomic model of brome mosaic virus using direct electron detection and real-space optimization. *Nature Communications, 5*, 4808. https://doi.org/10.1038/ncomms5808.

Wang, R. Y., Song, Y., Barad, B. A., Cheng, Y., Fraser, J. S., & DiMaio, F. (2016). Automated structure refinement of macromolecular assemblies from cryo-EM maps using Rosetta. *eLife, 5*, 17219. https://doi.org/10.7554/eLife.17219.

Webb, B., & Sali, A. (2016). Comparative protein structure modeling using MODELLER. *Current Protocols in Bioinformatics, 54*, 5.6.1–5.6.37. https://doi.org/10.1002/cpbi.3.

Webb, B., & Sali, A. (2021). Protein structure modeling with MODELLER. *Methods in Molecular Biology, 2199*, 239–255. https://doi.org/10.1007/978-1-0716-0892-0_14.

Wilmsen, H. U., Leonard, K. R., Tichelaar, W., Buckley, J. T., & Pattus, F. (1992). The aerolysin membrane channel is formed by heptamerization of the monomer. *The EMBO Journal, 11*(7), 2457–2463.

Xu, X., Godoy-Ruiz, R., Adipietro, K. A., Peralta, C., Ben-Hail, D., Varney, K. M., et al. (2020). Structure of the cell-binding component of the Clostridium difficile binary toxin reveals a di-heptamer macromolecular assembly. *Proceedings of the National Academy of Sciences of the United States of America, 117*(2), 1049–1058. https://doi.org/10.1073/pnas.1919490117.

Yamada, T., Yoshida, T., Kawamoto, A., Mitsuoka, K., Iwasaki, K., & Tsuge, H. (2020). Cryo-EM structures reveal translocational unfolding in the clostridial binary iota toxin complex. *Nature Structural & Molecular Biology, 27*(3), 288–296. https://doi.org/10.1038/s41594-020-0388-6.

Yip, K. M., Fischer, N., Paknia, E., Chari, A., & Stark, H. (2020). Atomic-resolution protein structure determination by cryo-EM. *Nature, 587*(7832), 157–161. https://doi.org/10.1038/s41586-020-2833-4.

Zhao, Y., Chen, S., Swensen, A. C., Qian, W. J., & Gouaux, E. (2019). Architecture and subunit arrangement of native AMPA receptors elucidated by cryo-EM. *Science, 364*(6438), 355–362. https://doi.org/10.1126/science.aaw8250.

Zheng, S. Q., Palovcak, E., Armache, J. P., Verba, K. A., Cheng, Y., & Agard, D. A. (2017). MotionCor2: Anisotropic correction of beam-induced motion for improved cryo-electron microscopy. *Nature Methods*, *14*(4), 331–332. https://doi.org/10.1038/nmeth.4193.

Zhou, M., Morgner, N., Barrera, N. P., Politis, A., Isaacson, S. C., Matak-Vinkovic, D., et al. (2011). Mass spectrometry of intact V-type ATPases reveals bound lipids and the effects of nucleotide binding. *Science*, *334*(6054), 380–385. https://doi.org/10.1126/science.1210148.

Zivanov, J., Nakane, T., Forsberg, B. O., Kimanius, D., Hagen, W. J., Lindahl, E., et al. (2018). New tools for automated high-resolution cryo-EM structure determination in RELION-3. *eLife*, 7, 42166. https://doi.org/10.7554/eLife.42166.

CHAPTER FOUR

Going full circle: Determining the structures of complement component 9

Bradley A. Spicer* and **Michelle A. Dunstone***
Department of Biochemistry and Molecular Biology, Biomedicine Discovery Institute, Monash University, Clayton, VIC, Australia
*Corresponding authors: e-mail address: bradley.spicer@monash.edu; michelle.dunstone@monash.edu

Contents

Abstract

Pore forming proteins (PFPs) undergo dramatic conformational changes to punch holes in the target membrane. These PFPs have the ability to self-assemble, by way of oligomerization, and have the capacity to transform from a water soluble state (commonly referred to as fluid phase) to a membrane adhered form. Accordingly, PFPs are metastable, that is they are inert until the right conditions cause the release of potential energy stored in the conformational fold leading to a vast structural rearrangement into a membrane-inserted oligomeric form. However, the metastable state of PFPs poses a problem of leading to aggregation and precipitation in conditions typically required for structural biology techniques. Here, we discuss the protein chemistry of the MACPF protein complement component 9 (C9). C9 is part of a larger complex assembly known as the membrane attack complex (MAC) that has been studied extensively for its ability to form pores in bacteria. An unusual artifact of human C9 is the ability to form a soluble oligomeric state of the channel portion of the MAC, called polyC9. PolyC9 formation does not require the presence of membranes or other complement factors. It is only

Methods in Enzymology, Volume 649
ISSN 0076-6879
https://doi.org/10.1016/bs.mie.2021.01.020

in recent years that structural studies of the MAC have become successful owing to improved recombinant DNA expression systems and the improvement of high-resolution techniques (both X-ray crystallography and single particle cryo-EM). We discuss the expression and purification of recombinant C9, crystallization of the soluble monomeric form of C9 and the preparation of the oligomeric polyC9.

1. Introduction

One research area that has benefitted from the use of recombinant expression is the area of complement biology, particularly the membrane attack complex (MAC). The assembly of the MAC represents the terminal complement pathway which acts by directly lysing foreign organisms but is also postulated to cause indirect cell death by the translocation of lysozyme (Martinez & Carroll, 1980). The MAC forms a 1.6 MDa complex that is a membrane spanning β-barrel pore. The earliest direct observation of MAC pores was on red blood cell membranes using electron microscopy in the 1960s (Borsos, Dourmashkin, & Humphrey, 1964). Following these observation, the MAC was determined to be an ultrastructure containing an asymmetric pore (Podack, Tschopp, & Müller-Eberhard, 1982; Tschopp, Müller-Eberhard, & Podack, 1982; Tschopp, Podack, & Müller-Eberhard, 1982). However, the long posed question in complement biology was; how does the MAC assemble into a pore? The assembly sequence of MAC formation was determined using the MAC components purified from human serum (Fig. 1). This research established that there is sequential

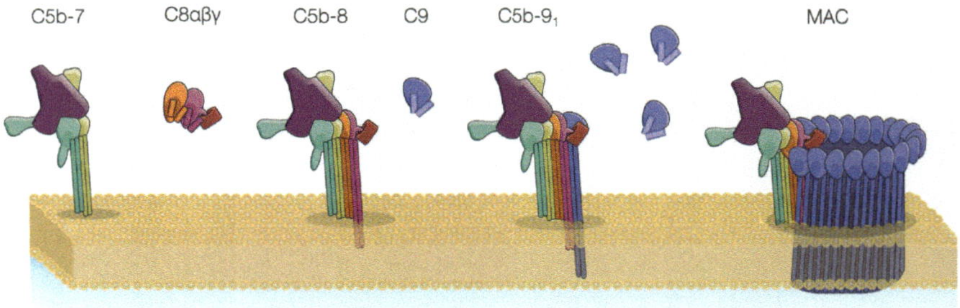

Fig. 1 Membrane attack complex (MAC) assembly pathway. Schematic representation of MAC addition to the final assembly on a membrane. The initial membrane binding occurs with the formation of C5b-7 which then recruits heterotrimeric C8 (αβγ). Next, up to 18 copies of C9 oligomerize to form the final membrane spanning pore. As the MAC assembly occurs, the membrane spanning regions progressively span further into the membrane as represented by the sticks.

formation of the following intermediates: C5b → C5b-6 → C5b-7 → C5b-8 and C5b-9_1 (Podack, Biesecker, & Müller-Eberhard, 1979; Podack & Müller-Eberhard, 1978). The C5b9_1 forms the nascent MAC that is a scaffold for the oligomerization of a further 15–17 C9 proteins to form the final MAC. Overall, C9 contributes to a majority of the membrane spanning portion of the MAC.

The first crystal structure of a MAC protein, C8α (Hadders, Beringer, & Gros, 2007) and a bacterial complement homologue, Plu-MACPF (Rosado et al., 2007) uncovered the evolutionary link between the MAC and bacterial pore forming toxins, the cholesterol dependent cytolysins (CDCs). Since this initial discovery, improvements in molecular biology and X-ray diffraction techniques have led to the determination of many structures of individual MAC proteins including C5, C6, C8, and C9, allowing a general molecular understanding of the pore forming mechanism (Aleshin, DiScipio, Stec, & Liddington, 2012; Aleshin, Schraufstatter, et al., 2012; Fredslund et al., 2008; Hadders et al., 2012; Lovelace, Cooper, Sodetz, & Lebioda, 2011; Spicer et al., 2018). Furthermore, the advancement in cryogenic electron microscopy has resolved the structures of the final MAC and of polyC9 (Dudkina et al., 2016; Menny et al., 2018; Serna, Giles, Morgan, & Bubeck, 2016; Spicer et al., 2018). Overall, these structures have allowed a deeper understanding of the assembly of the MAC and its concomitant pore forming activity.

Despite the vast number of crystal structures from serum derived protein, one of the major bottlenecks, with respect to studying MAC biology, has been the production of protein that can crystallize. Traditionally, MAC components were sourced from human plasma which precludes the study of mutants. This also means that there is a limit on the amount of MAC proteins that can be purified from small organisms (e.g., mouse). However, recombinant protein expression has advanced the structural biology studies of the individual MAC proteins and final complex by allowing for highly overexpressed samples to be purified. Desirable properties of the mammalian expression systems include native post-translational modifications such as disulfide bond formation and N-linked glycosylation that ensure the proper protein folding and function.

One of the more elusive targets for structural studies of the MAC is the final C9 protein. C9 is the only MAC protein with the propensity to self-oligomerize either within the context of the MAC or as the unusual polyC9 form (human C9 only). Moreover, C9 undergoes dramatic conformational changes that result in the final transmembrane pore, which is of major interest in structural biology studies.

1.1 Domain organization and properties of C9

Each of the MAC proteins contains a multidomain topology with a core set of domains conserved between C6, C7, C8α, C8β and C9. The C9 protein is the smallest protein of the MAC at 61 kDa and consists of four domains: an N-terminal thrombospondin-like (TSP) domain, a low density lipoprotein receptor A (LDLRA) domain, a membrane attack complex perforin-like (MACPF) domain and epidermal growth factor type 2 (EGF) domain (Fig. 2A). The sequence alignment of different species of C9 shows clear domain boundaries for each of these four domains (Fig. 2B). Indeed, within these discrete regions of the C9 protein, the sequence identity is relatively high (TSP 63%, LDLRA 56% and EGF 50%). Conversely, the TMH1 and TMH2 regions, which unfurl from the core of the MACPF domain and form the membrane spanning pore region, have the lowest level of sequence identity (14% and 22%, respectively). It is not known exactly why the TMH1 and TMH2 regions have a lower sequence conservation compared to the rest of the MACPF domain however it is speculated that the core of the domain is more important to conserve for correct folding and conservation of the oligomer interface. Conversely, the TMH1–2 regions can accommodate more change as long as they can span the target membrane using amphipathic β-strands. In addition, the distal N- and C-terminal regions show low conservation in both sequences and are predicted to be disordered. Further, the human C9 form contains a capped N-terminal glutamine (pyroglutamic acid) (Franc, Yang, & Heck, 2017).

C9 contains several post-translational modifications that are observed in several other MAC proteins. Notably, the ancillary domains (TSP1, LDLRA and EGF) contain three conserved intradomain disulfide bonds while the MACPF domain has two conserved disulfide bonds. The TSP1 contains a C-mannosylation on the first tryptophan of the WXXW motif (Hofsteenge, Blommers, Hess, Furmanek, & Miroshnichenko, 1999), although recent mass spectroscopy analysis has observed a proportion of mannan on the last tryptophan (Franc et al., 2017). Furthermore, in human C9, there are two N-linked glycans, one located on TMH1 and another on TMH2 (Fig. 3). In bovine and equine C9, there is an N-glycan located in a loop that separates the two α-helices of the HTH region. The mouse form contains three predicted N-linked glycans where the third N-glycan is co-located within the C-mannosylation motif (WSNWS). Each of the three N-glycan motifs were mutated (asparagine → aspartic acid) to produce an aglycosylated form of mouse C9 that crystallized (Fig. 3A). In the polyC9 structure the N-linked glycans are arranged such that they face the inside

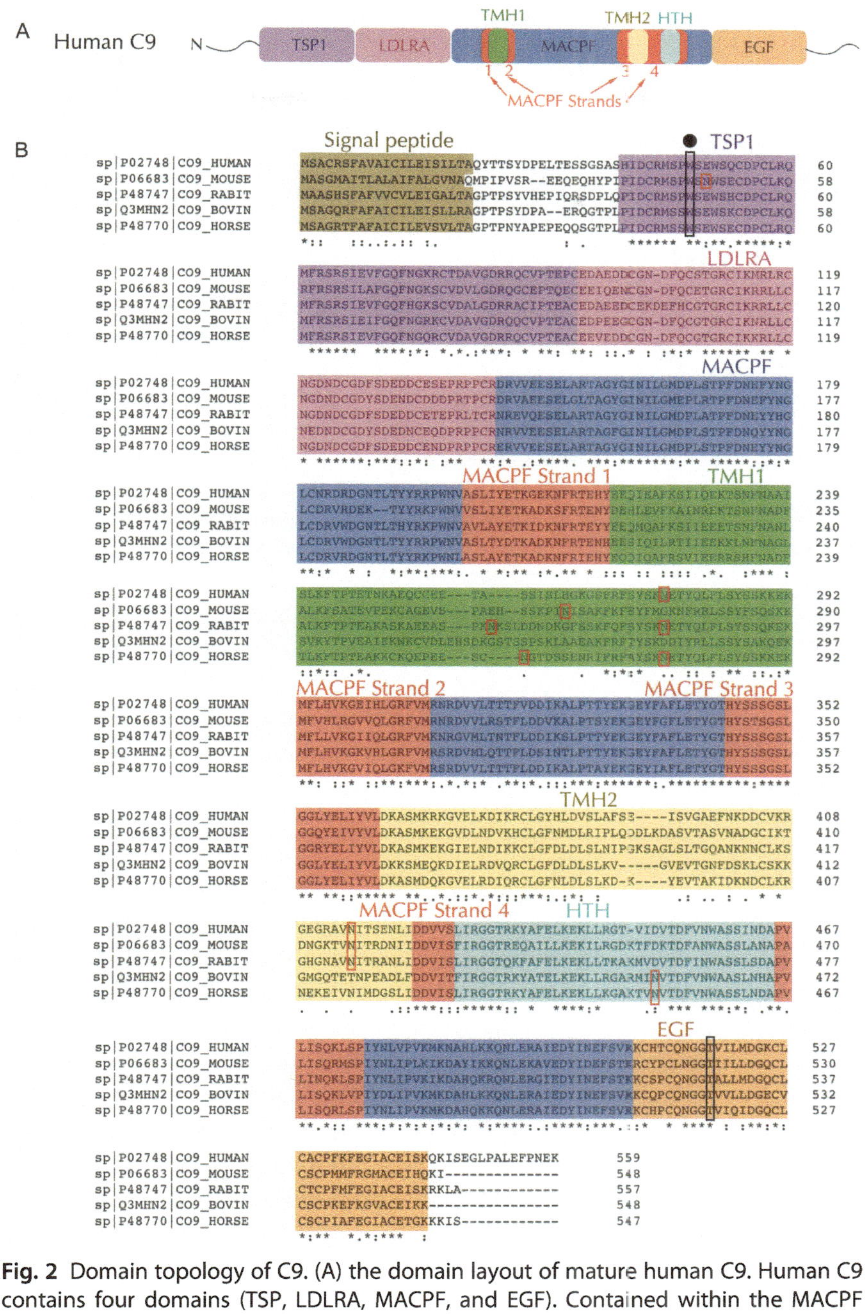

Fig. 2 Domain topology of C9. (A) the domain layout of mature human C9. Human C9 contains four domains (TSP, LDLRA, MACPF, and EGF). Contained within the MACPF domain are TMH1 and TMH2 which form the membrane spanning pore component of the MAC. (B) Sequence alignment between human, mouse, rabbit, bovine and horse C9 proteins. The regions of highest identity are in the ancillary domains (TSP, LDLRA, and EGF). Positions for the N-linked (red boxes) and O-linked glycans (black boxes) and C-mannosylation glycan (●) are shown.

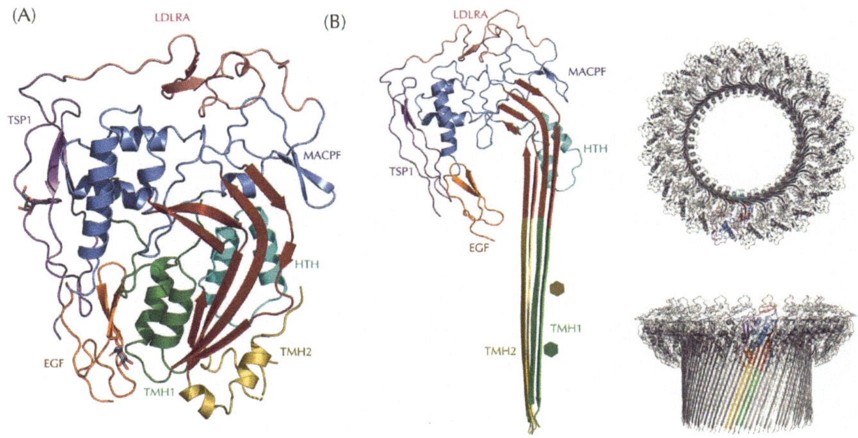

Fig. 3 Models of C9 monomer and polyC9. (A) The monomer structure of murine C9 (PDB ID: 6CXO). (B) The structure of polyC9 (PDB ID: 6DLW). A single unit of polyC9 is shown as well as the full pore from a top or front view. The relative positions of the N-linked glycans for TMH1 and TMH2 are shown on the polyC9 model (green and yellow hexagons).

of the lumen (Fig. 3B). A final O-linked glycan has been observed in the EGF domain in human (Shah et al., 2018) and mouse C9 (Fig. 3B) though it is not clear if there is a consensus motif to recognize this region. Overall, some of the properties of C9, such as the N-glycans and the variable termini and TMH regions, have the potential to introduce unwanted structural heterogeneity.

2. Sources of C9

Complement C9 has been studied extensively for its property to undergo conformational change from a soluble protein to polymerization of C9 to make membrane attack complex (MAC) pores. Several groups have attempted to crystallize C9 purified from human plasma with no success (personal correspondence). Accordingly, it was postulated that recombinant material could improve the success of structural studies by improving the quality of the purification, allowing the production of mutant variants and allowing the production of C9 from alternative species. This strategy was postulated to create more favorable properties for structural studies.

There are only a few published studies of full-length recombinant C9 using either the yeast expression system, the baculovirus expression system or CHO cells (Rossi, Wang, & Esser, 2010; Taylor, Trimby, & Campbell, 1997; Tomlinson et al., 1993; Tomlinson, Wang, Ueda, & Esser, 1995).

The C9 proteins expressed using the baculovirus and mammalian systems were both active in lytic assays indicating that the protein was folded, whereas yeast C9 was substantially less active (Tomlinson et al., 1993). In these regards, insect cell or mammalian cell expression are more appropriate in the study of C9.

Here, we describe the design of the plasmid and protein expression in HEK293F cells, a type of mammalian cell line. We further describe the protein purification, assays and structural techniques. The protein purification will include material expressed by both human and mouse C9 genes. In both cases the protein is secreted into the conditioned media of the HEK293F expression system (transient transfection method). The advantage of this approach over other mammalian cell lines and insect cell expression is the relatively high yield of purified protein. This system can yield milligram quantities of C9/100 mL volumes due to the improvement of commercial HEK cell lines. Moreover, the recombinant material retains the ability to produce folded protein with near native post-translational modifications (e.g., C-mannosylation, O- and N-linked glycosylation and disulfide bond formation). In our studies, recombinant forms of C9 (mouse and human) are hemolytic to the same extent as human plasma-derived C9, which is an overall indication of protein folding and assessment of quality.

2.1 Gene design for mammalian expression of C9 protein

The choice of leader peptide was important to the success of this research. All MAC proteins are secreted by the liver with the exception of C7, which is primarily produced by monocytes and macrophages (Würzner, Joysey, & Lachmann, 1994). Accordingly, these cells and tissues produce a form of the protein that is secreted into blood plasma by way of a secretion peptide. The secretion peptide is used to target the protein to the endoplasmic reticulum (ER) and into the trans Golgi network, allowing it to then be exported outside of the cell. On the path through the ER, the protein is targeted for post-translational modification such as N-linked glycosylation and formation of disulfide bonds, which all MAC proteins contain.

A typical strategy for the production of secreted protein is to use the mouse IgK leader peptide sequence (METDTLLLWVLLLWVPGSTG$^{\downarrow}$D [where $^{\downarrow}$ denotes the cleavage site]) to substitute the native peptide. This strategy has been effective at enhancing protein secretion in general (Elliger, Elliger, Lang, & Watson, 2002). As the peptide sequence is removed during translocation, it is imperative that any tag used for protein purification is introduced after the leader peptide sequence to avoid cleavage. There is a range of available plasmid vectors that can be utilized for transient transfections; we have tested pcDNA3.1(hygro +) and pSecTag2a. In our case, the C9 gene with the IgK leader peptide was synthesized and cloned by Genscript in a state that is ready for transfection. Additionally, the Genscript DNA sequence is optimized for RNA stability, which may improve yield. All subsequent genetic manipulation techniques can conveniently be performed using this construct. It is likely that the choice of both the signal peptide and codon optimization led to optimal expression.

2.2 Procedure for the transient transfection of C9 (human and mouse)

Here we detail the methods for expression of secreted C9 protein from mammalian cells in the untagged form (for both human and mouse C9). Notably, we have observed that His-tagged forms of C9 (e.g., 6 × histidine tag) are also capable of being secreted and may be convenient for purification, however, these forms may affect the downstream application, such as crystallization by interfering with the crystal contacts. These tagged forms of C9 are not described here. Typically, quantities of ~1 mg purified protein/ 100 mL of mammalian expression media can be obtained particularly in cells that have undergone a low number of passages (<30 passages). This expression system is also suitable for the expression of aglycosylated forms of C9 where expression yields are expected to be lower.

2.2.1 *Materials*

DNA preparation kit, e.g., Maxi prep or Giga prep (Qiagen)
2 × yeast extract-tryptone (YT) media or similar (e.g., Luria Broth)
Mammalian media
Tissue culture grade PBS
Tissue culture grade polyethylenimine (PEI) (25kDa, linear)
Lupin peptone (enzymatic hydrolysis of lupin flour protein providing a source of low molecular weight peptides and used as a cell culture supplement)
HEK293F cells or equivalent mammalian expression system for secretion of protein
Protease inhibitor tablets: cOmplete™, EDTA-free protease inhibitor (prepared by dissolving 1 tablet in 1 mL distilled H_2O) (Roche)
Mammalian cell incubators with appropriate gas fitting and rotating platform (recommended)
Benchtop centrifuge

2.2.2 Procedures

1. Perform a large-scale DNA preparation by Maxi or Giga prep. Both pSecTag2a and pcDNA3.1 are high copy vectors and should be prepared using an *E. coli* strain such as DH5α or Top10.
2. Prepare HEK293F cells for transient transfection by incubating at 37 °C with 5% (v/v) CO_2 in a shaking incubator. Set the platform oscillation rate to 110–140 rpm. The day before transfections, dilute the HEK293F cells to a cell density of $1.0–1.2 \times 10^6$ cell/mL in 90% of the final desired transfection volume (typically 90 mL cell suspension for a final volume of 100 mL).
3. The following day, prepare a transfection solution which is 10% of the final transfection volume (i.e., 10 mL for a final volume of 100 mL). The transfection solution is composed of PBS, pre-warmed to room temperature (19–23 °C), with the addition of transfection DNA (1 μg/mL of final transfection volume, e.g., 100 μg for a final 100 mL transfection volume) and PEI (4 μg per 1 μg DNA) in a sterile Falcon tube.
4. Incubate the DNA/PEI suspension for 15–20 min, then add dropwise with constant swirling to the HEK293F cells.
5. Measure the glucose concentration of the transfection culture using an Accu-Chek meter and adjust to 33 mM before placing in the shaking CO_2 incubator.
6. After 18 h, supplement the cells with lupin to a final concentration of 5 mg/mL and continue growing for a total of 4 days.
7. After transfection, harvest the HEK293F cell suspension at 2000 rcf, 4 °C.
8. Filter the conditioned media through consecutively smaller pore size from 0.8 μm, 0.45 μm to 0.22 μm filters with a vacuum trap manifold. To this filtrate add 200 μL aliquot of cOmplete, EDTA-free protease inhibitor per 100 mL of filtered media.

A time course of expression is recommended when testing mutant constructs as they may express poorly compared to wild-type. Typically, this can be performed by taking a small sample (e.g., 1 mL) each day after the initial transfection. The cells can be centrifuged at 2000 rcf and a TCA precipitation can be performed using the supernatant through a standard protocol. The secreted protein can be observed by SDS-PAGE with Coomassie staining/Western blot analysis.

2.3 Experimental procedure for the purification of recombinant C9 from mammalian media

Here we detail the purification protocol for human and mouse C9 (see also Figs. 4 and 5, respectively).

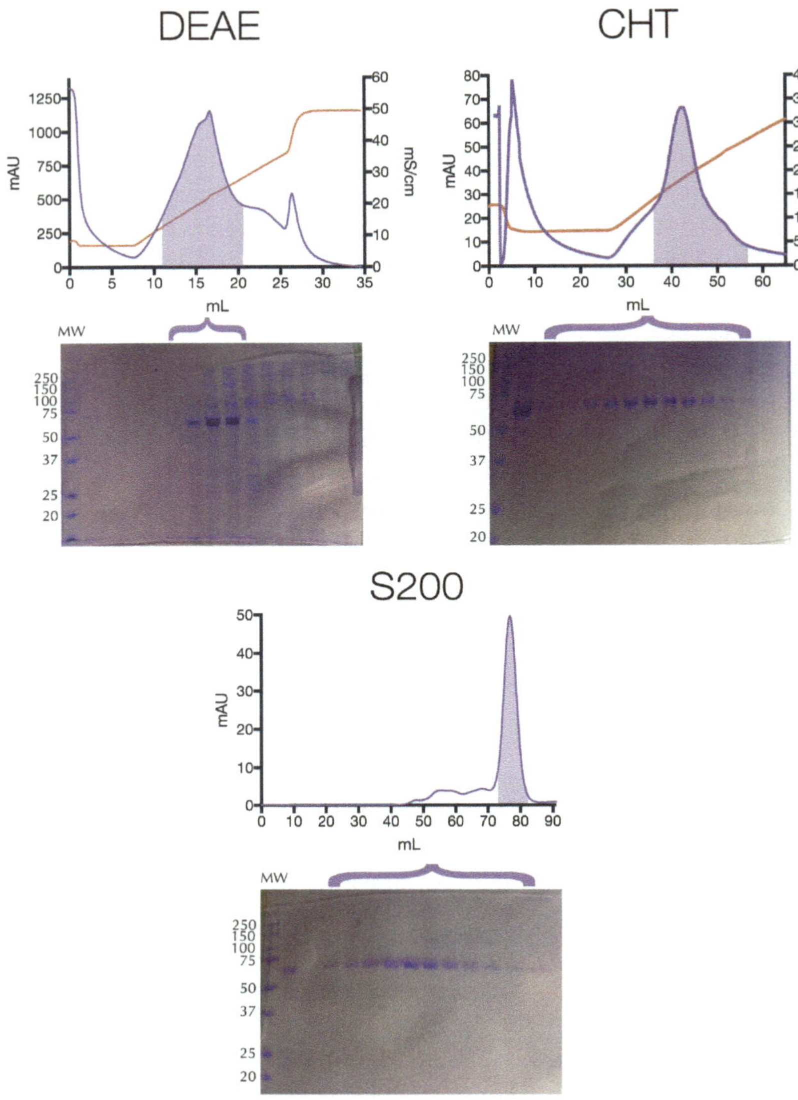

Fig. 4 Protein purification of human C9. Representative purification of untagged human C9 from transient transfection of HEK293F. Sequential purification by DEAE (top left); CHT (top right); and S200 (bottom) are shown. The chromatograms with 12% (w/v) non-reducing SDS-PAGE are shown (beneath). The pooled fractions for the next step are shown with brackets.

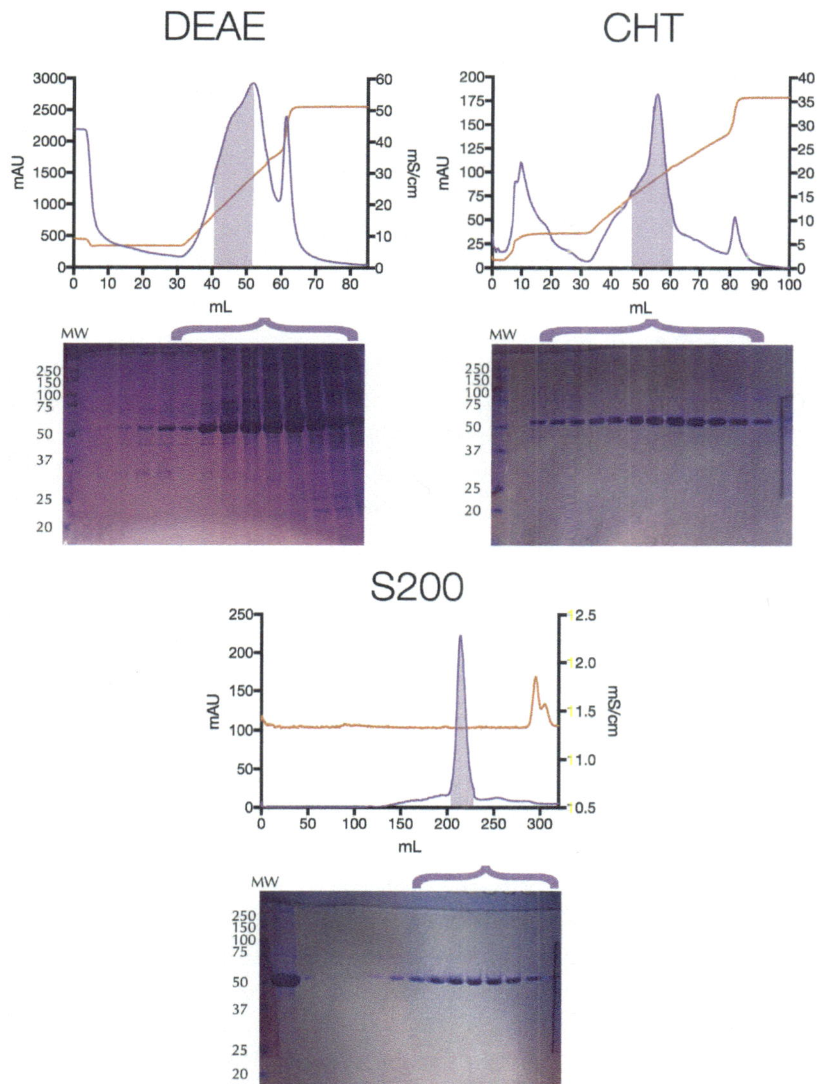

Fig. 5 Protein purification of mouse C9. Representative purification of untagged muC9 [aglyco] from transient transfection of HEK293F. Sequential purification by DEAE (top left); CHT (top right); and S200 (bottom) are shown. The chromatograms with 12% (w/v) non-reducing SDS-PAGE are shown (beneath). The pooled fractions for the next step are shown with brackets. Control reactions were performed containing C9-depleted serum.

2.3.1 Materials and equipment

FPLC and low-pressure pump (e.g., P-1 peristaltic pump (Cytiva); recommended for loading of columns)

Columns: HiTrap DEAE columns; Type 1 ceramic hydroxyapatite (CHT) prepacked columns and Superdex 200 column (16 mm diameter ×60 cm length or 26 mm diameter × 60 cm length)

30 kDa MWCO Amicon® concentrator (Millipore)

Standard SDS-PAGE equipment (MiniProtean 3 system or equivalent)

DEAE phosphate buffer: 10 mM Na_2HPO_4/NaH_2PO_4 pH 7.4, 20 mM NaCl

DEAE buffer A: 10 mM Na_2HPO_4/NaH_2PO_4 pH 7.4, 45 mM NaCl

DEAE buffer B: 10 mM Na_2HPO_4/NaH_2PO_4 pH 7.4, 500 mM NaCl

CHT phosphate buffer: 10 mM Na_2HPO_4/NaH_2PO_4 pH 7.0

CHT wash buffer: 10 mM Na_2HPO_4/NaH_2PO_4 pH 7.0, 100 mM NaCl

CHT buffer A: 45 mM Na_2HPO_4/NaH_2PO_4 pH 8.1

CHT buffer B: 350 mM Na_2HPO_4/NaH_2PO_4 pH 8.1

Human C9 size exclusion chromatography (SEC) buffer: 10 mM HEPES pH 7.2, 200 mM NaCl

Murine C9 SEC buffer: 10 mM HEPES pH 7.2, 100 mM NaCl

1. Dilute the filtrate with an equal volume of the DEAE phosphate buffer. Check the pH is 7.4 with a pH test strip, adjust if necessary with extra buffer.
2. Load the filtrate onto a HiTrap DEAE column equilibrated in DEAE buffer A with a peristaltic pump. Typically, a 1 mL column is sufficient to capture 100 mL of starting media but a 5 mL column can also be used with some modifications to the protocol.
3. Connect the HiTrap column to an FPLC and wash with at least 4 CV of DEAE buffer A. Elute the protein over a linear gradient to 70% with DEAE buffer B (over 20 CV with a 1 mL column or 6 CV with a 5 mL column).
4. The C9-containing fractions can be identified using a non-reducing Coomassie stained SDS-PAGE and/or through a hemolytic assay (Section 2.4). Pool the C9-containing fractions for CHT chromatography.
5. Dilute the pooled DEAE fractions with an equal volume of CHT phosphate buffer. Check that the pH is ~7.0 with a pH test strip, if not, adjust with additional buffer. Then bind to a prepacked CHT type 1 column equilibrated in CHT wash buffer.
6. Wash the protein bound CHT column with 4 CV of CHT buffer A prior to eluting. Elute with a linear gradient over 10 CV with

CHT buffer B. Note: this CHT step utilizes a mixed approach by simultaneously changing pH (from 7.0→8.1) and increasing conductivity.

7. The C9-containing fractions are identified as per step 4. Pool C9-containing fractions from CHT and concentrate protein in a 30 kDa MWCO Amicon® concentrator. Concentrate down to 1 mL for the 16 mm diameter × 60 cm length column or 2 mL for the 26 mm diameter × 60 cm length column.
8. Perform SEC with S200 into the respective SEC buffer for either human or mouse.
9. Assess the C9 purity as per step 4.

2.4 Hemolytic assays of recombinant C9 proteins: Titration-based method

Sheep red blood cells (sRBCs) are used for testing complement activity (through activation of the classical pathway) where the loss of hemoglobin from cells is used to infer lysis due to MAC formation. The hemoglobin is released into the supernatant from hemolytic reactions and subsequently detected by absorbance at 405 nm.

2.4.1 Materials

Sheep red blood cells
DHB++ buffer: 2.5% (w/v) D-glucose, 5 mM HEPES, 71 mM NaCl, 0.15 mM $CaCl_2$, 0.5 mM $MgCl_2$, pH 7.4
Rabbit anti-sheep IgG (Rockland Immunochemicals)
Celpresol (red blood cell preservative solution from bioCSL)
C9-depleted serum (Complement Technology, Inc)
Control C9 protein (Complement Technology, Inc)
BMG absorbance plate reader or equivalent
NanoDrop

1. Fresh sRBCs are collected into tubes with heparin, EDTA or citric acid to prevent coagulation. For long term storage (~1 month) the sRBCs can be washed three to four times in HBS pH 7.4 and resuspended to their original collection volume with Celpresol and kept at 4 °C.
2. On the day of the hemolytic assay, prepare the cells by washing 0.5 mL of sRBCs three times with 2.5 mL of ice-cold DHB++ buffer and centrifuge (3220 rcf, 4 °C, 2 min). Each time, gently remove the supernatant after centrifugation without disrupting the cells. The supernatant should be mostly clear with fresh cells.

3. Measure the cell density with a hemocytometer. Prepare a stock solution of sRBCs containing 6.5×10^8 cells/mL in DHB++ buffer. Sensitize the sRBCs by addition of anti-sheep antibody (to 0.75 mg/mL) and incubation at 30 °C for 30 min.
4. Wash the sensitized sRBCs to remove excess antibody (3220 rcf, 2 min) and resuspend with fresh DHB++.
5. Prepare sample dilutions of C9 protein (32 μg/mL), and 8–10 twofold serial dilutions of protein using the SEC buffer. This step is ideally performed in triplicate from independently prepared C9 samples for obtaining statistical error.
6. Prepare lysis reactions in a final volume of 220 μL by addition of 3.75×10^6 sensitized sRBCs to PCR tubes containing dilutions of C9 (15 μL) and C9-depleted serum (1 μL).
7. Incubate the reactions on a PCR thermocycler block heated to 37 °C for 30 min.
8. Stop the assay reactions by cooling to 4 °C and promptly centrifuging in a room temperature Qik Spin centrifuge equipped with a PCR strip tube rotor for 20 s at 6500 rpm.
9. Carefully extract 150 μL of the supernatant without disrupting the cell pellet and transfer to a 96-well round bottom plate.
10. Measure the absorbance at 405 nm to correlate the activity of C9 function.
11. Analyze the results by plotting the protein concentration from [pMol] to a logarithmic scale: X=Log(X). A negative (no C9) and positive control (0.5% (v/v) Triton-X) can be used to normalize activity from 0% to 100%, respectively.

2.5 Hemolytic assays of recombinant C9 proteins: Turbidity time-based method

A turbidity time-based measurement is recommended when testing C9 mutants that may have different activity compared to wild-type C9. This variation of the assay can be performed using erythrocytes coated with antibody and exposed to factors C1–C8 (EAC1–8), which can be prepared by following steps 1–5 from Section 2.4. In this case the erythrocytes used are sheep red blood cells (sRBC). It is recommended that a baseline measurement of turbidity is trialed with control C9 to determine how frequently the measurements should be acquired prior to testing several samples.

Follow steps 1–5 from Section 2.4 and then proceed as follows:

6. To the washed sensitized sRBCs, add C9-depleted serum in batch with 1 μL per 3.75×10^6 cells, mix the suspension by inverting several times. Incubate at 37 °C for 30 min. This renders the sRBCs as EAC1–8 which will be hemolytic in the presence of C9.
7. Add dilutions of C9 to a 96-well round bottom plate. To these C9 dilutions add the EAC1–8. Immediately start measuring the absorbance at 620 nm continuously for the desired length of time in a plate reader heated to 37 °C. It is best to have orbital shaking between acquisition cycles as the sRBCs will settle to the bottom of the wells

3. Structural applications of purified C9 protein

Next, we will detail the procedures for producing crystals of muC9 [aglyco] and human polyC9. This polyC9 form is convenient for studying the oligomerization process of the assembly. However, there is a degree of heterogeneity in the stoichiometry (i.e., rings of 21-, 22- or 23-C9 or incomplete arcs).

3.1 Preparation of muC9[aglyco] crystals

3.1.1 Materials

24-well hanging drop plates (22 mm diameter)
Cover slips and Vaseline petroleum jelly
Polyethylene glycol 3350
Disodium malonate (Hampton) (various pH buffers are prepared by mixing different ratios of 3.4 M sodium malonate stock solutions pH 4 (HR2–747) and pH 8 (HR2–807) to the desired pH. The stock solutions are commercially titrated with sodium hydroxide to the appropriate pH)
Additive screen reagents zinc chloride and benzamidine
Hampton PTFE (polytetrafluoroethylene) seed beads

3.1.2 Procedure

1. Pool purified muC9[aglyco] protein from pure fractions of size exclusion chromatography.
2. Concentrate the muC9[aglyco] protein in a 30 kDa MWCO concentrator to 9 mg/mL (~150 μM; measured by NanoDrop absorbance A280 nm with a calculated molar extinction coefficient (molar attenuation coefficient) of 0.885 M^{-1} cm^{-1}). Filter the protein using a 0.2 μm spin filter, centrifuge 16,000 rcf for 1 min at 4 °C.

3. A hanging drop vapor diffusion grid is prepared with different precipitant concentrations versus different pH by making reservoir liquors containing 16–20% (w/v) PEG 3350 against 0.2 M disodium malonate pH 6.5–8.0 (values expressed as final concentrations).
4. Set drops containing 1 μL of the reservoir liquor + 1 μL of protein. Seal the drops, incubate in a temperature-controlled environment at 22 °C. Initial crystals should appear within ~1 week as clusters of plates, however these are not recommended for beamline collection (Fig. 6A). Recommended additional step: wash the crystals several times using a nylon crystal loop to move into fresh reservoir liquor, and then resuspend in SEC buffer for SDS-PAGE analysis to ensure the crystals are composed of C9 (Fig. 6B).
5. Harvest several crystal plate clusters from the hanging drop using a nylon crystal loop and place into a volume of the reservoir liquor (~20 μL). Add a PFTE seed bead (Hampton) to the crystal suspension and vortex vigorously as per the manufacturer's instruction. Make 5–6 tenfold dilutions of the seed stock by diluting with the reservoir liquor. Test the seed stock by preparing a crystal tray in a single condition that worked above with 0.9 μL reservoir liquor + 0.9 μL muC9[aglyco] (9 mg/mL) + 0.2 μL various seed stock dilution (compare results to a non-seeded control). Store the seed stock at −80 °C.
6. Prepare a fine screen of crystals, using the appropriate dilution of seed stock C9 crystals that includes either of the two following additives; zinc chloride or benzamidine · HCl (step 5). The recommended screening condition is: 18% (w/v) PEG 3350, 0.1 M di-sodium malonate containing either 10 mM $ZnCl_2$ or 10 mM benzamidine · HCl. Set multiple drops per well, with different drop ratios of reservoir liquor::protein (e.g., 1.4 μL::0.4 μL; 0.9 μL::0.9 μL; or 0.4 μL::1.4 μL) each containing 0.2 μL of seed stock. Monitor daily for optimal crystals over several weeks until crystals are ready for harvest. Detach single crystals from clusters by prodding with a thin acupuncture needle (Fig. 6C and D).
7. Crystals can be cryo-preserved for X-ray detection by transferring individual crystals with a nylon crystal loop to a cryo-protectant of 25% (v/v) glycerol, 75% reservoir liquor, and then flash cooled in liquid nitrogen (Fig. 6E).

3.2 Preparation of polyC9

1. In order to study the structural changes C9 undergoes to form the MAC pore we chose to study the formation of the spontaneously formed

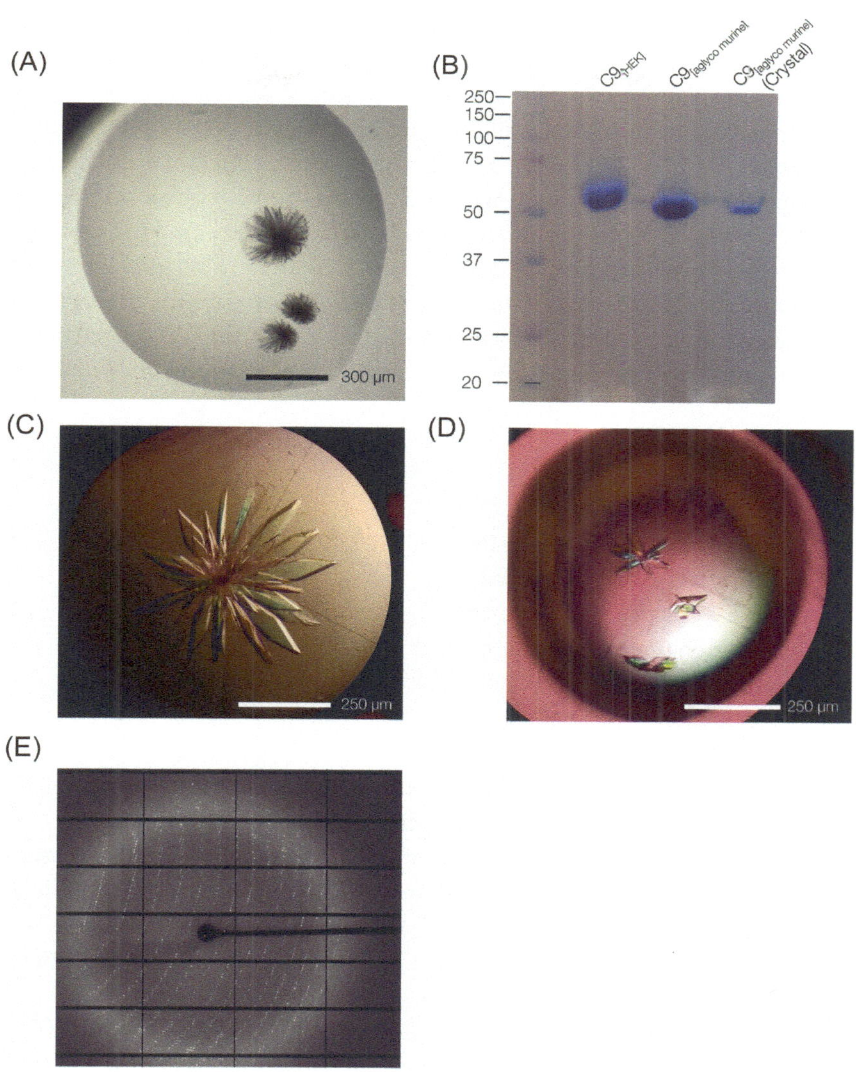

Fig. 6 Crystallization of muC9[aglyco]. (A) Initial C9 crystal plates can be formed using the hanging drop vapor diffusion method. (B) muC9[aglyco] crystals resolved by a 12% (w/v) non-reducing SDS-PAGE. (C) and (D) Optimized crystal growth with seeding and additive screen (zinc chloride, C) and (benzamidine HCl, D). (E) Typical X-ray diffraction of optimized C9 crystals. The X-ray diffraction image was taken at MX2, Australian Synchrotron, Melbourne, AU with a 0.25° oscillation range. The detector distance is 290.0 mm with the X-ray wavelength of 1.204 Å. The C9 crystal space group was determined as $P2_122_1$ with a unit cell constant of *a, b, c* (Å) 52.91, 148.87, 165.78; α, β, (°) 90.0, 90.0, 90.0.

polyC9. Although it has not been detected to occur in vivo, polyC9 represents a way of yielding high-resolution information not possible with the full MAC. Human C9 is the only MAC protein with the propensity to spontaneously self-oligomerize and only requires incubation at 37 °C at ~1–1.5 mg/mL to trigger formation. It is postulated that only the thermal change is required for this spontaneous form to occur. This protocol describes the method used to optimize polyC9 formation for the purposes of collecting cryo-EM datasets. The presence of the amphipol yields better particles by reducing aggregation and improving the distribution of the particles on the cryo-EM grid. We postulate that the amphipols shielded the hydrophobic patches normally interacting with membrane fatty acids however the amphipols were not observed in the maps.

3.2.1 Materials

PolyC9 buffer: 10 mM HEPES, 50 mM NaCl pH 7.5
Amphipol (A8–35; Anatrace)
Cryo-EM grids (e.g., Quantifoil R1.2/1.3 carbon support grids)
Glow discharger or plasma cleaner
Heating block with separately heated lid

3.2.2 Procedures

1. Pool the pure fractions of huC9 from SEC chromatography. Dialyze against the polyC9 buffer overnight at 4 °C.
2. Concentrate the dialyzed protein using a 30 kDa MWCO concentrator to approx. 2.0 mg/mL (NanoDrop A280 nm extinction coefficient 0.899 M^{-1} cm^{-1}).
3. Prepare a stock solution of amphipol at 0.02 mg/mL in the polyC9 buffer.
4. Prepare a range of C9 dilutions with the PolyC9 buffer (e.g., 1.1–1.5 mg/mL). Add 9 μL of protein to a thin-walled PCR tube + 1 μL of amphipol solution to a final concentration of 1.0–1.35 mg/mL.
5. Incubate protein overnight in a PCR machine with the heating block set to 37 °C and lid temperature set to 45 °C. This is the ideal method to prevent condensation of buffer inside the lid of the tube, where condensation can lead to an unwanted increase in protein concentration.
6. Prepare cryo-EM grids by applying a negative glow discharge or plasma clean.
7. Apply 2.5 μL of the polyC9 sample to the carbon side of the grid.

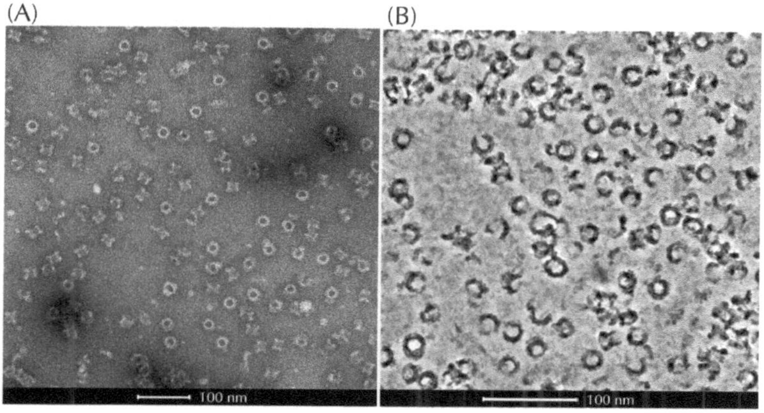

Fig. 7 Formation of polyC9 tubules. (A) Negative stain TEM micrograph of 0.25 μg sample applied to a carbon coated grid stained with 2% UA. (B) Cryo-EM micrograph applied to a holey Quantifoil grid (collected on a Titan Krios Microscope with a denoising filter applied).

8. Alternatively, negative stain grids can be prepared using carbon coated film support with negative glow discharge. A dilution of the sample (e.g., 100 μg/mL) may be necessary to achieve the desired negative stain, which will contain many rings compared to cryo-EM grids (Fig. 7). After application, excess solution can be wicked by touching the edge of the grid with Whatman paper. Then a solution of 2% (w/v) uranyl acetate is applied to the grid for 1 min and excess solution can be wicked by touching the edge of the grid with Whatman paper.

4. Summary

Complement C9 has been studied in its pivotal role in forming the channel part of the membrane attack complex (MAC). The majority of this research has been performed using C9 sourced from serum, which limits the types of experiments that can be performed, especially mutagenesis studies. In this chapter we describe how to make recombinant C9 using the HEK293 expression system in order to improve the success of structural and functional studies of C9. There were improvements in the purity/yield of the final C9 as well as the freedom to produce mutant variants of C9 and C9 from alternative species. We postulate that this recombinant expression and purification was key to the breakthroughs in structural studies, both crystallization and cryo-EM (Spicer et al., 2018). We hope that it will lead to further breakthroughs in other studies of C9. One outstanding area that we believe will benefit from this technology will be single molecule experiments on MAC assembly that rely on specific labeling of C9.

References

Aleshin, A. E., DiScipio, R. G., Stec, B., & Liddington, R. C. (2012). Crystal structure of C5b-6 suggests structural basis for priming assembly of the membrane attack complex. *The Journal of Biological Chemistry*, *287*(23), 19642–19652. https://doi.org/10.1074/jbc.M112.361121.

Aleshin, A. E., Schraufstatter, I. U., Stec, B., Bankston, L. A., Liddington, R. C., & DiScipio, R. G. (2012). Structure of complement C6 suggests a mechanism for initiation and unidirectional, sequential assembly of membrane attack complex (MAC). *The Journal of Biological Chemistry*, *287*(13), 10210–10222. https://doi.org/10.1074/jbc.M111.327809.

Borsos, T., Dourmashkin, R. R., & Humphrey, J. H. (1964). Lesions in erythrocyte membranes caused by immune hæmolysis. *Nature*, *202*(4929), 251–252. https://doi.org/10.1038/202251a0.

Dudkina, N. V., Spicer, B. A., Reboul, C. F., Conroy, P. J., Lukoyanova, N., Elmlund, H., et al. (2016). Structure of the poly-C9 component of the complement membrane attack complex. *Nature Communications*, 7, 10588. https://doi.org/10.1038/ncomms10588.

Elliger, S. S., Elliger, C. A., Lang, C., & Watson, G. L. (2002). Enhanced secretion and uptake of β-glucuronidase improves adeno-associated viral-mediated gene therapy of mucopolysaccharidosis type VII mice. *Molecular Therapy*, *5*, 617–626. https://doi.org/10.1006/mthe.2002.0594.

Franc, V., Yang, Y., & Heck, A. J. R. (2017). Proteoform profile mapping of the human serum complement component C9 reveals unexpected new features of N-, O- and C-glycosylation. *Analytical Chemistry*, *89*, 3483–3491. https://doi.org/10.1021/acs.analchem.6b04527.

Fredslund, F., Laursen, N. S., Roversi, P., Jenner, L., Oliveira, C. L. P., Pedersen, J. S., et al. (2008). Structure of and influence of a tick complement inhibitor on human complement component 5. *Nature Immunology*, *9*(7), 753–760. https://doi.org/10.1038/ni.1625.

Hadders, M. A., Beringer, D. X., & Gros, P. (2007). Structure of C8alpha-MACPF reveals mechanism of membrane attack in complement immune defense. *Science (New York, N.Y.)*, *317*(5844), 1552–1554. https://doi.org/10.1126/science.1147103.

Hadders, M. A., Bubeck, D., Roversi, P., Hakobyan, S., Forneris, F., Morgan, B. P., et al. (2012). Assembly and regulation of the membrane attack complex based on structures of C5b6 and sC5b9. *Cell Reports*, *1*(3), 200–207. https://doi.org/10.1016/j.celrep.2012.02.003.

Hofsteenge, J., Blommers, M., Hess, D., Furmanek, A., & Miroshnichenko, O. (1999). The four terminal components of the complement system are C-mannosylated on multiple tryptophan residues. *Journal of Biological Chemistry*, *274*(46), 32786–32794. https://doi.org/10.1074/jbc.274.46.32786.

Lovelace, L. L., Cooper, C. L., Sodetz, J. M., & Lebioda, L. (2011). Structure of human C8 protein provides mechanistic insight into membrane pore formation by complement. *The Journal of Biological Chemistry*, *286*(20), 17585–17592. https://doi.org/10.1074/jbc.M111.219766.

Martinez, R. J., & Carroll, S. F. (1980). Sequential metabolic expressions of the lethal process in human serum-treated Escherichia coli: Role of lysozyme. *Infection and Immunity*, *28*, 735–745.

Menny, A., Serna, M., Boyd, C. M., Gardner, S., Joseph, A. P., Morgan, B. P., et al. (2018). CryoEM reveals how the complement membrane attack complex ruptures lipid bilayers. *Nature Communications*, *9*(1), 5316. https://doi.org/10.1038/s41467-018-07653-5.

Podack, E. R., Biesecker, G., & Müller-Eberhard, H. J. (1979). Membrane attack complex of complement: Generation of high-affinity phospholipid binding sites by fusion of five hydrophilic plasma proteins. *Proceedings of the National Academy of Sciences of the United States of America*, *76*, 897–901. https://doi.org/10.1073/pnas.76.2.897.

Podack, E. R., & Müller-Eberhard, H. J. (1978). Binding of desoxycholate, phosphatidylcholine vesicles, lipoprotein and of the S-protein to complexes of terminal complement components. *Journal of Immunology*, *121*(3), 1025–1030.

Podack, E. R., Tschopp, J., & Müller-Eberhard, H. J. (1982). Molecular organization of C9 within the membrane attack complex of complement. Induction of circular C9 polymerization by the C5b-8 assembly. *Journal of Experimental Medicine*, *156*(1), 268–282. https://doi.org/10.1084/jem.156.1.268.

Rosado, C. J., Buckle, A. M., Law, R. H. P., Butcher, R. E., Kan, W.-T. W.-T., Bird, C. H., et al. (2007). A common fold mediates vertebrate defense and bacterial attack. *Science (New York, N.Y.)*, *317*(5844), 1548–1551. https://doi.org/10.1126/science.1144706.

Rossi, V., Wang, Y., & Esser, A. F. (2010). Topology of the membrane-bound form of complement protein C9 probed by glycosylation mapping, anti-peptide antibody binding, and disulfide modification. *Molecular Immunology*, *47*(7–8), 1553–1560. https://doi.org/10.1016/j.molimm.2010.01.013.

Serna, M., Giles, J. L., Morgan, B. P., & Bubeck, D. (2016). Structural basis of complement membrane attack complex formation. *Nature Communications*, *7*, 10587. https://doi.org/10.1038/ncomms10587.

Shah, A. K., Hartel, G., Brown, I., Winterford, C., Na, R., Lê Cao, K. A., et al. (2018). Evaluation of serum glycoprotein biomarker candidates for detection of esophageal adenocarcinoma and surveillance of Barrett's esophagus. *Molecular & Cellular Proteomics*, *17*, 2324–2334. https://doi.org/10.1074/mcp.RA118.000734.

Spicer, B. A., Law, R. H. P., Caradoc-Davies, T. T., Ekkel, S. M., Bayly-Jones, C., Pang, S.-S., et al. (2018). The first transmembrane region of complement component-9 acts as a brake on its self-assembly. *Nature Communications*, *9*(1), 3266. https://doi.org/10.1038/s41467-018-05717-0.

Taylor, K. M., Trimby, A. R., & Campbell, A. K. (1997). Mutation of recombinant complement component C9 reveals the significance of the N-terminal region for polymerization. *Immunology*, *91*(1), 20–27.

Tomlinson, S., Ueda, E., Maruniak, J. E., Garciacanedo, A., Bjes, E. S., & Esser, A. F. (1993). The expression of hemolytically active human complement protein C9 in mammalian, insect, and yeast cells. *Protein Expression and Purification*, *4*(2), 141–148. https://doi.org/10.1006/prep.1993.1020.

Tomlinson, S., Wang, Y., Ueda, E., & Esser, A. F. (1995). Chimeric horse/human recombinant C9 proteins identify the amino acid sequence in horse C9 responsible for restriction of hemolysis. *Journal of Immunology*, *155*(1), 436–444.

Tschopp, J., Müller-Eberhard, H. J., & Podack, E. R. (1982). Formation of transmembrane tubules by spontaneous polymerization of the hydrophilic complement protein C9. *Nature*, *298*(5874), 534–538.

Tschopp, J., Podack, E. R., & Müller Eberhard, H. J. (1982). Ultrastructure of the membrane attack complex of complement: Detection of the tetramolecular C9-polymerizing complex C5b-8. *Proceedings of the National Academy of Sciences of the United States of America*, *79*(23 I), 7474–7478.

Würzner, R., Joysey, V. C., & Lachmann, P. J. (1994). Complement component C7. Assessment of in vivo synthesis after liver transplantation reveals that hepatocytes do not synthesize the majority of human C7. *Journal of Immunology (Baltimore, Md.: 1950)*, *152*, 4624–4629.

CHAPTER FIVE

Preparation of *Clostridium perfringens* binary iota-toxin pore complex for structural analysis using cryo-EM

Tomohito Yamada[a,*] and Hideaki Tsuge[a,b,c,*]
[a]Faculty of Life Sciences, Kyoto Sangyo University, Kyoto, Japan
[b]Institute for Protein Dynamics, Kyoto Sangyo University, Kyoto, Japan
[c]Center for Molecular Research in Infectious Diseases, Kyoto Sangyo University, Kyoto, Japan
*Corresponding authors: e-mail address: i2087159@cc.kyoto-su.ac.jp; tsuge@cc.kyoto-su.ac.jp

Contents

Methods in Enzymology, Volume 649
ISSN 0076-6879
https://doi.org/10.1016/bs.mie.2021.01.032

Abstract

Iota toxin, a type of A-B toxin produced by *Clostridium perfringens*, comprises an enzymatic component (Ia) and a membrane-binding component (Ib). The translocation of Ia to the target cell via the pore formed by Ib allows it to function as an ADP-ribosyltransferase that inhibits actin polymerization in the host cell. The structure of Ia-bound Ib-pore has been determined using cryo-electron microscopy (cryo-EM), thereby elucidating the mechanism of the initial Ia translocation; however, open questions regarding Ia translocation still exist. In this chapter, we describe a new method of preparing Ia-bound Ib-pore complex samples for structural analysis at high resolution using cryo-EM. This method is different from previously reported methods for other A-B toxins. Consequently, it produces Ib-pore with two different states with short and long membrane-spanning β-barrel stem. We expect that this method will be useful in functional and structural studies of iota toxin and other binary toxins.

1. Introduction

A-B toxin family includes toxins composed of an enzymatic A component and a membrane-binding B component; these toxins are produced by various bacteria (Barth, Aktories, Popoff, & Stiles, 2004). First, the B component is produced as a soluble monomer. Before or after binding to the receptor on a target cell, the N-terminal propeptide of the B component is cleaved off by a protease, forming a ring-shaped heptameric or octameric oligomer (prepore) (Gibert, Petit, Raffestin, Okabe, & Popoff, 2000; Kintzer et al., 2009; Lacy, Wigelsworth, Melnyk, Harrison, & Collier, 2004). The A component then binds to the B component oligomer and the resultant complex is taken into the endosome (Mogridge, Cunningham, Lacy, Mourez, & Collier, 2002). In the acidified endosome, the prepore turns into a pore with the insertion of membrane-spanning β-barrel (Miller, Elliott, & Collier, 1999). Finally, the A component passes through this pore to enter the cell and exert toxicity (Fig. 1).

Clostridium perfringens, *C. difficile*, *C. Spiroforme*, and *C. botulinum* produce A-B toxins (iota toxin (Simpson, Stiles, Zepeda, & Wilkins, 1987), CDT (Perelle, Gibert, Bourlioux, Corthier, & Popoff, 1997), CST (Popoff & Boquet, 1988), and C2 (Aktories et al., 1986), respectively), which are called clostridial binary toxins because they are composed of two components, A and B. The A components (Ia, CDTa, CSTa, and C2I) function as enzymes that transfer ADP-ribose to G-actin of the host cell and inhibit polymerization, causing cell death (Vandekerckhove, Schering, Bärmann, & Aktories, 1987). To enter the cells, A components must be translocated via the pore formed by the respective B components (Ib, CDTb, CSTb, and C2II) in the acidified endosomes.

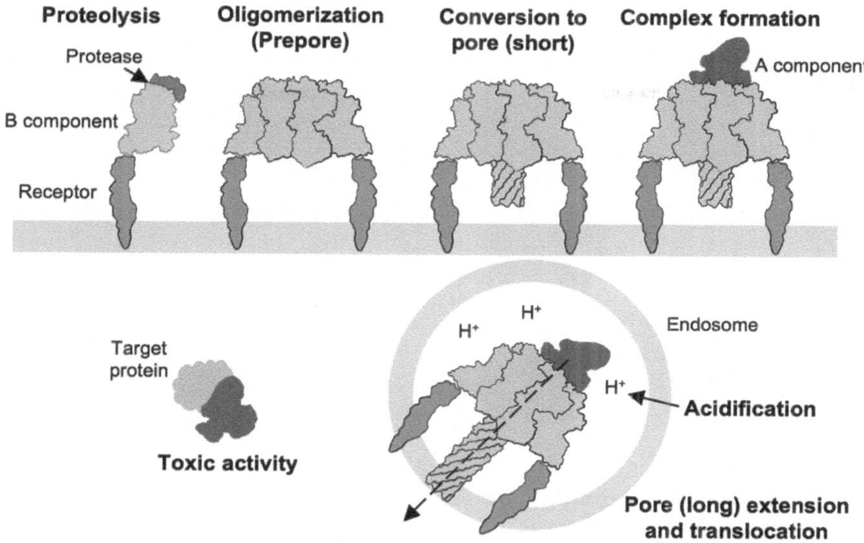

Fig. 1 Cellular uptake of clostridial A-B toxin. N-terminal pro-sequence B component is cleaved off by protease before or after binding to receptor on target cell membrane. Activated B component forms soluble heptameric or octameric oligomer (prepore). The membrane spanning β-barrel is partially formed. A component binds to the oligomer, followed by uptake to endosome. In acidified endosome, the perfect β-barrel is formed to make a path into the cytosol. Translocated A component causes the damage to the cell.

The amino acid sequences of the clostridial binary toxin A components have high similarity. Previous structural studies of A components revealed that they consist of two domains (Han, Craig, Putnam, Carozzi, & Tainer, 1999; Tsuge et al., 2003). Mutational, biochemical, and structural studies of A components further revealed that their N-terminal domains are necessary for binding to the respective B component oligomers, and their C-terminal domains possess ADP-ribosyltransferase activity (Marvaud et al., 2002; Tsuge et al., 2008; Tsurumura et al., 2013). Furthermore, in the case of *C. perfringens* iota toxin, the complex structure formed by Ia with the target G-actin has been revealed using X-ray crystallography, showing that the C-terminal domain of Ia is the enzymatic domain that recognizes actin (Tsuge et al., 2008). However, studies on the B component did not progress until recently because it forms large oligomers with membrane-spanning hydrophobic regions, making its crystallization for X-ray crystallography challenging. However, recent progress in cryo-EM has facilitated studies on the structures of Ib-pores and the Ia-bound Ib-pore complex (Yamada et al., 2020) (Fig. 2A–C), revealing that a single Ia binds to the Ib-pore via the N-terminal domain.

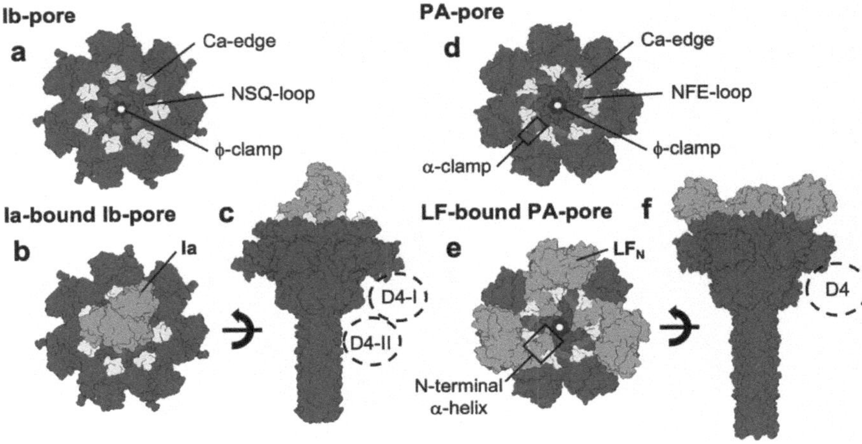

Fig. 2 Complex structure comparison between anthrax toxin and iota toxin. (A–C) Revealed model of iota toxin complex (PDBID: 6KLW). (A) Top view of Ib-pore. Three impressive bottlenecks are colored in different gray level. (B) Top view of iota toxin complex. (C) Side view of iota toxin complex. The model corresponding to D4-I and D4-II (dotted circle) is not shown because the signal was too weak to build the model. (D)–(F) The model was constructed based on PDBID: 3KWV and 6PSN. LF is shown only as N-terminal domain (LF_N). (D) Top view of PA-pore. Three impressive bottlenecks are colored in different gray level. The α-clamp is located on the cleft between two PA protomers. (E) Top view of anthrax toxin complex. (F) Side view of anthrax toxin complex. D4 receptor binding domain is shown as dotted circle.

On the other hand, *Bacillus anthracis* also produces an A-B binary toxin called anthrax toxin (Collier & Young, 2003), which is composed of either edema factor (EF) or lethal factor (LF) as the A component and protective antigen (PA) as the B component; thus, the functional binary toxin is either EF-PA or LF-PA. EF is a calmodulin-dependent adenyl cyclase, while LF is a Zn^{2+}-dependent protease that cleaves mitogen-activated-protein kinase (Tang & Guo, 2009; Tonello & Montecucco, 2009). As described in the clostridial binary toxin section, EF and LF are translocated into the target cell via the pore formed by PA and causes damage to the target cells.

The EF and LF structures determined by X-ray crystallography show that they differ from that of the clostridial binary toxin A component (Pannifer et al., 2001; Shen, Zhukovskaya, Guo, Florián, & Tang, 2005). In contrast, the amino acid sequences of all the B components (Ib, CDTb, C2II, and PA), except for the receptor binding domain, are well conserved (Barth et al., 2004). The similarity among the B components implies that translocation mechanisms are common to both clostridial binary toxins and anthrax toxin, despite the difference between their A components. In 2015, the

PA-pore structure was determined as the first structure of the heptameric B component-pore (Jiang, Pentelute, Collier, & Zhou, 2015). Interestingly, although the pore functions to translocate the A component, it was revealed that there is a narrow constriction site (ϕ-clamp) with a diameter of 6 Å in the translocation pathway (Krantz et al., 2005; Sun, Lang, Aktories, & Collier, 2008), which is too narrow for the translocation of the A component with a secondary or tertiary structure. Thus, to complete the translocation, unfolding of the A component is necessary; however, the precise mechanism of this process is not yet known.

Because of its sensitivity and toxicity to humans, anthrax toxin has been studied more extensively than other A-B toxins. Early studies on binary toxins revealed the complex structure of the LF-bound PA-prepore (Feld et al., 2010) (Fig. 2D–F). In the complex state, the N-terminal α-helix binds to a hydrophobic clamp called α-clamp, formed by two adjacent protomers (Cunningham, Lacy, Mogridge, & Collier, 2002). The N-terminal α-helix of LF moves to adjust to the α-clamp. Further studies on LF/EF-bound PA-pore structure showed that LF binds to both prepores and pores in the same binding mode (Hardenbrook et al., 2020). Notably, the tertiary structure of the N-terminal of LF/EF is unfolded for binding to the PA-prepore, and further structural change does not occur until the translocation.

In the case of clostridial binary toxin, the structural information has been limited only for the A component, and there have been no structural studies of the B component. However, in 2019 and early 2020, the structures of CDTb- and Ib-pores were determined, paving the way for studies on the translocation mechanisms in clostridial binary toxins (Anderson, Sheedlo, Jensen, & Lacy, 2020; Xu et al., 2020; Yamada et al., 2020). Interestingly, the Ia-bound Ib-pore structure provided new findings that was not revealed in studies on anthrax toxin. While three LF binds to the heptameric PA-pore, a single Ia binds to heptameric Ib-pores because of the difference in the A component (Fig. 2B, C, E, and F). The N-terminal of Ia does not bind to α-clamp of Ib-pore. Furthermore, the N-terminal α-helix is partially unfolded and continues to the ϕ-clmap.

In addition, the development of devices including camera and field emission electron gun and some software of cryo-EM have promoted structural studies of A-B toxin. However, high resolution structural information regarding A-B toxin complex are still limited to anthrax toxin and iota toxin because of difficulties in sample preparation. In this paper, we focus on the preparation of iota toxin sample. Furthermore, we discuss the oligomerization of CDTb because Ib and PA do not readily oligomerize, while CDTb easily forms a di-heptameric oligomer in vitro. We report here for the first time the method of Ib-pore sample preparation for structural analysis. This method is particularly useful, because the method used for sample preparation of anthrax toxin is not applicable to Ib-pores. This new method allowed the determination of the pore structure at high resolution using cryo-EM, and we expect that it will facilitate further advances in the study of the translocation mechanism of other binary toxins.

2. Sample preparation for single-particle analysis

Single-particle analysis is a variant of cryo-EM that is used to determine the structure of proteins and does not require crystals but rather a small amount of frozen protein sample with well-dispersed and various-oriented 10,000 to millions particle particles. Thus, it is an appropriate method for the analysis of samples, like membrane proteins which are difficult to crystalize. However, in order to achieve a high resolution, challenges concerning sample purification and solubilization of the membrane protein in the solution must be overcome.

If the molecule is too small (~100 kDa), the signal-to-noise ratio in the images captured by cryo-EM is too low for analysis. In case of membrane proteins, the sample is solubilized to avoid aggregation due to the use of lipid membranes using nanodisc or detergents (Mio & Sato, 2017; Rouck, Krapf, Roy, Huff, & Das, 2017). The use of liposomes to solubilize membrane proteins for cryo-EM is not preferred, because liposomes increase the thickness of ice on the grid. The excess thickness of ice resulting in a low signal to noise ratio. In contrast, nanodiscs, which are flat circulated lipid membranes formed by membrane scaffold proteins, are small and monodispersed, and are therefore preferred as solubilization agents for lipid membranes. Furthermore, detergents, such as LMNG (Lauryl Maltose Neopentyl Glycol) are also powerful tools for solubilization (Chae et al., 2010). LMNG has a strong binding affinity for the protein and can be used at remarkably low concentrations. This reduces the noise produced by unbound detergent molecular in cryo-EM images. When choosing the solubilization method, it is important to consider the cost of purification and solubilization with detergents. This challenge can be overcome by replacing the cheap detergent with nanodisc or LMNG at the last purification step. Furthermore, to obtain good quality cryo-EM images, it is important to remove the excess lipid or detergent by exchanging the buffer.

3. Character of membrane-spanning B-component pore

In the case of binary toxin B component pores, B component monomer purification is followed by oligomerization, purification of the oligomer, and solubilization. In the preparation of PA oligomers, which are well studied, it is known that acidic pH condition or the addition of urea promotes the conversion from prepore to pore (Akkaladevi et al., 2013; Vernier et al., 2009). In contrast, we found that the pore formation of Ib occurred at the same time upon oligomerization. Therefore, in the case of Ib, we need to perform solubilization simultaneously with oligomerization; otherwise, Ib oligomer aggregates due to the hydrophobicity of the membrane-spanning pore. We summarized the optimized process including some tips to be refined as described in this chapter (Fig. 3).

3.1 Oligomerization

B component oligomerization is regulated by the N-terminal propeptide, which is cleaved off by a protease before or after the B component binds to the receptor on the target cell membrane in vivo. CDTb oligomerizes efficiently in vitro. However, CDTb form a di-heptameric pore, which is stacked symmetrically via the receptor-binding domain.

Contrarily, anthrax PA does not oligomerize in vitro after proteolysis and the PA prepore is obtained by loading the monomer onto an anion exchange column (Christensen, Krantz, & Collier, 2006; Cunningham et al., 2002). It has also been reported that incubation of PA in acidic pH promotes the oligomerization, but also shows aggregation (Vernier et al., 2009). Thus, oligomerization using an anion exchange column is preferred for the preparation of PA-prepores.

3.1.1 Efficient Ib oligomerization with ethanol

Similar to PA, Ib rarely oligomerizes (~10%) after proteolysis by chymotrypsin. Because the oligomerization protocol for PA is not applicable to Ib, we evaluated stable and efficient conditions for Ib oligomerization and found that ethanol efficiently promoted Ib oligomerization (Fig. 4A). Ib oligomerizes with approximately 100% efficiency in the presence of 10% (v/v) ethanol, but excess ethanol could promote further aggregation of Ib. Therefore, we used 10% (v/v) ethanol to induce Ib oligomerization for cryo-EM sample preparation.

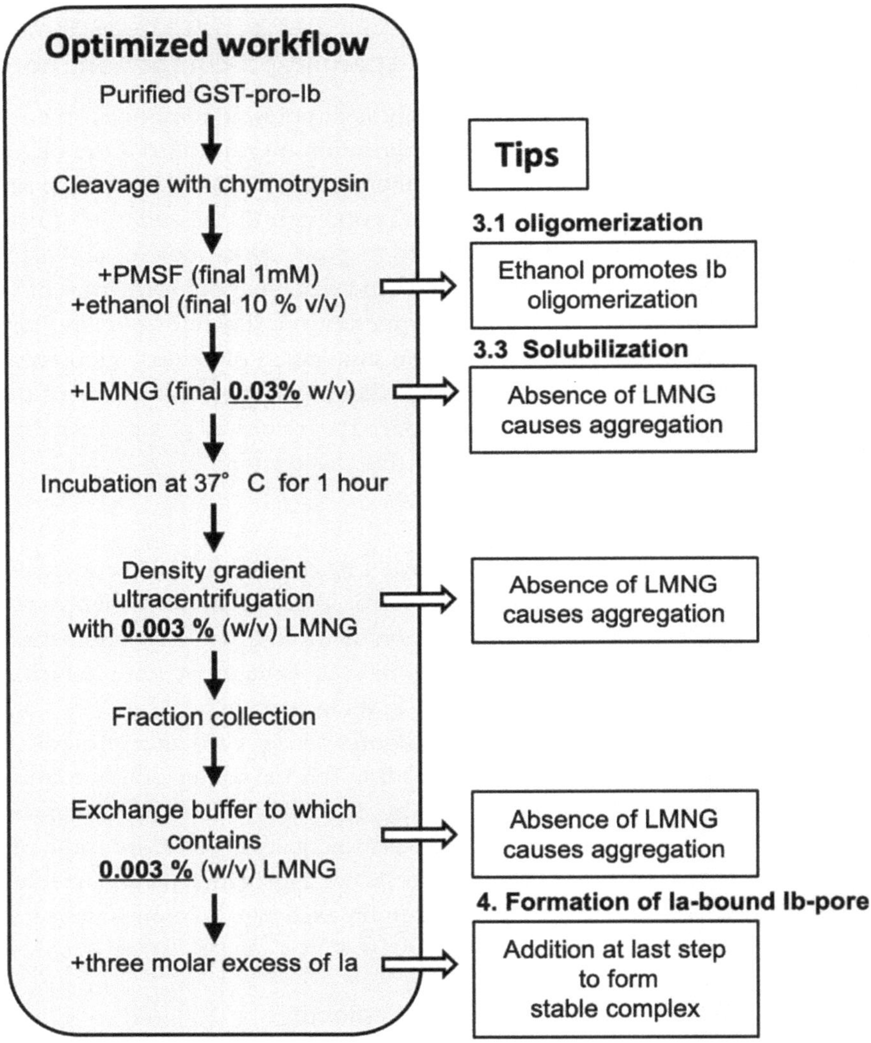

Fig. 3 Optimized workflow to prepare Ia-bound Ib-pore. Optimized workflow after purification of monomer Ib (gray box) and the tips which we found through sample preparation trials. Remarkable LMNG concentrations is underlined. The tips are described in each attached section with more details.

3.1.2 Efficient Ib oligomerization without ethanol

Although the amino acid sequence of the receptor-binding domain is well conserved between CDTb and Ib, the latter forms a heptameric physiological pore. However, the mechanisms behind the formation of di-heptamer

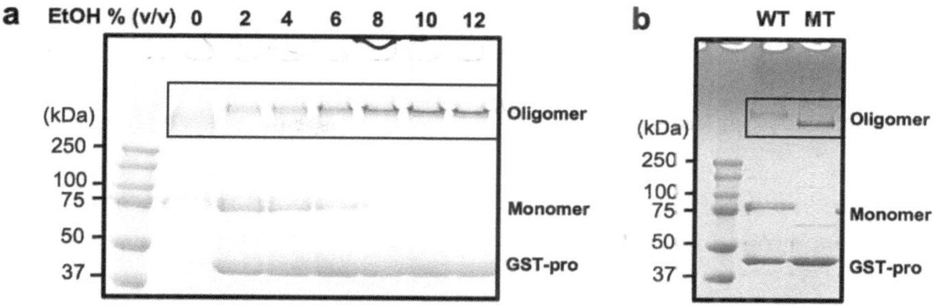

Fig. 4 Efficient oligomerization of propeptide-cleaved Ib. (A, B) The pictures are results of non-heating SDS-PAGE. (A) Cleaved Ib oligomerizes efficiently in the presence of ethanol. Because the protease inhibitor is diluted with ethanol, no inhibitor is added to the sample with 0% ethanol, allowing the proteolysis to undergo. While the monomer is disappeared along with oligomer formation, each band of cleaved off GST-pro (GST with pro region) looks same. The efficiency reaches approximately 100% in the presence of 10% ethanol. (B) The wild type Ib (WT) rarely oligomerizes without addition of excess ethanol. In contrast, the mutant Ib which lacks receptor binding D4-II (MT) oligomerizes efficiently.

and heptamer in these toxins are not yet fully understood. We hypothesized that the receptor-binding domain is also involved in determining the oligomerization efficiency. The receptor-binding domains of Ib and CDTb are composed of two subdomains, D4-I and D4-II (Fig. 2C).

Thus, we created a mutant Ib which lacks one of the two subdomains (D4-II) involved in the formation of the CDTb di-heptameric pore. Interestingly, this mutant Ib exhibited approximately 100% oligomerization at pH 8.0, without the addition of excess ethanol (Fig. 4B), indicating that B component oligomerization is regulated not only by the N-terminal propeptide but also the C-terminal receptor binding domain. Although the receptor-binding domain is not globally conserved among B components, the deletion of the receptor-binding domain may promote oligomerization, even in other B components.

3.2 Purification of B component oligomer

Following the oligomerization step, B component sample for cryo-EM was prepared using size exclusion chromatography or anion exchange chromatography in general, which facilitated the removal of the propeptide region cut off by proteolysis. By using this method, the B component sample did not require extensive purification prior to oligomerization.

We purified the Ib oligomer using density gradient ultracentrifugation, which enables the separation of molecules according to molecular weight. This method is advantageous for samples that cannot be purified using other chromatography systems. In density gradient ultracentrifugation, there is no stacking risk of the columns in chromatography system by aggregated proteins. In addition, the required volume of the buffer is smaller than that of the chromatography system, allowing the use of less expensive detergents. The amount of protein sample loaded should be sufficient for detecting the concentration of the fraction, which is measured using a nanodrop. However, excessive sample loading may decrease the purity. Therefore, it is necessary to optimize the ultracentrifugation speed and the amount of sample at the beginning of the process.

We prepared two buffers with different glycerol concentrations to form a glycerol density gradient, which can be prepared easily without machines. The low glycerol concentration buffer was carefully stacked onto the high concentration buffer in the ultracentrifugation tube. If both buffers are carefully stacked, the borderline between the two buffers is visible. Then, lay down the tube for 2 h to allow the diffusion sufficiently at the border between two buffer containing different glycerol concentration. Raising the tube again allows to form the density gradient from the bottom (high) to the top (low) (Fig. 5).

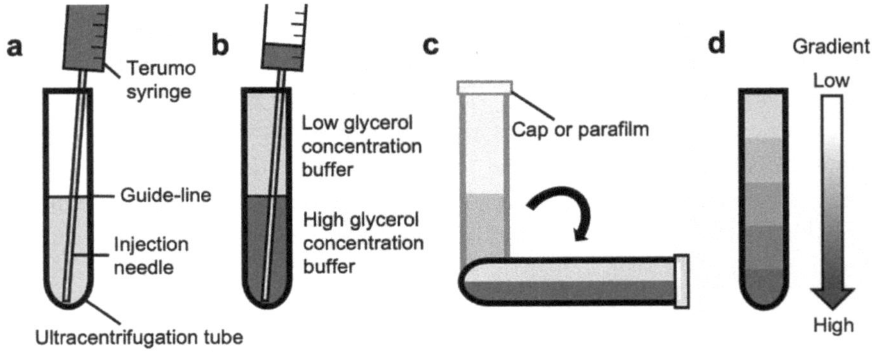

Fig. 5 Preparation of density gradient bed. (A) Draw a guideline on the tube and inject buffer containing low-concentration glycerol to the guideline. (B) Inject buffer containing high concentration glycerol slowly so that the buffers do not mix. A borderline should be seen between the two solutions. (C) Seal the tube with a cap or parafilm. The tube is laid down and left for 2 h at 4 °C. (D) Place the tube upright and remove the lid. If the empty volume is not sufficient to load the sample, remove some volume from the top of the gradient bed.

Before placing the sample in the buffer bed, it is necessary to remove excess volume from the gradient bed so that the sample (1 mL) can be accommodated; the weights of the two tubes must be then carefully adjusted. Next, 1 mL of sample was added to both tubes and the ultracentrifugation was performed at 230,139 × *g* for 16 h with a swing rotor P40ST (Hitachi). After ultracentrifugation, a hole was made at the bottom of the tube using a sharp needle to allow sample flow at a rate of 50 μL per drop. Thus, we collected 250 μL of each fraction and measured the concentration to generate the chromatogram (Fig. 6).

The presence of glycerol in the sample from the density gradient ultracentrifugation method affects the quality of cryo-EM images. Therefore, it is necessary to remove the glycerol by exchanging the buffer, which was archived using both PD10 (GE healthcare) and amicon.

3.3 Solubilization

The B-component pore functions as the translocation pathway by forming a membrane-spanning β-barrel that has a hydrophilic inner surface and a hydrophobic outer surface. Thus, it is essential to cover the hydrophobic outer surface with detergent micelles, liposomes, or nanodiscs to solubilize the pore.

As described in Section 3.1, PA-pore formation from prepore is promoted by the addition of urea or lowering the pH (Akkaladevi et al., 2013; Vernier et al., 2009). In contrast to PA, Ib does not require prepore to the pore conversion, because pore formation undergoes simultaneously

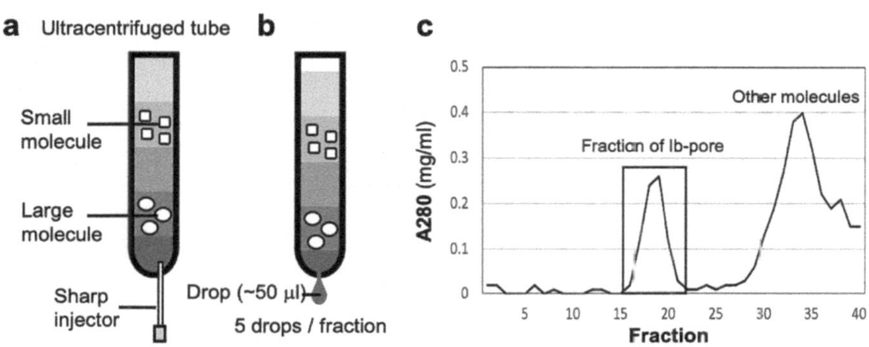

Fig. 6 Fraction collection after ultracentrifugation. (A) After density gradient ultracentrifugation, make a hole using a sharp injector on the bottom of the tube. (B) Collect 5 drops as 1 fraction until the entire solution comes out. If the drops are too small, expand the hole with needle for rapid collection. (C) Measure the protein concentration of each fraction and draw the chromatogram. We measured the Abs 280 nm using nanodrop (Thermo Fisher).

with the oligomerization. Thus, the Ib oligomer is formed as an insoluble pore due to the addition of ethanol, so the Ib-pores require solubilization during the oligomerization step. To solubilize the Ib pores, we added LMNG and ethanol after cleavage of the N-terminal propeptide.

Because excess detergent appears as noise in cryo-EM image, we tested the conditions for adding LMNG, including the step of its addition and its concentration. First, we performed Ib oligomerization with and without LMNG and subjected the Ib oligomers to ultracentrifugation. Second, we tested oligomerization with LMNG, followed by ultracentrifugation with and without LMNG. Third, we tested oligomerization and ultracentrifugation with LMNG, and exchange buffer with and without LMNG. The results revealed that LMNG was necessary for all the steps to inhibit the Ib oligomer aggregation.

Since Ib oligomer aggregates when there is no LMNG, 0.03% (w/v) LMNG was used for Ib solubilization. We also tested buffers containing lower concentrations of LMNG. Single particle analysis of the Ib oligomer sample with 0.01% (w/v) LMNG could not be performed initially, because there was too much noise when this sample was used for the initial grid screening using cryo-EM. The noise due to LMNG disappeared when the sample was diluted with the buffer containing no LMNG to one-fifth of the original concentration. However, dilution decreases the number of Ib-pore particles per image. To avoid this situation, we finally prepared the sample with 0.003% (w/v) LMNG, in which Ib-pore did not aggregate and there was almost no noise in the cryo-EM images.

Buffer exchange with amicon requires careful use of large filter size because detergents form micelles that cannot be removed with very small filter sizes.

4. Formation of Ia-bound Ib-pore

In previous studies, LF and EF were added to the PA-prepore to form the complex, followed by size exclusion chromatography to remove excess LF/EF (Hardenbrook et al., 2020). The sample was also treated with urea and nanodisc (pMSP1D1) to convert EF/LF-bound PA-prepore to EF/LF-bound PA-pore.

We have added Ia to Ib-pore and purified the complex by ultracentrifugation. But most of the particles are not Ia-bound Ib-pore, but Ib-pore. We hypothesized Ia dissociated from Ib-pore during the ultracentrifugation because the volume increase and the sample concentration decrease in this step. To analyze the structure at high resolution, the purification of Ia-bound Ib-pore particles was essential, and it was better that the particles are homogeneous. Hence, in this protocol, Ia was added after the purification of Ib-pore. Therefore, we mixed the excess Ia with purified Ib-pore, leading to increased molar concentration of Ia-bound Ib-pore.

In brief, Ia-bound Ib-pore complexes were prepared with a three-fold molar excess of Ia compared to the purified Ib-pore. Cryo-EM image of this sample revealed no obvious noise due to unbound Ia, indicating that further purification after Ia-Ib mixing is not necessary. In addition, the results of 3D classification during single particle analysis showed that almost all the classes are Ia-bound Ib-pores. Therefore, it can be said that the mixing ratio of Ia used in this protocol (three-fold molar excess of Ia to Ib-pore) is ideal. For the purification of Ia, we used a 6His-tag located at the C-terminus to avoid any interference with the translocation process.

Reconstitution of Ia with the Ib-pore could be also achieved with a purified Ia with N-terminal His-tag if the tag is cleaved before mixing it with Ib-pore. There was no problem in forming the complex and the binding mode was not noticeably different compared to the sample using Ia with C-terminal His-tag. However, there is a possibility that the N-terminal leftover may interfere with the translocation process.

5. Preparation of iota-toxin pore complex

5.1 Equipment

1. Bath culture shaker for 15 mL tube
2. Culture shaker for 5 L flask
3. Turbidimeter
4. Autoclave
5. Sonicator
6. Nanodrop (Thermo Fisher)
7. Empty column with volume of 120 mL
8. Chromatography system with UV detector
9. HiTrap Q HP 5 mL column (GE healthcare)
10. Centrifugal 30 K and 100 K filters (Amicon)
11. Ultracentrifugation tube (polypropylene) with volume of 13 mL (HITACHI)
12. Syringe with volume of 10 mL (TERUMO)
13. Injecting needle (TERUMO)
14. PD-10 (GE healthcare)

5.2 Chemical

1. IPTG (isopropyl β-D-1-thiogalactopyranoside)
2. LB liquid medium
3. Super Broth liquid medium
4. cOmplete (Protease inhibitor cocktail; Roche)
5. Ni-NTA agarose (GE Healthcare)
6. Glutathione sepharose 4B resin (GE Healthcare)
7. Sodium chloride
8. Calcium chloride
9. Tris (hydroxymethyl) aminomethane
10. HEPES
11. Imidazole
12. Dithiothreitol
13. Reduced glutathione
14. Glycerol
15. Chymotrypsin (SIGMA)
16. PMSF (Wako)
17. LMNG (Anatrace)

6. Protocol

6.1 Expression of Ia

1.1 Prepare *E. coli* BL21 star (DE3), transformed with pET-21b encoding Ia gene, as a colony or frozen glycerol stock.

1.2 Prepare 50 mg/mL ampicillin stock in 80% (v/v) ethanol, 500 mM IPTG stock in MilliQ, and 1 L of general LB liquid medium. LB liquid medium must be autoclaved.

1.3 Add 3 μL of ampicillin stock to 3 mL LB liquid medium so that the final concentration of ampicillin in the medium is 50 μg/mL. Similarly, add 1 mL ampicillin to 1 L LB liquid medium.

1.4 Transfer a single colony or small amount of glycerol stock into the medium containing ampicillin.

1.5 Incubate the culture at 37 °C for 16 h.

1.6 Add the entire pre-culture to 1 L fresh LB medium containing ampicillin.

1.7 Incubate the culture again at 37 °C until the OD_{600} is 1.5 (AU: Absorbance units).

1.8 Add 0.5 mL IPTG solution to the liquid culture so that the final concentration of IPTG is approximately 0.5 mM.

1.9 Incubate the culture again at 37 °C for 16 h.

1.10 Centrifuge at 12,092 × *g* for 5 min at 4 °C.

1.11 Remove the supernatant and collect the pellet. Wash the cell pellet with PBS.

1.12 Store the cells at −80 °C.

Tip: We used pET-21b to express Ia with a C-terminal his-tag. This method is also useful in obtaining Ia with an N-terminal his-tag.

Tip: We recommend preparing frozen glycerol stock for longer storage duration. Glycerol stock is prepared by mixing equal volumes of autoclaved 50% (v/v) glycerol and liquid culture. New stock should be prepared once every few months to retain the freshness of the cells.

Tip: In case of Ia with N-terminal his-tag, the culture should be incubated at 37 °C, but not 20 °C, after the addition of IPTG to avoid decrease in expression efficiency. For Ia with a C-terminal His-tag, the incubation temperature does not affect the expression efficiency.

6.2 Purification of Ia

2.1 Prepare the following buffers:

Buffer A. 20 mM Tris pH 8.0, 300 mM NaCl, and 20 mM imidazole

Buffer B. 20 mM Tris pH 8.0, 300 mM NaCl, and 500 mM imidazole

Buffer C. 20 mM Tris pH 8.0 and 2 mM $CaCl_2$

Buffer D. 20 mM Tris pH 8.0, 1 M NaCl, and 2 mM $CaCl_2$

Buffer E. 10 mM Tris pH 8.0 and 100 mM NaCl

2.2 Equilibrate 3 mL Ni-NTA agarose by loading three column volumes of buffer A.

2.3 Resuspend the frozen cells in buffer A with 1 tablet of cOmplete.

2.4 Sonicate the lysate (Branson Digital Sonifier set at 50% amplitude) on ice for 10 min, cooling down the suspension for 1 min after every 2 min of sonication.

2.5 Centrifuge at 180,000 × *g* for 40 min at 4 °C in ultracentrifuge tube.

2.6 Load the supernatant onto the equilibrated Ni-NTA agarose column.

2.7 Incubate the Ni-NTA agarose column at 4 °C for 1 h.

2.8 Open the plug and elute unbound proteins as a flow-through.

2.9 Wash the Ni-NTA agarose by loading buffer A.
2.10 Elute the Ni-NTA agarose by loading buffer B.
2.11 Collect the elution fractions and exchange the buffer with buffer C using amicon.
2.12 Load the sample onto the anion exchange chromatography column (HiTrap Q HP 5 mL column).
2.13 Elute the sample with a linear salt gradient (12 column volume) using buffer D.
2.14 Mix the fractions containing Ia and exchange the buffer with buffer E using Amicon centrifugal filter (30 K), followed by concentration of the sample.
2.15 Freeze the sample rapidly in liquid nitrogen and store at −80 °C.

In case of Ia with an N-terminal his-tag, the following steps are required after 2.14:

2.16 Equilibrate 3 mL Ni-NTA agarose by loading 3 column volumes of buffer A.
2.17 Incubate Ia with thrombin (0.001 unit per μg Ia) for 16 h at 20–23 °C.
2.18 Load the thrombin-treated Ia onto the Ni-NTA agarose column.
2.19 Incubate the Ni-NTA agarose column at 4 °C for 1 h.
2.20 Open the plug and elute unbound proteins as flow-through.
2.21 Wash the Ni-NTA agarose with buffer A and assemble the wash flow-through fractions.
2.22 Exchange the buffer with buffer E using Amicon centrifugal filter (30 K), followed by the concentration of the sample.
2.23 Freeze the sample rapidly in liquid nitrogen and store at −80 °C.

Tip: Except for treatment with thrombin, all the steps are performed at 4 °C or on ice. In particular, during sonication, the heat may induce aggregation.

6.3 Expression of Ib

3.1 Prepare *E. coli* origami, transformed with pGEX4T-1 encoding Ib gene, as a colony or frozen glycerol stock.
3.2 Prepare 50 mg/mL ampicillin stock in 80% (v/v) ethanol, 12.5 mg/mL tetracycline stock in 80% (v/v) ethanol, 15 mg/ml kanamycin in milliQ, 500 mM IPTG stock in milliQ, and 1 L Super Broth liquid medium containing 1 mL 5 N NaOH. The medium should be autoclaved.
3.3 Add 3 μL each of ampicillin, tetracycline, and kanamycin stocks into 3 mL Super Broth liquid medium. Similarly, 1 mL of the antibiotics was added to 1 L LB liquid medium.

3.4 Add one colony or a small amount of glycerol stock to the medium containing ampicillin.
3.5 Incubate the culture at 37 °C for 16 h.
3.6 Add the entire pre-culture into 1 L fresh Super Broth medium containing antibiotics.
3.7 Incubate the culture again at 37 °C until the OD_{600} reaches 0.6 (AU).
3.8 Place the culture flask on ice water.
3.9 Add 1 mL IPTG solution to the liquid culture so that the final concentration of IPTG will be approximately 1 mM.
3.10 Incubate the culture at 23 °C for 16 h for.
3.11 Centrifuge at 12,092 × *g* for 5 min at 4 °C.
3.12 Remove the supernatant, collect the pellet, and wash with PBS.
3.13 Store the cells at −80 °C.

Tip: Because Ib is a large protein that eventually converts to a membrane-binding protein, we selected GST-tag to increase its solubility. However, in case of PA, it has been reported that PA can be expressed with a His-tag and purified without any problem. Thus, it seems that his-tag could be a possible choice for Ib too.

Tip: The GST-tag is attached at the N-terminal upstream of the propeptide region. In the following step, we cleave the GST-tag together with the propeptide region; so, Ib needs no cleaving site between the GST-tag and propeptide.

Tip: Although the culture incubation methods are mostly similar to that in case of Ia, there are differences with respect to the antibiotics used and the temperature following the addition of IPTG.

6.4 Purification of Ib

4.1 Prepare the following buffers

Buffer F: 20 mM Tris pH 8.0, 150 mM NaCl, 2 mM $CaCl_2$, and 5 mM dithiothreitol.

Buffer G: 20 mM Tris pH 8.0, 150 mM NaCl, 2 mM $CaCl_2$, and 10 mM reduced glutathione.

Buffer H: 20 mM Tris pH 8.0, and 50 mL NaCl and 2.5 mM $CaCl_2$.

4.2 Equilibrate 3 mL glutathione sepharose 4B resin (GE Healthcare) by loading 3 column volumes of buffer F.
4.3 Resuspend the frozen cells in buffer F with 1 tablet of complete.

4.4 Sonicate the lysate (Branson Digital Sonifier set at 50% amplitude) on ice for 10 min, cooling down the suspension for 1 min after every 2 min of sonication.

4.5 Centrifuge at 180,000 × *g* for 40 min at 4 °C using ultracentrifuge tube.

4.6 Load the supernatant onto the equilibrated glutathione sepharose 4B resin column.

4.7 Incubate the glutathione sepharose 4B resin column at 4 °C for 1 h.

4.8 Open the plug and elute unbound proteins as flow-through.

4.9 Wash the glutathione sepharose 4B resin with loading buffer F until the concentration of the latest fraction is <0.1 mg/mL.

4.10 Elute bound-proteins to agarose by loading buffer G.

4.11 Collect the elution fractions and exchange the buffer with buffer H using Amicon centrifugal filter.

4.12 Freeze the sample rapidly in liquid nitrogen and store at −80 °C.

Tip: Because of the high specificity of the GST-tag, the purification of Ib is based only on the GST-tag method.

Tip: The reducing agent, DTT, of buffer F should be added just before use. Because the amount of Buffer G used for the elution is small, we aliquoted buffer G for and stored at −20 °C.

6.5 Preparation of density gradient ultracentrifugation (Fig. 5)

5.1 Prepare the following buffers

Buffer I: 10% (v/v) glycerol, 50 mM HEPES pH 7.5, 100 mM NaCl, 1 mM $CaCl_2$, and 0.003% (w/v) LMNG.

Buffer J: 30% (v/v) glycerol, 50 mM HEPES pH 7.5, 100 mM NaCl, 1 mM $CaCl_2$, and 0.003% (w/v) LMNG.

5.2 Place 5.25 mL of water in the ultracentrifugation tube and draw a line tracing the water level on the outside of the tube.

5.3 Remove the water from the tube.

5.4 Load 5.25 mL buffer I into the tube.

5.5 Load 5.25 mL buffer J into the tube from the bottom, using a needle-shaped injector, per the drawn line of 5.25 mL volume.

5.6 Seal the tube with a tube cap or parafilm and lay down the tube.

5.7 Keep the tube laid down at 4 °C for 2 h.

5.8 Raise the tube slowly and remove the seal.

5.9 Adjust the weight of the couple of tubes.

Tip: Prepare a couple of gradient beds in order to balance ultracentrifugation.

Tip: Carry out all the steps carefully and do not unsettle the gradient bed, as this would disrupt the density gradient.

Tip: Prepare the gradient bed such that sample preparation is complete at almost the same time, so that ultracentrifugation with a fresh density gradient can be performed.

6.6 Oligomerization and solubilization of Ib

6.1 Prepare following solutions

1 mg/mL chymotrypsin in water.

1% (w/v) LMNG in water.

100 mM PMSF in ethanol (dissolved later).

6.2 Add chymotrypsin to purified Ib with GST-tag (1/1000 quantity of chymotrypsin to Ib).

6.3 Incubate at 20–23 °C for 1 h.

6.4 Dissolve the PMSF in ethanol and add it to the Ib treated with chymotrypsin, such that the final PMSF concentration is 1 mM (final concentration of ethanol is 1% if 100 mM PMSF is used).

6.5 Add LMNG solution to Ib such that its final concentration is 0.03% (w/v).

6.6 Add ethanol to Ib such that its final concentration is 10% (v/v), including ethanol already added as PMSF solution.

6.7 Incubate at 37 °C for 1 h.

Tip: The concentration of LMNG for solubilization (0.03% w/v) is higher than that for the density gradient ultracentrifugation step (0.003% w/v).

Tip: Because Ib does not oligomerize efficiently without high concentrations of ethanol and the Ib oligomer is an insoluble pore, it is essential to add ethanol and LMNG.

6.7 Purification of Ib-pore

7.1 Prepare the following buffers

Buffer K: 10 mM HEPES (pH 7.5), 1 mM $CaCl_2$, and 0.003% (v/v) LMNG.

7.2 Load the sample (~1 mL) prepared in 6.6 onto the gradient bed prepared in 6.5.

7.3 Centrifuge at 230,139 × *g* for 16 h at 4 °C for the density gradient centrifugation.

7.4 Make a hole at the bottom of the tube using a needle, so that solution produces drops (Fig. 6).

7.5 Collect 5 drops as 1 fraction, repeating until all the solution in the tube comes out.

7.6 Measure the Abs 280 nm in each fraction to determine the fraction containing Ib-pore.

7.7 Assemble the fractions containing Ib-pore.

7.8 Exchange the buffer with buffer K using PD10, twice.

7.9 Further exchange the buffer with buffer K and concentrate the Ib-pore containing fraction to more than 1 mg/mL using Amicon centrifugal filter.

7.10 Freeze 50 μL drops of the sample in liquid nitrogen and store at −80 °C.

Tip: The loading amount of Ib-pore for ultracentrifugation should be approximately 1 mg per tube because of the limits in detection and separation.

Tip: If the sample volume is too much to load onto the tube, some volume need be removed from the top of the gradient bed, followed by loading of the sample.

Tip: Buffer exchanging to remove glycerol is crucial for high-resolution analysis. Thus, it should be performed carefully.

Tip: The filter size of the Amicon centrifugal filter should be 100 K for fast buffer exchange and to avoid stacking of LMNG on the filter.

Tip: The freezing step is not necessary if cryo-EM data are collected within a few days. Rapid freezing as droplets is essential. The volume of each droplet is approximately 50 μL, which is frozen by dropping into a cryo-tube cooled in liquid nitrogen.

6.8 Ia-bound Ib-pore complex formation

8. Add Ia prepared in 6.2 to Ib-pore prepared in 6.7 with three-fold molar excess

6.9 Cryo-EM grid preparation

9.1 Prepare the following buffer

Buffer L. 10 mM HEPES (pH 7.5) and 1 mM $CaCl_2$.

9.2 Dilute the prepared sample with buffer L.

9.3 Apply 2.6 μL of the diluted sample onto glow discharged Quantifoil R1.2/1.3 at 100% humidity.

9.4 Blot the sample grid with filter paper for 4.5 s.

9.5 Freeze the sample grid rapidly with liquid ethane.

Tip: Buffer L should not contain LMNG to reduce noise in cryo-EM images. In our experience, 5 times dilution of buffer L (final LMNG concentration = 0.0006%) just before grid preparation did not cause obvious aggregation.

Tip: The sample was applied under humid condition and blotting and freezing were carried out using Vitrobot.

Tip: The sample volume applied to the grid and blotting time are crucial for the thickness of the ice. We succeeded in preparing an ideal reproducible sample grid with the described conditions.

7. Summary

Single particle analysis is a well-established method that does not require crystals. Thus, it is an advantageous protocol compared with X-ray crystallography for analyzing large membrane proteins such as Ib-pores. It is not necessary to purify the samples completely, although it is better to maintain homogeneity. Consequently, we can observe the physiological motions of the sample. However, as the membrane protein sample is used for cryo-EM, mono-dispersibility and noise due to the sample content must be considered. Moreover, in case of the B-component pore, efficient formation of the oligomer is challenging.

In 2020, almost all the reported structures of A-B toxins are from *Bacillus anthracis*. Interestingly, although the amino acid sequences of B components have similarity, there is a significant difference between the amino acid sequences of the A components in anthrax toxin and clostridial binary toxin. Specifically, clostridial binary toxin studies will provide new knowledge regarding the translocation mechanism, which has not been revealed in anthrax toxin studies. Thus, a comparison of both binary toxins is essential to understand the dynamics of A-B toxin cytotoxicity.

In this chapter, we reported a new method wherein ethanol-induces Ib-pore formation and LMNG solubilizes these Ib pores, thus providing a high-resolution structure using cryo-EM. In addition, we deleted of the receptor-binding domain, which allowed efficient Ib oligomerization. We expect that these methods will contribute to both structural and functional studies of A-B toxins.

Acknowledgments

We appreciate K. Mitsuoka for working together with searching for conditions to prepare the sample at the beginning, JI. Kishikawa for teaching us techniques for single particle analysis, N. Nagahama and M. Oda for commenting on the research, H. Murata for laying background of purifying sample initially, A. Kawamoto and K. Iwasakai for working on Cryo-EM data acquisition and data processing, T. Yoshida for the model building, structure refinement, and analysis.

This work was supported by JSPS KAKENHI grant numbers 18K06170 and 17K15095. This research was partially supported by the Platform Project for Supporting Drug Discovery and Life Science Research (Basis for Supporting Innovative Drug Discovery and Life Science Research (BINDS)) from the Japan Agency for Medical Research and Development (AMED) under grant number JP19am0101072 (support number 1232). This work was partially supported by grants-in-aid from "Nanotechnology Platform" of the Ministry of Education, Culture, Sports, Science and Technology (MEXT), Japan.

References

Akkaladevi, N., Hinton-Chollet, L., Katayama, H., Mitchell, J., Szerszen, L., Mukherjee, S., et al. (2013). Assembly of anthrax toxin pore: Lethal-factor complexes into lipid nanodiscs. *Protein Science, 22*(4), 492–501.

Aktories, K., Bärmann, M., Ohishi, I., Tsuyama, S., Jakobs, K. H., & Habermann, E. (1986). Botulinum C2 toxin ADP-ribosylates actin. *Nature, 322*(6077), 390–392.

Anderson, D. M., Sheedlo, M. J., Jensen, J. L., & Lacy, D. B. (2020). Structural insights into the transition of Clostridioides difficile binary toxin from prepore to pore. *Nature Microbiology, 5*(1), 102–107.

Barth, H., Aktories, K., Popoff, M. R., & Stiles, B. G. (2004). Binary bacterial toxins: Biochemistry, biology, and applications of common Clostridium and Bacillus proteins. *Microbiology and Molecular Biology Reviews: MMBR, 68*(3), 373–402. table of contents.

Chae, P. S., Rasmussen, S. G. F., Rana, R. R., Gotfryd, K., Chandra, R., Goren, M. A., et al. (2010). Maltose–neopentyl glycol (MNG) amphiphiles for solubilization, stabilization and crystallization of membrane proteins. *Nature Methods, 7*(12), 1003–1008.

Christensen, K. A., Krantz, B. A., & Collier, R. J. (2006). The assembly and disassembly kinetics of anthrax toxin complexes. *Biochemistry, 45*(7), 2380–2386.

Collier, R. J., & Young, J. A. T. (2003). Anthrax toxin. *Annual Review of Cell and Developmental Biology, 19*(1), 45–70.

Cunningham, K., Lacy, D. B., Mogridge, J., & Collier, R. J. (2002). Mapping the lethal factor and edema factor binding sites on oligomeric anthrax protective antigen. *Proceedings of the National Academy of Sciences of the United States of America, 99*(10), 7049–7053.

Feld, G. K., Thoren, K. L., Kintzer, A. F., Sterling, H. J., Tang, I. I., Greenberg, S. G., et al. (2010). Structural basis for the unfolding of anthrax lethal factor by protective antigen oligomers. *Nature Structural & Molecular Biology, 17*(11), 1383–1390.

Gibert, M., Petit, L., Raffestin, S., Okabe, A., & Popoff, M. R. (2000). Clostridium perfringens iota-toxin requires activation of both binding and enzymatic components for cytopathic activity. *Infection and Immunity, 68*(7), 3848–3853.

Han, S., Craig, J. A., Putnam, C. D., Carozzi, N. B., & Tainer, J. A. (1999). Evolution and mechanism from structures of an ADP-ribosylating toxin and NAD complex. *Nature Structural Biology, 6*(10), 932–936.

Hardenbrook, N. J., Liu, S., Zhou, K., Ghosal, K., Hong Zhou, Z., & Krantz, B. A. (2020). Atomic structures of anthrax toxin protective antigen channels bound to partially unfolded lethal and edema factors. *Nature Communications, 11*(1), 840.

Jiang, J., Pentelute, B. L., Collier, R. J., & Zhou, Z. H. (2015). Atomic structure of anthrax protective antigen pore elucidates toxin translocation. *Nature, 521*(7553), 545–549.

Kintzer, A. F., Thoren, K. L., Sterling, H. J., Dong, K. C., Feld, G. K., Tang, I. I., et al. (2009). The protective antigen component of anthrax toxin forms functional octameric complexes. *Journal of Molecular Biology*, *392*(3), 614–629.

Krantz, B. A., Melnyk, R. A., Zhang, S., Juris, S. J., Lacy, D. B., Wu, Z., et al. (2005). A phenylalanine clamp catalyzes protein translocation through the anthrax toxin pore. *Science (New York, N.Y.)*, *309*(5735), 777–781.

Lacy, D. B., Wigelsworth, D. J., Melnyk, R. A., Harrison, S. C., & Collier, R. J. (2004). Structure of heptameric protective antigen bound to an anthrax toxin receptor: A role for receptor in pH-dependent pore formation. *Proceedings of the National Academy of Sciences of the United States of America*, *101*(36), 13147–13151.

Marvaud, J. C., Stiles, B. G., Chenal, A., Gillet, D., Gibert, M., Smith, L. A., et al. (2002). Clostridium perfringens iota toxin. Mapping of the Ia domain involved in docking with Ib and cellular internalization. *The Journal of Biological Chemistry*, *277*(46), 43659–43666.

Miller, C. J., Elliott, J. L., & Collier, R. J. (1999). Anthrax protective antigen: Prepore-to--pore conversion. *Biochemistry*, *38*(32), 10432–10441.

Mio, K., & Sato, C. (2017). Lipid environment of membrane proteins in cryo-EM based structural analysis. *Biophysical Reviews*, *10*(2), 307–316.

Mogridge, J., Cunningham, K., Lacy, D. B., Mourez, M., & Collier, R. J. (2002). The lethal and edema factors of anthrax toxin bind only to oligomeric forms of the protective antigen. *Proceedings of the National Academy of Sciences of the United States of America*, *99*(10), 7045–7048.

Pannifer, A. D., Wong, T. Y., Schwarzenbacher, R., Renatus, M., Petosa, C., Bienkowska, J., et al. (2001). Crystal structure of the anthrax lethal factor. *Nature*, *414*(6860), 229–233.

Perelle, S., Gibert, M., Bourlioux, P., Corthier, G., & Popoff, M. R. (1997). Production of a complete binary toxin (actin-specific ADP-ribosyltransferase) by Clostridium difficile CD196. *Infection and Immunity*, *65*(4), 1402–1407.

Popoff, M. R., & Boquet, P. (1988). Clostridium spiroforme toxin is a binary toxin which ADP-ribosylates cellular actin. *Biochemical and Biophysical Research Communications*, *152*(3), 1361–1368.

Rouck, J. E., Krapf, J. E., Roy, J., Huff, H. C., & Das, A. (2017). Recent advances in nanodisc technology for membrane protein studies (2012-2017). *FEBS Letters*, *591*(14), 2057–2088.

Shen, Y., Zhukovskaya, N. L., Guo, Q., Florián, J., & Tang, W. J. (2005). Calcium-independent calmodulin binding and two-metal-ion catalytic mechanism of anthrax edema factor. *The EMBO Journal*, *24*(5), 929–941.

Simpson, L. L., Stiles, B. G., Zepeda, H. H., & Wilkins, T. D. (1987). Molecular basis for the pathological actions of Clostridium perfringens iota toxin. *Infection and Immunity*, *55*(1), 118–122.

Sun, J., Lang, A. E., Aktories, K., & Collier, R. J. (2008). Phenylalanine-427 of anthrax protective antigen functions in both pore formation and protein translocation. *Proceedings of the National Academy of Sciences of the United States of America*, *105*(11), 4346–4351.

Tang, W. J., & Guo, Q. (2009). The adenylyl cyclase activity of anthrax edema factor. *Molecular Aspects of Medicine*, *30*(6), 423–430.

Tonello, F., & Montecucco, C. (2009). The anthrax lethal factor and its MAPK kinase-specific metalloprotease activity. *Molecular Aspects of Medicine*, *30*(6), 431–438.

Tsuge, H., Nagahama, M., Nishimura, H., Hisatsune, J., Sakaguchi, Y., Itogawa, Y., et al. (2003). Crystal structure and site-directed mutagenesis of enzymatic components from Clostridium perfringens iota-toxin. *Journal of Molecular Biology*, *325*(3), 471–483.

Tsuge, H., Nagahama, M., Oda, M., Iwamoto, S., Utsunomiya, H., Marquez, V. E., et al. (2008). Structural basis of actin recognition and arginine ADP-ribosylation by

Clostridium perfringens iota-toxin. *Proceedings of the National Academy of Sciences of the United States of America, 105*(21), 7399–7404.

Tsurumura, T., Tsumori, Y., Qiu, H., Oda, M., Sakurai, J., Nagahama, M., et al. (2013). Arginine ADP-ribosylation mechanism based on structural snapshots of iota-toxin and actin complex. *Proceedings of the National Academy of Sciences of the United States of America, 110*(11), 4267–4272.

Vandekerckhove, J., Schering, B., Bärmann, M., & Aktories, K. (1987). Clostridium perfringens iota toxin ADP-ribosylates skeletal muscle actin in Arg-177. *FEBS Letters, 225*(1–2), 48–52.

Vernier, G., Wang, J., Jennings, L. D., Sun, J., Fischer, A., Song, L., et al. (2009). Solubilization and characterization of the anthrax toxin pore in detergent micelles. *Protein Science: A Publication of the Protein Society, 18*(9), 1882–1895.

Xu, X., Godoy-Ruiz, R., Adipietro, K. A., Peralta, C., Ben-Hail, D., Varney, K. M., et al. (2020). Structure of the cell-binding component of the Clostridium difficile binary toxin reveals a di-heptamer macromolecular assembly. *Proceedings of the National Academy of Sciences of the United States of America, 117*, 1049–1058.

Yamada, T., Yoshida, T., Kawamoto, A., Mitsuoka, K., Iwasaki, K., & Tsuge, H. (2020). Cryo-EM structures reveal translocational unfolding in the clostridial binary iota toxin complex. *Nature Structural & Molecular Biology, 27*(3), 288–296.

CHAPTER SIX

AFM imaging of pore forming proteins

Adrian W. Hodel[a], Katharine Hammond[b,c,d], and Bart W. Hoogenboom[c,d,*]
[a]Killer Cell Biology Laboratory, Peter MacCallum Cancer Centre, Melbourne, VIC, Australia
[b]National Physical Laboratory, Teddington, United Kingdom
[c]London Centre for Nanotechnology, University College London, London, United Kingdom
[d]Department of Physics & Astronomy, University College London, London, United Kingdom
*Corresponding author: e-mail address: b.hoogenboom@ucl.ac.uk

Contents

Methods in Enzymology, Volume 649
ISSN 0076-6879
https://doi.org/10.1016/bs.mie.2021.01.002

Abstract

Pore forming proteins are released as water-soluble monomers that form—mostly oligomeric—pores in target membranes. Our understanding of such pore formation relies in part on the direct visualization of their assemblies on and in the membrane. Here, we discuss the application of atomic force microscopy (AFM) to visualize and understand membrane pore formation, illustrated specifically by studies of proteins of the MACPF/CDC superfamily on supported lipid bilayers. Besides detailed protocols, we also point out common imaging artefacts and strategies to avoid them, and briefly outline how AFM can be effectively used in conjunction with other methods.

1. Introduction

1.1 Atomic force microscopy

Atomic force microscopy (AFM) is a technique that has been extensively used to study pore forming proteins. In contrast to many other types of microscopy, AFM relies on a mechanical interaction with the sample, using a sharp tip to trace the surface topography of the sample at the nanometer scale. This surface sensitivity makes it exquisitely suited to visualize how pore forming toxins assemble on and insert into target membranes. AFM can be done in liquid, also enabling real-time imaging of protein assembly and membrane pore formation. Although AFM can be performed on live cell membranes, highest resolution has so far mostly been obtained on target membranes that consist of lipid bilayers on a solid support.

Here, we provide protocols, guidelines, and relevant literature on how AFM samples can be prepared and measured, how the data is processed, and what insights can be gathered about key aspects of membrane pore formation. This is illustrated with examples of pore forming proteins of the membrane attack complex/perforin/cholesterol-dependent cytolysin

(MACPF/CDC) superfamily, which play important roles in both bacterial attack and immune defence in all domains of life (Lukoyanova, Hoogenboom, & Saibil, 2016; Reboul, Whisstock, & Dunstone, 2016).

1.2 Key concepts of assembly and pore formation by pore forming proteins

To appreciate the use of AFM for studying pore forming proteins and MACPF/CDC proteins in particular, it is useful to briefly introduce the family and its mechanisms of assembly and pore formation. The members of the MACPF/CDC superfamily are a heterogeneous group of proteins with different functional pathways. For example, while the vertebrate immune protein macrophage expressed gene-1 (MPEG-1) pre-assembles into a pore before binding target membranes (Ni et al., 2020; Pang et al., 2019), the protein astrotactin-2, likely plays its role in neurodevelopment without forming pores at all (Ni, Harlos, & Gilbert, 2016). Generally, however, most MACPF/CDCs are expressed as water-soluble monomers that oligomerize and form transmembrane pores upon membrane binding (Lukoyanova et al., 2016).

1.2.1 Growing pore vs growing prepore pathways

In some cases, these proteins may assemble directly into the membrane to form transmembrane pores, where initial insertion of a single protein or small assembly induces a small pore, which is next increased by the addition of further assembly subunits. Alternatively, and quite commonly, they assemble on the membrane prior to pore formation, leading to so-called prepore assemblies or prepores, which next undergo a collective transition to insert into the membrane to form pores. These two basic pathways, which involve growing pores and transient, growing prepores, respectively, are schematically illustrated in Fig. 1A and B.

Examples for both pathways can be found in nature. In pore formation by the membrane attack complex (MAC) of the complement, multiple copies of the C9 subunit are added to the already inserted C5b-8 initiator complex, thus expanding the pore size in a growing pore fashion (Parsons et al., 2019). Suilysin and other CDCs fully determine their oligomeric states as prepores, and thus follow a growing prepore pathway. Cryo-electron microscopy (cryo-EM) reconstructions of suilysin prepores and pores are shown in Fig. 1C and D. Similar data is available for another CDC, pneumolysin (van Pee et al., 2017) and the MACPF pleurotolysin (Lukoyanova et al., 2015). Hybrid pathways are also possible as shown by the vertebrate immune protein perforin, which first oligomerizes into

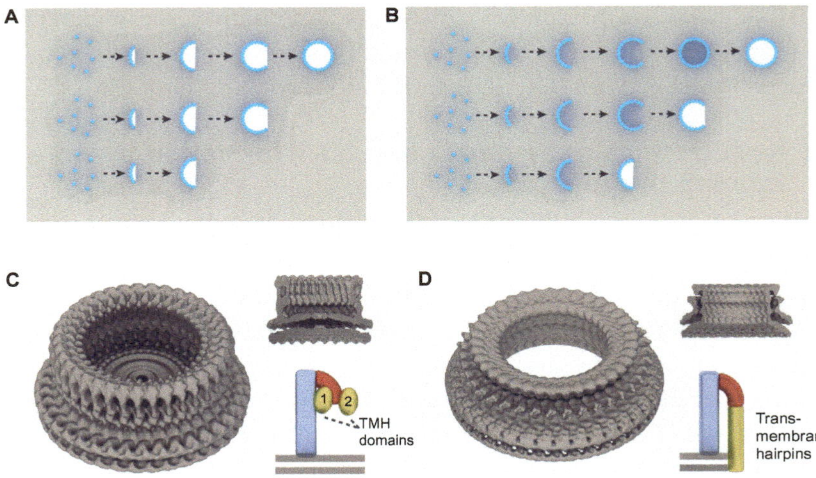

Fig. 1 Pores vs prepores in assembly pathways and structures. (A) Schematic of growing pores, in which initially small pores grow by addition of further subunits. (B) Schematic of growing prepores. The membrane is only perforated in the last step, after which the pore morphology is fixed. (C) left: Cryo-EM reconstruction of the suilysin prepore on a lipid vesicle. Top right: quarter slice of the EM reconstruction, showing the intact curved bilayer at the bottom of the prepore. Bottom right: schematic of the protein configuration in the prepore, where the two transmembrane helices (TMH, yellow) are retracted from the (gray) lipid bilayer. A central β-sheet domain and the protein backbone are depicted in red and blue, respectively. (D) as (C), for the suilysin pore. The EM reconstruction reveals the transmembrane channel at the center of the pore. The schematic depicts the former TMH domains after their transition into transmembrane hairpins that slice through the membrane. Note that the characteristic height collapse of CDCs upon membrane insertion (Czajkowsky, Hotze, Shao, & Tweten, 2004) is not included in the schematics. *Panels (A) and (B) are based on a concept by Gilbert, R. J. C. (2005). Inactivation and activity of cholesterol-dependent cytolysins: What structural studies tell us. Structure, 13(8), 1097–1106. https://doi.org/10.1016/j.str.2005.04.019, adapted with permission. 3D cryo-EM maps were retrieved from the electron microscopy databank (*ebi.ac.uk/pdbe/emdb*), EMD-2979 and EMD-2983 Leung, C., Dudkina, N. V., Lukoyanova, N., Hodel, A. W., Farabella, I., Pandurangan, A. P., … Hoogenboom, B. W. (2014). Stepwise visualization of membrane pore formation by suilysin, a bacterial cholesterol-dependent cytolysin. eLife, 3(5), e04247. https://doi.org/10.7554/eLife.04247.*

incomplete prepores that can insert the membrane and then continue oligomerization in the pore state (Leung et al., 2017).

1.2.2 Membrane insertion and disulfide locked mutants

The structural features and changes responsible for membrane insertion are well conserved within the MACPF/CDC family. They involve a central β-sheet domain that opens to release two α-helical bundles termed transmembrane helixes 1 and 2 (TMH1 and TMH2, see bottom right of Fig. 1C).

Upon release, the helixes unfurl and slice through the membrane as individual β-hairpins (bottom right of Fig. 1D) that, collectively, encase the pore lumen in a β-barrel structure.

Some proteins such as MPEG-1 or the bacterial toxin listeriolysin O may require sub-physiological pH to trigger membrane insertion, which can be exploited to study their prepore and pore forms separately (Mulvihill et al., 2015; Podobnik et al., 2015; Ruan, Rezelj, Bedina Zavec, Anderluh, & Scheuring, 2016). To trap pore formation in otherwise transient prepore states, membrane insertion can also be prevented by engineering a disulfide bond that locks one of the TMH domains to a more rigid part of the protein (Heuck, Hotze, Tweten, & Johnson, 2000; Hotze et al., 2001; Lukoyanova et al., 2015). The lock can be removed by reducing the disulfide bond, restoring pore formation of the protein, as illustrated in Fig. 2. The development of disulfide locked mutants has greatly aided the understanding of MACPF/CDCs and their characterization by AFM. For example, disulfide locked mutants of the mushroom protein pleurotolysin allowed in-depth study of how MACPF/CDCs unfold and insert into the membrane (Lukoyanova et al., 2015) and are now routinely used to study prepore structures and determine the oligomerization pathways of proteins that have no inherent trigger for insertion.

1.2.3 Arc- and ring-shaped pores

The final shapes of transmembrane pores can be heterogeneous due to different numbers of protein subunits present in individual assemblies. The resulting morphologies can be broadly divided into arc- and ring-shaped.

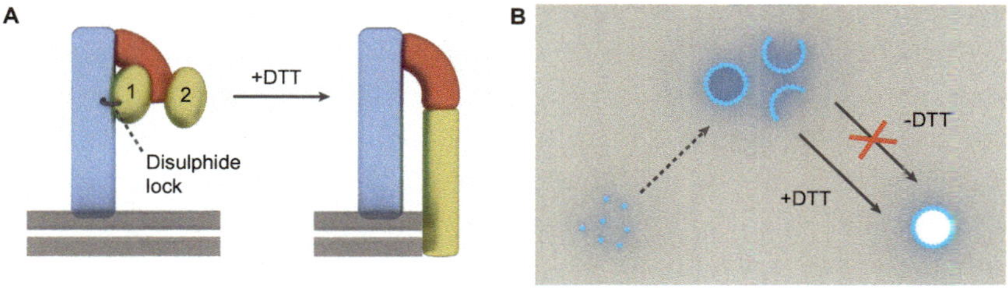

Fig. 2 Disulfide locked mutants to reversibly trap proteins in a prepore state. (A) Schematic example of a disulfide locked mutant, where the disulfide lock is situated between TMH1 (yellow, 1) and the protein backbone (blue). The lock prevents membrane insertion, until it is reduced by dithiothreitol (DTT). Addition of DTT restores the capacity of the protein to insert the membrane by unfurling both TMH domains into transmembrane β-hairpins. (B) Disulfide locked mutants are trapped in prepore assemblies unless DTT is added, which facilitates the study of these otherwise transient intermediates.

Ring-shaped pores form a β-barrel around the complete circumference of the pore lumen, although subunit count may still vary to some extent, giving rise to different sized rings. Arc-shaped pores can form functional transmembrane channels, but only partially seam the pore lumen, leaving an unsealed lipid edge (Leung et al., 2014; Sonnen, Plitzko, & Gilbert, 2014). The ratio between arc- and ring-shaped pores formed is influenced by the type and amount of protein used, and the composition of the host membrane (Rojko & Anderluh, 2015).

1.2.4 Lineaction and long-term plasticity of pores

Fully oligomerized and membrane inserted arc-shaped assemblies can diffuse in the membrane plane and alter pore morphology retrospectively. In a process called lineaction, arcs can align such that they encase a larger pore lumen than they would otherwise. Lineaction differs from oligomerization in that proteins do not interface with each other directly. Instead, the arc assemblies remain separate, and the alignment is driven by line tension along membrane edges. The idea of lineaction and its biological relevance largely originates from a study on listeriolysin O, which can create membrane lesions much larger than by the individual pore assemblies and is used by its host pathogen to escape cell compartments after phagocytosis (Ruan et al., 2016).

2. Protein requirements

The purity of the protein can be assessed via SDS page (Laemmli, 1970). The protein should leave a single band located at the molecular mass of the soluble monomer. The concentration of the purified protein can be measured by A280 absorbance on a NanoDrop (Thermo Fisher Scientific) or similar. The concentrations at which the pore forming proteins are deployed on their targets are often unknown in vivo. An exception is the MAC, where human serum contains between 350 and 900 nM of each MAC component (Heesterbeek et al., 2019). For practical purposes, protein stock solutions used for AFM sample preparations should contain enough protein to yield reasonable pore coverage and dilute enough to not risk spontaneous oligomerization in solution. We have used stock solutions with protein concentrations ranging from 0.1 to 1 mg/mL (e.g., ca. 2–20 μM for perforin or suilysin), which allows for further dilution during sample preparation.

To store protein stock solutions, we followed collaborator's or manufacturer's protocols. For example, CDCs were mixed with up to 10% w/w glycerine and aliquoted. The aliquots were then shock-frozen in liquid nitrogen and stored at −80 °C for long-term storage. When needed for experiments, only one aliquot was defrosted and stored at 4 °C. By contrast. Perforin was always kept at 4 °C and replaced regularly (especially the disulfide locked mutants) by freshly expressed batches of protein. To detect deterioration of the protein during storage, its activity can be monitored by lysis experiments (see Section 8.1) with a suitable, protein specific target, e.g., red blood cells. A change in protein activity also influences the concentration needed to achieve a useful pore coverage in AFM samples. Thus, AFM data of a protein batch or aliquot obtained at different timepoints can be used to detect protein deterioration as well, granted similar sample preparations were compared. In AFM data, loss of protein activity may or may not be accompanied by protein aggregation on the sample surface.

3. Supported lipid bilayer preparation

A range of model membrane systems can be used to study pore forming proteins, including lipid vesicles, lipid monolayers, and floating and tethered lipid bilayers (Pomorski, Nylander, & Cárdenas, 2014). For AFM experiments, model membranes are typically deposited on a solid, flat substrate. The solid support can be adapted according the type of membrane needed. Hydrophobic substrates such as highly ordered pyrolytic graphite is used to deposit supported monolayers. Mica or silicon oxide surfaces are suitable for supported lipid bilayers (SLBs). The experimental procedures presented here can all be realized with SLBs on mica, which provides an atomically flat substrate.

3.1 Lipid composition

Pore forming proteins can require specific lipid (and sometimes protein Boyd et al., 2016; Giddings, Zhao, Sims, & Tweten, 2004) receptors to bind a membrane surface and to form pores. As their name implies, cholesterol dependent cytolysins (CDCs) require some amount of cholesterol in the membrane to bind, typically >20%. Proteins with bactericidal properties like the MAC and MPEG-1 prefer bacterial membrane substrates or mimics that contain negatively charged lipids (Ni et al., 2020; Pang et al., 2019; Parsons et al., 2019). Similar preferences are observed with the recently discovered gasdermins (which are, perhaps, distant relatives of MACPF/CDCs Ruan, Xia, Liu, Lieberman, & Wu, 2018), corresponding to their role to bind mitochondrial membranes and the inner leaflets of plasma membranes

(Broz, Pelegrín, & Shao, 2020). Perforin can bind various types of lipids but requires liquid-disordered and uncharged membranes to form transmembrane pores (Hodel, Rudd-Schmidt, Trapani, Voskoboinik, & Hoogenboom, 2020; Rudd-Schmidt et al., 2019).

Lipid extracts from different tissues and from bacteria are commercially available and might be used to form suitable substrates for a given protein. Beyond that, the use of mixtures of purified or synthetic lipid species offers more freedom regarding properties that an SLB could display. The common basis of a lipid mixture is a bilayer forming phospho- or phosphosphingolipid. Such lipids each have a unique transition temperature T_m, below which they assemble into solid ordered bilayers (S_o) and above which they assemble into liquid disordered bilayers (L_d). An intermediate phase can be formed by incorporating hydrophobic molecules such as cholesterol. Cholesterol is not a bilayer-forming lipid on its own but intercalates with other lipid molecules, at higher concentrations leading to a liquid ordered (L_o) phase. The L_o phase shows increased lateral order compared to the L_d phase while still retaining its fluidity. A schematic of how the lipids are packed in the L_d, L_o, and S_o phase is shown in Fig. 3A.

The different phases can coexist within the same SLB: when lipids of different T_m are combined, phase separated SLBs containing regions of both

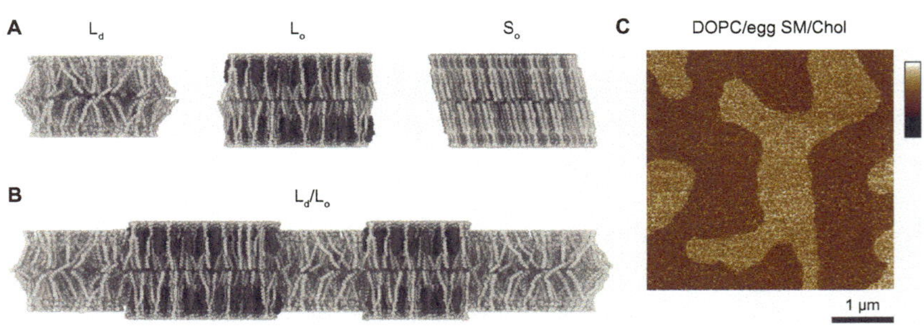

Fig. 3 Illustrations of different membrane phases and phase coexistence. (A) Different packing depending on the lipid phase. The more ordered, cholesterol (dark gray) containing L_o or S_o membranes exhibit a higher packing density of individual lipids compared to L_d membranes. Because of their lateral condensation, their thickness is increased. (B) Phase co-existence and separation of L_d and L_o domains in the same bilayer. The difference in thickness leads to step-edged phase boundaries. (C) AFM image of a roughly equimolar DOPC/egg SM/cholesterol (Chol) bilayer at ca. 30 °C. The different phase domains are clearly visible, as their difference in thickness is perceived as a change in height of ca. 1 nm between the lower (darker) L_d phase and the higher (brighter) L_o phase. The color scale ranges from −2 to 2 nm, with the surface of the L_d domains set to 0 nm. The full color scale is indicated on the right side.

S_o and L_d can be formed. When cholesterol is added to lipids with different T_m, SLBs can separate into L_d and L_o domains (schematic in Fig. 3B). Phase separated $L_d + L_o$ SLBs can be used to approximate the eukaryotic plasma membrane. Such mimics use a low T_m phosphatidylcholine species (e.g., DOPC or POPC) and high T_m sphingomyelin (e.g., purified egg SM) besides cholesterol, corresponding to the dominant types of lipids found in the outer leaflet of the eukaryotic membrane (van Meer, Voelker, & Feigenson, 2008). Conveniently, phase boundaries can be resolved with the AFM by a height difference between phases, typically of a few Ångströms to a nanometre (Fig. 3C) (Connell & Smith, 2006).

Lipid mixtures can also be complemented with lipids such as phosphatidylglycerol or phosphatidylserine to alter the surface charge of SLBs (Hodel et al., 2020; Rudd-Schmidt et al., 2019). Furthermore, incorporation of lipids with small headgroups like phosphatidylethanolamine, or lyso-lipids with only one acyl chain can change the tension stored in the bilayer, mimicking effects of mechanical stress to a membrane.

3.2 SLB formation

SLBs can be formed on a solid substrate like mica by the vesicle fusion method, which is schematically outlined in Fig. 4. The following subsections provide protocols for substrate and lipid vesicle preparation, and SLB deposition. Further recommendations on sample parameters and considerations on the limits of the method are provided thereafter.

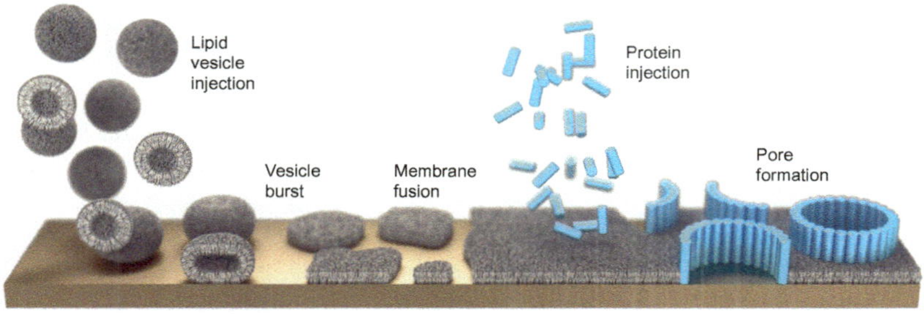

Fig. 4 Schematic illustration of SLB formation by vesicle fusion on a mica substrate in buffer. Unilamellar vesicles of the desired lipid composition are injected onto a mica substrate covered in adsorption buffer. When adsorbing to the mica surface, the vesicles burst, and at sufficient concentration fuse together to form an extended SLB covering the mica surface. Pore forming proteins can then be injected to use the SLB as a substrate for pore formation.

3.2.1 Preparation of mica supports

Mica provides an atomically flat hydrophilic surface that is compatible with lipid vesicle fusion. It delivers a clean background for high resolution imaging of pore forming proteins. The sample supports described in the following protocol have a metallic base that fits the magnetic sample holders of many AFM machines. A hydrophobic fluorinated ethylene propylene (FEP) layer provides a hydrophobic barrier around the mica disc. Mica discs of various sizes can be used and the protocols adjusted accordingly. The assembly of sample supports is further illustrated in Fig. 5.

1. Cut a ⌀ 15 mm circle of self-adhesive, vinyl backed FEP Bytac® laminate (Saint-Gobain Performance Plastics Corp.)
2. Attach the self-adhesive FEP to a ⌀ 15 mm steel disc (Agar scientific)
3. Glue a ⌀ 9.9 mm mica disc (Agar scientific) to the FEP surface with Araldite® 2-part epoxy resin (Huntsman Advanced Materials)
4. Leave to cure as per manufacturer's recommendations or longer

3.2.2 Considerations for optimal SLB formation

Vesicle adsorption and rupture are highly dependent on electrostatic interactions between the vesicles and the solid support (Richter, Bérat, & Brisson, 2006). Optimal bilayer formation can therefore be controlled by adjusting the pH and the ionic strength of solution. The conditions that best promote bilayer formation vary with lipid composition and charge, but quite generally, it is preferable for AFM studies on pore forming toxins to be conducted close to physiological salt conditions. We find that a buffer solution of 150 mM NaCl, 20 mM HEPES, pH 7.4 is compatible with a wide range of SLBs.

Besides tuning the electrostatic interactions between vesicles and solid support, the presence of divalent cations (Mg^{2+}, Ca^{2+}) has a direct effect on promoting SLB formation. This is particularly significant for mica substrates

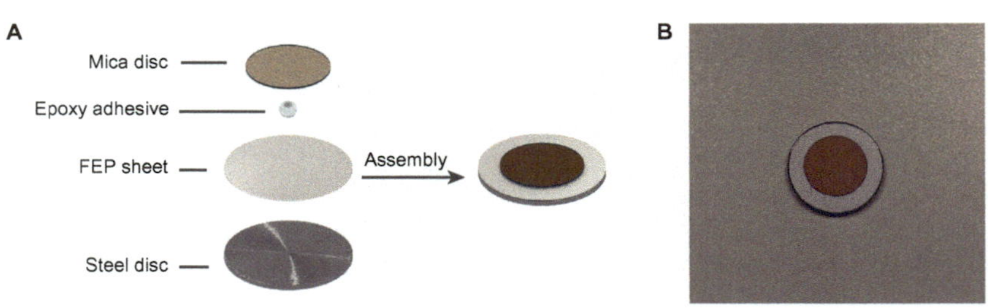

Fig. 5 Mica sample supports for AFM. (A) Schematic assembly of sample supports. (B) Photograph of an assembled sample support.

where the divalent cations act as a bridge between the negative mica substrate and the phospholipid head groups. Again, the optimal concentration varies with lipid composition, but in our lab, 10–30 mM total divalent cation concentration results in successful SLB formation. Note that Mg^{2+} and Ca^{2+} can have adverse effects on subsequent protein binding and assembly and, depending on the protein, may need to be diluted after SLB formation.

Vesicle size also influences bilayer formation. Larger vesicles can lead to SLBs containing discontinuities resulting from trapped intact vesicles (Richter et al., 2006). This effect is less likely for smaller vesicles; unilamellar vesicles that measure ≤100 nm in diameter can reliably lead to continuous SLBs. Finally, the vesicle concentration must be high enough to enable formation of a continuous SLB. 3–5 μg of lipid is sufficient for ⌀ 9.9 mm mica substrates.

3.2.3 Preparation of unilamellar vesicles

1. Lipids purchased in powder form are weighed and placed in a 4 mL PTFE capped glass vial (Agilent Technologies). Dissolve them in a suitable solvent (e.g., chloroform for the lipids outlined in these protocols). Depending on how much powder is available, final concentrations of lipid stocks can reach ca. 0.25–10 mg/mL. Repeat this step to generate stock solutions for all desired lipids
2. Using gastight syringes (Hamilton) or similar, solvent compatible equipment, extract the required amounts of each lipid stock solution and combine them in a separate glass vial to produce a lipid mixture of desired molar ratio and a total lipid weight of 0.25–1 mg
3. Evaporate the solvent in the lipid mixture under a gentle stream of nitrogen or argon gas to yield a dry lipid film
4. Re-suspend the lipid film in 0.5–1 mL buffer solution containing 150 mM NaCl and 20 mM HEPES, pH 7.4. Vortex until a turbid dispersion is formed in the vial. The turbidity is indicative of large multilamellar vesicles, with a diameter greater than the wavelength of visible light (i.e., >300–700 nm).
5. Sonicate the suspension for 30 min in a Fisherbrand FB11201 bath sonicator (Fisher Scientific) or similar at 37 kHz and full power for 30 min. The water bath temperature is set above the highest main transition temperature of all constituent lipids in the mixture. During the sonication process, the multilamellar vesicles are broken down, leading to reduced turbidity

6. The sonicated lipid suspension is aspirated into the syringe of a mini extruder kit (Avanti Polar Lipids, Inc.) equipped with a 50–100 nm polycarbonate membrane (GE Healthcare Lifesciences). The extrusion equipment is kept above the highest lipid transition temperature as well (room temperature or heated, as required). The suspension is pushed through the extruder membrane 15–31 times. The thus generated unilamellar vesicles (Hope, Bally, Webb, & Cullis, 1985) can be used for immediate SLB deposition

3.2.4 Deposition of SLBs

7. Cleave the mica surface of a sample support by placing a stripe of Scotch magic tape (3 M) on the mica disc and remove it again. The process is repeated until a complete layer of mica is removed from the disc. The freshly cleaved surface is free of contaminations and sufficiently clean for further sample preparation
8. On the freshly cleaved mica substrate, 80 μL of adsorption buffer containing 150 mM NaCl, 20 mM HEPES, 5 mM $CaCl_2$, and 20 mM $MgCl_2$, pH 7.4 are ejected. As outlined earlier, the ideal amount and type of divalent ions in the buffer can vary depending on the experiment. For example, pure DOPG bilayers adsorb fine at 10 mM $CaCl_2$ and in the absence of $MgCl_2$ (Rudd-Schmidt et al., 2019).
9. Using a micro-pipette, add 3–5 μg of the extruded lipid solution (e.g., 3–5 μL of a lipid suspension containing 1 mg/mL lipid) to the buffer on the freshly cleaved mica disc. Mix well by aspirating and re-suspending the buffer and lipids, and incubate for 30 min to allow for SLB formation. Incubation needs to take place above the highest main transition temperature of constituent lipids; if required the samples can be placed on a dedicated heater or directly on an AFM scanner with temperature control. During the incubation, samples should be covered to avoid contamination and buffer evaporation
10. After incubation, the newly formed SLB is washed by adding and removing 80 μL of adsorption buffer (150 mM NaCl, 20 mM HEPES, 5 mM $CaCl_2$, and 20 mM $MgCl_2$, pH 7.4) around 12 consecutive times. The washing steps remove surplus lipid vesicles from the supernatant buffer solution and leave a clean SLB that can be used to inject pore forming proteins, as outlined in Section 5

3.2.5 Further considerations

The vesicle fusion method does not always produce symmetric bilayers; where phospholipid mixtures contain negatively charged phospholipids, these preferentially accumulate on the proximal leaflet of the SLB due to stronger interactions with the negatively charged mica surface via divalent cations (Richter, Maury, & Brisson, 2005). Consequently, the accessible negatively charged lipid in the distal leaflet will be less than the nominal amount. Where phospholipid mixtures contain phospholipids with smaller headgroups (e.g., phosphatidylethanolamine) these preferentially accumulate in the inner leaflet of lipid vesicles and asymmetry in the adsorbing vesicles will likely propagate to asymmetry in the resulting bilayer. SLB preparation via the Langmuir-Blodgett technique allows more precise control over leaflet asymmetry than vesicle fusion, as the two leaflets are deposited separately (Mingeot-Leclercq, Deleu, Brasseur, & Dufrêne, 2008). For especially challenging cases where vesicle fusion fails, solvent assisted bilayer deposition may present another useful alternative (Ferhan et al., 2019).

4. The atomic force microscope

Several different manufacturers offer AFM models that are distinctive regarding the possible sample and scan sizes, scan speeds, and operation modes they offer. The AFMs used to record the data shown here were Bruker MultiMode machines capable of force-distance curve-based imaging (in their proprietary PeakForce tapping). The machines were further equipped with *E*-scanners offering $10 \times 10\,\mu m$ lateral and $2.5\,\mu m$ vertical scanning range or J-scanners (both Bruker) offering $125 \times 125\,\mu m$ lateral and $5\,\mu m$ vertical scanning range. Temperature control is possible by using a scanner mounted Bruker MultiMode sample heater stage, allowing temperatures up to $60\,^{\circ}C$ (at elevated temperatures, the fluid cell needs regular top-ups with milliQ water to prevent the sample from drying). The AFM specific guides and protocols apply to this setup. Of note, for real-time imaging of pore assembly at second time scales or faster, more specialized equipment may be used (Parsons et al., 2019; Ruan et al., 2016), although the overall principles remain the same.

4.1 Setup

The Bruker MultiMode follows a build-up that is similar to many other commercial AFMs and is schematically shown and described in Fig. 6. To measure in fluid, a transparent fluid-cell is used to mount the probe and retain the sample fluid in a flat interface. The probe that is mounted in the fluid cell consists of a sharp tip with a nanometre sized apex, mounted on the free end of a flexible cantilever. The sample to be analysed is mounted on a piezoelectric tube scanner which can move in three spatial dimensions with

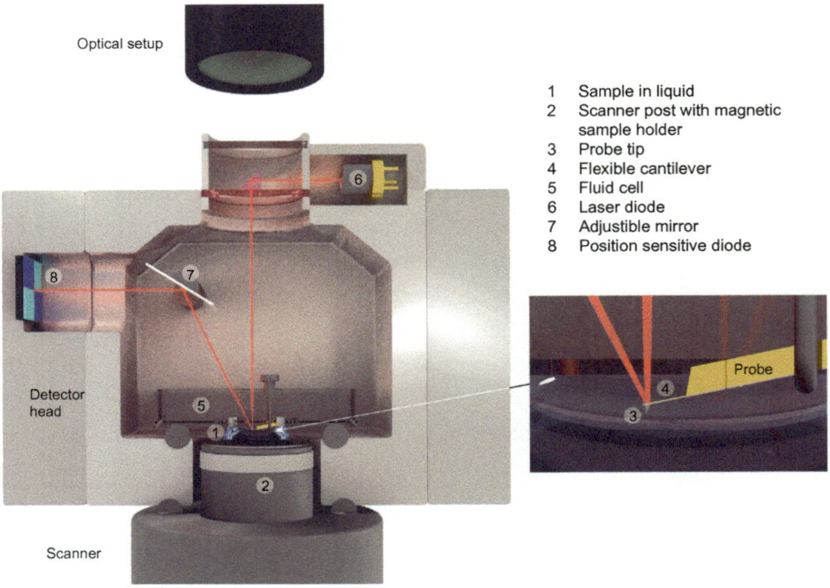

Fig. 6 Schematic setup of a MultiMode type AFM. Its three main parts are the scanner (mounted on a microscope base), the detector head, and an optical setup that aids alignment of, e.g., the laser and the sample. The sample (1) is mounted on the magnetic holder of the scanner post (2). In the liquid above the sample, a probe chip with the tip (3) and cantilever (4) facing the sample surface is placed via a fluid cell (5). The laser emitted from a diode (6) is reflected off the back of the cantilever, and, deflected by an adjustable mirror (7), aligned on a position sensitive diode (8). Drawing not to scale.

sub-nanometre precision. During imaging, the sample is raster-scanned relative to the probe tip. The choice of probe is discussed in more detail in Section 4.3 and typical imaging parameters are given in Section 4.4.

Before the AFM can be operated, probe, sample, fluid cell, and scan head are mounted, and the laser adjusted manually. To this end, the following steps should be taken.

1. Place a freshly cleaved mica disc hydrated with 100 µL buffer solution (150 mM NaCl, 20 mM HEPES, pH 7.4) onto the sample holder
2. Clean the AFM fluid cell using a detergent such as 1% MP Biomedicals™ 7X™-PF Cleaning Solution or similar to remove possible lipid contamination from previous experiments
3. Generously rinse the fluid cell with milliQ water to remove the detergent. The clean can be supplemented by a subsequential rinse with isopropanol and drying under a stream of nitrogen
4. Mount a probe into the AFM fluid cell using a dedicated set of tweezers
5. Place the fluid cell in the detector head and fix it in the clamping mechanism

6. Align the laser with the dedicated alignment knobs (and possibly a deflective mirror, see no. 7 in Fig. 6) on the detector head, such that it reflects off the back of the cantilever and onto the position sensitive diode. Alignment is optimized when the maximum sum signal is obtained
7. Minimize the lateral and vertical voltage readings by aligning the center of the position sensitive diode to the laser spot (as in Fig. 6, no. 8).
8. Allow the system to thermally equilibrate for 30 min
9. Exchange the mica disc for the sample to by analysed. If needed, re-align the position sensitive diode to the laser spot
10. Engage the cantilever and begin imaging

To finish a session, the fluid cell and sample is removed. The fluid cell and tweezer equipment are cleaned as outlined in step no. 2. Used probes are best stored separately to avoid mix-ups.

4.2 Measurement principles and imaging modes

When the probe tip establishes contact with the sample surface, the occurring forces will cause a proportional bending of the cantilever (Fig. 7A). By detecting the deflection of the laser beam that is reflected off the back of the cantilever, the tip-sample interaction forces can be recorded over time, by acquiring force-distance curves during sample engage and retraction from the sample (Fig. 7B). When imaging, the sample surface is scanned in a user defined pixel raster. A typical scanning pattern for a small pixel array is shown in Fig. 7C. The raster is scanned line-by-line, creating a fast and a slow image axis. By default, every line is scanned twice but in the opposite

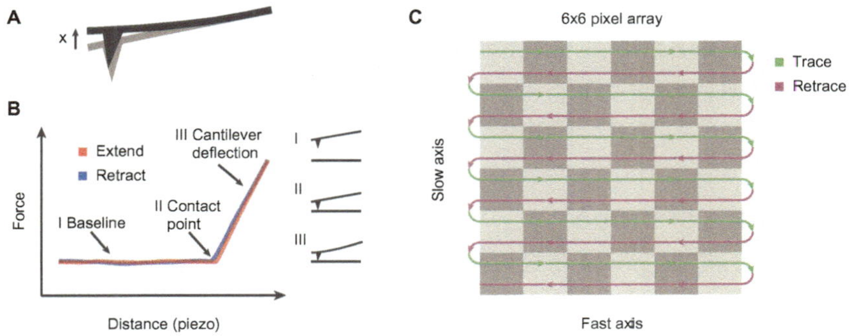

Fig. 7 AFM force-distance curves and scanning pattern. (A) The deflection x of a cantilever is proportional to the force exerted on the probe tip (Hooke's law). (B) Force-distance curves that are recorded by an AFM when a probe is extended toward or retracted from a solid surface. (C) Typical pattern of how a pixel array is raster-scanned during AFM imaging.

direction, generating two separate trace and retrace images. These two images can be used to improve imaging parameters (trace and retrace images should look as similar as possible) or to identify imaging artefacts (more on that in Section 7).

An image is recorded at a user-defined tip-sample interaction quantified by a value called setpoint. The setpoint can either be entered as a relative deflection value in Volts or, when the spring constant and deflection sensitivity is known (e.g., through calibration), as an absolute force (depending on the imaging mode). Different imaging modes can be chosen to define the motion of the probe during the scan. In brief, the main choices are between contact mode, tapping or AC mode, and force-distance curve-based mode. When imaging in contact mode, the cantilever is permanently engaged with the sample surface, with the interaction force kept constant at the setpoint as the tip scans over the sample. In tapping mode, the cantilever is oscillated at its resonance frequency. Here, the setpoint defines a decrease in amplitude compared to the free oscillation as measured just above the surface, and the probe is only in intermittent contact with the sample surface. In force-distance curve-based imaging, the probe is engaged at a frequency that is significantly lower than the resonance frequency of the cantilever. Like in tapping mode, the probe is only in intermittent contact with the sample. However, since the cantilever is not resonating, meaningful force-distance curves can still be recorded and thus the absolute imaging force evaluated at each measurement point, properly referenced to the cantilever deflection at some distance away from the sample surface. Consequently, force-distance curve-based imaging is less prone to drifts in the imaging force, which can be preferable for long experiments, e.g., recording changes on the membrane due to pore forming proteins. In all modes, the electronic feedback compares the measured force or amplitude with the setpoint value and compensates for any changes by moving the piezoelectric scanner closer or further apart from the AFM probe. This vertical scanner movement then represents the local sample height and is used to generate the sample topography in an AFM image.

4.3 Choice of probe

A wide variety of probes for different purposes can be purchased commercially. Some probes can only be used in certain machines or imaging modes. Generally, silicon nitride cantilevers with silicon tips offer the low spring constants and sharp tips needed to image MACPF/CDCs. Good results were obtained with commercial AC-40 (Olympus) or MSNL (Bruker) probes, the latter one being more economical due to its multiple levers available per chip, of which levers E and F could be used consecutively. The parameters of these probes are listed in Table 1.

4.4 Imaging parameters for force-distance curve-based imaging on Bruker MultiMode machines

Table 2 provides basic parameters for imaging profiles to perform PeakForce tapping at 2 kHz with MSNL-E and -F probes. The profile for MSNL-F can also be used on AC-40 probes.

Table 1 Parameters of some common probes used in the study of pore forming proteins with AFM.

Probe	Resonance freq. (liquid) [kHz]	K [N/m]	Tip radius [nm]
MSNL-E	9–16[a]/8–10[b]	0.05–0.20	2–12
MSNL-F	30–53[a]/30–35[b]	0.3–1.4	2–12
AC40	17–45	0.07–0.14	<15

[a]Resonance frequency in liquid estimated as ⅓ of the nominal value in air.
[b]Resonance frequencies determined by in-house calibrations of 8 individual cantilevers from 3 different wafer batches.

Table 2 Basic parameters used to perform PeakForce tapping on pore forming proteins on MultiMode 8 AFM machines, using MSNL-E and -F cantilevers.

Parameter	MSNL-E	MSNL-F
Engage Parameters		
PeakForce Engage Amplitude	10 nm	10 nm
PeakForce Engage Setpoint[a,b]	55 mV	35 mV
Engage int. gain	8.00	8.00
Imaging Parameters		
Scan size	800–2000 nm	800–2000 nm
Scan Rate	0.704 Hz	0.704 Hz
Feedback Gain[a]	25	35
PeakForce Setpoint[a,b]	14 mV	12 mV
PeakForce Amplitude	10 nm	10 nm
PeakForce Frequency	2 kHz	2 kHz
Lift Height	20 nm	20 nm

[a]On a MultiMode 8 AFM system; values may differ in other setups.
[b]On our Multimode 8 system, these force values (in mV), correspond to forces of ca. 200 pN (55 mV) and 50 pN (14 mV) for MSNL-E cantilevers and ca. 300 pN (35 mV) and 100 pN (12 mV) for MSNL-F.

5. Imaging of pore forming proteins

Bare SLBs can be imaged by AFM to check for lipid coverage or phase separation. Pore forming protein is then injected onto an SLB to form the prepore and pore structures that are subject to study.

5.1 Protein injection on SLBs

To make SLBs that contain prepore or pore features, pore forming proteins can be injected either onto the sample mounted in the AFM setup or, for more reproducible results, on an external hot plate using the following steps:

1. Heat the SLB to 37 °C and allow to equilibrate for 5–10 min
2. Dilute the stock protein in adsorption buffer to a volume of 20–40 μL, containing enough protein to yield the desired final concentration
3. Using a micropipette, inject the diluted protein into the supernatant above the SLB, increasing its volume to ca. 100–120 μL. For reproducible results, hold the micropipette upright, inject onto the center of the sample, and do not aspirate liquid after injection
4. Incubate for ca. 15 min at 37 °C to allow the protein to carry out its pore forming function. Incubation times may vary for certain proteins or experiments and are subject to optimization
5. The resulting sample can next be imaged by AFM

5.2 Appearance of pores and imaging of the lumen

Typical AFM images of MACPF/CDC pores show arc- and ring-shaped assemblies scattered on a flat background. Exemplary images for suilysin, perforin, and the MAC are shown in Fig. 8. The pores these proteins form are different in (average) diameter and height above the membrane. Upon closer inspection, differences in the ratio of arcs and rings being formed, arcs occurring in isolation or interlocked with each other, or variabilities in assembly curvatures can be observed. These subtle differences may already hint to differences in the assembly pathways and can be quantified for further discussion (more in Section 6). Indeed, all three proteins shown in Fig. 8 follow fundamentally different assembly pathways (Leung et al., 2014, 2017; Parsons et al., 2019).

In terms of maximum spatial resolution, the individual protein subunits of pore structures formed by different MACPF/CDCs and gasdermin D have been resolved (Czajkowsky et al., 2004; Mulvihill et al., 2018, 2015; Ni et al., 2020). Besides, an AFM tip with a sufficiently high aspect ratio

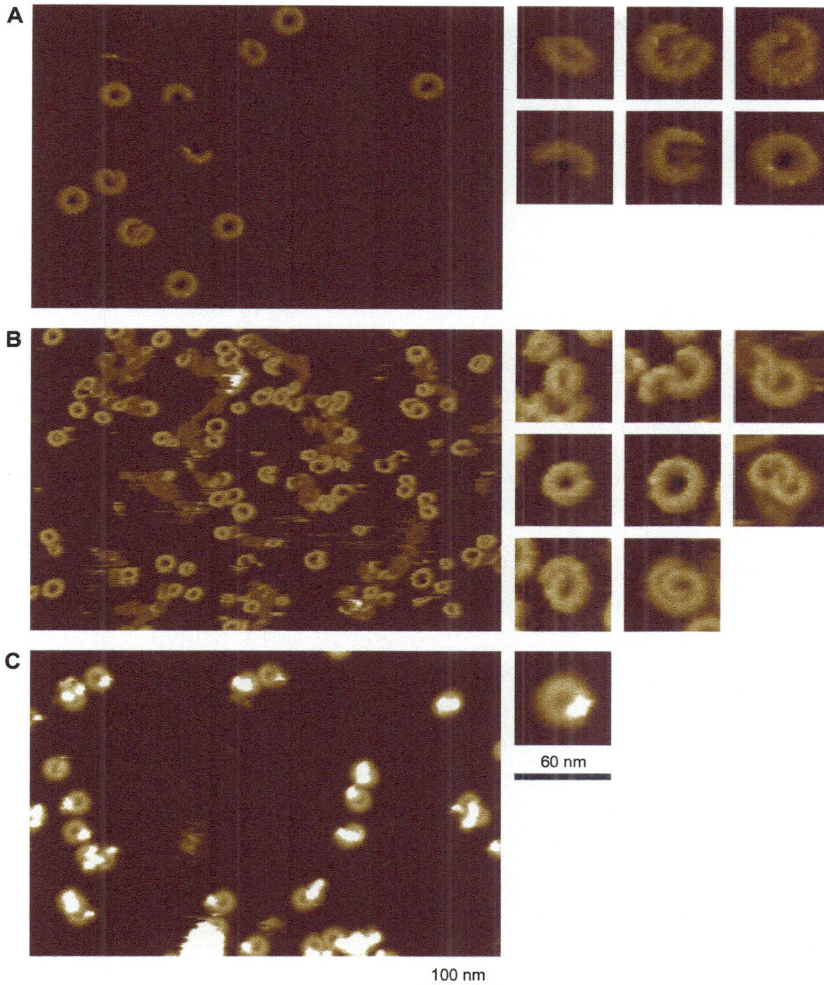

Fig. 8 AFM images of pores formed by three different proteins. (A) Pores formed by the CDC suilysin. A selection of arc- and ring-shaped features formed by suilysin are showed enlarged in separate tiles on the right side. (B) As (A), but for perforin. (C) As (A) and (B), but for the complement components that for the MAC. The color scale (color scale bar as in Fig. 3C) for all images ranges from −9 to 16 nm with the membrane surface set to 0 nm. *The data in panel (C) was reproduced with permission Parsons, E. S., Stanley, G. J., Pyne, A. L. B., Hodel, A. W., Nievergelt, A. P., Menny, A., … Hoogenboom, B. W. (2019). Single-molecule kinetics of pore assembly by the membrane attack complex. Nature Communications, 10, 2066. https://doi.org/10.1038/s41467-019-10058-7.*

Fig. 9 Imaging of the lumen of suilysin pores. (A) AFM image of arc- and ring-shaped pores. In this high contrast color scale (fully represented in the color bar on the top right), the pore lumen is visible in black at the center of a pore. The color scale was set from −5 to 10 nm, with the membrane surface fixed to 0 nm. Excerpt taken from the same raw data as Fig. 8A. (B) Cross section of a 3D render of a suilysin pore exposing its lumen, recorded by AFM. The image is overlaid with an atomic model representation of the suilysin pore. Color scale as in (A). (C) Height profile extracted along the cross-section in (B). *Figure panels (B) and (C) were reproduced with permission Hodel, A. W., Leung, C., Dudkina, N. V., Saibil, H. R., & Hoogenboom, B. W. (2016). Atomic force microscopy of membrane pore formation by cholesterol dependent cytolysins. Current Opinion in Structural Biology, 39, 8–15. https://doi.org/10.1016/j.sbi.2016.03.005.*

can probe the pore lumen and thus confirm the presence of a transmembrane channel (Fig. 9). In real world settings this is possible with large and/or low-profile pores such as CDCs (Leung et al., 2014; Mulvihill et al., 2015; van Pee, Mulvihill, Müller, & Yildiz, 2016), gasdermin D (Mulvihill et al., 2018; Sborgi et al., 2016), or perforin (Leung et al., 2017); the narrower and higher the channels, the more difficult it is to probe the presence/absence of the membrane in their center. High-aspect probes are advertised by several manufacturers but can also be found when screening different batches of standard probes.

5.3 Imaging prepores

Transient prepore assemblies can be observed if the injected protein is prevented from inserting the membrane, for example by using disulfide locked mutants or a pH trigger where applicable. Temperature sensitive proteins such as perforin can, to some extent, also be held in a prepore state at low temperatures (Leung et al., 2017). Otherwise, the protein injection is as described in Section 5.1.

The main difficulty when imaging prepores is their rapid lateral movement and sensitivity to lateral forces exerted by the scanning AFM tip, since they are not stabilized by the contact—across the supported lipid bilayer—with the solid support as pores are. Prepores may thus appear as diffuse streaks in AFM images (Fig. 10A) and for certain proteins visible as plateaus

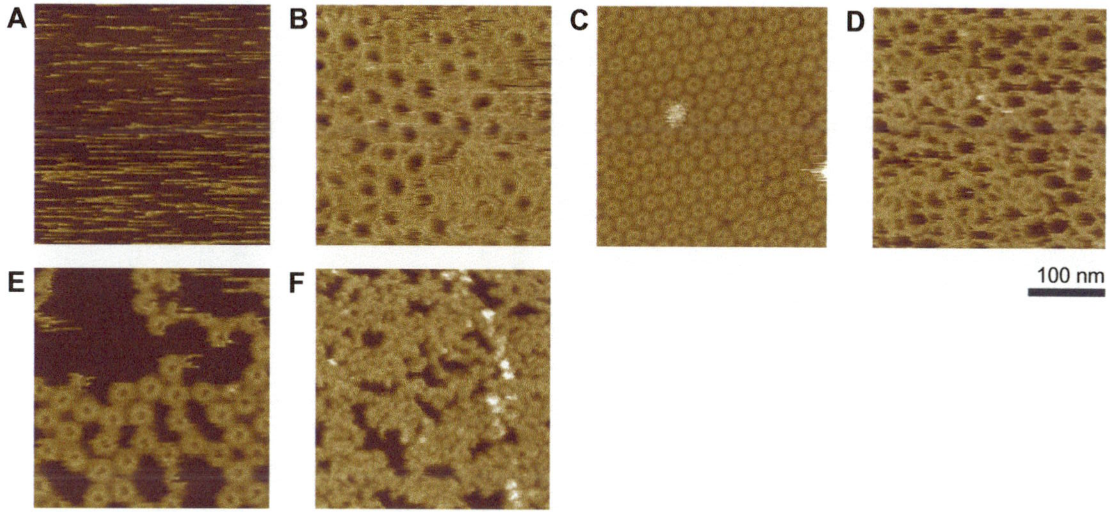

Fig. 10 AFM images of prepores from different proteins and under different conditions. (A) Diffuse streaks caused by mobile suilysin prepores. (B) Suilysin prepore rings and arcs immobilized by dense packing. (C) Soluble, pre-assembled MPEG-1 construct (Pang et al., 2019) bound to a lipid membrane and immobilized by dense packing. (D) Suilysin prepores immobilized by glutaraldehyde fixation. (E) MPEG-1 constructs used in (C), immobilized by glutaraldehyde fixation. (F) Small perforin prepores immobilized by glutaraldehyde fixation. The color scale (color scale bar as in Fig. 3C) for all images ranges from −9 to 16 nm with the membrane surface set to 0 nm. *Figure panels (B, C, E, and F) reproduced with permission Leung, C., Dudkina, N. V., Lukoyanova, N., Hodel, A. W., Farabella, I., Pandurangan, A. P., … Hoogenboom, B. W. (2014). Stepwise visualization of membrane pore formation by suilysin, a bacterial cholesterol-dependent cytolysin. eLife, 3(5), e04247. https://doi.org/10.7554/eLife.04247; Leung, C., Hodel, A. W., Brennan, A. J., Lukoyanova, N., Tran, S., House, C. M., … Hoogenboom, B. W. (2017). Real-time visualization of perforin nanopore assembly. Nature Nanotechnology, 12(5), 467–473. https://doi.org/10.1038/nnano.2016.303; Pang, S. S., Bayly-Jones, C., Radjainia, M., Spicer, B. A., Law, R. H. P., Hodel, A. W., … Whisstock, J. C. (2019). The cryo-EM structure of the acid activatable pore-forming immune effector macrophage-expressed gene 1. Nature Communications, 10, 4288. https://doi.org/10.1038/s41467-019-12279-2.*

that are contained within the phase boundaries of different lipid domains (Hodel et al., 2020; Leung et al., 2014, 2017). There are three established methods to counteract the (effects of) rapid movement of prepores and to enable their visualization. First, the protein can be injected at a high density, such that subsequently formed prepores are densely packed and confined to their location (Fig. 10B and C). Second, the AFM scan speed can be increased on dedicated high-speed AFM systems, sometimes combined with active cooling systems to lower protein diffusivity in addition (Leung et al., 2014). Filmlike data produced by consecutive high-speed AFM imaging is available for some CDC prepores, showing their morphology and diffusivity, including the first real-time visualization of prepore assembly (Ruan et al., 2016). Third, prepores can be cross-linked with a suitable agent like glutaraldehyde to form larger plaques of interconnected assemblies that can be imaged (Fig. 10D and E). The method offers fast results practically independent of protein concentration and aided the interpretations of small prepore assemblies formed by perforin (Fig. 10F). To cross-link pores with glutaraldehyde, the following steps should be considered:

1. Use a fresh vial of 8% glutaraldehyde EM fixative (TAAB Laboratories) or comparable
2. At room temperature (20–25 °C), inject ca. 0.04%–0.12% v/v of glutaraldehyde into the supernatant of a sample containing prepores
3. Incubate for 10 min and wash the supernatant with excess adsorption buffer
4. Image using a new and soft AFM probe such as MSNL-E (Bruker).

5.4 Real-time observation of assembly and pore formation

Transition of the protein into the pore state can be imaged by AFM in real-time and provide further details on membrane insertion and potential growing pore assembly mechanisms and lipid re-arrangements. Crucial to such experiments is that the protein can be triggered to convert to the pore state during imaging, or that pore formation is slow enough to set-up imaging beforehand (Leung et al., 2017; Mulvihill et al., 2015; van Pee et al., 2016). Triggers can be a change in pH by acid injection (Liu et al., 2016; Ni et al., 2020), addition of DTT to disulfide locked mutants (Czajkowsky et al., 2004; Leung et al., 2014, 2017), or injection of protein components required for pore formation (Parsons et al., 2019). The following steps provide a guideline to trigger pore formation by addition of DTT during imaging:

1. Set up imaging of a sample containing a disulfide locked MACPF/CDC. The experiment benefits from fast imaging—optimize scan speed, resolution, scanning area and PeakForce frequency to maximize data output
2. Prepare a 200 mM stock solution of DTT from powder
3. Using a micropipette, immediately inject 2–8 μL of the DTT stock solution into the liquid filled channel in the AFM fluid cell. Physical contact between micropipette and AFM can disturb imaging temporarily
4. Keep imaging until pore formation is complete or the AFM probe is not suitable for imaging anymore

The amount of data that can be captured in real-time depends on the balancing between image resolution and scanning speed. Using a MultiMode AFM and 2 kHz PeakForce tapping, images can be captured within the order of minutes (Fig. 11A). Such scan speeds may suffice to follow major structural transitions of assemblies in larger populations, and extensive lipid rearrangements. However, there is a clear trend in research toward faster imaging to resolve dynamic events within single assemblies. For example, by recording subsequent frames at second time scale, it is possible to follow the growth of the MAC pore as additional copies of C9 insert into the membrane (Fig. 11B). Imaging at second or sub-second timeframes requires specialized AFM machines capable to scan at high frequencies, and cantilevers that can be stably operated under these conditions.

6. Data analysis

Raw AFM data is post-processed to produce representable images and extract quantitative data. Processing can be performed using manufacturer supplied software or open-source software such as Gwyddion (gwyddion.net) (Nečas & Klapetek, 2012). The following examples of the most important steps of image processing are produced using the NanoScope Analysis software version 2.0 (Bruker). More intricate analysis may require specialized tools beyond standard software. As an example, manual tracing and extraction of pore features is discussed at the end of this section.

6.1 Image flattening

Raw AFM data is often not leveled; during scanning the sample may be tilted, and the scanner movement induces a bow distortion that needs correction. Furthermore, a height reference point is to be defined and set to zero. AFM data can be flattened accordingly, with respect to a flat reference

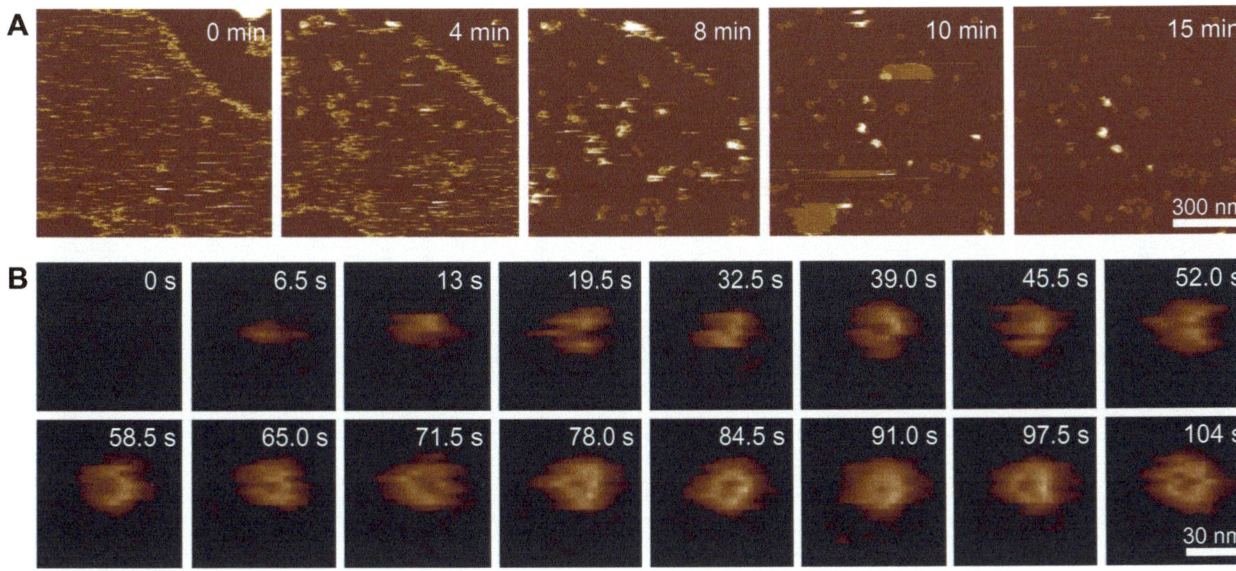

Fig. 11 Real-time AFM data of pore formation at different temporal resolutions. (A) Image sequence depicting disulfide locked suilysin prepore to pore conversion after addition of DTT. Note that the protein is confined to a cholesterol-rich membrane domain. (B) Gradual growth of a MAC pore shown in real time with high-speed AFM. The full range of the color scale (color scale bar as in Fig. 3B) is set to 35 nm in A and 16 nm in B. *Both image sequences were reproduced with permission Leung, C., Dudkina, N. V., Lukoyanova, N., Hodel, A. W., Farabella, I., Pandurangan, A. P., … Hoogenboom, B. W. (2014). Stepwise visualization of membrane pore formation by suilysin, a bacterial cholesterol-dependent cytolysin. eLife, 3(5), e04247. https://doi.org/10.7554/eLife.04247; Parsons, E. S., Stanley, G. J., Pyne, A. L. B., Hodel, A. W., Nievergelt, A. P., Menny, A., … Hoogenboom, B. W. (2019). Single-molecule kinetics of pore assembly by the membrane attack complex. Nature Communications, 10, 2066. https://doi.org/10.1038/s41467-019-10058-7.*

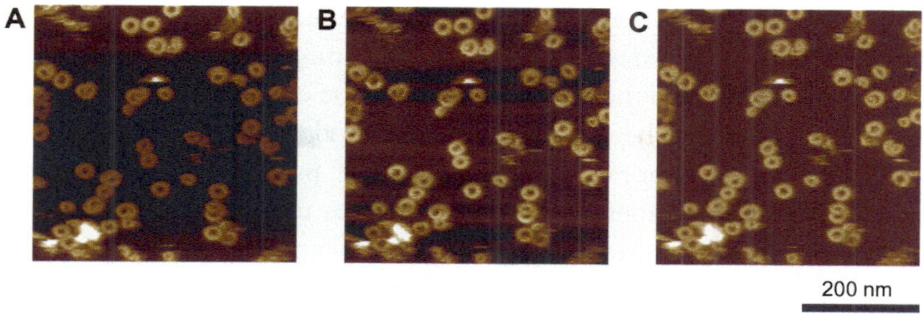

Fig. 12 AFM image transformation during the flattening process. (A) Raw AFM image without flattening applied. (B) AFM image flattened by applying a first order polynomial fit to each line, without using masks. (C) AFM image flattened additionally by applying a second order polynomial fit to each line, and pore features masked out for the fits by a height threshold.

Table 3 Steps and parameters suggested to flatten AFM images containing pore assemblies.

Parameter	1st step	2nd/3rd steps
Flatten Order	1st or 2nd	2nd
Flatten Z Thresholding Direction	No Thresholding	Use Z <
Find Threshold for	The whole image	The whole image
Mark Excluded Data	Yes	Yes
Number Histogram Binds	512	512
Threshold Height	0 nm[a]	2 nm
Use Histogram	Off	Off

[a]Protein aggregations or membrane discontinuities may need to be masked manually by tracing a rectangle on the image display.

surface, commonly the lipid substrate. A good strategy to flatten AFM images with pores is shown in Fig. 12, with software parameters listed in Table 3. First, a first or second order polynomial fit is applied to each line of the scanned topography image. The thus initially generated image overall looks flatter than the raw data but shows artefactual depressions (darker areas) in the membrane background. These arise when the protruding protein assemblies are taken into account for estimating the mean background profile (Fig. 12A and B). To avoid such artefacts, the flattening procedure can be repeated twice with a 2 nm height threshold when estimating the to-be-subtracted background. With the height threshold enabled, the protein features are ignored when applying the polynomial fit, and only the flat reference surface is considered. In the resulting image, the membrane surface appears uniform in height/color (Fig. 12C). The outlined flattening strategy works well on data where there is abundant and flat membrane. Flattening procedures can be more complex in case of protein aggregations or of discontinuities in the membrane, and of high or irregular protein coverage. In such cases, the masking of features for the background fitting can be done manually, and further adjustments of the flattening parameters may be required.

6.2 Color scale

In AFM images, local sample height is represented by a color scale. To facilitate comparison between different images, the color scale should incorporate a fixed height range and ideally also a reference height, commonly the membrane surface. The height range and the colors used in the color scale should be selected such that they enable clear visualization of the surface features. A commonly used color scale for AFM data, as in Fig. 3C, is well suitable for pore forming proteins: the shades of yellow and red can be used to represent the protein and membrane surface, while potential defects or aggregations at the extreme ends or outside the color scale appear black and white, respectively. However, to emphasize on different topographical features, it can be useful to change the color scale such as in Fig. 9. Overall, it is common practice to use a nonlinear and saturated color scale to best maximize the color contrast for features of interest at the sample surface. This nonlinearity should be taken into account when interpreting AFM images.

6.3 Height profiles

Protein height is a key characteristic that AFM can record at high (Ångström) resolution and can be represented in height profiles. Height profiles are extracted from (flattened) AFM data by taking a cross-section across the recorded topography. Ideally, the cross-section is taken along the fast scan axis as lines can display (small) height shifts caused by drift during scanning or inadequate flattening. Besides protein features, a height profile should include sufficient data from the membrane surface, as a suitable reference height. An example of a height profile can be found in Fig. 9C.

6.4 Feature tracing

To quantify the sizes of protein assemblies and the amount of proteins at the surface, it is important to systematically identify and trace relevant features in AFM data. This can be done, e.g., manually in the IMOD software v4.9.12 (BL3DEMC &Regents of the University of Colorado) (Kremer, Mastronarde, & McIntosh, 1996). Single images or image stacks can be loaded, and the features traced on a separate image layer. Different types of traces allow to distinguish between arcs and rings, features apparent in different images, or other properties. The length of the traces can be exported for further analysis and presented, for example, in assembly size histogram. An example of traced arcs and rings in AFM data is provided in Fig. 13.

Where possible, automated tracing approaches are favored over manual tracing, or more advanced image processing methods (e.g., artificial intelligence) may be used. However, the required software often needs adaptation to recognize and trace specific features appropriately. An example of a customizable toolkit for automated tracing is the recently introduced python toolkit TopoStats (Beton et al., 2020).

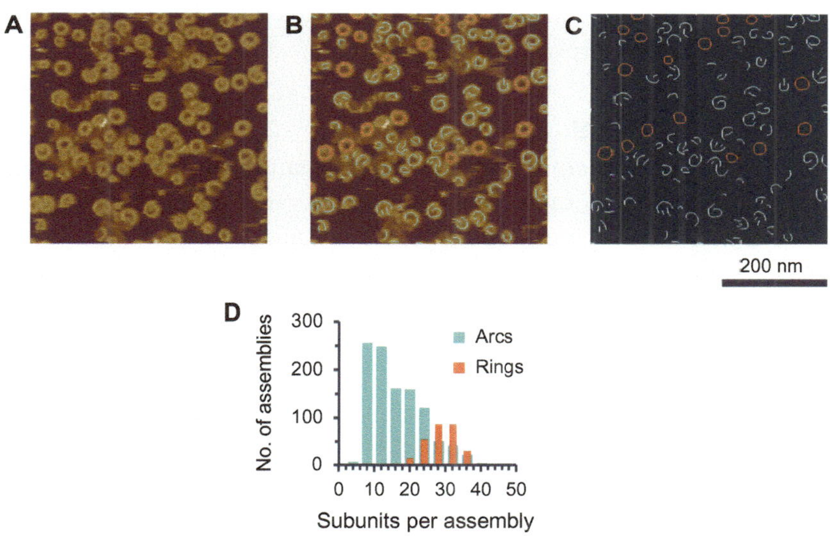

Fig. 13 Tracing of arc- and ring-shaped pores. (A) AFM image showing perforin pore features to be traced. (B) AFM image overlaid with manual traces. Arcs are traced in light blue, and rings in orange. (C) Image showing traces only. (D) Histogram of assembly sizes. The histogram is made from a larger dataset that includes the traces shown in (B and C).

7. Imaging artefacts and sample defects

Imaging artefacts and sample defects can lower the quality of AFM images and compromise their interpretation. Below is a compilation of commonly occurring artefacts when imaging pore forming proteins, and how they can be avoided, illustrated by the example of perforin.

7.1 Probe contamination

The tips of AFM probes can get contaminated, e.g., by protein sticking to it, sometimes within seconds after starting a scan. Small additional protrusions or contaminations can act like a secondary probe, causing sample features to be artefactually repeated in the images (see Fig. 14). Because this is determined by the geometry of the contaminated tip, such artefacts are readily detected by noting how sample features always repeat in the same direction; for example, this could be the appearance of a shadow image at the same side of each sample feature. When such tip contamination occurs, the imaging can be interrupted to remove the fluid cell with the cantilever, rinse it in milliQ water, and wash the sample with imaging buffer. If that is to no avail, the cantilever needs replacement.

7.2 Tip broadening

In addition (and related) to the possible contamination issues mentioned above, the inherent geometry of AFM probes is variable at nanometre length scales. Even for the same probe type and specification, some tips may be broader or otherwise differently shaped than others, due to variability in the manufacturing process or—though good practice should avoid this—due to damage incurred during the experiment. The tip shape affects the apparent dimensions of pores in images (as any AFM image represents a convolution of the sample surface with the tip geometry), and the same pores may therefore appear different when imaged with different tips (Fig. 15). Of note, AFM probes produced from the same wafer usually show good homogeneity and using AFM probes with the same wafer number can help to minimize variability in the results due to tip broadening.

7.3 Parachuting, snap-in, and adhesion

Some artefacts are caused by unstable imaging conditions. Parachuting is an artefact that emerges when the electronic or mechanical feedback is not fast

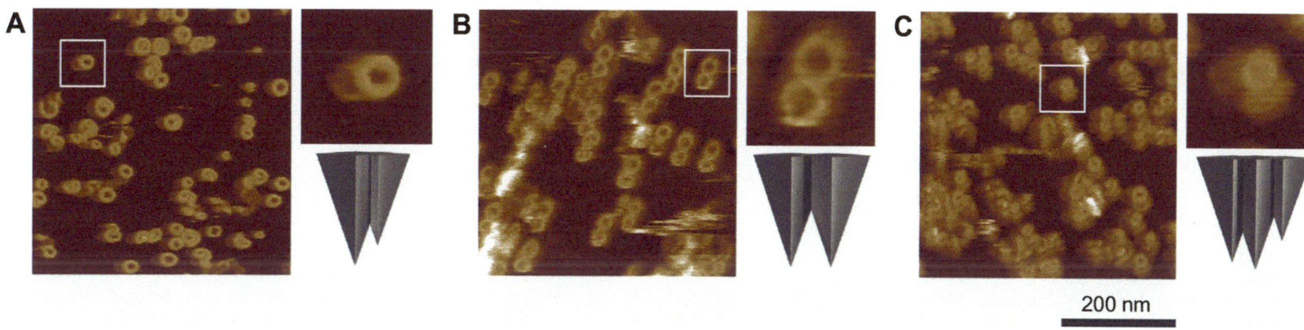

Fig. 14 Tip artefacts can occur when the tip of the AFM probe is contaminated. (A) Slight double tip causing shadow-like features on the left side of pore structures. A magnified view of the outlined area and schematic representation of the tip is provided on the right side of the AFM image. (B) Severe case of a double tip, leading to all pore structures appearing doubled in the same direction. (C) Severe case of a multi-tip, where the pore structures can hardly be distinguished due to the convolution with the contaminated probe tip.

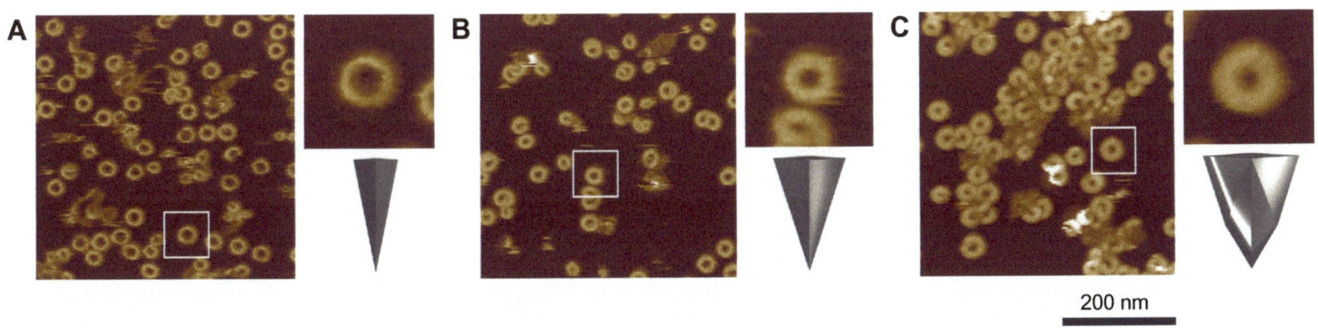

Fig. 15 Tip broadening artefacts. Tip broadening is visible in increased magnitude from (A) to (C). A detail image of the outlined square and a schematic illustration of the tip shape is provided on the right side of the AFM images in each Figure panel.

enough to adjust the tip-sample distance as the probe scans the sample: after passing an elevated obstacle, the AFM tip will lose surface contact for a limited time before the feedback loop brings it back into contact. By default, most AFM controllers record every scan line twice, from left to right (trace) and from right to left (retrace), which acts as a control for artefacts. Since parachuting effects depend on the scan direction, they appear at opposite sides of the sample features in trace and retrace images (Fig. 16A). They can be avoided by increasing the electronic gain levels in the AFM software or, if increased gains lead to instabilities, by decreasing the scan speed.

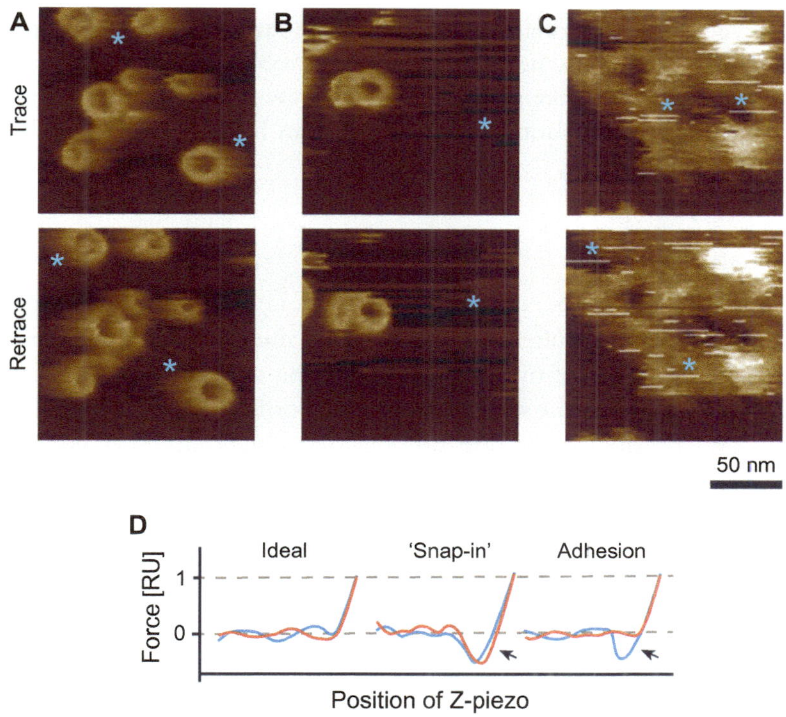

Fig. 16 Parachuting, snap-in, and adhesion artefacts. (A) Images of AFM pores showing parachuting artefacts. The artefacts point in different directions in trace and retrace images related to the direction of scanning. (B) Snap-in artefacts visible as dark streaks in a bilayer, caused by the probe tip making inadvertent contact with the solid substrate. (C) Adhesion artefacts on fixed prepore structures, visible as tall (white), horizontal streaks that change their location between Trace and Retrace images. Asterisks in (A–C) locate examples of the respective artefacts. (D) Force curves (colored red during extend and blue during retract phase, as in Fig. 7B) exemplifying the presence of snap-in and adhesion artefacts during imaging, compared to an ideal force curve.

Snap-in occurs when the probe tip experiences a strong adhesion to the surface (locally or generally). This makes it harder to control the tip-sample interaction, compromising the image quality and risking degradation of sample and/or tip. For example, the tip may cut through the SLB and image the underlying mica substrate instead. This behavior is recognized when defects appear in the lipid surface and the force-distance curves show a negative forces in during imaging (Fig. 16B and D). Re-engaging the probe and restarting the scan might alleviate the problem and replacing the probe may be necessary. Excessive tip-sample adhesion causes artefacts that appear as tall features on AFM recordings that do not produce a continuous shape from one line of the scan to the next, and that may not be locally conserved, e.g., between trace and retrace images (Fig. 16C and D). Adhesion is common when a contaminated probe is used and, for example, a cross-linking agent like glutaraldehyde is present in the sample. Repeated washing steps and mounting a new probe can help to minimize adhesion effects.

7.4 Sample drift

If the microscope is not equilibrated properly, i.e., still changing its temperature, if the sample buffer is being depleted due to evaporation, or if the scanner shows signs of hysteresis or creep, the sample area of interest can gradually drift out of the scanning area. Such drift can cause sample features to appear elongated in the direction of drift (Fig. 17A). To avoid this, it is important to allow the microscope/scanner to adjust to changing temperatures prior to imaging, to always ensure that enough buffer is present, especially when measuring at elevated temperatures, and, depending on scanner properties, to let the scanner repeatedly scan the same area.

7.5 Protein aggregation

Protein aggregations of the sample surface are common and can hardly be avoided for some pore forming proteins, while others yield cleaner samples. Aggregates may contain proteins or protein/lipid mixtures (e.g., protein-covered vesicles), and an example image is provided in Fig. 17B. Washing steps can be repeated, and incubation times shortened to reduce the amount of aggregated protein at the surface. The protein may have also degraded and need to be replaced.

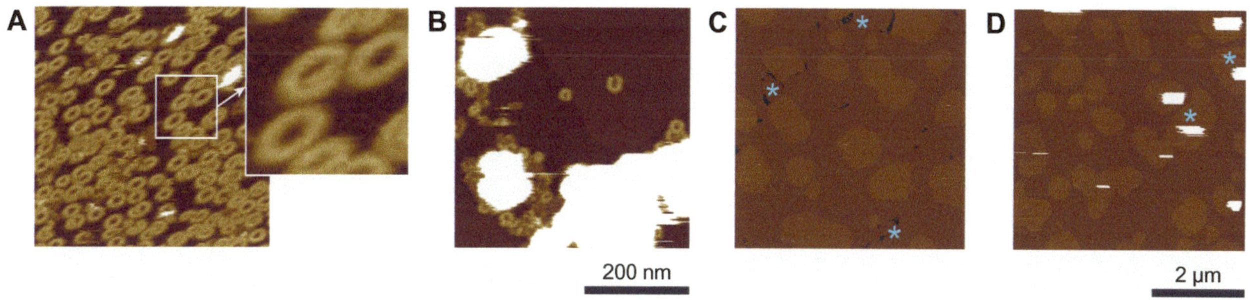

Fig. 17 Drift, protein, and SLB artefacts and defects. (A) Drift visible on an AFM image of pore structures, including a magnified view of the outlined square area. The pores appear elongated from the bottom left to the top right corner. (B) Aggregation on top of pore structures. Their height is beyond the color scale used here; hence they appear as white areas without any features. (C) SLB showing phase separation and incomplete coverage of the underlying substrate, visible as gaps in the lower (darker) phase domains. (D) SLB showing lipid condensation on top of the higher (lighter) phase domains. Blue asterisks point out some of the defects in panels (C and D).

7.6 SLB defects

Depositions of lipid vesicles at inadequate amounts or under unsuitable conditions may lead to artefacts in the SLB. Insufficient amounts of lipid added on a mica substrate leads to incomplete coverage with an SLB (Fig. 17C), and pore forming proteins tend to sequester at lipid boundaries. Excess lipid may lead to transient condensation of lipid layers on an existing SLB (Fig. 17D). SLB defects can also develop over time, for example, at low divalent ion concentrations. If the defects are present all over the sample surface, new samples need to be prepared under optimized conditions. Otherwise, changing the scanning area can help to avoid them.

8. Complementarity to other methods

Like any technique, AFM has its strengths and limitations. Hence, the study of new proteins may pose challenges that can be solved by combining data obtained by different techniques. The following is a brief list of some methods that have been used alongside AFM in the study of MACPF/CDCs, briefly outlining the complementary information they can provide.

8.1 Lysis and membrane perforation assays

Lysis assays can measure and quantify pore formation and other defects that void the integrity of free-standing (i.e., unlike AFM experiments, unsupported) model membranes and live cell membranes over time. Lysis can be detected via the diffusion of a marker through cellular or vesicular membranes or by cell death or vesicle bursting. Lysis assays are usually bulk measurements involving large numbers of pores and can be used to test the lytic function of a protein or its specificity to different targets. Somewhat similarly, membrane perforation can also be studied by electrophoresis, where measurements on a smaller number of assemblies and at high temporal resolution provided early indications of arc-shaped pores (El-Rachkidy, Davies, & Andrew, 2008; Praper, Sonnen, Viero, et al., 2011). Lysis assays can complement AFM studies, for example, with bulk measurements of pore forming activity and data gathered on live cells.

8.2 Quartz-crystal microbalance with dissipation monitoring

Quartz-crystal microbalance with dissipation monitoring (QCM-D) is used to measure deposition of lipids and proteins averaged over the surface of a quartz crystal oscillator. In brief, such deposition causes a shift in the oscillator resonance frequency that increases with the deposited mass (including hydrodynamically coupled solvents) as well as a dampening of the oscillation due to energy dissipation into the deposited layer that indicates changes in viscoelastic properties. The method is useful to identify binding and pore forming steps, and has been employed, for example, in the study of the heterogeneous MAC pore (Parsons et al., 2019). Silicon oxide coated QCM-D sensors are compatible with bilayer deposition via vesicle fusion allowing matching experimental conditions with AFM and, therefore, direct comparison of results. QCM-D can pick up events that are difficult to capture or interpret with AFM alone, because the AFM tip perturbs weakly-bound matter, such as bound monomers or transiently bound proteins or lipid vesicles.

8.3 Labeled MACPF/CDCs

Protein labels can be used to visualize MACPF/CDCs in complex environments such as (live) cells. Fluorescent tags are useful to co-localize the proteins with potential membrane targets (Johnson et al., 2012; Praper, Sonnen, Kladnik, et al., 2011; Rudd-Schmidt et al., 2019), or to monitor protein binding and cell morphology in flow cytometry experiments (Heesterbeek et al., 2019; Rudd-Schmidt et al., 2019). Isotope labels have been employed to study the specificity of CDCs to different pools of cholesterol available in the cell membrane (Das, Brown, Anderson, Goldstein, & Radhakrishnan, 2014). Compared with AFM, the advantages are enhanced temporal resolution, chemically specific detection, and better detection of matter that is otherwise easily perturbed by the interaction with the scanning AFM probe.

8.4 Negative stain electron microscopy of lipid monolayers

Lipid monolayer preparations containing MACPF/CDCs at various stages of pore formation can be transferred to carbon grids for electron microscopy (EM) and negatively stained. EM images of such samples routinely resolve the individual subunits of prepore and pore assemblies; see, for example, (Boyd et al., 2016; Law et al., 2010; Leung et al., 2014; Vadia et al., 2011). Furthermore, the absence of a solid support may be favorable when studying kinetic properties of a protein, as diffusion is not limited through physical contact with the substrate after inserting into the membrane. Traces and subunit counts of fixed prepore and pore assemblies can thus add quantitative data to complement AFM experiments recorded on bilayers in solution and help to control for dependence of observed features on the exact sample preparation method.

8.5 Cryo-EM reconstructions

Cryo-EM is used to build 3D structural models of protein assemblies that are associated with lipid vesicles, spontaneously formed in solution, or resolubilized after assembly on the membrane. Reconstructions are predominantly done for ring-shaped prepores (Leung et al., 2014; Lukoyanova et al., 2015; Ni et al., 2020; Pang et al., 2019; Tilley, Orlova, Gilbert, Andrew, & Saibil, 2005; van Pee et al., 2017) and pores (Dudkina et al., 2016; Law et al., 2010; Leung et al., 2014; Lukoyanova et al., 2015; Ni et al., 2018; Pang et al., 2019; Serna, Giles, Morgan, & Bubeck, 2016; van Pee et al., 2017), although arcs have been reconstructed too (Sonnen et al., 2014). AFM data on (pre)pores can be corroborated by direct comparison of AFM topography with 3D structural models (see Fig. 9B).

8.6 Computational and theoretical modeling of assembly

Theoretical or computational models of the pore assembly can predict experimentally measured pore size distributions (Leung et al., 2014) and can offer details on individual assembly steps that remain unresolved in AFM data. It can be used, for example, to determine how many subunits are added at a time during assembly, and to help identifying the critical kinetic steps involved (Parsons et al., 2019). There are several such computational models available in literature (Agrawal, Apoorva, & Ayappa, 2017; Lee, Senior, Wallace, Woolley, & Griffiths, 2016; Leung et al., 2014). Computational approaches have also been used to simulate the lipid behavior in the lumen of newly formed pores (Desikan, Padmanabhan, & Ganapathy Ayappa, 2020; Vögele et al., 2019).

Acknowledgments

This work was supported by the Biotechnology and Biological Sciences Research Council (BBSRC, BB/N015487/1, to B.W.H.) and the Swiss National Science Foundation (p2skp3_187634 to A.W.H.). We thank R. Thorogate for help in preparing Fig. 5.

References

Agrawal, A., Apoorva, K., & Ayappa, K. G. G. (2017). Transmembrane oligomeric intermediates of pore forming toxin Cytolysin a determine leakage kinetics. *RSC Advances*, 7(82), 51750–51762. https://doi.org/10.1039/c7ra07304f.

Beton, J. G., Moorehead, R., Helfmann, L., Gray, R., Hoogenboom, B. W., Praveen Joseph, A., et al. (2020). TopoStats—An automated tracing program for AFM images. *BioRxiv*, 2020.09.23.309609. https://doi.org/10.1101/2020.09.23.309609.

Boyd, C. M., Parsons, E. S., Smith, R. A. G., Seddon, J. M., Ces, O., & Bubeck, D. (2016). Disentangling the roles of cholesterol and CD59 in intermedilysin pore formation. *Scientific Reports*, *6*, 38446. https://doi.org/10.1038/srep38446.

Broz, P., Pelegrín, P., & Shao, F. (2020). The gasdermins, a protein family executing cell death and inflammation. *Nature Reviews Immunology*, *20*(3), 143–157. https://doi.org/10.1038/s41577-019-0228-2.

Connell, S. D., & Smith, D. A. (2006). The atomic force microscope as a tool for studying phase separation in lipid membranes (review). *Molecular Membrane Biology*, *23*(1), 17–28. https://doi.org/10.1080/09687860500501158.

Czajkowsky, D. M., Hotze, E. M., Shao, Z., & Tweten, R. K. (2004). Vertical collapse of a cytolysin prepore moves its transmembrane β-hairpins to the membrane. *The EMBO Journal*, *23*(16), 3206–3215. https://doi.org/10.1038/sj.emboj.7600350.

Das, A., Brown, M. S., Anderson, D. D., Goldstein, J. L., & Radhakrishnan, A. (2014). Three pools of plasma membrane cholesterol and their relation to cholesterol homeostasis. *eLife*, *3*, e02882. https://doi.org/10.7554/eLife.02882.

Desikan, R., Padmanabhan, P., & Ganapathy Ayappa, K. (2020). Opening of smaller toxin pores by lipid micelle formation. *Proceedings of the National Academy of Sciences of the United States of America*, *117*(10), 5107–5108. https://doi.org/10.1073/pnas.1921467117.

Dudkina, N. V., Spicer, B. A., Reboul, C. F., Conroy, P. J., Lukoyanova, N., Elmlund, H., et al. (2016). Structure of the poly-C9 component of the complement membrane attack complex. *Nature Communications*, 7, 10588. https://doi.org/10.1038/ncomms10588.

El-Rachkidy, R. G., Davies, N. W., & Andrew, P. W. (2008). Pneumolysin generates multiple conductance pores in the membrane of nucleated cells. *Biochemical and Biophysical Research Communications*, *368*(3), 786–792. https://doi.org/10.1016/j.bbrc.2008.01.151.

Ferhan, A. R., Yoon, B. K., Park, S., Sut, T. N., Chin, H., Park, J. H., et al. (2019). Solvent-assisted preparation of supported lipid bilayers. *Nature Protocols*, *14*(7), 2091–2118. https://doi.org/10.1038/s41596-019-0174-2.

Giddings, K. S. K., Zhao, J., Sims, P. P. J., & Tweten, R. K. R. (2004). Human CD59 is a receptor for the cholesterol-dependent cytolysin intermedilysin. *Nature Structural & Molecular Biology*, *11*(12), 1173–1178. https://doi.org/10.1038/nsmb862.

Heesterbeek, D. A., Bardoel, B. W., Parsons, E. S., Bennett, I., Ruyken, M., Doorduijn, D. J., et al. (2019). Bacterial killing by complement requires membrane attack complex formation via surface-bound C5 convertases. *The EMBO Journal*, *38*, e99852. https://doi.org/10.15252/embj.201899852.

Heuck, A. P., Hotze, E. M., Tweten, R. K., & Johnson, A. E. (2000). Mechanism of membrane insertion of a multimeric β-barrel protein: Perfringolysin O creates a pore using ordered and coupled conformational changes. *Molecular Cell*, *6*(5), 1233–1242. https://doi.org/10.1016/S1097-2765(00)00119-2.

Hodel, A. W., Rudd-Schmidt, J. A., Trapani, J. A., Voskoboinik, I., & Hoogenboom, B. W. (2020). Lipid specificity of the immune effector perforin. *Faraday Discussions*. https://doi.org/10.1039/d0fd00043d.

Hope, M. J., Bally, M. B., Webb, G., & Cullis, P. R. (1985). Production of large unilamellar vesicles by a rapid extrusion procedure. Characterization of size distribution, trapped volume and ability to maintain a membrane potential. *Biophysica et Biophysica Acta*, *812*(1), 55–65. https://doi.org/10.1016/0005-2736(85)90521-8.

Hotze, E. M., Wilson-Kubalek, E. M., Rossjohn, J., Parker, M. W., Johnson, A. E., & Tweten, R. K. (2001). Arresting pore formation of a cholesterol-dependent cytolysin by disulfide trapping synchronizes the insertion of the transmembrane β-sheet from a prepore intermediate. *The Journal of Biological Chemistry*, *276*(11), 8261–8268. https://doi.org/10.1074/jbc.M009865200.

Johnson, B. B., Moe, P. C., Wang, D., Rossi, K., Trigatti, B. L., & Heuck, A. P. (2012). Modifications in perfringolysin O domain 4 alter the cholesterol concentration threshold required for binding. *Biochemistry*, *51*(16), 3373–3382. https://doi.org/10.1021/bi3003132.

Kremer, J. R., Mastronarde, D. N., & McIntosh, J. R. (1996). Computer visualization of three-dimensional image data using IMOD. *Journal of Structural Biology*, *116*, 71–76.

Laemmli, U. K. (1970). Cleavage of structural proteins during the assembly of the head of bacteriophage T4. *Nature*, *227*(5259), 680–685. https://doi.org/10.1038/227680a0.

Law, R. H. P., Lukoyanova, N., Voskoboinik, I., Caradoc-Davies, T. T., Baran, K., Dunstone, M. A., et al. (2010). The structural basis for membrane binding and pore formation by lymphocyte perforin. *Nature*, *468*(7322), 447–451. https://doi.org/10.1038/nature09518.

Lee, A. A., Senior, M. J., Wallace, M. I., Woolley, T. E., & Griffiths, I. M. (2016). Dissecting the self-assembly kinetics of multimeric pore-forming toxins. *Interface*, *13*(114), 20150762. https://doi.org/10.1098/rsif.2015.0762.

Leung, C., Dudkina, N. V., Lukoyanova, N., Hodel, A. W., Farabella, I., Pandurangan, A. P., et al. (2014). Stepwise visualization of membrane pore formation by suilysin, a bacterial cholesterol-dependent cytolysin. *eLife*, *3*(5), e04247. https://doi.org/10.7554/eLife.04247.

Leung, C., Hodel, A. W., Brennan, A. J., Lukoyanova, N., Tran, S., House, C. M., et al. (2017). Real-time visualization of perforin nanopore assembly. *Nature Nanotechnology*, *12*(5), 467–473. https://doi.org/10.1038/nnano.2016.303.

Liu, X., Zhang, Z., Ruan, J., Pan, Y., Magupalli, V. G., Wu, H., et al. (2016). Inflammasome-activated gasdermin D causes pyroptosis by forming membrane pores. *Nature*, *535*(7610), 153–158. https://doi.org/10.1038/nature18629.

Lukoyanova, N., Hoogenboom, B. W., & Saibil, H. R. (2016). The membrane attack complex, perforin and cholesterol-dependent cytolysin superfamily of pore-forming proteins. *Journal of Cell Science*, *129*(11), 2125–2133. https://doi.org/10.1242/jcs.182741.

Lukoyanova, N., Kondos, S. C., Farabella, I., Law, R. H. P., Reboul, C. F., Caradoc-Davies, T. T., et al. (2015). Conformational changes during pore formation by the perforin-related protein pleurotolysin. *PLoS Biology*, *13*(2), e1002049. https://doi.org/10.1371/journal.pbio.1002049.

Mingeot-Leclercq, M.-P., Deleu, M., Brasseur, R., & Dufrêne, Y. F. (2008). Atomic force microscopy of supported lipid bilayers. *Nature Protocols*, *3*(10), 1654–1659. https://doi.org/10.1038/nprot.2008.149.

Mulvihill, E., Sborgi, L., Mari, S. A., Pfreundschuh, M., Hiller, S., & Müller, D. J. (2018). Mechanism of membrane pore formation by human gasdermin-D. *The EMBO Journal*, *37*, e98321. https://doi.org/10.15252/embj.201798321.

Mulvihill, E., van Pee, K., Mari, S. A., Müller, D. J., Yildiz, Ö., Van Pee, K., et al. (2015). Directly observing the lipid-dependent self-assembly and pore-forming mechanism of the cytolytic toxin listeriolysin O. *Nano Letters*, *15*(10), 6965–6973. https://doi.org/10.1021/acs.nanolett.5b02963.

Nečas, D., & Klapetek, P. (2012). Gwyddion: An open-source software for SPM data analysis. *Central European Journal of Physics*, *10*(1), 181–188. https://doi.org/10.2478/s11534-011-0096-2.

Ni, T., Harlos, K., & Gilbert, R. (2016). Structure of astrotactin-2: A conserved membrane protein involved in neuronal development. *Open Biology*, *6*, 160053. https://doi.org/10.1098/rsob.160053.

Ni, T., Jiao, F., Yu, X., Aden, S., Ginger, L., Williams, S. I., et al. (2020). Structure and mechanism of bactericidal mammalian perforin-2, an ancient agent of innate immunity. *Science Advances*, *6*(5), eaax8286. https://doi.org/10.1126/sciadv.aax8286.

Ni, T., Williams, S. I., Rezelj, S., Anderluh, G., Harlos, K., Stansfeld, P. J., et al. (2018). Structures of monomeric and oligomeric forms of the toxoplasma gondii perforin-like protein 1. *Science Advances*, *4*(3), eaaq0762. https://doi.org/10.1126/sciadv.aaq0762.

Pang, S. S., Bayly-Jones, C., Radjainia, M., Spicer, B. A., Law, R. H. P., Hodel, A. W., et al. (2019). The cryo-EM structure of the acid activatable pore-forming immune effector macrophage-expressed gene 1. *Nature Communications*, *10*, 4288. https://doi.org/10.1038/s41467-019-12279-2.

Parsons, E. S., Stanley, G. J., Pyne, A. L. B., Hodel, A. W., Nievergelt, A. P., Menny, A., et al. (2019). Single-molecule kinetics of pore assembly by the membrane attack complex. *Nature Communications*, *10*, 2066. https://doi.org/10.1038/s41467-019-10058-7.

Podobnik, M., Marchioretto, M., Zanetti, M., Bavdek, A., Kisovec, M., Cajnko, M. M., et al. (2015). Plasticity of Lysteriolysin O pores and its regulation by pH and unique histidine. *Scientific Reports*, *5*, 9623. https://doi.org/10.1038/srep09623.

Pomorski, T. G., Nylander, T., & Cárdenas, M. (2014). Model cell membranes: Discerning lipid and protein contributions in shaping the cell. *Advances in Colloid and Interface Science*, *205*, 207–220. https://doi.org/10.1016/j.cis.2013.10.028.

Praper, T., Sonnen, A. F.-P., Kladnik, A., Andrighetti, A. O., Viero, G., Morris, K. J., et al. (2011). Perforin activity at membranes leads to invaginations and vesicle formation. *Proceedings of the National Academy of Sciences*, *108*(52), 21016–21021. https://doi.org/10.1073/pnas.1107473108.

Praper, T., Sonnen, A., Viero, G., Kladnik, A., Froelich, C. J., Anderluh, G., et al. (2011). Human perforin employs different avenues to damage membranes. *Journal of Biological Chemistry*, *286*(4), 2946–2955. https://doi.org/10.1074/jbc.M110.169417.

Reboul, C. F., Whisstock, J. C., & Dunstone, M. A. (2016). Giant MACPF/CDC pore forming toxins: A class of their own. *Biochimica et Biophysica Acta - Biomembranes*, *1858*(3), 475–486. https://doi.org/10.1016/j.bbamem.2015.11.017.

Richter, R. P., Bérat, R., & Brisson, A. R. (2006). Formation of solid-supported lipid bilayers: An integrated view. *Langmuir*, *22*(8), 3497–3505. https://doi.org/10.1021/la052687c.

Richter, R. P., Maury, N., & Brisson, A. R. (2005). On the effect of the solid support on the interleaflet distribution of lipids in supported lipid bilayers. *Langmuir*, *21*(1), 299–304. https://doi.org/10.1021/la0478402.

Rojko, N., & Anderluh, G. (2015). How lipid membranes affect pore forming toxin activity. *Accounts of Chemical Research*, *48*(12), 3073–3079. https://doi.org/10.1021/acs.accounts.5b00403.

Ruan, Y., Rezelj, S., Bedina Zavec, A., Anderluh, G., & Scheuring, S. (2016). Listeriolysin O membrane damaging activity involves arc formation and lineaction—Implication for *Listeria monocytogenes* escape from phagocytic vacuole. *PLoS Pathogens*, *12*(4), e1005597. https://doi.org/10.1371/journal.ppat.1005597.

Ruan, J., Xia, S., Liu, X., Lieberman, J., & Wu, H. (2018). Cryo-EM structure of the gasdermin A3 membrane pore. *Nature*, *557*(7703), 62–67. https://doi.org/10.1038/s41586-018-0058-6.

Rudd-Schmidt, J. A., Hodel, A. W., Noori, T., Lopez, J. A., Cho, H. J., Verschoor, S., et al. (2019). Lipid order and charge protect killer T cells from accidental death. *Nature Communications*, *10*(1), 5396. https://doi.org/10.1038/s41467-019-13385-x.

Sborgi, L., Rühl, S., Mulvihill, E., Pipercevic, J., Heilig, R., Stahlberg, H., et al. (2016). GSDMD membrane pore formation constitutes the mechanism of pyroptotic cell death. *The EMBO Journal*, *35*(16), 1766–1778. https://doi.org/10.15252/embj.201694696.

Serna, M., Giles, J. L., Morgan, B. P., & Bubeck, D. (2016). Structural basis of complement membrane attack complex formation. *Nature Communications*, 7, 10587. https://doi.org/10.1038/ncomms10587.

Sonnen, A. F.-P., Plitzko, J. M., & Gilbert, R. J. C. (2014). Incomplete pneumolysin oligomers form membrane pores. *Open Biology*, *4*, 140044. https://doi.org/10.1098/rsob.140044.

Tilley, S. J., Orlova, E. V., Gilbert, R. J. C., Andrew, P. W., & Saibil, H. R. (2005). Structural basis of pore formation by the bacterial toxin pneumolysin. *Cell*, *121*(2), 247–256. https://doi.org/10.1016/j.cell.2005.02.033.

Vadia, S., Arnett, E., Haghighat, A. C., Wilson-Kubalek, E. M., Tweten, R. K., & Seveau, S. (2011). The pore-forming toxin listeriolysin o mediates a novel entry pathway of *L. monocytogenes* into human hepatocytes. *PLoS Pathogens*, 7(11), 1002356. https://doi.org/10.1371/journal.ppat.1002356.

van Meer, G., Voelker, D. R., & Feigenson, G. W. (2008). Membrane lipids: Where they are and how they behave. *Nature Reviews Molecular Cell Biology*, *9*(2), 112–124. https://doi.org/10.1038/nrm2330.

van Pee, K., Mulvihill, E., Müller, D. J., & Yildiz, Ö. (2016). Unraveling the pore-forming steps of pneumolysin from *Streptococcus pneumoniae*. *Nano Letters*, *16*(12), 7915–7924. https://doi.org/10.1021/acs.nanolett.6b04219.

van Pee, K., Neuhaus, A., D'Imprima, E., Mills, D. J., Kühlbrandt, W., & Yildiz, Ö. (2017). CryoEM structures of membrane pore and prepore complex reveal cytolytic mechanism of pneumolysin. *eLife*, *6*, e23644. https://doi.org/10.7554/eLife.23644.

Vögele, M., Bhaskara, R. M., Mulvihill, E., van Pee, K., Yildiz, Ö., Kühlbrandt, W., et al. (2019). Membrane perforation by the pore-forming toxin pneumolysin. *Proceedings of the National Academy of Sciences of the United States of America*, *116*(27), 13352–13357. https://doi.org/10.1073/pnas.1904304116.

CHAPTER SEVEN

High-speed atomic force microscopy to study pore-forming proteins

Fang Jiao[a,b], Yi Ruan[c], and Simon Scheuring[a,b,*]
[a]Department of Anesthesiology, Weill Cornell Medicine, New York, NY, United States
[b]Department of Physiology and Biophysics, Weill Cornell Medicine, New York, NY, United States
[c]Zhejiang Provincial Key Laboratory & Collaborative Innovation Center for Quantum Precision Measurement, College of Science, Zhejiang University of Technology, Hangzhou, China
*Corresponding author: e-mail address: sis2019@med.cornell.edu

Contents

Abstract

Pore-forming proteins (PFPs) include virulence factors that are produced by many pathogenic bacteria. However, PFPs also comprise non-virulence factors, such as apoptotic Bcl2-like proteins, and also occur in non-pathogenic bacteria and indeed in all kingdoms of life. Pore-forming proteins are an ancient class of proteins, which are tremendously powerful in damaging cell membranes. In general, upon binding to lipid membranes, they convert from the soluble monomeric form into an oligomeric state, and then undergo a dramatic conformational change to form transmembrane pores. Thus, PFPs render the plasma membrane of their target cells permeable to solutes, potentially leading to cell death, or to more subtle manipulations of cellular functions.

Methods in Enzymology, Volume 649
ISSN 0076-6879
https://doi.org/10.1016/bs.mie.2021.01.033

Recent cryo-EM and X-ray crystallography studies revealed high-resolution structures of several PFPs in their pre-pore and pore states, however many aspects regarding the cues that induce pore formation, the pre-pore to pore conformational transition, the mechanism of membrane permeation and associated dynamics are still less well understood, and direct visualization of the dynamics of these transitions are missing. Using high-speed atomic force microscopy (HS-AFM), the kinetics of oligomerization and the pre-pore to pore transition dynamics of various PFPs, such as Listeriolysin O (LLO), lysenin, and Perforin-2 (PFN2), could be studied. These studies revealed that LLO does not form pores of regular shape or size, but rather forms membrane inserted arcs that propagate and damage lipid membranes as lineactants. In contrast, lysenin forms stable pre-pore and pore nonameric rings and HS-AFM allowed to study their diffusion on and the pH-dependent insertion into the membrane. Similarly, PFN2 underwent pre-pore to pore transition upon acidification. The openness of the HS-AFM system allowed the transition to be imaged in real time and revealed that all observed molecules transitioned into the pore state within 3 s. In this chapter, we detail protocols to prepare lipids, form supported lipid bilayers, and provide guidelines for real-time, real-space HS-AFM observations of PFPs in action.

Abbreviations

AFM	atomic force microscopy
HS-AFM	high-speed atomic force microscopy
LLO	listeriolysin O
PFN2	perforin-2
PFPs	pore-forming proteins
PFTs	pore-forming toxins
SLB	supported lipid bilayers

1. Introduction

Cellular membranes consist of a multitude of different lipids and proteins. They are a vital part of all cells, delineating intracellular compartments or forming a protective barrier for the cells, such as the plasma membrane, in eukaryotic cells. Many effective strategies have evolved to destroy their integrity for the purpose of cellular defense, attack, hunting or digestion. One of the most ancient mechanisms to affect membrane integrity is to punch holes into them by means of pore-forming proteins (PFPs). PFPs, an ancient class of protein found in both pathogenic and nonpathogenic bacteria and indeed in all kingdoms of life (Alouf, 2001; Ni et al., 2020; Pipkin & Lieberman, 2007), are remarkably efficient to interfere with membrane integrity and thus jeopardize the very compartmentalization function of biological membranes.

Among the PFPs, pore-forming toxins (PFTs) are the most common class of bacterial secreted proteins, and well studied (Alouf, 2001; Bashford, 2001; Dal Peraro & van der Goot, 2016). PFTs act by damaging the plasma membrane of the target host cell. PFTs are the crucial virulence determinants during bacterial infections and diseases (Gilbert, 2002). They are divided into two categories based on the secondary structure of the regions that line the transmembrane pore channel, namely the α-PFTs and the β-PFTs (Iacovache, Bischofberger, & van der Goot, 2010; Lesieur, Vecsey-Semjen, Abrami, Fivaz, & van der Goot, 1997). Most PFTs have an intriguing, and for proteins quite unique, triple-life, explicitly they exist as water-soluble monomeric proteins, these monomers interact with membranes by recognizing specific receptors (*i.e.* lipids, proteins), where they commonly oligomerize and form a membrane-associated pre-pore, which then undergoes a dramatic conformational change to form stable pores. Thus, PFTs are water-soluble, membrane-associated and transmembrane proteins at times. In the latter stage, they disrupt the membrane integrity, which provides access to host cells, host cell contents, or potentially kills the host cells (Czajkowsky, Hotze, Shao, & Tweten, 2004; Podobnik et al., 2015; Ramachandran, Tweten, & Johnson, 2005; Shatursky et al., 1999; Shepard, Shatursky, Johnson, & Tweten, 2000; Tilley, Orlova, Gilbert, Andrew, & Saibil, 2005).

Besides PFTs, perforins also belong to the family of PFPs. Perforins are mammalian proteins, where they are expressed and secreted by cytotoxic lymphocytes. Perforins play crucial roles in the immune defense against pathogens and in the removal of cancer cells. For example, perforin is the only molecule of the human immune system that allows entry of granzyme into the target cells, leading to the initiation of apoptosis. The loss of perforin expression or function is catastrophic and leads to fatal immune dysregulation (Voskoboinik, Whisstock, & Trapani, 2015). Perforin is part of the membrane attack complex/perforin/cholesterol-dependent cytolysin (MACPF/CDC) superfamily (Reboul, Whisstock, & Dunstone, 2016). The pore-formation mechanism of this superfamily involves the unraveling of two α-helical regions that refold into transmembrane hairpins (TMH1 and TMH2) and create giant β-barrel transmembrane pores (Gilbert, 2005; Law et al., 2010; Ni et al., 2020). It should be noted that perforins show remarkable structural reminiscence to β-PFTs.

The main principle of PFP action is that they are secreted as water-soluble (generally monomeric) proteins, which upon recognition of (specific) receptors bind to the membrane surface and oligomerize, and finally transfer

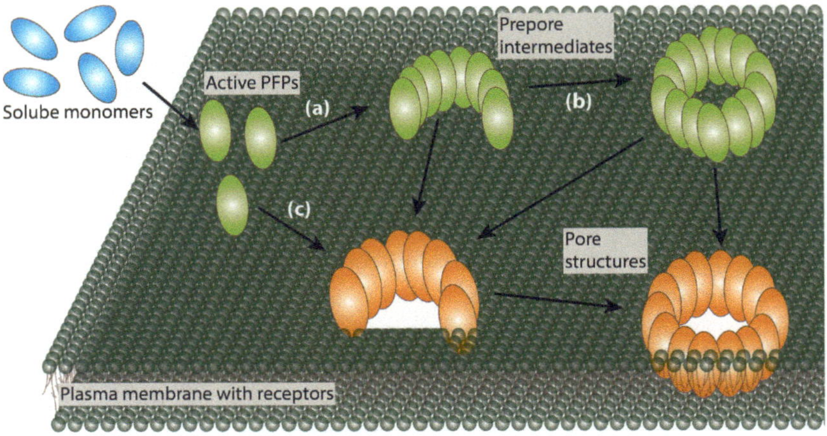

Fig. 1 Schematic representation of pore formation mechanisms. Soluble pore-forming proteins (PFPs) are recruited to the membrane surface by specific interactions with receptors (protein receptors or lipid receptors). Upon membrane binding, active PFPs concentrate and oligomerize into arc-, and/or ring-shaped pre-pore and pore structures. There are several pathways that PFPs can form pores: (A) PFPs assemble into pre-pore arc structures. Following, pre-pore arcs undergo conformational transitions to pore arcs. (B) PFPs assemble into full ring pre-pore structures that then undergo conformational transition to form either arc- and/or full ring-shaped pore structures. (C) Concomitant with oligomerization, PFPs insert the membrane to form arc-shaped pore structures, that may close into ring pores. Soluble PFPs are represented in blue, pre-pore PFPs in green, and pore PFPs in orange, respectively.

parts of their polypeptide chain into and across the lipid bilayer to form functional transmembrane pores (Fig. 1) (Dal Peraro & van der Goot, 2016; Ni et al., 2020; Ruan, Rezelj, Zavec, Anderluh, & Scheuring, 2016). Dramatic structural changes accompany the transition process. X-ray crystallography and electron cryo microscopy (cryo-EM) have provided high-resolution structures for soluble states, oligomeric membrane-associated complexes, and pore complexes (Bokori-Brown et al., 2016; Iacovache et al., 2016; Jiang, Pentelute, Collier, & Zhou, 2015; Katayama et al., 2010; Law et al., 2010; Leung et al., 2014; Lukoyanova et al., 2015; Serna, Giles, Morgan, & Bubeck, 2016; Tilley et al., 2005; van Pee et al., 2017). On the other hand, much less is known about the cues, environmental determinants of modulation, pathways and kinetics of the pore-formation mechanism of PFPs.

X-ray crystallography and cryo-EM are most powerful structural techniques, and provided atomic scale structures of PFPs. However, they often operate under vacuum and low temperature conditions or require specific non-native sample preparation, such as crystallization in the case of X-ray diffraction, which might influence the nativeness of the protein under investigation. Furthermore, these techniques rely on averaging and might therefore miss rare states. Conventional Atomic Force Microscopy (AFM) provided topography maps of PFTs oligomerization on membrane and under close-to-native conditions and contributed complementary information to EM and X-ray crystallography structures of PFTs. However, conventional AFM image acquisition rates are in the minutes time range, and thus only provided static snapshots of PFPs or documented very slow dynamic process. Fluorescence-based optical microscopy techniques have high sensitivity, capable of tracking single molecules, and good temporal resolution and are therefore likely the most popular methods to study the function of biological samples, but they rely on the availability of specific fluorophores to label the molecules under study, and the data depicts the fluorophore rather than the PFP itself. Moreover, the labeling procedure may change the properties and behavior of the molecules within the membrane or affect their organization. Thus, a direct dynamic visualization of the PFPs' nano-scale mechanism of action was lacking due to unavailability of suitable experimental tools.

High-Speed Atomic Force Microscopy (HS-AFM) (Ando et al., 2001), a unique tool for simultaneously studying the structure and dynamics of proteins, combines all advantages of conventional AFM but with high temporal resolution in the sub-second range (Ando, Uchihashi, & Scheuring, 2014). Therefore, HS-AFM, owing to its high spatio-temporal resolution, represents itself as a powerful technique to investigate PFPs on membranes in action. Observation of PFP diffusion on membranes, pre-pore to pore transition and the assembly kinetics of PFP oligomers in real-time can be provided by HS-AFM. Here, we review and provide methodological protocols how the performance of HS-AFM can be leveraged to assess dynamics at biologically relevant time scales of PFPs, namely for the analysis of Listeriolysin O (LLO), lysenin, and perforin-2 (PFN2) (Ni et al., 2020; Ruan et al., 2018). However, from our experiences with PFPs, β-PFT and α-PFTs such as actinoporin (Morante et al., 2016), the methodological protocols discussed here should be adaptable with minor modifications to most PFPs.

Listeriolysin O (LLO) is *Listeria monocytogenes*' powerful molecular weapon in host cell invasion, to cause the disease listeriosis (Koster et al., 2014). Until now, *Listeria* infection is treated by antibiotics, but as the development of novel antibiotics is a serious bottleneck, an improved understanding of LLO action may provide novel approaches to fight the disease.

Previous structural and functional analysis (Koster et al., 2014; Moors, Levitt, Youngman, & Portnoy, 1999; Schuerch, Wilson-Kubalek, & Tweten, 2005) showed that LLO is initially a water-soluble protein of 58 kDa molecular weight, is a β-PFT and belongs to the CDC family, and its activity is cholesterol- and pH-dependent. After binding to a cholesterol-rich cell membrane, LLO monomers oligomerize and undergo a major conformational change, that allows the oligomeric assemblies to penetrate the membrane and form pore complexes. This membrane insertion of LLO oligomers permeabilizes the membrane of the *listeria*-containing vacuole and enables escape of *Listeria* from the phagosome into the infected cells. Cholesterol-dependence, endosomolytic poreformation, and pH-dependent membrane insertion characteristics of LLO have been revealed in the last decades, recent works reported structural details of its monomers and oligomeric complexes, such as the structural depiction of LLO by EM and the pores observation by conventional AFM (Koster et al., 2014; Podobnik et al., 2015; Vadia et al., 2011). Nevertheless, the slow dynamics of conventional AFM and sample preparation in EM prohibited a detailed understanding of the dynamic action of LLO.

Lysenin, belonging to the family of eukaryotic β-PFTs, is a 33 kDa protein that is found in the coelomic fluid of the earthworm *Eisenia fetida* (De Colibus et al., 2012; Yoshiyuki, Kenichi, Terumi, & Hideshi, 1996). Lysenin specifically binds to sphingomyelin (SM), a major sphingolipid species in animal cells and a major lipid constituent of their plasma membrane, but not to other phospholipids in cell membranes. Lysenin induces erythrocyte lysis and is cytotoxic to vertebrate spermatozoa and amphibian larvae as well as cultured mammalian cells (Ishitsuka & Kobayashi, 2004; Kobayashi & Yamaji-Hasegawa, 2006). Upon binding to SM-containing membranes of target cells, lysenin assembles into oligomers in an SM-dependent manner, with the subsequent formation of pores with a diameter of ~3 nm in the target membrane (Bokori-Brown et al., 2016; De Colibus et al., 2012). Despite the wealth of structural information of lysenin, little was known about the assembly dynamics, pre-pore to pore state transition and associated activation mechanism.

Perforin-2 (PFN2, also known as MPEG-1) has recently been identified as a novel and critically important component of innate immunity conserved throughout the animal kingdom (Benard et al., 2015; D'Angelo, Dunstone, Whisstock, Trapani, & Bird, 2012; McCormack & Podack, 2015; Wiens et al., 2005), and is evolutionarily close to the last common ancestor of the MAC and perforin (McCormack, de Armas, Shiratsuchi,

Ramos, & Podack, 2013; McCormack & Podack, 2015). PFN2 is constitutively expressed in macrophages, however its expression can be induced in many different cell types by interferons (α, β and γ), bacterial infection, and pathogen-associated molecular patterns such as lipopolysaccharide (LPS) (Botto et al., 2009; Fields, McCormack, de Armas, & Podack, 2013; McCormack et al., 2013, 2015; McCormack, Lyapichev, Olsson, Podack, & Munson, 2015). Knockdown or deletion of PFN2 renders isolated cells and whole animals susceptible to infection, leads to increased morbidity and in many cases to death (McCormack et al., 2013; McCormack, de Armas, et al., 2015; McCormack, Lyapichev, et al., 2015). In human patients with pulmonary nontuberculous mycobacteria, four point mutations of PFN2 have been identified (McCormack et al., 2017), and macrophages expressing these mutant versions of PFN2 are unable to control *S. typhimurium* or *S. aureus* infection (McCormack et al., 2017). Thus, PFN2 plays a vital role in the defense against intracellular bacteria. It assembles into a pre-pore as a ring of 16 subunits, the pore forming transmembrane hairpins (TMHs) of which point in the opposite direction with respect to the membrane-binding domains. The reconfiguration of the MACPF and membrane-binding P2 domains triggered by acidification then results in pore formation (Ni et al., 2020).

HS-AFM successfully visualized the LLO action cycle from LLO oligomerization on the membrane surface to pre-pore to pore transition and finally to the processive membrane lysis, at high spatio-temporal resolution (Ruan et al., 2016). HS-AFM also revealed that the pore formation process of lysenin was pH-dependent but independent of neighboring lysenins despite their crowding in SM-containing membrane domains (Munguira, Takahashi, Casuso, & Scheuring, 2017). Moreover, fundamentals of lysenin membrane diffusion and aggregation were extracted from HS-AFM movies (Munguira et al., 2016). HS-AFM also revealed the PFN2 pre-pore oligomer assembly and acidic pH triggered PFN2 pre-pore to pore transition process (Ni et al., 2020). Thus, HS-AFM is a powerful tool for investigating PFP dynamics, such as the oligomerization, pore formation, and membrane destabilization. To mimic the native-like environment, when using AFM imaging the PFPs dynamics, PFPs are mostly embedded in supported lipid bilayers (SLBs). SLBs are biological model membranes deposited on top of a solid support (Castellana & Cremer, 2006). In this chapter, we describe in detail the protocols to form supported SLBs through vesicle fusion on small mica disk HS-AFM sample stages. The SLBs serve as a mimic of the target membrane to investigate PFP action. We also provide the guidelines for

HS-AFM imaging of PFPs (LLO, lysenin and PFN2) on such model membranes, methods that should be readily applicable or adaptable for the investigation of PFPs in general.

2. Materials and equipment

2.1 Reagents and PFPs

- 1,2-Dioleoyl-*sn*-glycero-3-phosphocholine (DOPC), highest purity, from Avanti Polar Lipids.
- 100% *E. coli* total phospholipid extract (composed of PE: PG: CA: unknown = 57.5: 15.1: 9.8: 17.6), from Avanti Polar Lipids.
- Egg Sphingomyelin (SM), from Avanti Polar Lipids.
- Cholesterol: ovine wool >98% pure, from Avanti Polar Lipids. All lipid stock solutions (in chloroform) are stored at −20 °C.
- Chloroform: analytical standard, ≥99.9%, from Supelco.
- Methanol: analytical standard, ≥99.9%, from Supelco.
- HEPES: ≥99.5% (titration), from Sigma-Aldrich.
- Sodium hydroxide (NaOH): BioXtra, ≥98% (acidimetric), pellets (anhydrous), from Sigma-Aldrich.
- Sodium chloride (NaCl): BioXtra, ≥99.5%, from Sigma-Aldrich.
- 2-(*N*-Morpholino)ethanesulfonic acid hydrate (MES): low moisture content, ≥99% (titration), from Sigma-Aldrich.
- Magnesium chloride ($MgCl_2$): anhydrous, ≥98%, from Sigma-Aldrich.
- Calcium chloride ($CaCl_2$): BioXtra, ≥99.5%, from Sigma-Aldrich.
- Hydrochloric acid (HCl): ACS reagent, 37%, from Sigma-Aldrich.
- Ethylenediaminetetraacetic acid (EDTA): BioUltra, anhydrous, ≥99% (titration), from Sigma-Aldrich.
- Milli-Q water.
- LLO is prepared as described in Podobnik et al. (Podobnik et al., 2015).
- Lysenin is prepared as described in Munguira et al. (Munguira et al., 2016).
- PFN2 is prepared as described in Ni et al. (Ni et al., 2020).

2.2 Supplies

- Mica disk (1.5 mm in diameter), from RIBM Co, or cut in-lab from a larger mica sheet.
- Glass rod (diameter: 1.5 mm, height: 3 mm), from RIBM Co.
- Ultra-short AFM tips: USC-F1.2-k0.15, from Nano World.
- Glass vial: screw top, amber glass, volume 2 mL, from Supelco.

- Nitrogen: gas, 99.99 atom % ^{14}N, from Sigma-Aldrich.
- Hamilton glass syringes: 10 μL, 25 μL, 50 μL, from Sigma-Aldrich.
- Scotch tape.
- Kimtech wipes, from Kimberly-Clark Professional.
- Clear nail polish.
- Hydrophobic barrier pen, from Thermo Fisher Scientific.
- Cotton bud.
- 100 mL, 250 mL beaker.
- Fast glass glue, from Loctite.
- Humid container, in-lab made from plastic petri dishes.

2.3 Equipment

- pH meter.
- Chemical fume hood.
- Vacuum desiccator.
- Water bath sonicator, from VWR.
- Tip sonicator, from Qsonica.
- HS-AFM: SS-NEX, from RIBM.

3. Methods

Supported lipid bilayers (SLBs) represent an attractive way to mimic biological membranes, especially for the investigation of membrane protein assembly and dynamics. Various different methods have been developed to prepare SLBs, the most common way remains the fusion of small unilamellar lipid vesicles (SUVs) on hydrophilic solid surfaces (alternatively, one can deposit giant unilamellar vesicles (GUVs) (Angelova & Dimitrov, 1986), or form SLBs through Langmuir transfer (Grandbois, Clausen-Schaumann, & Gaub, 1998; Ross, Steinem, Galla, & Janshoff, 2001; Tamm & Mcconnell, 1985)). The SUVs are relatively unstable in solution due to their high membrane curvature (10–50 nm in diameter). Upon addition of such SUVs onto the sample support, SUVs adsorb on the substrate, then rupture and fuse to form a continuous supported bilayer. Numerous approached can be used to prepare SUVs, including extrusions, sonication, and freeze-thaw cycles. The method we usually use to prepare SUVs is sonication. To reproducibly form SUVs using sonication, both bath sonication and tip sonication are introduced in this chapter. We recommend bath sonication for lipid mixtures of DOPC and cholesterol, and tip sonication for *E. coli* total phospholipid extract lipids, as the convenient ways to prepare

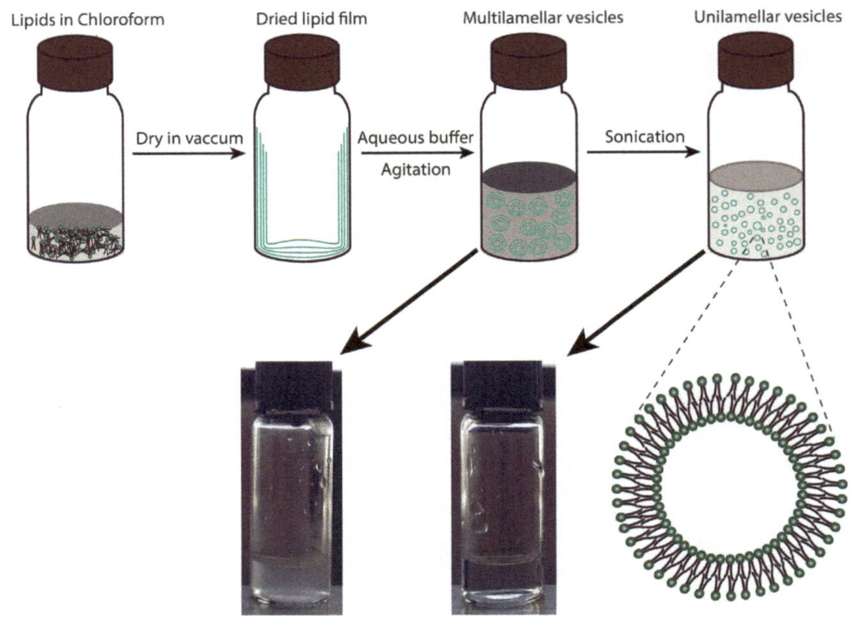

Fig. 2 Preparation of small unilamellar vesicles (SUVs). Lipids are mixed at proposed weight or molar ratios in chloroform; then dried in a N_2-gas stream and left in a vacuum chamber for more than 2 h to dry. The dried lipid film is hydrated in aqueous buffer and agitated to give a cloudy resuspension solution with multilamellar vesicles. Sonication is used to produce clear solutions of small unilamellar vesicles (SUVs) of desired lipid composition.

SUVs (Fig. 2). Afterwards, lipid solutions with well dispersed SUVs are incubated on the negative charged mica surface to form a continuous lipid membrane. Finally, monomers of PFPs are deposited on the lipid membrane. The processes and dynamics of monomer oligomerization and prepore-to-pore transition are revealed by HS-AFM. The main steps of the protocol described below are for HS-AFM experiments.

3.1 Liposome preparation

1. Pre-rinse glass syringes with organic solution (Chloroform: Methanol, 3:1, v:v) 15 times. Remember to rinse the syringes both pre- and post-use.
2. Mix the (DOPC: Cholesterol, 4:1, v:v)—or any other lipid mixture specific for the PFP under investigation - in organic solution (Chloroform:

Methanol, 3:1, v:v) in a glass vial, final concentration ~3 mM. For native mixed lipids, *e.g.* *E. coli* total phospholipid extract, drop 10 μL 25 mg/mL lipid in Chloroform in glass vial (Fig. 2).

3. Evaporate the organic solvent to dry lipids using a very light stream of clean nitrogen gas. Keep the glass vial slowly rotating during the gas flow, to produce a well-spread lipid film on the glass vial interior wall surface.
4. Transfer the glass vial containing the lipid film to a vacuum desiccator and keep it there for >2 h (we usually keep it overnight). This will evacuate the remaining chloroform in the lipid film after gas drying.
5. Hydrate the lipid mixture with desired buffer or milli-Q water (case examples: DOPC/cholesterol mixtures in milli-Q water; *E. coli* total phospholipid extract lipids in 10 mM Hepes-NaOH, pH 7.5, 200 mM NaCl) to desired lipid concentration (DOPC/cholesterol mixtures at 500 μM; *E. coli* total phospholipid extract lipids at 0.5 mg/mL).
6. Place the resuspend lipids into a water bath for 20 s, where the water is heated to ~70 °C, then agitate for 40 s. Repeat 5 cycles until all lipids are released from the inside wall surface and resuspend in the solution (Fig. 2).
7. First, seal the glass vial with parafilm to prevent lipid loss during the sonication steps; the sonication force may lead to the agitation of the small volume (Fig. 2). Sonicate with bath sonicator: to prepare SUVs of DOPC/cholesterol mixtures use bath sonication of the lipid suspension in water for ~40 min until the solution is completely clear (the sonication time depends on the lipid composition, concentration, and volume). Add some ice in the bath to prevent the water from heating. To optimize sonication, find the ideal water level where the sonication power transfer is the strongest.
8. Sonicate with tip sonicator: to prepare the SUVs of *E. coli* total phospholipid extract lipids, use tip sonication of the lipid suspension for ~5 min until the solution totally clears (the sonication time depends on lipid composition, concentration, and volume). To do this well, clean the sonicator tip with ethanol, and water, and wipe off the water with paper wipes. Immerse the sonicator tip in the center of the lipid suspension. Avoid that the sonicator tip touches the bottle neck and the bottom substrate. Optimize the sonication power, to maximize sonication while avoiding that the solution jolts out of the glass vial. To prevent overheating of the solution during sonication, pulse the sonicator 1 s on and 1 s off, and keep the glass vial in a beaker with ice.

9. Once the lipids solution clears, store lipid SUVs keep at ~4 °C. The maximal storage time is around 10 days. It is recommended to sonicate (bath sonication for 20 min or tip sonication for 3 min) again before use.
10. SUVs may also be flash frozen in liquid nitrogen and stored at −80 °C for as long as a month without noticeable change in properties.

3.2 Supported lipid bilayer preparation

The method is adapted from ref. Uchihashi, Kodera, and Ando (2012).

1. Clean the *Z*-piezo surface with paper wipes wetted with acetone to remove any dust and to confirm the flatness of the Z-piezo surface.
2. Glue a glass rod on top of the Z-piezo of the HS-AFM scanner with nail polish and incubate overnight to let the nail polish cure.
3. Glue a 1.5 mm diameter mica disk with fast glass glue on top of the glass rod, and let it cure for over 30 min (Fig. 3A).
4. Use scotch tape to cleave the mica and produce a clean and atomically flat mica surface.
5. Drop immediately after cleavage 1 μL incubation buffer (LLO experiment: 20 mM MES, pH 5.6, 100 mM NaCl, 5 mM $MgCl_2$; Lysenin experiment: 20 mM Hepes-NaOH, pH 7.5, 120 mM NaCl; PFN2 experiment: 10 mM Hepes-NaOH, pH 7.5, 200 mM NaCl, 2 mM $CaCl_2$) onto the mica disk. Divalent ions are added in the incubation buffer to promote vesicle fusion and bonding to the negative charged mica surface.
6. Deposit 4 μL lipid stock solution (LLO experiment: DOPC/cholesterol mixture at 500 μM; Lysenin experiment: Egg sphingomyelin/cholesterol; PFN2 experiment: *E. coli* total phospholipid extract lipids at 0.5 mg/mL) on mica the disk, mix with the incubation buffer, and incubate for ~30 min in the humid chamber. A continuous SLB with homogeneous surface will form (Fig. 3B, right). The incubation time depends on the lipid components and concentration. Incubation time for over or equal 30 min is recommended, as short incubation times may result in small lipid membrane islands, although they might be desirable for certain experiments (Fig. 3B, left).
7. Rinse 5 times with 200 μL incubation buffer (LLO experiment: 20 mM MES, pH 5.6, 100 mM NaCl, 5 mM $MgCl_2$; Lysenin experiment: 20 mM Hepes-NaOH, pH 7.5, 100 mM NaCl; PFN2 experiment: 10 mM Hepes-NaOH, pH 7.5, 200 mM NaCl, 2 mM $CaCl_2$). This will remove free SUVs in solution. Keep the mica substrate immersed at all time

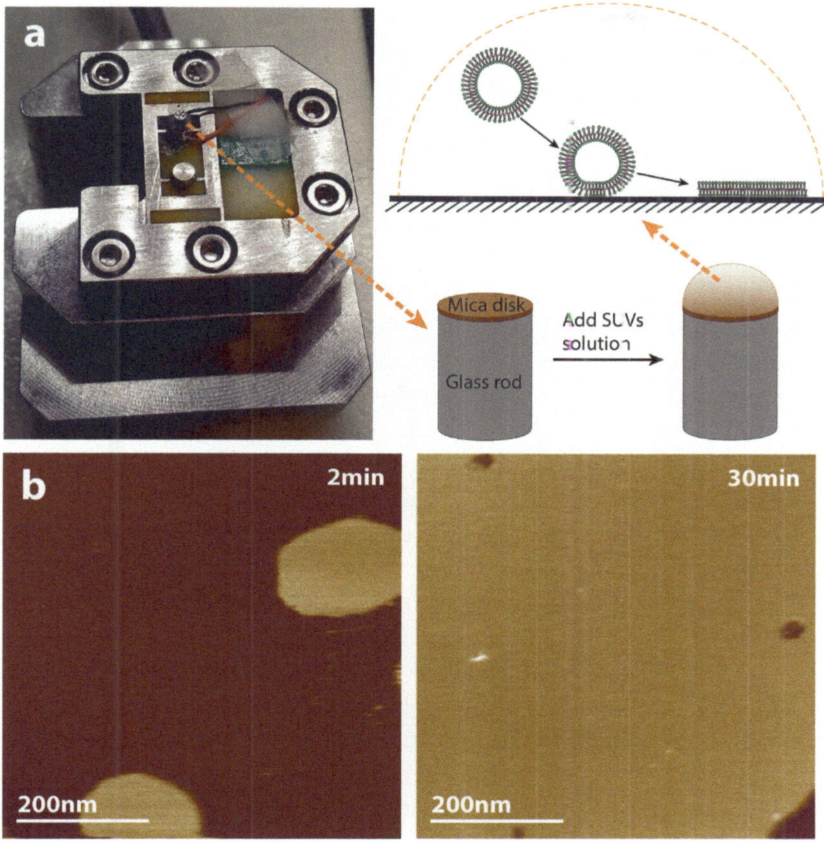

Fig. 3 Formation of supported lipid bilayers (SLBs) through small unilamellar vesicles (SUVs) fusion. (A) A mica disk with 1.5 mm diameter is glued on the glass rod. The lipid vesicle solution is deposited on the freshly cleaved mica disk, incubated in a sealed container with water-saturated atmosphere by wetted paper wipes. After incubation, lipid solution is thoroughly rinsed with buffer to remove redundant vesicles. The glass rod is glued on top of the z-scanner of the HS-AFM scanner (left). (B) HS-AFM images show lipid membrane islands (left) and a continuous lipid bilayer (right) formed when incubating *E. coli* SUVs on mica for 2 min and 30 min, respectively.

3.3 HS-AFM imaging of lipid membrane

1. Clean the HS-AFM liquid cell and the cantilever holder using a cotton bud wetted with mild detergent, separately. Afterward, rinse them with Milli-Q water thoroughly. Sonicate them with the cantilever holder in a beaker filled with Milli-Q water for 5 min (if needed, add small amounts of mild detergent). Then rinse them again with abundant Milli-Q water and dry them using clean nitrogen flow.

2. Equip HS-AFM with a clean ultra-short (8 μm) HS-AFM cantilever, from Nanoworld or Olympus, with nominal spring constant of 0.15 N/m, resonance frequency of ~0.6 MHz and a quality factor of ~1.5 in buffer. The cantilever should be firmly clamped in the holder, avoiding however cantilever fracture. Fill 100 μL imaging buffer into the HS-AFM liquid cell (LLO experiment: 20 mM MES, pH 5.6, 100 mM NaCl, 1 mM EDTA; Lysenin experiment: 20 mM Hepes-NaOH, pH 7.5, 120 mM NaCl; PFN2 experiment: 10 mM Hepes-NaOH, pH 7.5, 200 mM NaCl, 2 mM $CaCl_2$) and align the laser to the backside of the cantilever at the extreme position where the tip is situated, align the laser try to get the maximum readout sum signal and minimize the subtract signal value by alignment of the photo detector.
3. Place the rinsed mica disk with the pre-incubated lipid bilayer into the HS-AFM liquid cell and immerse it into the imaging buffer. Move and approach the mica disk close to the cantilever. Keep enough distance between the mica surface and the cantilever to avoid the tip from crashing with the sample stage during the alignment. If the tip crashes, the cantilever needs to be replaced and all previous procedures must be repeated.
4. Wait for about 10 min to allow the whole system to stabilize and equilibrate, to minimize drift while imaging.
5. Operate the HS-AFM in oscillating mode (also known as amplitude modulation mode, tapping mode). Small oscillation free and set point amplitude of ~1 nm and ~0.9 nm, respectively, are used, to achieve minimum tip sample interaction and best possible resolution. Scanning different areas to find continuous lipid membrane without defects and readjusting imaging parameters (cantilever oscillation amplitude, excitation frequency and set point voltage) is highly recommended until images with high contrast are recorded (Fig. 3B). If despite optimization of the imaging parameters it is impossible to get high-resolution images (or double-tip image artifacts are apparent), the tip might be contaminated, blunt or not ending in a single apex. Clean or replace the cantilever and repeat the procedure again.
6. Identify regions where a continuous lipid membrane SLB formed on the mica substrate (Fig. 3B, right).

3.4 HS-AFM imaging of LLO dynamics

1. Find a good scan area of continuous lipid membrane (DOPC/Cholesterol 4:1, v:v).

2. Deposit 3 μL of 17 μM LLO soluble monomers into the imaging buffer to a final concentration of ~500 nM. In case the operator continues real-time HS-AFM imaging during injection of the LLO solution into the HS-AFM liquid cell, the image quality likely gets disturbed for ~10 s due to the perturbation. It is typically recommended to briefly withdraw the tip from the surface during these initial moments upon protein addition to avoid HS-AFM tip contamination.
3. Optimize the imaging parameters (cantilever oscillation amplitude, excitation frequency, scanning speed and setpoint value) until high resolution images are achieved.
4. Manually compensate for sample stage drift, in case it occurs during imaging, through readjustment of the X-Y-piezo offset voltage.
5. Use ImageJ software with a dedicated Plugins developed for HS-AFM image (Husain, Boudier, Paul-Gilloteaux, Casuso, & Scheuring, 2012) or your own image processing software to perform data processing (image flattening, frame alignment, particle analysis). Further, project specific analysis of the data can be performed in Matlab or Origin.

Using these experimental methods, we have directly imaged the assembly of LLO into arc-shaped oligomers on the membrane surface (Fig. 4A). We reasoned that the membrane disturbance of a monomer penetrating a membrane leaflet is energetically costly and that bringing several units together might minimize this energy cost, driving oligomerization. This process occurred on time scales as short as 10 s at 500 nM LLO concentration. Once arcs have reached maturity, *i.e.* assembly of ~20 monomers in an arc of about 50 nm length, assembly stalled and further length analysis over 60 s revealed length fluctuations of a few nanometers, the size of single subunits (Fig. 4B). The formation of complete circular ring-shaped complexes from arc-shaped LLO was never observed. Closest to this, annealing of neighboring arcs was regularly observed. In this case, arc-shaped oligomers were interlocked to exhibit a rather ellipsoidal surface contour that did not further evolve as a function of time (Fig. 4C). Membrane insertion only occurred when rather advanced arc assemblies had been formed, indicative that the pre-pore to pore transition needed an advanced state of oligomerization and might therefore be cooperative (Fig. 4D). Quantitatively, single molecule analysis showed varying residence lifetime of pre-pore complexes on the membrane of up to minutes, followed by insertion that was completed within seconds. This more rapid membrane insertion compared to the residence lifetime of pre-pores is further evidence for a cooperative conformational change within the units of the pre-pore arcs for membrane insertion (Fig. 4E). Further oligomerization of already membrane embedded

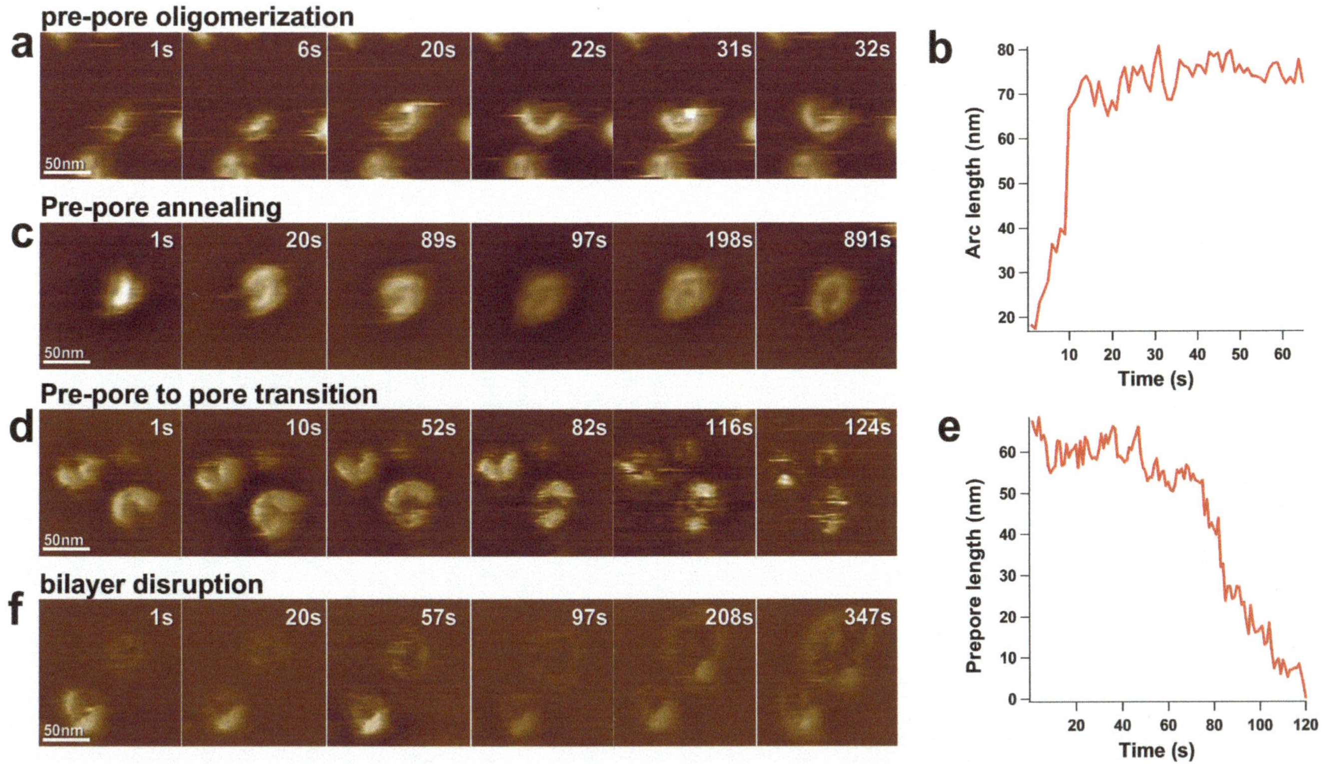

Fig. 4 Dynamic characterization of LLO assembly and action by HS-AFM. Buffer condition: 20 mM MES, pH 5.6, 100 mM NaCl, 1 mM EDTA, LLO final concentration 500 nM. (A) Pre-pore oligomerization process. (B) The oligomerization rate is about 5 nm/s (~2 subunits/s) until arc maturation after ~10 s. (C) Arc annealing: arc-shaped oligomers interlock to form a final ellipsoid LLO assembly that does not further evolve. (D) Membrane insertion process, *i.e.* pre-pore to pore transition. (E) The pre-pore is stable on the bilayer for 60s and then inserts rapidly and entirely with ~2 nm/s (about 1 subunits/s). (F) Following oligomerization and pre-pore to pore transition, bilayer disruption occurs at ~600 nm^2/s. *Reproduced with permission, under the Creative Commons Attribution (CC BY) license, from Ruan, Y., Rezelj, S., Zavec, A. B., Anderluh, G., & Scheuring, S. (2016). Listeriolysin O membrane damaging activity involves arc formation and lineaction—Implication for Listeria monocytogenes escape from phagocytic vacuole.* PLoS Pathogens, *12(4), e1005597.*

arc-shaped oligomers with other existing LLO arcs in the insertion state was never observed. These results indicated that the oligomerization and membrane insertion actions of LLO were spatiotemporally uncoupled. Once LLO was in the membrane, the dynamic destruction of the bilayer could directly be observed: membrane defects occurred inside the arc and grew by lateral propagation of the toxins in the membrane (Fig. 4F). It is notable that the perimeter edged by LLO remaining in the pre-pore state did not propagate in membrane disruption (Fig. 4F). The LLO membrane disruption modality, efficiency and dynamics depended strongly on the cholesterol content of the membrane. Therefore, all the above described experiments were performed with 0%, 10%, 20%, 30% and 40% cholesterol content in the membrane (Ruan et al., 2016).

3.5 HS-AFM imaging of lysenin dynamics

1. Mix Egg Sphingomyelin (SM) and cholesterol (chol) (Avanti Polar Lipids) at SM:chol, 1:1 (molar ratio) and form giant unilamellar vesicles (GUVs) through electroswelling.
2. For GUV formation, deposit 10 μL at 3 mM of each lipid dissolved in chloroform: methanol 3:1 on two glass plates coated with indium tin oxide with 70–100 Ω resistivity (Sigma-Aldrich).
3. Place the glass plates covered with the solution 60 min in the desiccator for complete solvent evaporation.
4. A U-shaped rubber piece of ~1 mm thickness was sandwiched between the two indium tin oxide coated slides. The so-formed chamber was filled with ~400 μL of 200 mM sucrose solution and exposed to 1.5 V sinusoidal 10 Hz AC current for 3 h followed by squared 5 Hz AC current for 15 min, at 55 °C.
5. GUVs were harvested from the chamber.
6. Place 1 μL of GUV solution on a 1.5 mm diameter freshly cleaved mica disk covered with 1 μL of phosphate buffer saline (PBS) and incubated for 30 min to form the supported lipid bilayers (SLBs). Rinse with PBS 10 times.
7. Incubate 1 μL of purified lysenin for 15 min. Rinse with PBS 10 times.
8. Image the lysenin on the SM:chol SLB using HS-AFM as described above for LLO. To change the pH, to observe pre-pore to pore transition, add 3 mL of 5 mM HCl solution. The pH in the fluid cell can be determined before and after the HS-AFM experiments or by scaling up the volumes 10-fold and measuring the pH with a pH meter (HANNA instruments). Alternatively, a pump (Harvard Instruments)

can be connected by two silicone tubes to the fluid cell pool cantilever holder for injection and extraction of buffer of different pH at a rate of 0.17 mL/s, *e.g.* to exchange from a buffer 20 mM HEPES, pH 7.5,120 mM NaCl to a buffer 20 mM MES, pH 4.0, 120 mM NaCl. Again, determine the pH by scaling up the experimental volumes 10-fold and measuring the pH with a pH meter.

Using these experimental methods, the diffusion dynamics of lysenin on the bilayer were investigated using HS-AFM. Not only stable hexagonal arrays on SM-enriched domains were observed, but also highly dynamic lysenin was observed around these domains (Munguira et al., 2016). In experiments, where the pH in the liquid chamber was changed, pH-dependent membrane insertion of lysenin toxin was directly visualized in time-lapse movies (Fig. 5) (Munguira et al., 2017). Initially, under physiological conditions, only ~15% of the lysenin nonamers were in the pore state, protruding ~2.5 nm from the membrane, whereas ~85% formed a hexagonal array of pre-pores, protruding ~5.0 nm. Decreasing the pH from 7.5 to <6.0 during the HS-AFM movie acquisition induced transition from the pre-pore to the pore state in many molecules, as detected in HS-AFM frames by oligomer rings that appear darker due to their ~2.5 nm decreased topography. Statistical assessment, using a protrusion threshold at ~3.75 nm (between the pre-pore and pore heights), of the total areas occupied by pre-pores and pores, combined with cross-correlation-based particle counting revealed that the percentage of pores increased from ~15 to ~47% (Fig. 5B, black circles). As a result of the pH change and the consecutive increase in the number of pore state lysenins, the hexagonal array lost order and the oligomers displayed increased lateral dynamics. The change of topography and increase in lateral dynamics could be assessed by the calculation of standard deviation maps (SD maps) during topography averaging or as a function of time in image series. Calculating SD maps over five subsequent frames (Fig. 5C) at the times corresponding to the topography maps shown in Fig. 5A and plotting the average SD value as a function of time highlighted the increase of lysenin mobility with pH drop and pre-pore to pore transition of about half of the proteins (Fig. 5B, gray triangles).

3.6 Dynamics of PFN2 pre-pore assembly and pre-pore to pore transition

1. Form a continuous *E. coli* (PE: PG: CA: unknown = 57.5: 15.1: 9.8: 17.6) lipid membrane on the mica substrate through SUV or GUV fusion as described in the experiments above. Verify the formation of a continuous SLB using HS-AFM imaging.

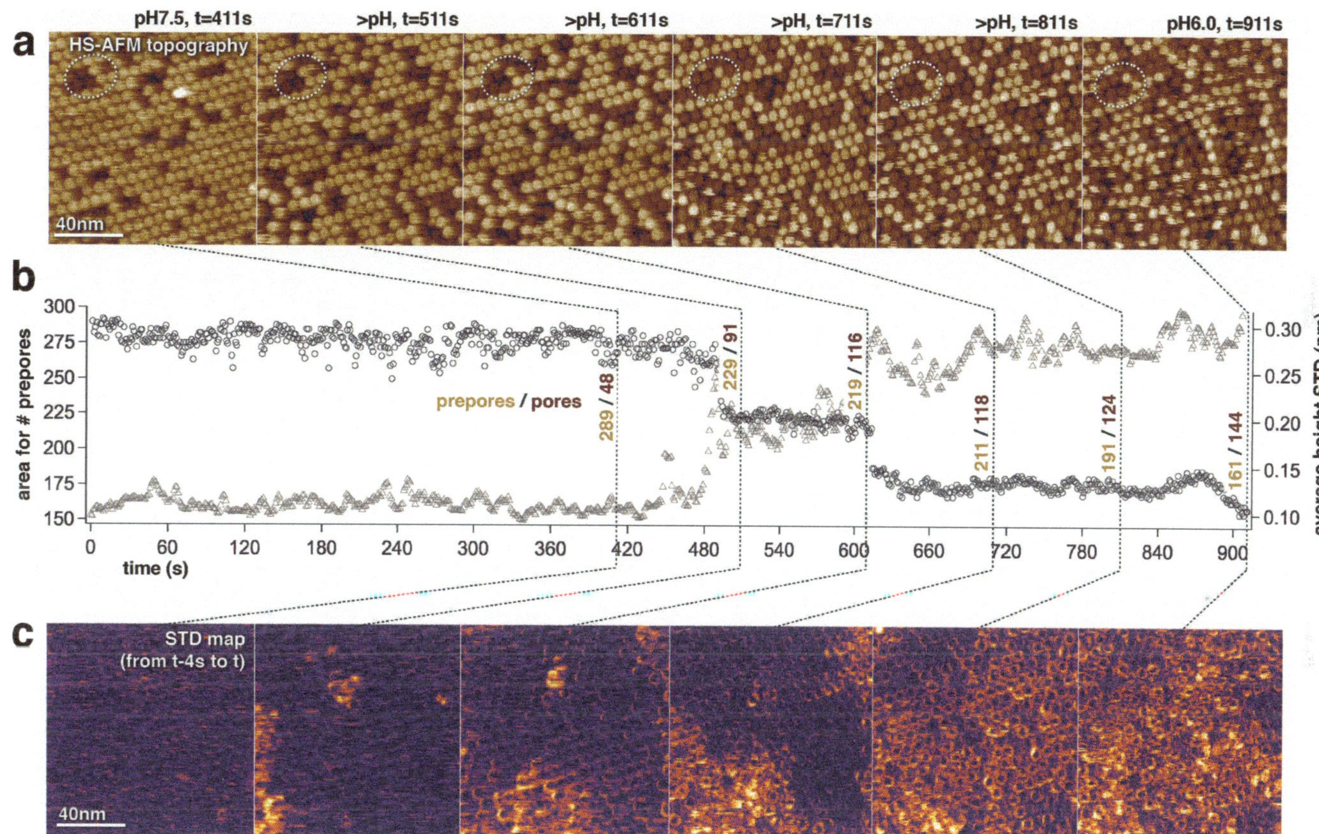

Fig. 5 See figure legend on next page.

2. Lift the HS-AFM head from the liquid cell, while always maintaining ~2 μL imaging buffer 10 mM Hepes-NaOH, pH 7.5, 200 mM NaCl, 2 mM $CaCl_2$ on the mica disk support.
3. Deposit 0.5 μL of 0.215 mg/mL PFN2 monomer solution onto the lipid bilayer surface (bare mica) to a final concentration of ~0.043 mg/mL. Incubate for 10 min in the humid container. Rinse the mica disk with abundant imaging buffer.
4. Image the PFN2 covered SLB (or bare mica) with the HS-AFM in imaging buffer. Optimize the imaging parameters (cantilever oscillation amplitude, excitation frequency, scanning speed and setpoint value) until high resolution images are achieved.
5. Change the ratio of set point and free amplitude (A_{set}/A_{free}) from ~0.9 to ~0.6 and back to ~0.9 to study the reversible lateral displacement of PFN2 prepores on the membrane. Under such conditions, the average estimated applied force on the HS-AFM tip changes from ~30 to ~200 pN and back to ~30 pN (Miyagi & Scheuring, 2016).
6. To study the pre-pore to pore transition, maintain *in-situ* imaging of the prepore ring structures. At the same time, deposit ~15 μL of 0.1 M HCl (through passive diffusion) in the fluid cell containing ~120 μL of imaging buffer to decrease the pH of the buffer solution to about pH 4.0. The change in pH is estimated by large-scale mixing 150 μL of 0.1 M HCl into 1.20 mL of the imaging buffer. Alternatively, a similar approach to acidify the environment as used for LLO can be used.
7. Use ImageJ software with a dedicated Plugins developed for HS-AFM image (Husain et al., 2012) or your own image processing software to perform data processing (image flattening, frame alignment, particle analysis). Further analysis of the data can be performed in Matlab.

Fig. 5 Lysenin toxin membrane insertion upon acidification. (A) HS-AFM movie frames of lysenin on an SM/chol (1:1) bilayer. During HS-AFM movie acquisition, the imaging solution is acidified, leading to a decrease in the number of pre-pore lysenins (bright circles, protruding 5 nm) that insert into the membrane to yield an increased number of pore lysenins (dark circles, protruding ~2.5 nm). (B) Graph of the number of pre-pores (*i.e.*, image area occupied by pixels higher than the 3.75 nm threshold) as a function of movie acquisition time (black circles). The precise number of pre-pores and pores has been analyzed by cross-correlation searches at six specific time points (dashed lines) corresponding to the frames shown in (A). The average value of the SD maps shown in (C) reports about the topographical changes, *i.e.*, the lateral mobility between subsequent frames, throughout movie acquisition. (C) SD maps over five consecutive frames (t-4 to t, where t is the time of the recording of the topography frames shown in A). *Reproduced with permission, Cell Press/Elsevier, from Munguira, I. L. B., Takahashi, H., Casuso, I., & Scheuring, S. (2017). Lysenin toxin membrane insertion is pH-dependent but independent of neighboring lysenins.* Biophysical Journal, 113*(9), 20292036.*

Using these experimental methods, PFN2 pre-pore oligomeric assemblies were studied using HS-AFM (Fig. 6). When PFN2 monomers were simply added directly to a freshly cleaved atomically-flat mica surface (Fig. 6A), HS-AFM found a wide distribution of arcs as well as rings with sizes ranging from 1 to 16 subunits (16 subunits corresponding to the complete ring) (Fig. 6C, green). In these experiments only 7% of the assemblies overall formed complete rings (16 subunits). In contrast, on lipid bilayers (Fig. 6B) the majority of membrane-bound oligomers formed complete rings (Fig. 6C, magenta). The complete rings were in total $\geq$80% of the assemblies found on lipid bilayers, arcs only occurred on the competing borders of adjacent patches of rings arranged in hexagonal arrays. From this, we concluded that PFN2 existed as monomers or small-sized oligomers in solution and that their direct adsorption to the mica where lateral diffusion is hindered did not allow the protein to form complete rings. In contrast, when incubated on a membrane, where lateral 2D-diffusion is allowed, PFN2 formed stable rings with 24 nm in diameter, which furthermore assembled into a hexagonal, *i.e.* the densest, molecular packing (Fig. 6C, inset). The occasional observation of incomplete PFN2 rings on membranes allows the determination of subunit association and dissociation rates, which were similar, 2.4 subunits s^{-1} and 2.3 subunits s^{-1} respectively, in this crowded environment where many arcs competed for subunits (Fig. 6D). To further explore whether the oligomers observed using HS-AFM were membrane-bound PFN2 pre-pores, we performed a force-sweep experiment to clear them from the area of observation: the distribution of oligomeric assemblies recovered by diffusion within ~10 s after minimizing the force applied to the tip, as expected for membrane associated prepore complexes (Fig. 6E). Finally, we performed direct pre-pore to pore activation experiments in the HS-AFM. Upon buffer exchange and exposure to a low pH shock, the PFN2 pre-pore rings began to form pores after 14 s and all ~70 pre-pores being imaged transited to the pore state within 3 s (Fig. 6F). The resulting structures were higher (raised by ~4 nm) compared to the pre-pore state, in agreement with the pores imaged by negative stain EM (Ni et al., 2020).

4. Summary

In this methods chapter we summarize the general approaches to achieve dynamic HS-AFM imaging of PFPs on SLBs and showcase these methods on three examples (Munguira et al., 2016, 2017; Ni et al., 2020; Ruan et al., 2016). Similar approaches, although not at high temporal

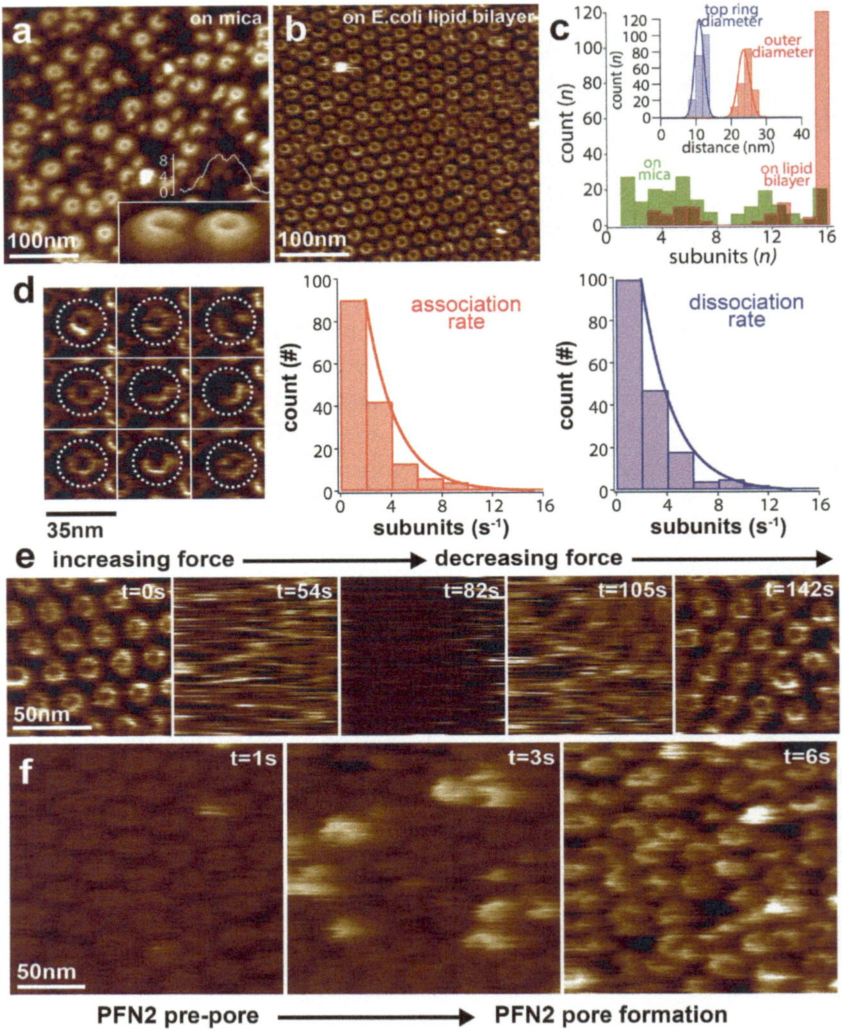

Fig. 6 HS-AFM of PFN2 pre-pore structure, assembly, and pre-pore to pore transition. (A) HS-AFM high-resolution image of PFN2 on mica. Most of the PFN2 oligomers are incomplete, arc-shaped assemblies. The individual subunits can be distinguished (inset). The inset section profile shows that pre-pore rings have a height of ~8.5 nm. (B) HS-AFM image of PFN2 on an *E. coli* lipid bilayer. Essentially all PFN2 prepore oligomers are complete rings that assemble into a hexagonal lattice. (C) Histogram analysis of PFN2 assembly on mica and on an *E. coli* lipid SLB. Inset: top ring diameter, 11.2 ± 0.9 nm, and outer diameter, 23.8 ± 1.2 nm, respectively. (D) HS-AFM images (frame acquisition speed: 500 ms) and histogram analysis of PFN2 subunit association ($2.4\,s^{-1}$) and dissociation ($2.3\,s^{-1}$) rates. Histograms of association and dissociation rates were fit to a single exponential decay function. Full rings, where all subunits have neighbors on both sides, were stable over extended periods of time. (E) HS-AFM images of PFN2 oligomers on an *E. coli* lipid bilayer ($t = 0$ s and

(Continued)

resolution, have been successfully used by us and others to image a larger variety of PFPs (Leung et al., 2014, 2017; Mulvihill et al., 2018; Mulvihill, van Pee, Mari, Muller, & Yildiz, 2015; Pang et al., 2019; Parsons et al., 2019; van Pee, Mulvihill, Muller, & Yildiz, 2016).

Taken together, the first step of the experimental procedure consists of the formation of a supported lipid bilayer (SLB) membrane on the mica surface, preferably through fusion of SUVs. SUVs are prepared as described in the methods implicating a bath sonication or tip sonication step, to get homogeneous SUVs and reproducible experimental results. SUVs can be formed from a wide variety of lipid mixtures depending on the specificity of the PFP under investigation. Using HS-AFM, the various stages of the membrane complex formation and supramolecular assembly in a native-like environment could be visualized at high spatial and temporal resolution, which provided new insights in the assembly process of pore-forming proteins. The spatial resolution is typically good enough that in best resolved images subunits can be counted and structural states assessed, meanwhile, we the temporal resolution can allow the dynamics of the subunit association to be studied. Finally, real-time imaging of the pre-pore to pore transition dynamics can be achieved and thus, if unknown, biochemical cues to induce these transitions, can be tested. Thus, HS-AFM is an outstanding technique to probe both structure and dynamics of PFPs assemblies in real time, and offers a powerful complementary approach to high-resolution structural investigation by X-ray crystallography and cryo-EM, notably to characterize parameters that cannot be contributed by these techniques, such as dynamics, transitions, *etc.* In the future, the kinetics of oligomerization and the pre-pore to pore transition dynamics of more PFPs will be revealed by using HS-AFM, which will ultimately provide a better understanding of the mode of action of PFPs.

Fig. 6—Cont'd $t=142$ s) exposed to HS-AFM tip force induced removal ($t=54$ s), giving access to an empty bilayer ($t=82$ s) followed by assembly recovery ($t=105$ s), in agreement with bilayer surface associated prepores. (F) HS-AFM image (left) of PFN2 on an *E. coli* lipid bilayer as in (B), then (middle and right) images of the same area upon acidification to about pH 4.0. The pre-pore to pore transition was accompanied by an increase in height and sometimes by splitting the ring-shaped pre-pores into arcs. *Reproduced with permission, under the Creative Commons Attribution (CC BY) license, from Ni, T., Jiao, F., Yu, X. L., Aden, S., Ginger, L., Williams, S. I., et al. (2020). Structure and mechanism of bactericidal mammalian perforin-2, an ancient agent of innate immunity.* Science Advances, 6*(5), eaax8286.*

5. Notes

- In Section 3.1, in all processes the lipid samples described should be protected from light using aluminum foil wrapping of the vials.
- In Section 3.1, step 2, the composition and proportion of the lipids in the mixture should be adjusted depending on the experimental requirement.
- In Section 3.1, step 3, the application of a very light stream of clean nitrogen is necessary. If the stream is too strong, the solution may splash out of the glass vial or dry in spaces where it is difficult to be resolved, leading to a loss of lipids and a misjudgment of the concentration.
- In Section 3.1, step 8, optimize the sonicator working power and tip position in solution to avoid loss of the lipid solution.
- In Section 3.2, step 4, after cleaving the mica with scotch tape, double check the presence of a freshly cleaved mica on the glass rod. If the whole mica sheet is cleaved off, repeat from Section 3.2, step 3.
- In Section 3.2, step 5, the divalent ion added with 5 mM $MgCl_2$ in the incubation buffer can be replaced by $MgSO_4$, or $CaCl_2$. If there is phosphatidylinositol in the lipid mixture, avoid to use divalent ions during SLB preparation to avoid phosphatidylinositol aggregation. Because of the absence of divalent ions, lipid mixtures containing phosphatidylinositol are preferably incubated for longer time on the mica surface ($\geq$1 h) in Section 3.2, step 6.
- In Section 3.2, step 6, when using the SUVs that were stored at ~4 °C, sonication (bath sonication for 20 min or tip sonicating for 3 min) right before use is recommended. This will promote the formation of a continuous SLB. Be careful that the mica disk never dries during SUV incubation. In case of problems to keep the lipids solution on top of the mica disk on the glass rod, use a hydrophobic pen and paint a hydrophobic barrier around the glass rod.
- In Section 3.3, step 2, avoid breaking or contaminating the HS-AFM tip.
- In Section 3.4, step 4, stage drift occurs frequently. Small drift is acceptable and can be corrected during data analysis. Large drift is indicative that a longer equilibration time is needed or that the scanner must be repositioned.
- The imaging buffer can be adjusted according to experiment requirements. Case example protocols are provided here.
- A schematic summary of the workflow for a HS-AFM imaging study of PFPs is shown in Table 1

Table 1 Stepwise generalized protocol of a PFP HS-AFM imaging experiment.

Step	Experiment
1	*Liposome preparation (SUV, GUV, etc)* • Mixture, evaporation and sonication • Store in the fridge until use
2	*Supported lipid bilayer (SLB) preparation* • Freshly cleaved mica surface on glass rod • Deposition of SUVs on mica • Incubation in humid chamber
3	*HS-AFM imaging of SLB membrane* • Equip HS-AFM with cantilever • Set HS-AFM in in oscillating mode
4	*Identification of a continuous SLB* • Image SLB with HS-AFM after different SUV incubation times • Identify conditions where a continuous SLB is reproducibly formed (if this step fails, go back to step 2)
5	*HS-AFM imaging of PFP assembly dynamics* • Inject PFP (soluble form) into the HS-AFM liquid cell • Optimize the imaging parameters • Observe PFP adsorption on SLB, oligomerization, diffusion, and supramolecular packing
6	*HS-AFM imaging of PFP state transition dynamics* • Optimize the imaging parameters • Decrease the buffer pH through HCl injection or slow controlled buffer exchange while HS-AFM imaging (if other experimental cues are needed to induce pre-pore to pore transition, adapt the protocol accordingly)

Acknowledgments

S.S. acknowledges support from National Institute of Health (NIH): NCCIH DP1AT010874 and NIGMS R01NS110790.

Y.R. acknowledges support from National Nature Science Foundation of China Grants, NSFC [61805213] and the Fundamental Research Funds for the Provincial Universities of Zhejiang.

References

Alouf, J. E. (2001). Pore-forming bacterial protein toxins: An overview. *Pore-Forming Toxins*, *257*, 1–14.

Ando, T., Kodera, N., Takai, E., Maruyama, D., Saito, K., & Toda, A. (2001). A high-speed atomic force microscope for studying biological macromolecules. *Proceedings of the National Academy of Sciences of the United States of America*, *98*(22), 12468–12472.

Ando, T., Uchihashi, T., & Scheuring, S. (2014). Filming biomolecular processes by high-speed atomic force microscopy. *Chemical Reviews*, *114*(6), 3120–3188.
Angelova, M. I., & Dimitrov, D. S. (1986). Liposome electroformation. *Faraday Discussions*, *81*, 303–311.
Bashford, C. L. (2001). Pore-forming toxins: Attack and defence at the cell surface. *Cellular & Molecular Biology Letters*, *6*(2a), 328–333.
Benard, E. L., Racz, P. I., Rougeot, J., Nezhinsky, A. E., Verbeek, F. J., Spaink, H. P., et al. (2015). Macrophage-expressed perforins mpeg1 and mpeg1.2 have an anti-bacterial function in zebrafish. *Journal of Innate Immunity*, 7(2), 136–152.
Bokori-Brown, M., Martin, T. G., Naylor, C. E., Basak, A. K., Titball, R. W., & Savva, C. G. (2016). Cryo-EM structure of lysenin pore elucidates membrane insertion by an aerolysin family protein. *Nature Communications*, 7, 11293.
Botto, M., Kirschfink, M., Macor, P., Pickering, M. C., Wurzner, R., & Tedesco, F. (2009). Complement in human diseases: Lessons from complement deficiencies. *Molecular Immunology*, *46*(14), 2774–2783.
Castellana, E. T., & Cremer, P. S. (2006). Solid supported lipid bilayers: From biophysical studies to sensor design. *Surface Science Reports*, *61*(10), 429–444.
Czajkowsky, D. M., Hotze, E. M., Shao, Z. F., & Tweten, R. K. (2004). Vertical collapse of a cytolysin prepore moves its transmembrane beta-hairpins to the membrane. *EMBO Journal*, *23*(16), 3206–3215.
Dal Peraro, M., & van der Goot, F. G. (2016). Pore-forming toxins: Ancient, but never really out of fashion. *Nature Reviews Microbiology*, *14*(2), 77–92.
D'Angelo, M. E., Dunstone, M. A., Whisstock, J. C., Trapani, J. A., & Bird, P. I. (2012). Perforin evolved from a gene duplication of MPEG1, followed by a complex pattern of gene gain and loss within Euteleostomi. *BMC Evolutionary Biology*, *12*, 59.
De Colibus, L., Sonnen, A. F. P., Morris, K. J., Siebert, C. A., Abrusci, P., Plitzko, J., et al. (2012). Structures of lysenin reveal a shared evolutionary origin for pore-forming proteins and its mode of sphingomyelin recognition. *Structure*, *20*(9), 1498–1507.
Fields, K. A., McCormack, R., de Armas, L. R., & Podack, E. R. (2013). Perforin-2 restricts growth of chlamydia trachomatis in macrophages. *Infection and Immunity*, *81*(8), 3045–3054.
Gilbert, R. J. C. (2002). Pore-forming toxins. *Cellular and Molecular Life Sciences*, *59*(5), 832–844.
Gilbert, R. J. C. (2005). Inactivation and activity of cholesterol-dependent cytolysins: What structural studies tell us. *Structure*, *13*(8), 1097–1106.
Grandbois, M., Clausen-Schaumann, H., & Gaub, H. (1998). Atomic force microscope imaging of phospholipid bilayer degradation by phospholipase A(2). *Biophysical Journal*, *74*(5), 2398–2404.
Husain, M., Boudier, T., Paul-Gilloteaux, P., Casuso, I., & Scheuring, S. (2012). Software for drift compensation, particle tracking and particle analysis of high-speed atomic force microscopy image series. *Journal of Molecular Recognition*, *25*(5), 292–298.
Iacovache, I., Bischofberger, M., & van der Goot, F. G. (2010). Structure and assembly of pore-forming proteins. *Current Opinion in Structural Biology*, *20*(2), 241–246.
Iacovache, I., De Carlo, S., Cirauqui, N., Dal Peraro, M., van der Goot, F. G., & Zuber, B. (2016). Cryo-EM structure of aerolysin variants reveals a novel protein fold and the pore-formation process. *Nature Communications*, 7, 12062.
Ishitsuka, R., & Kobayashi, T. (2004). Lysenin: A new tool for investigating membrane lipid organization. *Anatomical Science International*, *79*(4), 184–190.
Jiang, J. S., Pentelute, B. L., Collier, R. J., & Zhou, Z. H. (2015). Atomic structure of anthrax protective antigen pore elucidates toxin translocation. *Nature*, *521*(7553), 545–U323.

Katayama, H., Wang, J., Tama, F., Chollet, L., Gogol, E. P., Collier, R. J., et al. (2010). Threedimensional structure of the anthrax toxin pore inserted into lipid nanodiscs and lipid vesicles. *Proceedings of the National Academy of Sciences of the United States of America*, *107*(8), 3453–3457.

Kobayashi, T., & Yamaji-Hasegawa, A. (2006). Lysenin: A new probe for sphingomyelin. In Y. Hirabayashi, Y. Igarashi, & A. H. Merrill (Eds.), *Sphingolipid biology* (pp. 475–482). Tokyo: Springer Japan.

Koster, S., van Pee, K., Hudel, M., Leustik, M., Rhinow, D., Kuhlbrandt, W., et al. (2014). Crystal structure of listeriolysin O reveals molecular details of oligomerization and pore formation. *Nature Communications*, *5*, 3690.

Law, R. H. P., Lukoyanova, N., Voskoboinik, I., Caradoc-Davies, T. T., Baran, K., Dunstone, M. A., et al. (2010). The structural basis for membrane binding and pore formation by lymphocyte perforin. *Nature*, *468*(7322), 447–U277.

Lesieur, C., Vecsey-Semjen, B., Abrami, L., Fivaz, M., & van der Goot, F. G. (1997). Membrane insertion: The strategies of toxins. *Molecular Membrane Biology*, *14*(2), 45–64.

Leung, C., Dudkina, N. V., Lukoyanova, N., Hodel, A. W., Farabella, I., Pandurangan, A. P., et al. (2014). Stepwise visualization of membrane pore formation by suilysin, a bacterial cholesterol-dependent cytolysin. *eLife*, *3*, e04247.

Leung, C., Hodel, A. W., Brennan, A. J., Lukoyanova, N., Tran, S., House, C. M., et al. (2017). Real-time visualization of perforin nanopore assembly. *Nature Nanotechnology*, *12*(5), 467–473.

Lukoyanova, N., Kondos, S. C., Farabella, I., Law, R. H. P., Reboul, C. F., Caradoc-Davies, T. T., et al. (2015). Conformational changes during pore formation by the perforin related protein pleurotolysin. *PLoS Biology*, *13*(2), e1002049.

McCormack, R. M., de Armas, L. R., Shiratsuchi, M., Fiorentino, D. G., Olsson, M. L., Lichtenheld, M. G., et al. (2015). Perforin-2 is essential for intracellular defense of parenchymal cells and phagocytes against pathogenic bacteria. *eLife*, *4*, e06508.

McCormack, R., de Armas, L. R., Shiratsuchi, M., Ramos, J. E., & Podack, E. R. (2013). Inhibition of intracellular bacterial replication in fibroblasts is dependent on the perforin-like protein (perforin2) encoded by macrophage-expressed gene 1. *Journal of Innate Immunity*, *5*(2), 185–194.

McCormack, R. M., Lyapichev, K., Olsson, M. L., Podack, E. R., & Munson, G. P. (2015). Enteric pathogens deploy cell cycle inhibiting factors to block the bactericidal activity of Perforin-2. *eLife*, *4*, e06505.

McCormack, R., & Podack, E. R. (2015). Perforin-2/Mpeg1 and other pore-forming proteins throughout evolution. *Journal of Leukocyte Biology*, *98*(5), 761–768.

McCormack, R. M., Szymanski, E. P., Hsu, A. P., Perez, E., Olivier, K. N., Fisher, E., et al. (2017). MPEG1/perforin-2 mutations in human pulmonary nontuberculous mycobacterial infections. *JCI Insight*, *2*(8), e89635.

Miyagi, A., & Scheuring, S. (2016). Automated force controller for amplitude modulation atomic force microscopy. *Review of Scientific Instruments*, *87*(5), 053705.

Moors, M. A., Levitt, B., Youngman, P., & Portnoy, D. A. (1999). Expression of listeriolysin O and ActA by intracellular and extracellular Listeria monocytogenes. *Infection and Immunity*, *67*(1), 131–139.

Morante, K., Bellomio, A., Gil-Carton, D., Redondo-Morata, L., Sot, J., Scheuring, S., et al. (2016). Identification of a membrane-bound prepore species clarifies the lytic mechanism of actinoporins. *Journal of Biological Chemistry*, *291*(37), 19210–19219.

Mulvihill, E., Sborgi, L., Mari, S. A., Pfreundschuh, M., Hiller, S., & Muller, D. J. (2018). Mechanism of membrane pore formation by human gasdermin-D. *EMBO Journal*, *37*(14), e98321.

Mulvihill, E., van Pee, K., Mari, S. A., Muller, D. J., & Yildiz, O. (2015). Directly observing the lipid-dependent self-assembly and pore-forming mechanism of the cytolytic toxin listeriolysin O. *Nano Letters*, *15*(10), 6965–6973.

Munguira, I., Casuso, I., Takahashi, H., Rico, F., Miyagi, A., Chami, M., et al. (2016). Glasslike membrane protein diffusion in a crowded membrane. *ACS Nano*, *10*(2), 2584–2590.

Munguira, I. L. B., Takahashi, H., Casuso, I., & Scheuring, S. (2017). Lysenin toxin membrane insertion is pH-dependent but independent of neighboring lysenins. *Biophysical Journal*, *113*(9), 20292036.

Ni, T., Jiao, F., Yu, X. L., Aden, S., Ginger, L., Williams, S. I., et al. (2020). Structure and mechanism of bactericidal mammalian perforin-2, an ancient agent of innate immunity. *Science Advances*, *6*(5), eaax8286.

Pang, S. S., Bayly-Jones, C., Radjainia, M., Spicer, B. A., Law, R. H. P., Hodel, A. W., et al. (2019). The cryo-EM structure of the acid activatable pore-forming immune effector macrophage-expressed gene 1. *Nature Communications*, *10*, 4288.

Parsons, E. S., Stanley, G. J., Pyne, A. L. B., Hodel, A. W., Nievergelt, A. P., Menny, A., et al. (2019). Single-molecule kinetics of pore assembly by the membrane attack complex. *Nature Communications*, *10*, 2066.

Pipkin, M. E., & Lieberman, J. (2007). Delivering the kiss of death: Progress on understanding how perforin works. *Current Opinion in Immunology*, *19*(3), 301–308.

Podobnik, M., Marchioretto, M., Zanetti, M., Bavdek, A., Kisovec, M., Cajnko, M. M., et al. (2015). Plasticity of listeriolysin O pores and its regulation by pH and unique histidine. *Scientific Reports*, *5*, 9623.

Ramachandran, R., Tweten, R. K., & Johnson, A. E. (2005). The domains of a cholesterol-dependent cytolysin undergo a major FRET-detected rearrangement during pore formation. *Proceedings of the National Academy of Sciences of the United States of America*, *102*(20), 7139–7144.

Reboul, C. F., Whisstock, J. C., & Dunstone, M. A. (2016). Giant MACPF/CDC pore forming toxins: A class of their own. *Biochimica et Biophysica Acta-Biomembranes*, *1858*(3), 475–486.

Ross, M., Steinem, C., Galla, H. J., & Janshoff, A. (2001). Visualization of chemical and physical properties of calcium-induced domains in DPPC/DPPS Langmuir-Blodgett layers. *Langmuir*, *17*(8), 2437–2445.

Ruan, Y., Kao, K., Lefebvre, S., Marchesi, A., Corringer, P. J., Hite, R. K., et al. (2018). Structural titration of receptor ion channel GLIC gating by HS-AFM. *Proceedings of the National Academy of Sciences of the United States of America*, *115*(41), 10333–10338.

Ruan, Y., Rezelj, S., Zavec, A. B., Anderluh, G., & Scheuring, S. (2016). Listeriolysin O membrane damaging activity involves arc formation and lineaction—Implication for Listeria monocytogenes escape from phagocytic vacuole. *PLoS Pathogens*, *12*(4), e1005597.

Schuerch, D. W., Wilson-Kubalek, E. M., & Tweten, R. K. (2005). Molecular basis of listeriolysin O pH dependence. *Proceedings of the National Academy of Sciences of the United States of America*, *102*(35), 12537–12542.

Serna, M., Giles, J. L., Morgan, B. P., & Bubeck, D. (2016). Structural basis of complement membrane attack complex formation. *Nature Communications*, 7, 10587.

Shatursky, O., Heuck, A. P., Shepard, L. A., Rossjohn, J., Parker, M. W., Johnson, A. E., et al. (1999). The mechanism of membrane insertion for a cholesterol-dependent cytolysin: A novel paradigm for pore-forming toxins. *Cell*, *99*(3), 293–299.

Shepard, L. A., Shatursky, O., Johnson, A. E., & Tweten, R. K. (2000). The mechanism of pore assembly for a cholesterol-dependent cytolysin: Formation of a large prepore complex precedes the insertion of the transmembrane beta-hairpins. *Biochemistry*, *39*(33), 10284–10293.

Tamm, L. K., & Mcconnell, H. M. (1985). Supported phospholipid-bilayers. *Biophysical Journal*, *47*(1), 105–113.
Tilley, S. J., Orlova, E. V., Gilbert, R. J. C., Andrew, P. W., & Saibil, H. R. (2005). Structural basis of pore formation by the bacterial toxin pneumolysin. *Cell*, *121*(2), 247–256.
Uchihashi, T., Kodera, N., & Ando, T. (2012). Guide to video recording of structure dynamics and dynamic processes of proteins by high-speed atomic force microscopy. *Nature Protocols*, 7(6), 1193–1206.
Vadia, S., Arnett, E., Haghighat, A. C., Wilson-Kubalek, E. M., Tweten, R. K., & Seveau, S. (2011). The pore-forming toxin listeriolysin O mediates a novel entry pathway of L. monocytogenes into human hepatocytes. *PLoS Pathogens*, 7(11), e1002356.
van Pee, K., Mulvihill, E., Muller, D. J., & Yildiz, O. (2016). Unraveling the pore-forming steps of pneumolysin from Streptococcus pneumoniae. *Nano Letters*, *16*(12), 7915–7924.
van Pee, K., Neuhaus, A., D'Imprima, E., Mills, D. J., Kuhlbrandt, W., & Yildiz, O. (2017). CryoEM structures of membrane pore and prepore complex reveal cytolytic mechanism of pneumolysin. *eLife*, *6*, e23644.
Voskoboinik, I., Whisstock, J. C., & Trapani, J. A. (2015). Perforin and granzymes: Function, dysfunction and human pathology. *Nature Reviews Immunology*, *15*(6), 388–400.
Wiens, M., Korzhev, M., Krasko, A., Thakur, N. L., Perovic-Ottstadt, S., Breter, H. J., et al. (2005). Innate immune defense of the sponge Suberites domuncula against bacteria involves a MyD88-dependent signaling pathway. Induction of a perforin-like molecule. *Journal of Biological Chemistry*, *280*(30), 27949–27959.
Yoshiyuki, S., Kenichi, H., Terumi, N., & Hideshi, K. (1996). A novel protein, lysenin, that causes contraction of the isolated rat aorta: Its purification from the coelomic fluid of the earthworm, eisenia foetida. *Biomedical Research*, *17*(3), 197–203.

CHAPTER EIGHT

The use of giant unilamellar vesicles to study functional properties of pore-forming toxins

Saša Aden, Tina Snoj, and Gregor Anderluh*
Department for Molecular Biology and Nanobiotechnology, National Institute of Chemistry, Ljubljana, Slovenia
*Corresponding author: e-mail address: gregor.anderluh@ki.si

Contents

Methods in Enzymology, Volume 649
ISSN 0076-6879
https://doi.org/10.1016/bs.mie.2021.01.016

Abstract

Pore-forming toxins (PFTs) act upon lipid membranes and appropriate model systems are of great importance in researching these proteins. Giant unilamellar vesicles (GUVs) are an excellent model membrane system to study interactions between lipids and proteins. Their main advantage is the size comparable to cells, which means that GUVs can be observed directly under the light microscope. Many PFTs properties can be studied by using GUVs, such as binding specificity, membrane reorganization upon protein binding and oligomerization, pore properties and mechanism of pore formation. GUVs also represent a good model for biotechnological approaches, e.g., in applications in synthetic biology and medicine. Each research area has its own demands for GUVs properties, so several different approaches for GUVs preparations have been developed and will be discussed in this chapter.

1. Giant unilamellar vesicles (GUVs) as a membrane model system

Pore Forming Toxins (PFTs) act at the level of cell membranes and therefore appropriate lipid model systems are needed to study their interactions with membranes *in vitro*. Many biophysical approaches and corresponding model membrane systems have been used to study PFTs, e.g., supported lipid bilayers, vesicles, nanodiscs, planar lipid bilayers, etc. (Anderluh & Lakey, 2008; Beseničar, Maček, Lakey, & Anderluh, 2006; Marchioretto, Podobnik, Dalla Serra, & Anderluh, 2013; Podobnik, Kraševec, Zavec, et al., 2016). Vesicles resemble the basic compartment structure of biological cells in that they consist of a self-enclosed lipid bilayer that mimics the plasma membrane, and are therefore a very popular model system to study membrane-interacting proteins. Various experimental approaches can be used to produce vesicles of different sizes (from nanometers to tens of microns) (Dimova, Aranda, Bezlyepkina, et al., 2006). GUVs are the largest among vesicle membrane systems and have high impact in membrane biophysics research. Due to their size (from 1 to several tens of micrometers), they can be observed under the light microscope, allowing direct observation and manipulation of membrane interactions. On the other hand, studying interactions with smaller vesicles usually requires the use of indirect methods, and in addition, due to their small size, i.e., in sonicated vesicles, they exhibit higher membrane curvature, which may affect membrane properties and consequently protein interactions. Cell-sized GUVs therefore much better reflect the properties and behavior of the cell membrane (Dimova et al., 2006; Oropeza-Guzman, Ríos-Ramírez, & Ruiz-Suárez, 2019; Walde, Cosentino, Engel, & Stano, 2010). However, since GUVs lack the cytoskeleton to offer support from the inside, they are extremely fragile and tend to be easily damaged by mechanical or osmotic stress. Fragility is the major drawback of GUVs, and common manipulation techniques that work for smaller vesicles (such as washing by centrifugation or gel filtration) are usually ineffective for GUVs.

One of the most important advantages of membrane systems, such as GUVs, over cellular systems is the precise control of lipid membrane composition. Very often the lipid composition is simple, so that the unknown factors in the studied system are reduced. For example, lipids commonly found in membranes of animals, plants, and bacteria, such as phosphatidylcholine, phosphatidylserine, phosphatidylethanolamine (PE), cholesterol, and sphingomyelin (SM), are often used because of their affordable commercial availability. On the other hand, lipids extracted from natural sources such as plants (Buré, Cacas, Wang, et al., 2011) or archaea (Morii, Yagi, Akutsu, et al., 1999) are much more difficult to obtain in high purity and large quantities. These lipids are crucial for the functioning of host organisms and important for the interaction of virulence factors with plant membranes (Lenarčič, Albert, Böhm, et al., 2017; Mamode Cassim, Gouguet, & Gronnier, 2019). They are also interesting for their beneficial properties, e.g., it has been shown that archaeal lipids can form very large and stable GUVs (Rezelj, Kozorog, Švigelj, et al., 2018). GUVs can also be formed from erythrocyte ghost membranes (Mikelj, Praper, Demič, et al., 2013; Montes, Alonso, Goñi, & Bagatolli, 2007) or from native pulmonary surfactant membranes (Bernardino de la Serna et al., 2004, 2009). Finally, synthetic amphiphilic diblock copolymers can also be used to make giant vesicular structures called polymerosomes, which can serve as capsules for drug delivery (Discher, Won, Ege, et al., 1999).

GUVs are a very practical tool for studying a wide range of membrane properties and membrane-protein interactions. Examples of the use of GUVs for such studies are following diffusion of lipids and integral membrane proteins (Wang, Ingram, & Weisshaar, 2010), lipid membrane viscosity (Tabaei, Gillissen, Kim, et al., 2016), curvature and elasticity of lipid membrane (Olbrich, Rawicz, Needham, & Evans, 2000; Rawicz, Olbrich, McIntosh, et al., 2000), lipid-lipid, lipid-protein and protein-protein interactions (Kahya, 2010), imaging of lipid phase separation (Cans, Andes-Koback, & Keating, 2008; Zhao, Wu, Shao, et al., 2007), arrangement and packing of lipids in microdomains (Kahya, 2010; Kahya, Scherfeld, & Schwille, 2005), lipid dynamics and organization in rafts (Kahya, Scherfeld, Bacia, & Schwille, 2004; Sezgin, Levental, Mayor, & Eggeling, 2017), etc.

In addition, GUVs are very often used for synthetic biology applications, such as building sensors and engineered surfaces (Chung, Lowe, Chan, et al., 2009; Zhou, Loftus, Mulley, & Jenkins, 2010), and constructing synthetic cells (Noireaux, Maeda, & Libchaber, 2011). There is still quite a long way to construct a complete minimal synthetic cell that would be capable of self-regulation and self-replication; however, some important steps have been made to achieve this goal using GUVs. For example, cell-free protein synthesis within GUVs (Nishimura, Matsuura, Nishimura, et al., 2012), the design of a vesicle bioreactor (Noireaux & Libchaber, 2004), the design of the light-guided motility of a minimal synthetic cell (Bartelt, Steinkühler, Dimova, & Wegner, 2018), insertion of basic biological functions inside GUVs such as replication (Kita, Matsuura, Sunami, et al., 2008), lipid synthesis (Exterkate, Caforio, Stuart, & Driessen, 2018), reconstruction of cytoskeleton (Hürtgen, Vogel, & Schwille, 2019; Pontani, van der Gucht, Salbreux, et al., 2009), signaling pathways (Hui & Vale, 2014; Li, Banjade, Cheng, et al., 2012), cell division (Furusato, Horie, Matsubayashi, et al., 2018; Kretschmer, Ganzinger, Franquelim, & Schwille, 2019), and self-replication (Kurihara, Tamura, Shohda, et al., 2011). Compartmentalization (Elani, Law, & Ces, 2015) and communication between synthetic cells (Aufinger & Simmel, 2019; Lentini, Santero, Chizzolini, et al., 2014) have also been successfully applied to synthetic cells.

2. Methods for GUVs formation

Various methods have been developed to form GUVs, such as gentle hydration, electroformation, droplet-transfer method, or by using microfluidics, etc. Using appropriate methods, GUVs can be prepared from quite simple to more complex binary or ternary lipid mixtures, as well as from natural extracts.

2.1 Gentle hydration method

First records of GUVs preparations date back to 1969, when it was observed that phospholipid membranes can swell in sucrose solution (Reeves & Dowben, 1969). The swelling method (Faucon, Mitov, Méléard, et al., 1989) is still in use nowadays. In the spontaneous swelling method, also named gentle hydration, the thin lipid film is deposited on a glass substrate, where lipids swell into vesicles by rehydration in aqueous solutions (Méléard, Bagatolli, & Pott, 2009). It was shown that the negative charges on the lipids do not affect the lipid composition of the GUVs, formed by gentle hydration, with regard to the initial lipid mixture. On the other hand, the incorporation of charged lipids is one of the limitations of the electroformation of GUVs (described in Section 2.2) and despite the recent improvements in electroformation protocols, gentle hydration remains the preferred method when the lipid mixture contains a higher percentage of charged lipids or is carried out under physiological salt conditions (Morales-Penningston, Wu, Farkas, et al., 2010; Walde et al., 2010). However, spontaneously swelled GUVs show more defects (encapsulated vesicles, different lipid structures attached to the inner or outer membrane leaflet), their yield is variable and low, formation usually takes several days and works well only with restricted lipid compositions and ionic conditions (Rodriguez, Pincet, & Cribier, 2005). The hydration method was improved by gel-assisted GUVs formation, which allowed to use a wider range of lipid compositions, buffer conditions and can even encapsulate biomolecules (Horger, Estes, Capone, & Mayer, 2009; Weinberger, Tsai, Koenderink, et al., 2013).

2.2 Electroformation method

Electroformation has been the preferred method to produce GUVs because of its high yields of spherical giant vesicles with a narrower size distribution, high unilamellarity and fewer defect structures (Bagatolli, Parasassi, & Gratton, 2000; Rodriguez et al., 2005). Electroformation was initially proposed in 1986 (Angelova & Dimitrov, 1986) and is an alternative method to spontaneous swelling with enhancement by applied electric voltage. Electroformation is a two-step procedure, where lipid film is firstly deposited on the electrode (e.g., glass slide with conductive coating of indium tin oxide (ITO) or platinum wire) (Fig. 1A and B). The second step is electroformation process itself, where lipid film is rehydrated by aqueous solution and applied electric current (Walde et al., 2010). The swelling of the lipid film to form vesicles (Fig. 1A) is poorly understood from a theoretical point of view, although osmotic and electrostatic forces appear to play some role. One possible mechanism is that the bilayers should be separated and destabilized to bend and fuse to form vesicles. External electric fields can facilitate both by creating a gentle mechanical agitation (Angelova & Dimitrov, 1986). Success in producing relatively homogeneous GUVs depends on finding a set of optimal experimental conditions, i.e., type, composition, and amount of lipids used, thickness of the film and its mode of deposition, temperature (preferably above the transition temperature for

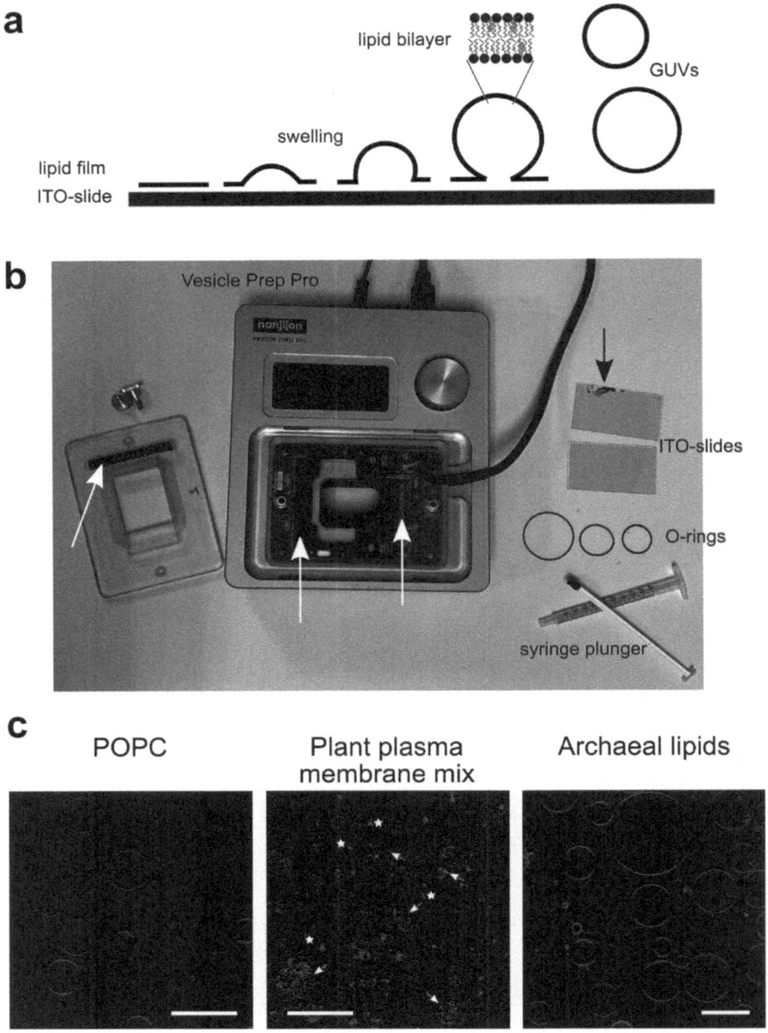

Fig. 1 GUVs electroformation. (A) Schematic representation of GUVs electroformation on ITO-slides. First the lipid film is deposited on the conductive site of the slide and after the addition of aqueous solution and voltage application the lipids swell in the solution, forming closed bilayer, GUVs. (B) Photography of the Vesicle Prep Pro apparatus (Nanion, Germany) for GUVs electroformation. White arrows mark the electrodes, which are touching conductive sites of the ITO-slides. ITO-slides are on the right sides. Black arrow marks the burn of the rehydrating glucose solution on the glass after electroformation. O-rings on the photo are of sizes 28×1 mm, 18×1 mm and 16×1 mm from left to right. The black rubber part of the syringe plunger is used to apply silicon grease on the O-rings. (C) Examples of GUVs prepared by electroformation. GUVs prepared from a single lipid (1-palmitoyl-2-oleoyl-*sn*-glycero-3-phosphocholine, POPC), from the mixture of POPC, glycosyl inositol phosphoryl ceramides (GIPC), plant-specific

all lipids), parameters of the applied electric field, and hydration medium (Angelova & Dimitrov, 1986; Oropeza-Guzman et al., 2019; Walde et al., 2010). Electroformation has been optimized for many different lipid mixtures (Fig. 1C) (e.g., to improve compatibility with charged lipids (Steinkühler, De Tillieux, Knorr, et al., 2018)) and aqueous environments, even high salinity conditions with physiologically relevant electrolyte concentrations (Méléard et al., 2009; Mikelj et al., 2013; Montes et al., 2007; Pott, Bouvrais, & Méléard, 2008). However, the yield is optimal for uncharged lipids and low salt conditions, as charges usually interfere with the process (Göpfrich, Haller, Staufer, et al., 2019).

The most basic electroformation protocol is divided into three steps (Méléard et al., 2009). In the first step, which starts as soon as the electroformation chamber is filled with the desired solution, the electrodes are connected to a function generator and an alternating voltage is applied. It is steadily increased at a fixed frequency up to its maximum value at the end of step one. The GUVs begin to bud from the swelling lipid film and gradually increase in size. At this stage, the electric field must be strong enough to maintain a globular shape of the growing vesicles. The second step or the swelling period, during which the electric field parameters are constant, is crucial for controlling the final size of the GUVs. This step can be omitted if the electric field amplitude slowly increases to its maximum (at low ionic strengths). The third step is the "rebounding" period, where the frequency is reduced to promote sealing of the vesicles and further detachment into solution. Vesicles bounce back and forth in the low frequency electric field oscillations like a rebounding ball (Méléard et al., 2009).

2.3 Droplet transfer method

In recent years, the aim to construct synthetic cells has led to increased interest in methods that allow the preparation of vesicles with different inner and

---

sphingolipids purified from tobacco leaves, and plant-based sterols (campesterol:β-Sitosterol:stigmasterol; 1:1:1 (n:n:n)) (POPC:GIPC:sterols 47:23:30 mol%), plant plasma membrane mix), and from lipids purified from hyperthermophilic archaeon *Aeropyrum pernix* K1 consisting of 91 mol% of 2,3-di-*O*-sesterterpanyl-*sn*-glycerol-1-phospho-1′-(2′-*O*-α-d-glucosyl)-*myo*-inositol ($C_{25,25}$-archaetidyl[glucosyl]inositol) and 9 mol% of 2,3-di-*O*-sesterterpanyl-*sn*-glycerol-1-phospho-*myo*-inositol ($C_{25,25}$-archaetidylinositol). In all cases rhodamine-DHPE was used to stain lipid membranes. When lipids are purified from natural sources or when complex lipid mixtures are used, the yield of large spherical GUVs is not so high and, additionally, many defects (labeled in plant plasma membrane mix GUVs with white arrows) can be observed, together with clumps of residual lipid material (labeled with stars). The scale bar is 50 μm.

outer solutions. A widely used method for this purpose is the droplet transfer method, which is also known by other names, such as phase-transfer method (Karamdad, Hindley, Bolognesi, et al., 2018), water-in-oil method (Elani, Gee, Law, & Ces, 2013), and emulsion-based method (Huang, Kim, & Arriaga, 2017). One of the main advantages of this method is that it allows the preparation of asymmetric membranes of GUVs (different inner and outer lipid composition of the membrane bilayer) (Pautot, Frisken, & Weitz, 2003) and GUVs with inner solution that differ from the surroundings. Thus, the droplet transfer method is the preferred approach to prepare, for example, GUVs with encapsulated *in vitro* transcription and translation (IVTT) reagents mix, which is one of the steps toward formation of synthetic cells. By using microfluidic techniques (described in Section 2.4), this approach can further increase reproducibility and allow better control of GUVs production, such as better manipulation, production of monodisperse GUVs, and increased efficiency of encapsulation (Karamdad, Law, Seddon, et al., 2014; Paterson, Reboud, Wilson, et al., 2014; van Swaay & deMello, 2013).

The basic principle of GUVs preparation by droplet transfer method lies in transferring the emulsion droplets (water/oil interface) through an aqueous phase (oil/water interface). First, the aqueous solution that we want to encapsulate inside GUVs is placed on the oil phase containing dissolved lipids. With vigorous stirring, an emulsion is formed in which the water droplets disperse in the oil and are enclosed by the lipid monolayer intended for the inner leaflet (Fig. 2A and B). The emulsion is then placed over the second oil/water interface, which contains the lipid monolayer on its surface, with the aqueous solution representing the outer GUV solution (Fig. 2C). The lipids deposited on the surface of the oil phase may have a different composition than the inner leaflet. Sedimentation causes the emulsion to sink through this second monolayer of lipids due to the heavier water droplets (Fig. 2D). As a result of sedimentation, the emulsion pulls the lipids of the second monolayer to complete the bilayer, and the GUVs are formed in the aqueous phase (Fig. 2D) (Pautot et al., 2003).

2.4 Microfluidic methods

Robust microfluidics-based approaches to the formation of GUV have been introduced as an alternative to electroformation. Several such techniques have achieved the formation of GUVs with controlled contents in a repeatable manner, using oil-water interfaces to define lipid membranes (similar

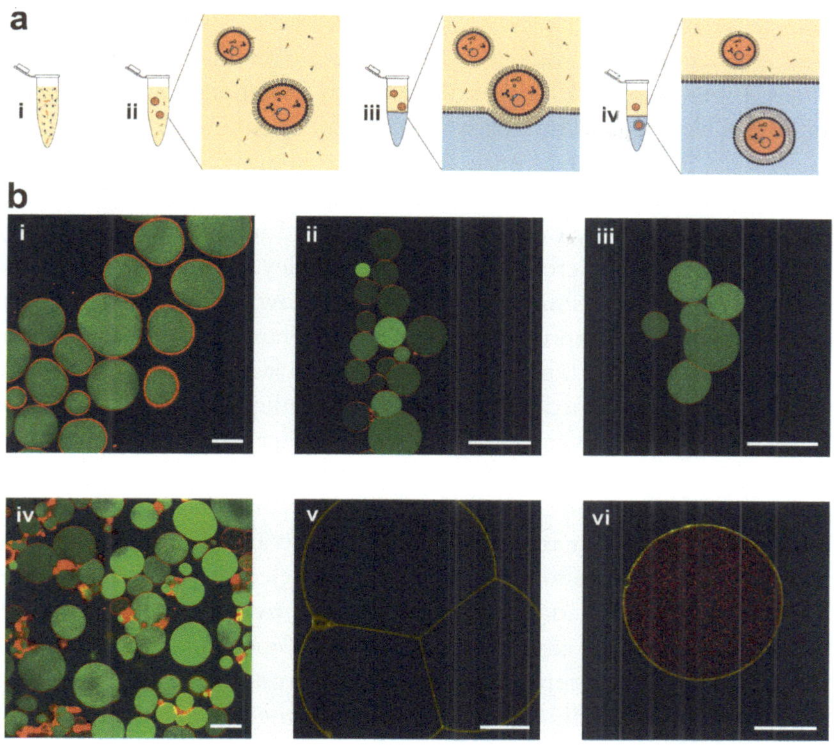

Fig. 2 (A) GUVs preparation by the droplet transfer method. i: Lipids are dissolved in the oil phase (yellow), ii: Emulsion is formed by adding the inner solution (orange; e.g., fluorescent probes or IVTT components such as amino acids, tRNA, *E. coli* S30 extract and plasmid with gene encoding desired protein for production inside GUVs or other special molecules for encapsulation inside GUVs) and stirring or mixing, iii: Emulsion is placed over the second monolayer of lipids on the oil/water interface, iv: Emulsion is sedimented over the outer aqueous solution (blue) by passing through second lipids monolayer and enclosing the bilayer to form GUVs. (B) Examples of GUVs prepared by droplet transfer method. i: GUVs with lipid composition 1,2-dioleoyl-*sn*-glycero-3-phosphocholine (DOPC): cholesterol in molar ratio 4:1 (1 mM lipids in mineral oil) and inner sucrose solution with fluorescein isothiocyanate (FITC)-dextran 70 kDa, ii: GUVs with lipid composition DOPC: cholesterol in molar ratio 1:1 (1 mM lipids in squalane) and inner sucrose solution with FITC-dextran 4 kDa. Prepared by centrifugation at 1500 × *g* in the microtiter plate, iii: the same as in ii, but prepared by centrifugation at 1000 × *g* in microtiter plate, iv: GUVs with lipid composition DOPC:cholesterol in molar ratio 4:1 (1 mM lipids in mineral oil) and inner sucrose solution with FITC-dextran 150 kDa, prepared by centrifugation at 300 × *g* in a microtube, v: GUVs with lipid composition archaeal lipids: cholesterol in molar ratio 1:1 (1 mM lipids in mineral oil) and inner solution with IVTT reaction components, including mCherry gene, prepared by centrifugation at 2000 × *g* in a microtube. There is no detectable fluorescent signal from mCherry after 7 h, vi: GUVs prepared like in v, but observed 70 h later. Fluorescent signal from mCherry is apparent. i–iv: Green fluorescent signal represents fluorescence of FITC-dextran and red fluorescent signal fluorescence of rhodamine-DHPE in the GUVs' membrane. v and vi: Yellow fluorescent signal represents fluorescence of NBD-PE in the GUVs' membrane and red fluorescent signal fluorescence of mCherry. Scale bar is 50 μm in all panels.

principle as described in Section 2.3). Vesicles are produced either by continuous flows using microdevices that hydrodynamically focus fluid streams (Utada, Lorenceau, Link, et al., 2005) or more commonly by pulsed microfluidic jets that deform interfaces (Funakoshi, Suzuki, & Takeuchi, 2007; Stachowiak, Richmond, Li, et al., 2008, 2009). In such an approach, a planar lipid bilayer is formed at the interface of two aqueous droplets that are initially surrounded by oil (containing dissolved lipids) and then brought into contact in the microfluidic chamber (Bayley, Cronin, Heron, et al., 2008). By applying a short pulsed liquid jet flow ejected from a fine jet nozzle brought near the membrane, the membrane deforms and stretches into an outward bud. The vesicle is generated when its neck is pinched off. One GUV is formed for each pulse of the liquid jet against the lipid bilayer, which remains intact after the vesicle is formed. Repeated pulses against the same lipid bilayer trigger the formation of multiple, monodisperse vesicles. Jet of fluid can contain the desired components (e.g., proteins, fluorescent probes) that are trapped inside the vesicles. Since bilayer deformation is large (relative to molecular radii), any ejected materials, regardless of their size, concentration or chemical properties (e.g., charge), will be packed. Another great advantage of such direct encapsulation is the elimination of the washing step of the fragile GUVs from the excess components in the surrounding medium, which is needed in the case of electroformation or gentle hydration. At last, because the vesicle is created from a single bilayer, only unilamellar GUVs are expected to form that is not always the case in before mentioned lipid film hydration methods (Funakoshi et al., 2007; Stachowiak et al., 2008, 2009).

This method was further improved to avoid the use of oil for the delivery of lipids, because some lipids have poor solubility in oil solutions and in addition, oil residues may contaminate the generated GUVs. In this approach, planar bilayers were formed by delivering lipids through an aqueous phase in the form of small unilamellar vesicles (SUVs). Two aqueous droplets were kept separated by a thin divider in a chamber filled with oil (but without dissolved lipids), and the SUVs prepared with the oil-insoluble lipids of interest were loaded into the aqueous droplets. The SUVs gradually fused with the oil-water interface of the droplet, forming a lipid monolayer around each droplet. When the divider was removed, droplets came into contact and lipid bilayer was formed. GUVs with asymmetric bilayers can be generated that way (Richmond, Schmid, Martens, et al., 2011).

Electroformation may also be carried out in the microfluidic device. Paterson et al. created mammalian biomimetic vesicles with a membrane composition representative of typical mammalian membranes by on chip electroformation. Special polydimethylsiloxane cast with electroformation chamber and microtrap array chamber, connected with multiple channels, was placed between two ITO-coated slides. After being electroformed, the GUVs are immediately positioned using laminar flow into the trap microarrays for single vesicle analysis (Paterson et al., 2014).

3. Electroformation of GUVs

We will describe the electroformation of GUVs by using Vesicle Prep Pro apparatus (Nanion, Germany). It is a relatively inexpensive, commercially available apparatus for simple and automated preparation of GUVs (Fig. 1B). Flexible protocol design, temperature control and other adjustable parameters (current amplitude, frequency) allow the formation of GUVs from a wide variety of lipids (e.g., highly charged lipids or lipids with high phase transition temperatures) (Fig. 1C).

3.1 Equipment and materials

1. Lipid mixtures in organic solvents (e.g., chloroform, chloroform: methanol)
2. 2mL glass vial (Supelco, Cat. No. 27134)
3. Sucrose buffer (290mM sucrose, 1mM MES, pH 5.7)
4. Glucose buffer (290mM glucose, 1mM MES, pH 5.7)
5. Protein buffer (20mM MES, 150mM NaCl, pH 5.7)
6. Sterile filters (0.22μM size) and syringes
7. Osmometer (Osmomat 3000, Gonotec)
8. Vesicle Prep Pro station with electroformation chamber, ITO slides and O-ring
9. Voltmeter
10. Silicon grease (e.g., Rotisilon, Roth: Cat. No. 8537.1) and syringe plunger with rubber stopper
11. Conical 15mL Falcon centrifuge tube
12. Vacuum concentrator (e.g., Eppendorf Concentrator plus)

3.2 Procedure

1. Prepare sucrose buffer, glucose buffer, and protein buffer. Use chemicals of the highest purity. See also Note 1

2. All solutions must be isotonic to prevent fragile GUVs from collapsing (e.g., bursting or shrinking due to the osmotic difference between the inner and outer solutions). Measure the osmolarity of all solutions using an osmometer. Adjust the osmolarity by adding more water or NaCl (the maximum difference allowed is ± 1 mOsmol/kg).
3. Filter the buffers in a sterile environment using sterile syringes and 0.22 μM filters
4. Aliquot the solutions (sucrose buffer at least 800 μL, glucose buffer at least 550 μL, protein buffer at least 3 mL) and store them at −20 °C to prevent bacterial contamination. See also Note 2
5. Prepare desired lipid mixture (e.g., 5 mM POPC:cholesterol:fluorescent membrane probe Rhodamine-DHPE in a molar ratio of 4:1:0.5, dissolved in chloroform) in the glass vial. Allow the lipids to warm up at room temperature (21 °C) for approximately 30 min before handling. See also Notes 3 and 4
6. Create the appropriate electroformation protocol and upload it onto the Vesicle Prep Pro station using the VesicleControl software (e.g., Step 1: Increase the amplitude from 0 to 3 V at a frequency of 5 Hz and a temperature of 36 °C in 15 min. Step 2: Decrease the amplitude from 3 to 1.6 V and the frequency from 5 to 1 Hz in 4 h. Step 3: Turn the voltage to 0 V). See also Note 5
7. Take out a pair of ITO-slides from the package and make sure that there is no protective folium attached to their surface
8. Identify and mark the conductive sides of the ITO-slides by using a common voltmeter
9. Using the Hamilton syringe place 10 μL of lipid mixture on the conductive side in the center of one of the ITO-slides. Draw a spiral with the lipid mixture. Pay attention that there is always a drop of solution at the Hamilton syringe's tip to avoid scratching of the ITO-slide when dispersing lipids. See also Note 6
10. Dry the ITO-slide with lipid deposit for at least 30 min under the reduced pressure in the vacuum concentrator
11. Preheat sucrose solution to the starting temperature and turn on the Vesicle Prep Pro apparatus. Select the appropriate protocol but do not start it yet. Preheat the apparatus first and wait until the starting temperature is reached
12. Place the ITO-slide in the bottom part of the Vesicle Prep Pro chamber with the lipid film facing up

13. Prepare the rubber O-ring (28 × 1 mm) by coating it with silicone grease on one side with the help of syringe plunger with rubber stopper. This should be done carefully, taking care not to use the excess of grease. Place the O-ring around the dried lipids
14. Add 750 μL of the preheated sucrose buffer with a pipette. Start on one side of the O-ring and slowly move toward the middle. Hydrating should be done extremely carefully to keep the lipid film attached to the ITO-slide. See also Notes 7 and 8
15. Align the other ITO-slide at one shorter edge so that it sits on top of the conductive pad of the bottom of the Vesicle Prep Pro Chamber with conductive side facing down. Slowly lower the other side until the slide rests on the O-ring. In this way it can be avoided having air bubbles trapped between the two ITO-slides
16. Put the top part of the Vesicle Prep Pro chamber on the station, so that the conductive pads are at different ends of the slides. The conductive pad of the top part needs to make contact with the bottom slide
17. Place the screws in the designated holes. Tighten them so the O-ring gets slightly squeezed and forms a tight seal with the slides. See also Note 8
18. Run the protocol. Don't forget to remove light sources when using fluorescent probes: cover the monitor with a pad (the blue lights are steadily on while the protocol is running) and put a dark box over the entire station
19. Remove the vesicles by pipette immediately as soon as the protocol is finished, as they are not stable on ITO-slides. Slowly release them on the wall of the conical 15 mL Falcon tube and let them acclimate to room temperature for 15 min. If the air bubble is captured at the bottom of the tube, it should be removed by pipette
20. Add 500 μL of the glucose buffer. Do it drop by drop with automated pipette just above the surface of the sucrose solution containing GUVs, i.e., one drop every 3 s. Sucrose buffer should be equilibrated at room temperature. Leave GUVs to sediment for 15 min. See also Note 9
21. Add 1 mL of the protein buffer drop by drop, i.e., one drop every 2 s. Wait 5 min and add another 2 mL of the protein buffer
22. Leave the GUVs at room temperature or at 4 °C overnight. Put the Falcon tube in a safe place without any disturbances, wrapped in an aluminum foil

23. Carefully pipette away the solution above the sedimented GUVs, leaving about 200 μL at the bottom. Transfer the solution with GUVs to 1.5 mL Eppendorf tube. Use the GUVs the same day. Store them at 4 °C if necessary, but preferably use them within 4 days

3.3 Notes

1. Concentrations of sucrose and glucose depend on the osmolarity of the salt solution in the experiments in which GUVs are used and can be adapted based on the used protein buffer. Typical sugar concentration is between 100 and 300 mM. Other sugars may also be employed (e.g., 210 mM to 1 M D-sorbitol as an intra-GUV solution (e.g., Barthmes, Liao, Jiang, et al., 2016). Different buffers and pH ranges, or even MiliQ water, may be employed to match the experimental conditions, where GUVs are used
2. It is important not to reuse the solutions once opened and to take fresh solutions for each experiment because of the high sugar concentration and potential contamination with microorganisms
3. We observed that choice of the right highly volatile organic solvent may be important for GUV production from specific lipid mixtures. Commonly used are pure chloroform or a mixture of chloroform and more polar methanol (e.g., in 9:1 volume ratio). Diethyl eter or trichloromethan may also be used (e.g., Barthmes et al., 2016). All lipid compounds should have about the same solubility in the chosen solvent, otherwise solvent evaporation from the lipid film will lead to the precipitation of the less soluble lipid and thus possibly influencing the exact composition of the final GUV membrane
4. It is important to protect the lipid mixture with the fluorescent probe from the light (e.g., turn the light off in a fume hood, cover a vial in aluminum foil, etc.).
5. Depending on the specific lipid mixture and ionic conditions the following parameters can be varied: temperature, voltage and frequency and their time course. Generally, with higher electrolyte concentrations in the medium, higher voltages and frequencies should be applied (Méléard et al., 2009). However, electroformation conditions at very high voltages may lead to lipid oxidation (Breton, Amirkavei, & Mir, 2015). If the lipid mixture contains too many charged lipids, electroformation may not succeed and some other method of GUVs preparation should be employed (e.g., gentle hydration, for which the presence of charged lipids is even beneficial) (Hishida, Seto, & Yoshikawa, 2005; Rodriguez et al., 2005).

6. Thickness of the deposited lipid film and its quality are important for the successful GUVs production. The lipids should be evenly spread in the center of the ITO-slide. The volume of lipids solution and lipid concentration should be adjusted to get optimal results. Typically, concentrations of 2.5–10 mM and volumes of 5–20 μL are appropriate for 28 × 1 mm O-ring
7. Different sizes of O-rings are available, e.g., 16 × 1 mm (for 250 μL of hydration solution), 18 × 1 mm (for 275 μL of hydration solution) and 28 × 1 mm (for 750 μL of hydration solution).
8. Sugar solutions may get burned into the ITO-slides at high temperatures. If the burned parts are localized only at the edges of the slide, such a slide can still remain functional (Fig. 1B). Burns are more likely to occur if the screws are overtightened and hydrating solution squeezed outside the O-ring. However, ITO-coated glass slides age with each subsequent use and this may lead to degradation of the quality and size of GUVs from some cationic lipid mixtures (Herold, Chwastek, Schwille, & Petrov, 2012).
9. Steps 20 and 21 are optional. The GUVs are transferred into the glucose solution to enhance their sedimentation to the bottom of the Falcon tube due to the density differences between the entrapped sucrose and the outer glucose solution. If needed, this also provides better optical contrast in the light microscope (Mally, Majhenc, Svetina, & Žekš, 2002).

4. Droplet transfer method

Protocol for the formation of GUVs from archaea-isolated lipids by droplet transfer method, is presented here. It is specific for this special lipid mixture, but can be optimized for other purposes.

4.1 Equipment and materials

1. Mineral oil (Sigma-Aldrich: Cat. No. M5904)
2. Lipids (e.g., archaeal lipids:cholesterol:fluorescent membrane probe NBD-PE in molar ratio 50:49.5:0.5 dissolved in chloroform; archaeal lipids purified from from hyperthermophilic archaeon *Aeropyrum pernix* K1 consisting of 91 mol% of $C_{25,25}$-archaetidyl[glucosyl]inositol and 9 mol% of $C_{25,25}$-archaetidylinositol)

3. 2 mL glass vial (Supelco, Cat. No. 27134)
4. Vacuum concentrator (Eppendorf Concentrator plus)
5. Water bath sonicator (Elma Transsonic 460)
6. Osmometer (Osmomat 3000, Gonotec)
7. Needle (BD Microlance 3, 0.45 × 16 mm)

4.2 Procedure

1. Place 85 μL of 8.5 mM lipids with a certain composition (i.e., in this case composed of archaeal lipids:cholesterol:fluorescent membrane probe NBD-PE in molar ratio 50:49.5:0.5; i.e., Rezelj et al. (2018) dissolved in chloroform in the glass vial
2. Place the vial in the vacuum concentrator and allow the chloroform to evaporate completely, i.e., for at least 2 h
3. Add 700 μL of mineral oil to the dried lipid film so that the final lipid concentration in mineral oil is approximately 1 mM. See also Note 1
4. Sonicate the solution of lipids in mineral oil in a water bath for 1 h at 70 °C
5. After sonication, mix the solution with a vortex and store in the dark at room temperature until further use
6. Prepare the inner GUVs solution by mixing the components that should be trapped inside the GUVs (e.g., desired chemical agents and buffers, biological molecules, fluorescent probes, IVTT reaction components (see also Notes 2–6)) to 11 μL final volume
7. Prepare the outer GUVs solution to 55 μL final volume. The outer GUVs solution can be different from the inner GUVs solution, but both must be osmotically balanced. Use an osmometer. See also Notes 6 and 7
8. Place 10 μL of the inner GUVs solution on top of 200 μL of the lipids dissolved in the mineral oil and vortex for 20 s, so that an emulsion forms
9. Place all of the prepared emulsion on top of 55 μL outer GUVs solution and incubate for 10 min. See also Note 8
10. Centrifuge at 2000 × g for 15 min. See also Note 9
11. GUVs sediment in the pellet. Collect them by puncturing the bottom of the tube with a needle where pellet is located. While puncturing, keep the tube lid open. Place the new tube under it and close the lid of the upper tube so that the flow of solution of GUVs through to hole to the new tube is promoted

12. Optionally, incubate the collected GUVs at the required temperature and time to promote protein expression in case of IVTT or other reactions that you wish to conduct inside GUVs, before observing the sample under the microscope
13. Place the sample on the cover glass and observe it under the microscope as described in Section 5

4.3 Notes

1. In case the vesicles do not form when lipids are dissolved in the mineral oil, one can also consider other non-polar solvents (e.g., hexane, dodecane, squalene, squalane).
2. Inner GUVs solution may be of desired composition and can contain various compounds, i.e., fluorescent probes at high concentration (i.e., several mM) or can be quite complex and can contain a lot of components that can even allow IVTT to be conducted in the interior of GUVs after formation. Inner solution in case of IVTT inside GUVs can be very complex and composed of the following components with final concentration: 50 mM HEPES, 0.2 mg/mL tRNA *E. coli* MRE600, 2.3 mM DTT, 1.5 mM ATP pH 8,0, 1.5 mM GTP pH 8,0, 0.9 mM CTP pH 8,0, 0.9 mM UTP, 34 μg/mL folinic acid, 0.75 mM cAMP, 27.6 mM NH_4OAc, 3 mM amino acid mix (all amino acids dissolved and mixed together as described in Caschera and Noireaux (2015)), 30 mM 3-PGA, 0.26 mM CoA, 0.33 mM NAD, 500 mM sucrose, 200 mM K-glutamate, 13 mM magnesium acetate, 5 nM DNA construct (fluorescent protein (e.g., mCherry) or fluorescent fusion protein (e.g., mCherry-PFT), T7 RNA Polymerase (200 U), RNase inhibitor (10 U), 33% S30 extract and DEPC water should be added to reach the final volume of the inner solution (Rezelj et al., 2018).
3. The S30 extract (ribosomes with transcription and translation factors from *E. coli*) can be prepared as described in Dreier and Plückthun (2011).
4. Preparation of IVTT components should be done in an RNase free environment
5. Before preparation of IVTT components check the solubility and pH stability of the components. Preferably, make aliquots for all stock solutions and use them only one when preparing the final IVTT reaction mix. Do not thaw stock solutions many times, due to instability of some components, especially some amino acids, tRNA and folinic acid

6. Outer reaction mix can be very similar to inner IVTT reaction mix, so that the concentrations are the same to adjust osmotic balance and to have feeding solution outside. It is important that the outer solution does not contain components which would promote gene expression (tRNA, DNA, T7 RNA polymerase, S30 extract).
7. To promote sedimentation of the vesicles, use sucrose in the inner solution and the same concentration of glucose in the outer solution. The final concentration of sucrose and glucose should be determined individually, depending on the lipid composition and other components in the system. In most of the cases 300 mM concentration could be sufficient, but it can go even up to 1 M
8. Incubation of the emulsion on the surface of the outer solution is an important step to allow the formation of the lipid monolayer on the oil/water interface. In the case vesicles do not form after centrifugation, optimization of the incubation time must be considered (10 min to several hours, depending on the lipid composition and non-polar solvents).
9. The centrifugation length and force should be optimized in case there is no vesicle formation and other different lipid compositions are used

5. Monitoring binding of PFTs to GUVs

Many studies have used GUVs to elucidate the interactions between PFTs and membrane, binding, oligomerization, pore formation mechanism and pore properties. Monitoring of PFTs binding to the membrane is used to study lipid specificity (Fig. 3A), domain specificity, lipid reorganization upon binding, effects of proteins on membranes upon binding, etc. Following the binding of PFTs to membranes is also important to determine the functionality of the proteins, e.g., after cell-free expression inside GUVs (Fig. 3B). The binding of proteins to GUVs membranes can be directly observed by employing fluorescent probes. PFTs can be fluorescently labeled by fusing them with fluorescent proteins such as green fluorescent protein or mCherry (Rezelj et al., 2018; Skočaj, Resnik, Grundner, et al., 2014). PFTs can also be labeled with fluorescently labeled antibodies that specifically bind to PFTs or their domains (Ni, Williams, Rezelj, et al., 2018). Another possibility is to label PFTs with fluorescent probes, such as nitrobenzoxadiazole (NBD), Alexa probes, etc., by covalently binding them to engineered amino acid side chains (Anderluh, Hong, Boetzel, et al., 2003; Malovrh, Viero, Serra, et al., 2003; Rojko, Kristan, Viero, et al., 2013).

There is also a large variety of fluorescent lipids that allow staining of lipid membrane. Basically, there are two possibilities for lipid labeling: chain-labeled or head group labeled lipid analogs. The use depends on the experiment. For studying the labeled lipid on the membrane surface, the labeling on the head group could have an important impact, so in this case chain-labeled lipids (e.g., BODIPY® Glycerophospholipids, NBD-labeled acyl chains) are preferred. On the other hand, head group labeled lipids can be used in cases, where properties of the chain regions are important, e.g., for phase determination in biological membranes. For head group labeling, phosphatidylethanolamine has been most commonly used as an anchor lipid for such probes (e.g., Rhodamine-PE, NBD-PE, Dansyl-PE, Texas Red-PE) (Klymchenko & Kreder, 2014; Maier, Oberle, & Hoekstra, 2002). One can also use fluorescently labeled sphingolipids, steroids and lipopolysaccharides. For membrane domains and lipid monolayers imaging dialkylcarbocyanine probes (DiI, DiO, DiD, DiR) are often used (Kahya et al., 2004). Fluorescent quenchers can be used to measure the density of the membrane packing, e.g., the quenching of NBD in the chain region of lipid probes by the aqueous phase quencher sodium dithionite (Huster, Müller, Arnold, & Herrmann, 2001). Fluorescent dyes can be employed also for tracking the changes in the environment, such as pH and changes of the solvent surrounding the dye (Demchenko, Mély, Duportail, & Klymchenko, 2009).

Many PFTs are specific for a particular membrane lipid. GUVs are an excellent system for studying the lipid specificity of PFTs. For example, *Toxoplasma gondii* perforin-like protein 1 (TgPLP1) has been shown to preferentially bind to PE-containing GUVs membrane (Ni et al., 2018). Another example is listeriolysin O, a member of the cholesterol dependent cytolysins (CDCs) produced by the pathogenic bacteria *Listeria monocytogenes*, which are specific for cholesterol. By using GUVs made from archaeal lipids, that contain myo-inostiol and glucose in their lipid headgroup, it was revealed that listeriolysin O can additionally bind to membranes in the absence of cholesterol (Fig. 3A) (Rezelj et al., 2018). Additionally, listeriolysin O was produced inside GUVs using IVTT and its functionality was monitored by binding to the GUVs membrane (Fig. 3B) (Rezelj et al., 2018). Equinatoxin II, a pore forming toxin from sea anemones, lipid specificity for SM was also studied by using GUVs (Schön, García-Sáez, Malovrh, et al., 2008). Furthermore, lipid domains can be directly observed in GUVs with fluorescent microscopy. In this way, the association of perfringolysin O, another CDC from bacteria

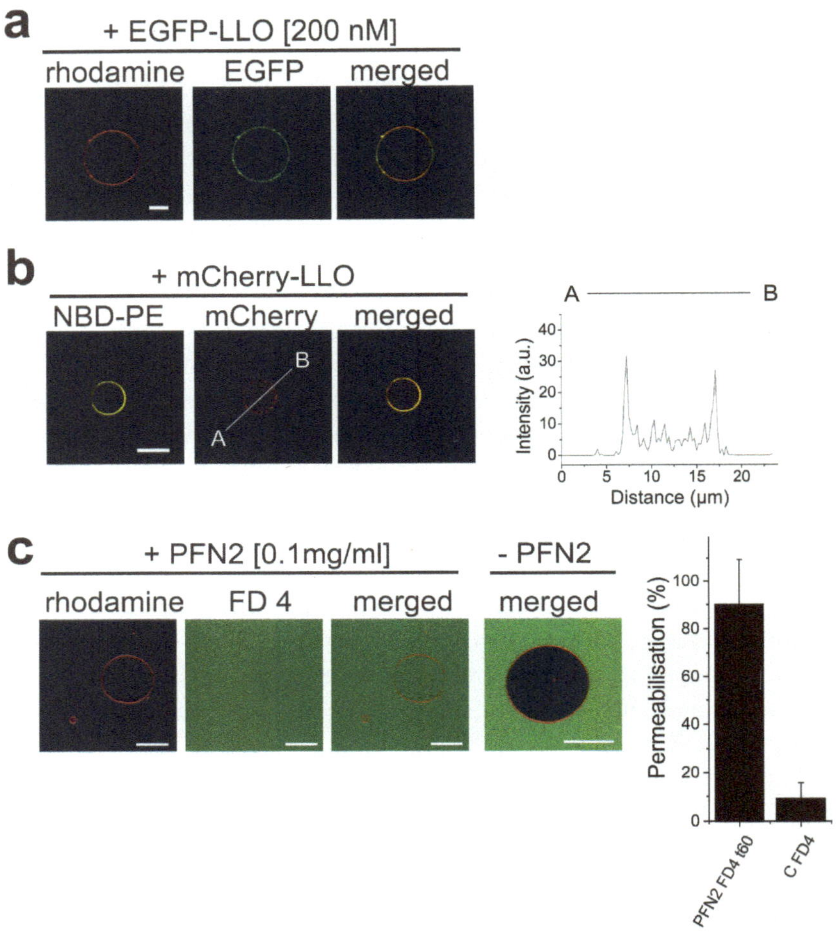

Fig. 3 GUVs as a membrane system for studying functional properties of PFTs. (A) Binding of LLO fused with enhanced green fluorescent protein (EGFP) to GUVs composed of archaeal lipids. The lipid membrane is visualized by using fluorescent membrane probe Lissamine Rhodamine B 1,2-Dihexadecanoyl-*sn*-Glycero-3-Phosphoethanolamine (rhodamine-DHPE). Scale bar is 10 μm. (B) IVTT of mCherry-LLO inside archaeal GUVs (+0.5 mol % fluorescent membrane probe 1,2-dipalmitoyl-sn-glycero-3-phosphoethanolamine-N-(7-nitro-2-1,3-benzoxadiazol-4-yl (NBD-PE)) produced by the droplet transfer method. On the right side of images are corresponding plots of mCherry protein fluorescence intensity (calculated in red channel as denoted on the image along the white lines from A to B). Scale bar is 10 μm. (C) Permeabilization of GUVs composed of POPC:POPS:*E. coli* lipid extract (80:10:10; *w/w*) (+1 mol% rhodamine-DHPE) by addition of perforin-2 (PFN2; 0.1 mg/mL) and fluorescent dextran of 4 kDa (FD 4) at pH 5.7 after 60-min incubation. Scale bar is 20 μm. *Examples in panels (A and B) are reproduced from Rezelj, S., Kozorog, M., Švigelj, T., et al. (2018). Cholesterol enriched archaeosomes as a molecular system for studying interactions of cholesterol-dependent cytolysins with membranes.* The Journal of Membrane Biology, 251, *1–15. https://doi.org/10.1007/s00232-018-0018-y, in panel (C) from Ni, T., Jiao, F., Yu, X., et al. (2020). Structure and mechanism of bactericidal mammalian perforin-2, an ancient agent of innate immunity.* Science Advances, 6, *eaax8286. https://doi.org/10.1126/sciadv.aax8286 by permission.*

Clostridium perfringens, with lipid ordered domains has been elucidated (Nelson, Chiantia, & London, 2010). It was also demonstrated that equinatoxin II, was able to form pores in SM-containing membranes when liquid-ordered and -disordered phases coexisted (Schön, García-Sáez, Malovrh, et al., 2008). And finally, it was shown by using GUVs that lysenin, a sphingomyelin-specific PFT from the earthworm, induced pore formation in membranes containing SM (Alam, Kobayashi, & Yamazaki, 2012) and that it is able to recognize the membrane domains containing SM (Ikenouchi, Suzuki, Umeda, et al., 2012).

5.1 Equipment and materials

1. Micro cover glasses (24 × 60 mm, thickness No. 1.5; Marienfeld Superior: Cat. No. 0107242)
2. Micro cover glasses (18 × 18 mm, thickness No. 1, Brand: Cat. No. 470819)
3. Aluminum foil
4. Bovine Serum Albumin (BSA, CAS: 9048-46-8, Serva: Cat. No. 11920.06)
5. MilliQ water prepared by the Millipore water system
6. Protein buffer (e.g., 20 mM MES, 150 mM NaCl, pH 5.7)
7. Electroformed and sedimented GUVs from Section 3
8. Fluorescently labeled or fluorescent fusion proteins
9. Leica TCS SP5 laser-scanning microscope

5.2 Procedure

1. Prepare 10 mL of 10 mg/mL BSA solution in MilliQ water
2. Coat cover glasses with BSA by spreading the 10 mg/mL BSA solution on top of the cover glasses (24 × 60 mm). After 5 min, wash off with MilliQ water and dry under the air stream. See also Note 1
3. Fold an aluminum foil and cut it into approximately 0.5 × 0.5 mm squares, which will be used as spacers between two cover glasses. See also Note 2
4. Mix electroformed and sedimented GUVs (Section 3) with fluorescent fusion proteins, e.g., prepare 10 μL final mixture with 9 μL GUV solution and 1 μL protein solution (e.g., 200 nM final concentration of fluorescent fusion protein. As a negative control, use protein buffer or fluorescent probe/protein instead. See also Notes 3–5

5. Incubate the mixture for 30 min at room temperature in the dark. See also Note 6
6. Place 5 μL of the prepared sample on top of BSA coated cover glass
7. Add 3 pre-cut squares of aluminum foil around the 5 μL drop on top of cover glass and then carefully place a smaller cover glass (18 × 18 mm) on top of that. In this way, the thickness of the solution is reduced and, consequently, the movement of the GUVs. See also Note 7
8. Place the slide with the sample on top of 40 × or 63 × oil-immersion objective of the laser-scanning microscope. Avoid any compression or distortion of the sample. See also Note 8
9. Use the appropriate lasers and detector wavelength according to the fluorescent probes used (e.g., for EGFP excitation use 488 nm and set emission detection from 505 to 530 nm. For rhodamine containing GUV membranes use laser at 543 nm and set emission detection from 570 to 620 nm). See also Note 9
10. Acquire images and optionally analyze binding using ImageJ software and documentation (http://rsb.info.nih.gov/ij/index.html) as described in Section 6.2

5.3 Notes

1. It is important to incubate the cover glass with BSA before adding and imaging the GUVs, otherwise the GUVs are prone to burst when in contact with glass
2. Aluminum foil may be used as a spacer between two cover glasses due to its affordable availability and because the exact distance between the cover glasses is not needed. In case the exact spacing is needed, there are commercially available fixed size spacers
3. When mixing GUVs with proteins, make sure to use pipettes with tips for 20–200 μL and not the smaller ones (1–10 μL), since the small tips can rapture the GUVs while pipetting through its small hole
4. As GUVs quickly sink to the bottom of the tube, GUVs should be evenly spread in the solution. Before pipetting to the final mixture with the proteins, the GUVs solutions should be mixed gently
5. If the GUVs are too dilute, they can be left to sediment again to the bottom of the tube and then part of the solution above removed and mixed again. In this way, a more concentrated GUVs solution can be obtained. This is possible when GUVs are produced in sucrose and sedimented in glucose, due to their density difference. In case GUVs are produced

in pure MilliQ water, the process of sedimentation might not be as effective or would take a long time (several days instead of several hours).

6. Depending on the protein used, longer or shorter incubation times and temperatures can be used
7. To minimize the movement of GUVs during microscopy, it is better to prepare GUVs in sucrose solution and let them sediment in glucose solution, as such GUVs also tend to sediment on the cover glass and thus movement of GUVs is further reduced
8. Immersion oil is toxic, so gloves should be used when handling. When using for fluorescence microscopy, it is advised to use special low-auto fluorescence oil, as many general oils will fluoresce in certain environments
9. Choose the appropriate combination of fluorescent probes on the protein and in the membrane to avoid spectral overlap. Also, be sure to use separate channels for excitation with each laser and fluorescence emission detection to avoid possible crosstalk

6. Analyzing permeabilization activity of PFTs by using GUvs

Permeabilization studies on GUVs are widely used to investigate the properties of transmembrane pores. The size of the pores formed can be estimated from the size of fluorescently labeled molecules that can pass through the pores from the outside of the vesicles, where they are added to GUVs, to the inside. In this way, pore properties of listeriolysin O and its pH dependent mutant (Kisovec, Rezelj, Knap, et al., 2017; Ruan, Rezelj, Zavec, et al., 2016), of pore forming proteins from immune system of vertebrates, such as perforin (Praper, Beseničar, Istinič, et al., 2010; Praper, Sonnen, Viero, et al., 2011) and perforin-2 (Ni, Jiao, Yu, et al., 2020) (Fig. 3C), etc., were studied. GUVs also allowed studying the mechanism of sugar transport (such as glucose) across the vesicle membrane by, e.g., equinatoxin II-induced pores (Mally et al., 2002). Similarly, transport kinetics through α-hemolysin pores were revealed by following calcein efflux from GUVs after addition of protein (Robinson, Kuhn, Eyer, & Dittrich, 2013).

Environmental conditions that allow optimal pore formation can also be studied employing GUVs, e.g., pH of the solution, presence of Ca^{2+} or other ions, certain inhibitors of binding and pore formation, etc. (Omersa, Aden, Kisovec, et al., 2020; Praper et al., 2010). Since GUVs can be observed under the microscope, it is easy to follow the changes in the membrane shape and morphology of GUVs, which can provide clues to the mechanism of protein-membrane interaction. For example, it has been shown that perforin is able to induce membrane invaginations (Praper, Sonnen, Kladnik, et al., 2011) that presumably arise from insertion of protein into one monolayer, creating a tension that is released by membrane bending (Zimmerberg & Kozlov, 2006). Recently membrane deformation with wrinkling and layer-by-layer peeling of multilamellar giant vesicles was shown for pneumolysin, a CDC from bacterium *Streptococcus pneumonia* (Drücker, Iacovache, Bachler, et al., 2019).

The permeabilization activity of PFTs can be directly monitored by optical microscopy due to the cell size of GUVs. This can be achieved by simply placing the vesicles on the glass slides or in a more sophisticated way by micromanipulating and trapping them in a microfluidic device (Mally et al., 2002). The first option is simple, it allows the analysis of many GUVs at the same time, and no special equipment other than the confocal microscope is needed. However, it is difficult to exchange solutes in the exterior of the GUVs. On the other hand, microfluidic approaches are associated with low throughput but allow precise control over membrane exposure to solvent and facile exchange of solutes (Drücker et al., 2019; Mally et al., 2002).

One of the most commonly used transport molecules are fluorescently labeled dextrans of different molecular weights (Praper, Sonnen, Viero, et al., 2011; Praper, Sonnen, Kladnik, et al., 2011; Ruan et al., 2016). The protocol for GUVs observation on glass slides, following the transport of dextrans into the vesicles as a consequence of pore formation by PFTs, will be described.

6.1 Equipment and materials

1. Micro cover glasses (24 × 60 mm, thickness No. 1.5; Marienfeld Superior: Cat. No. 0107242)
2. Micro cover glasses (18 × 18 mm, thickness No. 1, Brand: Cat. No. 470819)
3. Aluminum foil
4. Bovine Serum Albumin (BSA, CAS: 9048-46-8, Serva: Cat. No. 11920.06)
5. MilliQ water purified by the Millipore water system
6. Protein buffer (i.e., 20 mM MES, 150 mM NaCl, pH 5.7)
7. Electroformed and sedimented GUVs from Section 3

8. 10 mg/mL fluorescein isothiocyanate (FITC)-dextran solutions
9. Solution of pore-forming protein
10. Leica TCS SP5 laser-scanning microscope

6.2 Procedure

1. Prepare BSA coated cover glasses and aluminum foil squares as described in Section 5.2 (steps 1–3). See also Notes 1 and 2 in Section 5.3
2. Dissolve FITC-dextrans of different molecular weights (e.g., 4 kDa, 20 kDa, 70 kDa, etc.) in protein buffer to a the final concentration of 10 mg/mL. See also Note 1
3. Sonicate the FITC-dextran solutions in bath sonicator for 10 min to disrupt possible aggregates. See also Note 2
4. Mix electroformed and sedimented GUVs (Section 3) with FITC-dextran and pore-forming protein, e.g., prepare 10 μL final mixture with 8 μL GUV solution, 1 μL FITC-dextran solution (1 mg/mL final concentration), and 1 μL of PFT (500 nM final concentration). As a negative control, add protein buffer instead of the protein. See also Note 3 and Notes 3–5 in Section 5.3
5. Incubate the mixture for the required time (e.g., 30 min) at room temperature in the dark. See also Note 6 in Section 5.3
6. Place 5 μL of the prepared sample on top of BSA coated cover glass (24 × 60 mm).
7. Add 3 pre-cut squares of aluminum foil around the 5 μL drop on top of cover glass (24 × 60 mm) and then carefully place a smaller cover glass (18 × 18 mm) on top of that
8. Place the slide with the sample on top of 40 × or 63 × oil-immersion objective of laser-scanning microscope. See also Note 8 in Section 5.3
9. Use the appropriate lasers and detector wavelength according to the fluorescent probes used. For FITC-dextran excitation use laser with excitation at 488 nm and set emission detection from 497 to 534 nm (green channel). For rhodamine containing GUV membranes use laser at 543 nm and set emission detection from 570 to 620 nm (red channel). See also Notes 4 and 5
10. Capture images and analyze permeabilization of the GUVs using ImageJ software (Fig. 3C). Open the image you would like to analyze in ImageJ and split it into red, green and blue channels (go to Image → Color → Split channels). Now open your "red channel"

image and overlay it on top of "green channel" image (go to Image → Overlay → Add image and set opacity to 60%). Use the oval selection tool to select the desired area in the image and measure dextran fluorescence at this spot. Percentage of permeabilization of individual vesicle may be calculated by the following equation:

$$\%permeabilisation = \frac{GUVin}{GUVout}\ x\ 100$$

Value of GUV*out* represents average value of FITC-dextran fluorescence intensity from 5 different points outside GUVs on the individual image. GUV*in* is the FITC-dextran fluorescence intensity inside individual GUV on the same image.

6.3 Notes

1. Dextrans are hygroscopic, take care when handling. They should be acclimated for approximately 30 min before use after taken out of the fridge. Once opened, the package should be sealed with Parafilm for further storage
2. Dextrans can be passed across the gel filtration column to assure homogeneity. Fractions can be collected and analyzed for size by dynamic light scattering and for permeability of GUVs
3. 1 mg/mL final concentration of dextran is optimal; the image is bright enough at moderate laser intensities without the need to raise gain (voltage) on the photomultiplier tube (PMT) detector too high. Higher laser intensities needed at lower dextran concentrations may damage the GUVs, and raising gain on PMT can amplify weak signal but also amplifies the background noise, so it is advised to stay in the middle range
4. Dextrans with fluorescent probes other than fluorescein isothiocyanate (FITC) are available (e.g., Texas Red, Cascade Blue, different Alexa Fluor and Rhodamine dyes). Appropriate one should be selected according to the fluorescently labeled lipid in GUVs' membrane to avoid spectral overlap. Furthermore, be careful to use separate channels for excitation with each laser and fluorescence emission detection to certainly avoid possible crosstalk
5. The addition of dextran dissolved in protein buffer to GUVs does not normally trigger osmotic changes. However, GUVs from each lipid mixture are unique in their sensitivity, so attention should be paid if well-known morphological signs of osmotic imbalance occur (e.g., dumbbell or pear

shape, hollow inner circles, small buddings, tubulation) (as evidenced in Okano, Inoue, Koseki, and Suzuki (2018)). GUVs prepared in pure MilliQ water are usually more sensitive to the addition of dextran

7. Summary and conclusions

GUVs proved to be an excellent model for studying membrane-interacting proteins. GUVs have many advantages over smaller model membranes, such as their membrane curvature that is similar to cells, direct observation of PFT-membrane interaction, asymmetric bilayers can be formed, etc. On the other hand, GUVs can sometimes be more difficult to produce or require special equipment (electroformation chamber, microfluidic device) for production. Moreover, they are very fragile and must be handled with great care. There are a variety of methods for the preparation of GUVs. Electroforming is the most commonly used. For other special purposes, such as building artificial cells or entrapping special molecules, the droplet transfer method or other methods that use microfluidics are better choices. Successful formation of GUVs for a plethora of purposes is possible by choosing the right preparation method and adapting the protocol to the specific lipid mixture and ionic environment. Among the many possible biophysical characterizations of PFTs with GUVs, GUVs are mainly used to elucidate binding specificity and membrane permeabilization. GUVs are also very promising in biotechnological applications, such as in synthetic biology for the construction of artificial cells or as large scale membrane models of delivery systems.

Acknowledgment

The research in our laboratory is supported by the Slovenian Research Agency (program grant Molecular interactions P1-0391 and projects J7-1819 and J4-8225).

References

Alam, J. M., Kobayashi, T., & Yamazaki, M. (2012). The single-giant unilamellar vesicle method reveals lysenin-induced pore formation in lipid membranes containing sphingomyelin. *Biochemistry*, *51*, 5160–5172. https://doi.org/10.1021/bi300448g.

Anderluh, G., Hong, Q., Boetzel, R., et al. (2003). Concerted folding and binding of a flexible colicin domain to its periplasmic receptor TolA. *The Journal of Biological Chemistry*, *278*, 21860–21868. https://doi.org/10.1074/jbc.M300411200.

Anderluh, G., & Lakey, J. H. (2008). Disparate proteins use similar architectures to damage membranes. *Trends in Biochemical Sciences*, *33*, 482–490. https://doi.org/10.1016/j.tibs.2008.07.004.

Angelova, M. I., & Dimitrov, D. S. (1986). Liposome electroformation. *Faraday Discussions of the Chemical Society*, *81*, 303–311. https://doi.org/10.1039/dc9868100303.

Aufinger, L., & Simmel, F. C. (2019). Establishing communication between artificial cells. *Chemistry - A European Journal, 25*, 12659–12670. https://doi.org/10.1002/chem.201901726.

Bagatolli, L. A., Parasassi, T., & Gratton, E. (2000). Giant phospholipid vesicles: Comparison among the whole lipid sample characteristics using different preparation methods: A two photon fluorescence microscopy study. *Chemistry and Physics of Lipids, 105*, 135–147. https://doi.org/10.1016/S0009-3084(00)00118-3.

Bartelt, S. M., Steinkühler, J., Dimova, R., & Wegner, S. V. (2018). Light-guided motility of a minimal synthetic cell. *Nano Letters, 18*, 7268–7274. https://doi.org/10.1021/acs.nanolett.8b03469.

Barthmes, M., Liao, J., Jiang, Y., et al. (2016). Electrophysiological characterization of the archaeal transporter NCX_Mj using solid supported membrane technology. *The Journal of General Physiology, 147*, 485–496. https://doi.org/10.1085/jgp.201611587.

Bayley, H., Cronin, B., Heron, A., et al. (2008). Droplet interface bilayers. *Molecular BioSystems, 4*, 1191–1208. https://doi.org/10.1039/b808893d.

Bernardino de la Serna, J., Perez-Gil, J., Simonsen, A. C., et al. (2004). Cholesterol rules: Direct observation of the coexistence of two fluid phases in native pulmonary surfactant membranes at physiological temperatures. *Journal of Biological Chemistry, 279*(39), 40715–40722. https://doi.org/10.1074/jbc.M404648200.

Beseničar, M., Maček, P., Lakey, J. H., & Anderluh, G. (2006). Surface plasmon resonance in protein–membrane interactions. *Chemistry and Physics of Lipids, 141*, 169–178. https://doi.org/10.1016/j.chemphyslip.2006.02.010.

Breton, M., Amirkavei, M., & Mir, L. M. (2015). Optimization of the electroformation of giant unilamellar vesicles (GUVs) with unsaturated phospholipids. *The Journal of Membrane Biology, 248*, 1–9. https://doi.org/10.1007/s00232-015-9828-3.

Buré, C., Cacas, J.-L., Wang, F., et al. (2011). Fast screening of highly glycosylated plant sphingolipids by tandem mass spectrometry. *Rapid Communications in Mass Spectrometry, 25*, 3131–3145. https://doi.org/10.1002/rcm.5206.

Cans, A.-S., Andes-Koback, M., & Keating, C. D. (2008). Positioning lipid membrane domains in giant vesicles by micro-organization of aqueous cytoplasm mimic. *Journal of the American Chemical Society, 130*, 7400–7406. https://doi.org/10.1021/ja710746d.

Caschera, F., & Noireaux, V. (2015). Preparation of amino acid mixtures for cell-free expression systems. *BioTechniques, 58*, 40–43.

Chung, M., Lowe, R. D., Chan, Y.-H. M., et al. (2009). DNA-tethered membranes formed by giant vesicle rupture. *Journal of Structural Biology, 168*, 190–199. https://doi.org/10.1016/j.jsb.2009.06.015.

Bernardino de la Serna, J., Hansen, S., Hannibal-Bach, H. K., et al. (2009). Biophysical, structural and compositional characterization at the molecular level of native pulmonary surfactant membranes directly isolated from mice wild-type and knocked-out protein D bronco-alveolar lavage fluid. *Biophysical Journal, 96*, 451a. https://doi.org/10.1016/j.bpj.2008.12.2319.

Demchenko, A. P., Mély, Y., Duportail, G., & Klymchenko, A. S. (2009). Monitoring biophysical properties of lipid membranes by environment-sensitive fluorescent probes. *Biophysical Journal, 96*, 3461–3470. https://doi.org/10.1016/j.bpj.2009.02.012.

Dimova, R., Aranda, S., Bezlyepkina, N., et al. (2006). A practical guide to giant vesicles. Probing the membrane nanoregime via optical microscopy. *Journal of Physics. Condensed Matter, 18*, S1151–S1176. https://doi.org/10.1088/0953-8984/18/28/S04.

Discher, B. M., Won, Y.-Y., Ege, D. S., et al. (1999). Polymersomes: Tough vesicles made from diblock copolymers. *Science, 284*, 1143–1146. https://doi.org/10.1126/science.284.5417.1143.

Dreier, B., & Plückthun, A. (2011). Ribosome display: a technology for selecting and evolving proteins from large libraries. In D. J. Park (Ed.), *PCR Protocols* (pp. 283–306). Totowa, NJ: Humana Press.

Drücker, P., Iacovache, I., Bachler, S., et al. (2019). Membrane deformation and layer-by-layer peeling of giant vesicles induced by the pore-forming toxin pneumolysin. *Biomaterials Science*, 7, 3693–3705. https://doi.org/10.1039/C9BM00134D.

Elani, Y., Gee, A., Law, R. V., & Ces, O. (2013). Engineering multi-compartment vesicle networks. *Chemical Science*, *4*, 3332–3338. https://doi.org/10.1039/C3SC51164B.

Elani, Y., Law, R. V., & Ces, O. (2015). Protein synthesis in artificial cells: Using compartmentalisation for spatial organisation in vesicle bioreactors. *Physical Chemistry Chemical Physics*, *17*, 15534–15537. https://doi.org/10.1039/C4CP05933F.

Exterkate, M., Caforio, A., Stuart, M. C. A., & Driessen, A. J. M. (2018). Growing membranes in vitro by continuous phospholipid biosynthesis from free fatty acids. *ACS Synthetic Biology*, 7, 153–165. https://doi.org/10.1021/acssynbio.7b00265.

Faucon, J. F., Mitov, M. D., Méléard, P., et al. (1989). Bending elasticity and thermal fluctuations of lipid membranes. Theoretical and experimental requirements. *Journal de Physique*, *50*, 2389–2414. https://doi.org/10.1051/jphys:0198900500170238900.

Funakoshi, K., Suzuki, H., & Takeuchi, S. (2007). Formation of giant lipid vesicle like compartments from a planar lipid membrane by a pulsed jet flow. *Journal of the American Chemical Society*, *129*, 12608–12609. https://doi.org/10.1021/ja074029f.

Furusato, T., Horie, F., Matsubayashi, H. T., et al. (2018). De novo synthesis of basal bacterial cell division proteins FtsZ, FtsA, and ZipA inside giant vesicles. *ACS Synthetic Biology*, 7, 953–961. https://doi.org/10.1021/acssynbio.7b00350.

Göpfrich, K., Haller, B., Staufer, O., et al. (2019). One-pot assembly of complex giant unilamellar vesicle-based synthetic cells. *ACS Synthetic Biology*, *8*, 937–947. https://doi.org/10.1021/acssynbio.9b00034.

Herold, C., Chwastek, G., Schwille, P., & Petrov, E. P. (2012). Efficient electroformation of supergiant unilamellar vesicles containing cationic lipids on ITO-coated electrodes. *Langmuir*, *28*, 5518–5521. https://doi.org/10.1021/la3005807.

Hishida, M., Seto, H., & Yoshikawa, K. (2005). Smooth/rough layering in liquid-crystalline/gel state of dry phospholipid film, in relation to its ability to generate giant vesicles. *Chemical Physics Letters*, *411*, 267–272. https://doi.org/10.1016/j.cplett.2005.06.045.

Horger, K. S., Estes, D. J., Capone, R., & Mayer, M. (2009). Films of agarose enable rapid formation of giant liposomes in solutions of physiologic ionic strength. *Journal of the American Chemical Society*, *131*, 1810–1819. https://doi.org/10.1021/ja805625u.

Huang, Y., Kim, S.-H., & Arriaga, L. R. (2017). Emulsion templated vesicles with symmetric or asymmetric membranes. *Advances in Colloid and Interface Science*, *247*, 413–425. https://doi.org/10.1016/j.cis.2017.07.013.

Hui, E., & Vale, R. D. (2014). In vitro membrane reconstitution of the T-cell receptor proximal signaling network. *Nature Structural & Molecular Biology*, *21*, 133–142. https://doi.org/10.1038/nsmb.2762.

Hürtgen, D., Vogel, S. K., & Schwille, P. (2019). Cytoskeletal and actin-based polymerization motors and their role in minimal cell design. *Advanced Biosystems*, *3*, 1800311. https://doi.org/10.1002/adbi.201800311.

Huster, D., Müller, P., Arnold, K., & Herrmann, A. (2001). Dynamics of membrane penetration of the fluorescent 7-nitrobenz-2-oxa-1,3-diazol-4-yl (NBD) group attached to an acyl chain of phosphatidylcholine. *Biophysical Journal*, *80*, 822–831.

Ikenouchi, J., Suzuki, M., Umeda, K., et al. (2012). Lipid polarity is maintained in absence of tight junctions. *The Journal of Biological Chemistry*, *287*, 9525–9533. https://doi.org/10.1074/jbc.M111.327064.

Kahya, N. (2010). Protein–protein and protein–lipid interactions in domain-assembly: Lessons from giant unilamellar vesicles. *Biochimica et Biophysica Acta (BBA) - Biomembranes*, *1798*, 1392–1398. https://doi.org/10.1016/j.bbamem.2010.02.028.

Kahya, N., Scherfeld, D., Bacia, K., & Schwille, P. (2004). Lipid domain formation and dynamics in giant unilamellar vesicles explored by fluorescence correlation spectroscopy. *Journal of Structural Biology*, *147*, 77–89. https://doi.org/10.1016/j.jsb.2003.09.021.

Kahya, N., Scherfeld, D., & Schwille, P. (2005). Differential lipid packing abilities and dynamics in giant unilamellar vesicles composed of short-chain saturated glycerol-phospholipids, sphingomyelin and cholesterol. *Chemistry and Physics of Lipids, 135*, 169–180. https://doi.org/10.1016/j.chemphyslip.2005.02.013.

Karamdad, K., Hindley, J. W., Bolognesi, G., et al. (2018). Engineering thermoresponsive phase separated vesicles formed via emulsion phase transfer as a content-release platform. *Chemical Science*, ***9***, 4851–4858. https://doi.org/10.1039/C7SC04309K.

Karamdad, K., Law, R. V., Seddon, J. M., et al. (2014). Preparation and mechanical characterisation of giant unilamellar vesicles by a microfluidic method. *Lab on a Chip, 15*, 557–562. https://doi.org/10.1039/C4LC01277A.

Kisovec, M., Rezelj, S., Knap, P., et al. (2017). Engineering a pH responsive pore forming protein. *Scientific Reports*, 7, 42231. https://doi.org/10.1038/srep42231.

Kita, H., Matsuura, T., Sunami, T., et al. (2008). Replication of genetic information with self-encoded replicase in liposomes. *Chembiochem, 9*, 2403–2410. https://doi.org/10.1002/cbic.200800360.

Klymchenko, A. S., & Kreder, R. (2014). Fluorescent probes for lipid rafts: From model membranes to living cells. *Chemistry & Biology, 21*, 97–113. https://doi.org/10.1016/j.chembiol.2013.11.009.

Kretschmer, S., Ganzinger, K. A., Franquelim, H. G., & Schwille, P. (2019). Synthetic cell division via membrane-transforming molecular assemblies. *BMC Biology, 17*, 43. https://doi.org/10.1186/s12915-019-0665-1.

Kurihara, K., Tamura, M., Shohda, K., et al. (2011). Self-reproduction of supramolecular giant vesicles combined with the amplification of encapsulated DNA. *Nature Chemistry, 3*, 775–781. https://doi.org/10.1038/nchem.1127.

Lenarčič, T., Albert, I., Böhm, H., et al. (2017). Eudicot plant-specific sphingolipids determine host selectivity of microbial NLP cytolysins. *Science, 358*, 1431–1434. https://doi.org/10.1126/science.aan6874.

Lentini, R., Santero, S. P., Chizzolini, F., et al. (2014). Integrating artificial with natural cells to translate chemical messages that direct E. coli behaviour. *Nature Communications, 5*, 4012. https://doi.org/10.1038/ncomms5012.

Li, P., Banjade, S., Cheng, H.-C., et al. (2012). Phase transitions in the assembly of multivalent signalling proteins. *Nature, 483*, 336–340. https://doi.org/10.1038/nature10879.

Maier, O., Oberle, V., & Hoekstra, D. (2002). Fluorescent lipid probes: Some properties and applications (a review). *Chemistry and Physics of Lipids, 116*, 3–18. https://doi.org/10.1016/S0009-3084(02)00017-8.

Mally, M., Majhenc, J., Svetina, S., & Žekš, B. (2002). Mechanisms of equinatoxin II-induced transport through the membrane of a giant phospholipid vesicle. *Biophysical Journal, 83*, 944–953. https://doi.org/10.1016/S0006-3495(02)75220-X.

Malovrh, P., Viero, G., Serra, M. D., et al. (2003). A novel mechanism of pore formation membrane penetration by the n-terminal amphipathic region of equinatoxin. *The Journal of Biological Chemistry, 278*, 22678–22685. https://doi.org/10.1074/jbc.M300622200.

Mamode Cassim, A., Gouguet, P., Gronnier, J., et al. (2019). Plant lipids: Key players of plasma membrane organization and function. *Progress in Lipid Research, 73*, 1–27. doi:10.1016./j.plipres.2018.11.002.

Marchioretto, M., Podobnik, M., Dalla Serra, M., & Anderluh, G. (2013). What planar lipid membranes tell us about the pore-forming activity of cholesterol-dependent cytolysins. *Biophysical Chemistry, 182*, 64–70. https://doi.org/10.1016/j.bpc.2013.06.015.

Méléard, P., Bagatolli, L. A., & Pott, T. (2009). Giant unilamellar vesicle electroformation: From lipid mixtures to native membranes under physiological conditions. In *Methods in Enzymology* (pp. 161–176). In: Academic Press.

Mikelj, M., Praper, T., Demič, R., et al. (2013). Electroformation of giant unilamellar vesicles from erythrocyte membranes under low-salt conditions. *Analytical Biochemistry, 435*, 174–180. https://doi.org/10.1016/j.ab.2013.01.001.

Montes, L.-R., Alonso, A., Goñi, F. M., & Bagatolli, L. A. (2007). Giant unilamellar vesicles electroformed from native membranes and organic lipid mixtures under physiological conditions. *Biophysical Journal, 93*, 3548–3554. https://doi.org/10.1529/biophysj.107.116228.

Morales-Penningston, N. F., Wu, J., Farkas, E. R., et al. (2010). GUV preparation and imaging: Minimizing artifacts. *Biochimica et Biophysica Acta (BBA) - Biomembranes, 1798*, 1324–1332. https://doi.org/10.1016/j.bbamem.2010.03.011.

Morii, H., Yagi, H., Akutsu, H., et al. (1999). A novel phosphoglycolipid archaetidyl(glucosyl) inositol with two sesterterpanyl chains from the aerobic hyperthermophilic archaeon Aeropyrum pernix K1. *Biochimica et Biophysica Acta—Molecular and Cell Biology of Lipids, 1436*, 426–436. https://doi.org/10.1016/S0005-2760(98)00157-X.

Nelson, L. D., Chiantia, S., & London, E. (2010). Perfringolysin O association with ordered lipid domains: Implications for transmembrane protein raft affinity. *Biophysical Journal, 99*, 3255–3263. https://doi.org/10.1016/j.bpj.2010.09.028.

Ni, T., Jiao, F., Yu, X., et al. (2020). Structure and mechanism of bactericidal mammalian perforin-2, an ancient agent of innate immunity. *Science Advances, 6*, eaax8286. https://doi.org/10.1126/sciadv.aax8286.

Ni, T., Williams, S. I., Rezelj, S., et al. (2018). Structures of monomeric and oligomeric forms of the toxoplasma gondii perforin-like protein 1. *Science Advances, 4*, eaaq0762. https://doi.org/10.1126/sciadv.aaq0762.

Nishimura, K., Matsuura, T., Nishimura, K., et al. (2012). Cell-free protein synthesis inside giant unilamellar vesicles analyzed by flow cytometry. *Langmuir, 28*, 8426–8432. https://doi.org/10.1021/la3001703.

Noireaux, V., & Libchaber, A. (2004). A vesicle bioreactor as a step toward an artificial cell assembly. *Proceedings of the National Academy of Sciences of the United States of America, 101*, 17669–17674. https://doi.org/10.1073/pnas.0408236101.

Noireaux, V., Maeda, Y. T., & Libchaber, A. (2011). Development of an artificial cell, from self-organization to computation and self-reproduction. *Proceedings of the National Academy of Sciences of the United States of America, 108*, 3473–3480. https://doi.org/10.1073/pnas.1017075108.

Okano, T., Inoue, K., Koseki, K., & Suzuki, H. (2018). Deformation modes of giant unilamellar vesicles encapsulating biopolymers. *ACS Synthetic Biology, 7*, 739–747. https://doi.org/10.1021/acssynbio.7b00460.

Olbrich, K., Rawicz, W., Needham, D., & Evans, E. (2000). Water permeability and mechanical strength of polyunsaturated lipid bilayers. *Biophysical Journal, 79*, 321–327.

Omersa, N., Aden, S., Kisovec, M., et al. (2020). Design of protein logic gate system operating on lipid membranes. *ACS Synthetic Biology, 9*, 316–328. https://doi.org/10.1021/acssynbio.9b00340.

Oropeza-Guzman, E., Ríos-Ramírez, M., & Ruiz-Suárez, J. C. (2019). Leveraging the coffee ring effect for a defect-free electroformation of giant unilamellar vesicles. *Langmuir, 35*, 16528–16535. https://doi.org/10.1021/acs.langmuir.9b02488.

Paterson, D. J., Reboud, J., Wilson, R., et al. (2014). Integrating microfluidic generation, handling and analysis of biomimetic giant unilamellar vesicles. *Lab on a Chip, 14*, 1806–1810. https://doi.org/10.1039/C4LC00199K.

Pautot, S., Frisken, B. J., & Weitz, D. A. (2003). Engineering asymmetric vesicles. *Proceedings of the National Academy of Sciences, 100*, 10718–10721. https://doi.org/10.1073/pnas.1931005100.

Podobnik, M., Kraševec, N., Zavec, A. B., et al. (2016). How to study protein-protein interactions. *Acta Chimica Slovenica, 63*, 424–439. https://doi.org/10.17344/acsi.2016.2419.

Pontani, L.-L., van der Gucht, J., Salbreux, G., et al. (2009). Reconstitution of an actin cortex inside a liposome. *Biophysical Journal, 96*, 192–198. https://doi.org/10.1016/j.bpj.2008.09.029.

Pott, T., Bouvrais, H., & Méléard, P. (2008). Giant unilamellar vesicle formation under physiologically relevant conditions. *Chemistry and Physics of Lipids*, *154*, 115–119. https://doi.org/10.1016/j.chemphyslip.2008.03.008.

Praper, T., Beseničar, M. P., Istinič, H., et al. (2010). Human perforin permeabilizing activity, but not binding to lipid membranes, is affected by pH. *Molecular Immunology*, *47*, 2492–2504. https://doi.org/10.1016/j.molimm.2010.06.001.

Praper, T., Sonnen, A. F.-P., Kladnik, A., et al. (2011). Perforin activity at membranes leads to invaginations and vesicle formation. *Proceedings of the National Academy of Sciences*, *108*, 21016–21021. https://doi.org/10.1073/pnas.1107473108.

Praper, T., Sonnen, A., Viero, G., et al. (2011). Human perforin employs different avenues to damage membranes. *The Journal of Biological Chemistry*, *286*, 2946–2955. https://doi.org/10.1074/jbc.M110.169417.

Rawicz, W., Olbrich, K. C., McIntosh, T., et al. (2000). Effect of chain length and unsaturation on elasticity of lipid bilayers. *Biophysical Journal*, *79*, 328–339.

Reeves, J. P., & Dowben, R. M. (1969). Formation and properties of thin-walled phospholipid vesicles. *Journal of Cellular Physiology*, *73*, 49–60. https://doi.org/10.1002/jcp.1040730108.

Rezelj, S., Kozorog, M., Švigelj, T., Ulrih, N. P., Žnidaršič, N., Podobnik, M., et al. (2018). Cholesterol enriched archaeosomes as a molecular system for studying interactions of cholesterol-dependent cytolysins with membranes. *The Journal of Membrane Biology*, *251*(3), 491–505. https://doi.org/10.1007/s00232-018-0018-y.

Richmond, D. L., Schmid, E. M., Martens, S., et al. (2011). Forming giant vesicles with controlled membrane composition, asymmetry, and contents. *Proceedings of the National Academy of Sciences*, *108*, 9431–9436.

Robinson, T., Kuhn, P., Eyer, K., & Dittrich, P. S. (2013). Microfluidic trapping of giant unilamellar vesicles to study transport through a membrane pore. *Biomicrofluidics*, 7(4), 044105. https://doi.org/10.1063/1.4816712.

Rodriguez, N., Pincet, F., & Cribier, S. (2005). Giant vesicles formed by gentle hydration and electroformation: A comparison by fluorescence microscopy. *Colloids and Surfaces. B, Biointerfaces*, *42*, 125–130. https://doi.org/10.1016/j.colsurfb.2005.01.010.

Rojko, N., Kristan, K.Č., Viero, G., et al. (2013). Membrane damage by an α-helical pore-forming protein, equinatoxin II, proceeds through a succession of ordered steps. *The Journal of Biological Chemistry*, *288*, 23704–23715. https://doi.org/10.1074/jbc.M113.481572.

Ruan, Y., Rezelj, S., Zavec, A. B., et al. (2016). Listeriolysin O membrane damaging activity involves arc formation and lineaction—Implication for listeria monocytogenes escape from phagocytic vacuole. *PLoS Pathogens*, *12*, e1005597. https://doi.org/10.1371/journal.ppat.1005597.

Schön, P., García-Sáez, A. J., Malovrh, P., et al. (2008). Equinatoxin II permeabilizing activity depends on the presence of sphingomyelin and lipid phase coexistence. *Biophysical Journal*, *95*, 691–698. https://doi.org/10.1529/biophysj.108.129981.

Sezgin, E., Levental, I., Mayor, S., & Eggeling, C. (2017). The mystery of membrane organization: Composition, regulation and roles of lipid rafts. *Nature Reviews. Molecular Cell Biology*, *18*, 361–374. https://doi.org/10.1038/nrm.2017.16.

Skočaj, M., Resnik, N., Grundner, M., et al. (2014). Tracking cholesterol/sphingomyelin-rich membrane domains with the ostreolysin A-mCherry protein. *PLoS One*, *9*, e92783. https://doi.org/10.1371/journal.pone.0092783.

Stachowiak, J. C., Richmond, D. L., Li, T. H., et al. (2008). Unilamellar vesicle formation and encapsulation by microfluidic jetting. *Proceedings of the National Academy of Sciences*, *105*, 4697–4702. https://doi.org/10.1073/pnas.0710875105.

Stachowiak, J. C., Richmond, D. L., Li, T. H., et al. (2009). Inkjet formation of unilamellar lipid vesicles for cell-like encapsulation. *Lab on a Chip*, *9*, 2003–2009. https://doi.org/10.1039/B904984C.

Steinkühler, J., De Tillieux, P., Knorr, R. L., et al. (2018). Charged giant unilamellar vesicles prepared by electroformation exhibit nanotubes and transbilayer lipid asymmetry. *Scientific Reports*, *8*, 11838. https://doi.org/10.1038/s41598-018-30286-z.

Tabaei, S. R., Gillissen, J. J. J., Kim, M. C., et al. (2016). Brownian dynamics of electrostatically adhering small vesicles to a membrane surface induces domains and probes viscosity. *Langmuir*, *32*, 5445–5450. https://doi.org/10.1021/acs.langmuir.6b00985.

Utada, A. S., Lorenceau, E., Link, D. R., et al. (2005). Monodisperse double emulsions generated from a microcapillary device. *Science*, *308*, 537–541. https://doi.org/10.1126/science.1109164.

van Swaay, D., & deMello, A. (2013). Microfluidic methods for forming liposomes. *Lab on a Chip*, *13*, 752–767. https://doi.org/10.1039/C2LC41121K.

Walde, P., Cosentino, K., Engel, H., & Stano, P. (2010). Giant vesicles: Preparations and applications. *Chembiochem*, *11*, 848–865. https://doi.org/10.1002/cbic.201000010.

Wang, T., Ingram, C., & Weisshaar, J. C. (2010). Model lipid bilayer with facile diffusion of lipids and integral membrane proteins. *Langmuir*, *26*, 11157–11164. https://doi.org/10.1021/la101046r.

Weinberger, A., Tsai, F.-C., Koenderink, G. H., et al. (2013). Gel-assisted formation of giant unilamellar vesicles. *Biophysical Journal*, *105*, 154–164. https://doi.org/10.1016/j.bpj.2013.05.024.

Zhao, J., Wu, J., Shao, H., et al. (2007). Phase studies of model biomembranes: Macroscopic coexistence of Lα + Lβ, with light-induced coexistence of Lα + Lo Phases. *Biochimica et Biophysica Acta (BBA) - Biomembranes*, *1768*, 2777–2786. https://doi.org/10.1016/j.bbamem.2007.07.009.

Zhou, J., Loftus, A. L., Mulley, G., & Jenkins, A. T. A. (2010). A thin film detection/response system for pathogenic bacteria. *Journal of the American Chemical Society*, *132*, 6566–6570. https://doi.org/10.1021/ja101554a.

Zimmerberg, J., & Kozlov, M. M. (2006). How proteins produce cellular membrane curvature. *Nature Reviews. Molecular Cell Biology*, 7, 9–19. https://doi.org/10.1038/nrm1784.

CHAPTER NINE

Preparation and utility of asymmetric lipid vesicles for studies of perfringolysin O-lipid interactions

Shinako Kakuda, Bingchen Li, and Erwin London*

Department of Biochemistry and Cell Biology, Stony Brook University, Stony Brook, NY, United States

*Corresponding author: e-mail address: erwin.london@stonybrook.edu

Contents

Abstract

Studying the interaction of pore-forming toxins, including perfringolysin O (PFO), with lipid is crucial to understanding how they insert into membranes, assemble and associate with membrane domains. In almost all past studies, symmetric lipid bilayers, i.e., bilayers having the same lipid composition in each monolayer (leaflet), have been used to study this process. However, practical methods to make asymmetric lipid vesicles have now been developed. These involve a cyclodextrin-catalyzed lipid exchange process in which the outer leaflet lipids are switched between two lipid vesicle populations with different lipid compositions. By use of alpha class cyclodextrins, it is practical to include a wide range of sterol concentrations in asymmetric vesicles. In this article, protocols for preparing asymmetric lipid vesicles are described, and to illustrate how they may be applied to studies of pore-forming toxin behavior, we summarize what has been learned about PFO conformation and its lipid interaction in symmetric and in asymmetric artificial lipid vesicles.

Methods in Enzymology, Volume 649
ISSN 0076-6879
https://doi.org/10.1016/bs.mie.2021.01.005

1. Introduction

For membrane proteins, including cholesterol-dependent cytolysins, an important approach to investigate protein-lipid interactions and their impact upon structure and function are studies using purified protein mixed with artificial lipid membranes. To date, these studies have generally used symmetric lipid bilayers, in which the inner and outer lipid monolayer (leaflet) of the lipid bilayer have the same lipid composition. However, many natural membranes have lipid asymmetry, in which the lipid composition differs in the two leaflets. This is certainly true for the plasma membrane of eukaryotic cells, in which the outer leaflet is rich in sphingolipids (SM) and phosphatidylcholine (PC), while the inner leaflet is relatively enriched in phosphatidylethanolamine (PE) and anionic phospholipids such as phosphatidylserine (PS) (Lorent et al., 2020). The level of cholesterol asymmetry remains controversial (Courtney, Fung, Maxfield, Fairn, & Zha, 2018; Liu et al., 2017). The extent to which lipid asymmetry affects membrane protein behavior is little studied. This is a question of importance for PFO, which acts at the plasma membrane, and perhaps internal membranes that also possess lipid asymmetry.

2. Studies of the effect of lipid composition upon PFO binding and PFO domain localization using symmetric vesicles

Much information about the impact of lipid composition upon PFO behavior has been gained from studies using symmetric lipid vesicles. This includes a basic picture of the steps involved in membrane insertion (Johnson & Heuck, 2014). These steps include lipid binding, which requires interaction of residues at the tip of domain 4 with cholesterol, then oligomerization into a 30–50-mer, followed by a large structural rearrangement in domains 2 and 3 in which two sets of helix-containing loops on domain 3 insert into the membrane, forming two hairpins with two transmembrane (TM) beta-strands each. The final structure of the membrane-inserted segments is an oligomeric TM beta-barrel with a large aqueous pore. Based on mutational and kinetics studies, the events during the membrane insertion process in symmetric vesicles are believed to take place in the order membrane binding, oligomerization, insertion, and pore-forming beta-barrel assembly, with the last two steps perhaps being nearly simultaneous.

The nature of PFO interaction with lipid has been examined in studies in which lipid composition was systematically varied. These studies have examined the effect of lipid structure upon both insertion into membranes, and the degree of association of PFO with membrane domains (Flanagan, Tweten, Johnson, & Heuck, 2009; Lin & London, 2013a, 2013b, 2014a; Nelson, Chiantia, & London, 2010; Nelson, Johnson, & London, 2008; Sokolov & Radhakrishnan, 2010). In both artificial lipid vesicles and plasma membranes, liquid ordered (Lo) domains (rich in sphingolipids which tend to have saturated acyl chains and cholesterol) can co-exist with liquid disordered (Ld) domains (which are more enriched in phospholipids having one unsaturated acyl chain) (Brown & London, 2000).

The requirement of cholesterol in order that PFO can bind to membranes is absolute. In membranes containing a mixture of sterol and phospholipids, a stronger PFO interaction with sterol results in PFO binding to the membrane at a lower sterol concentration. The 3 beta OH group of the sterol is essential, but the details of the structure of sterol rings and aliphatic tail often have only a modest effect on binding (Nelson et al., 2008). This indicates that PFO mainly binds to the polar end of the sterol, and the remainder for the sterol remains in contact with the lipid bilayer rather than bound in a pocket on the PFO molecule. This model is also consistent with the effect of sterol structure on domain localization (see below).

It appears that, as a function of the structure of the phospholipids in a symmetric lipid vesicle, PFO binding to cholesterol increases with the degree to which cholesterol is exposed to aqueous solution (Flanagan et al., 2009; Nelson et al., 2008; Sokolov & Radhakrishnan, 2010). The partial exposure of membrane-inserted sterol to aqueous solution corresponds to a high free energy state because small OH group cannot fully shield the hydrophobic sterol rings from water (Huang & Feigenson, 1999). As a consequence, for two different membranes with equal cholesterol concentrations, but different physical state or phospholipid composition, PFO preferentially binds to membranes in which there is loose packing between sterol and phospholipid, i.e., the lipids are in the Ld state. In such membranes the headgroups of the lipids cannot come close enough to the sterol to fully shield cholesterol from water. Differences in binding to different sterols may often reflect differences in how closely they pack with membrane phospholipids (Nelson et al., 2008). Relative to membranes containing PC, PFO binding to and insertion into membranes is also promoted by anionic lipid (PS) and by PE (Nelson et al., 2008). This presumably reflects favorable Coulombic electrostatic interactions with anionic lipids in the case of PS, and the fact that PE has a small headgroup that is even less effective in shielding cholesterol from water than that of PC.

The discussion above might suggest that PFO would favor greater association with Ld domains than with Lo domains. However, this ignores the fact that the cholesterol concentration in Lo domains is higher than in Ld domains, a factor that should tend to compensate for differences in sterol exposure to water. Studies in symmetric lipid vesicles have examined PFO domain localization directly. It has been found that PFO can localize in both Lo and Ld domains, with localization highly dependent upon the degree of mismatch between the length of the TM beta-strands and the width of the hydrophobic core of the lipid bilayer in the Lo and Ld domains. The less the degree of mismatch, the stronger the association with that type of domain (Lin & London, 2013a). It should be noted that the possibility that PFO in Lo domains is surrounded by a thin ring of lipids in the Ld state cannot be ruled out (Nelson et al., 2010). Such nanodomains are below the resolution of microscopy. It should also be noted that the final domain localization of PFO is not necessarily the same as that to which PFO monomers initially bind.

The type of sterol bound to PFO also influences its affinity for Lo and Ld domains. Cholesterol partitions preferentially into Lo domains, and this promotes localization of PFO bound to cholesterol in Lo domains. In contrast, PFO binding to a sterol that favors Ld domains has been found to result in localization of PFO in Ld domains (Lin & London, 2013b). The dependence of PFO domain localization upon the preference of a sterol to localize in Lo or Ld domains suggests that the PFO-bound sterol remains largely in contact with the lipid bilayer. As noted above, exposure to the bilayer rather than localization in a deep pocket within the PFO molecule is also consistent with the modest effect of sterol ring and aliphatic tail structure upon binding to PFO. Sterol bound to a deep cleft within the PFO would lose contact with lipid and be unlikely to directly influence PFO domain localization.

3. Methods for preparing asymmetric vesicles using cyclodextrins

One barrier to studies of membrane proteins in a natural lipid environment has been the inability to easily prepare stable artificial asymmetric lipid bilayers to mimic natural membranes. Methods have now been developed to prepare asymmetric lipid vesicles. At present, the most versatile approach to prepare such vesicles involves lipid exchange using a cyclodextrin (CD), as will be described below (Cheng & London, 2011; Cheng & Megha, 2009;

Doktorova et al., 2018; Heberle et al., 2016; Li et al., 2016; Li & London, 2020; St. Clair & London, 2019). The basic principles are straightforward (Fig. 1). In this method, two populations of lipid vesicles with different lipid compositions are mixed. The population that will provide the outer leaflet (the donor population, which donates what is the outer leaflet lipid in the asymmetric vesicle) is used in excess over the acceptor population which provides what is the inner leaflet lipid of the asymmetric vesicle (it is the population that accepts the outer leaflet lipid from the donor).

CD is added at the appropriate time (see below). The CD pulls an individual lipid molecule out of the outer leaflet of a donor vesicle (the donor) and then deposits it in the outer leaflet of another vesicle (an acceptor). This leads to a continuous exchange of lipids that results in equilibration of the outer leaflet lipid compositions. As a result of this and the use of excess donor, the desired acceptor population ends up with an outer leaflet lipid composition close to that of the donor before exchange. The exact composition of the outer leaflet after exchange depends in a complex fashion on the

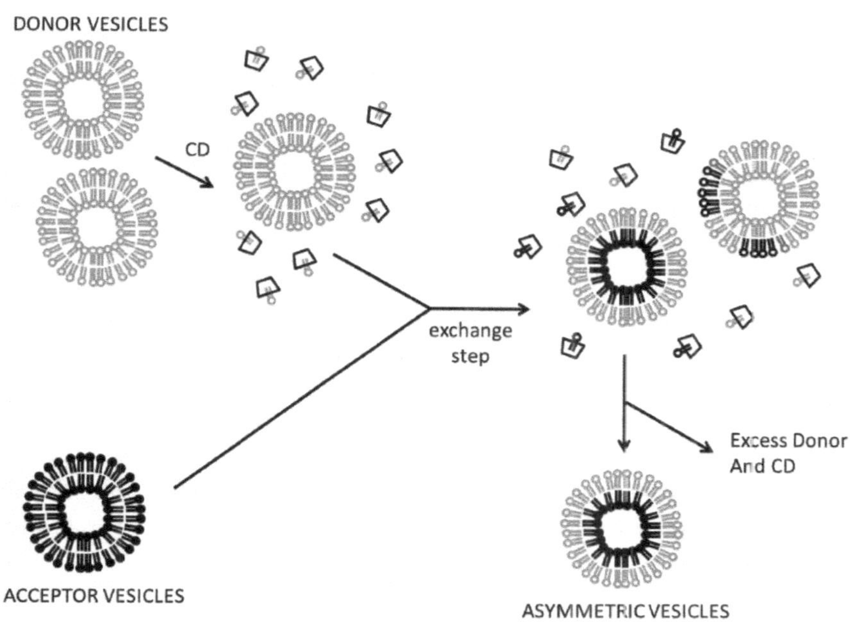

Fig. 1 Schematic illustration of preparation of asymmetric vesicles. *From London, E. (2019). Membrane structure-function insights from asymmetric lipid vesicles.* Accounts of Chemical Research, 52, *2382–2391. doi: https://doi.org/10.1021/acs.accounts.9b00300. © American Chemical Society.*

values of the many equilibria involved. These values can be dependent on the lipid composition of the donor and acceptor, which change during the experiment, as well as the lipid-CD binding constants. Under the usual experimental conditions, the composition of the outer leaflet of the asymmetric vesicle is not exactly the same as one would predict if the lipids in the donor and acceptor were randomized during exchange, although it is often close. Thus, especially using donor vesicles with a mixture of lipids, it is necessary to empirically determine what ratio of lipids must be used in the donor to obtain the desired outer leaflet composition in the asymmetric vesicles after exchange. In some cases, two rounds of exchange can be useful to achieve nearly complete exchange (see below).

It should be noted that CD binding to lipid is a cooperative process (Anderson, Tan, Ganz, & Seelig, 2004), and so the concentration of CD used is critical. In addition, some CDs will dissolve lipid vesicles that is completely destroy the vesicles at the high CD concentrations that are most useful for exchange (MβCD and MαCD). To minimize problems, donor lipid in excess of that which can be dissolved by the CDs is mixed with the CD in a first step. This saturates the CD with lipid and avoids or minimizes subsequent solubilization of acceptor vesicles by the CD. It may also increase the extent of exchange because lipids originally on the inner leaflet of donor vesicles load onto the CD.

The final step is the isolation of the lipid-exchanged acceptor vesicles, which are the desired asymmetric vesicles. This is carried out by centrifugation either initially preparing donor or acceptor (usually the acceptor vesicles) with trapped sucrose to impart different densities to them (Cheng & London, 2011; Doktorova et al., 2018; Heberle et al., 2016; Lin & London, 2014b). It is important to note that the vesicle yield is very sensitive to the density of the vesicles and the media in which they are centrifuged. Generally, we can achieve about 10% yields relative to the input of acceptor vesicles. What limits yield is unclear. In some cases, it may reflect incomplete pelleting. In other cases, there may be imbalanced exchange in which the difference in the number of lipids delivered to or extracted from the acceptor vesicles is large enough to cause vesicle lysis with loss of trapped sucrose and/or formation of smaller vesicles that are difficult to pellet.

For inclusion of cholesterol in asymmetric vesicles, at least two approaches are possible. One is to add cholesterol complexed with MβCD to asymmetric vesicles lacking cholesterol, using a low enough MβCD concentration that other lipids do not bind to the MβCD. However, a better method involves carrying out lipid exchange with alpha class CDs (MαCD or HPαCD).

These CDs have a cavity too small to accommodate cholesterol, but can bind phospholipids and sphingolipids because they can accommodate acyl chains. To obtain asymmetric vesicles with the desired amount of cholesterol, one only has to include the desired amount of cholesterol in the acceptor vesicles. No cholesterol is incorporated into the donor vesicles. In fact, if cholesterol were present in the donor vesicles and a CD.

that dissolves up lipid vesicles were used, the free cholesterol released after solubilization of donor vesicles might well stick to the acceptor vesicles, which would be difficult to control and thus very undesirable.

Which alpha CD is best for preparing asymmetric vesicles with cholesterol, MαCD or HPαCD? We have used both, and different researchers in the lab have different personal preferences. Protocols for both will be described below. Considerations include efficiency of exchange, amount of CD needed, and lipid specificity. In artificial lipid vesicles efficient exchange is possible with both HPαCD and MαCD. The ability of these two CDs to exchange lipids has not be compared in great detail to date. The amount of MαCD needed is less than the amount of HPαCD. With HPαCD we found we can increase the extent of exchange with two rounds of exchange, although yield decreases (Lin & London, 2015). In some cases, the fact that HPαCD does not dissolve up donor vesicles may be an advantage. MαCD has been used to prepare asymmetric vesicles with various sterols (Lorent et al., 2020; St. Clair & London, 2019). This has not been attempted yet with HPαCD.

4. Differences in lipid binding and conformation of PFO in symmetric and asymmetric vesicles

In one study our group took advantage of the ability to prepare asymmetric vesicles to define the effect of lipid asymmetry upon PFO conformation (Lin & London, 2014a). We summarize that work as an example to show how asymmetric vesicles can be used to increase the understanding of pore-forming toxin behavior. Vesicles that mimic the regions of plasma membranes in the Ld state were prepared, with an outer leaflet composed of PC and an inner leaflet of PS and PE. Cholesterol levels were varied. PFO was then added to the external aqueous solution. Five parameters that respond to PFO conformation and/or lipid interaction were measured as a function of cholesterol concentration. The first was Trp fluorescence, which increases when PFO binds to membranes. The second was FRET using FRET donor and acceptor attached to point mutant A215C of

PFO. The third was insertion into the bilayer, as measured by the change in fluorescence at acrylodan-labeled residue 215, a residue on a membrane-inserting segment forming a TM beta hairpin. The fourth parameter was formation of a complete beta-barrel, which results in formation of an SDS-resistant oligomer which can be detected on SDS gels. The last parameter measured was pore formation, measured by release of biocytin trapped within the vesicles by PFO (11).

These parameters were measured for PFO added to asymmetric vesicles, and compared to those for PFO added to three types of symmetric vesicles, in each case as a function of cholesterol concentration. The three types of symmetric vesicles were: (1) Scrambled, with the same overall composition as the asymmetric vesicles, (2) With a composition equivalent to that in the inner leaflet of the asymmetric vesicles, and (3) Equivalent to that in the outer leaflet of the asymmetric vesicles. As shown in Fig. 2, the behavior of PFO is different in each type of vesicle. In scrambled and outer leaflet compositions, all five changes occur to a similar extent at each cholesterol concentration as cholesterol concentration is increased (Fig. 2B and D). This indicates once the protein binds to membranes it tends to undergo all the changes that convert it to the pore-forming state. In contrast, in the presence of asymmetric vesicles or symmetric vesicles with the composition of the inner leaflet, a structural intermediate was found to dominate at intermediate cholesterol concentrations (Fig. 2A and C). This intermediate had undergone oligomerization and deep insertion, but did not form the SDS-resistant pore-forming beta-barrel state. Additional experiments confirmed that the intermediate conformation was TM, as determined by the observation that PFO biotinylated at a residue at the distal tip of the beta hairpin that forms upon TM insertion reached the solution on the *trans* side of the membrane (opposite the side of the membrane from which PFO inserted).

In terms of formation of a stable TM intermediate that did not convert to the pore-forming state, the behavior of PFO in the inner leaflet lipid mixture was similar to that in the asymmetric vesicles. However, both the insertion to form the TM intermediate, and conversion to the pore-forming state required lower concentrations of cholesterol than in the asymmetric vesicles. This may reflect the increased exposure of cholesterol due to their high concentration of PS and PE, consistent with the effect of PS and PE in symmetric vesicles (Nelson et al., 2008). This would promote interaction of PFO with the inner leaflet vesicles at lower cholesterol concentrations.

The difference of PFO behavior in symmetric outer leaflet vesicles and asymmetric vesicles which have the same outer leaflet but a different inner leaflet raises the question: how can the headgroups of the inner leaflet lipids

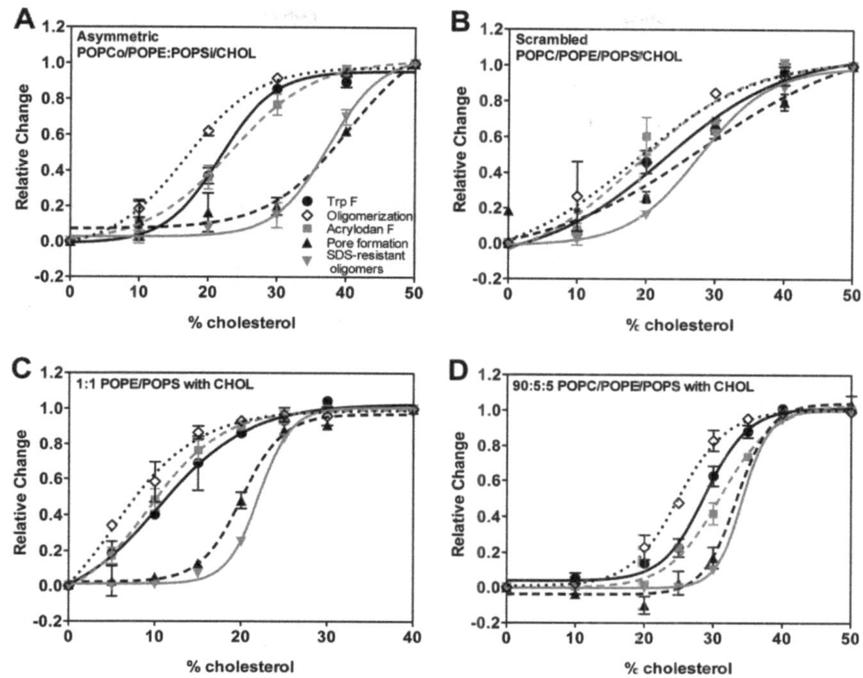

Fig. 2 Change in PFO (PFO A215C) behavior as a function of cholesterol concentration in asymmetric vesicles and in symmetric vesicles at room temperature. (A) POPCo/POPE: POPSi/cholesterol asymmetric vesicles. o = outer leaflet lipid, i = inner leaflet lipid. (B) Scrambled vesicles. (C) 1:1 POPE/POPS symmetric vesicles (mimicking the inner leaflet of asymmetric vesicles). (D) 90:5:5 POPC/POPE/POPS symmetric vesicles (mimicking the outer leaflet of asymmetric vesicles). The parameters measured are shown in the key to the figure (within panel A). Trp fluorescence (Trp F) detects binding to lipid vesicles and acrylodan fluorescence (of labeled residue 215) detects insertion of the beta hairpin. Oligomerization was detected by FRET, and is distinguished from SDS-resistant oligomers, which are believed to reflect formation of a complete beta-barrel. *This research and figure was adapted from one originally published in the Journal of Biological Chemistry Lin, Q., & London, E. (2014a). The influence of natural lipid asymmetry upon the conformation of a membrane-inserted protein (perfringolysin O).* The Journal of Biological Chemistry, *289(9), 5467–78.*

influence insertion when PFO only can sense them once TM sequences reach the distal surface in TM insertion is irreversible? The likely answer is that formation of the intermediate TM inserted state is not irreversible, and it is in equilibrium with non-inserted state. This is supported by the observation that removal of cholesterol from PFO forming the TM intermediate reversed PFO conformational changes and even binding to vesicles. In contrast, formation of the pore-forming state was irreversible.

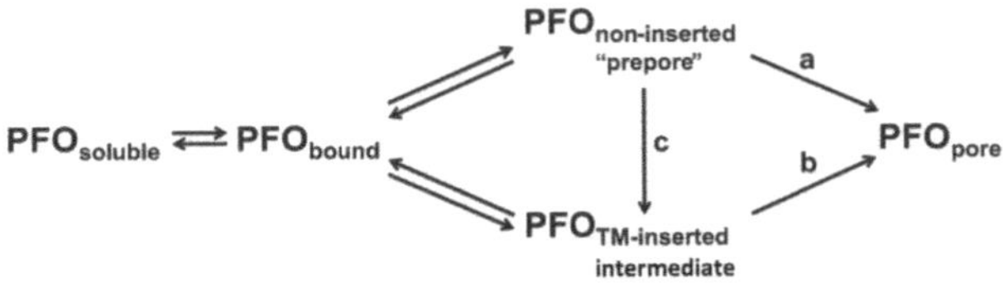

Fig. 3 Schematic diagram of the possible steps in PFO pore formation. Pathways a, b, and c represent different possible steps in the assembly process. *This research and figure was adapted from one originally published in the Journal of Biological Chemistry. Lin, Q., & London, E., (2014a). The influence of natural lipid asymmetry upon the conformation of a membrane-inserted protein (perfringolysin O).* The Journal of Biological Chemistry, 289*(9), 5467–78, © the American Society for Biochemistry and Molecular Biology.*

Fig. 3 summarizes potential pathways for PFO membrane insertion and assembly. The upper pathway from the aqueous monomer to final pore-forming PFO structure in membranes (via "a") shows the species detected in symmetric vesicles with a composition similar to the outer leaflet of plasma membranes, and is the traditional pathway in which any TM inserted intermediate is too transient or at too low a concentration to easily detect. The lower pathway (via "b") includes the TM intermediate. A hybrid process (via "c") is also possible. It should be noted that the latter pathways are not kinetically-defined, but rather involve structural intermediates that can be trapped lipid compositions unable to carry out the complete insertion process. There is always the potential that structural intermediates are off-pathway in terms of the kinetic events that occur during insertion in vivo. Nevertheless, the formation of structural intermediates shows they are sufficiently low free energy to form under reasonable conditions, and might do so during insertion in vivo.

One important general lesson of these experiments is clear. Lipid asymmetry can influence the conformational behavior of a membrane protein, and studies that take this into account may give a more complete picture of the influence of lipid upon membrane protein structure and function.

5. Remaining questions about the behavior of PFO in membranes

Our understanding of PFO behavior in asymmetric membranes is by no means complete. First, as noted above, the experiments in asymmetric membranes only examined conformations that represent structural intermediates.

The evolution of changes in the conformation of PFO vs. time after addition to lipid vesicles is needed to detect if the structural intermediates in asymmetric vesicles are also kinetic intermediates that form during the assembly of the pore-forming structure. A second unanswered question concerns the structure of the TM intermediate. Spectroscopically, it appears identical to the pore-forming structure in terms of how it is inserted in the bilayer. Does it contain a partially formed beta-barrel with an incomplete ring of beta-strands? If so, it is difficult to visualize what type of interactions form between the hydrophilic face of the beta sheet, which normally faces an internal aqueous pore, and the lipid bilayer. Contact between such a hydrophilic surface and the acyl chains of lipids seems improbable. C-shaped partial rings piled against one another back to back are one possibility. Third, the lipid compositions studied to date only roughly mimic plasma membranes. It is possible the behavior of PFO could be substantially different in an asymmetric membrane with physiological levels of sphingolipids included in the outer leaflet. In this regard, whether PFO localizes in Lo or Ld domains in asymmetric membranes is also unanswered.

Finally, the effect of PFO upon lipid asymmetry is unknown. That is, we do not know if lipid asymmetry changes after PFO inserted into the membrane. This is a particularly difficult question to answer because pore formation makes it impossible to use many of the simpler methods to assess lipid asymmetry, such as whether a labeled lipid incorporated into just one leaflet of an asymmetric vesicle is or is not restricted to that leaflet after PFO insertion. The problem is that pore formation allows reagents added externally (to define lipid sidedness) to pass through the lipid bilayer and interact with lipids on the inner leaflet as well as the outer leaflet. Nevertheless, it seems unlikely that there is complete loss of lipid asymmetry because the behavior of PFO added to asymmetric membranes is so different from that of PFO added to symmetric membranes with the same lipid composition.

Given the many remaining questions, and the fact that other cholesterol-dependent cytolysins have not been studied in asymmetric membranes, it is clear that asymmetric membranes should have an important role in future studies of this class of proteins. It is hoped that the detailed protocols below will help the interested researcher prepare asymmetric lipid vesicles for such studies. The first protocol shown uses HPαCD. The second uses MαCD. Note that we recently extended the latter method to use of cationic lipids, which although unnatural, should allow a wider range of asymmetries to be prepared (Li & London, 2020).

6. Asymmetric large unilamellar vesicle (aLUVs) preparation procedure using HPαCD

6.1. Materials

6.1.1. Palmitoyl-2-oleoyl-*sn*-glycero-3-phosphocholine (POPC), palmitoyl-2-oleoyl-*sn*-glycero-3-phosphoethanolamine (POPE), palmitoyl-2-oleoyl-*sn*-glycero-3-phosphoserine (POPS), cholesterol (CHOL), egg sphingomyelin (egg SM), purchased from Avanti Polar Lipids (Alabaster, AL).

6.1.2. Randomly 2-hydroxypropylated-α-cyclodextrin also called (2-hydroxypropyl)-α-cyclodextrin (HPαCD), purchased from Sigma-Aldrich (St. Louis, MO). Average molecular weight is ~1180 Da.

6.1.3. 10 × phosphate-buffered saline (PBS), purchased from Bio-RAD (Hercules CA). When diluted to 1 × concentration is 10 mM sodium phosphate, 150 mM NaCl (pH 7.8 ± 0.2).

6.1.4. Thin-layer chromatography (HP-TLC) plates (Silica Gel 60), purchased from VWR International (Batavia, IL).

6.1.5. Sucrose, crystal, purchased from J.T Baker, Avantor (Center Valley, PA).

6.1.6. Dry ice or liquid nitrogen.

6.1.7. Acetone, if dry ice is used, purchased from Fisher Chemical (Fair Lawn, NJ).

6.1.8. 100 nm-pore polycarbonate membrane, purchased from Whatman.

6.2. Equipment

6.2.1. Microdispensers, 25 μL, 50 μL, and 100 μL, and corresponding glass bores, purchased from Drummond Scientific Co, (Broomall, PA) (see Note 8.1).

6.2.2. Water bath.

6.2.3. Beckman 640 spectrophotometer or suitable alternative spectrophotometer and quartz cuvettes (Beckman Instruments, Fullerton, CA).

6.2.4. Beckman L8–80M Ultracentrifuge and SW 60 rotor (Beckman Instruments, Fullerton, CA).

6.2.5. Multitube vortexer (VWR, Westchester, PA). (see Note 8.2).

6.2.6. Convection oven (GCA Corp, Precision Scientific, Chicago, IL).

6.2.7. 1 mL syringe equipped Lipid Extruder (Avanti Polar Lipids, Alabaster, AL).
6.2.8. Epson 1640XL (Epson America, Long Beach, CA), or similar scanner.
6.2.9. ImageJ 1.48v software (Wayne Rasband, National Institutes of Health, Bethesda, MD).
6.2.10. Cahn C-33 Microbalance (Thermo Electron Corporation) or other microbalance.
6.2.11. Vacuum Pump and Vacuum Desiccator.
6.2.12. Hot plate or high temperature oven for heating TLC plates.
6.2.13. Low speed centrifuge.
6.2.14. TLC tank.
6.2.15. Glass tubes.
6.2.16. Nitrogen gas.

6.3. aLUVs preparation protocol using HPαCD.

6.3.1. Stock solution preparation.

6.3.1.1. Non-fluorescent Lipids are dissolved in chloroform and stored at −20 °C. The concentrations of lipids are measured by dry weight.

6.3.1.2. 385 mM HPαCD is dissolved in distilled water and stored at 4 °C. At this high concentration the HPαCD powder greatly affects the volume of the solution. The total volume of the solution after the powder dissolves has to be measured carefully. (see Note 8.3).

6.3.1.3. 10–15 and 25–28% (w/w) sucrose dissolved in distilled water or 1 × PBS.

6.3.2. Preparation of asymmetric LUVs (aLUVs) using lipid exchange catalyzed by HPαCD.

6.3.2.1. Donor multilamellar vesicles (MLVs) preparation and mixture with HPαCD. The desired amounts of phospholipids (total 10.4 μmol) are taken from the appropriate stock solutions using the micro-dispenser and combined in a glass tube. For example, consider preparation of aLUV with egg SM and POPC in the outside leaflet, POPC in the inside leaflet, and ~25 mol% cholesterol in both leaflets. In this case, Donor MLVs might be made with total 10.4 μmol lipid having the desired mole ratio of egg

SM: POPC. (see Note 8.4). There is no need to include cholesterol in the donor lipid mixture as it is not transferred between vesicles by HPαCD. The lipids are dried under nitrogen until visually dry. The lipid should be dispersed as a thin film on the walls of the glass tube. Gently warming the sample helps dry the solvent more quickly. The dried lipids are further dried by placing them in a vacuum desiccator, and subjecting them to a high vacuum for 1 h. The dried lipid mixtures in glass tubes are placed in a 70 °C water bath for hydration by adding 177 μL of 385 mM HPαCD. The lipid-HPαCD mixture is incubated at 70 °C for 2–3 min and vortexed several times until most lipids are evenly mixed. Then 473 μL 1 × PBS is added. The final MLV lipid concentration is 16 mM, HPαCD concentration is 105 mM and the volume is 0.65 mL. The lipid-HPαCD mixture is incubated in the water bath at 70 °C for another 2–3 min and vortexed until the dried lipid film dispersed evenly (a few times over a period of 2–3 min). The dispersed lipid-HPαCD mixture is further vortexed for 2 h at 55 °C in a multitube vortexer placed in an oven (see Note 2).

6.3.2.2. Acceptor large unilamellar vesicles (LUVs) preparation. Acceptor LUVs are made with 5.5 μmol of lipid. For the example noted above, they would contain POPC: CHOL = 75: 25. The desired lipid mixture are dried in the same fashion as for the donor lipid samples. (see Note 8.5) Dried lipid mixtures in glass tubes are placed in a 70 °C water bath and hydrated by adding 550 μL of 28% (w/w) sucrose (see list 6.3.2.4), dispersing the lipids 70 °C and vortexing several times over 5 min to form crude MLVs. The samples are then vortexed using a multitube vortexer placed in an oven at 55 °C for 15 min. To prepare LUVs from MLVs, the MLVs dispersions are subjected to seven cycles of freezing-thaw using a mixture of dry ice and acetone, or liquid nitrogen, to freeze the MLVs, and then warm water to thaw them, followed by extrusion 11 times through using a 1 mL extruder equipped with 100 nm-pore polycarbonate membranes. To wash away the untrapped sucrose, the resulting ~0.45 mL of acceptor LUVs (~0.1 mL is lost during extrusion) in 28 (w/w) % sucrose is mixed with 3.3 mL 1 × PBS and the LUVs pelleted by ultracentrifugation at 190,000 × *g* for 30 min at 23 °C (using a Beckman L8–80 M ultracentrifuge with an SW 60 rotor). LUV pellets are resuspended to 0.45 mL in 1 × PBS.

6.3.2.3. Lipid exchange step. Lipid exchange step is performed at a concentration of 8 mM donor lipid and 4 mM acceptor lipid in a total volume of 1 mL. To do this, 500 μL of the donor MLVs-HPαCD mixture is combined with 400 μL of acceptor LUV and 100 μL of 1 × PBS and then vortexed for 30 min at 55 °C using the multitube vortexer. During this step, lipid exchange occurs and the acceptors are converted into asymmetric vesicles. Of course, donor vesicles also pick up some acceptor vesicle lipid. After cooling to near room temperature, the donor-acceptor vesicle mixtures were overlaid on 2.8 mL 15 (w/w) % sucrose (see list 6.3.2.4) and then centrifuged at 190,000 × *g* for 50 min at 23 °C After centrifugation the resulting pellet is loose, and to avoid disturbing it remove about 3.3 mL of supernatant carefully, leaving the lowest ~400–500 μL of the supernatant, and then add 3.3 mL of 1 × PBS to the centrifuge tube, mix, and centrifuge again at 190,000 × *g* for 30 min at 23 °C. This time the pellet is tight, and the entire supernatant can now be carefully removed. The final pellet is resuspended in 500 μL of 1 × PBS. Lipid yield determined by HP-TLC is typically 10% of the theoretical based on the initial amount of acceptor lipid used (see below).

6.3.2.4. Optimizing sucrose concentrations needed. The sucrose concentration inside the vesicles is chosen to maximize yield while minimizing contamination of the asymmetric vesicles with donor lipids. For vesicles with an outer leaflet composed of egg SM and POPC, an inner leaflet composed of POPC, and 25 mol% CHOL we load the vesicles with 28 mol% sucrose and centrifuge over 15 (w/w) % sucrose, as described above. However, the amount of sucrose needed can depend upon lipid composition. For some cases we trapped 25% (w/v) sucrose/ 1 × PBS and centrifuged over 10 (w/v) % sucrose (11). To optimize sucrose concentrations, the protocol described above is carried out in samples lacking either donor lipid or acceptor lipid. Conditions in which there is no lipid pellet in the sample lacking acceptor lipid and a maximal pellet in the sample lacking donor lipid are chosen for the lipid exchange experiments.

6.4. Analysis of asymmetric LUV lipid composition and yield by HP-TLC. The lipids from 200 μL of asymmetric vesicle sample is extracted by methanol and chloroform in a ratio of 0.5:1:1 aLUVs preparation: methanol: chloroform (v/v). The methanol is added to the vesicles first, vortexed briefly to homogenize, and then the chloroform is added. The sample, depending on lipids used, can be milky at this point, and can be centrifuged at low speed for a few min to help the upper and lower layers separate. The lower chloroform layer contains the lipids. Carefully remove the upper aqueous layer by pipette and discard. Most sucrose is in the upper layer. The lower layer is then dried under a nitrogen stream. In the meantime, an HP-TLC plate is warmed at 90–100 °C for 20 min and then cooled to room temperature (about 23 °C) before use. This helps to activate the silica gel by driving off water and gives more reproducible TLC lipid mobilities. The dried asymmetric vesicle lipids are re-dissolved in 40–50 μL of 1:1 chloroform: methanol (v/v) and 20–30 μL are loaded on the TLC plate. Pure lipid standards are combined and various amounts (2–10 μg ~ 3–14 nmol of each lipid) are loaded onto the same HP-TLC plate as the asymmetric vesicle samples to be quantified. For vesicles containing egg SM, POPC and CHOL, the plates were developed in the solvent system 6:8:2:2:1 chloroform/acetone/methanol/acetic acid/water (v/v) to separate the lipids. Samples containing POPC, POPE, POPS and cholesterol were generally chromatographed in two solvents. The first solvent system was 50:38:8:4 chloroform: methanol: water: acetic acid (v/v). After the solvent front migrated about halfway up the plate, the plate was air-dried for 5 min. Then the plate was rechromatographed in 1:1 hexane: ethyl acetate (v/v) until the solvent front migrated to near the top of the plate. When the solvent migrates to ~80% of the full height of the plate, the plate is removed and dried for at least 30 min in a fume hood. To visualize lipid bands, TLC plates are evenly sprayed with 3% (w/v) cupric acetate and 8% (v/v) phosphoric acid solution, dried for at least 30 min, and charred on a hot plate or in a high temperature oven at 180–200 °C for 5–10 min. For best charring, the charring time should be maximized to get the darkest lipid bands *before* the remainder of the plate starts to char and darken. Charred HP-TLC plates are scanned and the intensity of lipids bands are quantified by ImageJ software. The amount of each lipid in the asymmetric vesicle samples is calculated by comparing band intensity to that in the standard bands. To quantify the lipids the bands of interest are highlighted, and their intensity is measured. A background intensity of a region without lipid on the same plate, but of the same size as the bands being quantified should be subtracted. Alternatively, a single size can be used for all the bands of interest. In this case, thin bands will have more background area than thicker bands, but this does not affect the quantitation. A third method is to choose one lane of the TLC with the lipid of interest, and quantify using the WAND tool.

7. aLUVs preparation procedure using MαCD

7.1. Materials

7.1.1. Palmitoyl-2-oleoyl-*sn*-glycero-3-phosphocholine (POPC), 1-palmitoyl-2-oleoyl-*sn*-glycero-3-ethylphosphocholine (chloride salt) (POePC), 1,2-dioleoyl-3-trimethylammonium-propane (chloride salt) (DOTAP), 1-palmitoyl-2-oleoyl-*sn*-glycero-3-phospho-(1′-rac-glycerol) (sodium salt) (POPG), 1-palmitoyl-2-oleoyl-*sn*-glycero-3-phospho-L-serine (sodium salt) (POPS), 1-palmitoyl-2-oleoyl-*sn*-glycero-3-phosphate (sodium salt) (POPA), and cholesterol (CHOL), purchased from Avanti Polar Lipids (Alabaster, AL).

7.1.2. Randomly methylated-α-cyclodextrin, also called methyl-α-cyclodextrin (MαCD), purchased from AraChem Cyclodextrin Shop (Tilburg, the Netherlands). The degree of substitution (number of methyl groups) is 10–12.

7.1.3. Other materials listed in 6.1.

7.2. Equipment

7.2.1. Equipment listed in 6.2.

7.2.2. Refractometer.

7.3. Asymmetric large unilamellar vesicle (aLUVs) preparation procedure using MαCD.

7.3.1. Stock solution preparation.

7.3.1.1. Lipids are dissolved in chloroform and stored at −20 °C. The concentrations of lipids are measured by dry weight.

7.3.1.2. 285 mM MαCD is dissolved in distilled water and stored at 4 °C. If a standard curve of refractive index vs. concentration is constructed first, the concentration of the MαCD solution can be determined subsequently by comparison of its refractive index to the standard curve. But see Note 8.3 for correctly preparing standard samples to account for volume changes when CD is added to water.

7.3.1.3. 10 × phosphate-buffered saline (PBS) diluted to 1 × or 4 ×. The final concentration at 1 × is 10 mM sodium phosphate, 150 mM NaCl (pH 7.8 ± 0.2).

7.3.1.4. 7 and 23% (w/w) sucrose dissolved in 4 × PBS and 1 × PBS, respectively.

7.3.2. Preparation of aLUVs using lipid exchange catalyzed by MαCD.

7.3.2.1. Donor vesicles (MLVs) preparation and mixture with MαCD. The desired amounts of lipids are taken out of the appropriate stock solutions using the microdispenser and combined in a glass tube. Total 8 μmol lipid. (see Note 8.6). CHOL should NOT be included in the donor lipid mixture with this CD as it is not transferred between vesicles by MαCD and after donor vesicles are partly solubilized the cholesterol could stick to acceptor vesicles. The lipids are dried under nitrogen until visually dry. The lipid should be dispersed as a thin film on the walls of the glass tube. Gently warming the sample helps dry the solvent more quickly. To remove residual chloroform the dried lipids are placed in a vacuum desiccator, and subjected to a high vacuum for 1 h. The dried lipids mixtures in glass tubes are placed in a 70 °C water bath for hydration by adding 430 μL of 1 × PBS. The sample is incubated at 70 °C for 2–3 min and vortexed several times until the dried lipid film disperses evenly. Then 70 μL of 285 mM MαCD is added and the mixture will be vortexed several times for 2–3 min. The final MLVs lipid concentration is 16 mM, MαCD concentration is 40 mM and the volume is 0.5 mL. The dispersed lipid-MαCD mixture is further vortexed for 2 h at 55 °C in a multitube vortexer placed in an oven (see Note 2).

7.3.2.2. Acceptor large unilamellar vesicles (LUVs) preparation. 4 μmol of lipid containing the desired lipid mixture are dried in the same fashion as for the donor lipid samples. (see Note 8.7). Dried lipid mixtures in glass tubes are placed in a 70 °C water bath and hydrated by adding 500 μL of 23% (w/w) sucrose, dispersing the lipids 70 °C and vortexing several times over 5 min to form crude MLVs. The samples are then vortexed using a multitube vortexer placed in an oven at 55 °C for 15 min. To prepare LUVs from MLVs, the MLVs dispersions are subjected to seven cycles of freezing-thaw using a mixture of dry ice and acetone or liquid nitrogen to freeze the samples, and then warm water to thaw them, followed by extrusion 11 times through using a 1 mL extruder equipped with 100 nm-pore polycarbonate membranes. To wash away the untrapped sucrose, the resulting ~0.5 mL (0.01–0.03 mL is lost during extrusion) of acceptor LUVs is mixed with 3.5 mL 1 × PBS and the LUVs pelleted by ultracentrifugation at 190,000 × *g* for 30 min at 23 °C (using a Beckman L8–80 M ultracentrifuge with an SW 60 rotor). LUV pellets are resuspended to 0.5 mL in 1 × PBS.

7.3.2.3. Lipid exchange step. Lipid exchange step is performed at a concentration of 8 mM donor lipid and 4 mM acceptor lipid in a total volume of 1 mL. To do this, 500 μL of the donor MLVs-MαCD mixture is combined with 500 μL of acceptor LUV then incubated for 45 min at 37 °C using the incubator with shaking. During this step, lipid exchange occurs and the acceptors become converted to asymmetric vesicles. Of course, donor vesicles also pick up some acceptor vesicle lipid. After cooling, the donor-acceptor vesicle mixture is overlaid on 3 mL of 7 (w/w) % sucrose in 4 × PBS (see list 7.3.2.4) and then centrifuged at 190,000 × *g* for 45 min at 23 °C. After centrifugation the resulting pellet is loose. Remove about 3.5 mL of supernatant carefully, leaving the lowest ~400–500 μL of the supernatant, and then add 3.5 mL of 1 × PBS to the centrifuge tube, mix, and centrifuge again at 190,000 × *g* for 30 min at 23 °C. This time the pellet is tight, and the entire supernatant can be carefully removed. The final pellet is resuspended in 500 μL of 1 × PBS. Lipid yield determined by HP-TLC is typically 10% of the theoretical based on the initial amount of acceptor lipid used (see below). The extent of exchange, is determined by HP-TLC (see below) and can vary significantly depending on the lipids used (Li & London, 2020).

7.3.2.4. Optimizing sucrose concentrations needed. The sucrose concentration inside the vesicles is chosen to maximize yield while minimizing contamination of the asymmetric vesicles with donor lipids. This can depend on lipids chosen. To optimize sucrose concentrations, the protocol described above is carried out in samples lacking either donor lipid or acceptor lipid. Conditions in which there is no lipid pellet in the sample lacking acceptor lipid and a maximal pellet in the sample lacking donor lipid are chosen for the lipid exchange experiments.

7.4. Analysis of asymmetric LUV lipid composition and yield by HP-TLC. The lipids from 200 μL of the asymmetric vesicle sample is extracted with 1:2:2 of deionized water: methanol: chloroform (v/v). The 1 mL of deionized water is added to the vesicles solution first, vortexed briefly to homogenize, and then methanol is added, vortexed briefly to homogenize. Lastly, the chloroform is added, and the mixture is vortexed until milky. It can be centrifuged at low speed for a few min to help the separation of upper and lower layers. The lower chloroform layer contains the lipids. Carefully remove the upper aqueous layer by pipette and discard. (To get more complete extraction, a second round of extraction can be carried out, the upper layer solution can be moved to a tube and then dried by nitrogen flow with gentle heating. After extracting as described above, the lower layer can be combined with the lower layer solution from the first round.) The lower layer is then dried under a nitrogen stream. In the meantime, an HP-TLC plate is warmed at 90–100 °C for 20 min and then cooled to room temperature before use. This helps activate the silica gel by driving off water and gives more reproducible TLC lipid mobilities. The dried aLUVs lipids are re-dissolved 50–100 μL of 1:1 chloroform: methanol (v/v) and 10–40 μL are loaded on the TLC plate. Pure lipid standards are combined and various amounts ((2–10 μg ∼ 3–14 nmol) of each lipid) are loaded onto the same HP-TLC plate as the asymmetric vesicle samples to be quantified. A typical solvent system we used for TLC was 50:18:18 chloroform: methanol: acetic acid (v/v). After the solvent front migrated almost to the top of the plate, the plate removed from the TLC tank, and air-dried for at least 30 min in a fume hood. TLC bands were visualized as described in list 6.4.

8. Notes

8.1. It is essential to use glass bore positive displacement (solid plunger) dispensers/pipettors. You cannot accurately pipet with and ordinary air displacement pipettors using plastic tips, because of volatility of the solvent will cause inaccurate measurements and components in the plastic may contaminate the sample.

8.2. A multitube vortexer is very convenient as it can take multiple samples and set to always on position, avoiding having to hold the sample being vortexed in place by hand for long periods.

8.3. If the volume is large it can be measured in a graduated cylinder, but if it is small, we use an adjustable air displacement pipette to measure the total volume of the solution.

8.4. The investigator may want to use different donor lipids than the POPC used as donor in reference (Lin & London, 2014b). If a lipid mixture is used, the ratio of lipids in the donor mixture and the ratio of what is exchanged into the acceptor will not be identical and the outer leaflet composition must be calculated from the final lipid composition of the asymmetric vesicles, which can be determined from HP-TLC as described here and references (Li & London, 2020; Lin & London, 2014b). Also, in addition to the unlabeled lipids, we often wish to include a fluorescent lipid. Details are given in (St. Clair, Kakuda, & London, 2020).

8.5. The investigator may want to use different acceptor lipids than the 1:1 mol:mol mixture of POPE, POPS plus cholesterol, used in (Lin & London, 2014b).

8.6. The investigator may want to use different donor lipids than the SM, POPC, and the 3:1 mixtures of POPC with DOTAP, POPG, POPA, POPS, POePC, used as donor in references (Li & London, 2020; Lin & London, 2014b; St. Clair & London, 2019). If a lipid mixture is used, the ratio of lipids in the donor mixture and the ratio of what is exchanged into the acceptor will not be identical and the outer leaflet composition must be calculated from the final lipid composition of the asymmetric vesicles, which can be determined from HP-TLC as described here and reference (Li & London, 2020; Lin & London, 2014b).

8.7. The investigator may want to use different donor lipids than in (Li & London, 2020; St. Clair et al., 2020). In these papers, the acceptors included dioleoylphosphatidylcholine (DOPC), or POPC, or the 3:1 mixtures of POPC with the same lipids noted in the donor samples, as well as CHOL, generally 25–40 mol%. Note that we found yield of aLUVs often decreased at lower CHOL concentrations using our protocols.

Acknowledgments

This work was supported by NIH grant GM 122493 and NSF grant DMR 1709035.

References

Anderson, T. G., Tan, A., Ganz, P., & Seelig, J. (2004). Calorimetric measurement of phospholipid interaction with methyl-beta-cyclodextrin. *Biochemistry*, *43*(8), 2251–2261.

Brown, D. A., & London, E. (2000). Structure and function of sphingolipid- and cholesterol-rich membrane rafts. *The Journal of Biological Chemistry*, *275*(23), 17221–17224.

Cheng, H. T., & London, E. (2011). Preparation and properties of asymmetric large unilamellar vesicles: Interleaflet coupling in asymmetric vesicles is dependent on temperature but not curvature. *Biophysical Journal, 100*(11), 2671–2678.
Cheng, H. T., & Megha, E. (2009). London, Preparation and properties of asymmetric vesicles that mimic cell membranes: Effect upon lipid raft formation and transmembrane helix orientation. *The Journal of Biological Chemistry, 284*(10), 6079–6092.
Courtney, K. C., Fung, K. Y., Maxfield, F. R., Fairn, G. D., & Zha, X. (2018). Comment on 'Orthogonal lipid sensors identify transbilayer asymmetry of plasma membrane cholesterol'. *eLife*, 7, e38493.
Doktorova, M., Heberle, F. A., Eicher, B., Standaert, R. F., Katsaras, J., London, E., et al. (2018). Preparation of asymmetric phospholipid vesicles for use as cell membrane models. *Nature Protocols, 13*(9), 2086–2101.
Flanagan, J. J., Tweten, R. K., Johnson, A. E., & Heuck, A. P. (2009). Cholesterol exposure at the membrane surface is necessary and sufficient to trigger perfringolysin O binding. *Biochemistry, 48*(18), 3977–3987.
Heberle, F. A., Marquardt, D., Doktorova, M., Geier, B., Standaert, R. F., Heftberger, P., et al. (2016). Subnanometer structure of an asymmetric model membrane: Interleaflet coupling influences domain properties. *Langmuir, 32*(20), 5195–5200.
Huang, J., & Feigenson, G. W. (1999). A microscopic interaction model of maximum solubility of cholesterol in lipid bilayers. *Biophysical Journal, 76*(4), 2142–2157.
Johnson, B. B., & Heuck, A. P. (2014). Perfringolysin O structure and mechanism of pore formation as a paradigm for cholesterol-dependent cytolysins. *Sub-Cellular Biochemistry, 80*, 63–81.
Li, G., Kim, J., Huang, Z., St Clair, J. R., Brown, D. A., & London, E. (2016). Efficient replacement of plasma membrane outer leaflet phospholipids and sphingolipids in cells with exogenous lipids. *Proceedings of the National Academy of Sciences of the United States of America, 113*(49), 14025–14030.
Li, B., & London, E. (2020). Preparation and drug entrapment properties of asymmetric liposomes containing cationic and anionic lipids. *Langmuir, 36*, 12521–12531.
Lin, Q., & London, E. (2013a). Altering hydrophobic sequence lengths shows that hydrophobic mismatch controls affinity for ordered lipid domains (rafts) in the multitransmembrane strand protein perfringolysin O. *The Journal of Biological Chemistry, 288*(2), 1340–1352.
Lin, Q., & London, E. (2013b). Transmembrane protein (perfringolysin o) association with ordered membrane domains (rafts) depends upon the raft-associating properties of protein-bound sterol. *Biophysical Journal, 105*(12), 2733–2742.
Lin, Q., & London, E. (2014a). The influence of natural lipid asymmetry upon the conformation of a membrane-inserted protein (perfringolysin O). *The Journal of Biological Chemistry, 289*(9), 5467–5478.
Lin, Q., & London, E. (2014b). Preparation of artificial plasma membrane mimicking vesicles with lipid asymmetry. *PLoS One, 9*, e87903.
Lin, Q., & London, E. (2015). Ordered raft domains induced by outer leaflet sphingomyelin in cholesterol-rich asymmetric vesicles. *Biophysical Journal, 108*(9), 2212–2222.
Liu, S. L., Sheng, R., Jung, J. H., Wang, L., Stec, E., O'Connor, M. J., et al. (2017). Orthogonal lipid sensors identify transbilayer asymmetry of plasma membrane cholesterol. *Nature Chemical Biology, 13*(3), 268–274.
Lorent, J. H., Levental, K. R., Ganesan, L., Rivera-Longsworth, G., Sezgin, E., Doktorova, M., et al. (2020). Plasma membranes are asymmetric in lipid unsaturation, packing and protein shape. *Nature Chemical Biology, 16*(6), 644–652.
Nelson, L. D., Chiantia, S., & London, E. (2010). Perfringolysin O association with ordered lipid domains: Implications for transmembrane protein raft affinity. *Biophysical Journal, 99*(10), 3255–3263.

Nelson, L. D., Johnson, A. E., & London, E. (2008). How interaction of perfringolysin O with membranes is controlled by sterol structure, lipid structure, and physiological low pH: Insights into the origin of perfringolysin O-lipid raft interaction. *The Journal of Biological Chemistry*, *283*(8), 4632–4642.

Sokolov, A., & Radhakrishnan, A. (2010). Accessibility of cholesterol in endoplasmic reticulum membranes and activation of SREBP-2 switch abruptly at a common cholesterol threshold. *The Journal of Biological Chemistry*, *285*(38), 29480–29490.

St. Clair, J. W., Kakuda, S., & London, E. (2020). Induction of ordered lipid raft domain formation by loss of lipid asymmetry. *Biophysical Journal*, *119*(3), 483–492.

St. Clair, J. W., & London, E. (2019). Effect of sterol structure on ordered membrane domain (raft) stability in symmetric and asymmetric vesicles. *Biochimica et Biophysica Acta - Biomembranes*, *1861*(6), 1112–1122.

CHAPTER TEN

Molecular basis for the activation of actinoporins by lipids

Jose M.M. Caaveiro[a,*] and Kouhei Tsumoto[b,c,d,*]

[a]Department of Global Healthcare, Graduate School of Pharmaceutical Sciences, Kyushu University, Fukuoka-shi, Fukuoka, Japan
[b]Department of Bioengineering, Graduate School of Engineering, The University of Tokyo, Bunkyo-ku, Tokyo, Japan
[c]Institute of Medical Science, The University of Tokyo, Minato-ku, Tokyo, Japan
[d]Department of Chemistry and Biotechnology, School of Engineering, The University of Tokyo, Bunkyo-ku, Tokyo, Japan
[*]Corresponding authors: e-mail address: jose@phar.kyushu-u.ac.jp; tsumoto@bioeng.t.u-tokyo.ac.jp

Contents

Abstract

Actinoporins are a family of homologous pore forming proteins from sea anemones. They are one of the few families of eukaryotic toxins that have been characterized in depth. Actinoporins are activated by lipids in the context of bilayers, especially in cell and in model membranes containing the lipid sphingomyelin. These proteins must undergo conformational changes induced upon interaction with lipids in the membrane, where they form cytotoxic pores causing cell death and lethality. Herein we

Methods in Enzymology, Volume 649
ISSN 0076-6879
https://doi.org/10.1016/bs.mie.2021.01.008

review a list of procedures and techniques to study this family of toxins, with the goal of elucidating the physicochemical, thermodynamic and structural basis for their activation by lipids. The emerging picture indicates that actinoporins undergo a stepwise process that includes binding to the membrane, oligomerization, and pore formation, in this order. The key transformation from the inactive oligomer to the active pore is catalyzed by sphingomyelin, explaining the key role of this lipid in the function of actinoporins.

1. Introduction

Actinoporins are a group of cnidarian pore-forming toxins (PFT) produced by sea anemones intensively studied in the context of basic science and biotechnological applications. PFT comprise a large group of proteins produced in all kingdoms of life, including bacteria, plants, and animals (Dal Peraro & van der Goot, 2016). PFTs are part of the pathogenic as well as the defensive arsenals of these organisms, inducing lesions in the plasma membrane conducive to cell lysis. These lesions are caused by the opening of a pore traversing the membrane, of which two structural classes have been identified so far: the alpha-barrel pores and the beta-barrel pores, each giving name to the α-PFTs and β-PFTs families of PFTs (Anderluh & Lakey, 2008).

Cnidaria is an ancient phylum of marine animals, of which sea anemones are one of its most representative groups. The venom of sea anemones is a very complex liquid, containing numerous compounds such as biogenic amines, neurotoxic phospholipases, metalloproteinases, and PFTs (Podobnik & Anderluh, 2017). This venom is discharged at very high speed (few microseconds) into their preys or other target organisms by specialized cells present in the tentacles of sea anemones known as nematocytes, using a sophisticated secretory organella termed nematocyst. The PFTs present in the venom of sea anemones are called actinoporins (Actiniaria refers to the order to which sea anemones belong), although actinoporins are also found in other Cnidaria organisms (Podobnik & Anderluh, 2017).

Actinoporins are a multigenic family of cytolysins of roughly 20 kDa, positively charged proteins displaying high pI, and generally lacking cysteine residues (Kristan, Viero, Dalla Serra, Macek, & Anderluh, 2009). Although the IUBMB introduced a systematic nomenclature for this family of PFTs, by which they are classified as transmembrane solute transporters with the code 1.C.38 (Saier, 2000), the vast majority of studies still refers to them as actinoporins. Actinoporins form pores in the plasma membrane of eukaryotic cells giving rise to a number of noxious biological effects (see below).

The comparison of the primary structure of actinoporins indicates that this family PFTs are well conserved and display a notable degree of sequence similarity (Fig. 1). The similarity is equally high when comparing the three-dimensional structures of actinoporins whose crystal structures have been determined (Garcia-Linares et al., 2016; Podobnik & Anderluh, 2017). Actinoporins have been extensively studied since the isolation of the first member of the family, equinatoxin, nearly 50 years ago (Ferlan & Lebez, 1974), and numerous toxic and cytolytic activities have been reported at the cellular- and organism-level (Anderluh & Macek, 2002). These effects include high lethality in mammals attributed to cardiorespiratory arrest and coronary vasospasm, high cytotoxicity to a variety of cells affecting their ultrastructure, hemolysis (lysis of RBC) elicited by colloid-osmotic shock, cell swelling, interference with synaptosomal release or uptake of neurotransmitters, modification of GABA and choline uptake, release of glutamate from rat cortical synaptosomes, and others. However, the ultimate biological role of actinoporins for sea anemones has not been unambiguously clarified. Many carnivorous sea animals produce toxins functioning as biological weapons to paralyze preys such as crabs, worms, or small fish (Garcia-Ortega et al., 2011). It is thus possible that actinoporins have evolved to support different feeding demands, and/or for defense purposes (Rojko, Dalla Serra, Macek, & Anderluh, 2016).

A significant effort has been made by researchers interested in actinoporins, and PFTs in general, to understand the molecular and physicochemical basis underlying the cytolytic and biological effects described above. This effort has required the coalescence of numerous techniques, some of which will be explained in more detail throughout this review, and has greatly advanced our knowledge on how these family of cytolysins work. But above all, the dissection of the mechanism of action of actinoporins is an ongoing endeavor, that reflects intense technological progress and effective cooperation among fellow researchers. The aforementioned efforts have been focused on answering a simple question: How a small, monomeric, and water-soluble protein is capable of producing lytic transmembrane pores on cells? The process of answering this simple question challenges established principles in protein science, and defies our perceptions of how cell membranes are organized.

The mechanistic lessons (discussed in greater detail in Sections 2 and beyond) together with the pioneering efforts of scientists and engineers have

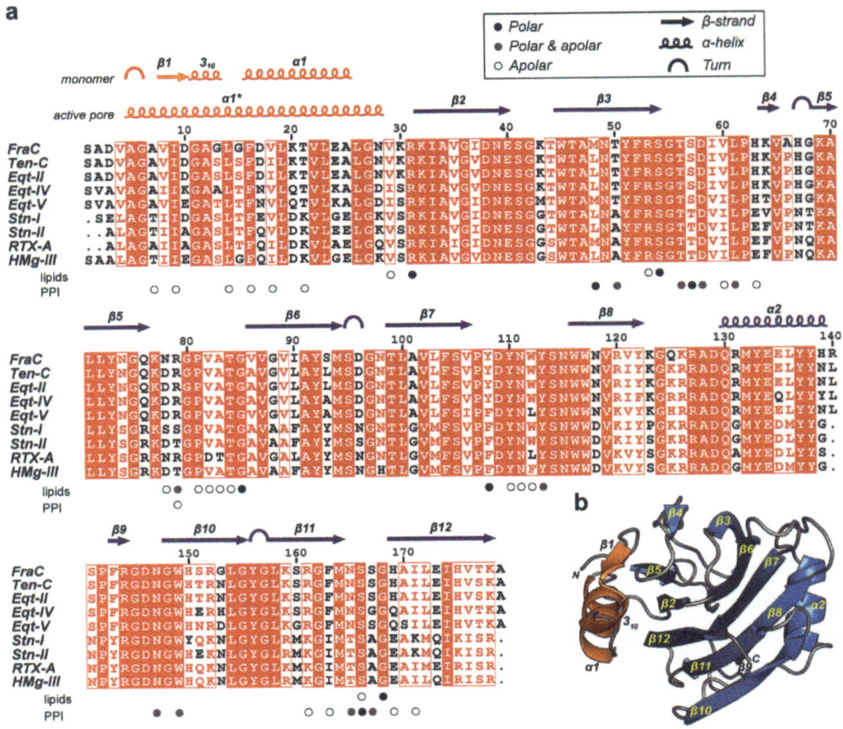

Fig. 1 Sequence alignment of actinoporins. (A) Sequence alignment of representative members of the actinoporin family of PFTs. The following actinoporins are shown: FraC from *Actinia fragacea*; tenebrosin-C (Ten-C) from *Actinia tenebrosa*; equinatoxin II (Eqt-II), equinatoxin IV (Eqt-IV), and equinatoxin V (Eqt-V) from *Actinia equina*; sticholysin I (Stn-) and sticholysin II (Stn-II) from *Stichodactyla helianthus*; cytolysin RTX-A (RTX-A) from *Radianthus macrodactylus*; and magnificalysin III (HMg-III) from *Heteractis magnifica*. In the alignment, the red background, the red text and the black text are indicative of the degree of conservation of the residues in descending order. In addition, the secondary structure elements of the N-terminal region of FraC (corresponding to the monomer and pore conformations only) are shown immediately above the amino acid sequence. The symbols employed are helices (representing α-helix or 3_{10}-helix), arrows (for β-strands) and inverted U symbols (for turns). The secondary structure of FraC is well conserved after residue 30 in the two types of structures of FraC considered, so only one representation is shown. To prepare the sequence alignment the software ESPript was employed (Gouet, Robert, & Courcelle, 2003). The figure also identifies the residues involved in lipid-protein and protein-protein (PPI) interactions. The black solid, gray solid, and empty circles correspond to residues engaging in polar, apolar, or both polar and apolar interactions, respectively. (B) The structure of the monomeric, water-soluble actinoporin FraC is shown to illustrate the secondary structure elements of the toxin. The N-terminal and the β-strand regions are depicted in blue and orange, respectively. Each element is numbered in the order it appears in the sequence, starting from the N-terminus. *This figure was adapted Tanaka, K., Caaveiro, J. M. M., Morante, K., Gonzalez-Manas, J. M., & Tsumoto, K. (2015). Structural basis for self-assembly of a cytolytic pore lined by protein and lipid.* Nature Communications, 6, *6337. doi:10.1038/Ncomms7337.*

also advance the practical use of actinoporins in several key technical fields (Ramirez-Carreto, Miranda-Zaragoza, & Rodriguez-Almazan, 2020). In some respects, this process could be considered a model showcasing how basic science may contribute to the advancement of biotechnology and biomedicine. First, actinoporins have been employed as immunotoxins for cancer therapy. In this approach, actinoporins are attached to and antibody or another carrier protein in order to specifically target various types of cancer cells (Lv et al., 2016; Tejuca, Anderluh, & Dalla Serra, 2009; Tejuca et al., 2004). Although the approach successfully inactivates cancer cells, problems related to its specificity and safety must be addressed before this method can be brought to clinical trials.

A second technological application consists in designing actinoporin-based nanopore biosensors. This class of stochastic biosensors detect individual molecules, such as nucleic acids, based on the gradual blockage of a steady electrical current as these individual molecules traverse trough the nanopore (Maglia, Heron, Stoddart, Japrung, & Bayley, 2010). Although a device designed to sequence DNA has already been commercialized (Goldstein, Beka, Graf, & Klassen, 2019), other areas such the monitoring of enzymatic reactions, or the identification of peptides and proteins, are highly desirable. In this respect, the actinoporin FraC has been successfully employed in laboratory experiments to detect posttranslational modifications (Restrepo-Perez, Wong, Maglia, Dekker, & Joo, 2019), discriminate peptide and protein biomarkers (Huang, Willems, Soskine, Wloka, & Maglia, 2017), and detect peptide with a resolution of up to 44 Da (Huang, Voet, & Maglia, 2019), among others. Third, another technological area where actinoporins are being actively investigated include the development of sphingomyelin (SM) probes to identify and characterize its distribution on the plasma membrane of eukaryotic cells (Makino et al., 2015). This application is based on the observation that actinoporins show specificity for this lipid as profusely discussed throughout this review. And fourth, actinoporins are also explored as adjuvants for vaccine design to improve the immune response, facilitating the entry of antigens in antigen presenting cells (Laborde et al., 2017).

In summary, actinoporins are a fascinating group of proteins that have evolved to recognize specific elements on the plasma membrane of eukaryotic cells where they undergo a number of transformations leading to the formation of a lytic pore. Herein, we will review the key elements that explain the specificity of actinoporins for certain lipids and membranes, and the concrete changes triggered by these lipids in actinoporins from structural and thermodynamic points of view. Finally, we will go one step further to show that, in the absence of lipids, the remarkable transformations leading to an active pore can be reversed in the presence of certain detergents.

2. Purification and hemolysis

The purification method of actinoporins has undergone significant changes since they were identified for the first time, reflecting technological advances in the field of protein cloning and purification. Initially, and for a significant amount of time, actinoporins were obtained in good yield and sufficient purity from the exuded liquid of the animal itself (Anderluh et al., 2003; Bellomio et al., 2009; Caaveiro et al., 2001; de los Rios et al., 1999; Ferlan & Lebez, 1974). This methodology employed fractional precipitation and chromatographic techniques, leading to the separation of several fractions each one containing a different isoform of the actinoporin under study. For example, in the species *Actinia fragacea* (strawberry anemone) several isoforms of fragaceatoxin A–E (FraA–E) were isolated, exhibiting slightly different properties from each other such as the hemolytic activity and protein stability (Morante et al., 2019). The development of techniques of recombinant protein expression has modified previous protocols employing natural sources. It is now more common to produce the protein in the host *Escherichia coli*, enabling site-directed mutagenesis and other protein engineering approaches. For example, the protein may be expressed lacking any tag, and purified by a two-step procedure that includes cation exchange chromatography (taking advantage of the high pI of actinoporins), and size exclusion chromatography (Anderluh, Pungercar, Strukelj, Macek, & Gubensek, 1996; Bellomio et al., 2009). This simple approach yields sufficient amount of protein of adequate purity for structural studies (Fig. 2). Alternatively, the protein is expressed with a hexahistidine tag that is subjected to immobilized-metal affinity chromatography (de los Rios et al., 2000).

Hemolytic activity is a particularly convenient and inexpensive approach to verify the presence of actinoporins and to estimate the potency of the various isoforms. The assay is based on the change of turbidity of a suspension of red blood cells (RBC) in the presence of a hemolytic substance (Louw & Visser, 1977). From the kinetic profiles it is possible to extract several parameters related to the hemolytic potency of the solution, such as the lag time, the time to reach 50% hemolysis, and the maximal hemolysis rate (V_{max}).

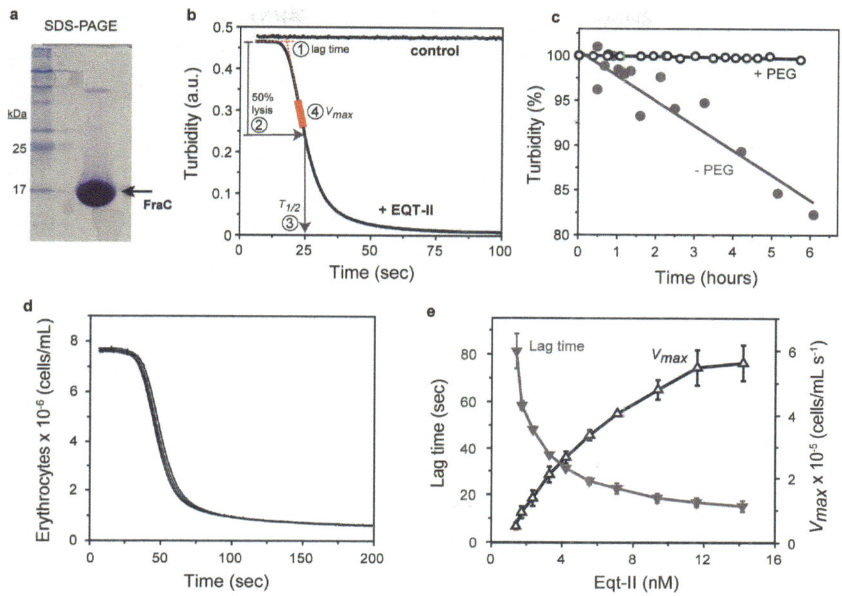

Fig. 2 Purification and hemolytic activity of actinoporins. (A) SDS-PAGE of recombinant actinoporin FraC expressed in *E. coli*. The lane containing FraC was loaded with a large amount of protein (~100 μg) to better observe the presence of contaminants. In addition to FraC, only a faint band at higher molecular weight was observed. (B) Kinetics of hemolysis of horse erythrocytes in the presence of the actinoporin Equinatoxin-II (Eqt-II). The data shows a clear contrast between the turbidity signal of a control sample (untreated) and a sample treated with 11.5 nM EQT-II. In each kinetic profile four different elements can be distinguished: (1) the lag time, representing the period passed between the addition of the toxin and the appearance of clear signs of hemolysis, (2) the value of 50% hemolysis, (3) the time at which 50% hemolysis is achieved ($T_{1/2}$), and (4) the maximum hemolysis rate (V_{max}) calculated at the maximum change of turbidity per unit of time. Hemolysis generally ends with the complete disappearance of the RBC. The concentration at the beginning of the experiment was $\sim 7.5 \times 10^7$ erythrocytes/mL. (C) Protective effect of polyethylene glycol (PEG) 10,000 in a solution of RBC. The presence of 2 mM PEG 10,000 (+PEG) decreases the spontaneous hemolysis compared with the sample without PEG 10,000 (−PEG), greatly improving de reproducibility of the kinetic studies for at least 5 h (Caaveiro, Gutierrez-Aguirre, Tribout, Paredes, & Gonzalez-Manas, 2000). The black open symbol and the gray closed circles corresponded to the turbidity of solutions of RBC in the presence or absence of PEG 10,000, respectively. No toxin is present in this experiment. Lines correspond to the least square fittings. (D) Reproducibility of the kinetics of hemolysis by Eqt-II when the RBC solution is supplemented with PEG 10,000. A total of seven kinetic traces are shown. (E) Parameters of hemolysis lag time (gray solid symbols) and V_{max} (black open symbol) as a function of the concentration of Eqt-II.

The greater the potency of the toxin solution the shorter the lag time and $T_{1/2}$, and greater the value of V_{max}. The reproducibility of the kinetic profiles benefits from the presence of polyethylene glycol (PEG) 10,000 in the solution of RBC. The presence of PEG at a concentration of 2 mM decreases the spontaneous hemolysis of RBC from 69 erythrocytes/ $mL\,s^{-1}$, to 1 erythrocyte/$mL\,s^{-1}$ at room temperature (20–25 °C) for at least 5 h, dramatically improving de reproducibility of the kinetic parameters inferred form the hemolysis profiles (Caaveiro et al., 2000).

Alternatively, the hemolytic activity of actinoporins can be monitored by the appearance of hemoglobin in the supernatant of a solution of RBC treated with actinoporins. This assay is even simpler than the kinetic studies, and can be easily set-up in 96-well plate format. In this modality, the key parameter is HC_{50}, representing the concentration of toxin producing 50% hemolysis (Alegre-Cebollada, Cunietti, Herrero-Galan, Gavilanes, & Martinez-del-Pozo, 2008).

2.1 Equipment

1. Orbital shaker with temperature control
2. Emulsiflex C5 (cell homogenizer) or other cell-disruption equipment
3. AKTA chromatographic system with UV detection
4. Resource S column (cation exchange)
5. HiLoad 16/60 Superdex 75 pg column (size exclusion chromatography)
6. Amicon concentrators (molecular weight cut-off 10 kDa) and centrifuge
7. Nanodrop microvolume spectrophotometer
8. SDS-PAGE gel electrophoresis apparatus
9. UV–visible spectrophotometer equipped with stirring unit
10. Cell counter, for example, Coulter Counter ZF

2.2 Key reagents and biologicals

1. RBC from horse (or from other species)
2. pBAT-4 expression vector (other bacterial vectors are also suitable, e.g., pET series)
3. *Escherichia coli* (e.g., BL21 (DE3)).
4. Ampicillin (or other antibiotic selection marker)
5. Isopropyl β-D-1-thiogalactopyranoside (IPTG)
6. Standard buffers
7. PEG 10,000

2.3 Protocol

(1) In this protocol we describe the expression and purification of the recombinant actinoporin FraC. The DNA encoding the sequence of FraC was cloned into pBAT-4 expression vector as described in detail in a previous report (Bellomio et al., 2009) and is considered the starting point of this protocol.

(2) The expression host is *E. coli* BL21 (DE3) cells. The cells are transformed with the expression vector and grown at 37 °C in LB culture supplemented with ampicillin at 0.1 mg/mL. Generally 4–8 L of medium are prepared. Protein expression was induced with IPTG (1 mM) when OD reached a value of 0.5–0.6. Cells are harvested after 4 h by centrifugation at 6000 × *g* for 10 min at 4 °C.

(3) The harvested cells are resuspended in a volume approximately six times greater than the mass of wet cell pellet in a buffer containing 50 mM Tris-HCl (pH 7.5) and subjected to disruption in an Emulsiflex C5 homogenizer (other disruption method such as sonication is also adequate).

(4) The lysate is centrifuged at 40,000 × *g* for 30 min at 4 °C and the supernatant collected.

(5) After filtration with a Millex GP 0.22 μm unit, the sample was applied to a Resource S cationic-exchange chromatography column (6 mL) equilibrated with lysis buffer. FraC was eluted with a gradient of buffer with salt (50 mM Tris, and 1 M NaCl at pH 7.4). Samples containing FraC are pooled together and concentrated to less than 2 mL.

(6) The concentrated sample is further purified by size exchange chromatography in a HiLoad 16/60 Superdex 75 pg column equilibrated with 50 mM Tris, 200 mM NaCl at pH 7.4. This step is not mandatory, although it is strongly recommended when purifying protein for crystallization experiments.

(7) The fractions containing the protein are pooled together and concentrated to the desired protein concentration with an Amicon filter (MWCO 10 kDa). Protein concentration is monitored with a nanodrop microvolume spectrophotometer at 280 nm using a molar absorptivity value of 44,000 M^{-1} cm^{-1}.

(8) The protein solution is frozen by submersion in liquid N_2, and stored at −80 °C until use.

(9) RBC (horse, rat, sheep, etc....) are repeatedly washed in assay buffer (10 mM HEPES, 150 mM NaCl, pH 7.5) by centrifugation (2000 × *g*,

5 min) and resuspended until a clear supernatant is obtained. When precise data is sought after, the RBC suspension is stabilized by supplementing the solution with 2 mM PEG 10,000.

(10) The solution of RBC was adjusted at a concentration of $\sim 7 \times 10^7$ erythrocytes/mL.

To measure the kinetics of hemolysis, the RBC are placed in a cuvette and the turbidity at 700 nm monitored constantly during the experiment. The appropriate amount of actinoporin is added and the signal recorded until hemolysis is completed. Instead of using the volume of the pipette as an indication, the mass of the solution of RBC and the mass of solution of actinoporin are measured in an analytical balance to calculate the precise amount of toxin added to the solution of RBC. In that case it is recommended to use a cuvette with a lid to minimize the evaporation of solvent when the rate of hemolysis is very slow.

(11) The maximal rate of hemolysis for each curve (V_{max}) is calculated by a regression of the segment where the rate of turbidity change is greatest. The time necessary to reach half hemolysis $T_{1/2}$, is calculated at the point where the total change of turbidity has decreased to half that value. The lag time is estimated as the intersection of the regression line employed to calculate V_{max} and the initial turbidity value. Since the stock solution of blood decays we employ FraC WT as a control.

3. Interaction of actinoporins with liposomes

Actinoporins are activated by lipids forming bilayers, and not by detergents or short lipids forming micelles. Maximal activation is achieved when the membranes contain a class of lipids termed sphingomyelin (SM), a lipid that is present in abundance in eukaryotic cells. Soon after the first reports researchers realized that SM could be a key lipid for the function of actinoporins, since liposomes containing this lipid (made by the injection method) strongly inhibited the hemolytic activity of actinoporins, whereas liposomes containing other classes of lipids such as phosphatidylcholine (PC), phosphatidylcholine (PE) or phosphatidylserine (PS) did not achieve such effect even at mg quantities (Macek, Sencic, & Lebez, 1982). Numerous studies so far have shown that actinoporins preferentially permeabilize lipid membranes, and interact with other model systems such as monolayers and supported lipid bilayers, containing the lipid SM (Garcia-Ortega et al., 2011; Rojko et al., 2016). The specificity of actinoporins for SM is actually relevant from a biological point of view, because the cell membranes of sea anemones do not contain SM. In sea anemones the headgroup of SM is replaced by a phosphono derivative, greatly differing from that of SM. This derivative is critical to make sea anemones resistant to their own toxin (Meinardi, Florinchristensen, Paratcha, Azcurra, & Florinchristensen, 1995). A protein receptor that facilitates the action of actinoporins has not been found to date.

Although the general consensus among researchers is that SM is a special lipid for actinoporins, the picture is a bit more complex. In the standard assay to monitor the lytic activity of actinoporins in lipid membranes (liposomes), it is clearly observed that liposomes composed of only SM or only the synthetic lipid 1,2-dioleoyl-*sn*-glycero-3-phosphocholine (DOPC) are not susceptible to the action of actinoporins (Fig. 3). This seems to contradict the established principle that SM activates actinoporins. The greatest activity is observed when the concentration of SM and DOPC in the liposomes are equal (i.e., 50% each), a result described for Eqt-II from *Actinia equina* (Caaveiro et al., 2001) and later confirmed for other actinoporins. The oligomerization state of the protein after contacting these three different types of liposomes, and the degree of protection from a protease are also essentially identical in liposomes composed of SM or in liposomes composed of DOPC. The protein only oligomerizes and is fully protected when the lipid composition contains the same number of molecules of SM and DOPC (equimolar mixtures). Thus, the activity of actinoporins must depend on some additional factor that is maximized in conditions of maximum heterogeneity (1:1 mixtures).

These and other types of evidence resulted in a number of reports in which the list of lipids and lipids mixtures activating actinoporins was expanded. For example, the presence of cholesterol in the liposomes increased the pore-forming activity, especially if SM was also present as shown, for example, with sticholysin II from *Stichodactyla helianthus* (de los Rios, Mancheno, Lanio, Onaderra, & Gavilanes, 1998). Also, under conditions of low ionic strength, electrostatic interactions between actinoporins (positively charged, high pI) and liposomes containing the negatively charged lipid phosphatidylglycerol (PG) promoted significant levels of lytic activity (Caaveiro et al., 2001).

Collectively, these data indicate that SM is the most efficient activator of actinoporins, but not the only factor to consider. Further research has convincingly shown that the activity of actinoporins are maximized in membranes containing SM in mixtures with other lipids that organize in lipid domains. Lipid domains are caused by phase separation in the plane of

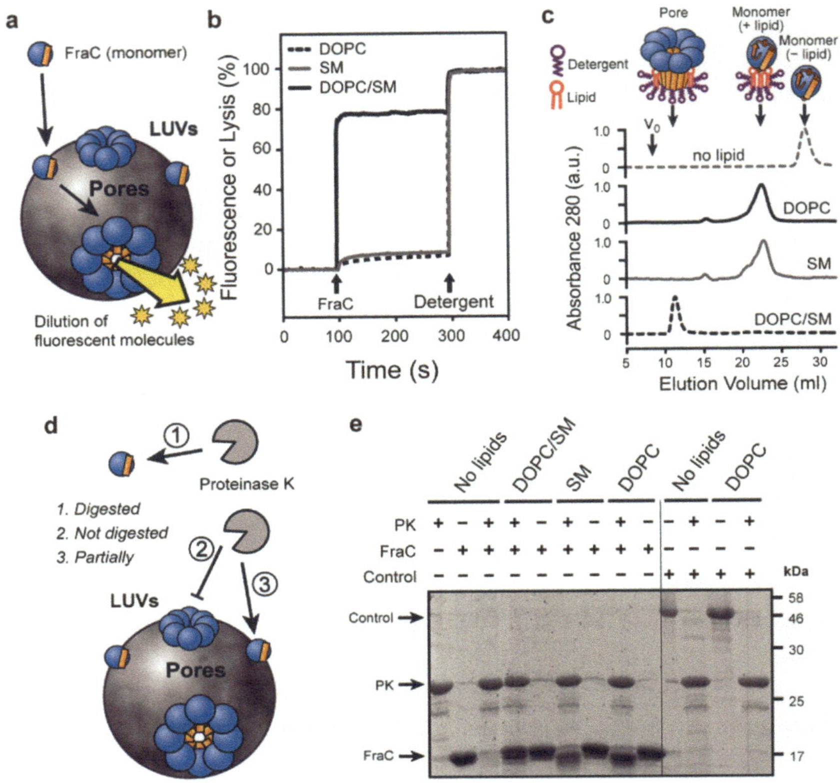

Fig. 3 Interaction of actinoporins with lipid membranes. (A) Scheme illustrating the lytic (leakage) assay. LUVs liposomes encapsulating a pair of fluorescent and quencher molecules at high concentration are prepared with a defined lipid composition. When water-soluble actinoporin (in this case FraC) is activated and forms a lytic pore, the encapsulated molecules exit the liposomes and are diluted in the medium, where the quencher now is no longer effective, and the fluorescence signal increases. (B) Typical leakage assay. The treatment of liposomes with FraC results in the leakage of encapsulated solutes only in certain lipid compositions. In this example, only liposomes composed of an equimolar mixture of the lipids DOPC and SM are susceptible to the presence of FraC. (C) Oligomerization of FraC after incubation with LUVs is monitored by size exclusion chromatography in the presence of a detergent that solubilizes the liposomes. Three different sizes are observed. The smaller size corresponded to that of the monomer (no incubation with liposomes), followed by an intermediate state of monomers containing lipids and/or detergents. The largest particles corresponded to the pore form of FraC, composed of several units of protein, and appearing near the exclusion volume. (E) Schematic view of the protease protection assay. In this experiment the protective effect of the liposomes is evaluated. When water-soluble FraC encounters the unspecific proteinase K, it is digested. However, when FraC is bound to the liposomes, proteinase K does not completely digest the protein. In the case of a bound monomer, only the N-terminal is suitable to digestion by the protease. When the molecules of FraC participate assemble as a pore, the protein is not digested. (F) Typical results from a protease protection assay. It is shown that the degree of protection depends on the lipid composition of the liposomes. In the figure control refers to BSA, a protein that does not interact with liposomes. PK, proteinase K. *This figure has been adapted Tanaka, K., Caaveiro, J. M. M., Morante, K., Gonzalez-Manas, J. M., & Tsumoto, K. (2015). Structural basis for self-assembly of a cytolytic pore lined by protein and lipid.* Nature Communications, 6, *6337. doi:10.1038/Ncomms7337.*

the membrane between lipids of different nature (Heberle & Feigenson, 2011). This concept was incorporated in the lexicon of cell membranes as lipid rafts, to explain protein dynamics and cell signaling (Simons & Ikonen, 1997; Simons & Vaz, 2004). Lipid rafts are "islands" or clusters rich in SM and cholesterol that favor the activity of protein receptors and other membrane proteins. The selection of lipid compositions that maximize the presence of lipid domains is clearly correlated with the activity of actinoporins, a circumstance that facilitates their oligomerization and/or insertion in the hydrophobic core of the membrane (Barlic et al., 2004; Morante et al., 2016; Schon et al., 2008).

Isothermal titration calorimetry (ITC) is the premier technique in interaction analysis and it has been also applied to the study of lipid-actinoporin interactions since the pioneering work of Alegre-Cebollada et al. (2008). When employing this technique, useful parameters such as the apparent dissociation constant and the enthalpy of the interaction can be determined. This information allows quantitative comparisons of the interaction of FraC between vesicles of different lipid compositions (see Protocol 3.3.3). For example, in Fig. 4 it is observed that the binding profile of FraC to liposomes that favor the assembly of the active pore is different to that of liposomes where the actinoporin is inactive: both the enthalpy and the apparent binding affinity are substantially greater in the liposomes where the pore is assembled (Tanaka, Caaveiro, Morante, Gonzalez-Manas, & Tsumoto, 2015). The technique may also be used to determine the role of certain residues in the oligomerization and lytic process or to assess their role in direct protein-lipid interactions (Alegre-Cebollada et al., 2008; Morante, Caaveiro, Tanaka, Gonzalez-Manas, & Tsumoto, 2015; Morante, Caaveiro, Viguera, Tsumoto, & Gonzalez-Manas, 2015).

Herein we describe general protocols to examine the lytic activity of actinoporins in liposomes using classical approaches and a high-resolution thermodynamic technique employed to evaluate their mutual interaction.

3.1 Equipment

1. Rotary evaporator and pump
2. Heating block
3. Water bath
4. UV–visible spectrophotometer
5. Vortex mixer
6. Avanti mini-extruder

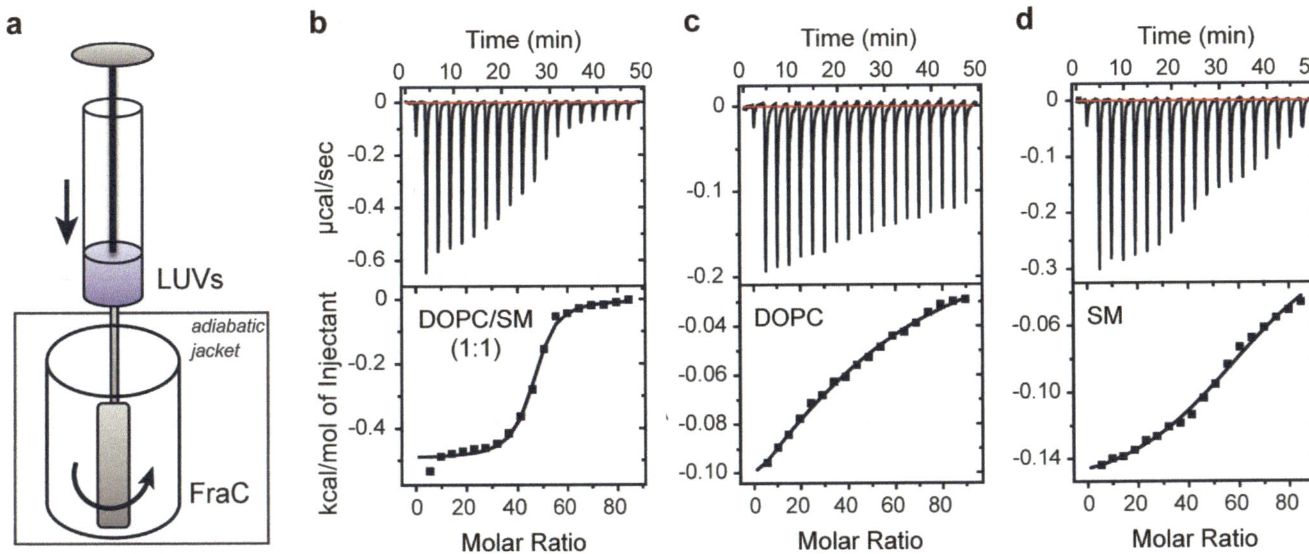

Fig. 4 Lipid-protein interaction monitored by ITC. (A) Experimental set-up. Protein sample is placed in the cell compartment under intense stirring, whereas the liposomes are injected from a highly precise syringe in a stepwise fashion, acting as the titrant. The cell compartment is placed within an adiabatic jacket to eliminate the exchange of heat with the surroundings. (B–D) Examples of titration of FraC with three different populations of liposomes each having different lipid composition as shown in each panel. The upper portion of the panels corresponds to the titration kinetics. The bottom portion of the panels shows the binding isotherm obtained from the titration from above. In the bottom portion of the panels, the solid line corresponds to the non-linear regression of the integrated heat (black solid squares) to a binding model, obtaining various relevant thermodynamic parameters. *This figure has been partially adapted from Tanaka, K., Caaveiro, J. M. M., Morante, K., Gonzalez-Manas, J. M., & Tsumoto, K. (2015). Structural basis for self-assembly of a cytolytic pore lined by protein and lipid.* Nature Communications, 6, *6337. doi:10.1038/Ncomms7337.*

7. Desalting column (PD-10)
8. Spectrofluorometer
9. Protein purification system (AKTA)
10. Size exclusion chromatography column
11. SDS-PAGE gel electrophoresis apparatus
12. Degassing device
13. ITC microcalorimeter such as VP-ITC or iTC200
14. Dialysis device for small volumes (<500 μL)

3.2 Reagents

1. Purified protein, for example, FraC
2. Sphingomyelin (SM) from porcine brain
3. 1,2-Di-(9*Z*-octadecenoyl)-*sn*-glycero-3-phosphocholine (DOPC)
4. 8-Aminonaphthalene-1,3,6-trisulfonic acid (ANTS).
5. *N,N'*-*p*-xylenebispyridinium bromide (DPX).
6. Liquid nitrogen
7. Triton X-100 (a detergent)
8. Proteinase K

3.3 Protocols for the study of actinoporin-liposome interactions

3.3.1 *Protocol (leakage assay)*

(1) Preparation of protein was explained above (Protocol 2.3).

(2) Liposomes (large unilamellar vesicles, LUVs) were prepared by the extrusion method. The appropriate volume of lipids dissolved in chloroform/methanol (2:1) is evaporated in a rotary evaporator overnight. It is highly recommended to prepare a lipid stock (10–50 mM) and precisely determine its concentration. For phospholipids the method of Bartlett or any of their variants is adequate (Bartlett, 1959).

(3) The dried lipids are resuspended in buffer 10 mM HEPES, 50 mM NaCl, 25 mM ANTS, and 90 mM DPX at pH 7.5, by vigorous mixing in a Vortex mixer or equivalent to a lipid concentration of 5–10 mM. Other buffers are perfectly adequate, except buffers containing phosphate or compounds with phosphorous since it interferes with the analytical procedure to determine the lipid concentration. The suspension takes a dense white aspect reflecting the appearance of large multilamellar vesicle (MLVs).

(4) Subject the lipid sample to 10 cycles of freeze-thaw by subjecting the solution to alternative immersion in liquid N_2 and a water bath at a temperature above the melting temperature (T_m) of the lipids.

(5) Produce LUVs by using a Avanti mini-extruder or other equivalent device. The use of polycarbonate filters of 100 nm yields LUVs of a hydrodynamic diameter of approximately 115 nm and with an average number of lamellae of 1–1.5, which is adequate for most experiments described here. The sample is subjected to 10–20 cycles of extrusion at a temperature 5 °C above the T_m of the lipids composing the liposomes. When the T_m value is around or above room temperature, this can be achieved by placing the extruder on a hot plate. When a complex mixture of lipids is employed to prepare the liposomes, the T_m can be estimated as the weighted average of the T_m of the individual components. The presence of cholesterol tends to abolish the melting transition, specially at high concentrations. In case of doubt it is appropriate to use the T_m of the compound with the highest value.

(6) Pass the liposomes through a desalting column (or similar) to separate the solution not encapsulated inside the liposomes from the liposomes themselves. LUVs suspensions have a characteristic translucent appearance that can be used as a cue to visually monitor the elution of the liposomes from the desalting column.

(7) Determine the concentration of lipid in the liposomes using the same method as indicated in step 2.

(8) The leakage experiment is carried out in a spectrofluorometer with excitation and emission wavelengths set at 355 and 515 nm, respectively. It is recommended to place a cut-on 475 nm filter (or similar wavelength) between the sample and the emission monochromator to reduce the scattering from the liposomes.

(9) LUVs at 100 μM are placed in a fluorescence cuvette under constant stirring. Toxin was added at the desired concentration. Pore formation starts almost immediately, so it is important to minimize the time of addition of the toxin. The reaction continues for as long as necessary. To obtain the full release of the fluorescent probe (100% signal) each sample is treated with the detergent Triton X-100 at 0.1% (w/v) (final concentration). Because LUVs are not homogeneously unilamellar (despite their name), leakage is generally underestimated by ~20% (see, for example, Fig. 3B). The experiment can be carried out a room temperature.

(10) Data is analyzed according to the following formula.

Leakage (%) = $(F - F_0)/(F_{100} - F_0) \times 100$.
where F is the fluorescence at any moment, F_0, the fluorescence before addition of the protein, and F_{100} the signal obtained in the presence of Triton-X100.

3.3.2 Protocol (protection assay)

(1) Preparation of protein was explained above (Protocol 2.3).
(2) Preparation of LUVs as explained in Protocol 3.3.1. The buffer employed to resuspend the dry lipid (step 3) should not include the fluorescent molecules explained in the protocol of the leakage experiment. We generally employ 10 mM HEPES, 200 mM NaCl, at pH 7.5.
(3) LUVs of the desired composition are incubated with FraC (50 μM) for 30 min at room temperature in a buffer containing 10 mM HEPES, 200 mM NaCl at pH 7.5. A molar ratio lipid-to-protein of 50:1 to 100:1 is generally appropriate.
(4) Proteinase K, a broad-spectrum serine protease, at 50 μM is incubated with the protein/lipid mixture for 24 h at room temperature. The buffer is supplemented with 5 mM $CaCl_2$ to facilitate the activity of the protease. The reaction is stopped with 5 mM phenylmethylsulfonyl fluoride for 10 min, and the sample analyzed by SDS-PAGE.
(5) A schematic explanation and typical results are shown in Fig. 4D and E. Proteinase K completely digests FraC in the absence of liposomes, reflecting the accessibility of the enzyme to the actinoporin moiety. In the presence of liposomes, we have observed two types of cleavage patterns, both being different from the pattern observed in the liposome-free experiment. On the one hand, with liposomes where FraC is inactive, such as DOPC or SM, the accessibility of the protease decreases abruptly, and only a small portion of the protein (the N-terminal region) is cleaved, leaving a protein band that is ~3 kDa shorter than the untreated protein. On the other hand, in liposomes where FraC forms the active pore the major fraction of the protein is not cleaved, reflecting the inaccessibility of the protease to the N-terminal region since it forms a transmembrane pore (as it will be explained in Section 4).
(6) This protocol can be adapted to other proteases. For example, when liposomes composed of negatively charged residue are employed, the presence of Ca^{2+} (necessary for proteinase K) may induce liposome aggregation and ruin the experiment. Therefore, other proteases not requiring Ca^{2+} may better suited under these particular circumstances.

(7) Control experiments, for example, with a protein not known to bind liposomes such as bovine serum albumin, is recommended as shown in Fig. 3E.

3.3.3 *Protocol (ITC)*

(1) Protein and LUVs are prepared as explained in Protocols 2.3 and 3.3.2.
(2) Lipids and protein must be dialyzed in the same buffer prior to the experiment, and samples are degassed under a low-pressure device with constant stirring.
(3) The ITC experiment is carried out in a microcalorimeter, such as iTC200 or VP-ITC. The cell chamber is loaded with protein at 20–50 μM, and the syringe with a suspension of LUVs of the desired composition is injected in aliquots under constant stirring. A total of 15–20 injections is generally adequate.
(4) The heat released after each injection is determined from the raw data. Each exothermic or endothermic peak is integrated after subtraction of the baseline. And the binding isotherms were fitted to a model in which a molecule of protein binds to *n* molecules of lipid with the software ORIGIN 7. An excellent overview of the thermodynamic equations is given elsewhere (Alegre-Cebollada et al., 2008).
(5) In this experiment it is strongly recommended that the buffer employed during the preparation of the liposomes is also used during the titration to minimize spurious signal caused by the heat of dilution (buffer mismatch with respect to the buffer encapsulated inside the liposomes) specially under conditions where the protein forms pores.

4. Crystallization of the pore of FraC

Chronologically, Eqt-II was the first actinoporin to be crystallized, revealing a beta-sheet fold interrupted by a transversal alpha-helix, and a N-terminal region comprising the first 30 residues adopting a mixture of helix and irregular secondary structure (Athanasiadis, Anderluh, Macek, & Turk, 2001) (Fig. 1B). Subsequently, a crystal structure of Stn-II with a lipid fragment corresponding to the lipid headgroup (phosphorylcholine) was reported at 2.4 Å, revealing for the first time the position of a lipid binding site in the aromatic pocket of the protein (Mancheno, Martin-Benito, Martinez-Ripoll, Gavilanes, & Hermoso, 2003). A few years later, the structure of the actinoporin FraC was determined in a oligomeric state consisting of nine identical subunits of protein (Mechaly et al., 2011). The structure of each individual protein chain in the 9-mer is essentially unchanged with respect to the monomeric water-soluble form. The oligomer is arranged in a doughnut-shape entity, in which the phosphorylcholine binding site lays on a plane compatible with the position of the 9-mer on a lipid membrane. However, no portion of the protein seems to be embedded in the membrane, and therefore this structure does not explain pore formation by FraC.

More recently, the structure of the pore of FraC at 3.15 Å with several lipids bound, as well as some other accessory structures, were reported (Tanaka, Caaveiro, Morante, et al., 2015) (Fig. 5). The structure shows for the first time the molecular details of the active form of an actinoporin pore, such as the conformational change of the N-terminal region to a single alpha-helix spanning the entire membrane. The pore is composed by 8 identical protein subunits, in which the transmembrane portion is organized as an alpha-barrel. This type of pore architecture is different from that of the beta-barrels, as exemplified by the bacterial α-toxin heptamer from *Staphylococcus aureus* (Song et al., 1996).

The structure of the pore of actinoporins reveals several lipids bound, of which one of them "bridges" two consecutive protein subunits, occupying a crucial position. The lipid modeled in that crucial position was SM, although the resolution of the structure is not sufficiently high to unequivocally assign it to either one of the components of the liposomes, SM or DOPC. Based on the interactions established between lipid and protein, however, it is more likely to be SM than DOPC (Tanaka, Caaveiro, Morante, et al., 2015). This bridging lipid lines the walls of the pore, and although most of the pore is made or protein, the contribution of the lipid is sufficient to make it a hybrid protein-lipid pore. Two annular lipids bound to each protomer complete the structure, and when all the lipids are seen in the context of the pore they lay on the same plane of the membrane. In one of the accessory structures, the water-soluble form of FraC was crystalized in the presence of short-chain and water-soluble lipids, revealing up to four distinct lipid binding sites. Two of these lipids are bound to the same binding site than the annular lipids seen in the pore. The other two lipids bind to new binding sites, demonstrating the concept of lipid multivalency in actinoporins. Collectively, actinoporins have at least four lipid binding sites for each monomer needed for initial adhesion to the membrane, and a fifth one (bridging lipid) occupied by SM that is proposed to activate the transformation from the inactive oligomer to the active pore.

In the experimental section, the preparation of the crystal of the pore is explained in some detail, since it employs an elaborate procedure (Fig. 6).

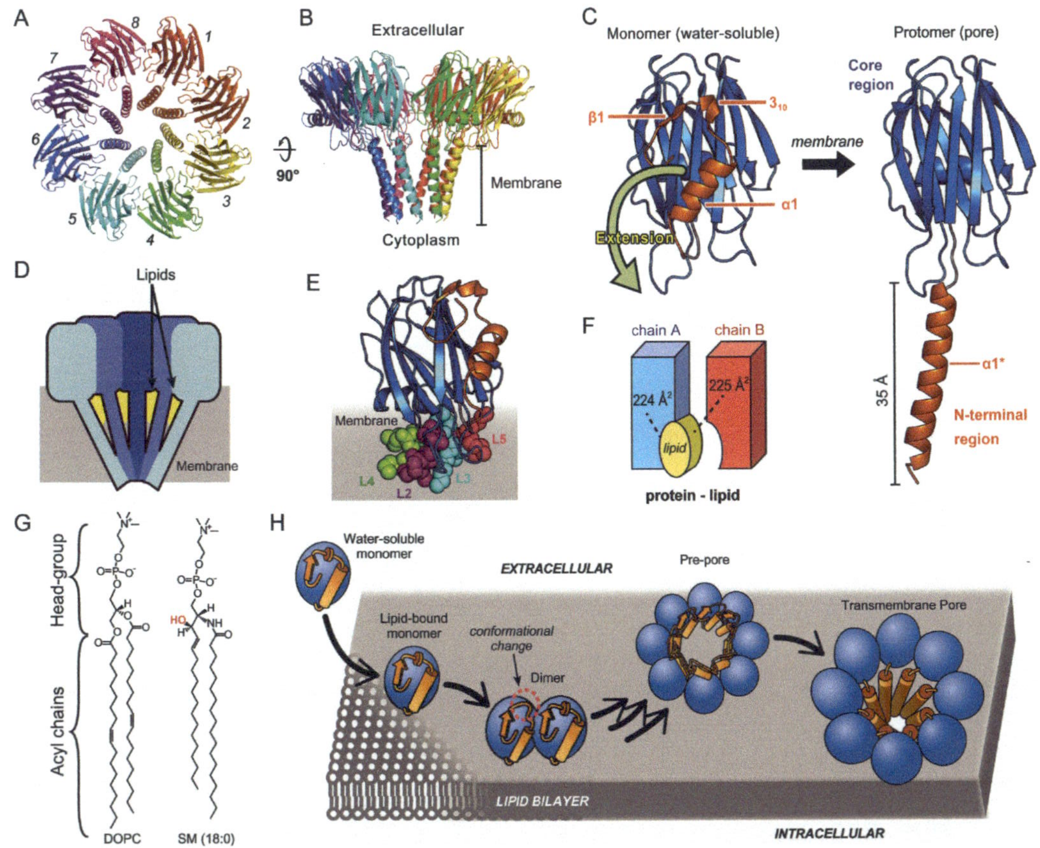

Fig. 5 See figure legend on opposite page.

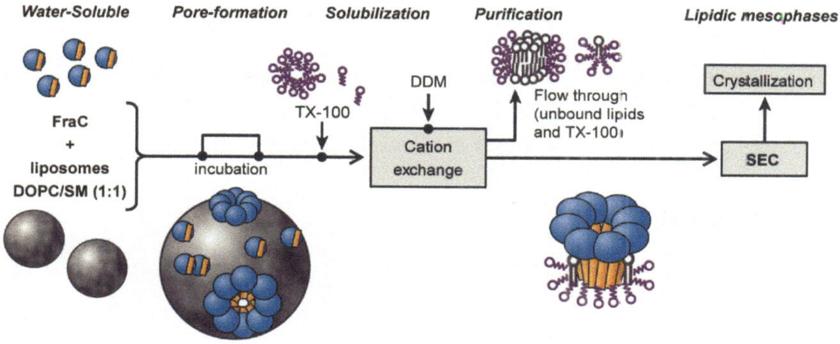

Fig. 6 Preparation of pore specimens for crystallization. To crystallize the active pore of actinoporins (in this case FraC) it is necessary to first generate a homogeneous sample of pores. For that, the monomeric protein and a suitable suspension of liposomes, such as LUVs composed of the equimolar mixture of DOPC and SM, were first incubated with each other. The incubation generates pores of FraC inserted in the LUVs. Prior to crystallization, the liposomes must be disintegrated (solubilized) because it is not possible to crystallize proteins embedded in liposomes. The detergent Triton X-100 was employed for that, since it combines a good solubilization efficacy without compromising the architecture of the pore. The pores were subsequently purified in a cation exchange chromatography column to remove detergent/lipid mixed micelles, and exchanged to the milder detergent DDM. In the absence of detergent, the protein aggregates, so it is necessary to keep a small concentration of detergent above the critical micellar concentration throughout the purification. The pore sample is applied to a size exclusion chromatography column to separate the pore specimens from a small fraction of monomeric protein. The sample is ready for the crystallization experiment after adjusting the protein concentration to the desired value. *This figure has been adapted from Tanaka, K., Caaveiro, J. M. M., Morante, K., Gonzalez-Manas, J. M., & Tsumoto, K. (2015). Structural basis for self-assembly of a cytolytic pore lined by protein and lipid.* Nature Communications, 6, *6337. doi:10.1038/Ncomms7337.*

Fig. 5 Crystal structure of the pore of the actinoporin FraC. (A, B) Top and side view of the pore. The pore is composed of eight protomers of FraC. In the pore, the beta-rich domain (core-region) is facing the extracellular side, and the N-terminal has undergone a dramatic conformational change, lining the walls of the pore in the transmembrane region. (C) Depiction of the conformational change in the N-terminal region (orange) upon pore formation. On the left, the structure of the water-soluble protein, and on the right the structure of one protomer in the pore. (D) Schematic view of the pore in which fenestrations filled-up with lipids are shown in yellow. (E) Structure of the water-soluble monomer with four DHPC lipids bound. The lipid positions are labeled L2-L5. (F) The position of lipid L1 is schematically shown. This critical lipid occupies a position between two protein chains in the pore, and could be identified as the yellow shade in panel (D). (G) Structures of typical lipids. On the left, DOPC, and on the right SM. (H) Model of activation of FraC by membranes. Based on the available data we propose the stepwise model of pore formation of actinoporins. This model is based on four different crystal structures (water-soluble, lipid-bound, dimer and pore) and a cryo-microscopy and AFM study of FraC on liposomes of DOPC (Morante et al., 2016). *The figure was adapted Tanaka, K., Caaveiro, J. M. M., Morante, K., Gonzalez-Manas, J. M., & Tsumoto, K. (2015). Structural basis for self-assembly of a cytolytic pore lined by protein and lipid.* Nature Communications, 6, *6337. doi:10.1038/Ncomms7337.*

The crystallization of the other structures follow standard procedures (Tanaka, Caaveiro, Morante, et al., 2015) and readers are directed to other excellent crystallization texts for a general overview (McPherson, 2016).

4.1 Equipment

1. Crystallization robot ORIX-8 (Douglas Instruments)
2. Incubator
3. Crystallization tools
4. Synchrotron radiation source (X-rays)

4.2 Reagents

1. Crystallization kits and reagents
2. *N*-dodecyl-beta-maltoside (DDM)
3. Triton X-100

4.3 Protocol (crystallization of pore)

(1) The protein and liposomes were prepared following the Protocols 2.3 and 3.3.2, respectively.

(2) Water-soluble, monomeric FraC at 50 μM was incubated with LUVs of equimolar mixture of SM and DOPC at a lipid-to-protein molar ratio of 200:1. The buffer was Tris 50 mM, NaCl 200 mM at pH 7.4. The mixture was incubated for 30 min at room temperature, followed by the complete solubilization of the liposomes with Triton X-100 at 1% (v:v) to separate the pores from the bilayer, since it is not possible to crystallize proteins embedded in membranes. The addition of excessive detergent could strip the pore particles from essential lipids attached to them, rendering them unstable.

(3) To separate the pore particles from mixed micelles of lipids and detergent Triton X-100, the solution was applied to a Resource S column equilibrated with a buffer composed of 50 mM Tris (pH 7.4) and DDM (0.015%). The pores of FraC were eluted with a buffer containing 50 mM Tris (pH 7.4), NaCl 1 M, and DDM (0.015%). The eluted samples containing FraC were pooled together and subsequently subjected to size exclusion chromatography in a Superdex 200 10/300 column equilibrated with 50 mM Tris (pH 7.4), NaCl 200 mM, and DDM (0.015%).

(4) The solution containing the pores of FraC are concentrated to ~10 mg/mL and subjected to extensive crystallization trials by the method of sitting drop with a crystallization robot, obtaining numerous conditions in which large and well-shaped crystals are formed. However, and despite intense optimization by the sitting and hanging drop, these crystals never diffracted beyond ~10 Å.

(5) The solubilized pores of FraC were crystallized by the hanging drop method using the lipidic-sponge phase screen (Wohri et al., 2008) by mixing FraC at 7 mg/mL and a lipidic solution containing Jeffamine M600, 1,2,3-heptanetriol, monoolein, ammonium sulfate and HEPES (pH 7.0). The reservoir solution was composed of 0.1 g/L of sodium/potassium phosphate, 550 mM sodium acetate and 750 mM HEPES (pH 6.2). Crystals developed to full size in about 6 months at 20 °C.

(6) Suitable crystals were harvested and frozen by direct immersion in liquid N_2. Incubation in a special cryo-solution prior to freezing was not necessary.

(7) Data collection, data processing, structure determination by the method of molecular replacement using the coordinates of the monomeric form, and refinement, were carried by standard procedures.

5. Regeneration of monomeric water-soluble protein from transmembrane pores

We have extensively discussed the transformation of the naturally produced protein in a lytic pore activated by lipids and specially by SM. We asked ourselves if it was possible to adjust the experimental conditions to produce the reverse transformation from the pore form to the water-soluble protein. Although the transformation to the membrane-anchored-pore is an irreversible reaction in physiological conditions, it was considered whether the opposite reaction (pore to monomer) could happen under a different environment. For that to happen, the pore should be purified as explained in Protocol 4.3. We hypothesized that treatment of the isolated pores with the adequate detergent could efficiently remove all the lipids (it was shown that at least three lipids strongly attach to the pore in the crystal structure), triggering the regeneration of the amphipathic, N-terminal transmembrane helix of the pore into the water-soluble conformation (Fig. 5), i.e., regenerating the monomeric protein. So we aimed to answer the question of whether it is possible to regenerate soluble monomeric FraC from the transmembrane oligomeric form or not. Our research has shown that the answer to that question is yes (Fig. 7). For that, the pore must be separated from the lipids by a stepwise exchange of detergents

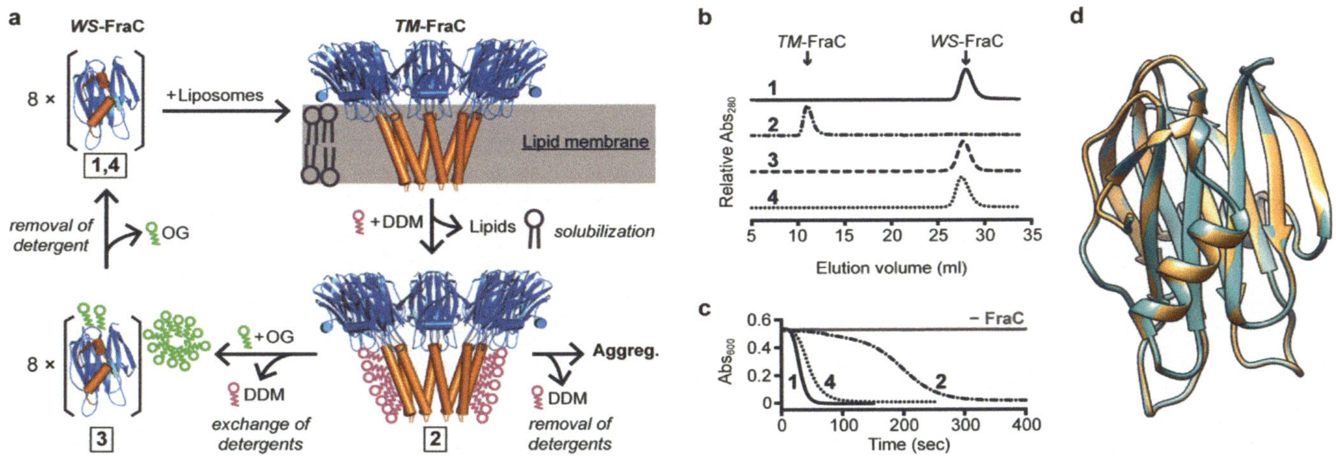

Fig. 7 Bidirectional transformation of actinoporins. (A) Scheme of the steps conducive to regenerate the water-soluble from of FraC after inducing pore formation in susceptible liposomes composed of SM/DOPC. The numbers within the boxes indicate different states where the protein can be isolated: 1 and 4 correspond to the water-soluble form. Although 1 and 4 are presumably identical, the process to obtain them is very different: 1 correspond to the purified protein and 4 is the monomeric protein obtained after treating the pore form with detergents. (B) Purification of FraC by size exclusion chromatography. The numbers 1–4 correspond to the same states of FraC explained in panel (A). In the panel TM refers to transmembrane and *WS* to water soluble. (C) Hemolysis of actinoporins purifies in different states. The numbers 1–4 correspond to the same states explained in panel (A). (D) X-ray crystal structure of water-soluble FraC before and after transformation to pore and regeneration. The two crystal structures are essentially identical, demonstrating that the regeneration of FraC successfully produces the same structural molecule before the transformation into pores.

as explained below. The use of octyl-glucoside (OG) or other detergents with a short chain generally led to the high recovery of the monomeric protein, and a X-ray crystal structure confirmed that the three-dimensional structure of the monomer before and after was identical (Tanaka, Caaveiro, & Tsumoto, 2015).

5.1 Equipment

1. Protein purification system (AKTA)
2. Size exclusion chromatography column
3. Resource S cation exchange column
4. Synchrotron X-ray radiation source

5.2 Reagents

1. Detergents (a selection of ~20 detergents)

5.3 Protocol

(1) Preparation of pores as explained in the protocol 4.3 (steps 1–3).

(2) FraC pores in detergent (68 or 30 μg) solubilized in 50 mM Tris (pH 7.4), 200 mM NaCl, and 0.3 mM DDM were loaded in a column containing 150 μL of SP Sepharose Fast Flow cationic exchange resin. The column was washed first with 20 column volumes (CV) of 50 mM Tris (pH 7.4), 200 mM NaCl with various detergents (wash 1), followed by 20 CV of 50 mM Tris (pH 7.4), 200 mM NaCl supplemented with 0.3 mM DDM (wash 2). FraC was eluted from the column with 3 CV of 50 mM Tris-HCl, 500 mM NaCl, and 0.3 mM DDM at pH 7.4. A fraction (250 μL) of the eluted protein was examined by analytical size exclusion chromatography in a Superdex 200 10/300 GL column (GE Healthcare) at 4 °C and the elution peaks monitored at 280 nm.

(3) Prior to quantify the transformation from pore to monomer, it was necessary to run a control experiment. For that purpose, the profile of untreated pores of FraC (68 μg) was monitored. The peak area of the control sample was set as 100%. Peaks in experimental samples were quantified relative to the control peak.

(4) To isolate the regenerated water-soluble FraC, the pores of FraC were first produced as above. The suspension of FraC pores was loaded onto a Resource S cationic exchange column equilibrated within 50 mM Tris (pH 7.4), 200 mM NaCl, and 0.3 mM DDM (buffer A). The column was then washed with buffer A replacing DDM with 50 mM OG

(20 CV) and eluted in a gradient with a second buffer composed of 50 mM Tris (pH 7.4), 1 M NaCl, and 50 mM OG to obtain the regenerated water-soluble protein.

(5) Optional. We verified that the structure of the regenerated water-soluble form of FraC was identical to that of the water-soluble form before addition of the liposomes. The protein (1 μL) at 7 mg/mL and 1 μL of crystallization solution composed of 24% PEG 8000, 200 mM $(NH_4)_2SO_4$ and 100 mM sodium cacodylate (pH 7.4) were mixed in a hanging drop crystallization tray to obtain a stock of crystal seeds. These crystal seeds were used to produced larger rod-shaped single crystals in a crystallization solution containing 1 μL of protein a t 7 mg/mL, 1 μL of a crystallization solution of 8% PEG 8000 and 100 mM Tris-HCl (pH 8.5) and 0.2 μL solution of micro-seeds.

(6) Suitable crystals were identified, harvested, immersed in a solution of mother liquor supplemented with 20% glycerol and subsequently frozen by immersion in liquid N_2 until data collection at the synchrotron.

(7) Data collection, data processing, structure determination by the method of molecular replacement using the coordinates of the monomeric form, and refinement, were carried by standard procedures.

6. Summary

The understanding of the mechanism of activation of actinoporins in membranes has greatly advanced since their discovery more than 40 years ago. In particular, this chapter describes some of the most recent molecular-level-insights on the stepwise events that lead to pore formation of the model actinoporin FraC, and the techniques employed to study them. The structural, kinetic and thermodynamic characterization of the lytic mechanism of FraC in membranes is one of the most detailed not only for actinoporins, but for pore-forming toxins in general. Looking ahead, two key question remain unanswered. First, determining the mechanism of insertion of the amphipathic portion of the protein into the membrane, and in particular distinguishing between the two extreme possibilities: an individual (one molecule at a time) insertion, or a cooperative one. And second, how the lipid controls the detachment of the N-terminal amphipathic helix from the body of the toxin. A detail understanding of this process could facilitate the design of engineered toxins with desired lipid specificities different from the natural variant, and facilitate the generation of immunotoxins and other therapeutic modalities exploiting the extraordinary abilities of actinoporins.

References

Alegre-Cebollada, J., Cunietti, M., Herrero-Galan, E., Gavilanes, J. G., & Martinez-del-Pozo, A. (2008). Calorimetric scrutiny of lipid binding by sticholysin II toxin mutants. *Journal of Molecular Biology*, *382*(4), 920–930.

Anderluh, G., Dalla Serra, M., Viero, G., Guella, G., Macek, P., & Menestrina, G. (2003). Pore formation by equinatoxin II, a eukaryotic protein toxin, occurs by induction of nonlamellar lipid structures. *Journal of Biological Chemistry*, *278*(46), 45216–45223. https://doi.org/10.1074/jbc.M305916200.

Anderluh, G., & Lakey, J. H. (2008). Disparate proteins use similar architectures to damage membranes. *Trends in Biochemical Sciences*, *33*(10), 482–490. https://doi.org/10.1016/J.Tibs.2008.07.004.

Anderluh, G., & Macek, P. (2002). Cytolytic peptide and protein toxins from sea anemones (Anthozoa : Actiniaria). *Toxicon*, *40*(2), 111–124. https://doi.org/10.1016/S0041-0101(01)00191-X.

Anderluh, G., Pungercar, J., Strukelj, B., Macek, P., & Gubensek, F. (1996). Cloning, sequencing, and expression of equinatoxin II. *Biochemical and Biophysical Research Communications*, *220*(2), 437–442. https://doi.org/10.1006/bbrc.1996.0391.

Athanasiadis, A., Anderluh, G., Macek, P., & Turk, D. (2001). Crystal structure of the soluble form of equinatoxin II, a pore-forming toxin from the sea anemone Actinia equina. *Structure*, *9*(4), 341–346.

Barlic, A., Gutierrez-Aguirre, I., Caaveiro, J. M., Cruz, A., Ruiz-Arguello, M. B., Perez-Gil, J., et al. (2004). Lipid phase coexistence favors membrane insertion of equinatoxin-II, a pore-forming toxin from Actinia equina. *Journal of Biological Chemistry*, *279*(33), 34209–34216.

Bartlett, G. R. (1959). Phosphorus assay in column chromatography. *Journal of Biological Chemistry*, *234*(3), 466–468.

Bellomio, A., Morante, K., Barlic, A., Gutierrez-Aguirre, I., Viguera, A. R., & Gonzalez-Manas, J. M. (2009). Purification, cloning and characterization of fragaceatoxin C, a novel actinoporin from the sea anemone *Actinia fragacea*. *Toxicon*, *54*(6), 869–880.

Caaveiro, J. M. M., Echabe, I., Gutierrez-Aguirre, I., Nieva, J. L., Arrondo, J. L., & Gonzalez-Manas, J. M. (2001). Differential interaction of equinatoxin II with model membranes in response to lipid composition. *Biophysical Journal*, *80*(3), 1343–1353.

Caaveiro, J. M. M., Gutierrez-Aguirre, I., Tribout, M., Paredes, S., & Gonzalez-Manas, J. M. (2000). An evaluation of some models systems commonly used in the study of the sea anemone toxin equinatoxin II with membranes. *Acta Biologica Slovenica*, *43*, 93–97.

Dal Peraro, M., & van der Goot, F. G. (2016). Pore-forming toxins: Ancient, but never really out of fashion. *Nature Reviews Microbiology*, *14*(2), 77–92. https://doi.org/10.1038/nrmicro.2015.3.

de los Rios, V., Mancheno, J. M., del Pozo, A. M., Alfonso, C., Rivas, G., Onaderra, M., et al. (1999). Sticholysin II, a cytolysin from the sea anemone *Stichodactyla helianthus*, is a monomer-tetramer associating protein. *FEBS Letters*, *455*(1–2), 27–30. https://doi.org/10.1016/S0014-5793(99)00846-7.

de los Rios, V., Mancheno, J. M., Lanio, M. E., Onaderra, M., & Gavilanes, J. G. (1998). Mechanism of the leakage induced on lipid model membranes by the hemolytic protein sticholysin II from the sea anemone Stichodactyla helianthus. *European Journal of Biochemistry*, *252*(2), 284–289. https://doi.org/10.1046/j.1432-1327.1998.2520284.x.

de los Rios, V., Onaderra, M., Martinez-Ruiz, A., Lacadena, J., Mancheno, J. M., del Pozo, A. M., et al. (2000). Overproduction in Escherichia coli and purification of the hemolytic protein sticholysin II from the sea anemone Stichodactyla helianthus. *Protein Expression and Purification*, *18*(1), 71–76. https://doi.org/10.1006/prep.1999.1174.

Ferlan, I., & Lebez, D. (1974). Equinatoxin, a lethal protein from actinia-equina .1. Purification and characterization. *Toxicon*, *12*(1), 57–61. https://doi.org/10.1016/0041-0101(74)90099-3.

Garcia-Linares, S., Rivera-de-Torre, E., Morante, K., Tsumoto, K., Caaveiro, J. M. M., Gavilanes, J. G., et al. (2016). Differential effect of membrane composition on the pore-forming ability of four Different Sea Anemone Actinoporins. *Biochemistry, 55*(48), 6630–6641. https://doi.org/10.1021/acs.biochem.6b01007.

Garcia-Ortega, L., Alegre-Cebollada, J., Garcia-Linares, S., Bruix, M., Martinez-Del-Pozo, A., & Gavilanes, J. G. (2011). The behavior of sea anemone actinoporins at the water-membrane interface. *Biochimica et Biophysica Acta, 1808*(9), 2275–2288.

Goldstein, S., Beka, L., Graf, J., & Klassen, J. L. (2019). Evaluation of strategies for the assembly of diverse bacterial genomes using MinION long-read sequencing. *BMC Genomics, 20*, 23. https://doi.org/10.1186/s12864-018-5381-7.

Gouet, P., Robert, X., & Courcelle, E. (2003). ESPript/ENDscript: Extracting and rendering sequence and 3D information from atomic structures of proteins. *Nucleic Acids Research, 31*(13), 3320–3323. https://doi.org/10.1093/Nar/Gkg556.

Heberle, F. A., & Feigenson, G. W. (2011). Phase separation in lipid membranes. *Cold Spring Harbor Perspectives in Biology, 3*(4). https://doi.org/10.1101/cshperspect.a004630, a004630.

Huang, G., Voet, A., & Maglia, G. (2019). FraC nanopores with adjustable diameter identify the mass of opposite-charge peptides with 44 Dalton resolution. *Nature Communications, 10*, 837. https://doi.org/10.1038/s41467-019-08761-6.

Huang, G., Willems, K., Soskine, M., Wloka, C., & Maglia, G. (2017). Electro-osmotic capture and ionic discrimination of peptide and protein biomarkers with FraC nanopores. *Nature Communications, 8*, 935. https://doi.org/10.1038/s41467-017-01006-4.

Kristan, K. C., Viero, G., Dalla Serra, M., Macek, P., & Anderluh, G. (2009). Molecular mechanism of pore formation by actinoporins. *Toxicon, 54*(8), 1125–1134.

Laborde, R. J., Sanchez-Ferras, O., Luzardo, M. C., Cruz-Leal, Y., Fernandez, A., Mesa, C., et al. (2017). Novel adjuvant based on the pore-forming protein Sticholysin II encapsulated into liposomes effectively enhances the antigen-specific CTL-mediated immune response. *Journal of Immunology, 198*(7), 2772–2784. https://doi.org/10.4049/jimmunol.1600310.

Louw, A. I., & Visser, L. (1977). Kinetics of erythrocyte lysis by snake-venom cardiotoxins. *Biochimica et Biophysica Acta, 498*(1), 143–153. https://doi.org/10.1016/0304-4165(77)90095-2.

Lv, X. X., Zhang, J., Xu, R., Dong, Y. G., Sun, A. Y., Shen, Y. L., et al. (2016). Gigantoxin-4-4D5 scFv is a novel recombinant immunotoxin with specific toxicity against HER2/neu-positive ovarian carcinoma cells. *Applied Microbiology and Biotechnology, 100*(14), 6403–6413. https://doi.org/10.1007/s00253-016-7487-7.

Macek, P., Sencic, L., & Lebez, D. (1982). Isolation and partial characterization of three lethal and hemolytic toxins from the sea anemone Actinia cari. *Toxicon, 20*(1), 181–185. https://doi.org/10.1016/0041-0101(82)90189-1.

Maglia, G., Heron, A. J., Stoddart, D., Japrung, D., & Bayley, H. (2010). Analysis of single nucleic acid molecules with protein Nanopores. *Methods in Enzymology, 475*, 591–623. https://doi.org/10.1016/S0076-6879(10)75022-9.

Makino, A., Abe, M., Murate, M., Inaba, T., Yilmaz, N., Hullin-Matsuda, F., et al. (2015). Visualization of the heterogeneous membrane distribution of sphingomyelin associated with cytokinesis, cell polarity, and sphingolipidosis. *The FASEB Journal, 29*(2), 477–493. https://doi.org/10.1096/fj.13-247585.

Mancheno, J. M., Martin-Benito, J., Martinez-Ripoll, M., Gavilanes, J. G., & Hermoso, J. A. (2003). Crystal and electron microscopy structures of sticholysin II actinoporin reveal insights into the mechanism of membrane pore formation. *Structure, 11*(11), 1319–1328.

McPherson, A. (2016). *Introduction to macromolecular crystallography*. Wiley-Blackwell.

Mechaly, A. E., Bellomio, A., Gil-Carton, D., Morante, K., Valle, M., Gonzalez-Manas, J. M., et al. (2011). Structural insights into the oligomerization and architecture of eukaryotic membrane pore-forming toxins. *Structure, 19*(2), 181–191.

Meinardi, E., Florinchristensen, M., Paratcha, G., Azcurra, J. M., & Florinchristensen, J. (1995). The molecular-basis of the self non-self selectivity of a coelenterate toxin. *Biochemical and Biophysical Research Communications, 216*(1), 348–354. https://doi.org/10.1006/bbrc.1995.2630.
Morante, K., Bellomio, A., Gil-Carton, D., Redondo-Morata, L., Sot, J., Scheuring, S., et al. (2016). Identification of a membrane-bound Prepore species clarifies the lytic mechanism of Actinoporins. *Journal of Biological Chemistry, 291*(37), 19210–19219. https://doi.org/10.1074/jbc.M116.734053.
Morante, K., Bellomio, A., Viguera, A. R., Gonzalez-Manas, J. M., Tsumoto, K., & Caaveiro, J. M. M. (2019). The isolation of new pore-forming toxins from the sea Anemone *Actinia fragacea* provides insights into the mechanisms of Actinoporin evolution. *Toxins, 11*(7), 401. https://doi.org/10.3390/toxins11070401.
Morante, K., Caaveiro, J. M. M., Tanaka, K., Gonzalez-Manas, J. M., & Tsumoto, K. (2015). A pore-forming toxin requires a specific residue for its activity in membranes with particular physicochemical properties. *Journal of Biological Chemistry, 290*(17), 10850–10861. https://doi.org/10.1074/jbc.M114.615211.
Morante, K., Caaveiro, J. M. M., Viguera, A. R., Tsumoto, K., & Gonzalez-Manas, J. M. (2015). Functional characterization of Val60, a key residue involved in the membrane-oligomerization of fragaceatoxin C, an actinoporin from *Actinia fragacea*. *FEBS Letters, 589*(15), 1840–1846. https://doi.org/10.1016/j.febslet.2015.06.012.
Podobnik, M., & Anderluh, G. (2017). Pore-forming toxins in Cnidaria. *Seminars in Cell & Developmental Biology, 72*, 133–141. https://doi.org/10.1016/j.semcdb.2017.07.026.
Ramirez-Carreto, S., Miranda-Zaragoza, B., & Rodriguez-Almazan, C. (2020). Actinoporins: From the structure and function to the generation of biotechnological and therapeutic tools. *Biomolecules, 10*(4), 539. https://doi.org/10.3390/biom10040539.
Restrepo-Perez, L., Wong, C. H., Maglia, G., Dekker, C., & Joo, C. (2019). Label-free detection of post-translational modifications with a Nanopore. *Nano Letters, 19*(11), 7957–7964. https://doi.org/10.1021/acs.nanolett.9b03134.
Rojko, N., Dalla Serra, M., Macek, P., & Anderluh, G. (2016). Pore formation by actinoporins, cytolysins from sea anemones. *Biochimica et Biophysica Acta, 1858*(3), 446–456. https://doi.org/10.1016/j.bbamem.2015.09.007.
Saier, M. H. (2000). A functional-phylogenetic classification system for transmembrane solute transporters. *Microbiology and Molecular Biology Reviews, 64*(2), 354–411. https://doi.org/10.1128/Mmbr.64.2.354-411.2000.
Schon, P., Garcia-Saez, A. J., Malovrh, P., Bacia, K., Anderluh, G., & Schwille, P. (2008). Equinatoxin II permeabilizing activity depends on the presence of sphingomyelin and lipid phase coexistence. *Biophysical Journal, 95*(2), 691–698.
Simons, K., & Ikonen, E. (1997). Functional rafts in cell membranes. *Nature, 387*(6633), 569–572.
Simons, K., & Vaz, W. L. (2004). Model systems, lipid rafts, and cell membranes. *Annual Review of Biophysics and Biomolecular Structure, 33*, 269–295.
Song, L., Hobaugh, M. R., Shustak, C., Cheley, S., Bayley, H., & Gouaux, J. E. (1996). Structure of staphylococcal alpha-hemolysin, a heptameric transmembrane pore. *Science, 274*(5294), 1859–1866.
Tanaka, K., Caaveiro, J. M. M., Morante, K., Gonzalez-Manas, J. M., & Tsumoto, K. (2015). Structural basis for self-assembly of a cytolytic pore lined by protein and lipid. *Nature Communications, 6*, 6337. https://doi.org/10.1038/Ncomms7337.
Tanaka, K., Caaveiro, J. M. M., & Tsumoto, K. (2015). Bidirectional transformation of a metamorphic protein between the water-soluble and transmembrane native states. *Biochemistry, 54*(46), 6863–6866. https://doi.org/10.1021/acs.biochem.5b01112.

Tejuca, M., Anderluh, G., & Dalla Serra, M. (2009). Sea anemone cytolysins as toxic components of immunotoxins. *Toxicon, 54*(8), 1206–1214. https://doi.org/10.1016/j.toxicon.2009.02.025.

Tejuca, M., Diaz, I., Figueredo, R., Roque, L., Pazos, F., Martinez, D., et al. (2004). Construction of an immunotoxin with the pore forming protein St1 and ior C5, a monoclonal antibody against a colon cancer cell line. *International Immunopharmacology, 4*(6), 731–744. https://doi.org/10.1016/j.intimp.2004.02.010.

Wohri, A. B., Johansson, L. C., Wadsten-Hindrichsen, P., Wahlgren, W. Y., Fischer, G., Horsefield, R., et al. (2008). A lipidic-sponge phase screen for membrane protein crystallization. *Structure, 16*(7), 1003–1009. https://doi.org/10.1016/j.str.2008.06.003.

CHAPTER ELEVEN

Biophysical approaches to study actinoporin-lipid interactions

Juan Palacios-Ortega[a,b], Esperanza Rivera-de-Torre[a,c], José G. Gavilanes[a], J. Peter Slotte[b], Álvaro Martínez-del-Pozo[a], and Sara García-Linares[a,*]

[a]Department of Biochemistry and Molecular Biology, Complutense University, Madrid, Spain
[b]Biochemistry, Faculty of Science and Engineering, Åbo Akademi University, Turku, Finland
[c]Department of Biotechnology and Biomedicine, Technical University of Denmark, Kongens Lyngby, Denmark
*Corresponding author: e-mail address: sglinares@ucm.es

Contents

Methods in Enzymology, Volume 649
ISSN 0076-6879
https://doi.org/10.1016/bs.mie.2020.12.017

Abstract

Protein-lipid interactions are crucial events from a biochemical point of view, like the interaction of proteins with the cell plasma membrane, and their study is of great importance. Actinoporins are a very powerful tool to study this kind of interactions, since they are soluble proteins in an aqueous environment, capable of inserting into membranes when they have the adequate composition. In fact, actinoporins have been used to study protein-lipid interactions for many years now. Sometimes it is not possible to use real biological membranes in the experiments, so model membranes need to be used. This article aims to give a thorough description of many of the techniques used to study actinoporin-lipid interactions, using both biological and model membranes: Hemolysis, release of vesicles content, surface plasmon resonance, isothermal titration calorimetry, fluorescence-based measurements, etc. Some of these techniques measure the actinoporins activity and some measure their binding properties. The combination of all the techniques described can offer valuable information about the thermodynamics and the kinetics of the actinoporin-lipid interaction.

1. Introduction

Proteins can be classified based on many different principles: Folding patterns, structural domains, sequence motifs, function, shape, or enzymatic activity, to name a few. One of the most general categorizations, based on a rather broad criterion, sorts proteins according to whether they are soluble or reside at the membrane. While this method can comprise most proteins, there are some that elude this categorization. Those are the proteins that can change their partition between aqueous environments and membrane phases depending on environmental features, which would need a category of their own.

A family of proteins that can be included in that separate group are pore-forming toxins (PFTs) (Dal Peraro & van der Goot, 2016). These toxins remain stably folded in solution. That conformation is also maintained when they encounter a membrane that lacks the receptor required. However, if the specific molecule recognized by the toxin is accessible in that membrane, binding occurs, triggering a conformational change that leads, ultimately, to oligomerization and the formation of pores, driving the death of the targeted cell by osmotic shock.

Some of the best known PFTs are actinoporins. These are small, basic, single-polypeptide proteins, produced by sea anemones which specifically recognize sphingomyelin (SM) (Alegre-Cebollada, Oñaderra, Gavilanes, & Martínez-del-Pozo, 2007; Anderluh & Maček, 2002; Maček, 1992). There are four actinoporins that have been studied in detail: Sticholysins I and II (StnI and StnII) produced by *Stichodactyla helianthus*, equinatoxin II (EqtII) produced by *Actinia equina*, and fragaceatoxin C (FraC) produced by *Actinia fragacea*. The three-dimensional soluble structures of these four proteins have been solved by crystallographic and/or nuclear magnetic resonance methods (Athanasiadis, Anderluh, Maček, & Turk, 2001; García-Linares et al., 2013; Hinds, Zhang, Anderluh, Hansen, & Norton, 2002; Mancheño, Martín-Benito, Martínez-Ripoll, Gavilanes, & Hermoso, 2003; Mechaly et al., 2011; Mechaly, Bellomio, Morante, González-Mañas, & Guerin, 2009). Over the years, the pore complex of FraC could be isolated, and its structure resolved using X-ray crystallography (Tanaka, Caaveiro, Morante, González-Mañas, & Tsumoto, 2015).

These structures have provided great insight for those of us investigating actinoporin behavior. However, they only represent the initial and final states of the process. And even if, in principle, most of the information regarding the process of pore formation is contained in their structure, it cannot be inferred completely directly and solely from the structures, since other aspects, especially lipids, play an essential role in this process. That is to say, the action mechanism of these proteins cannot be understood without studying their membrane interactions.

Therefore, understanding the behavior of these toxins demands a thorough characterization of their interaction with the membranes. This is accomplished with the combination of a great variety of methods, of which we take advantage of their respective strengths. The use of this assortment of techniques is essential, since each one reveals only a few aspects of the interaction. It is only when put together that the results obtained really become meaningful.

The first step to characterize actinoporin-lipid interactions is, obviously, having availability of such protein, purified to homogeneity. This process shall not be covered in this chapter. Nevertheless, the reader is encouraged to consult the following references (Alegre-Cebollada et al., 2007; Anderluh, Pungerear, Štrukelj, Maček, & Gubenšek, 1996; Bellomio et al., 2009). Next, the membranes that the toxin of interest interacts with need to be prepared. This is covered in the section regarding vesicle preparation if model membranes are to be used. This process, quite straight forward, is a common place for most of the subsequent techniques, sometimes requiring some experiment-specific adjustments. In case the activity of actinoporins is to be tested with biological membranes, erythrocytes are usually chosen, due to their easy availability. Preparation of the erythrocytes is covered in the section regarding hemolysis assays.

As for the assays, they can be classified in two categories: those that report on the activity of the toxins, and those that do not. It should be noted that, when it comes to actinoporins, and PFTs in general, the term activity refers to their capacity to form pores and induce cell/vesicle damage. No enzymatic activity is carried out by these toxins, i.e., pore formation entails no change whatsoever on the covalent bonds in neither of the intervening parts and the toxins are nor recovered after pore formation to damage other suitable target membranes.

The usual activity experiments performed with actinoporins are the hemolysis and the aqueous contents release assays. Other procedures reported in this article give information about the actinoporins binding capability to membranes. The information provided by all other most commonly performed approaches does not directly inform about the activity. In some cases, though, the results can correlate since the parameters reported do affect the activity.

There are two techniques that can be used to obtain information regarding the overall process of pore formation. On the one hand, surface plasmon resonance (SPR) can be used, in principle, to measure the affinity of a toxin for a given membrane composition. Nevertheless, this is often not possible with actinoporins since their binding to membranes is largely irreversible. Still, SPR can be used to compare different membrane compositions in a qualitative manner. If the mass of vesicles attached to the chip is registered, one can also estimate how many lipids per bound protein are for each combination. On the other hand, isothermal titration calorimetry (ITC) can provide a complete thermodynamic characterization of the interaction between a selected actinoporin and the chosen membrane composition, including the affinity constant, and other thermodynamic data such as the enthalpy of the interaction.

Finally, a series of fluorescence experiments can be used to gather information regarding details of the interaction between the toxins and specific lipid species. The intrinsic fluorescence of the proteins, and the use of specific labels and fluorescent lipid analogs can provide great insight regarding specific interactions, protein oligomerization, and the effects toxin binding can have on the distribution of a given lipid in the membrane.

2. Hemolysis

It is common to use erythrocytes as a substrate to measure actinoporin's activity. Red blood cells can be easily obtained from a wide variety of animal sources and are stably stored at 4 °C. Each species shows a specific lipid composition. However, they share common features that make them suitable to study actinoporins' activity: The presence of sphingomyelin and a high cholesterol content (around 26%) (Nelson, 1967). One should be careful, nevertheless, since some animal red blood cells have a remarkably low SM content. Eventually, they can be used as controls as well. Actinoporins form pores in the cell membrane of erythrocytes, leading to cell burst by osmotic shock. It is therefore possible to measure actinoporins' activity by following the turbidity decrease of a red blood cell suspension upon interaction with actinoporin solutions.

2.1 Materials and equipment

- Buffer: 10 mM Tris–HCl, pH 7.4, 145 mM NaCl
- 0.1% (*w/v*) Na_2CO_3
- Defibrinated/heparinized blood
- Multichannel pipettor
- 96-multiwell plate
- Centrifuge suitable for standard 15 mL tube centrifugation
- 96-multiwell plate reader equipped with 620–655 nm filter

2.2 Step-by-step method details

2.2.1 Preparation of erythrocytes suspensions

Timing: 1 h

This step is intended to remove plasma proteins. Actinoporins could avidly bind to lipoproteins.

1. 10 mL of blood are centrifuged at 700 × *g* for 20 min at 25 °C
2. Supernatant is carefully discarded with a Pasteur pipette, avoiding disturbing the sediment. Then, the sediment is gently resuspended in Tris–HCl, NaCl buffer by inverting the tube. Cells are washed by centrifugation, again, at 700 × *g* for 10 min
3. Repeat step 2 at least four more times or until the supernatant is clear
4. To ensure that the suspension's turbidity is within the sensitivity range of the 96-multiwell plate reader, it is necessary to test erythrocyte dilutions between 1/10 to 1/100 dilution range, in identical conditions as the ones that will be used in the assay: 100 μL of erythrocytes suspension dilution mixed with 100 μL of buffer (that will be substituted by protein solution in the assay) have to display an optical density value at 620–655 nm lower than 1.0
5. Washed erythrocytes can be used for hemolysis assays within 24 h, although it is recommendable to check suspension dilution every 8 h

Note: While resuspending sediment or diluting erythrocytes it is not recommended to shake or hit the tube, since erythrocytes would burst due to mechanical stress. It is possible that, when blood is stored in the fridge for a long time, erythrocytes need more washes since they become more fragile because of aging.

2.2.2 Hemolysis assay

Timing: 50 min/plate

1. Twofold serial dilutions are prepared in 96-multiwell plates. This format allows for assaying two proteins with triplicates or three proteins in duplicates simultaneously, which is especially useful when comparing different wild-type or modified versions of the protein. The maximum concentration of protein employed will depend on the specific toxin used. A good orientation is the already determined HC_{50} value for well-known actinoporins (García-Linares et al., 2016). The final volume on each well should be 100 μL
2. Every plate must include a control for 0% hemolysis, obtained with 100 μL of buffer that will allow to monitor spontaneous hemolysis during the kinetic assay and a control for 100% hemolysis, obtained with 100 μL of 0.1% (*w/v*) Na_2CO_3
3. Upon addition of 100 μL of erythrocytes dilution to each well, the $OD_{620–655}$ is registered every 20 s for 20 min

Note: avoid shaking steps when measuring.

2.2.3 Hemolysis assay to test synergy

Actinoporins appear as multigene families in the sea anemone genomes. Typically, a single sea anemone species has a battery of independent genes encoding highly similar actinoporins, each displaying different pore-forming activity (García-Linares et al., 2016). It has been previously described that some of these actinoporins can act in synergy when mixed, displaying higher activity than expected if both isoforms would act independently. To test synergy between two actinoporin isoforms using a hemolysis assay, it is necessary to test the relative concentration of each isoform separately and in combination within the same plate. For example, when testing a final concentration of 10 nM, at a 20%–80% proportion, it is necessary to assay 2 and 8 nM of the isoforms representing the 20% and 80% fractions of the combined sample, respectively. Such strategy allows to compare experimental kinetic values with the theoretical activity obtained by the simple addition of the activity values displayed by the relative concentrations of the isolated proteins.

2.3 Expected outcomes

For each concentration assayed, it will be possible to build a kinetic curve by representing the percentage of hemolysis as a function of time (Fig. 1A).

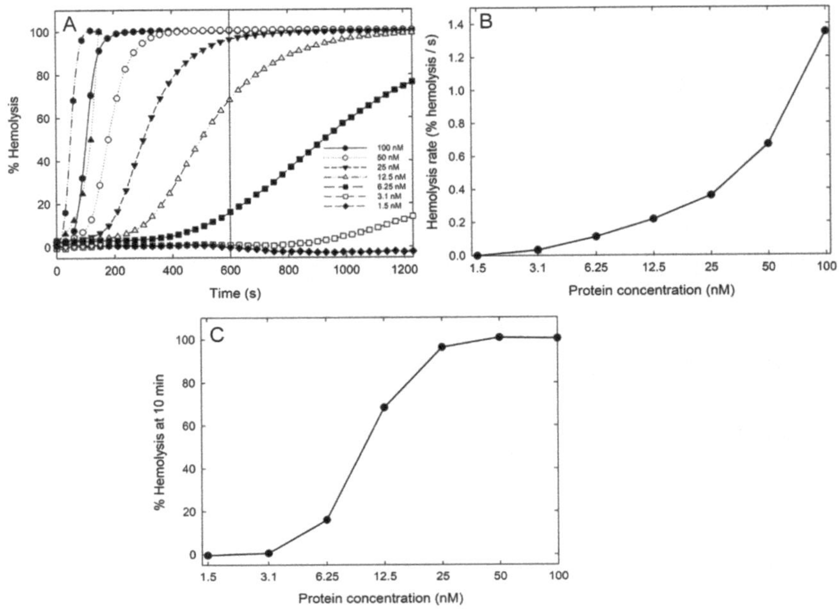

Fig. 1 *Hemolysis.* (A) Curves represent hemolytic activity exerted by a range of concentration of the actinoporin StnI. (B) and (C) are dose-response curves built from graph A data. (B) represents the maximum hemolytic rate values (expressed as percentage of hemolysis/s) vs the logarithm of total protein concentration. (C) represents the % of hemolysis reached after 10 min of StnI addition (vertical line in (A) indicates this time).

From each %hemolysis/time curve, the initial hemolytic rate is calculated as the slope of the linear increment range of the graph (Fig. 1B). In addition, it is also possible to determine the maximum hemolysis obtained for each protein concentration after 10 min (Fig. 1C). Both approximations allow for the construction of dose-response curves, enabling for different subsequent interpretations.

2.4 Quantification and analysis

Initial hemolytic rate curves allow for the evaluation of the activity in terms of binding and pore formation steps, while maximum hemolysis values are related to the number of pores that can be formed and are therefore related to the stoichiometry.

Every experiment should be tested at least by duplicate on each plate and it should be repeated three times with different blood batches.

2.5 Advantages

Hemolysis assays are simple, cheap, accessible, and versatile experiments that allow for a fast interpretation of actinoporins' activity.

2.6 Limitations

Unlike model vesicles, that have a well-defined composition, erythrocytes are complex membranes that, while better mimicking physiological activity of actinoporins, yield results which are also more difficult to interpret.

3. Preparation of vesicles

The preparation of vesicles is a common procedure for many of the experiments discussed in this article and for protein-lipid interactions studies in general. This section will cover the main aspects of the vesicles preparation process to obtain a population of mainly large unilamellar vesicles of a certain diameter.

3.1 Materials and equipment

- Lipids
- Methanol or chloroform:methanol 2:1 (*v*/v) solution
- System for drying lipids under nitrogen flow (it is ideal to also have a separate system that can generate vacuum to ensure complete removal of the organic solvent)
- Graduated glass syringes
- Disposable glass tubes
- Vesicle hydration buffer: Tris–HCl, NaCl
- Water bath
- Vortexer
- Mini extruder
- Polycarbonate filters (Whatman - Millipore), typically with diameters between 100 and 200 nm

3.2 Step-by-step method details

3.2.1 Preparation of lipid films

Timing: 30 min to 2 h

The commercial presentation of lipids is usually a solid powder, although it is possible to purchase them already dissolved in organic solvents such us methanol or mixtures containing chloroform. It is *extremely important* to consider that most plastic consumables in the laboratory can be affected by organic solvents. Thus, to prepare lipid films it is necessary to work with glass graduated pipettes and disposable glass tubes. Methanol or chloroform: methanol (2:1) solvents are used to dissolve pure lipids at a known concentration. Lipids can be then mixed to obtain the desired lipid mixture for the assay, in the wished molar ratio. Typical mixtures used when assaying actinoporins are PC:SM:Chol (1:1:1), PC:SM:Chol (80:20:10), and PC: SM (80:20), from 0.5 to 10 mM.

The lipid solutions are dried out under a stream of nitrogen. Ideally, vacuum is applied to the dried films for 2–3 h to ensure the evaporation of all the organic solvent. Any solvent left in the film might affect the formation and behavior of vesicles or damage the extrusion filter. Once the solvent is completely evaporated, aliquots might be gassed with nitrogen or argon to avoid oxidation of certain lipids, like polyunsaturated fatty acids, which are especially prone to oxidation. Dried lipid films aliquots can be stored for long term at −20 °C until use.

3.2.2 Preparation of large unilamellar vesicles (LUVs) from lipid films

1. Lipid films are hydrated with the amount of hydration buffer necessary to achieve the desired final lipid concentration. This will be the buffer that will define the inner content of the vesicles
2. The mixture is vortexed intensely. The lipid films, in contact with an aqueous solution, will form multilamellar vesicles (MLVs). The suspension is then hydrated at a temperature above the melting temperature of the lipid mixture for 30 min to 1 h. It is recommendable to vortex every 10 min to ensure the homogeneity of the suspension
3. The MLVs preparation is intensely vortexed one more time, and then the suspension is passed through a 100 or 200 nm polypropylene filter with a mini extruder device 15 times, to ensure a homogeneous population of LUVs. An odd number of passes are used to retain any large contaminant on one side of the extruder

3.3 Expected outcomes

After extrusion, you will end up with a vesicle preparation in which the majority of the particles will be large unilamellar vesicles of the specific diameter you use to extrude them.

3.4 Quantification and analysis

Take aliquots of your LUV preparation after it has been extruded to determine the actual concentration of lipids by means of phosphorus quantification methods like (Bartlett, 1959) or (Rouser, Fkeischer, & Yamamoto, 1970). This calculation will not be possible if the buffer used was phosphate based, such as PBS.

3.5 Advantages

This technique allows to prepare lipid vesicles of the desire composition and obtain a large population of the specific size we need.

3.6 Limitations

A small proportion of the vesicles may still have a different diameter or be multilamellar. Homogeneity of the LUVs can be checked by means of laser scattering measurements, for example.

4. Release of aqueous contents from model lipid vesicles

Actinoporins bind to SM-containing membranes forming stable pores that are selectively permeable to some cations. According to electrophysiology studies and leakage of PEGs of different molecular mass, the hydrodynamic size of the pore ranges between 1 and 2 nm (Casallanovo et al., 2006; Tejuca, Serra, Ferreras, Lanio, & Menestrina, 1996).

In comparison with hemolysis assays, leakage of soluble contents from model vesicles allows controlling lipid composition while also giving insight of the primary steps of pore formation. Release of aqueous contents from model lipid vesicles is a standard procedure that has been used to evaluate pore-formation efficiency of actinoporins. However, regardless of the probe of choice, actinoporins action does not empty vesicles completely, at least with the probes so far employed. Calcein and carboxifluorescein have been for decades the standard ones in the field (Alm, García-Linares, Gavilanes, Martínez-del-Pozo, & Slotte, 2015; Carretero et al., 2018; García-Linares et al., 2015, 2016; Maula et al., 2013; Palacios-Ortega et al., 2019; Rivera-de-Torre et al., 2016; Ros et al., 2019). However, recent studies remarked that both fluorescent probes were indeed negatively charged and too big to leak through the pore (Palacios-Ortega, Rivera-de-Torre, Gavilanes, Slotte, & Martínez-del-Pozo, 2020). This same study revealed that positively charge fluorescent probes such as rhodamine 6G, though still too big, seem to be a better choice for performing this type of experiments. Irrespective of the probe chosen, once the final pores are stably formed on the membrane, the release through the pore seems impossible, since they are too small for the probes. It is for this reason that leakage of fluorescent probes, which are generally larger than the pore of actinoporins, report transient permeation of the membrane as consequence of the assembly into the final oligomeric pore structures, rather than the passage of solutes through the stably assembled pores. Leakage assays demonstrate then that, besides the proper pore, the first stages of pore-formation would inflict serious damages to living cells as well, thus being an important feature to study about the activity of actinoporins.

4.1 Materials and equipment

- Material and chemicals for LUVs preparation (see Section 3)
- Vesicle hydration buffer: Typically, 10 mM Tris–HCl, pH 7.4, 140 mM NaCl
- Actinoporins (sticholysins)
- Fluorescent probe, such as calcein or rhodamine 6G
- Sephacryl S200HR (Sigma-Merck)
- Triton X-100 10% (*v*/v)
- $C_{12}E_8$ (octaethylene glycol monododecyl ether) 10% (v/v)
- Liquid nitrogen
- Empty Poly-Prep chromatography columns (BioRad)
- Fluorimeter
- Fluorescence quartz cuvette

4.2 Step-by-step method details

4.2.1 Calcein

If calcein is dissolved at high enough concentrations, it becomes self-quenched. Hence, its fluorescence emission is negligible when encapsulated in liposomes at such concentrations. Release of calcein from LUVs leads to a dramatic increase in fluorescence emission due to the dilution of the probe in the total volume of the spectrophotometric cell. Fluorescence emission is proportional to the fraction of dye leaked from the model vesicles.

Preparation of calcein loaded vesicles
Timing: 2 h

- 250–500 nmol lipid films are hydrated with 200 μL of 100 mM calcein dissolved in Tris–HCl, NaCl hydration buffer
- After hydration, and before extrusion, the MLVs are subjected to 10 freezing/thawing cycles with liquid nitrogen to improve encapsulation efficiency. Thawing is done at the same temperature as hydration
- Once extruded, LUVs are separated from non-entrapped calcein by gel filtration on Sephacryl S200HR. Briefly, plastic columns are filled with the resin and equilibrated with the vesicle hydration buffer. Then, the 200 μL LUVs suspension is applied in the column, 1 mL of hydration buffer is added, and vesicles are eluted by centrifugation in that unique 1 mL elution fraction
- Calcein-loaded vesicles are used for permeabilization studies within 8 h from preparation

Calcein leakage assay
Timing: 10 min/assay

- Lipid concentration during the assay should be about 2.5 μM
- Emission is recorded at 550 nm as a function of time, using excitation at 480 nm. Temperature must be constantly controlled. Typically, experiments are performed at a fixed temperature between 23 °C and 25 °C, depending of the lipid composition of the vesicles used
- To ensure that non-spontaneous leakage does not take place, fluorescence emission should be recorded for a few minutes before toxin addition. A steady signal should be obtained for stable vesicles
- Protein is then added in a volume less than 2% of the total volume of the cuvette to avoid dilution effects. Different concentrations (typically ranging from 1 to 100 nM) should be tested to build dose-response curves in terms of calcein leakage rates
- Once the signal has reached a steady level after the addition of the toxin and has maintained it for some time, 10 μL of 10% (*v*/v) Triton X-100 is added to the cuvette to induce LUVs disintegration. The fluorescence value obtained after total release of the probe makes possible to calculate the fraction of calcein released by action of the toxin. The following expression is used to that end:

$$\%\ \text{leakage} = \frac{(F_{obs}(t) - F_0)}{(F_{max} - F_0)} \times 100$$

where F_{obs} (t) is the emission at a given time, F_0 is the basal emission prior to protein addition, and F_{max} is the maximum emission value upon lysis of the vesicles by the addition of the detergent.

4.2.2 Rhodamine 6G (R6G)

R6G, in contrast with calcein, is a cationic probe. Therefore, it is a more suitable molecule to test leakage upon actinoporins' action, since they form cation-selective pores (Varanda & Finkelstein, 1980). R6G release rates are faster and much more sensitive than any other tested probe, including calcein. Therefore, charge is a critical factor to measure permeation, even considering that the probes do not leak through the actual final stable pores. Consequently, given that neither calcein nor R6G would have the right size to cross the pore lumen, the final leakage value reaches a plateau below 100%. Nevertheless, the faster release of R6G suggests that it can escape from vesicles easier than calcein through transient pores because of its positive charge (Palacios-Ortega et al., 2020).

Preparation of R6G loaded vesicles

Timing: 2 h

- R6G is dissolved in distilled water to saturation and then a five times concentrated version of the hydration buffer is added to the R6G solution. Insoluble fraction of the probe is discarded by centrifugation. The approximately concentration of R6G in the hydration buffer should be 5 mM
- R6G loaded vesicles are then prepared as detailed for calcein loaded vesicles

R6G leakage assay

Timing: 10 min/assay

- Lipid concentration during the assay should be around 2.5 μM as in calcein assays
- Emission is followed at 555 nm as a function of time upon excitation at 525 nm. Temperature must be constantly controlled between 23 °C and 25 °C
- As in calcein assays it is necessary to control spontaneous leakage before adding different protein concentrations
- Once the signal is steady after the addition of the lytic protein, 10 μL of 10% (*v*/v) $C_{12}E_8$ is added to the cuvette to induce the disintegration of LUVs. Triton X-100 should not be used for R6G assays since it quenches the probe. Percentage of released probe is calculated as described above

4.3 Expected outcomes

For each concentration assayed, it will be possible to build a kinetic curve by representing the percentage of leakage as a function of time (Fig. 2A). From each %leakage/time curve, the initial lytic rate is calculated as the slope of the linear increment range of the graph (Fig. 2B). Experimental data can be fitted in a model for peptide-induced membrane perturbation defined by (Andersson, Danielsson, Gräslund, & Mäler, 2007):

$$\frac{F_{obs}(t)}{F_{max}} = 1 - \exp\left(J_2\left(e^{-t \cdot v_{relax}} - 1\right)\right)$$

The presented model describes release curves resulting from transient perturbations affecting the membrane, where F_{obs} (t) is the fluorescence registered at a certain time and F_{max} is the maximum fluorescence registered in

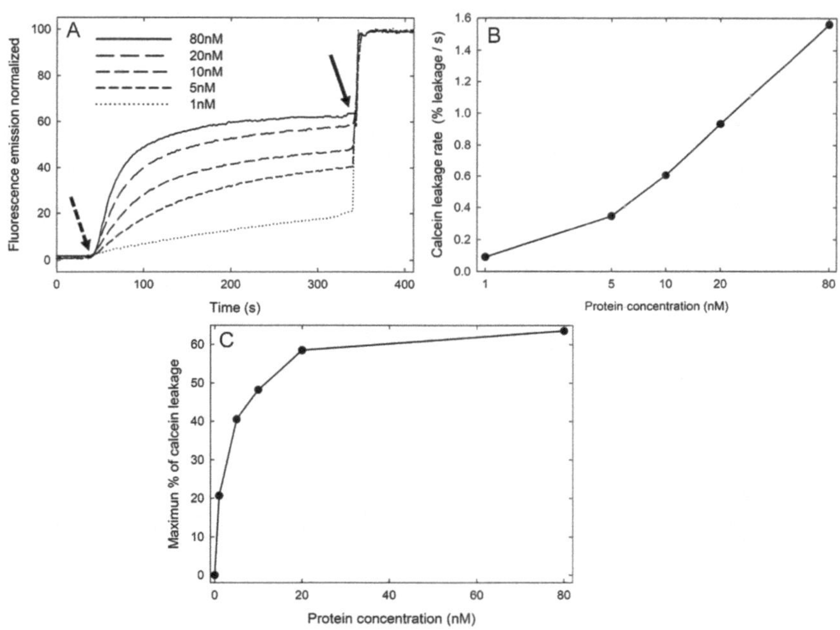

Fig. 2 *Calcein leakage.* (A) Curves represent calcein release activity exerted by a range of concentration of the actinoporin StnI from DOPC:SM:Chol (1:1:1) model vesicles. Dashed arrow indicates the addition of the lytic protein and solid arrow indicates de addition of detergent to lysate vesicles. (B) and (C) are dose-response curves built from graph A data. (B) represents the maximum calcein release rate values (expressed as percentage of leakage/s) vs the logarithm of total protein concentration. (C) represents the maximum steady percentage of leakage reached before adding detergent to obtain 100% release control.

the system after addition of detergent. Once the system is stable and there are no more perturbations, the leak ceases. The model has two parameters that defines the behavior of the system. J_2 is proportional to the magnitude of the perturbations and if the model fits the experimental data, it can be calculated as:

$$J_2 = -\ln(1 - L_{max})$$

where L_{max} is the value of the final, maximum release induced by the toxin. The term v_{relax} is related to the duration of the perturbations. The product $J_2 \cdot v_{relax}$ is equal to the slope of the release curves at $t = 0$. In addition, it is also possible to determine the maximum release rate obtained for each protein concentration after as the steady value obtained before the addition of the detergent (Fig. 2C). Both approximations allow for the construction of dose-response curves that lead to different interpretations.

4.4 Quantification and analysis

Due to their particular features, actinoporins are not able to induce 100% leakage from vesicles. The size of the actinoporin pore does not enable the pass of the probe through the lumen, as explained. Therefore, it is usually considered that final release values are not highly representative of the process of pore formation. Instead, it is a common practice to calculate dose-response curves in terms of leakage rate to evaluate the permeabilization activity of these proteins.

Every experiment should be tested at least by duplicate with different LUVs batches.

4.5 Advantages

Release of soluble contents from vesicles is an easy way to test actinoporins' activity in terms of permeabilization due to membrane disruption/perturbations prior to the formation of the final stable pores (once the pores are formed, the probe is too large to escape through them and leakage reaches a plateau). The development and interpretation of the results are usually simple, and the use of model vesicles allows to select the specific composition desired to test the activity.

4.6 Limitations

Release of the soluble probes presented does not account for leakage through the stable pores.

5. Surface plasmon resonance (SPR)

Surface plasmon resonance is a highly sensitive and label-free method used to study surface changes and it is a well-established method utilized to measure binding affinity and kinetics. Lipid vesicle layers are spread on a sensor surface and actinoporins interaction on lipids is determined by measuring the change of the angle of reflection of an incident beam due to the adsorption of actinoporins to the vesicle layer.

5.1 Materials and equipment

- Material and chemicals for LUVs preparation (see Section 3)
- Buffer: 10 mM Tris–HCl, pH 7.4, 140 mM NaCl
- Binding agent: Actinoporins (sticholysins) or membrane binding agent (MBA) of interest
- Detergents: 3-[(3-Cholamidopropyl)dimethyl-ammonio]-1-propanesulfonate (CHAPS)
- Bovine serum albumin (BSA)
- NaOH
- Isopropanol
- Bath sonicator
- SPR instrument (BioNavis)
- Sensor gold chip coated with a carboxymethylated dextran layer treated with N-hydroxysuccinimide and N-ethyl-N′-(dimethylaminopropyl)-carbodiimide (Löfås & Johnsson, 1990)

5.2 Step-by-step method details

Timing: 50 min; LUVs preparation, 1 h

1. Filter and degas by sonication (using an ultrasonic bath) the Tris–HCl, NaCl buffer
2. Prepare:

    50 mM NaOH in Tris–HCl, NaCl buffer
    50 mM NaOH:isopropanol (2,3)
    0.1 mg/mL BSA in Tris–HCl, NaCl buffer
    10 mM CHAPS in filtered and degassed water
    4 μM actinoporin in Tris–HCl, NaCl buffer
    0.5 mM LUVs of the desired composition (see Section 3)

3. Degas the protein preparation by centrifugation (12,000 g, 5 min, 25 °C).
4. Place the carboxymethyl dextran coated sensor chip in the instrument
5. Clean the sensor chip surface using the detergent CHAPS (2 × 2 min injection).
6. Apply the lipid vesicles (0.5 mM) to the dextran surface for 12 min
7. Unbound vesicles were removed using 50 mM NaOH (3 min injection), in order to make the experiment only with stably bound vesicles
8. Use bovine serum albumin (BSA) as a blocking molecule to prevent any non-specific binding to the substrate (3 min injection).
9. Apply actinoporins (4 μM) for 10 min on the lipid surface
10. Inject buffer for 5 min to study toxin dissociation
11. Regenerate the sensor surface using CHAPS. For vesicles containing cholesterol, 50 mM NaOH:isopropanol (2,3) gives better results than CHAPS

Notes

1. Temperature = 23 °C.
2. Flow = 10 μL/min.
3. To finish every injection of the different reactants, 200 μL of buffer are injected.
4. For a more detailed explanation of the protocol see (Cooper, Try, Carroll, Ellar, & Williams, 1998) and (Del Vecchio & Stahelin, 2016).

5.3 Expected outcomes

The instrument returns bond mass in ng/cm^2 which can be represented versus time to obtain a sensorgram that shows the progress of the interaction (Fig. 3).

5.4 Quantification and analysis

The relative molecular ratios can be calculated if the ng/cm^2 are recorded for both the lipids and the MBA, provided the molecular mases are known.

5.5 Advantages

SPR can be utilized to measure the binding of molecules in real-time without the use of labels.

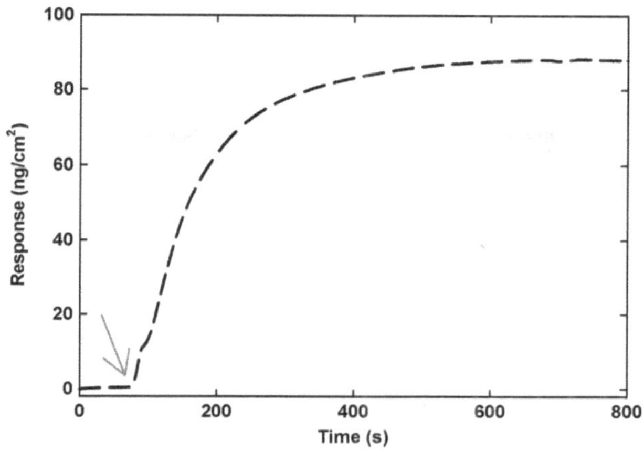

Fig. 3 *Surface plasmon resonance.* Typical sensorgram showing the interaction of the actinoporin FraC with supported DOPC:SM (80:20) model vesicles. Protein was added at time indicated by arrow, and the signal was recorded as a function of time.

5.6 Limitations

Mass transport limitations can occur if the actinoporin transfer to the vesicles on the sensor surface is not adequate, in which case the actinoporin concentration near the surface will be different from the bulk concentration. Increasing the flow rate is an easy way to limit mass transport effects.

6. Isothermal titration calorimetry (ITC)

Isothermal titration calorimetry is a highly suitable technique for studying the energetics of interactions between biological molecules (Velázquez-Campoy, Ohtaka, Nezami, Muzammil, & Freire, 2004). This technique has been widely used to study actinoporins' affinity for membranes of different compositions, by injecting aliquots of the lipid suspension into a protein solution continuously stirred. The heat released in each injection can be calculated from the raw data by integration of the peaks after subtraction of the baseline. Experimental data can be fitted to obtain the thermodynamic parameters of the reaction.

6.1 Materials and equipment

- Material and chemicals for LUVs preparation (see Section 3)
- Buffer: 10 mM Tris–HCl, pH 7.5, 100 mM NaCl, 1 mM EDTA
- Binding agent: Actinoporins (sticholysins) or MBA of interest

- Dialysis system
- Bath sonicator
- VP-ITC calorimeter (Malvern MicroCal)
- ThermoVac (Malvern MicroCal)
- Origin software (OriginLab) provided by MicroCal

6.2 Step-by-step method details

Timing: 3.5 h; MBA dialysis, 2–3 h; LUVs preparation, 1 h

1. Prepare the actinoporin at the proper concentration (see note 1) in Tris–HCl, NaCl buffer
2. Dialyze the actinoporin against 500 mL of buffer for 2–3 h at room temperature (20–25 °C) and keep the outer dialysis medium
3. Prepare LUVs of the desired composition (see Section 3) using the actinoporin dialysis media (see note 2) and let them stabilize to experimental temperature. The concentration of vesicles depends on the concentration of MBA and the stoichiometry of the reaction (see note 3).
4. Filter the dialysis media and degas by bath sonication. Degas both the membrane-binding agent and LUVs under constant stirring using a vacuum system
5. Fill the calorimeter cell using the proper Hamilton syringe. It is crucial not to introduce bubbles in the cell
6. Fill the injection syringe with the vesicle preparation following the calorimeter manufacturer instructions
7. Run the adequate protocol for your experiment. Some parameters (total number of injections, volume of injection, cell temperature, reference power, stirring speed, etc.) should be adjusted to better register the binding reaction depending on the characteristics of the reaction (stoichiometry, exothermic/endothermic reaction, etc.)
8. Injection of vesicles against buffer in the cell under the same experimental conditions must be done to eliminate dilution heat contribution

6.3 Expected outcomes

ITC experiments give a thermogram showing the characteristic sequence of peaks corresponding to each ligand injection into the sample cell (Fig. 4A). A binding isotherm is obtained by integrating each injection peak as a function of lipid/protein molar ratio in the calorimetric cell (Fig. 4B).

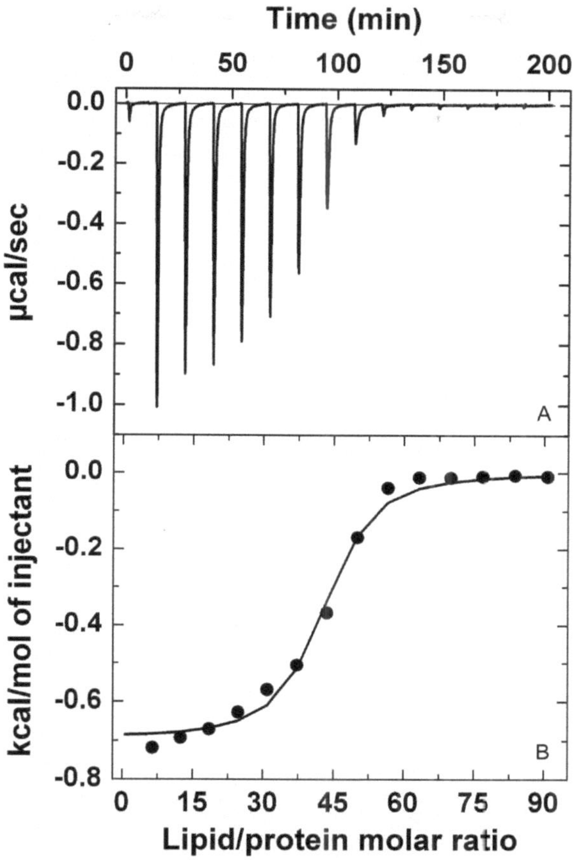

Fig. 4 *Isothermal titration calorimetry.* Panel (A) shows the characteristic heat peaks of an exothermic reaction, in this case the interaction of StnI with DOPC:SM (80:20) model vesicles. Panel (B) represents the binding isotherm obtained by integration of the peaks as a function of lipid/protein molar ratio.

Notes

1. The concentration of the MBA should be adjusted so the c value ($c = K_b \times N \times [MBA]$) is between 10 and 1000 (preferably 500). In our case, we used actinoporins at 1–10 μM, depending on the value of the binding constant (K_b) and the stoichiometry of the reaction (N) (García-Linares et al., 2016). There are alternate ways of analyzing thermodynamics of membrane interactions for cases that cannot be described by stochiometric binding but are consistent with phase partitioning instead (White, Wimley, Ladokhin, & Hristova, 1998).

2. It is imperative that the buffer matches exactly between the actinoporin and vesicles preparations to avoid heat contributions from buffer mixing and dilution.
3. In our case, lipid vesicles concentration was between 1 and 10 mM.
4. The volume of samples depends on the specific model of the calorimeter.
5. For extraordinarily high binding constants (i.e., nano-picomolar range) in combination with high heat release, a good experimental approximation to obtain a curve that can be fitted in the correct c interval (note 1) is reducing the concentration of cell and titration components and increasing the number of injections while reducing their volume (Rivera-de-Torre et al., 2016).

6.4 Quantification and analysis

(Alegre-Cebollada, Cunietti, Herrero-Galán, Gavilanes, & Martínez-del-Pozo, 2008):

1. Use the Origin software to open the raw data
2. Follow the software instructions to calculate the heat released in each injection from the raw data by integration of the peaks after subtraction of the baseline
3. The binding isotherms are adjusted to a model in which a molecule of protein binds the membrane involving N lipid molecules
4. Obtain K_b, N, and ΔH values from the Origin integration
5. Obtain ΔG and ΔS values according to the expression: $\Delta G = -R \cdot T \cdot \ln(K_b/(N \cdot \nu_L)) = \Delta H - T \cdot \Delta S$ (where ν_L is the specific molar volume of lipids in L/mol).

6.5 Advantages

ITC gives a basic thermodynamic characterization of the actinoporin-lipid interaction in a single experiment (including association, stoichiometry, and binding enthalpy) without the need of labeling or immobilizing the reactants.

6.6 Limitations

ITC requires large amounts of samples and only one ligand can be tested at a time. ITC methodology requires heat is either consumed or generated by the reaction. The lower the changes in enthalpic component of binding, the lower is the signal, and the less reliable the measurements becomes.

7. Evaluating the effect of actinoporin binding on SM-clustering

The effect that actinoporin binding has on SM organization in the membrane can be studied using pyrene-labeled SM analogs. Pyrene moieties can form excited-state dimers, known as excimers, if two of them are in very close range of each other during the lifetime of their excited state (Lakowicz, 2006). Pyrene emission, which regularly displays three peaks, becomes unstructured and red-shifted when excimers are formed (Lakowicz, 2006).

Therefore, pyrene-labeled lipids can be used to detect lipid clustering and declustering by means of measuring the relative pyrene monomer (M) and excimer (E) emission, i.e., the E/M ratio (Hresko, Sugár, Barenholz, & Thompson, 1986, 1987; Jones & Lentz, 1986). This has been used, for example, to detect formation of domains. In our case, we have used it to monitor the effect of actinoporin membrane binding on the distribution of SM in the membrane, using SM analogs labeled in their acyl chain. In practice, this technique can be used to follow the effect of any MBA on any lipid that can be labeled with pyrene.

7.1 Materials and equipment

- Material and chemicals for LUVs preparation (see Section 3)
- Buffer: 10mM Tris–HCl, pH 7.4, 140mM NaCl
- Binding agent: Actinoporins (sticholysins) or MBA of interest
- Probes: Pyrene-labeled lipids
- Steady-state fluorimeter, preferably one that can operate in T-format (to measure simultaneously emission of the monomer at 392nm and the excimer at 480nm)
- Quartz cuvette

7.2 Step-by-step method details

Timing: 10–15min/assay (depending on the kinetics of the process); LUVs preparation, 1h

1. Prepare LUVs of the desired composition (see Section 3) and let them stabilize to experimental temperature (see note 1).
2. Set the excitation wavelength at 345nm. Monomer and excimer emission are recorded at 392 and 480nm respectively (see general note 1).
3. Record E/M emission for the vesicles alone (see notes 2–4).

4. Record E/M emission in the presence of the MBA (see general note 2). If the kinetics of the process are of interest, the MBA can be added to the measurement started on 3

7.3 Expected outcomes

Results allow to build dose-dependent curves to quantify the effect of a given MBA on the distribution of SM, with the possibility of also comparing different membrane systems.

7.4 Advantages

This is a straightforward system to evaluate effect of membrane binding on SM-clusters.

7.5 Limitations

Requires pyrene-labeled SM, which is not always easy to acquire, and might require self-manufacturing. The labeled compound may also behave slightly different from endogenous SM.

Notes

1. In our case, we used POPC:PSM:pyrSM:Chol in ratios 5:3:2:0 and 5:3:2:1, because we also wanted to see the effect of Chol presence (Palacios-Ortega et al., 2019). Depending on which MBA is tested, and the pyrene-labeled lipid, the ratio might need to be tuned for best sensitivity.
2. Vesicle concentration would need to be adjusted depending on total pyrene concentration, instrument sensitivity and, later, desired protein/lipid ratio. In our case, we used 1.0 μM lipids, with the protein lipid ratio being below 0.2 (maximum value assayed).
3. We recorded the E/M ratio in a time dependent manner, measuring emission at 392 and 480 nm, because initially we were interested in the kinetics of the process. Nevertheless, as long as all measurements are consistent and comparable, the ratio can also be calculated from the emission spectra, as long as the sample has reached equilibrium when they are recorded. Either way, light scattering due to vesicles needs to be evaluated and corrected for if required.
4. If the software of the fluorimeter allows for it, the E/M ratio can be calculated on the go. If else, E and M signals can be acquired independently and processed later.

8. Measuring the effect of actinoporin binding on lipid distribution

Actinoporin binding can affect lipid distribution on the membrane. In order to test that effect, we used a lipid-mimicking fluorescent probe in combination with a phase-selective quencher which is also embedded within the membrane as a lipid component. When the quencher is in contact with the fluorescent probe, it prevents the probe from emitting. In practice, we can consider that collisional quenching only takes place upon molecular contact (Lakowicz, 2006).

Thus, by measuring the relative quantum yield of the probe, we can assess its exposure to the quencher. Since the quencher is confined to one phase, this method allows to determine if the selected MBA affects the phase partitioning of the fluorescent probe. In our case, we used it to determine that StnII binding to the membrane alters the distribution of cholesterol in the membrane, using a fluorescent analog of cholesterol, cholestatrienol (CTL) (Palacios-Ortega et al., 2019).

8.1 Materials and equipment

- Material and chemicals for LUVs preparation (see Section 3)
- Buffer: 10 mM Tris–HCl, pH 7.4, 140 mM NaCl
- Binding agent: Actinoporins (sticholysins) or MBA of interest
- Probes: Fluorescent lipid analogs, such as CTL
- Phase selective quencher: 7-SLPC (1-palmitoyl-2-(7-doxyl)-stearoyl-phosphatidylcholine) or equivalent
- Steady-state fluorimeter
- Quartz cuvette

8.2 Step-by-step method details

Timing: 10–20 min/sample; LUVs preparation, 1 h

1. Prepare two sets of LUVs of the desired composition (see Section 3), differing only in the quencher. In one of them, the quencher should be absent and replaced by a regular lipid with similar physicochemical properties (see note 1).
2. Let the prepared vesicles stabilize to experimental temperature
3. Set the excitation and emission wavelengths according to the spectral properties of the probe of choice. For CTL, we used 310 and 400 nm for excitation and emission, respectively (see general note 1).

4. Starting with, say, the F_0 sample, record probe emission in absence of the MBA
5. Add the MBA and record probe emission until equilibrium is reached
6. Repeat points 4 and 5 for the F sample. If the kinetics of the process is of interest, add the MBA at the same time point as for the previous sample

8.3 Expected outcomes

Results (Fig. 5) allow to conclude if an MBA affects the micro-environment surrounding a given lipid species (in our case, CTL).

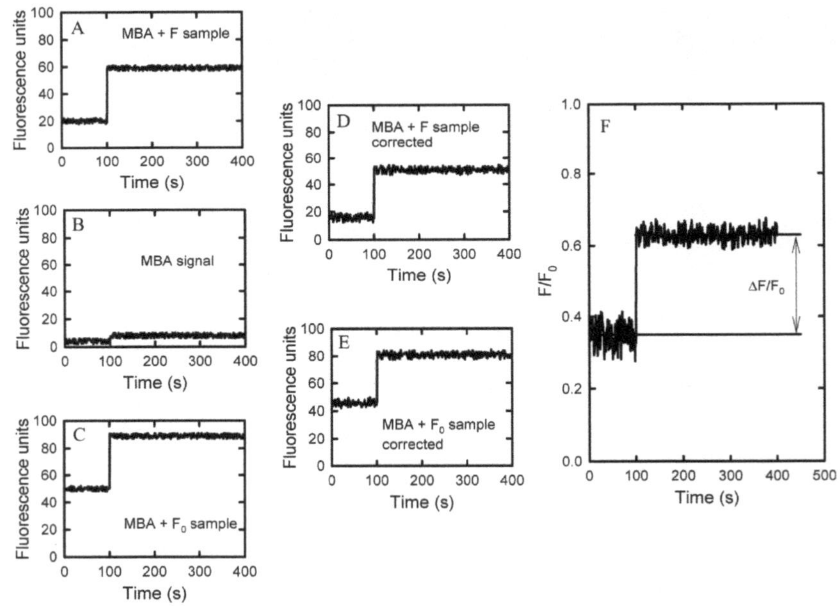

Fig. 5 *Lipid distribution.* Panels (A) to (C) represent the experimental results expected when adding a MBA to a quencher-containing sample (F sample; panel (A)), directly to plain buffer (panel (B)), and to a quencher-free sample (panel (C)). Data in panels (A) and (C) need to be corrected for MBA contribution by subtraction of data in (B), as shown in panels (D) (for data in (A)) and (E) (for data in (C)). Finally, the F/F_0 ratio can be calculated by division of data in (D) over (E), yielding the result shown in (F). In the example, the F/F_0 ratio before time 100 s corresponds to the quenching of the probe in absence of the MBA. Addition of the MBA, in the example, increases the value of the F/F_0 ratio (i.e., $\Delta F/F_0 > 0$). Since the probe is less exposed to the quencher than it was in absence of the MBA, it can be concluded that MBA binding hinders the probe-quencher interaction.

8.4 Quantification and analysis

1. If the MBA interferes with the signal of the probe, the final signal should be corrected for MBA effects (note 2).
2. Obtain the F/F_0 ratio, using the time point of MBA addition as reference to align the traces
3. If the F/F_0 ratio does not change due to MBA addition, it can be concluded that the MBA does not change the distribution of the probe
4. If the F/F_0 is affected by MBA addition, the $\Delta F/F_0$ can be calculated. If $\Delta F/F_0 > 0$, the MBA can be said to reduce the exposure of the probe to the quencher. If $\Delta F/F_0 < 0$, the MBA can be said to increase probe-quencher interactions. Elucidation of the mechanisms causing the changes in probe exposure to quencher requires further experiments

Notes

1. Since the intensity of the probe's emission is the parameter of interest, the concentration of the probe in the membrane has to be such that it does not quench itself. CTL, for example, has to be kept below 5 mol % (Schroeder, Nemecz, Gratton, Barenholz, & Thompson, 1988). The samples with and without quencher will be referred to as F and F_0, respectively.
2. For example, if the MBA is a protein, like sticholysins, and the probe used is CTL, excitation at 310 nm does excite Trp residues (though not much), which also emit at 400 nm. In that case, Trp emission has to be measured and subtracted from the final signal to obtain the contribution of CTL.

8.5 Advantages

The method is very precise in elucidating if a given MBA promotes or hinders the displacement of the probed lipid to the quencher-rich phase.

8.6 Limitations

Further experiments are required to gain mechanistical insight in the process.

9. Membrane binding of proteins based on Trp emission

Membrane binding of proteins can be observed monitoring Trp fluorescence. In order to do that, the protein of interest needs to have exposed Trp residues that change their exposure to the solvent upon membrane binding, either by going into the membrane, or by being shielded from the solvent by a neighboring monomer if the protein oligomerizes.

If the protein has multiple Trp residues, and Trp mutants are available, this approach also enables to find out which Trp residues are in close contact with the membrane or involved in protein-protein interactions as a consequence of oligomerization upon pore-formation. Furthermore, if a three-wdimensional structure of the soluble protein structure is available, this approach enables to estimate the orientation of the membrane-bound form of the protein relative to the bilayer based on the changes in the emission of each Trp residue. This is only possible, however, if the structural changes that occur concomitant to membrane binding do not significantly alter the soluble fold or the protein, or at least the regions that include the Trp residues. This latter scenario is the one observed for actinoporins.

9.1 Materials and equipment

- Material and chemicals for LUVs preparation (see Section 3)
- Buffer: 10 mM Tris–HCl, pH 7.4, 140 mM NaCl
- Binding agent: Actinoporins (sticholysins), or MBA of interest
- Steady-state fluorimeter
- Quartz cuvette

9.2 Step-by-step method details

Timing: 3–5 min/spectrum, i.e., 1 h/titration; LUVs preparation, 1 h

1. Prepare a set of LUVs of the desired composition (see Section 3) and let them stabilize to experimental temperature
2. Since emission of Trp is to be measured, excitation can be set to 295 nm to avoid contributions from Tyr and Phe. However, to determine the thermodynamics of binding via titration, the relevant observation is the net change in the spectroscopic signal, and for that, selective excitation of Trp residues is not required. In that case, excitation at 280 nm can be used. For more detailed suggestions see (Ladokhin, Jayasinghe, & White, 2000).
3. Record the emission spectra of the soluble protein from 305 to 475 nm
4. Titrate the sample with vesicles until saturation is reached. Record the emission spectra after each addition of vesicles
5. If Trp mutants are available, repeat steps 3 and 4 using those

9.3 Expected outcomes

Results (Fig. 6) allow to determine the lipid-protein ratio required to achieve complete binding of the available MBA in solution.

9.4 Quantification and analysis

1. The relative quantum yield of the protein after each addition of the vesicles can be calculated by integrating the emission spectra and comparing the result to that of the protein in solution
2. The calculated relative quantum yields can be plotted either as a function of lipid/protein molar ratio (for stoichiometric binding), or as a function of lipid concentration (for phase partitioning, as described in (Ladokhin et al., 2000)).

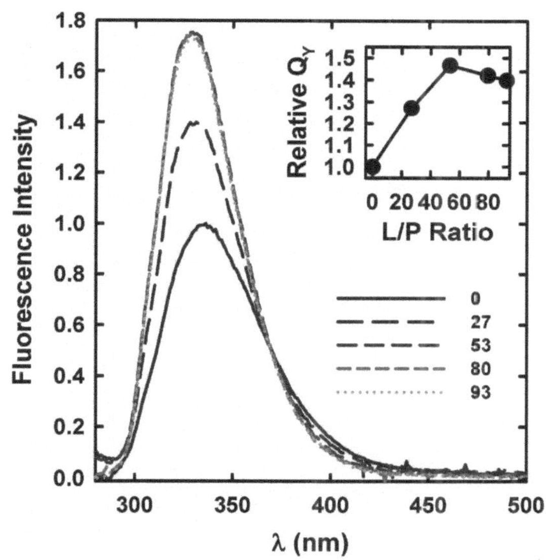

Fig. 6 *Trp emission.* Fluorescence emission of StnII under 260 nm excitation at increasing lipid/protein (L/P) molar ratios. Notice how the emission is greatly increased as the lipid amount is increased in the sample, while the position of the emission maximum is blue shifted. Insert: the emission of all spectra in the graph was quantified by numerical integration and normalized by the value obtained for the spectra obtained in absence of lipids. Plot of the quantum yield as a function of the L/P ratio in the sample clarifies that saturation is reached at ~55 L/P molar ratio. Increased resolution can be achieved by using smaller increments when adding lipids to the sample

3. If Trp mutants have been used, the relative quantum yields (calculated to each protein's emission in absence of lipids) can be used to see which Trp residues interact with the membrane. The details depend on the mutants that are available

9.5 Advantages

The only requirement is that the MBA has Trp residues, or that one can be introduced via mutagenesis.

9.6 Limitations

At least one of the Trp residues needs to change its environment as a consequence of binding, otherwise no change in signal is observed, unless specialized quencher-enhanced titration protocols are used (Posokhov, Gottlieb, & Ladokhin, 2007).

General notes:

1. In all fluorescence experiments it is good practice to first record an emission scan to assess the emission spectra of the sample. If this is the case, it should be remembered to perform all regular corrections, such as subtraction of the background, including the Raman peak of water
2. Make sure, before performing the experiment, that the spectral properties of the MBA do not interfere with those of the probe. If that is the case, the interference should be estimated, and the signals corrected accordingly

Funding

This research was supported by the Sigrid Jusélius Foundation, the Jane and Aatos Erkko Foundation, and the Magnus Ehrnrooth Foundation (to J.P.S.) and by UCM-Banco Santander grants PR75/18-21561 and PR87/19-22556 (to Á.M.-d.-P.). J.P.-O. has a funded doctoral student position from ISB/ÅA. A UCM-Banco Santander fellowship was granted to E.R.-d.-T. S.G.-L. is a Real Colegio Complutense-Harvard postdoctoral fellow.

References

Alegre-Cebollada, J., Clementi, G., Cunietti, M., Porres, C., Oñaderra, M., Gavilanes, J. G., et al. (2007). Silent mutations at the 5′-end of the cDNA of actinoporins from the sea anemone *Stichodactyla helianthus* allow their heterologous overproduction in *Escherichia coli*. *Journal of Biotechnology*, *127*, 211–221.

Alegre-Cebollada, J., Cunietti, M., Herrero-Galán, E., Gavilanes, J. G., & Martínez-del-Pozo, Á. (2008). Calorimetric scrutiny of lipid binding by sticholysin II toxin mutants. *Journal of Molecular Biology*, *382*, 920–930.

Alegre-Cebollada, J., Oñaderra, M., Gavilanes, J. G., & Martínez-del-Pozo, Á. (2007). Sea anemone actinoporins: The transition from a folded soluble state to a functionally active membrane-bound oligomeric pore. *Current Protein and Peptide Science*, *8*, 558–572.

Alm, I., García-Linares, S., Gavilanes, J. G., Martínez-del-Pozo, Á., & Slotte, J. P. (2015). Cholesterol stimulates and ceramide inhibits sticholysin II-induced pore formation in complex bilayer membranes. *Biochimica et Biophysica Acta - Biomembranes*, *1848*, 925–931.

Anderluh, G., & Maček, P. (2002). Cytolytic peptide and protein toxins from sea anemones (Anthozoa: Actiniaria). *Toxicon*, *40*, 111–124.

Anderluh, G., Pungerear, J., Štrukelj, B., Maček, P., & Gubenšek, F. (1996). Cloning, sequencing and expression of equinatoxin II. *Biochemical and Biophysical Research Communications*, *220*, 437–442.

Andersson, A., Danielsson, J., Gräslund, A., & Mäler, L. (2007). Kinetic models for peptide-induced leakage from vesicles and cells. *European Biophysics Journal*, *36*, 621–635.

Athanasiadis, A., Anderluh, G., Maček, P., & Turk, D. (2001). Crystal structure of the soluble form of equinatoxin II, a pore-forming toxin from the sea anemone *Actinia equina*. *Structure*, *9*, 341–346.

Bartlett, G. R. (1959). Colorimetric assay methods for free and phosphorylated glyceric acids. *Journal of Biological Chemistry*, *234*, 469–471.

Bellomio, A., Morante, K., Barlič, A., Gutiérrez-Aguirre, I., Viguera, A. R., & González-Mañas, J. M. (2009). Purification, cloning and characterization of fragaceatoxin C, a novel actinoporin from the sea anemone *Actinia fragacea*. *Toxicon*, *54*, 869–880.

Carretero, G. P. B., Vicente, E. F., Cilli, E. M., Alvarez, C. M., Jenssen, H., & Schreier, S. (2018). Dissecting the mechanism of action of actinoporins. Role of the N-terminal amphipathic alpha-helix in membrane binding and pore activity of sticholysins I and II. *PLoS One*, *13*, e0202981.

Casallanovo, F., de Oliveira, F. J., de Souza, F. C., Ros, U., Martínez, Y., Penton, D., et al. (2006). Model peptides mimic the structure and function of the N-terminus of the pore-forming toxin sticholysin II. *Biopolymers*, *84*, 169–180.

Cooper, M. A., Try, A. C., Carroll, J., Ellar, D. J., & Williams, D. H. (1998). Surface plasmon resonance analysis at a supported lipid monolayer. *Biochimica et Biophysica Acta - Biomembranes*, *1373*, 101–111.

Dal Peraro, M., & van der Goot, F. G. (2016). Pore-forming toxins: Ancient, but never really out of fashion. *Nature Reviews Microbiology*, *14*, 77–92.

Del Vecchio, K., & Stahelin, R. (2016). Using surface plasmon resonance to quantitatively assess lipid-protein interactions. *Methods in Molecular Biology*, *1376*, 141–153.

García-Linares, S., Alm, I., Maula, T., Gavilanes, J. G., Slotte, J. P., & Martínez-del-Pozo, Á. (2015). The effect of cholesterol on the long-range network of interactions established among sea anemone sticholysin II residues at the water-membrane interface. *Marine Drugs*, *13*, 1647–1665.

García-Linares, S., Castrillo, I., Bruix, M., Menéndez, M., Alegre-Cebollada, J., Martínez-del-Pozo, Á., et al. (2013). Three-dimensional structure of the actinoporin sticholysin I. influence of long-distance effects on protein function. *Archives of Biochemistry and Biophysics*, *532*, 39–45.

García-Linares, S., Rivera-de-Torre, E., Morante, K., Tsumoto, K., Caaveiro, J. M., Gavilanes, J. G., et al. (2016). Differential effect of membrane composition on the pore-forming ability of four different sea anemone actinoporins. *Biochemistry*, *55*, 6630–6641.

Hinds, M. G., Zhang, W., Anderluh, G., Hansen, P. E., & Norton, R. S. (2002). Solution structure of the eukaryotic pore-forming cytolysin equinatoxin. II: Implications for pore formation. *Journal of Molecular Biology*, *315*, 1219–1229.

Hresko, R. C., Sugár, I. P., Barenholz, Y., & Thompson, T. E. (1986). Lateral distribution of a pyrene-labeled phosphatidylcholine in phosphatidylcholine bilayers: Fluorescence phase and modulation study. *Biochemistry*, *25*, 3813–3823.

Hresko, R. C., Sugár, I. P., Barenholz, Y., & Thompson, T. E. (1987). The lateral distribution of pyrene-labeled sphingomyelin and glucosylceramide in phosphatidylcholine bilayers. *Biophysical Journal, 51*, 725–733.

Jones, M. E., & Lentz, B. R. (1986). Phospholipid lateral organization in synthetic membranes as monitored by pyrene-labeled phospholipids: Effects of temperature and prothrombin fragment 1 binding. *Biochemistry, 25*, 567–574.

Ladokhin, A. S., Jayasinghe, S., & White, S. H. (2000). How to measure and analyze tryptophan fluorescence in membranes properly, and why bother? *Analytical Biochemistry, 285*, 235–245.

Lakowicz, J. R. (2006). *Principles of fluorescence spectroscopy* (3rd ed.). Boston: Springer.

Löfås, S., & Johnsson, B. (1990). A novel hydrogel matrix on gold surfaces in surface plasmon resonance sensors for fast and efficient covalent immobilization of ligands. *Journal of the Chemical Society, Chemical Communications*, (21), 1526–1528.

Maček, P. (1992). Polypeptide cytolytic toxins from sea anemones (Actiniaria). *FEMS Microbiology Immunology, 5*, 121–129.

Mancheño, J. M., Martín-Benito, J., Martínez-Ripoll, M., Gavilanes, J. G., & Hermoso, J. A. (2003). Crystal and electron microscopy structures of sticholysin II actinoporin reveal insights into the mechanism of membrane pore formation. *Structure, 11*, 1319–1328.

Maula, T., Isaksson, Y. J., García-Linares, S., Niinivehmas, S., Pentikäinen, O. T., Kurita, M., et al. (2013). 2NH and 3OH are crucial structural requirements in sphingomyelin for sticholysin II binding and pore formation in bilayer membranes. *Biochimica et Biophysica Acta - Biomembranes, 1828*, 1390–1395.

Mechaly, A. E., Bellomio, A., Gil-Carton, D., Morante, K., Valle, M., González-Mañas, J. M., et al. (2011). Structural insights into the oligomerization and architecture of eukaryotic membrane pore-forming toxins. *Structure, 19*, 181–191.

Mechaly, A. E., Bellomio, A., Morante, K., González-Mañas, J. M., & Guerin, D. M. (2009). Crystallization and preliminary crystallographic analysis of fragaceatoxin C, a pore-forming toxin from the sea anemone *Actinia fragacea. Acta Crystallographica. Section F, Structural Biology and Crystallization Communications, 65*, 357–360.

Nelson, G. J. (1967). Lipid composition of erythrocytes in various mammalian species. *Biochimica et Biophysica Acta, 144*, 221–232.

Palacios-Ortega, J., García-Linares, S., Rivera-de-Torre, E., Gavilanes, J. G., Martínez-del-Pozo, Á., & Slotte, J. P. (2019). Sticholysin, sphingomyelin, and cholesterol: A closer look at a tripartite interaction. *Biophysical Journal, 116*, 2253–2265.

Palacios-Ortega, J., Rivera-de-Torre, E., Gavilanes, J. G., Slotte, J. P., & Martínez-del-Pozo, Á. (2020). Evaluation of different approaches used to study membrane permeabilization by actinoporins on model lipid vesicles. *Biochimica et Biophysica Acta - Biomembranes, 1862*, 183311.

Posokhov, Y. O., Gottlieb, P. A., & Ladokhin, A. S. (2007). Quenching-enhanced fluorescence titration protocol for accurate determination of free energy of membrane binding. *Analytical Biochemistry, 362*, 290–292.

Rivera-de-Torre, E., García-Linares, S., Alegre-Cebollada, J., Lacadena, J., Gavilanes, J. G., & Martínez-del-Pozo, Á. (2016). Synergistic action of actinoporin isoforms from the same sea anemone species assembled into functionally active heteropores. *Journal of Biological Chemistry, 291*, 14109–14119.

Ros, U., Carretero, G. P. B., Paulino, J., Crusca, E., Jr., Pazos, F., Cilli, E. M., et al. (2019). Self-association and folding in membrane determine the mode of action of peptides from the lytic segment of sticholysins. *Biochimie, 156*, 109–117.

Rouser, G., Fkeischer, S., & Yamamoto, A. (1970). Two dimensional thin layer chromatographic separation of polar lipids and determination of phospholipids by phosphorus analysis of spots. *Lipids, 5*, 494–496.

Schroeder, F., Nemecz, G., Gratton, E., Barenholz, Y., & Thompson, T. E. (1988). Fluorescence properties of cholestatrienol in phosphatidylcholine bilayer vesicles. *Biophysical Chemistry*, *32*, 57–72.

Tanaka, K., Caaveiro, J. M., Morante, K., González-Mañas, J. M., & Tsumoto, K. (2015). Structural basis for self-assembly of a cytolytic pore lined by protein and lipid. *Nature Communications*, *6*, 6337.

Tejuca, M., Serra, M. D., Ferreras, M., Lanio, M. E., & Menestrina, G. (1996). Mechanism of membrane permeabilization by sticholysin I, a cytolysin isolated from the venom of the sea anemone *Stichodactyla helianthus*. *Biochemistry*, *35*, 14947–14957.

Varanda, W., & Finkelstein, A. (1980). Ion and nonelectrolyte permeability properties of channels formed in planar lipid bilayer membranes by the cytolytic toxin from the sea anemone *Stoichactis helianthus*. *Journal of Membrane Biology*, *55*, 203–211.

Velázquez-Campoy, A., Ohtaka, H., Nezami, A., Muzammil, S., & Freire, E. (2004). Isothermal titration calorimetry. *Current Protocols in Cell Biology*. Supplement 23, 17.18.11-17.18.24.

White, S. H., Wimley, W. C., Ladokhin, A. S., & Hristova, K. (1998). Protein folding in membranes: Determining energetics of peptide-bilayer interactions. *Methods in Enzymology*, *295*, 62–87.

CHAPTER TWELVE

Conformational switching, refolding and membrane insertion of the diphtheria toxin translocation domain

Alexey S. Ladokhin[a,*], Alexander Kyrychenko[b], Mykola V. Rodnin[a], and Victor Vasquez-Montes[a]

[a]Department of Biochemistry and Molecular Biology, Kansas University Medical Center, Kansas City, KS, United States

[b]Institute of Chemistry and School of Chemistry, V. N. Karazin Kharkiv National University, Kharkiv, Ukraine

*Corresponding author: e-mail address: aladokhin@kumc.edu

Contents

Abstract

Diphtheria toxin is among many bacterial toxins that utilize the endosomal pathway of cellular entry, which is ensured by the bridging of the endosomal membrane by the toxin's translocation (T) domain. Endosomal acidification triggers a series of conformational changes of the T-domain, that take place first in aqueous and subsequently in

Methods in Enzymology, Volume 649
ISSN 0076-6879
https://doi.org/10.1016/bs.mie.2020.12.016

membranous milieu. These rearrangements ultimately result in establishing membrane-inserted conformation(s) and translocation of the catalytic moiety of the toxin into the cytoplasm. We discuss here the strategy for combining site-selective labeling with various spectroscopic methods to characterize structural and thermodynamic aspects of protonation-dependent conformational switching and membrane insertion of the diphtheria toxin T-domain. Among the discussed methods are FRET, FCS and depth-dependent fluorescence quenching with lipid-attached bromine atoms and spin probes. The membrane-insertion pathway of the T-domain contains multiple intermediates and is governed by staggered pH-dependent transitions involving protonation of histidines and acidic residues. Presented data demonstrate that the lipid bilayer plays an active part in T-domain functioning and that the so-called Open-Channel State does not constitute the translocation pathway, but is likely to be a byproduct of the translocation. The spectroscopic approaches presented here are broadly applicable to many other systems of physiological and biomedical interest for which conformational changes can lead to membrane insertion (e.g., other bacterial toxins, host defense peptides, tumor-targeting pHLIP peptides and members of Bcl-2 family of apoptotic regulators).

Abbreviations

DA	distribution analysis
FCS	fluorescence correlation spectroscopy
FRET	Förster resonance energy transfer
HDX-MS	hydrogen deuterium exchange mass spectrometry
IF	interfacial
LUV	large unilamellar lipid vesicles
LysoUB	FRET quencher for NBD fluorescence used to study membrane topology of membrane proteins
MD	molecular dynamics (simulations)
MP(s)	membrane protein(s)
NBD-PE	fluorescent lipid (1,2-dipalmitoyl-snglycero-3-phosphoethanolamine-N-(7-nitro-2-1,3-benzoxadiazol-4-yl)
OCS	Open-Channel State
POPC	1-palmitoyl-2-oleoyl-*sn*-glycero-3-phosphocholine
POPG	1-palmitoyl-2-oleoyl-*sn*-glycero-3-phospho-(1′-rac-glycerol)
QP	quenching profile
T-domain	diphtheria toxin translocation domain
TM	transmembrane

1. Introduction

Diphtheria toxin is an exotoxin secreted by *Corynebacterium diphtheriae* and causes disease in humans by inhibiting protein synthesis. It enters the cell *via* the endosomal pathway [Murphy, 2011], a feature shared by many other A-B toxins, including botulinum, tetanus, and anthrax [Collier & Young, 2003; Hoch et al., 1985; Koriazova & Montal, 2003; Neale, 2003]. Although cellular entry of these toxins comprises complex and not fully understood processes, they have certain similarities with the entry pathway of diphtheria toxin. That is, they involve receptor-mediated endocytosis followed by endosome acidification and pH-triggered conformational changes. This refolding results in membrane insertion of the transporting protein and formation of either a permanent or transient passageway, through which the toxic enzymatic components are transported into the cell. Diphtheria toxin T-domain implements its function—translocation of the catalytic domain across the endosomal membrane under acidic conditions—without the help of any additional protein components [Oh, Senzel, Collier, & Finkelstein, 1999], and apparently as a monomer [Gordon & Finkelstein, 2001; Palchevskyy et al., 2006]. The ability of the T-domain to translocate relatively large macromolecules [Ren et al., 1999] in a pH-dependent manner, makes it a potential candidate for targeting tumors, which produce slightly acidic environment. Here, we will review the results of structural and thermodynamic studies of T-domain refolding, membrane insertion and translocation obtained in our lab for the past decade. We also present recent advancements in fluorescence-based methodologies for characterizing membrane insertion pathways, which are applicable to a large class of membrane-binding proteins.

Diphtheria toxin T-domain is considered by many to be the archetypal representative of a specific class of proteins that insert into biological membranes in a translocon-independent manner after being refolded from their initial water-soluble structures [Antignani & Youle, 2006; Vargas-Uribe, Rodnin, & Ladokhin, 2013]. Such transitions are common in a number of biologically important processes, such as the cellular entry of bacterial toxins [Collier & Young, 2003; Hoch et al., 1985; Koriazova & Montal, 2003; Ladokhin, 2013; Murphy, 2011; Neale, 2003], colicins [Jakes & Cramer, 2012; Zakharov & Cramer, 2004], and viruses [Kielian & Rey, 2006; Neuman et al., 2013], complement-mediated cytolysis, and the regulation of apoptosis by the Bcl-2 family of proteins [Antignani & Youle, 2006; Bogner, Leber, & Andrews, 2010; Chi, Kale, Leber, & Andrews, 2014; Hsu, Wolter, & Youle, 1997; Leber, Lin, & Andrews, 2007; Youle & Strasser, 2008]. The unique characteristic of these proteins is their ability to undergo conformational switching when they move from the polar environment of the aqueous medium to the non-polar milieu of the lipid bilayer in response to a given cellular/chemical signal. In the case of diphtheria toxin T-domain, this conformational switching occurs in response to the acidic environment of the endosome [Ladokhin, 2013], and ensures the toxin's entry into the cell.

The crystallographic structure of the soluble form of the T-domain [Bennett & Eisenberg, 1994; Weiss, Blanke, Collier, & Eisenberg, 1995] (Fig. 1A) consists of nine helices of various lengths (TH1–9), eight of

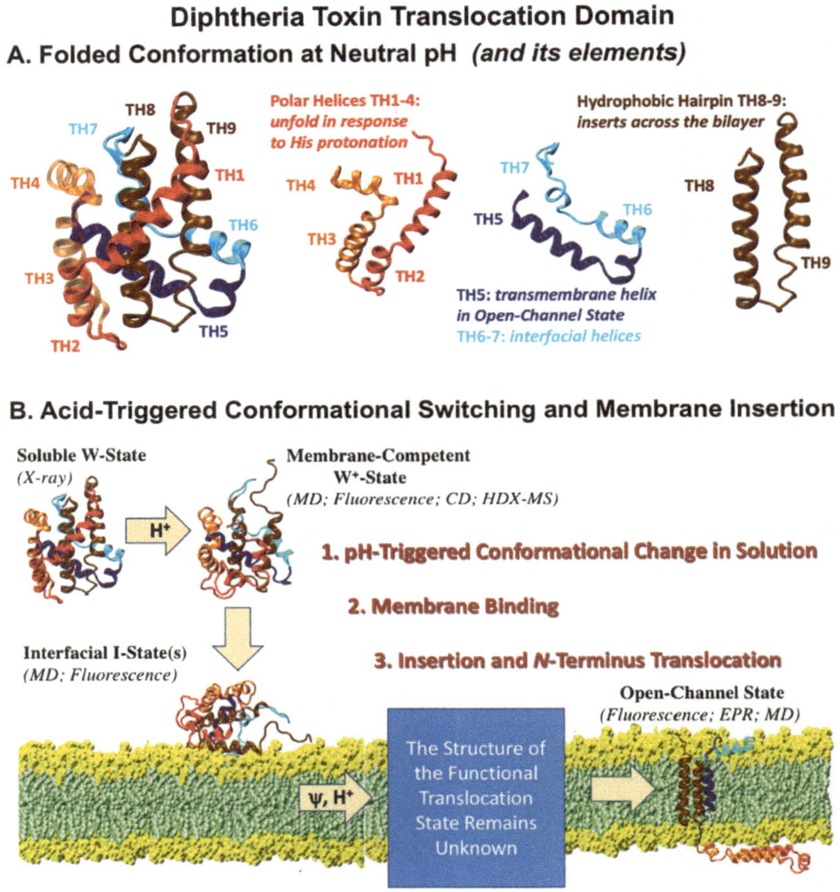

Fig. 1 (A) Crystallographic structure of the diphtheria toxin translocation domain [Bennett & Eisenberg, 1994; Weiss et al., 1995] consists of several structural elements that undergo conformational changes in response to protonation and membrane interactions. (B) pH-triggered membrane insertion pathway of the diphtheria toxin T-domain [Kurnikov et al., 2013; Kyrychenko, Posokhov, Rodnin, & Ladokhin, 2009], responsible for the cellular entry of the toxin.

which completely surround the most hydrophobic one, TH8. The N-terminal helices TH1–4 (red and orange), do not penetrate into the membrane and are translocated with the catalytic domain [Oh et al., 1999; Senzel, Huynh, Jakes, Collier, & Finkelstein, 1998] The core fragment consists of the TH8–9 hydrophobic hairpin that inserts first (brown helices in Fig. 1A), followed by the rest of the core sequence (TH5–7, blue and cyan).

A schematic representation of the pH-dependent membrane insertion pathway of the T-domain is shown in Fig. 1B [Kyrychenko et al., 2009]. Protonation of key histidine residues triggers unfolding of the polar helices and exposure of the hydrophobic core [Kurnikov et al., 2013; Rodnin et al., 2010; Rodnin, Li, Gross, & Ladokhin, 2016]. The resulting membrane-competent W^+-state (for protonated water-soluble) rapidly associates with the bilayer to form an interfacial intermediate I-state [Kyrychenko et al., 2009]. Subsequent insertion is facilitated by the presence of anionic lipids, which decrease the thermodynamic barrier for the insertion of the TH8–9 helical hairpin. The two protonation steps responsible for the formation of conformations capable of membrane association and insertion have overlapping pH ranges, suggesting that additional protonation can occur at the same pH value due to the shift of pK_a values of titratable residues after their partitioning into the interfacial zone of the lipid bilayer. While numerous studies had shown the co-existence of multiple insertion intermediates [Chenal et al., 2002; Kyrychenko et al., 2009; Ladokhin, Legmann, Collier, & White, 2004; Montagner et al., 2007; Palchevskyy et al., 2006; Rodnin et al., 2011; Vargas-Uribe, Rodnin, Kienker, Finkelstein, & Ladokhin, 2013; Wang et al., 1997], the structure of the functional state of the T-domain responsible for the translocation of its N-terminus along with the catalytic domain remains unknown. The putative structural model representing the core in the post-translocated state (Fig. 1B, bottom right structure) is based on the Open-Channel State (OCS) model [Senzel et al., 2000] (derived from conductivity measurements in planar bilayers [Huynh et al., 1997; Mindell, Silverman, Collier, & Finkelstein, 1994a, 1994b]) and is now confirmed by a combination of site-specific labeling and depth-dependent fluorescence quenching experiments [Kyrychenko et al., 2018].

One of the main reasons for the gaps in our understanding of the physiological action of the T-domain and other self-inserting proteins, is the shortage of appropriate experimental tools to study the protein folding inside the lipid bilayer. For example, the high-resolution methods (e.g., X-ray crystallography and NMR) are difficult to apply, because the refolding of the soluble conformation on membrane interface and subsequent transbilayer insertion produces multiple intermediate states. As a result, structural information has to be teased out by various types of low-resolution methods, like fluorescence and EPR [Heuck & Johnson, 2002; Kyrychenko et al., 2018, 2009; Ladokhin, Isas, Haigler, & White, 2002; Oh, Altenbach, Collier, & Hubbell, 2000; Oh et al., 1996; Perier et al., 2007; Vasquez-Montes et al., 2019; Wang et al., 1997], which can be complemented with Molecular Dynamics (MD) simulations to get atomistic presentation of the membrane-inserted conformations [Flores-Canales, Ladokhin, & Kurnikova, 2014; Flores-Canales, Vargas-Uribe, Ladokhin, & Kurnikova, 2015; Kyrychenko et al., 2018]. In the following sections, we will show how fluorescence-based approaches can be applied to monitor different aspects of protein membrane insertion, including kinetic pathways of membrane insertion and free energy landscapes, specific conformational changes, structure and topology of membrane-inserted states, as well as elucidating the role of lipids on the membrane protein insertion pathways.

2. Methods for studies of kinetics and thermodynamics of protein-membrane interactions

2.1 Application of environment-sensitive probes and FRET for membrane binding/insertion assays

The use of environment-sensitive fluorescence probes is a ubiquitous method to monitor partitioning of proteins into lipid bilayers. The examples of such probes include the natural amino acid tryptophan and the commercially available external probes NBD (derivatives of 7-nitro-2-1,3-benzoxadiazol-4-amine) and Bimane (derivative of 3-(bromomethyl)-2,5,6-trimethyl-1*H*,7*H*-pyrazolo[1,2-*a*]pyrazole-1,7-dione). When the protein-attached probe is transferred from aqueous to membranous environments, its quantum yield and fluorescence lifetime often increase and the position of emission maximum shifts toward shorter wavelengths (blue-shift). Thus, following these changes in spectroscopic signature allows monitoring the partitioning of a labeled protein between the aqueous solution and the lipid bilayer. The general principles and specific applications of spectroscopic titration for the quantitative analysis of thermodynamics of membrane partitioning, has been described in detail elsewhere [Ladokhin, 2009]. With respect to environment-sensitive probes, however, one should keep in mind that not all spectroscopic signals constitute linear-response functions, suitable for quantitative analysis of binding. Unlike the total intensity (or that measured at a constant wavelength), the peak intensity or the position of spectral maximum do not scale linearly with the fraction of bound species and therefore are not suitable for analysis of binding [Ladokhin, Jayasinghe, & White, 2000].

Intrinsic tryptophan fluorescence is a classical tool to monitor membrane partitioning and refolding of membrane-active peptides and proteins [Kyrychenko et al., 2014; Kyrychenko, Vasquez-Montes, Ulmschneider, & Ladokhin, 2015; Ladokhin et al., 2000; Reshetnyak, Segala, Andreev, & Engelman, 2007; Vargas-Uribe, Rodnin, Kienker, et al., 2013; Vasquez-Montes, Gerhart, King, Thevenin, & Ladokhin, 2018]. An important advantage of tryptophan fluorescence is that it does not require site-selective-labeling, because the amino acid is incorporated as part of the peptide sequence. However, tryptophan fluorescence has some disadvantages because proteins usually have multiple tryptophan residues, or the signal-to-noise ratio is often not optimal when proteins have a single tryptophan residue. In certain instances, this can be circumvented by the application of quenching-enhanced titration protocols, which utilize aqueous quenchers, such as I^- or Cs^+ ions. The latter would preferentially quench fluorescence coming from proteins in solution, rather than from membrane-bound proteins, thus increasing the difference in the signal upon binding [Posokhov, Gottlieb, & Ladokhin, 2007]. Additional issues regarding tryptophan fluorescence, such as how to deal with light-scattering effects arising from the presence of vesicles, are discussed elsewhere [Ladokhin et al., 2000]. Most of these challenges, however, can be bypassed by site-selective-labeling of the protein with environment-sensitive dyes, such as NBD, BODIPY or Bimane, which can substantially improve the signal-to-noise ratio [Kaback & Wu, 1999; Rosconi & London, 2002; Vargas-Uribe, Rodnin, & Ladokhin, 2013; Wang, Rosconi, & London, 2006].

Application of the Förster Resonance Energy Transfer (FRET) is a common strategy to detect binding between two fluorescence-labeled particles [May & Kühn, 2011]. In contrast to the use of environment-sensitive probes, this approach does not require the probe to become embedded into the lipid bilayer to yield a fluorescence response. Instead, there is a change in signal that can be experimentally monitored because a donor probe transfers energy to an acceptor probe when they are within the Förster's distance [Kyrychenko et al., 2009; Vargas-Uribe, Rodnin, & Ladokhin, 2013; Vasquez-Montes et al., 2018]. The experiment consists of labeling the protein and membrane with the donor/acceptor pair, so that the partitioning of the protein causes FRET between the probes. Below, we will discuss how the comparison of FRET-based membrane-association kinetics with to NBD-based membrane-insertion kinetics allowed for characterization of insertion intermediates in diphtheria toxin T-domain.

The formation of intermediates and subsequent transbilayer insertion is mediated by a subtle balance of hydrophobic and electrostatic interactions between proteins and the membrane interface. The applications of fluorescence spectroscopy for kinetic measurements of the T-domain insertion along the pathway from aqueous W-state to interfacial intermediate I-state and finally to transbilayer inserted T-state is summarized in Fig. 2. Kinetic measurements using the FRET signal between AlexaFluor488-labeled T-domain and Rhodamine-labeled LUV were used to study the time-dependent association of T-domain with the membrane. The validity of this FRET approach was verified by comparing equilibrium titration profiles with those obtained by FCS (Fig. 3A, *circles* and *diamonds*). In addition, FRET-based binding kinetics (Fig. 2A and B) were compared to those observed with the environmentally sensitive probe NBD attached to a single-Cys residue in the Q369C mutant of the T-domain (Fig. 2A and C). This residue is relatively solvent-exposed in the soluble T-domain at neutral pH (W-state) and according to our topology result (Fig. 2A) is in the middle of the TM segment in the inserted T-state. Such a transition is expected to result in a substantial increase of the fluorescence of NBD, which was indeed observed, as shown by the kinetics measurements in Fig. 2C. In contrast to the fast FRET-based binding kinetics (Fig. 2B), fluorescence kinetics of NBD was slow and dependent on the lipid composition (Fig. 2C). This indicates that there is a membrane-bound, but not fully inserted, kinetic intermediate (I-state) that can be kinetically trapped. Accurate measurements of pH-dependent binding and insertion of the T-domain revealed that the protonation transitions allowing membrane binding and TM insertion are staggered rather than consecutive [Kyrychenko et al., 2009]. This is likely to be related to an additional protonation event of the T-domain on the interface due to variations in pK_a between bulk solvent and the membrane interfacial zone. While the first transition (W-to-W^+) involves

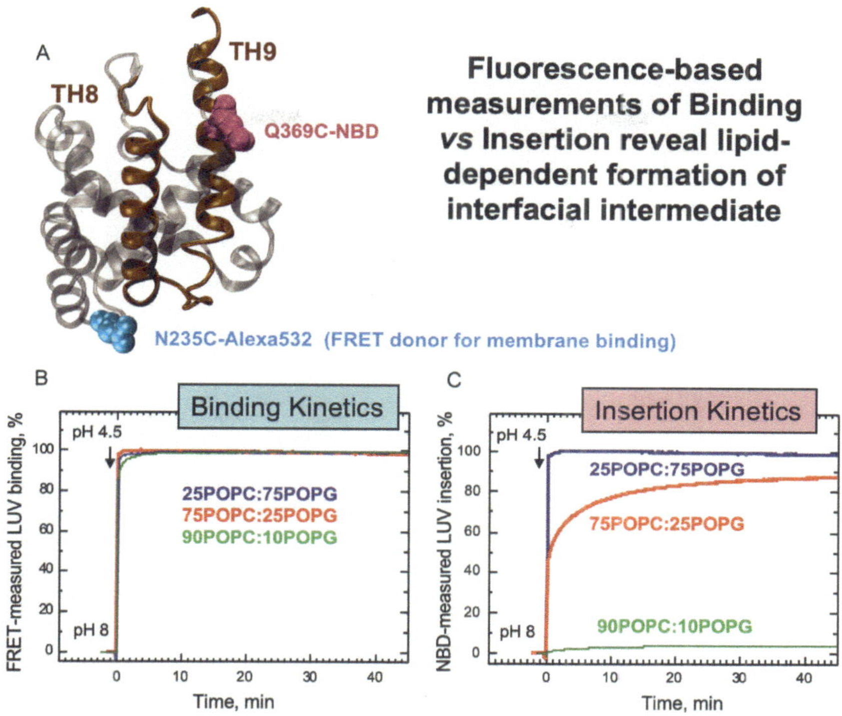

Fig. 2 (A) Crystallographic structure of the diphtheria toxin T-domain in soluble form at neutral pH [Bennett & Eisenberg, 1994]. The consensus insertion hairpin consisting of helices TH8 and TH9 is shown solid, while the rest of the structure is shown semitransparent. Two residues, N235 and Q369 (highlighted in CPK), along TH9 were replaced with cysteines (one at a time), providing unique labeling sites for fluorescence dye (NBD or AlexaFluor488) used in steady-state and kinetic measurements of insertion. (B and C) Membrane binding and insertion kinetics of the T-domain measured with LUV of specified lipid compositions (color-coded). (B) Binding kinetics were followed by the change in FRET signal between donor-labeled T-domain and acceptor-labeled LUV. (C) Insertion kinetics were followed by changes in fluorescence intensity of environment-sensitive probe NBD attached to the center region of the membrane-insertion domain (Q369C-NBD). Differences observed in the two types of kinetics reveal the insertion intermediate, which depends on pH and lipid composition. *Modified from Kyrychenko, A., Posokhov, Y.O., Rodnin, M.V., Ladokhin, A.S. (2009) Kinetic intermediate reveals staggered pH-dependent transitions along the membrane insertion pathway of the diphtheria toxin T-domain. Biochemistry, 48, 7584–7594.*

protonation of N-terminal histidines [Kurnikov et al., 2013; Rodnin et al., 2010, 2016], the second one (Interfacial-to-Transmembrane state) depends on the protonation of E362 and another, not yet identified, residue [Ghatak et al., 2015; Kyrychenko et al., 2018].

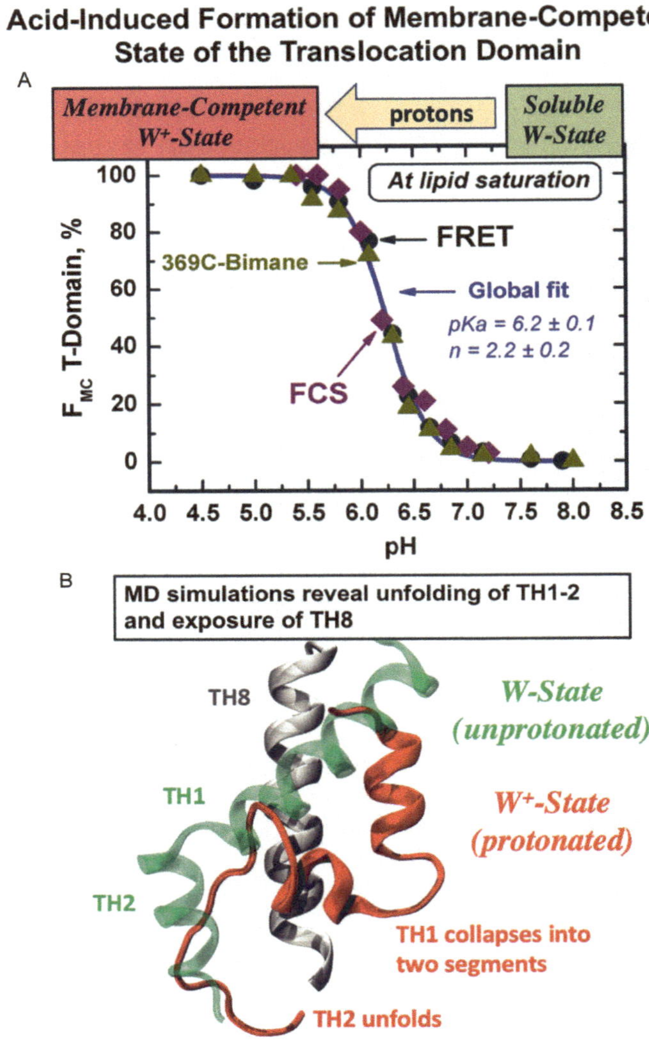

Fig. 3 pH-dependent conversion of the T-domain from the soluble W-state into the membrane-competent W^{+}-state. (A) Identification of W-to-W+ transition through membrane binding at lipid saturation. Mobility measurements based on FCS (*diamonds*) and measurements of FRET between donor-labeled T-domain and acceptor-labeled LUVs (*circles*) were performed as described in [Kyrychenko et al., 2009]. Triangles represent the increase in fluorescence intensity of environment-sensitive probe bimane attached to a single cysteine in position 369 (Fig. 2A). (B) Illustration of conformational changes associated with protonation of T-domain's six histidines studied by MD simulations [Kurnikov et al., 2013]. The collapse of TH1–2 and exposure of TH8 are also confirmed by HDX-MS [Li, Rodnin, Ladokhin, & Gross, 2014; Rodnin et al., 2016]. *Reprinted from Kyrychenko, A., Posokhov, Y.O., Rodnin, M.V., Ladokhin, A.S. (2009) Kinetic intermediate reveals staggered ph-dependent transitions along the membrane insertion pathway of the diphtheria toxin t-domain. Biochemistry, 48, 7584–7594; Copyright 2013, with permission from Elsevier.*

2.2 FCS-based methods to quantify membrane interactions in the presence of membrane-competent and incompetent protein conformations

Combining linear-response spectroscopic techniques with lipid titrations is a popular approach in the thermodynamic characterization of membrane binding of proteins and peptides [Ladokhin et al., 2000; White & Wimley, 1998]. Measurements of fluorescence intensity or those of quantum yield-weighted anisotropy provide sensitive and accurate measures of the fractional saturation of membrane partitioning. While all titration-based approaches can be extremely useful for simple binding studies, their application to complex multistate systems is more challenging. In contrast, Fluorescence Correlation Spectroscopy (FCS), which relies on diffusion measurements is well suited to tackle such systems and was successfully applied to thermodynamic measurements of membrane interactions of the T-domain [Kurnikov et al., 2013; Posokhov, Rodnin, Das, Pucci, & Ladokhin, 2008; Vargas-Uribe et al., 2015]. Below, we briefly summarize the formalism of FCS analysis as it applies to a complex binding equilibrium typical for the T-domain and other proteins that undergo pH-dependent membrane binding [Kurnikov et al., 2013; Ladokhin, 2009; Posokhov, Rodnin, Das, et al., 2008; Vargas-Uribe et al., 2015].

The autocorrelation function $G(\tau)$ for single diffusing species undergoing Brownian motion can be described with the Eq. (1) [Haustein & Schwille, 2007; Hess, Huang, Heikal, & Webb, 2002]:

$$
\begin{aligned}
G(\tau) &= \frac{1}{N} \times g(\tau) \\
&= \frac{1}{N} \cdot \left(1 + \frac{T}{1-T} \cdot e^{-\tau/\tau_{TR}}\right) \cdot \left(\frac{1}{1+\tau/\tau_D}\right) \cdot \left(\frac{1}{1+\tau/S^2\tau_D}\right)^{1/2}
\end{aligned} \tag{1}
$$

where N is the average number of fluorescent molecules in the focal volume and τ_D is the correlation time of the particles. The correlation time represents the diffusion time through the focus volume and equals $\tau_D = \omega^2/4D$, where ω^2 is the square of the radius of the laser focus and D is the diffusion constant. S is the ratio of the distances from the center of the laser beam focus in the radial and axial directions, respectively. T is the fraction of fluorophores in the triplet state and τ_{Tr} is the triplet lifetime.

The measured correlation function $G(\tau)$ of a multicomponent system is a weighted sum of the autocorrelation functions of each component $G_i(\tau)$ with amplitudes A_i [Ladokhin, 2009; Posokhov, Rodnin, Lu, & Ladokhin, 2008; Rhoades, Ramlall, Webb, & Eliezer, 2006; Rusu, Gambhir, McLaughlin, & Radler, 2004] as shown by Eq. (2).

$$G(\tau) = \sum_{i=1}^{M} q_i^2 N_i^2 G_i(\tau) / \left[\sum_{i=1}^{M} q_i N_i\right]^2 = \sum_{i=1}^{M} q_i^2 N_i g_i(\tau) / \left[\sum_{i=1}^{M} q_i N_i\right]^2 = \sum_{i=1}^{M} A_i \cdot g_i(\tau) \quad (2)$$

where N_i is the mean particle number and q_i is the ratio of the fluorescence yield of the i_{th} component to that of the first component. $q_i = 1$ if binding of dye-labeled membrane protein to vesicles does not change fluorescence intensity of fluorescent labels. For binding of membrane protein to vesicles only two diffusing species are considered, namely, the fluorescently labeled proteins (index P) and LUV with bound fluorescently labeled proteins (index V):

$$G(\tau) = A_P \cdot g_P(\tau) + A_V \cdot g_V(\tau) \quad (3)$$

If the quantum yield of a fluorophore is not environment-sensitive and does not change upon protein association with the membrane (which is often the case with AlexaFluor series of dyes), the amplitudes in Eq. 3 will depend only on the numbers of fast and slow moving particles in the focal volume (N_P and N_V, respectively).

$$A_P = \frac{N_P}{(N_V + N_P)^2}; \quad A_V = \frac{N_V}{(N_V + N_P)^2} \quad (4)$$

Under conditions of infinite dilution regime, the fraction of bound protein is equal [Ladokhin, 2009; Posokhov, Rodnin, Lu, & Ladokhin, 2008]:

$$f_B = \frac{A_V}{A_V + A_P} \quad (5)$$

The resulting titration isotherm is fitted to a mole fraction partitioning equation [Ladokhin et al., 2000; White, Wimley, Ladokhin, & Hristova, 1998], modified to account for the presence of membrane-incompetent protein species [Kyrychenko et al., 2009; Ladokhin, 2009; Posokhov, Rodnin, Lu, & Ladokhin, 2008]:

$$f_B = F_{MC} \cdot \frac{K_x \cdot [L]}{[W] + K_x \cdot [L]} \quad (6)$$

$$\Delta G = -RT \ln K_x \quad (7)$$

where F_{MC} is the membrane-competent fraction (determined at lipid saturation), $[L]$ is lipid concentration, [W] is water concentration (55.3 M), and K_x is a mole fraction partitioning coefficient which is used to determine the free energy ΔG of binding. Because FCS measurements are not dependent on the approximations of the signal saturation critical for other techniques, they often serve as a golden standard to which other methods, such as application of environment-sensitive probes and FRET, can be compared to (e.g., Fig. 3A, [Kurnikov et al., 2013]).

The described-above methodology was applied to determine both the pH-dependent formation of the membrane-competent form of the T-domain and the free energy of transfer of the T-domain from solution to the interfacial and membrane inserted states [Kurnikov et al., 2013; Posokhov, Rodnin, Das, et al., 2008; Vargas-Uribe et al., 2015]. The free energy measurements required application of fluorinated surfactants, that can chaperone membrane protein insertion and are particularly useful for controlling aggregation in solution, without compromising bilayer insertion [Kyrychenko et al., 2012; Palchevskyy et al., 2006; Rodnin et al., 2008]. The main gain in free energy was found to occur upon partitioning of the membrane-competent form of the T-domain to the interface (−8.2 kcal/mol), while the subsequent insertion was found to be only half as favorable (−3.0/−4.0 kcal/mol) [Posokhov, Rodnin, Das, et al., 2008; Vargas-Uribe et al., 2015]. These results confirm the conclusions of an earlier hydrophobicity analysis [Ladokhin et al., 2004] with regard to the critical role of the bilayer interface in modulating the insertion process.

2.3 pH-dependent formation of membrane-competent state of the T-domain

The formation of the membrane-competent form of the T-domain (W^+-state) is the first step along a complex pathway, leading from a soluble conformation with a known crystallographic structure [Bennett & Eisenberg, 1994] (W-state) to ultimately the membrane-inserted states (Fig. 1B) [Kyrychenko et al., 2018, 2009; Ladokhin, 2013; Palchevskyy et al., 2006; Rodnin et al., 2010; Vargas-Uribe et al., 2015]. Protonation of T-domain histidine sidechains has been recognized to play an important role in various stages of the membrane insertion process. In particular, H257 and H223 were implicated to act as a molecular switch that triggers disruption of the native structure of W-state of T-domain in low pH solutions [Kurnikov et al., 2013; Ladokhin, 2013; Rodnin et al., 2010, 2011]. The pH-dependent conformational transition W-to-W^+ is normally identified in a membrane-binding assay conducted under conditions of LUV saturation [Ladokhin, 2013; Vargas-Uribe et al., 2015]. The application of three independent fluorescence methods for the T-domain's interaction with LUV are summarized in Fig. 3: (*i*) diffusion measurements by FCS technique (Fig. 3A, *diamonds*), (*ii*) association of donor-labeled T-domain and acceptor-labeled vesicles by FRET (Fig. 3A, *circles*) [Kyrychenko et al., 2009], and (*iii*) changes in polarity of a specifically attached environment-sensitive bimane probe by increase in fluorescence intensity (Fig. 3A, *triangles*) [Kurnikov et al., 2013]. All combined datasets were fitted with Eq. 8 and could be adequately described by a single fitting curve with a pK_a of 6.2 and a Hill coefficient of about 2 (Fig. 3A, *line*). Thus, these measurements confirmed that W^+-state is an experimentally observable state of the protein with a conformation different from that of the W-state. The structure and dynamics of the pH-dependent conformational switching of the T-domain have been a subject of MD simulations reported elsewhere [Flores-Canales & Kurnikova, 2015; Flores-Canales et al., 2015; Ghatak et al., 2015; Kurnikov et al., 2013; Kyrychenko et al., 2018].

$$F_{MC} = \frac{1}{1 + 10^{n(pH - pKa)}} \tag{8}$$

where pKa is a negative logarithm of the dissociation constant, and n is the Hill coefficient.

3. Methods for studies of structure and topology of membrane-inserted proteins

3.1 FRET-based topology assay

Another application of FRET methodology is related to the determination of bilayer topology of membrane-inserted proteins (i.e., position of a particular site with respect to the two leaflets of the lipid bilayer). The insertion topology can be determined by combining site-selective labeling of the protein with a fluorescent probe and symmetric or asymmetric labeling of the membrane with a medium-range FRET-quencher [Everett, Zlotnick, Tennyson, & Holloway, 1986; Ladokhin, 2000; Ladokhin & White, 2004; Matsuzaki, Murase, Fujii, & Miyajima, 1995; Wimley & White, 2000]. The idea is that *cis*- or *trans*-positioning of the attached probe in the bilayer can be deduced by comparing the fluorescent response of the fluorophore upon interaction of the donor-labeled protein with membranes labeled symmetrically or asymmetrically with acceptors. Examples of these quenchers include Lyso-MC, a methylcoumarin derivative capable of quenching tryptophan fluorescence [Ladokhin & White, 2004; Wimley & White, 2000]; and Lyso-UB, a UniBlue chromophore derivative capable of quenching NBD fluorescence [Ladokhin et al., 2002; Posokhov & Ladokhin, 2006]. In both cases, the quenching occurs through non-radiative Förster type energy transfer (FRET). Because single-chain *lyso*-lipids have significant water solubility and form micelles, they can be incorporated easily into the outer leaflet when added to preformed bilayer vesicles. The bulky charged quencher headgroup and single-acyl chain act in concert to slow lipid flip-flop, resulting in an asymmetric distribution, which is stable for topology measurements under equilibrium conditions.

While Lyso-MC has been useful for determining the topology of relatively short tryptophan-containing peptides [Ladokhin & White, 2004; Wimley & White, 2000], several problems arise when it is applied to proteins (e.g., weak tryptophan fluorescence, poor spectral selectivity, parasitic Lyso-MC fluorescence). These problems can be bypassed by using Lyso-UB as quencher of NBD fluorescence [Ladokhin et al., 2002; Posokhov & Ladokhin, 2006]. Another advantage of Lyso-UB is that the Förster radius from NBD to Lyso-UB, which ranges from 30 to 35 Å depending on donor's quantum yield and spectral position, is about the same as the thickness of the bilayer hydrocarbon core. Thus, FRET efficiency will be the highest when the quencher and fluorophore are in the same leaflet and the lowest when they are separated by the bilayer. This can be measured by monitoring changes in donor intensity or lifetime [Posokhov & Ladokhin, 2006], where the latter is preferred because it is less susceptible to artifacts associated with partial loss of sample.

In order to test the insertion topology of the diphtheria toxin T-domain, the NBD probe was attached to the very tip of the insertion domain formed by helices TH8 and TH9 using the L350C and P378C mutants (Fig. 4) [Kyrychenko et al., 2009]. If these helices adopt a TM conformation upon insertion, the probe that is translocated across the bilayer becomes inaccessible to the externally added Lyso-UB quencher. If the helices remain interfacial, the quenching will be substantial for both mutants. Lyso-UB quenching experiments were performed after T-domain insertion had come to equilibrium at 1 h after mixing the T-domain and LUV at low pH. In order to describe Lyso-UB quenching, and the resulting topology, in a quantitative way, the fluorescence decay for all samples was analyzed and the ratio of the average lifetime upon addition of the quencher was calculated, as described in [Posokhov & Ladokhin, 2006]. We found that the degree of quenching depends on the lipid composition. The amplitude-averaged lifetime ratio τ_α/τ_0 of about 1.1 observed in 75POPC:25POPG

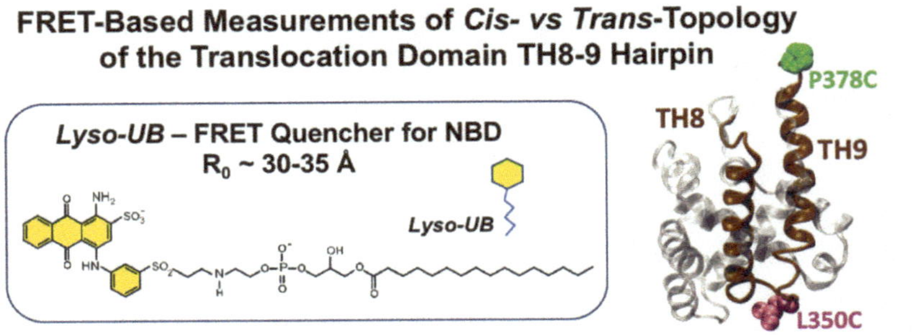

Fluorescence Lifetime Quenching of NBD-labeled Mutants with Lyso-UB
(Lyso-UB resides in the Cis-leaflet)

T-Domain labeling site	Membrane composition	Lyso-UB	τ_1, ns (α_1)	τ_2, ns (α_2)	τ_α, ns	τ_α/τ_0 (Topology)
L350C-NBD	25PC:75PG	none	1.50 (0.32)	7.90 (0.68)	5.85	1.12 (*trans*)
		+ 2%	1.42 (0.37)	7.47 (0.63)	5.25	
L350C-NBD	75PC:25PG	none	1.33 (0.35)	7.20 (0.65)	5.13	1.11 (*trans*)
		+ 2%	1.32 (0.40)	6.79 (0.60)	4.61	
L350C-NBD	90PC:10PG	none	1.18 (0.50)	6.27 (0.50)	3.72	1.70 (*cis*)
		+ 2%	0.94 (0.59)	3.97 (0.41)	2.19	
P378C-NBD	25PC:75PG	none	1.23 (0.51)	4.54 (0.49)	2.98	1.72 (*cis*)
		+ 2%	0.85 (0.67)	3.50 (0.33)	1.73	

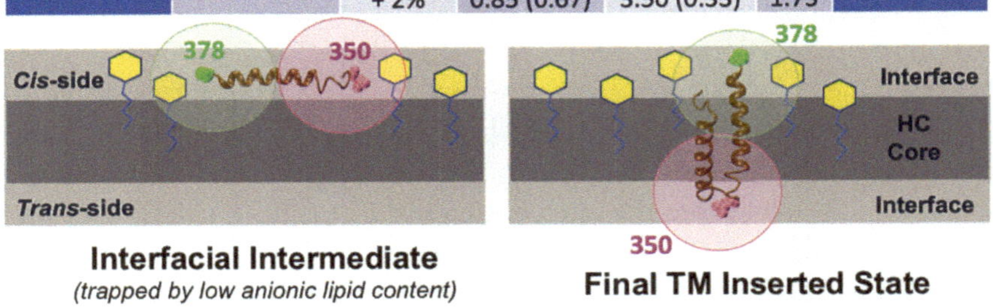

Fig. 4 (*Top left*) Chemical structure of LysoUB, a compound comprised of UniBlue chromophore attached to a *lyso*-lipid. (*Top right*) Structure of the T-domain with highlighted residues P378 and L350, flanking TH9. (*Middle*) Fluorescence lifetime quenching with Lyso-UB of the NBD-labeled P378C and L350C mutants in membranes containing various fractions of zwitterionic PC and anionic PG lipids [Kyrychenko et al., 2009]. (*Bottom*) Schematic representation of insertion pathway of TH8–9 of T-domain studied with LysoUB/NBD topology method. Shaded circles illustrate the approximate range of efficient quenching from an NBD-labeling site, consistent with the quenching results in the middle panel.

and 25POPC:75POPG is indicative of efficient translocation, while the ratio τ_α/τ_0 of 1.7 observed in 90POPC:10POPG confirms an interfacial topology (Fig. 4). In fact, the same quenching ratio was observed in a control experiment using the NBD-labeled P378C mutant, for which no translocation is expected during TH8-TH9 insertion (Fig. 4). These topology methods were critical for correct interpretation of measurements of lipid-dependent kinetics and thermodynamics [Kyrychenko et al., 2009; Posokhov, Rodnin, Das, et al., 2008; Vargas-Uribe et al., 2015].

3.2 Method of distribution analysis (DA) for depth-dependent fluorescence quenching by lipid-attached quenchers

Accurate determination of the penetration depth of proteins and peptides into lipid bilayers is an important step in studying of binding, topology and folding of membrane proteins. However, extracting of quantitative information about the membrane location of an intrinsic fluorophore or a site-selectively attached external probe from fluorescence studies is typically complicated by broad distributions of the transverse positions of both quenchers and fluorophores. Therefore, depth-dependent fluorescence quenching with a series of lipid-attached quenchers, such as bromine atoms or paramagnetic Tempo and Doxyl groups (Fig. 5A), has become an important spectroscopic tool for calculating the precise depth of a fluorescent moiety [Ladokhin, 1997, 1999, 2014; London & Ladokhin, 2002]. While bromolipids can be utilized as quenchers for tryptophan and pyrene fluorescence, the lipid-attached spin moieties, in addition to those, can also quench other organic fluorophores, including NBD, rhodamine, bimane and anthracene. The transverse position of bromine atoms has been determined by X-ray diffraction [McIntosh & Holloway, 1987], while the position of spin labels has been recently refined by MD simulations [Kyrychenko & Ladokhin, 2013].

The main assumption of the depth-dependent quenching methodology is that the quenching between a fluorescence dye and a quencher will be greatest when they are located at the same depth within the bilayer. To quantitate membrane penetration the quenching profiles are measured and fitted using Distribution Analysis (DA) method [Ladokhin, 1997, 2014]. DA reconstructs a depth-dependent quenching profile by using the quenching data collected from a series of quenchers, located at the known depths (h) within the membrane (Fig. 5). The DA methodology approximates the transverse quenching profile ($QP(h)$) of a fluorophore with a symmetrical twin Gaussian function (Fig. 5B), which has three fitting parameters: h_m = the center (mean) of the quenching profile, σ = the width of the distribution, and S = the area of the quenching profile. These parameters correspond to the most probable depth of penetration, fluctuations in the transverse position, and overall accessibility to quenching (i.e., quenching efficiency), respectively [Ladokhin, 1999]. Unlike in the previously described

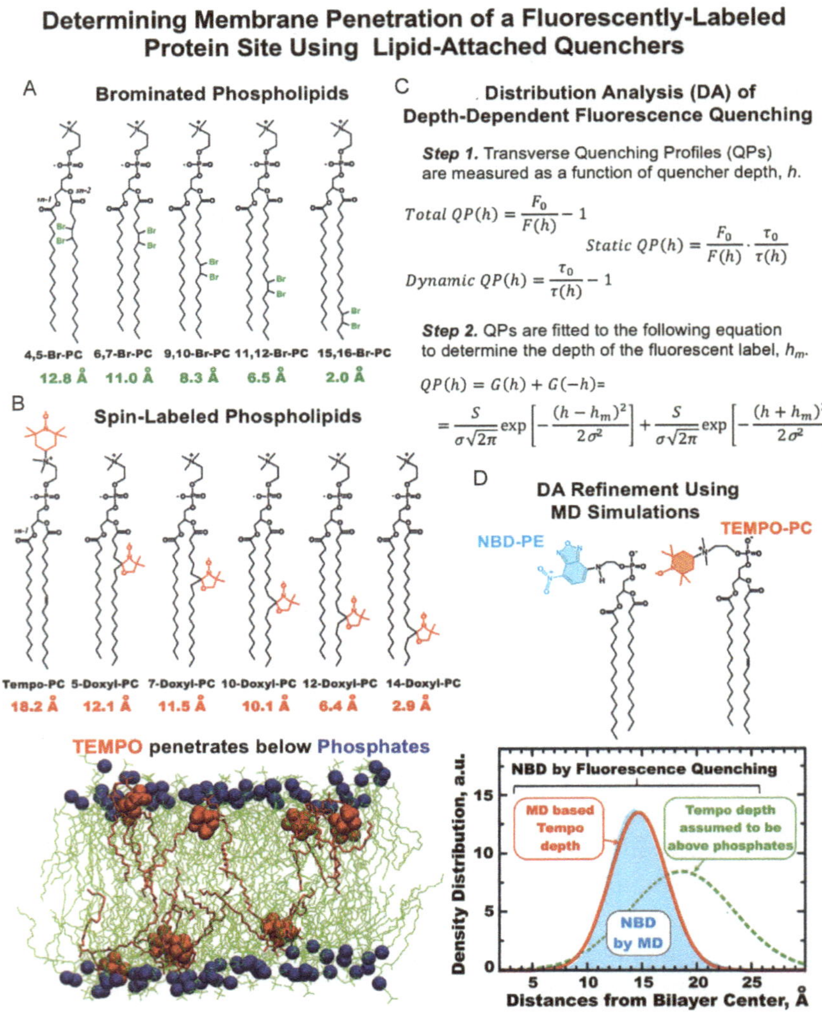

Fig. 5 (A and B) Lipid-attached fluorescence quenchers used for depth-dependent measurements and their depth (the distances from bilayer center to the quencher) [Kyrychenko & Ladokhin, 2013; McIntosh & Holloway, 1987]. (A) Brominated phospholipids: 1-palmitoyl-2-(*n*,*n*-dibromo)stearoyl-*sn*-glycero-3-phosphocholine (*n*,*n*-Br-PC). (B) Spin-labeled phospholipids: 1-palmitoyl-2-oleoyl-*sn*-glycero-3-phospho(TEMPO) choline (Tempo-PC), 1-palmitoyl-2-stearoyl-(*n*-doxyl)-*sn*-glycero-3-phosphocholine (*n*-doxyl-PC). (C) The DA methodology approximates the transverse quenching profile (*QP*(*h*)) with a sum of two mirror-image Gaussian functions (*G*(*h*)), where h_m—the center (mean) of the quenching profile, σ—the width of the distribution, and S—the area of the quenching profile [Ladokhin, 1997, 1999, 2014; London & Ladokhin, 2002]. (D) Top Right: Structure of NBD-PE and TEMPO-PC. Bottom Left: An MD snapshot of a lipid bilayer composed of Tempo-PC and POPC with a ratio of that correspond to 11 mol% of bilayer spin-labeling, respectively. The POPC phosphorus atoms are shown as *blue balls*. POPC and Tempo-PC are shown in stick representations in *olive* and *red*, respectively, with the Tempo moieties shown using van der Waals representation Bottom Right:

LysoUB quenching method (Section 3.1), the quenchers are distributed in both leaflets. In order to account for possible trans-leaflet quenching of deeply penetrating fluorophores, the mirror-image $G(-h)$ component is added to main Gaussian component $G(h)$ [Ladokhin, 1999].

The accuracy of membrane depth calculations is crucially dependent on the knowledge of the precise positioning of the lipid-attached spin-quenchers within the bilayer. Therefore, the average positions of the spin quencher calculated from the center of the lipid bilayer have recently been determined by molecular dynamics (MD) simulations of a series of spin-labeled lipids in the model membrane [Kyrychenko & Ladokhin, 2013]. A combination of MD simulations and depth-dependent fluorescence quenching was used to calibrate the methodology for extracting quantitative information on membrane penetration of fluorescent-labeled molecules [Kyrychenko, Rodnin, & Ladokhin, 2015]. One of such model systems, NBD-PE in a POPC bilayer, is illustrated in Fig. 5D. The MD-simulated depth profile of the lipid-attached NBD probe is shown as a blue distribution on the Bottom Right panel (Fig. 5D). The red and green lines correspond to two alternative application of the DA methodology to the same set of experimental data obtained with the full set of lipid-attached spin labels, including TEMPO (Fig. 5B). The difference between the two DA-generated profiles is in the transverse positioning of the TEMPO group. Assuming that TEMPO is located above phosphates, as sometimes suggested in the literature [Schrader & Han, 2017], results in large discrepancy between MD and experimental position of NBD (green line). In contrast, positioning of TEMPO below the level of phosphates during DA analysis (as suggested by MD simulations [Kyrychenko & Ladokhin, 2013]) and Fig. 5D, Bottom Left panel), results in excellent correlation between the reconstructed quenching profile (Fig. 5D, Bottom Right panel, red curve) and MD-generated [Kyrychenko, Rodnin, & Ladokhin, 2015] profile for the NBD (shaded blue area).

Comparison of the MD-simulated distribution of a NBD moiety [Kyrychenko, Rodnin, & Ladokhin, 2015] (*shaded area*) with those estimated from experimental data (*red* and *green lines*). *Green profile* corresponds to the result obtained when Tempo is assumed to reside above the phosphate level, as suggested by Schrader and Han [Schrader & Han, 2017], while *red profile* corresponds to the result obtained when position of Tempo is taken from our MD simulation ([Kyrychenko & Ladokhin, 2013] and central panel). The latter gives excellent correspondence to the MD-determined profile of the NBD (blue), thus validating the DA approach. *Reprinted by permission from [Kyrychenko, A., Ladokhin, A.S. (2020) Location of TEMPO-PC in lipid bilayers: Implications for fluorescence quenching. The Journal of Membrane Biology, 253, 73–77.], COPYRIGHT 2020 from Springer*

3.3 Application of DA method to structural studies of translocation domain in the lipid bilayer

The membrane depth of NBD probes attached at various positions in the diphtheria toxin T-domain were examined by depth-dependent fluorescence quenching using the DA methodology [Kyrychenko et al., 2018]. The steady-state and lifetime fluorescence of NBD-labeled T-domain inserted into lipid vesicles were examined in the absence and presence of spin-labeled lipids. Depth-dependent *QP*'s were constructed using measurements of fluorescence intensity, F, and lifetime, τ, as functions of quencher depth, h, to determine the transverse position of the NBD probes in the lipid bilayer (see Section 3.2 for details).

We use two limiting cases to illustrate the quantitative application of depth-dependent quenching with the DA methodology: The shallow penetrating residue P378C (Fig. 6) and the deeply penetrating residue A282C (Fig. 7). Panels A in Figs. 6 and 7 show the primary experimental data for the steady-state intensity (triangles) and lifetime quenching (squares) of these residues. The data is plotted as the fluorescence ratio between measurements in the absence (F_0 or τ_0) and presence (F or τ) of spin lipids. In both cases the "dynamic" quenching (triangles) is much weaker than the total quenching (squares). The former is likely influenced by experimental difficulties of lifetime measurements in scattering samples. As a result, the "dynamic" quenching refers to all events that occur on a timescale longer than 0.1 ns. The presented τ parameter represents the amplitude-weighted average of the two longest lifetime components.

Data in panels B in Figs. 6 and 7 was derived from the primary experimental data (panels A) and represent the depth-dependent static *QP*, calculated as $QP(h) = (F_0/F(h))/(\tau(h)/\tau_0)$. The datasets were fitted (solid lines) with DA equation from Fig. 5C to obtain the most probable depth of the NBD probe, h_m, and the dispersion in transverse position, σ. The colored profiles denote individual Gaussian functions, G(h); while the total *QP*'s are a sum of G(h) and its mirror image G($-h$), where both functions share the same parameters. The single gaussian fit is a simplified approach that only considers the quenching in a single leaflet, while the double Gaussian function used in DA analysis also accounts for transleaflet quenching. The improvement of depth determination by the double gaussian approach is more evident at deeply penetrating positions, such as in the case of A282C-NBD (Fig. 7) and is discussed in a greater detail below.

The robustness of the NBD penetration depths, h_m, obtained by DA were determined by subjecting the fits to support-plane analysis [Montgomery & Peck, 1982]. This statistical procedure relies on the χ^2 parameter of the

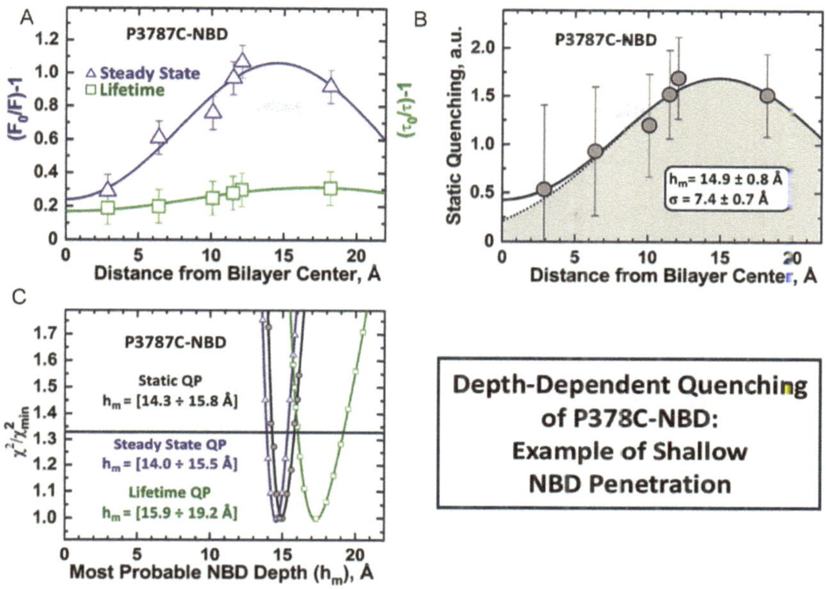

Fig. 6 Application of Distribution Analysis (DA) to depth-dependent fluorescence quenching of shallow penetrating probe in P378C-NBD mutant [Kyrychenko et al., 2018]. (A) Fluorescence steady-state (triangles) and time-resolved (squares) quenching profiles plotted against the depth of the following quenchers (from right to left): Tempo–PC and *n*-Doxyl–PCs with $n=5, 7, 10, 12$, and 14. (B) Static QP calculated from the data in panel A as $QP(h)=(F_0/F(h))/(\tau(h)/\tau_0)$. Solid lines represent the fit of the corresponding datasets using DA equation from Fig. 5C. The shaded areas represent a Gaussian function for the single leaflet quenching profiles. The sum of the shaded area, G(*h*), and its mirror image, G(−*h*), represents the total fitting function of DA, with the total of three fitting parameters: h_m, σ and *S*. (C) Support-plane analysis of the robustness of the determination of the parameter for the most probable depth of the NBD dye (see text for details).

goodness of the least-squares fit of a series of solutions where h_m was fixed at various values (in 0.2–0.5 Å increments) around the most optimal solution (χ^2_{min}). The support plane analysis of the quenching distributions obtained from the primary fluorescence data and DA analysis are plotted in Fig. 6C and 7C. A $\chi^2/\chi^2_{min}=1.33$ cut-off (horizontal line), corresponding to 1-standard deviation, was used to estimate the h_m ranges that produce statistically indistinguishable fits [Montgomery & Peck, 1982].

In the case of the shallow penetrating fluorophore P378C-NBD (Fig. 6C), the steady-state *QP* (blue) and static *QP* (black) produce similar results, with well-defined profiles centered at ∼15 Å from the center of the bilayer. The lack of definition in the lifetime QP suggests that the highest

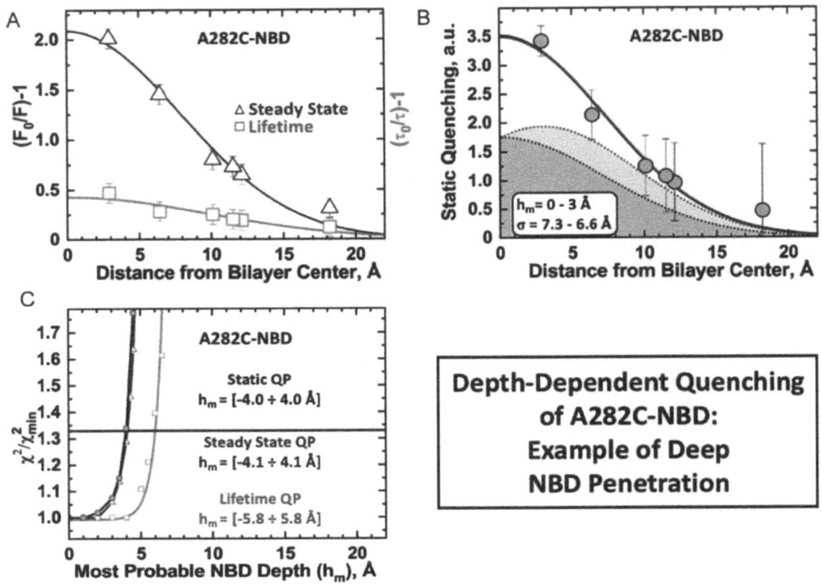

Fig. 7 Application of DA to depth-dependent fluorescence quenching of deep penetrating probe in A282C-NBD mutant [Kyrychenko et al., 2018]. The steady-state (panel A, triangles) and time-resolved data (panel A, squares), as well as the static (panel B) quenching profiles were obtained as described in Fig. 6. Solid shapes represent two examples of individual solutions for a single-leaflet quenching profile (G(*h*) in Eq. 1) with $h_m = 0$ Å (gray) and $h_m = 3$ Å (purple). Both solutions produce essentially overlapping two-leaflet quenching profiles (correspondingly colored solid lines that go through the data). (C) Support-plane analysis of the robustness of the determination of the parameter for the most probable depth of the NBD dye (see text for details).

contribution to the dynamic quenching component is present at depths ranging from 16 to 19 Å. All three *QP*'s of the deeply penetrating fluorophore A282C-NBD (Fig. 7) are less defined than those determined for P378C-NBD. This difference is primarily due to the uncertainty introduced by the cross-leaflet quenching near the middle of the bilayer. As a result, data fitting in this region is challenging and yields equivalent solutions for distributions centered between 0 and 3 Å (individual colored profiles in Fig. 7B). Consistently, the support plane analysis indicates that the NBD's most probable position for residue A282C ranges within the 4 Å range on either side of the bilayer center (Fig. 7C).

The exact positioning of the helix will allow us to generate realistic starting positions for future studies of the membrane-inserted T-domain by MD computer simulations.

3.4 Is the Open-Channel State needed for the translocation? (Positioning of TH5 helix in T-domain WT and H322Q mutant)

Recent results indicate that C-terminal histidines (Fig. 8), and especially H322, are critical for the insertion of the N-terminal part of the T-domain and formation of the OCS [Rodnin et al., 2011; Vargas-Uribe, Rodnin, Kienker, et al., 2013]. Replacements with Q, R or K, while not affecting folding in solution or proper insertion of TH8–9, lead to an alternative folding pathway and inhibit the channel activity in planar lipid bilayer. This alternative folding is revealed by the red-shifted tryptophan fluorescence, associated with changes in insertion of W281 located in TH5. The spectroscopic changes become even more pronounced in a single-Trp mutant W206F (Fig. 8, central panel and Ref. [Ladokhin et al., 2017]). This variation in the insertion mode is confirmed by the results on an independent depth-dependent quenching experiments using three commercially available bromolipids with differed depths, h: 6-7BrPC ($h=11$ Å), 9-10BrPC ($h=8.3$ Å) and 11-12BrPC ($h=6.5$ Å) (Fig. 5A). Each of the bromolipids was mixed with anionic POPG in 1:1 M ratio. The intensity of tryptophan fluorescence of the T-domain was measured in the presence of quenching vesicles, $F(h)$ and in quencher-free POPC-POPG LUV, F_0. The "total" quenching profile was generated as $QP(h)=F_0/F(h)-1$ and analyzed using the DA method [Ladokhin, 1997, 2014] by fitting to the twin Gaussian function (equation in Fig. 5B and C). Because the quenching profile for the mutant was poorly defined, the dispersion parameter σ representing the width of the distribution was fixed at 5, 6 or 7 Å. Regardless of the chosen value, the results indicate that W281 in the WT is located close to the center of the lipid bilayer, but remains close to the interface in the H223Q mutant (Fig. 8, central panel). Both spectroscopic experiments presented in Fig. 8 support the conclusion that the H322Q mutant does not form an OCS conformation [Ladokhin et al., 2017; Rodnin et al., 2011; Vargas-Uribe, Rodnin, Kienker, et al., 2013].

While the exact mechanism for how the replacement of H322 influences the topology of a segment some 40 residues away remains a mystery. This result, however, confirms that fluorescence measurements constitute an important tool for studies of membrane protein folding. It is also surprising that the strong inhibition of the channel activity, caused by replacement of C-terminal histidines, does not translate into comparable loss of the cytotoxic activity when the same mutations are introduced into the full-length toxin (Fig. 8, right panel and Ref. [Ladokhin et al., 2017]). This implies that OCS is not directly involved in the translocation pathway for the catalytic domain but is likely to be a byproduct of translocation process.

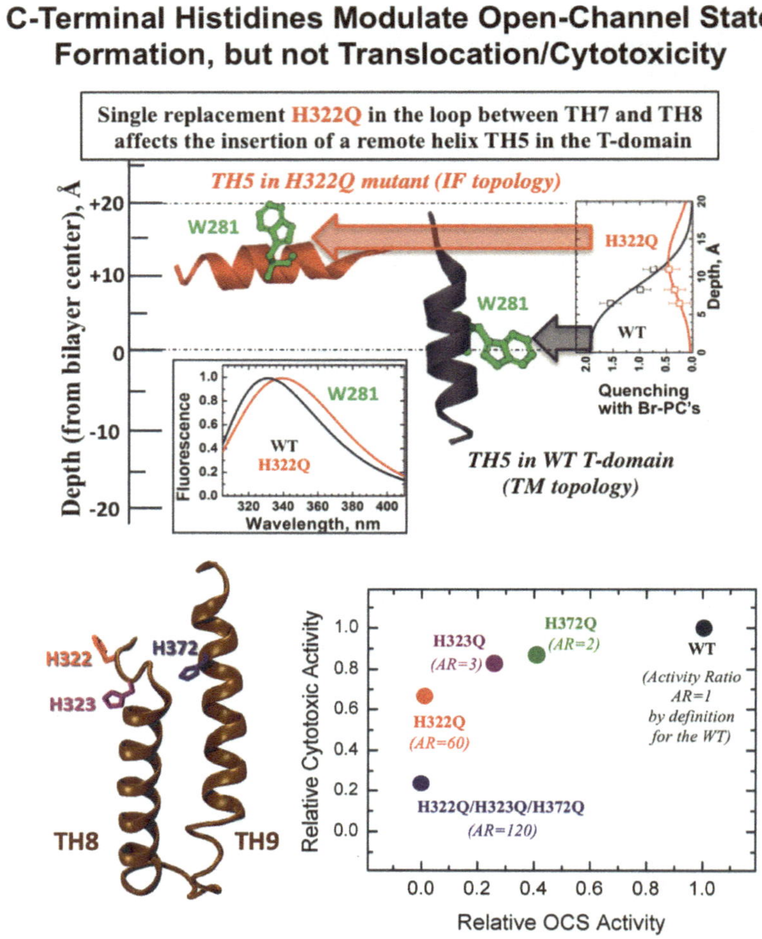

Fig. 8 *Left bottom panel*: TH8–9 hydrophobic hairpin with highlighted histidine residues. *Top panel*: A single replacement of H322 results in repositioning of the TH5 helix from TM state (WT, black) to IF state (H322Q mutant, red), associated with loss of channel activity [Rodnin et al., 2011; Vargas-Uribe, Rodnin, Kienker, et al., 2013]. This repositioning can be followed by changes in spectral position of W281 (lower insert) and it's quenching with brominated lipids (upper insert). *Right bottom panel*: Summary of the relative cellular toxicity in various mutants of diphtheria toxin and relative OCS activity of the T-domain with the same mutations in planar bilayers [Ladokhin, Vargas-Uribe, Rodnin, Ghatak, & Sharma, 2017]. The Activity Ratio (AR) between the two activities significantly deviates from one for the mutants with C-terminal histidine substitutions, indicating that channel activity is not required for the cytotoxicity associated with the translocation of the catalytic domain. *Reprinted by permission from Kyrychenko, A., Posokhov, Y.O., Vargas-Uribe, M., Ghatak, C., Rodnin, M.V., Ladokhin, A.S., Fluorescence applications for structural and thermodynamic studies of membrane protein insertion, in: C.D. Geddes (Ed.), Reviews in fluorescence 2016, Springer 2017, pp. 243–274; COPYRIGHT 2017 from Springer.*

4. Perspectives

Fluorescence spectroscopy is an important tool for studies of insertion of proteins into lipid bilayers. As demonstrated with the example of diphtheria toxin translocation domain, various fluorescence approaches can be applied to obtain critical structural and thermodynamic information enabling the deciphering of complex insertion/refolding pathways. Measurements of the insertion topology and bilayer penetration of protein sites, selectively labeled with appropriate probes, have a potential to fill an important knowledge gap in the structure-function relationship of a wide variety of proteins interacting with lipid bilayers. The methods described here are opening the doors for addressing fundamental questions related to lipid-dependent regulation of protein function *via* conformational switching and signaling on membrane interfaces. Future studies of the diphtheria toxin will be geared toward deciphering the elusive structure of the active conformation involved in the translocation (Fig. 1B); characterizing the protonation states of the titratable residues responsible for conformational switching in the lipid bilayer and ultimately visualizing and studying the infection process inside the cell.

Acknowledgments

Supported by NIH GM-126778 (A.S.L.). A.K. acknowledges the Grant 0119U002536 of Ministry of Education and Science of Ukraine.

References

Antignani, A., & Youle, R. J. (2006). How do Bax and Bak lead to permeabilization of the outer mitochondrial membrane? *Current Opinion in Cell Biology*, *18*, 685–689.

Bennett, M. J., & Eisenberg, D. (1994). Refined structure of monomeric diphtheria toxin at 2.3 å resolution. *Protein Science*, *3*, 1464–1475.

Bogner, C., Leber, B., & Andrews, D. W. (2010). Apoptosis: Embedded in membranes. *Current Opinion in Cell Biology*, *22*, 845–851.

Chenal, A., Savarin, P., Nizard, P., Guillain, F., Gillet, D., & Forge, V. (2002). Membrane protein insertion regulated by bringing electrostatic and hydrophobic interactions into play. A case study with the translocation domain of the diphtheria toxin *The Journal of Biological Chemistry*, *277*, 43425–43432.

Chi, X., Kale, J., Leber, B., & Andrews, D. W. (2014). Regulating cell death at, on, and in membranes. *Biochimica et Biophysica Acta*, *1843*, 2100–2113.

Collier, R. J., & Young, J. A. (2003). Anthrax toxin. *Annual Reviews in Cell Develepment Biology*, *19*, 45–70.

Everett, J., Zlotnick, A., Tennyson, J., & Holloway, P. W. (1986). Fluorescence quenching of cytochrome b_5 in vesicles with an asymmetric transbilayer distribution of brominated phosphatidylcholine. *The Journal of Biological Chemistry*, *261*, 6725–6729.

Flores-Canales, J. C., & Kurnikova, M. (2015). Microsecond simulations of the diphtheria toxin translocation domain in association with anionic lipid bilayers. *The Journal of Physical Chemistry. B*, *119*, 12074–12085.

Flores-Canales, J. C., Ladokhin, A. S., & Kurnikova, M. (2014). Multiscale simulations of diphtheria toxin T-domain membrane association. *Biophysical Journal*, *106*, 441a.

Flores-Canales, J. C., Vargas-Uribe, M., Ladokhin, A. S., & Kurnikova, M. (2015). Membrane association of the diphtheria toxin translocation domain studied by coarse-grained simulations and experiment. *The Journal of Membrane Biology*, *248*(3), 1–15.

Ghatak, C., Rodnin, M. V., Vargas-Uribe, M., McCluskey, A. J., Flores-Canales, J. C., Kurnikova, M., et al. (2015). Role of acidic residues in helices TH8–TH9 in membrane interactions of the diphtheria toxin T domain. *Toxins*, *7*, 1303–1323.

Gordon, M., & Finkelstein, A. (2001). The number of subunits comprising the channel formed by the T domain of diphtheria toxin. *The Journal of General Physiology*, *118*, 471–480.

Haustein, E., & Schwille, P. (2007). Fluorescence correlation spectroscopy: Novel variations of an established technique. *Annual Review of Biophysics and Biomolecular Structure*, *36*, 151–169.

Hess, S. T., Huang, S., Heikal, A. A., & Webb, W. W. (2002). Biological and chemical applications of fluorescence correlation spectroscopy: A review. *Biochemistry*, *41*, 697–705.

Heuck, A. P., & Johnson, A. E. (2002). Pore-forming protein structure analysis in membranes using multiple independent fluorescence techiniques. *Cell Biochemistry and Biophysics*, *36*, 89–101.

Hoch, D. H., Romero-Mira, M., Ehrlich, B. E., Finkelstein, A., DasGupta, B. R., & Simpson, L. L. (1985). Channels formed by botulinum, tetanus, and diphtheria toxins in planar lipid bilayers: Relevance to translocation of proteins across membranes. *Proceedings of the National Academy of Sciences*, *82*, 1692.

Hsu, Y. T., Wolter, K. G., & Youle, R. J. (1997). Cytosol-to-membrane redistribution of bax and Bcl-x(L) during apoptosis. *Proceedings of the National Academy of Sciences of the United States of America*, *94*, 3668–3672.

Huynh, P. D., Cui, C., Zhan, H. J., Oh, K. J., Collier, R. J., & Finkelstein, A. (1997). Probing the structure of the diphtheria toxin channel: Reactivity in planar lipid bilayer membranes of cysteine-substituted mutant channels with methanethiosulfonate derivatives. *The Journal of General Physiology*, *110*, 229–242.

Jakes, K. S., & Cramer, W. A. (2012). Border crossings: Colicins and transporters. *Annual Review of Genetics*, *46*, 209–231.

Kaback, H. R., & Wu, J. H. (1999). What to do while awaiting crystals of a membrane transport protein and thereafter. *Accounts of Chemical Research*, *32*, 805–813.

Kielian, M., & Rey, F. A. (2006). Virus membrane-fusion proteins: More than one way to make a hairpin. *Nature Reviews. Microbiology*, *4*, 67–76.

Koriazova, L. K., & Montal, M. (2003). Translocation of botulinum neurotoxin light chain protease through the heavy chain channel. *Nature Structural Biology*, *10*, 13–18.

Kurnikov, I. V., Kyrychenko, A., Flores-Canales, J. C., Rodnin, M. V., Simakov, N., Vargas-Uribe, M., et al. (2013). pH-triggered conformational switching of the diphtheria toxin T-domain: The roles of N-terminal histidines. *Journal of Molecular Biology*, *425*, 2752–2764.

Kyrychenko, A., Freites, J. A., He, J., Tobias, D. J., Wimley, W. C., & Ladokhin, A. S. (2014). Structural plasticity in the topology of the membrane-interacting domain of HIV-1 gp41. *Biophysical Journal*, *106*, 610–620.

Kyrychenko, A., & Ladokhin, A. S. (2013). Molecular dynamics simulations of depth distribution of spin-labeled phospholipids within lipid bilayer. *The Journal of Physical Chemistry. B*, *117*, 5875–5885.

Kyrychenko, A., Lim, N. M., Vasquez-Montes, V., Rodnin, M. V., Freites, J. A., Nguyen, L. P., et al. (2018). Refining protein penetration into the lipid bilayer using fluorescence quenching and molecular dynamics simulations: The case of diphtheria toxin translocation domain. *The Journal of Membrane Biology*, *251*, 379–391.

Kyrychenko, A., Posokhov, Y. O., Rodnin, M. V., & Ladokhin, A. S. (2009). Kinetic intermediate reveals staggered pH-dependent transitions along the membrane insertion pathway of the diphtheria toxin T-domain. *Biochemistry, 48*, 7584–7594.
Kyrychenko, A., Rodnin, M. V., & Ladokhin, A. S. (2015). Calibration of distribution analysis of the depth of membrane penetration using simulations and depth-dependent fluorescence quenching. *The Journal of Membrane Biology, 248*, 583–594.
Kyrychenko, A., Rodnin, M. V., Posokhov, Y. O., Holt, A., Pucci, B., Killian, J. A., et al. (2012). Thermodynamic measurements of bilayer insertion of a single transmembrane helix chaperoned by fluorinated surfactants. *Journal of Molecular Biology, 416*, 328–334.
Kyrychenko, A., Vasquez-Montes, V., Ulmschneider, M. B., & Ladokhin, A. S. (2015). Lipid headgroups modulate membrane insertion of pHLIP peptide. *Biophysical Journal, 108*, 791–794.
Ladokhin, A. S. (1997). Distribution analysis of depth-dependent fluorescence quenching in membranes: A practical guide. *Methods in Enzymology, 278*, 462–473.
Ladokhin, A. S. (1999). Analysis of protein and peptide penetration into membranes by depth-dependent fluorescence quenching: Theoretical considerations. *Biophysical Journal, 76*, 946–955.
Ladokhin, A. S. (2000). Fluorescence spectroscopy in peptide and protein analysis. In R. A. Meyers (Ed.), *Peptides and proteins* (pp. 5762–5779). Chichester: John Wiley & Sons, Ltd.
Ladokhin, A. S. (2009). Fluorescence spectroscopy in thermodynamic and kinetic analysis of pH-dependent membrane protein insertion. *Methods in Enzymology, 466*, 19–42.
Ladokhin, A. S. (2013). pH-triggered conformational switching along the membrane insertion pathway of the diphtheria toxin t-domain. *Toxins, 5*, 1362–1380.
Ladokhin, A. S. (2014). Measuring membrane penetration with depth-dependent fluorescence quenching: Distribution analysis is coming of age. *Biochimica et Biophysica Acta, 1838*, 2289–2295.
Ladokhin, A. S., Isas, J. M., Haigler, H. T., & White, S. H. (2002). Determining the membrane topology of proteins: Insertion pathway of a transmembrane helix of annexin 12. *Biochemistry, 41*, 13617–13626.
Ladokhin, A. S., Jayasinghe, S., & White, S. H. (2000). How to measure and analyze tryptophan fluorescence in membranes properly, and why bother? *Analytical Biochemistry, 285*, 235–245.
Ladokhin, A. S., Legmann, R., Collier, R. J., & White, S. H. (2004). Reversible refolding of the diphtheria toxin T-domain on lipid membranes. *Biochemistry, 43*, 7451–7458.
Ladokhin, A. S., Vargas-Uribe, M., Rodnin, M. V., Ghatak, C., & Sharma, O. (2017). Cellular entry of the diphtheria toxin does not require the formation of the open-channel state by its translocation domain. *Toxins, 9*, 299.
Ladokhin, A. S., & White, S. H. (2004). Interfacial folding and membrane insertion of a designed helical peptide. *Biochemistry, 43*, 5782–5791.
Leber, B., Lin, J., & Andrews, D. W. (2007). Embedded together: The life and death consequences of interaction of the Bcl-2 family with membranes. *Apoptosis: An International Journal on Programmed Cell Death, 12*, 897–911.
Li, J., Rodnin, M. V., Ladokhin, A. S., & Gross, M. L. (2014). Hydrogen deuterium exchange and mass spectrometry reveal the pH-dependent conformational changes of diphtheria toxin t domain. *Biochemistry, 53*, 6849–6856.
London, E., & Ladokhin, A. S. (2002). Measuring the depth of amino acid residues in membrane-inserted peptides by fluorescence quenching. *Current Topics in Membranes, 52*, 89–115.
Matsuzaki, K., Murase, O., Fujii, N., & Miyajima, K. (1995). Translocation of a channel-forming antimicrobial peptide, magainin 2, across lipid bilayers by forming a pore. *Biochemistry, 34*, 6521–6526.
May, V., & Kühn, O. (2011). *Excitation energy transfer, Charge and energy transfer dynamics in molecular systems* (pp. 467–558). Wiley-VCH Verlag GmbH & Co. KGaA.

McIntosh, T. J., & Holloway, P. W. (1987). Determination of the depth of bromine atoms in bilayers formed from bromolipid probes. *Biochemistry, 26*, 1783–1788.
Mindell, J. A., Silverman, J. A., Collier, R. J., & Finkelstein, A. (1994a). Structure function relationships in diphtheria toxin channels. II. A residue responsible for the channel's dependence on *trans* ph. *The Journal of Membrane Biology, 137*, 29–44.
Mindell, J. A., Silverman, J. A., Collier, R. J., & Finkelstein, A. (1994b). Structure-function relationships in diphtheria toxin channels. III. Residues which affect the *cis* pH dependence of channel conductance. *The Journal of Membrane Biology, 137*, 45–57.
Montagner, C., Perier, A., Pichard, S., Vernier, G., Menez, A., Gillet, D., et al. (2007). Behavior of the n-terminal helices of the diphtheria toxin T domain during the successive steps of membrane interaction. *Biochemistry, 46*, 1878–1887.
Montgomery, D. C., & Peck, E. A. (1982). *Introduction to linear regression analysis*. New York: Wiley.
Murphy, J. R. (2011). Mechanism of diphtheria toxin catalytic domain delivery to the eukaryotic cell cytosol and the cellular factors that directly participate in the process. *Toxins, 3*, 294.
Neale, E. A. (2003). Moving across membranes. *Nature Structural Biology, 10*, 2–3.
Neuman, B. W., Kiss, G., Al-Mulla, H. M., Dokland, T., Buchmeier, M. J., Weikl, T., et al. (2013). Direct observation of membrane insertion by enveloped virus matrix proteins by phosphate displacement. *PLoS One, 8*, e57916.
Oh, K. J., Altenbach, C., Collier, R. J., & Hubbell, W. L. (2000). Site-directed spin labeling of proteins. Applications to diphtheria toxin. *Methods in Molecular Biology, 145*, 147–169.
Oh, K. J., Senzel, L., Collier, R. J., & Finkelstein, A. (1999). Translocation of the catalytic domain of diphtheria toxin across planar phospholipid bilayers by its own T domain. *Proceedings of the National Academy of Sciences of the United States of America, 96*, 8467–8470.
Oh, K. J., Zhan, H., Cui, C., Hideg, K., Collier, R. J., & Hubbell, W. L. (1996). Organization of diphtheria toxin T domain in bilayers: A site-directed spin labeling study. *Science, 273*, 810–812.
Palchevskyy, S. S., Posokhov, Y. O., Olivier, B., Popot, J.-L., Pucci, B., & Ladokhin, A. S. (2006). Chaperoning of insertion of membrane proteins into lipid bilayers by hemifluorinated surfactants: Application to diphtheria toxin. *Biochemistry, 45*, 2629–2635.
Perier, A., Chassaing, A., Raffestin, S., Pichard, S., Masella, M., Menez, A., et al. (2007). Concerted protonation of key histidines triggers membrane interaction of the diphtheria toxin t domain. *The Journal of Biological Chemistry, 282*, 24239–24245.
Posokhov, Y. O., Gottlieb, P. A., & Ladokhin, A. S. (2007). Quenching-enhanced fluorescence titration protocol for accurate determination of free energy of membrane binding. *Analytical Biochemistry, 362*, 290–292.
Posokhov, Y. O., & Ladokhin, A. S. (2006). Lifetime fluorescence method for determining membrane topology of proteins. *Analytical Biochemistry, 348*, 87–93.
Posokhov, Y. O., Rodnin, M. V., Das, S. K., Pucci, B., & Ladokhin, A. S. (2008). FCS study of the thermodynamics of membrane protein insertion into the lipid bilayer chaperoned by fluorinated surfactants. *Biophysical Journal, 95*, L54–L56.
Posokhov, Y. O., Rodnin, M. V., Lu, L., & Ladokhin, A. S. (2008). Membrane insertion pathway of annexin B12: Thermodynamic and kinetic characterization by fluorescence correlation spectroscopy and fluorescence quenching. *Biochemistry, 47*, 5078–5087.
Ren, J., Kachel, K., Kim, H., Malenbaum, S. E., John, R., Collier, L., et al. (1999). Interaction of diphtheria toxin t domain with molten globule-like proteins and its implications for translocation. *Science, 284*, 955.
Reshetnyak, Y. K., Segala, M., Andreev, O. A., & Engelman, D. M. (2007). A monomeric membrane peptide that lives in three worlds: In solution, attached to, and inserted across lipid bilayers. *Biophysical Journal, 93*, 2363–2372.

Rhoades, E., Ramlall, T. F., Webb, W. W., & Eliezer, D. (2006). Quantification of alpha-synuclein binding to lipid vesicles using fluorescence correlation spectroscopy. *Biophysical Journal, 90*, 4692–4700.

Rodnin, M. V., Kyrychenko, A., Kienker, P., Sharma, O., Posokhov, Y. O., Collier, R. J., et al. (2010). Conformational switching of the diphtheria toxin T domain. *Journal of Molecular Biology, 402*, 1–7.

Rodnin, M. V., Kyrychenko, A., Kienker, P., Sharma, O., Vargas-Uribe, M., Collier, R. J., et al. (2011). Replacement of c-terminal histidines uncouples membrane insertion and translocation in diphtheria toxin t-domain. *Biophysical Journal, 101*, L41–L43.

Rodnin, M. V., Li, J., Gross, M. L., & Ladokhin, A. S. (2016). The pH-dependent trigger in diphtheria toxin T domain comes with a safety latch. *Biophysical Journal, 111*, 1946–1953.

Rodnin, M. V., Posokhov, Y. O., Contino-Pepin, C., Brettmann, J., Kyrychenko, A., Palchevskyy, S. S., et al. (2008). Interactions of fluorinated surfactants with diphtheria toxin T-domain: Testing new media for studies of membrane proteins. *Biophysical Journal, 94*, 4348–4357.

Rosconi, M. P., & London, E. (2002). Topography of helices 5-7 in membrane-inserted diphtheria toxin T domain: Identification and insertion boundaries of two hydrophobic sequences that do not form a stable transmembrane hairpin. *The Journal of Biological Chemistry, 277*, 16517–16527.

Rusu, L., Gambhir, A., McLaughlin, S., & Radler, J. (2004). Fluorescence correlation spectroscopy studies of peptide and protein binding to phospholipid vesicles. *Biophysical Journal, 87*, 1044–1053.

Schrader, A. M., & Han, S. (2017). Location of the tempo moiety of TEMPO-PC in lipid bilayers. *Biophysical Journal, 113*, 966–969.

Senzel, L., Gordon, M., Blaustein, R. O., Oh, K. J., Collier, R. J., & Finkelstein, A. (2000). Topography of diphtheria toxin's T domain in the open channel state. *The Journal of General Physiology, 115*, 421–434.

Senzel, L., Huynh, P. D., Jakes, K. S., Collier, R. J., & Finkelstein, A. (1998). The diphtheria toxin channel-forming t domain translocates its own NH_2-terminal region across planar bilayers. *The Journal of General Physiology, 112*, 317–324.

Vargas-Uribe, M., Rodnin, M. V., Kienker, P., Finkelstein, A., & Ladokhin, A. S. (2013). Crucial role of H322 in folding of the diphtheria toxin T-domain into the open-channel state. *Biochemistry, 52*, 3457–3463.

Vargas-Uribe, M., Rodnin, M. V., & Ladokhin, A. S. (2013). Comparison of membrane insertion pathways of the apoptotic regulator Bcl-xL and the diphtheria toxin translocation domain. *Biochemistry, 52*, 7901–7909.

Vargas-Uribe, M., Rodnin, M. V., Ojemalm, K., Holgado, A., Kyrychenko, A., Nilsson, I., et al. (2015). Thermodynamics of membrane insertion and refolding of the diphtheria toxin T-domain. *The Journal of Membrane Biology, 248*, 383–394.

Vasquez-Montes, V., Gerhart, J., King, K. E., Thevenin, D., & Ladokhin, A. S. (2018). Comparison of lipid-dependent bilayer insertion of pHLIP and its p20g variant. *Biochimica et Biophysica Acta, Biomembranes, 1860*, 534–543.

Vasquez-Montes, V., Vargas-Uribe, M., Pandey, N. K., Rodnin, M. V., Langen, R., & Ladokhin, A. S. (2019). Lipid-modulation of membrane insertion and refolding of the apoptotic inhibitor bcl-xl. *Biochimica et Biophysica Acta. Proteins and Proteomics, 1867*, 691–700.

Wang, Y., Malenbaum, S. E., Kachel, K., Zhan, H. J., Collier, R. J., & London, E. (1997). Identification of shallow and deep membrane-penetrating forms of diphtheria toxin T domain that are regulated by protein concentration and bilayer width. *The Journal of Biological Chemistry, 272*, 25091–25098.

Wang, J., Rosconi, M. P., & London, E. (2006). Topography of the hydrophilic helices of membrane-inserted diphtheria toxin T domain: TH1-TH3 as a hydrophilic tether. *Biochemistry, 45*, 8124–8134.

Weiss, M. S., Blanke, S. R., Collier, R. J., & Eisenberg, D. (1995). Structure of the isolated catalytic domain of diphtheria toxin. *Biochemistry*, *34*, 773–781.

White, S. H., & Wimley, W. C. (1998). Hydrophobic interactions of peptides with membrane interfaces. *Biochimica et Biophysica Acta*, *1376*, 339–352.

White, S. H., Wimley, W. C., Ladokhin, A. S., & Hristova, K. (1998). Protein folding in membranes: Determining the energetics of peptide-bilayer interactions. *Methods in Enzymology*, *295*, 62–87.

Wimley, W. C., & White, S. H. (2000). Determining the membrane topology of peptides by fluorescence quenching. *Biochemistry*, *39*, 161–170.

Youle, R. J., & Strasser, A. (2008). The BCL-2 protein family: Opposing activities that mediate cell death. *Nature Reviews Molecular Cell Biology*, *9*, 47–59.

Zakharov, S. D., & Cramer, W. A. (2004). On the mechanism and pathway of colicin import across the e. coli outer membrane. *Frontiers in Bioscience*, *9*, 1311–1317.

CHAPTER THIRTEEN

Cysteine-based crosslinking approach for characterization of oligomeric pore-forming proteins in the mitochondrial membranes

Zhi Zhang[a], Bo Huang[b], Xuejun C. Zhang[b], and Jialing Lin[a,c,*]

[a]Department of Biochemistry and Molecular Biology, University of Oklahoma Health Sciences Center, Oklahoma City, OK, United States

[b]Institute of Biophysics, Chinese Academy of Sciences, Beijing, China

[c]Stephenson Cancer Center, University of Oklahoma Health Sciences Center, Oklahoma City, OK, United States

*Corresponding author: e-mail address: jialing-lin@ouhsc.edu

Contents

Abstract

Mitochondria are important not only to healthy but also dying cells. In particular, apoptotic cell death initiates when the mitochondrial outer membrane is permeabilized by Bax, a protein of the Bcl-2 family. Bax shares a structural fold with some α-helical bacterial pore-forming toxins before these proteins actively engage membranes. Despite decades of intensive research, the structures of the pores formed by these proteins are mostly unknown, mainly because the pores are assembled by different numbers of the proteins whose conformation and interaction are highly dynamic. Site-specific crosslinking of the pore-forming proteins in cellular membranes where the pores are

Methods in Enzymology, Volume 649
ISSN 0076-6879
https://doi.org/10.1016/bs.mie.2021.01.023

assembled is a powerful approach to assess the biological pore structure, dynamics and function. In this chapter, we describe a cysteine-based site-specific crosslinking protocol for the Bax protein in the mitochondrial membrane. We discuss the expected results and the resulting structural-functional models for the pore-forming Bax oligomer, in comparison with other crosslinking approaches that have been used to study other mitochondrial protein complexes. At the end, we highlight the advantages of the crosslinking approaches as well as the limitations and alternative approaches.

1. Introduction

Mitochondria are the most important organelle in all eukaryotic cells. They are factories where many metabolic reactions take place, where many proteins are either produced within from the mitochondrial genome or imported from the cytosol after they are produced based on the cellular genome in the nucleus, where many lipids begin or end their synthesis, and where all the ATP is produced as the most used energy token for the cell. To produce ATP, an electrochemical gradient is generated across the inner membrane of mitochondria, resulting in a proton flow that powers the ATP synthase to covert ADP to ATP. Mitochondria also have an outer membrane. The double membranes segregate the mitochondria to two compartments, the matrix surrounded by the inner membrane and the intermembrane space between the inner and outer membranes, in which different reactions occur generating different products. The two membranes also carry out different reactions and have their own goods to make.

It is hence not a surprise that the permeability of both mitochondrial membranes is tightly controlled to allow certain, but not other, molecules to cross. The examples are voltage-dependent anion channel (VDAC) in the outer membrane and adenine nucleotide translocator (ANT) in the inner membrane that have regulated conductance for ions, metabolites and ATP-ADP exchange. The ATP synthase also has a channel to let protons flow through, producing a force that spins a transmembrane rotor to power the enzyme. In contrast to these integral membrane proteins that have a small channel to let ions or small molecules to cross the membranes in a controllable way, i.e., the voltage-gated ion channel of VDAC, there are large pores in the mitochondrial membranes. The translocases in the outer and inner membranes (TOM and TIM) contain pores that are large enough to allow proteins, at least in their extended unfolded state, to move from the cytoplasm to the mitochondrial intermembrane space or matrix. Comparing to these large pores that are part of the healthy cells, even larger pores are formed in the stressed cells. For example, the mysterious mitochondrial permeability transition pore, and the deadly pores formed by Bax and Bak proteins of the Bcl-2 family in the outer membrane that can be tens or even hundreds of nanometers wide in the tubular mitochondria whose dimension is in the order of micrometer (McArthur et al., 2018; Salvador-Gallego et al., 2016).

These mega pores are formed by oligomers of the same protein, i.e., Bax. Because the pores are oligomeric in nature, and heterogeneous in size, conventional structural biology approaches such as X-ray crystallography and nuclear magnetic resonance are not very useful to solve their three-dimensional structures in toto. The revolutionary cryo-electron microscopy is useful to review structures of large complexes that are homogeneous in size and shape. As an example, Gasdermin pore structure that may be formed in the mitochondrial membranes to release mitochondrial proteins and DNA to trigger inflammatory cytokine secretion and cell death was determined recently (Ruan, Xia, Liu, Lieberman, & Wu, 2018). In comparison to the Gasdermin pore that is formed by 27 or 28 Gasdermin molecules and has a diameter of 18 nm, Bax pore sizes are much larger and much more variable, perhaps formed by tens to hundreds of Bax proteins that interacted via variable regions for variable durations (Cosentino & Garcia-Saez, 2017). Thus, even cryo-electron microscopy may not be good enough to solve their structures.

Fortunately, crosslinking techniques come to the rescue, since they can detect the proximity between two sites in the same protein and hence the conformation and dynamics of the protein, or the proximity of two or more proteins in a complex and hence the protein-protein interaction and dynamics. More importantly, they can monitor the conformation and interaction as well as the dynamics of the proteins and complexes in situ, e.g., in the mitochondrial outer membrane where the conformationally dynamic Bax proteins form the structurally dynamic oligomeric pore. And, they are relatively cheaper, easy and fast to do as we and other have demonstrated. For example, we used site-specific photo and chemical crosslinking to investigate the homo-oligomeric Bax interactions in the mitochondrial membranes as well as the interactions of Bax with other proteins of the Bcl-2 family (Chi et al., 2020; Ding et al., 2014; Zhang et al., 2016). Kluck and colleagues analyzed the Bak oligomerization in the mitochondria isolated from cells using disulfide crosslinking (Dewson et al., 2008; Uren et al., 2017). Walensky laboratory coupled chemical crosslinking with mass spectrometry to study the Bax oligomers induced by detergent (Hauseman et al., 2020).

Here, we describe a cysteine-based site-specific crosslinking protocol that we have used to study the pore complex formed by active Bax proteins in the mitochondria isolated from animal tissue. We discuss the crosslinking data that we have used to model the structures of different dimeric interfaces in the oligomeric pore. We outline cautions that one must take when interpreting the crosslinking data in order to draw correct conclusions about the protein interaction detected by the crosslinking. Furthermore, we compare different crosslinking methods that have been used to study other mitochondrial protein complexes. At the end, we list the advantages and limitations of the crosslinking approach, and the alternative approaches that can also be used to study functional protein interactions in the mitochondrial membranes.

2. Materials and equipment

2.1 Materials

2.1.1. Reagents for molecular biology

- Vectors for construction of DNA plasmids for *in vitro* protein synthesis, e.g., pSPUTK (Stratagene) or pGEM series (Promega). These vectors contain promoters for T7 and SP6 RNA polymerases that flank a multiple cloning region where a protein-coding DNA sequence can be inserted. *In vitro* transcription of the resulting plasmid generates the mRNA. And in turn, *in vitro* translation of the mRNA generates the protein
- PCR reagents to produce the protein-coding DNA sequence with restriction endonuclease sites on both ends
- Restriction endonucleases, and DNA ligase to clone the protein-coding DNA sequence into the vector
- *E. coli* strains, growth media, antibiotics, transformation reagents, and plasmid DNA preparation kit (QIAprep Spin Miniprep kit, Qiagen) to produce the protein-coding plasmid

2.1.2. Reagents for protein synthesis

Note: Nuclease-free ultrapure water (NF H_2O) is obtained from Milli-Q Direct 8 water purification system (Millipore), autoclaved in bottles that are baked at 210 °C for overnight, and stored at room temperature (~22 °C), unless indicated otherwise. All solutions are prepared in NF H_2O with chemicals of the highest purity from Sigma or Thermo Fisher Scientific, filtered with 0.2 μm filter, and stored at 4 °C, unless indicated otherwise.

- Reagents for *in vitro* transcription to produce mRNAs, and *in vitro* translation to produce proteins are previously described (Lin, Johnson, & Zhang, 2019).
- Transcription/Translation-coupled (TNT) wheat germ extract system (Promega), see 3.1.3
- Cycloheximide, 200 μg/mL, prepare 1 mL, aliquot 10 μL for multiple uses, store at −20 °C

2.1.3. Reagents for activation and localization of proteins to mitochondria

- BH3 peptide: containing Bax residues 53–86 as described before (Tan et al., 2006), solubilized 10 mM in DMSO, dilute to 370 μM with NF H_2O, aliquot 30 μL for single use, store at −80 °C
- Mitochondria: isolated from *bak* knockout mouse liver as described (Pogmore, Pemberton, Chi, & Andrews, 2016), aliquot 10 μL for single use, stored at −80 °C
- AT-KCl buffer, prepare 50 mL according to Table 1, store at 4 °C
- N-ethylmaleimide (NEM), 1 and 0.4 M in 100% ethanol, prepare 1 mL in a 1.5-mL microfuge tube, wrap the tube with aluminum foil, store at 4 °C
- Note: Stop use the NEM solution when the color changes to green.
- Succinate, 1 M, prepare 10 mL, store at 4 °C
- AT-KCl-Succinate buffer, prepare 50 mL by mixing 0.25 mL 1 M succinate with 49.75 mL AT-KCL buffer so that [succinate] = 5 mM, store at 4 °C
- ATP, 0.1 M, prepare 1 mL, aliquots 10 μL for single use, store at −80 °C

Table 1 AT-KCl buffer.

Component	Volume (mL)	Final concentration
NF H_2O	29.8	
1 M Trehalose	15	300 mM
1 M HEPES, adjust to pH 7.5 with 5 M KOH	0.5	10 mM
1 M KCl	4	80 mM
0.5 M EGTA, adjust to pH 7.7 with 5 M NaOH	0.1	1 mM
0.5 M EDTA, adjust to pH 8.0 with 5 M NaOH	0.1	1 mM
10% (*w*/*v*) BSA	0.5	0.1% (w/v)
Final Volume	50	

- Phosphocreatine (CP), 4 mM, prepare 1 mL, aliquots 10 μL for single use, store at −80 °C
- Creatine phosphokinase (CPK), 4 mg/mL, prepare 1 mL, aliquots 10 μL for single use, store at −80 °C
- DTT, 50 mM, prepare 1 mL, aliquots 20 μL for single use, store at −20 °C
- Low Salt buffer, prepare 50 mL according to Table 2, filter with 0.2 μm filter, store at 4 °C

2.1.4. Reagents for protein crosslinking

- $NaAsO_2$, 0.2 M, prepare 1 mL, store at 4 °C
- Copper (II)(1,10-Phenanthroline)$_3$ (CuPhe) solution, prepare according to Table 3, aliquot 0.5 mL in microfuge tubes for multiple uses, wrap the tubes with aluminum foil, store at 4 °C
- Quench solution-I, freshly prepare according to Table 4, enough to stop 20 crosslinking reactions in 3.3.1, keep at room temperature before use
- Quench solution-II, prepare in a 15-mL tube according to Table 5, wrap the tube with aluminum foil, store at 4 °C
- Sample buffer, prepare according to Table 6, store at 22 °C
- CCl_3COOH, 25% (w/v), prepare 100 mL, store at 4 °C
- Acetone-HCl solution, mix 90 mL acetone and 1 mL 0.1 M HCl, store at 4 °C
- Reagents for SDS Polyacrylamide Gel Electrophoresis (SDS-PAGE) are previously described (Lin et al., 2019)
- [Methyl-^{14}C] methylated protein molecular weight markers (PerkinElmer)
- PageRuler prestained protein ladder (Thermo Fisher Scientific)

Table 2 Low salt buffer.

Component	Volume (mL)	Final concentration
NF H_2O	47.32	
1 M HEPES, adjust to pH 7.5 with 5 M KOH	1.25	25 mM
4 M CH_3COOK, adjust to pH 7.5 with 5 M KOH	1.38	110 mM
1 M $Mg(CH_3COO)_2$	0.05	1 mM
Final Volume	50	

Table 3 CuPhe solution.

Component	Quantity	Final concentration
NF H_2O	8 mL	
100% ethanol	2 mL	20% (v/v)
$CuSO_4$	0.075 g	30 mM
1,10-Phenanthroline	0.18 g	100 mM
Final Volume	10 mL	

Table 4 Quench solution-I.

Component	Volume (μL)	Final concentration (mM)
0.5 M EDTA, adjust to pH 8.0 with 5 M NaOH	216	390
0.4 M NEM	55.2	80
0.2 M $NaAsO_2$	5.5	4
Final Volume	276.7	

Table 5 Quench solution-II.

Component	Volume (mL)	Final concentration (mM)
NF H_2O	4.9	
0.5 M EDTA, adjust to pH 8.0 with 5 M NaOH	4	200
0.4 M NEM	1	40
0.2 M $NaAsO_2$	0.1	2
Final Volume	10	

2.2 Equipment

- Pipettes: Rainin Classic, 1000 μL, 200 μL, 20 μL, and 10 μL
- Pipet-Aid pipette controller: Drummond Scientific
- Centrifuge: Eppendorf, model 5415R, refrigerated with rotor for 1.5-mL microfuge tube
- Water bath incubator: Fisher Scientific, model Isotemp 215

Table 6 Sample buffer.

Component	Volume (mL)	Final concentration
NF H_2O	6.9	
1 M Tris	12	240 mM
0.5 M EDTA, adjust to pH 8.0 with 5 M NaOH	1.5	15 mM
100% (v/v) glycerol	15	30% (v/v)
25% (w/v) SDS	14.6	7.3% (w/v)
Add 2 mg bromophenol blue		0.004% (w/v)
Final Volume	50	

- Dry bath incubator: Fisher Scientific
- Vortex: Fisher Scientific, model Genie 2
- Vacuum concentrator: Thermo Fisher Scientific, model SpeedVAC
- Vertical gel electrophoresis apparatus: GibcoBRL, model V16
- Power supply: Bio-Rad, model PowerPac 3000
- Gyrotory shaker: New Brunswick, model G2
- Gel dryer: Bio-Rad, model 583
- Vacuum pump: Precision Scientific, model DD90
- Phosphor-imager: Fujifilm FLA-9000 multipurpose image scanner
- Phosphor-imaging plate, Fujifilm BAS storage phosphor screen, type MS, 20 × 25 cm
- UV/Vis spectrophotometer: Beckman, model DU520
- Benchtop pH meter: Orion, model Star A111
- Balances: Ohaus, models Standard and Explorer Precision
- Programmable thermal controller: MJ Research, model PTC-100
- Refrigerators: 4 °C, −20 °C, and −80 °C

3. Protocol

3.1 In vitro protein synthesis

Timing: Week 1 and Week 2, Day 1

3.1.1. Construction of DNA plasmids for *in vitro* synthesis of Bax proteins with no, single or double Cys residues

(a) The human Bax DNA was inserted between the *Nco*I and *Eco*RV sites in pSPUTK vector (previously available from Stratagene; now available from the corresponding author upon request; can be replaced by other vectors that are currently available, e.g., pGEM series from Promega). As we described previously (Ding et al., 2014; Zhang et al., 2016), the resulting plasmid with the start codon of the Bax DNA overlaps with the NcoI site (CCATGG) and the stop codon 5′ to the EcoRV site was used to generate Bax RNA *in vitro* through SP6 RNA polymerase catalyzed transcription. And in turn, the Bax mRNA was used to produce Bax protein *in vitro* through wheat germ extract or rabbit reticulocyte lysate-based translation system. Both translation systems are commercially available (Promega). Alternatively, we used the Bax plasmid to produce the Bax protein *in vitro* using transcription/translation coupled (TNT) system based on wheat germ extract or rabbit reticulocyte lysate (Promega) (Chi et al., 2020; Zhang et al., 2016). Since the TNT system is the simplest one to produce proteins, we describe it in 3.1.3

(b) To construct the plasmids coding for Bax mutants with no, single or double Cys residues, the Bax plasmid described in (a) was mutated, first by changing the two endogenous Cys codons to Ser codons to generate the Cys-null Bax plasmid (Bax C0), and second by changing codon(s) at specified position(s) to Cys codon(s) to generate the single or double-Cys Bax plasmid(s), e.g., the single-Cys Bax I31C, E69C, or E146C or double-Cys Bax L59C/M79C plasmid with the codon(s) for Ile^{31}, Glu^{69}, Glu^{146} or Leu^{59} and Met^{79} changed to the codon(s) for Cys, respectively. QuickChange kit (Agilent) was used to generate these Bax mutant plasmids according to the manufacturer's protocol. All the resulting plasmids were verified by DNA sequencing.

3.1.2. Expression and purification of DNA plasmids for the single or double-Cys, or Cys-null Bax

(a) The plasmids from 3.1.1 were amplified in *E. coli* DH5α or XL1-Blue cells and purified using QIAprep Spin Miniprep kit (Qiagen). The resulting plasmids in NF H_2O were quantified by measuring A_{260} of 50-fold diluted samples and calculating the concentrations of the undiluted stocks using the following equation:

$$[Plasmid\ Stock] = A_{260} \times 2.5\ (\mu g/\mu l)$$

(b) The plasmid stocks were then diluted so that the [plasmid] = 0.15 µg/µl; store at −20 °C for multiple uses.

Critical: The A_{260}/A_{280} ratio obtained from the diluted plasmids must be 1.8–2.0 so that the purity of the plasmids is sufficient for the TNT reaction in 3.1.3.

Pause Point: Pause for weekend.

Timing: Week 2, Day 1, Hour 1–2.

3.1.3. *In vitro* synthesis of single or double-Cys, or Cys-null Bax proteins using the transcription/translation-coupled (TNT) wheat germ extract system (Promega) by following the manufacturer's protocol. An example is given below

Table 7 Master TNT reaction.

Component[a]	Volume (µL)	Final concentration
NF H_2O	10.8	
TNT Wheat Germ Extract	25	1/2 dilution
TNT Reaction Buffer	2	1/25 dilution
TNT RNA Polymerase (SP6)	1	1/50 dilution
Amino Acid Mixture, Minus Methionine, 1 mM	1	20 µM
RNasin Ribonuclease Inhibitor, 40 U/µL	1	0.8 U/µL
[^{35}S]Methionine, 40.2 µCi/µL[b]	2.5	2 µCi/µL
Gently mix by pipetting, and centrifuge briefly		
Final Volume	43.3	For 10 5-µL TNT reaction

[a]*Note*: Add the components to the microfuge tube according to the order in Table 7.
[b]*Note*: [^{35}S]Methionine decays and the radioactivity at the time of TNT reaction is calculated using the following equation:

$$A_t = A_0 e^{\left(-0.693T/T_{1/2}\right)}$$

For this example: A_T = the radioactivity at the time of TNT reaction = 40.2 µCi/µL, A_0 = the radioactivity at the time specified by the manufacturer = 43.5 µCi/µL, T = the time between the manufacturer specified and the TNT reaction = 10 days, and $T_{1/2}$ = the time for the radioactivity of [^{35}S] isotope to decay to half of the manufacturer specified activity = 87.4 days.

a. Assemble a master TNT reaction in a 1.5-ml microfuge tube on ice for 10 5-μL individual reactions according to Table 7.

(b) Aliquot 4.3 μL of the master TNT reaction for each individual reaction to a 1.5-mL microfuge tube, and add 0.7 μL of 0.15 μg/μL plasmid for the single or double-Cys, or Cys-null Bax so that [plasmid] = 0.02 μg/μL.

(c) Conduct the TNT reaction in a 30 °C water bath for 90 min to produce the single or double-Cys, or Cys-null Bax protein.

(d) Stop the TNT reaction by adding 0.5 μL 200 μg/mL cycloheximide so that [cycloheximide] = 20 μg/mL, and incubating at 22 °C for 10 min.

Critical: Handling RNA: mRNAs and tRNAs for *in vitro* translation are very labile. Nucleases are secreted by humans, and transferred in our finger oil to any surface we touch. We therefore must avoid any contact between a mRNA or tRNA solution and any surface that may have been touched by us or any solution that may have been contaminated by nucleases. When transcriptions or translations do not work, it is invariably because of a nuclease contamination that degrades the mRNAs and tRNAs. Thus, stringent nuclease-free solutions and techniques are absolutely the most critical requirement for achieving successful transcription and translation.

3.2 *In vitro* targeting Bax proteins to mitochondria and fractionation

3.2.1. Isolation of mitochondria from *bak* knockout mouse liver as described before (Pogmore et al., 2016). The resulting mitochondria lack not only Bak protein due to the gene knockout but Bax protein due to its cytoplasmic localization in the liver cells. Therefore, there is no endogenous Bax and Bak in the mitochondria that can form complexes with the single or double-Cys Bax, and reduce the number of Bax complexes that can be crosslinked via the introduced Cys residues.

Timing: Week 2, Day 1, Hour 3–4.

3.2.2. Preparation of mitochondria before use for Bax targeting.

(a) 10 min before the TNT reactions complete, take one aliquot of frozen mitochondria in 1.5-mL microfuge tube (~10 μL, [protein] = 50 mg/mL, in AT-KCl buffer without KCl) from −80 °C; thaw the mitochondria by holding the tube between fingers; place on ice.

Note: 10 μL of the frozen mitochondria are enough for targeting of six different Bax proteins; increase accordingly for experiments with more proteins.

(b) Dilute the mitochondria by adding 490 μL AT-KCl buffer so that [protein] = 1 mg/mL; gently mix by pipetting; briefly centrifuge.

(c) Treat the mitochondria with NEM to block sulfhydryls in the mitochondrial proteins so they will not interfere with the crosslinking reaction via sulfhydryls of Bax proteins in 3.3: Add 9 μL 1 M NEM to the mitochondria so that [NEM] = 18 mM; incubate on ice for 10 min; centrifuge at 5000*g* and 4 °C for 2 min; wash the mitochondria pellet with 500 μL AT-KCl buffer by gently pipetting; centrifuge again to pellet the mitochondria.

(d) Resuspend the mitochondria pellet with 250 μL AT-KCl-Succinate buffer by gently pipetting until there are no visible clumps of mitochondria; briefly centrifuge; the resulting mitochondria's [protein] = 2 mg/mL.

(e) Add 5 μL 0.1 M ATP, 0.6 μL 4 mg/mL CPK, and 0.6 μL 4 mM CP to the mitochondria so that [ATP] = 2 mM, [CPK] = 10 μg/mL, and [CP] = 10 μM; gently mix by pipetting; briefly centrifuge; keep the mitochondria on ice and use within 2 h.

3.2.3. Activation and localization of Bax proteins to mitochondria

(a) To each TNT reaction from 3.1.3, add 35 μL of the mitochondria from 3.2.2, 0.9 μL 50 mM DTT, and 1.7 μL 370 μM BH3 peptide so that the mitochondrial [protein] = 1.6 mg/mL, [DTT] = 1 mM, and [BH3 peptide] = 15 μM; gently mix by tapping the tube; briefly centrifuge; incubate at 37 °C for 60 min to allow the BH3 peptide to activate the soluble Bax proteins resulting in integral Bax proteins that interact to form oligomeric pores in the mitochondrial outer membrane.

(b) Separate of the mitochondrial Bax proteins from the soluble proteins by centrifuging the samples from (a) at 13,000*g* and 4 °C for 5 min; use pipette to transfer each supernatant to a 1.5-mL microfuge tube for analysis of the soluble Bax proteins, if needed; see (Zhang et al., 2016) for an example.

(c) Resuspend each mitochondria pellet with the associated Bax proteins in 85 μL low salt buffer by gently pipetting.

(d) To each mitochondria sample, add 0.4 μL 0.2 M $NaAsO_2$ so that $[NaAsO_2]=1$ mM; gently mix by pipetting; incubate on ice for 5 min. This treatment decreases residual DTT that otherwise will inhibit the crosslinking in Section 3.3.

3.3 Crosslinking of mitochondrial Bax proteins via cysteine residues

Timing: Week 2, Day 1, Hour 5.

3.3.1. Disulfide crosslinking of the mitochondrial Bax proteins.

(a) Split each sample from 3.2.3 to two 40-μL aliquots in 1.5-mL microfuge tubes.

(b) To one aliquot from each sample, add 0.4 μL CuPhe solution so that $[CuSO_4]=0.3$ mM, and [1,10-phenanthroline] = 1 mM; incubate on ice for 30 min to let CuPhe, a redox catalyst, induce disulfide crosslinking of the Bax proteins.

(c) To stop each crosslinking reaction, add 10 μL of freshly prepared Quench solution-I so that [EDTA] = 77 mM, [NEM] = 15.5 mM, and $[NaAsO_2]=0.8$ mM; incubate at 37 °C for 10 min. The resulting samples contain Bax proteins that are crosslinked for "30 min."

(d) To the other aliquot from each sample, add 10 μL Quench solution-I to block the disulfide crosslinking before adding 0.4 μL of CuPhe solution; incubate on ice for 30 min and then at 37 °C for 10 min. The resulting samples contain Bax proteins that are crosslinked for "0 min," serving as the negative controls.

(e) Add 65 μL 25% (*w*/*v*) CCl_3COOH to each sample from (c) and (d) and incubate on ice to precipitate the proteins.

Critical: 3.3.1 must be done as soon as possible to avoid oxidation of the samples by oxygen in the solutions prior to the crosslinking reactions, which increases the background crosslinked protein bands in the "0 min" controls on the phosphor-images of SDS-PAGE gel.

Pause Point: Pause for overnight.

3.4 SDS-PAGE and phosphor-imaging of crosslinked Bax proteins

Timing: Week 2, Day 2, Hour 1–6.

3.4.1. Sample preparation for SDS-PAGE.

(a) Precipitate the proteins in the samples from 3.3 by centrifuging at 13,000*g* and 4 °C for 15 min; remove the supernatants by pipetting; wash each protein pellet with 350 μL Acetone-HCl solution by vortex; centrifuge again and remove the supernatants; dry the protein pellets in vacuum concentrator at 22 °C for 20 min.

(b) Mix 250 μL Quench solution-II with 250 μL Sample buffer at 22 °C.

(c) Resuspend each pellet in 25 μL of the Quench solution-II/ Sample buffer mixture by vortex; briefly centrifuge.

(d) Aliquot 10 μL from each "30 min" sample; add 1 μL 2-mercaptoethanol so that [2-mercaptoethanol] = 10% (*v*/v); vortex and briefly centrifuge, resulting in the reduced "30 min" samples.

(e) Mix 1 μL of [Methyl-^{14}C] methylated protein molecular weight markers with 9 μL of Quench-Sample Buffer.

(f) Incubate all the "30 min" crosslinked and reduced, "0 min" crosslinked samples, and the protein markers at 65 °C for 30 min; briefly centrifuge.

3.4.2. SDS-PAGE

(a) Set up SDS-PAGE apparatus according to laboratory or manufacture protocol.

Note: We make gels for SDS-PAGE according to Lin et al. (2019).

(b) Load 10 μL of the samples from 3.4.1 per well in gel.

Critical: Separate the reduced sample set from the nonreduced crosslinked sample set and the protein markers by a well loaded with some leftover mitochondria from 3.2.3 solubilized in Quench solution-II/Sample buffer mixture without 2-mercaptoethanol as described in 3.4.1. Otherwise, the crossover reducing agent from the reduced samples will affect the migration or integrity of the crosslinked samples during SDS-PAGE.

Alternative: Nonradioactive PageRuler prestained protein ladder (5 μL mixed with 5 μL of Quench solution-II/Sample buffer mixture) can be used instead of the radioactive protein molecular weight markers. Since the bands of these proteins are visible on the gel, their positions can be painted by radioactive ink made with the leftover samples from 3.4.8 that contains radioactive Bax proteins after the gel is dry on a filter paper. The radioactive paint allows the positions of the non-radioactive protein ladder to be detected by phosphor-imaging in 3.4.10.

(c) Perform SDS-PAGE according to laboratory or manufacture protocol to separate the crosslinked Bax proteins from the monomers.

Note: We perform SDS-PAGE according to Lin et al. (2019).

Timing: Week 2, Day 2, Hour 7–8; and Day 3–5.

3.4.3. Phosphor-imaging

(a) Perform phosphor-imaging according to Lin et al. (2019) to detect the radioactive Bax proteins and their crosslinked products. A phosphor-image acquired from Fujifilm FLA-9000 Multipurpose Image Scanner controlled by Fujifilm Image Reader FLA-9000 Program is shown in Fig. 1A.

(b) Analyze the phosphor-image using Fujifilm Multi Gauge Program. A structural model that fits the crosslinking data from the Bax I31C mutant is shown in Fig. 1B. Other structural models that fit the crosslinking data from Bax E69C and L59C/M79C mutants are described previously (Zhang et al., 2016).

4. Results and discussion

The most intuitive interpretations of the crosslinking data such as that shown in Fig. 1A are that, if a crosslinked band is detected for two proteins, the proteins form a dimer via the regions where the two Cys residues are located. The stronger the crosslinked band, the stronger the protein-protein interaction, and the more stable and longer lasting protein complex. If no crosslinking band is detected, the proteins do not interact via the Cys-containing regions. However, one should be cautious about these interpretations based on the following considerations.

The crosslinking between two proteins would occur when a Cys in one protein is close to a Cys in other protein in a complex formed by the two proteins. For example, a disulfide crosslinking occurs between two single-Cys Bax E69C proteins, each with a Cys replaced Glu^{69} in the sequence, as shown in Fig. 1A, lane 8. As expected, the apparent molecular weight of the crosslinked proteins is about twice of each protein's molecular weight. The crosslinked dimer shifts to the monomer when the sample is reduced as

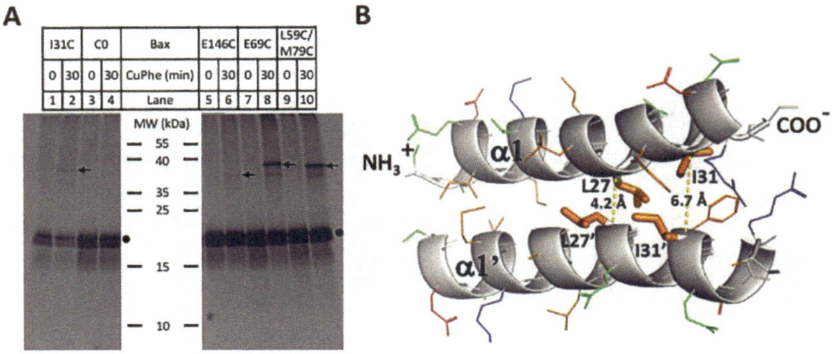

Fig. 1 Crosslinking of the mitochondrial Bax proteins and a dimer model for Bax α-helix 1. (A) The *in vitro* synthesized [^{35}S]Met-labeled single-Cys (I31C, E69C or E146C) or double-Cys (L59C/M79C), or Cys-null (C0) Bax protein was activated and localized to the mitochondria that were then isolated. CuPhe solution was added to the resulting mitochondria to induce disulfide crosslinking of the Bax proteins for 30 min, and Quench solution was then added to stop the crosslinking reactions, resulting in the "30 min" crosslinked samples. For the "0 min" crosslinked controls, Quench solution was added before the addition of CuPhe. All the samples and controls were analyzed by non-reducing SDS-PAGE and phosphor-imaging. In the resulting phosphor-images, the closed circles indicate Bax monomers, the arrows indicate disulfide-linked Bax dimers, and the molecular weights (MW) indicate the positions of prestained protein ladder. (B) A model for a dimer formed two parallel Bax α-helix 1 structures was generated computationally to fit the crosslinking data obtained from Bax I31C shown in panel A and other Bax mutants with Cys residues positioned in the α-helix 1 (data not shown) using the method we described before for modeling of another dimer formed by Bax α-helices 2, 3, and 4 to fit the crosslinking data from Bax E69C and other mutants; see fig. 3 in Zhang et al. (2016). PyMOL Molecular Graphics System (Schrödinger, LLC) was used to depict the model. The two helices are indicated by α1 and α1′. The termini of α1 are labeled by NH_3^+ and COO^-. The amino acid residues are shown as thin lines, except of the residues, I31 in α1 and I31′ in α1′, which formed a disulfide bond after they were replaced by Cys are shown as thick sticks with their α carbon atoms linked by a dashed line with the distance indicated. For Discussion below, the L27 and L27′ are depicted as the I31 and I31′.

described in 3.4.1 d) before SDS-PAGE as expected from a disulfide-linked dimer; see this data in fig. 3A and the related Appendix fig. S2B of Zhang et al. (2016). The crosslinking does not occur in the "0 min" crosslinked control (lane 7) so the oxidation of the sulfhydryls of the Cys residues is required for the disulfide bond to form. No crosslinking is detected with the Cys-null Bax C0 protein (lanes 3–4) so the crosslinking is dependent on the Cys residue. The disulfide crosslinking also occurs between two double-Cys Bax L59C/M79C proteins, each with two Cys substituted the Leu^{59} and Met^{79}, and requires the oxidation and dependent on the Cys residues as well (lanes 9–10). The other single-Cys mutant, Bax I31C or E146C, also generates a disulfide crosslinked dimer (lanes 1–2 or 5–6), albeit at much lower levels than the E69C dimer.

The negative "0 min" crosslinked control is necessary for one to claim the positive crosslinking detected in the respective "30 min" crosslinked sample. Occasionally, one may see a crosslinked protein band in the "0 min" control. An interpretation of this false positive data would be that the control sample is oxidized such that the proteins are crosslinked before the Quench solution is added in 3.3.1 d) to block the crosslinking. To avoid this, on should use the fresh-made sample under reducing condition, which is Critical as noted in 3.3.1. Alternatively, the Quench solution may deteriorate, thereby unable to block the crosslinking. To counter that, we prepare the Quench solution immediately prior to the use in 3.3.1 mainly because one of the components, NEM, is unstable in aqueous solution and sensitive to light, although the NEM stock in ethanol is stable at 4 °C without light.

We have never seen a crosslinked band from the Cys-null Bax. However, if a Cys-null protein generates a "crosslinked" band in the "30 min" lane of a non-reducing SDS-PAGE gel, one should first check whether the band disappears in a reducing SDS-PAGE gel. Second, one should check whether the band also appears in the "0 min" control of the Cys-null protein. Because if the answer is no from the first check, but yes from the second, the band is not generated by the disulfide crosslinking reaction, but rather by other means. For example, the TNT reaction may produce not only the Cys-null protein but a mysterious protein that has the same molecular weight as the crosslinked Cys-null proteins. The mysterious protein and other potential background proteins that the TNT may produce can be easily revealed by running a control TNT without the Cys-null mRNA.

Based on typical geometries of disulfide bonds determined from a survey of known protein structures (Ozhogina & Bominaar, 2009), one may conclude that the crosslinked Cys residues are less than 8 Å apart in terms of the distance between their α-carbons. For example, the distance between the α-carbons of two Cys residues that replace the I31 and I31′ would be 6.8 Å according to the structural model in Fig. 1B. As we documented in fig. 1 of Zhang et al. (2016), the crosslink data from the L59C/M79C and other Cys-containing Bax mutants could be fit by the structure of a symmetrical dimer formed by two Bax regions containing the α-helices 2 to 5, which was determined by crystallography (Czabotar et al., 2013). This led to the conclusion that two active Bax proteins in the mitochondrial outer membrane form a symmetrical dimer structure via their α-helices 2 to 5 regions, similar to the crystal structure (PDB ID: 4BDU). In contrast, the crosslinking via the two E69C residues is unexpected since their α-carbons are 12.2 Å apart in the crystal structure. This crosslinking data suggests that the conformation of the dimer formed by the α-helices 2 to 5 of full-length Bax protein at the mitochondrial membrane is more dynamic than the crystal structure such that the distance between the two E69C residues can be less than 8 Å at least for some time. Alternatively, the dimer may rearrange after the α-helix 5 is released from the dimer and inserts into the membrane, resulting in a dimer without the α-helix 5 in which the two E69C residues are closer than those in the α-helices 2 to 5 dimer. See fig. 3 of Zhang et al. (2016) for more evidence and logic for the dimer transition and a structural model for the dimer without the α-helix 5.

By comparing the arrow-indicated bands in lanes 8 and 10 of Fig. 1A, one may see that the two Bax E69C proteins that are linked by the single disulfide between two Cys^{69} residues migrate slower in the SDS-PAGE gel than the two Bax L59C/M79C proteins that are linked by the double disulfides, one between Cys^{59} and $Cys^{79'}$, and other between Cys^{79} and $Cys^{59'}$. Note that similar to the depiction of the residues in the two protomers of the α-helix 1 dimer in Fig. 1B, the two Cys residues without the prime symbol are of the first protomer in the α-helices 2 to 5 dimer, whereas the two Cys residues with the prime symbol are of the second protomer. These data are expected with two SDS-charged protein complexes that are crosslinked differently and hence have different shapes that run through the gel matrix with different speeds in an electrical field. These data further support the symmetrical dimer model in which the two α-helices 2 to 3 regions are antiparallel (Zhang et al., 2016). Thus, careful analysis of the crosslinking data from a carefully designed experiment can reveal more structural information about a protein complex.

Moreover, the model for another Bax dimer via α-helix 9 is shown in fig. 7 of Zhang et al. (2016). This model is not based a known structure but on the crosslinking data from Bax proteins with Cys residues in the α-helix 9. A computational transmembrane α-helical dimer model built on the conserved GxxxG/A dimerization motif in α-helix 9 fits the crosslinking data. Thus, crosslinking and computational modeling can generate a high-resolution dimer model without a known structure determined by X-ray crystallography, NMR or cryo-EM. With this said, atomic resolution structures of protein complexes determined by structural biologists are highly valuable to design of crosslinking experiments. For example, based on the structures one can position the Cys residues precisely to where the crosslinking can occur to bridge the interacting proteins. Even if a structure that is not of a protein complex but of the protein monomer that can form the complex, it is still useful since one can position the Cys residues on the surface that one suspects the binding would occur. Without any structures, one must fish out the interacting regions by position Cys or other reactive residues such as Lys throughout the interacting proteins, performing tedious crosslinking experiment, and hoping to detect the dimer from some Cys or Lys residues. Then one can structurally model the protein complex, and further test the structural model by more crosslinking with better positioned Cys or Lys residues guided by the model. On the other hand, the crosslinking approach can validate the crystal, NMR and cryo-EM structures to see if they represent the structures form by the full-length proteins and in the cognate membranes. Since most of the crystal, NMR and cryo-EM structures are determined without membranes and of truncated proteins, this validation is essential before one commits more resources to assess functional implications of the structures.

As discussed in context of Bax, crosslinking data can lead to structural models for dimers formed by different regions of an oligomerizing protein. To build a model for an oligomer from two or more dimer models, a first principle is that the oligomer is formed by a protein via multiple dimer interfaces that can be the same or different. For example, using A-B to represent a protein with two binding surfaces, A and B, an oligomer of A-B can be either represented by A-B:A-B:A-B... with the same B:A interface, or by A-B:B-A:A-B... with two different B:B and A:A interfaces. We have used this first principle to build an oligomer model for Bax pore in the mitochondrial membrane as shown in fig. 11 of Zhang et al. (2016). We have validated the oligomer model by crosslinking of Bax proteins via Cys residues located in both α-helices 2 to 5 and α-helix 9, and observing the oligomers crosslinked via both dimer interfaces. Importantly, we have demonstrated the functional relevance of the oligomer model by structure-guided mutagenesis that is expected to disrupt each dimeric interaction, followed by mitochondrial membrane permeabilization assay that reveals the mutations' impact on Bax pore formation.

The distances between the dimers in an oligomer can be estimated from the dimension of the protein monomer and the relative location of the binding surfaces (e.g., the distance between A and B in the protein A-B). If the monomer structure is not known or the structure changes greatly when the monomers form the dimer, one must determine the distance between the binding surfaces experimentally. One may use longer molecular rulers to measure this intramolecular distance, e.g., crosslinking two residues by a bifunctional crosslinker with long spacer arm up to 54 Å (i.e., PEG12-SPDP from Thermo Scientific Pierce), DEER between two paramagnetic centers (up to 60 Å for membrane proteins and 100 Å for soluble proteins (Jeschke, 2012)), and FRET from a donor to an acceptor (up to 170 Å using Thermo Fisher Scientific Alexa Fluor 594/647 donor-acceptor pair with the Forster distance $R_0 = 85$ Å (Johnson, 2005)).

As mentioned above, the crosslinking band generated by Bax I131C or E146C is weaker than that generated by Bax E69C. In general, reasons for "weak" vs "strong" crosslinking include:

(1) A smaller number of a mutant protein is localized to the mitochondria than another mutant. Since the number of protein complexes that are formed by the proteins are dependent on the number of the proteins, the fewer the proteins in the mitochondria, the fewer the protein complexes, and hence, the fewer of the proteins that are crosslinked. This is an interpretation of the data from Bax I131C since fewer proteins are in the mitochondria subjected to the crosslinking, judged by the lower intensity of the protein monomer band in the gel compared to the E69C monomer band. Normalization of the dimer band intensity by the total protein (dimer plus monomer) band intensity can only partially reduce the apparent difference between the crosslinking efficiencies of the two mutants but not between the monomer-dimer equilibria that are determined not only by the binding constant but the concentration of each mutant in the mitochondria.

(2) The interaction via an interface is weaker than that via another interface. This could be the interpretation of the data from Bax E146C because similar numbers of the E146C and E69C mutants are localized to the mitochondria, judged by the similar intensities of their monomer bands. Since the number of the E146C dimer is less than that of the E69C dimer as judged by the intensities of their dimer bands, the dimeric interaction via the E146C-containing regions would be weaker than that via the E69C-containing regions. A similar conclusion was drawn for the interactions of Bak, a structural and functional homolog of Bax, in part based on the crosslinking data (Uren et al., 2017). Thus, the interaction between the core regions of Bak is stronger than that between the other regions.

In general, a stable, and hence static, structure of a complex would increase the crosslinking between two proteins in the complex, resulting in more crosslinked dimers that can be detected by SDS-PAGE and phosphorimaging, western blot or dye staining. On the other hand, a less stable, and hence dynamic or transient, structure of a complex would decrease the crosslinking, resulting in fewer crosslinked dimers. This is how a crosslinking approach could distinguish these different protein complexes. However, whether one sees more or fewer crosslinked dimers on a gel would also depend on other factors, e.g., (1) above, and (3–4) below. Thus, to one extreme, the crosslinking would not distinguish the complexes. But to the other extreme, if all the other factors are the same, such that in a well-controlled experiment, a protein complex is more or less stable due to a point mutation that does not change any other properties of the interacting protein, the crosslinking would distinguish the wild-type complex from the mutated one.

(3) Similar to other point mutations, a Cys mutation in a region may weaken the protein interaction, whereas another Cys mutation in another region may not. To ascertain the Cys mutations do not adversely affect the protein function that is dependent on the protein interaction, one should assay the function of the mutants and compare to that of the wild-type protein. For example, we determined the function of Cys-containing Bax mutants in the mitochondrial outer membrane permeabilization before we used the mutants in the crosslinking experiments to determine the interaction of Bax proteins as shown in figure EV1 of Zhang et al. (2016).

(4) Two Cys residues are worse positioned than the other two Cys residues to detect the same interaction. Since the efficiency of a disulfide formation is dictated by the geometry of two Cys residues, a weaker crosslinking is expected when the two Cys residues are in a less favorable geometry for the disulfide formation. Using the structural model for Bax α-helix 1 dimer as an example, the I31C and I31C′ residues are 6.7 Å apart, perhaps too far away to form a disulfide bond efficiently. One may place the Cys residues to other locations that are closer, e.g., L27C and L27C′ that are 4.2 Å apart, and test if this increased the disulfide formation. The redox state of the environment for the Cys residues also dictates the efficiency of the disulfide formation. Thus, if two sulfhydryls are located in a reduced environment that cannot be oxidized, a disulfide will not form.

Knowing these caveats, one should not conclude whether the dimerization of one protein is of a higher affinity than the dimerization of other protein solely based on the crosslinking band intensity from the one is higher than the other. Since the two Cys-containing proteins are different, and they are both different from the wild-type protein, one should not conclude that the protein-protein interaction via one dimer interface is stronger than the interaction via other interface because the crosslinked band via the two Cys residues positioned in the one interface is stronger than the crosslinked band via the two Cys residues in the other interface.

With all the cautions that one must take to correctly interpret the crosslinking data, a crosslinking study will be useful in the following ways. As mentioned above, a crosslinking study can validate a high-resolution structure of a protein complex. In particular, if the structure is determined form part of the protein, the crosslinking study with the whole protein can determine whether the partial structure is used by the whole protein in the complex formation. More relevant to membrane proteins, such as the mitochondrial Bax protein, the structures are usually determined in solution or in crystals that lack membranes. These structures must be validated by experiments that use the proteins in the membranes, such as the crosslinking experiment with the Bax protein in the mitochondrial membrane. Moreover, there is no atomic resolution structure for any Bax pore in any membrane, so the crosslinking approach has been useful to generate structural models for the pores, which allow further elucidate the structure and function of the pore. Expanding the scope, various crosslinking approaches have been used to study other mitochondrial protein complexes and their dynamic structures and functions such as TIM (Banerjee, Gunsel, & Mokranjac, 2017), and TOM (Shiota et al., 2015; Shiota, Mabuchi, Tanaka-Yamano, Yamano, & Endo, 2011) with the results matched well with the later cryo-EM structures (Araiso et al., 2019; Araiso, Imai, & Endo, 2020; Bausewein et al., 2017; Tucker & Park, 2019). Even more powerful, when crosslinking approaches are combined with mass spectrometry, the mitochondrial protein interactome has been mapped at high structural resolutions, e.g., electron transport chain complexes and ATP synthase (Schweppe et al., 2017). Therefore, we are on shining crossroads of science toward full molecular and even atomic knowledge of a super important organelle that entered our cells as endosymbiotic bacteria 1.45 billion years ago (Martin & Mentel, 2010). No matter which road we forward and which vehicle we drive, fruitful reward will be ample. So, TAKE OFF and ENJOY!

5. Advantages

- Protein production by TNT is faster and cheaper than that by expression in cells
- The radioactive proteins produced from TNT can be detected by phosphor-imaging even at very low levels comparing to immunoblotting that is a less sensitive detection method and required specific antibodies
- Disulfide crosslinking uses "zero length" crosslinker. Longer crosslinkers that can link sulfhydryls are available (e.g., from Thermo Fisher Scientific) to crosslink Cys residues that are more distal in the interface. As an example, we use bismaleimidohexane (BMH), a homo-bifunctional crosslinker with two sulfhydryl-reactive moieties linked by a spacer arm of 13.8 Å, to crosslink Bax and Bim, a Bax activator, via Cys residues placed in the suspected binding surfaces (Chi et al., 2020).
- Hetero-bifunctional crosslinkers with a sulfhydryl-reactive group at one end and a photo-reactive group at the other end (available from Thermo Fisher Scientific) can be used to crosslink two interacting proteins of which one contains a Cys and the other does not. In addition, since photo-crosslinkers are less selective reacting with almost any atoms nearby as discussed above, they are useful to initially screen the binding regions before Cys-based crosslinking that can provide more structural information about the binding (Zhang et al., 2010).
- Distance constraints provided by various crosslinking methods are useful to model interface structures in protein complexes using inhouse or online structure computation programs (Ding et al., 2014; Zhang et al., 2016). The resulting interface structural models can feed molecular dynamics simulations that can refine the models or explore the interaction landscapes (Liao et al., 2016). For more information regarding computation and simulation of interacting protein structures, see (Ozdemir, Nussinov, Gursoy, & Keskin, 2019; Slater, Miller, & Kontoyianni, 2020).
- Cys serves as a specific labeling site in protein, useful for probing the protein conformation and environment with other techniques that requires site-specific labeling, e.g., fluorescence spectroscopy with fluorescent dyes (Johnson, 2005), EPR or NMR spectroscopy with paramagnetic probes, and accessibility to water soluble or lipophilic labeling agents. These techniques are complementary to the crosslinking and hence can verify and expand the research

6. Limitations and alternatives

- The phosphor-imaging requires expensive equipment
- Radioactive materials require license to use and cautions to handle
- The mitochondria from a specific gene-knockout mouse will be clean so one does not worry about the endogenous protein competing with the Cys mutant protein for complex formation, which would reduce the crosslinking via the Cys. One may use the wild-type mouse if the Cys mutant can still form enough complexes for the crosslinking. Or, one may use liposomes without any proteins, but they are not native membranes. If one wants to detect the crosslinking of the Cys mutant with the endogenous wild type protein that does not have any Cys or have Cys but it is not in the right position for the crosslinking, one may use a hetero-bifunctional crosslinker with a sulfhydryl reactive group at one end and a photo reactive group such as aryl azide (e.g., phenyl, hydroxyphenyl, or nitrophenyl azide) or diazirine at the other end, so that the two proteins can be linked by the crosslinker via the Cys in the mutant and an atom in the wild type protein that can be reached by the crosslinker. Note that when aryl azide is activated by ultraviolet photon, it forms a nitrene group that can initiate addition reactions with double bonds, insertion into C-H and N-H sites, or subsequent ring expansion to react with a nucleophile (e.g., primary amines). Photo-activation of diazirine creates reactive carbene intermediates that can form covalent bonds through addition reactions with any amino acid side chain or peptide backbone. These reagents are available from Thermo Fisher Scientific
- The crosslinking results are qualitative as discussed in Section 4. In comparison, FRET-based *in vitro* binding assay is quantitative (Lovell et al., 2008; Pogmore et al., 2016). FRET-based microscopy can even visualize and quantify protein interactions in live cells (Osterlund, Liu, & Andrews, 2015). However, it is difficult to distinguish the interactions via different binding interfaces in an oligomeric protein complex using FRET. Using the Bax oligomer as an example, when a FRET is detected between two fluorescent Bax proteins, is the binding mediated by the two α-helices 2 to 5 regions or the two α-helix 1 regions?

Acknowledgments

We thank Dr. David Andrews at Sunnybrook Research Institute for providing the mitochondria. This work was supported by a grant from National Institutes of Health grant (R01GM062964) to J.L., and by an Institutional Development Award from the National Institute of General Medical Sciences of National Institutes of Health (P20GM103640).

References

Araiso, Y., Imai, K., & Endo, T. (2020). Structural snapshot of the mitochondrial protein import gate. *The FEBS Journal*. https://doi.org/10.1111/febs.15661.

Araiso, Y., Tsutsumi, A., Qiu, J., Imai, K., Shiota, T., Song, J., et al. (2019). Structure of the mitochondrial import gate reveals distinct preprotein paths. *Nature*, *575*(7782), 395–401. https://doi.org/10.1038/s41586-019-1680-7.

Banerjee, R., Gunsel, U., & Mokranjac, D. (2017). Chemical crosslinking in intact mitochondria. *Methods in Molecular Biology*, *1567*, 139–154. https://doi.org/10.1007/978-1-4939-6824-4_9.

Bausewein, T., Mills, D. J., Langer, J. D., Nitschke, B., Nussberger, S., & Kuhlbrandt, W. (2017). Cryo-EM structure of the TOM core complex from Neurospora crassa. *Cell*, *170*(4), 693–700 e697. https://doi.org/10.1016/j.cell.2017.07.012.

Chi, X., Nguyen, D., Pemberton, J. M., Osterlund, E. J., Liu, Q., Brahmbhatt, H., et al. (2020). The carboxyl-terminal sequence of bim enables bax activation and killing of unprimed cells. *eLife*, *9*. https://doi.org/10.7554/eLife.44525.

Cosentino, K., & Garcia-Saez, A. J. (2017). Bax and Bak pores: Are we closing the circle? *Trends in Cell Biology*, *27*(4), 266–275. https://doi.org/10.1016/j.tcb.2016.11.004.

Czabotar, P. E., Westphal, D., Dewson, G., Ma, S., Hockings, C., Fairlie, W. D., et al. (2013). Bax crystal structures reveal how BH3 domains activate Bax and nucleate its oligomerization to induce apoptosis. *Cell*, *152*(3), 519–531. https://doi.org/10.1016/j.cell.2012.12.031.

Dewson, G., Kratina, T., Sim, H. W., Puthalakath, H., Adams, J. M., Colman, P. M., et al. (2008). To trigger apoptosis, Bak exposes its BH3 domain and homodimerizes via BH3: Groove interactions. *Molecular Cell*, *30*(3), 369–380. https://doi.org/10.1016/j.molcel.2008.04.005.

Ding, J., Mooers, B. H., Zhang, Z., Kale, J., Falcone, D., McNichol, J., et al. (2014). After embedding in membranes antiapoptotic Bcl-XL protein binds both Bcl-2 homology region 3 and helix 1 of proapoptotic Bax protein to inhibit apoptotic mitochondrial permeabilization. *The Journal of Biological Chemistry*, *289*(17), 11873–11896. https://doi.org/10.1074/jbc.M114.552562.

Hauseman, Z. J., Harvey, E. P., Newman, C. E., Wales, T. E., Bucci, J. C., Mintseris, J., et al. (2020). Homogeneous oligomers of pro-apoptotic BAX reveal structural determinants of mitochondrial membrane permeabilization. *Molecular Cell*, *79*(1), 68–83 e67. https://doi.org/10.1016/j.molcel.2020.05.029.

Jeschke, G. (2012). DEER distance measurements on proteins. *Annual Review of Physical Chemistry*, *63*, 419–446. https://doi.org/10.1146/annurev-physchem-032511-143716.

Johnson, A. E. (2005). Fluorescence approaches for determining protein conformations, interactions and mechanisms at membranes. *Traffic*, *6*(12), 1078–1092. https://doi.org/10.1111/j.1600-0854.2005.00340.x.

Liao, C., Zhang, Z., Kale, J., Andrews, D. W., Lin, J., & Li, J. (2016). Conformational heterogeneity of Bax helix 9 dimer for apoptotic pore formation. *Scientific Reports*, *6*, 29502. https://doi.org/10.1038/srep29502.

Lin, J., Johnson, A. E., & Zhang, Z. (2019). Photocrosslinking approach to investigate protein interactions in the BCL-2 family. *Methods in Molecular Biology*, *1877*, 131–149. https://doi.org/10.1007/978-1-4939-8861-7_9.

Lovell, J. F., Billen, L. P., Bindner, S., Shamas-Din, A., Fradin, C., Leber, B., et al. (2008). Membrane binding by tBid initiates an ordered series of events culminating in membrane permeabilization by Bax. *Cell*, *135*(6), 1074–1084. https://doi.org/10.1016/j.cell.2008.11.010.

Martin, W., & Mentel, M. (2010). The origin of mitochondria. *Nature Education*, *3*(9), 58. Retrieved from https://www.nature.com/scitable/topicpage/the-origin-of-mitochondria-14232356/.

McArthur, K., Whitehead, L. W., Heddleston, J. M., Li, L., Padman, B. S., Oorschot, V., et al. (2018). BAK/BAX macropores facilitate mitochondrial herniation and mtDNA efflux during apoptosis. *Science*, *359*(6378). https://doi.org/10.1126/science.aao6047.

Osterlund, E. J., Liu, Q., & Andrews, D. W. (2015). The use of FLIM-FRET for the detection of mitochondria-associated protein interactions. *Methods in Molecular Biology*, *1264*, 395–419. https://doi.org/10.1007/978-1-4939-2257-4_34.

Ozdemir, E. S., Nussinov, R., Gursoy, A., & Keskin, O. (2019). Developments in integrative modeling with dynamical interfaces. *Current Opinion in Structural Biology*, *56*, 11–17. https://doi.org/10.1016/j.sbi.2018.10.007.

Ozhogina, O. A., & Bominaar, E. L. (2009). Characterization of the kringle fold and identification of a ubiquitous new class of disulfide rotamers. *Journal of Structural Biology*, *168*(2), 223–233. https://doi.org/10.1016/j.jsb.2009.06.003.

Pogmore, J. P., Pemberton, J. M., Chi, X., & Andrews, D. W. (2016). Using forster-resonance energy transfer to measure protein interactions between Bcl-2 family proteins on mitochondrial membranes. *Methods in Molecular Biology*, *1419*, 197–212. https://doi.org/10.1007/978-1-4939-3581-9_15.

Ruan, J., Xia, S., Liu, X., Lieberman, J., & Wu, H. (2018). Cryo-EM structure of the gasdermin A3 membrane pore. *Nature*, *557*(7703), 62–67. https://doi.org/10.1038/s41586-018-0058-6.

Salvador-Gallego, R., Mund, M., Cosentino, K., Schneider, J., Unsay, J., Schraermeyer, U., et al. (2016). Bax assembly into rings and arcs in apoptotic mitochondria is linked to membrane pores. *The EMBO Journal*, *35*(4), 389–401. https://doi.org/10.15252/embj.201593384.

Schweppe, D. K., Chavez, J. D., Lee, C. F., Caudal, A., Kruse, S. E., Stuppard, R., et al. (2017). Mitochondrial protein interactome elucidated by chemical cross-linking mass spectrometry. *Proceedings of the National Academy of Sciences of the United States of America*, *114*(7), 1732–1737. https://doi.org/10.1073/pnas.1617220114.

Shiota, T., Imai, K., Qiu, J., Hewitt, V. L., Tan, K., Shen, H. H., et al. (2015). Molecular architecture of the active mitochondrial protein gate. *Science*, *349*(6255), 1544–1548. https://doi.org/10.1126/science.aac6428.

Shiota, T., Mabuchi, H., Tanaka-Yamano, S., Yamano, K., & Endo, T. (2011). In vivo protein-interaction mapping of a mitochondrial translocator protein Tom22 at work. *Proceedings of the National Academy of Sciences of the United States of America*, *108*(37), 15179–15183. https://doi.org/10.1073/pnas.1105921108.

Slater, O., Miller, B., & Kontoyianni, M. (2020). Decoding protein-protein interactions: An overview. *Current Topics in Medicinal Chemistry*, *20*(10), 855–882. https://doi.org/10.2174/1568026620666200226105312.

Tan, C., Dlugosz, P. J., Peng, J., Zhang, Z., Lapolla, S. M., Plafker, S. M., et al. (2006). Auto-activation of the apoptosis protein Bax increases mitochondrial membrane permeability and is inhibited by Bcl-2. *The Journal of Biological Chemistry*, *281*(21), 14764–14775. https://doi.org/10.1074/jbc.M602374200.

Tucker, K., & Park, E. (2019). Cryo-EM structure of the mitochondrial protein-import channel TOM complex at near-atomic resolution. *Nature Structural & Molecular Biology*, *26*(12), 1158–1166. https://doi.org/10.1038/s41594-019-0339-2.

Uren, R. T., O'Hely, M., Iyer, S., Bartolo, R., Shi, M. X., Brouwer, J. M., et al. (2017). Disordered clusters of Bak dimers rupture mitochondria during apoptosis. *eLife*, *6*. https://doi.org/10.7554/eLife.19944.

Zhang, Z., Subramaniam, S., Kale, J., Liao, C., Huang, B., Brahmbhatt, H., et al. (2016). BH3-in-groove dimerization initiates and helix 9 dimerization expands Bax pore assembly in membranes. *The EMBO Journal*, *35*(2), 208–236. https://doi.org/10.15252/embj.201591552.

Zhang, Z., Zhu, W., Lapolla, S. M., Miao, Y., Shao, Y., Falcone, M., et al. (2010). Bax forms an oligomer via separate, yet interdependent, surfaces. *The Journal of Biological Chemistry*, *285*(23), 17614–17627. https://doi.org/10.1074/jbc.M110.113456.

CHAPTER FOURTEEN

Topological analysis of type 3 secretion translocons in native membranes

Yuzhou Tang[a], Hanling Guo[a], Arjan J. Vermeulen[b], and Alejandro P. Heuck[b,*]

[a]Program in Molecular and Cellular Biology, University of Massachusetts, Amherst, MA, United States
[b]Department of Biochemistry and Molecular Biology, University of Massachusetts, Amherst, MA, United States
*Corresponding author: e-mail address: heuck@umass.edu

Contents

Abstract

PFPs (Pore-forming proteins) perforate cellular membranes to create an aqueous pore and allow the passage of ions and polar molecules. The molecular mechanisms for many of these PFPs have been elucidated by combining high resolution structural information of these proteins with biochemical and biophysical approaches. However, some PFPs do not adopt stable conformations and are difficult to study in vitro. An example of these proteins are the bacterial Type 3 Secretion (T3S) translocators. The translocators are secreted by the bacterium and insert into the target cell membrane to form a

Methods in Enzymology, Volume 649
ISSN 0076-6879
https://doi.org/10.1016/bs.mie.2021.01.036

translocon pore providing a portal for the passage of T3S toxins into eukaryotic cells. Given the important role that the T3S systems play in pathogenesis, methods to study these translocon pores in cellular membranes are needed. Using a combination of protein modifications and methods to selectively permeate and solubilized eukaryotic membranes, we have established an experimental procedure to analyze the topology of the *Pseudomonas aeruginosa* T3S translocon using *P. aeruginosa* strain variants and HeLa cell lines.

1. Introduction

PFPs (Pore-forming proteins) are widely spread in nature, present in bacteria, fungi, plants, and animals (Dal Peraro & van der Goot, 2016). The ultimate goal of PFPs is to perforate cellular membranes by insertion of polypeptide segments into the non-polar core of the lipid bilayers to create a transmembrane pore (Gilbert, Dalla Serra, Froelich, Wallace, & Anderluh, 2014). These segments are in general amphipathic, with the hydrophilic side facing the central aqueous pore and the hydrophobic side facing the lipid bilayer (Heuck & Johnson, 2005). Among PFPs, the bacterial pore-forming toxins (PFTs) are the largest known class of the protein family and the best characterized (Alouf & Freer, 1999). Their fundamental role in pathogenesis caught the attention of researchers early on, with the first reports dating from the early 1900s (Arrhenius, 1907). Molecular studies revealed later in the century that most PFTs are secreted as protein monomers that bind to the target cell membrane after recognizing a receptor molecule. This binding initiates a series of conformational changes that ultimately transform the water-soluble monomer into a membrane-inserted protein. While a single monomer may suffice to assist the translocation of the active portion of the toxin across the membrane (e.g., colicins and diphtheria toxin), the formation of a stable transmembrane pore requires the oligomerization of the monomers. Each monomer contributes one or more amphipathic segments that cooperatively insert forming the transmembrane pore.

Structural and functional characterization of PFPs revealed that the transmembrane segments that form the interface between the aqueous pore and the non-polar core of the membrane are typically α-helices or β-hairpins. Not surprisingly, PFTs were classified into two broad groups: α-PFTs and β-PFTs. In the α-PFTs group, we find colicins, ClyA, and actinoporins (Cosentino, Ros, & García-Sáez, 2016). The β-PFTs group includes *Staphyloccocus aureus* α-hemolysin and leucocidins, aerolysin, anthrax Protective Antigen, and the membrane attack complex/perforin/cholesterol-dependent cytolysin (MACPF/CDC) family (Boyd & Bubeck, 2018).

PFTs are produced as water-soluble monomers and therefore they can be expressed and purified in great quantities from heterologous organisms. Purified proteins combined with model membranes allowed the structural and functional characterization of intermediates (pre-pores) and final membrane-inserted complexes or pores. However, some PFPs like colicins, diphtheria toxin and the Bcl-2 related proteins are more difficult to study (Cosentino et al., 2016). These proteins are mostly α-helical and are involved in the translocation of other proteins. It is possible that the challenge of studying these proteins in vitro relies on the transient nature of the formed pore, or the need for lipid asymmetry, electrochemical gradients, and/or the assistance of other membrane components present only on native membranes.

Some PFPs, like the Type 3 Secretion (T3S) translocators, are not secreted as stable water-soluble monomers, and therefore are difficult to study in vitro. The T3S system is used by many pathogenic Gram-negative bacteria to inject proteins directly into the target cell (Deng et al., 2017; Vermeulen, Tang, & Heuck, 2016). The translocators (typically two different secreted proteins) are mostly α-helical but do not adopt a stable water-soluble monomeric state. Individually they adopt a molten globular conformation and have a tendency to aggregate in solution (Faudry, Job, Dessen, Attree, & Forge, 2007). In the cytosol of the bacterium, they exist as a complex with a cognate chaperone (Discola et al., 2014). This chaperone maintains the proteins in a translocation competent state. They are released from the chaperone while being secreted through the T3S system, to be released in close proximity of the target membrane where they insert, oligomerize, and form a pore by which other T3S secreted proteins are subsequently injected into the target cell cytosol (Fig. 1).

Given the molten globular nature of the translocators in aqueous solution and their tendency to aggregate, is not surprising that no high-resolution structures for isolated monomers or pores are available yet. The translocators (PopB and PopD in *Pseudomonas aeruginosa*) can be purified in heterologous systems in high quantities as a complex with their cognate chaperone (Faudry, Vernier, Neumann, Forge, & Attree, 2006; Romano et al., 2011). The structure of the water-soluble complex formed by the *Aeromonas hydrophila* AcrH chaperone and the translocator ApoB has

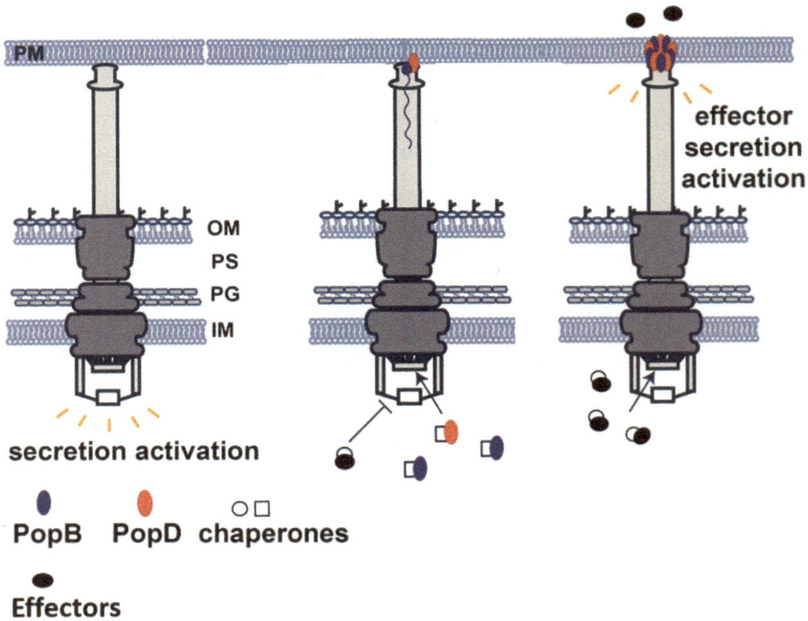

Fig. 1 Model of the T3S system and translocon assembly. T3S injectisomes are pre-assembled on the bacterial surface (OM, outer membrane; PS, periplasmic space; PG, peptidoglycan; IM, inner membrane). Contact with the host cell plasma membrane (PM) initiates the secretion of translocators and activates expression of T3S-related genes (left). PopB and PopD are secreted to form a hetero oligomer, presumably through the formation of hetero dimers (center). After assembly, the PopB-PopD translocon engages the needle and triggers the switch from secretion of translocators to secretion of effectors. Effectors are then injected into the host cell cytosol (right).

been solved (Nguyen et al., 2015). Unfortunately, as it is common for PFPs, predicting the structural characteristics of the pore complex by simply analyzing the structure of the water-soluble monomeric form is difficult (Rossjohn, Feil, McKinstry, Tweten, & Parker, 1997; Shatursky et al., 1999).

Efforts have been made to study the *P. aeruginosa* translocon assembly and structure using purified translocators and model membranes (Romano et al., 2011, 2016; Tang, Romano, Breña, & Heuck, 2018). However, the lack of a system to check the functionality of the assembled complexes in vitro make it difficult to verify the models built using studies based on purified components and model membranes. Methods that allow the study of the formed translocon pores in cellular membranes are needed.

Using a combination of protein modifications and methods to selectively permeate and solubilized eukaryotic membranes, we have established an experimental procedure to analyze the assembly and topology of the T3S translocons using a cellular infection system (Tang et al., 2018). In this chapter, we will describe the protocols employed to analyze the T3S translocon using *P. aeruginosa* strain variants and HeLa cells.

2. Topology analysis of the translocator PopD in cellular membranes: Experimental rationale

The T3S translocators PopB and PopD form a translocon channel in the target cell membrane through which T3S effector proteins gain access to the host cell cytosol (Goure et al., 2004). In model membranes, both translocators can also insert and form pores with discrete sizes (Faudry et al., 2006; Romano et al., 2011). This suggests that PopD and PopB insert one or more transmembrane segments across the membrane. To explore the topology of cell membrane-associated translocator, a tag that gets specifically modified when exposed to the cytosol of mammalian cells and not in the bacterial cytosol could be used to detect exposure of a protein segment to the target cell cytosol. Variants with the tag inserted at different locations in the translocator are produced in a strain of *P. aeruginosa* lacking that particular translocator. Exposure of the peptide-tag to the target cell cytosol will be determined by the detection of modified peptides using immunoblotting.

2.1 The bacterial strain

These studies required a bacterial strain with genetic modifications that (i) lack the protein being studied, (ii) maximize the expression of the T3S system and therefore, the amount of material for the analysis, and (iii) minimize the toxic effect the bacterium may have on the target cell to avoid target cell detachment from the culture flask. To maximize the expression of the T3S system and minimize toxicity, we obtained the *P. aeruginosa* PAKΔexsEΔexoSTY strain (Note 1). Genomic deletion mutants were obtained according to a two-step allelic exchange procedure described in Hmelo et al. (2015). Deletion of genes that encode the regulator ExsE exhibited increased assembly of T3S system on the bacterial surface (Lee, Zmina, Stopford, Toska, & Rietsch, 2014) Deletion of the three *P. aeruginosa* T3S secreted effectors ExoS, ExoT, and ExoY (Engel & Balachandran, 2009) minimize the detachment of the target cells used in the assay (see below).

Strains with deleted translocator were generated in the wild type PAK strain or the PAKΔexsEΔexoSTY (Lee, Smith, Tummler, & Lory, 2005). The gene encoding the translocator variant is then reintroduced into the strain on a plasmid. For example, when the translocator PopD is being analyzed, the PAKΔpopD strain is used to test the functionally of the PopD variant using a cytotoxicity assay (see below), and the PAKΔexsEΔexoSTY ΔpopD to study the topology of PopD. These strains are transformed with a plasmid encoding PopD variants. A list of strains used for these studies is shown in Table 1.

The pUCP18 plasmids containing the DNA fragment that encode the translocator WT or tagged-translocators were generated using Gibson assembly as described previously (Romano et al., 2016). The genomic expression of the translocators is regulated by the endogenous promoter region that was included in the cloned DNA fragment. This promoter regulates the expression of the plasmid encoded translocator and cognate chaperone (Fig. 2). Plasmids carrying the modified translocator gene were introduced into the indicated strains using electroporation, and positive clones were identified by their resistance to carbenicillin (Romano et al., 2016).

Note 1. Other toxins not related to the T3S system may affect the integrity of the target cell. This will be inherent in the pathogen being analyzed, the strain being used, and the experimental conditions.

2.2 The target cell line

The selection of the target cell line is important because the bacteria do not interact with every cell line in the same way. For example, the assay requires a target cell line capable of remaining attached to the culture flask after infection and withstand washes during the procedure is needed. For PAK infections, we found HeLa cells to be a good model system (Tang et al., 2018). However, case by case optimizations will be required to determine the optimal multiplicity of infection (MOI or bacterium/cell ratio) and incubation times for other systems. Examples of the optimization procedure are provided below.

2.3 HeLa cell cytotoxicity assay: Functionality test for modified translocators

The infection assay was adapted from Kaufman et al. (2000). Briefly, HeLa cells cultured in DMEM containing 10% Fetal Bovine Serum (FBS) were

Table 1 Strains and plasmids mentioned in this chapter.

Name	Short description	Source
Strain genotype		
PAK	Wild type *P. aeruginosa* PAK strain	Gift from Dr. S. Lory[a]
PAKΔpopD	PAK with genomic deletion of popD	Gift from Dr. S. Lory[a]
PAKΔexsEΔpopD	PAKΔpopD with genomic deletion of exsE	Tang et al. (2018)
PAKΔexoSTY	PAK with genomic deletion of exoS, exoT, and exoY	Gift from Dr. S. Lory[a]
PAKΔexsEΔexoSTY	PAKΔexoSTY with genomic deletion of exsE	Tang et al. (2018)
PAKΔexsEΔexoSTYΔpopD	PAKΔexsEΔexoSTY with genomic deletion of popD	Tang et al. (2018)
PAKΔexsEΔexoSTYΔpopB ΔpopD	PAKΔexsEΔexoSTY with genomic deletion of popB	Tang et al. (2018)
Plasmid		
pUCP18	Broad-host-range plasmid vector. Gift from Dr.Dara. Frank.	Romano et al. (2016)
pUCPGHD	pUCP18 plasmid encoding WT popD and WT pcrH under control of pcrG promoter	Romano et al. (2016)
pUCPGHD-Q40GSK	Change WT popD to popD with a single GSK-tag insertion after Gln40 on pUCPGHD plasmid	Tang et al. (2018)

[a]More details in (Lee et al., 2005).

maintained in a 37 °C incubator supplemented with 5% CO_2. Before infection, cells were washed once with pre-warmed Dullbeco PBS buffer (DPBS) and incubated in fresh DMEM without FBS to avoid FBS induced T3S system activation (Lee, Mazmanian, & Schneewind, 2001). Bacteria were grown overnight at 37 °C in Miller lysogeny broth (LB), the next day the

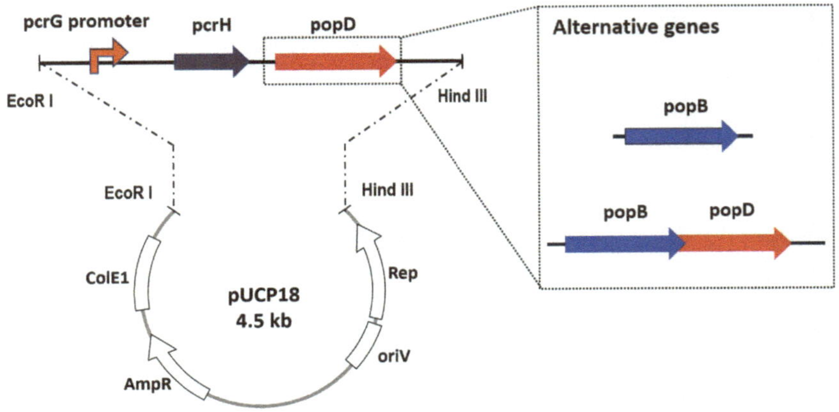

Fig. 2 Scheme of the plasmid employed to express different translocator variants in *P. aeruginosa* PAK strains (Romano et al., 2016). The promoter for the GVHBD operon regulates the expression of PopD (and/or PopB) and its cognate chaperone PcrH. The target fragment is cloned between EcoR I site and Hind III site on the *Escherichia coli-P. aeruginosa* shuttle vector pUCP18. The plasmid contains high-copy-number ColE1 origin of replication and broad-host-range origin of replication, oriV, which comes from *P. aeruginosa*. The plasmid also contains Ampicillin resistant gene (AmpR) for selection and replication protein gene (Rep) from *P. aeruginosa*.

cells were diluted to an optical density at 600 nm (OD_{600}) of 0.1 in fresh LB and grown until the OD_{600} reached a value of 1. Bacteria were spun down at 16,000 × *g* for 2 min and suspended in DMEM. Bacteria added to HeLa cells grown to confluence (1.2×10^6 cells/well) in 6-well plates and incubated at 37 °C. The rounding of HeLa cells was determined and compared with the control with PAK WT and no bacteria (Fig. 3).

2.4 Optimization of the experimental conditions used to isolate membrane-associated translocators

HeLa cell rounding is a typical morphology change caused by translocated T3S effector proteins that mediate cytoskeleton rearrangement. Cell morphology was examined using a Nikon phase contrast microscope. We arbitrarily categorized the cell rounding using a scale from 0 to 5, where (0) represented no cell rounding, (1) some cell rounding, (2) ~50% cell rounding, (3) all cell rounded but still attached, (4) some cells detachment from the flask, and (5) complete cell detachment.

To examine if the translocators per se have any effect on cell rounding, wild type PAK and effector-less PAK strains were incubated with HeLa cells and compared with the effector-less strain that does lack the translocators

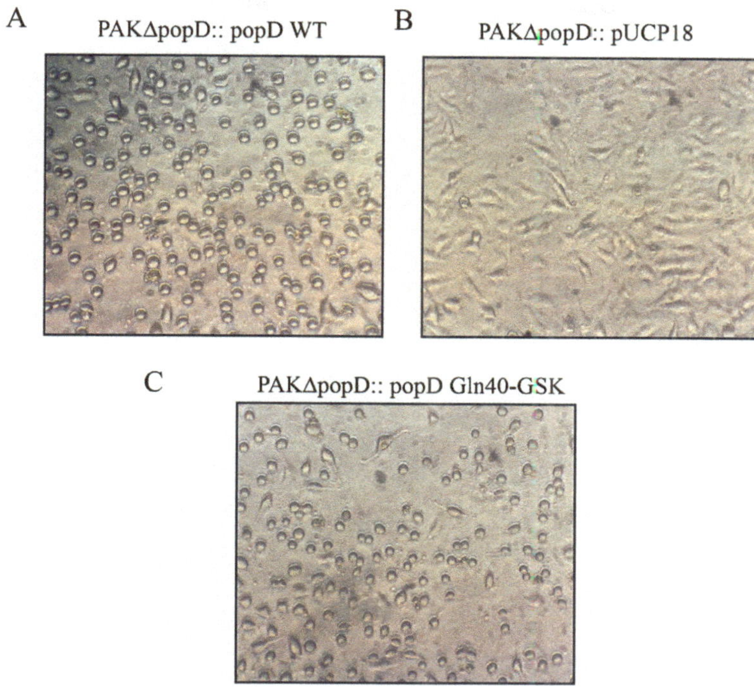

Fig. 3 Cell rounding assay for the functionality of modified translocators. To test the functionality of tagged PopD translocators, PAKΔpopD complemented with popD WT (A), empty vector (B), or tagged popD (C), were incubated with HeLa cells for 2 h with a MOI of 30. The functionality of the modified translocators is assessed by the rounding of the HeLa cell due to disrupted cytoskeleton caused by T3S injected effectors. *Adapted from Tang, Y., Romano, F.B., Breña, M., & Heuck, A.P. (2018). The Pseudomonas aeruginosa type III secretion translocator PopB assists the insertion of the PopD translocator into host cell membranes.* The Journal of Biological Chemistry 293, *8982–8993.*

(Fig. 4A–C). At MOI of 30 or higher, PAK WT caused severe cell death and fast cell detachment in less than 30 min. Effector-less PAK (ΔSTY) showed minimal cell rounding within 150 min, confirming that cell rounding is largely caused by T3S effectors. Elimination of the regulator ExsE in the effector-less PAK strain (ΔEΔSTY) expedited cell death. All cells were rounded after 90 min incubation at a MOI of 30 (Fig. 4B) and the time shortened to 60 min at a MOI of 100 (Fig. 4C). Further deletion of translocators PopB and PopD (ΔEΔSTYΔBD) slowed down the cell rounding process. This strain still caused severe cell death, but only over a long incubation time. These results suggested that the translocators alone contribute to cell

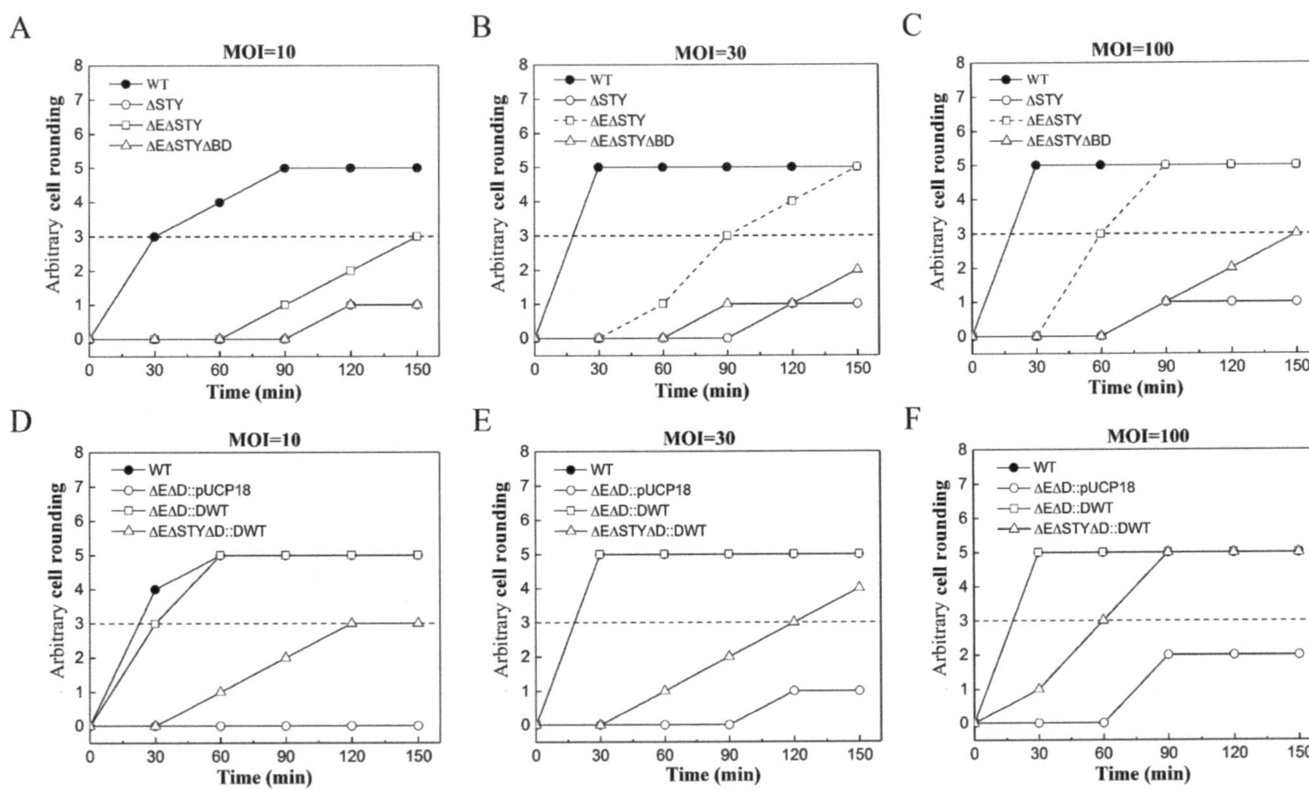

Fig. 4 See figure legend on opposite page.

death in HeLa cells infected with *P. aeruginosa*. Moreover, other factors caused cell death in the absence of the translocators and effectors, but only at longer incubation times.

Next, we determined the optimal incubation time for the bacterial infection assay (Fig. 4D–F). The HeLa cells need to remain firmly attached to the flask during washes to remove away the unattached bacteria. Therefore, a scale of 3 in Fig. 4 represents the limit for how long a particular strain could be incubated (cells start to detach if incubated longer). A strain that does not produce a functional translocon PAKΔexsEΔpopD (ΔEΔD) complemented with empty vector pUCP18 does not significantly round cells at low MOI, but some cell rounding was observed at MOI of 30 and higher. Complementing PAKΔEΔD with a plasmid carrying the gene codifying for wild type PopD (PopD WT) restored the ability of the strain to cause rapid cell death. In the absence of effectors, rounding was observed at longer times of incubation and higher MOI. In conclusion, an MOI of 30 or lower could be used with incubation times up to 2 h for the infection assay of *P. aeruginosa* PAKΔEΔSTY with HeLa cells. At higher MOI the incubation time needs to be reduced to avoid cell detachment.

2.4.1 Selective solubilization of HeLa cells

To remove any translocators that could be injected into HeLa cells after the infection and therefore not forming part of an active translocon complex, one needs a method that lyses HeLa cells but not bacteria. It has been reported that low concentrations of Triton X-100 lyse mammalian cells

Fig. 4 Examination of HeLa cell rounding during incubation with *P. aeruginosa*. Prior to infection, 0.2–0.4 × 10^6 HeLa cells were washed twice with pre-warmed DPBS and replenished with DMEM supplemented with 10% FBS. Indicated PAK strains grown overnight (~12 h) were diluted in fresh LB at an OD_{600} of 0.15, and allowed to grow until OD_{600} of 1. Aliquots containing 0.01–0.1 mL of bacterial culture corresponding to a MOI of 10, 30, or 100 were added to a monolayer of HeLa cells and incubated for a total of 2.5 1 h at 37 °C in a 5% CO_2 atmosphere. Cell morphology was examined every 30 min, and cell detachment was estimated by eye. Top panel: PAK WT, ΔexoSTY (ΔSTY), ΔexsEΔexoSTY (ΔEΔSTY), and ΔexsEΔexoSTYΔpopBpopD (ΔEΔSTYΔBD) were incubated with HeLa cells for a total of 2.5 h at a MOI of 10 (A), 30 (B), or 100 (C). Bottom panel: PAK WT, ΔEΔD::pUCP18 (empty vector), ΔEΔD::D WT, and ΔEΔSTYΔD::D WT was incubated with HeLa cells for a total of 2.5 h at a MOI of 10 (D), 30 (E), or 100 (F). The extent of cell rounding was visualized every 30 min, and categorized using a scale of 0–5 as indicated in the text. The dashed line at the y-scale of 3 indicates the optimal incubation conditions for isolation of membrane-associated translocon, where rounding is maximum, and cells are still attached.

while leaving *P. aeruginosa* intact (Bucior, Tran, & Engel, 2014; Ha & Jin, 2001; Urbanowski, Brutinel, & Yahr, 2007). We therefore treated *P. aeruginosa* PAK cells and HeLa cells with different amounts of Triton X-100 to determine the optimal concentration of detergent to be used.

2.4.1.1 Effect of triton X-100 on PAK

PAKΔexsEΔexoSTYΔpopD::popD in 4 mL of LB culture ($OD_{600}=0.62$) was equally divided into four aliquots and collected by centrifugation at 4 °C (3,000 × *g*, 20 min). Three tubes of bacteria pellets were resuspended in PBS (control), 0.05% or 0.1% Triton X-100 and incubated for 30 min on ice. A fourth tube of pelleted bacteria was resuspended in 80 μL sodium dodecyl sulfate (SDS) sample buffer at 20–23 °C and served as a control for total protein ("Total" in Fig. 5). After incubation, solubilized proteins were isolated by centrifugation at 4 °C (18,000 × *g*, 10 min) and precipitated with trichloro acetic acid (TCA) using 10 μg Bovine Serum Albumin as a carrier. The samples were analyzed by immunoblotting using anti-RpoA antibody (Neoclone). RpoA is a subunit of RNA polymerase in *P. aeruginosa* and serves as a marker ofbacterium cytosol. No significant amounts ofRpoA were detected in the control sample (PBS) or the samples incubated with 0.05% or 0.1% Triton X-100 (Fig. 5A).

2.4.1.2 Effect of triton X-100 on HeLa cells

A total of 6 × 10^6 HeLa cells grown in DMEM with 10% FBS were scraped in DPBS buffer, equally divided into 3 aliquots, and collected by centrifugation at 4 °C (2000 × *g*, 10 min). Two aliquots of cells were lysed in 0.05% or 0.1% Triton X-100 for 30 min on ice. The solubilized membrane and cytosolic proteins (S, supernatant) were separated from insoluble cell debris (P, pellet) by centrifugation (18,000 × *g*, 4 °C for 10 min), and precipitated with methanol/chloroform as described below in Section 3.3.3 (Note 2). The third aliquot of cells was resuspended directly in 80 μL SDS sample buffer and served as a control for total proteins (Total). All samples were analyzed by coomassie staining (Fig. 5B).

In summary, we confirmed here that 0.1% Triton X-100 can efficiently lyse HeLa cells without affecting the integrity of the *P. aeruginosa* membranes. However, we strongly recommend testing the reagents used to specifically lyse mammalian cells since we have observed lysis when using a slightly higher concentration of Triton X-100 (e.g., 0.5%, data not shown).

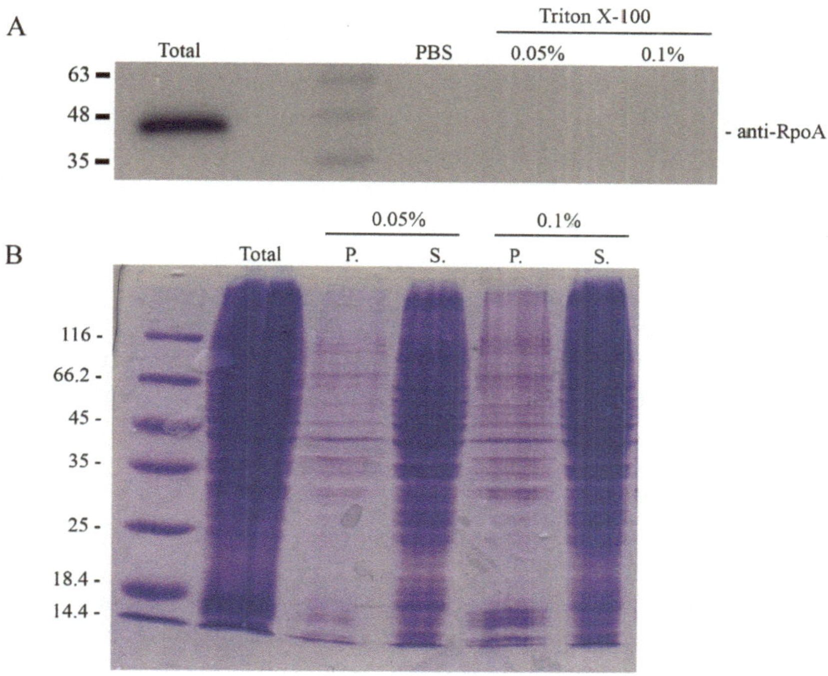

Fig. 5 0.1% TtritonX-100 efficiently lyses HeLa cells while leaving *P. aeruginosa* intact. (A) PAKΔexsEΔexoSTYΔpopD::popD WT in 4 mL of LB culture (OD_{600} = 0.62) was equally divided into four aliquots and collected by centrifugation. The bacteria pellet was resuspended in 80 μL SDS sample buffer (Total), 200 μL PBS (control no lysis), 0.05% or 0.1% Triton X-100. Solubilized proteins were analyzed by immunoblotting using anti-RpoA antibody as indicated in the text. (B) HeLa cells were resuspended in 80 μL SDS sample buffer (Total) or lysed with the indicated concentration of Triton X-100 as detailed in the text. The solubilized membrane and cytosolic proteins (S, supernatant) were separated from insoluble cell debris (P, pellet) and precipitated with methanol/chloroform. All samples were analyzed by SDS-PAGE and coomassie brilliant blue staining. Molecular mass markers are indicated in kDa.

Note 2. When samples contain membranes or detergents, we use the chloroform/methanol procedure to precipitate proteins rather than TCA precipitation.

2.4.2 Quantification of HeLa cell associated translocators

2.4.2.1 Effect of the incubation time

The optimal infection time was determined by incubating PAKΔexsEΔexoSTYΔpopD::popD WT with HeLa cells at a MOI of 100 in DMEM supplemented with 10% FBS for 30 min, 1 h, or 2 h

(Note 3). At the end of each incubation, cells were lysed by 0.18% Triton X-100, and the total amount of protein was determined using the Bicinchoninic Acid (BCA) assay. Aliquots containing 40 μg of total protein were analyzed for PopB and PopD by immunoblotting as described previously (Tang et al., 2018). The amount of HeLa cell associated translocators increased twofold after doubling the incubation time from 30 min to 1 h. However, a longer incubation did not show a further increase in the amount of recovered translocator (Fig. 6). Therefore, incubation times of 1 h or lower will be used below to isolate membrane associated translocators in this assay.

Note 3. When the analysis is performed using incubation times that exceed 1 h, the culture media should be supplemented with FBS to avoid cell death and detachment.

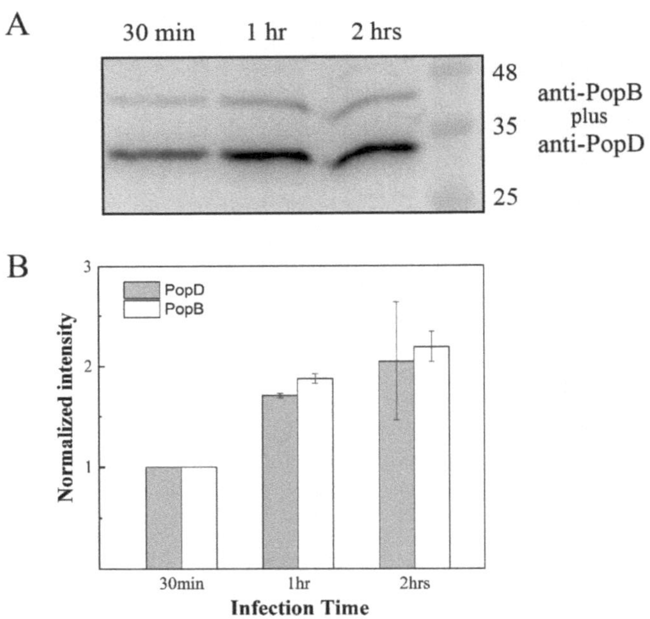

Fig. 6 Optimization of infection time for the detection of translocators. (A) PAK ΔexsEΔexoSTYΔpopD::popD WT was incubated with HeLa cells as indicated in the text. Cell lysates were collected by Triton X-100 solubilization and a total of 40 μg protein from each sample was analyzed for PopB and PopD by immunoblotting. A representative blot from two independent experiments is shown. (B) Band intensities were quantified using ImageJ. To compare results from different lanes, band intensities of PopB and PopD were normalized to the one observed for 30 min. Data shown are an average of quantifications from two blots and error bars correspond to the range.

2.4.2.2 Effect of the multiplicity of infection

HeLa cells were incubated with PAKΔexsEΔexoSTYΔpopD::popD WT at the indicated MOI (Fig. 7). Compared with the number of translocators detected using a MOI of 30, a higher number of bacteria did not show a significant increase in the number of translocators associated with HeLa cells. Since using a low number of bacteria minimize cell detachment and contamination with translocators that are not directly associated with target cell membranes (e.g., bacterial lysis or secretion into the media), we selected a MOI of 30 for the assay of membrane-associated translocators.

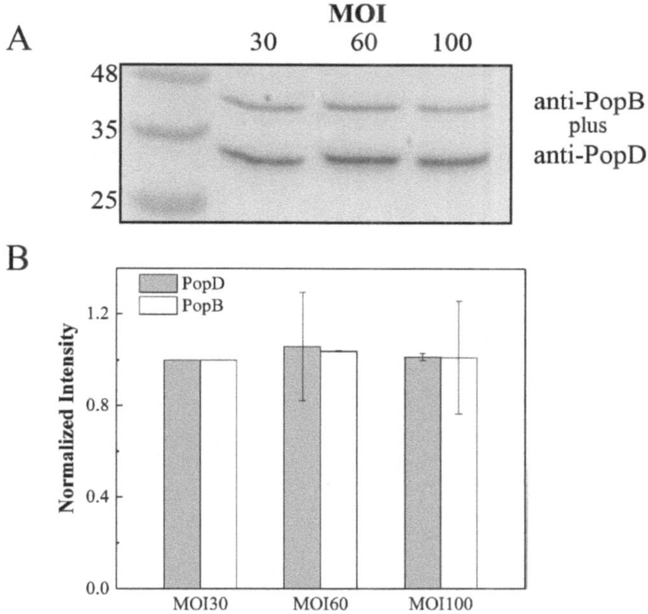

Fig. 7 Comparison of PopD and PopB detected on HeLa cells when incubated with *P. aeruginosa* using different MOI. (A) PAK ΔexsEΔexoSTYΔpopD::popD WT was incubated with HeLa cells at an MOI of 30, 60 or 100 in FBS-free DMEM for 1 h at 37 °C. Cell lysates were collected by Triton solubilization and quantified using BCA assay. A total of 40 µg protein from each sample was analyzed for PopB and PopD by immunoblotting. Representative blots from two independent experiments are shown. (B) Quantification of band intensities of blots. Band intensities were quantified using ImageJ. To compare results from different lanes, band intensities of PopB and PopD were normalized to the one observed for MOI 30. Data shown are an average of quantifications from two blots and error bars correspond to the range. *Adapted from Tang, Y., Romano, F.B., Breña, M., & Heuck, A.P. (2018). The Pseudomonas aeruginosa type III secretion translocator PopB assists the insertion of the PopD translocator into host cell membranes.* The Journal of Biological Chemistry 293, *8982–8993.*

We have established the optimal conditions to maximize the amount of PopB and PopD translocators isolated from HeLa cells after incubations with effector-less strains of *P. aeruginosa*. However, the topological analysis of the translocators required to isolate the membrane-inserted translocators and minimize any potential source of contaminants that could interfere with the analysis (Fig. 8). For example, when analyzing exposure of segments to HeLa cell cytosol (1) translocators secreted to the media could be associated with the HeLa cell membrane but do not form functional translocons (potential false negative); (2) some bacteria could remain associated with the HeLa cells after washes and therefore, the permeabilization and membrane

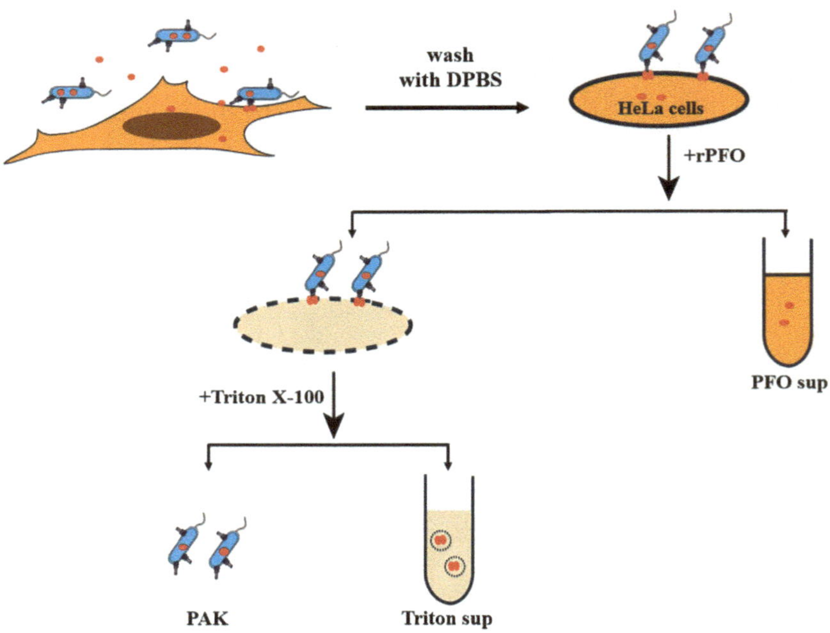

Fig. 8 Diagram of the translocon solubilization procedure. Infected cells are first washed with DPBS to remove free bacteria (marked blue) and free secreted PopD (marked red) in the media. Then the cells are permeabilized with rPFO and centrifuged to separate the released cytosolic components (PFO sup). The pellet containing permeabilized HeLa cells and attached bacteria was treated with 0.1% Triton X-100 to selectively solubilize HeLa cell membranes. Insoluble cell debris and bacteria were removed using centrifugation. The supernatant (Triton sup) contains the solubilized membrane proteins. *Adapted from Tang, Y., Romano, F.B., Breña, M., & Heuck, A.P. (2018). The Pseudomonas aeruginosa type III secretion translocator PopB assists the insertion of the PopD translocator into host cell membranes.* The Journal of Biological Chemistry 293, *8982–8993.*

solubilization procedure should not affect the integrity of the bacterial membranes (potential false negative); and (3) after assembly of a functional translocon, translocators being secreted could be injected into the target cell cytosol before the T3S system switch to effector secretion mode (potential false positive) (e.g., Francis & Wolf-Watz, 1998; Iwai et al., 2007). We have analyzed and found experimental conditions that minimize these potential problems, as described below.

2.4.2.3 Translocators secreted to the media do not stably associate with HeLa cells membranes

It is well known that PopB and PopD can be secreted into the culture media by chelating Ca^{2+} (e.g., with EGTA, Goure et al., 2004), or by the presence of FBS in the cell culture medium (Lee et al., 2001; Tang et al., 2018). HeLa cells were infected with PAKΔexsEΔexoSTYΔpop D::popD WT using a MOI of 30 in three different conditions: (i) in PBS, (ii) in DMEM, or (iii) in DMEM plus 10% FBS. Both PopB and PopD were found associated with HeLa cells ("Triton solubilized proteins" in Fig. 9A), but the translocators were found largely secreted into the infection media only when FBS was present (Fig. 9A). We incubated supernatants from infections done in culture media containing FBS ("secreted proteins" in Fig. 9B) with HeLa cells for 1 h at 37 °C and found that PopB and PopD were only associated with HeLa cell membranes if secreted by the bacteria (Fig. 9B), suggesting that secretion and membrane-insertion are coupled during translocon assembly at neutral pH.

2.4.3 HeLa cell permeabilization to release cytosolic content

To eliminate the potential false positive contribution of injected translocators into the target cell cytosol, we tested a few methods for the selective permeabilization of mammalian cells without affecting bacteria. The presence of HeLa cell cytosolic components and plasma membrane was detected by immunoblotting using antibodies against GAPDH (Glyceraldehyde 3-phosphate dehydrogenase) or against Na^{+}/K^{+} ATPase (a mammalian cell plasma membrane protein), respectively (Tang et al., 2018). When needed, antibodies against RpoA were used to detect bacteria lysis (Fig. 5).

2.4.3.1 Physical methods

We first tested the efficiency of passing the sample after infection repeatedly through a fine needle, or by exposing the sample to cycles of freeze-thaw.

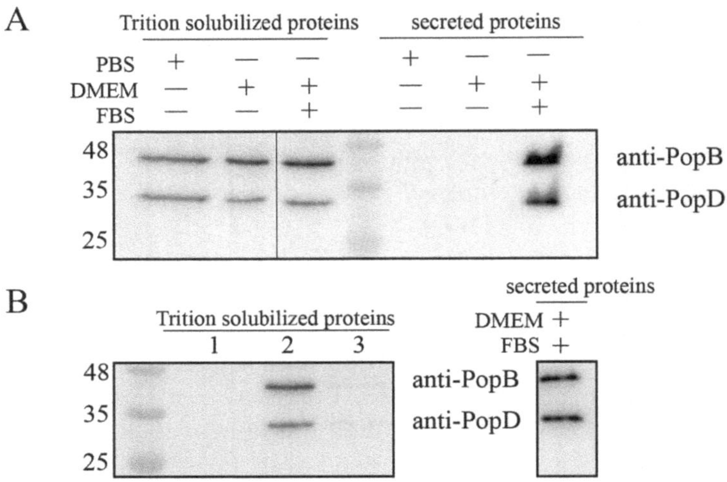

Fig. 9 Association of PopB and PopD into cell membranes occurs only in the presence of *P. aeruginosa*. (A) PopB and PopD were detected on HeLa cell membranes after incubations with PAK at physiological pH. PAK ΔexoEΔexoSTYΔpopD::popD WT strain was incubated with HeLa cells in the indicated medium for 1 h. After the incubation, secreted proteins in 1 mL of medium were collected by centrifugation, precipitated with TCA, and analyzed by immunoblotting using anti-PopB and anti-PopD antibodies. Infected HeLa cells were lysed with 0.1% Triton X-100, and Triton-solubilized proteins were separated from residual bacteria and insoluble debris using centrifugation. 20 μg of total protein in each Triton-solubilized protein sample was analyzed for the presence of PopB and PopD using immunoblotting. (B) translocators secreted to the medium did not bind to HeLa cells. HeLa cells alone (lane 1), incubated with PAK (lane 2), or incubated with the supernatant containing secreted proteins (lane 3) were collected and analyzed as described in A. The supernatant containing secreted PopB and PopD was obtained by centrifugation after culturing PAK with HeLa cells in DMEM plus FBS for 1 h. Representative blots from two independent experiments are shown. *Adapted from Tang, Y., Romano, F.B., Breña, M., & Heuck, A.P. (2018). The Pseudomonas aeruginosa type III secretion translocator PopB assists the insertion of the PopD translocator into host cell membranes.* The Journal of Biological Chemistry 293, *8982–8993.*

HeLa cells (7.2×10^6) incubated with the PAKΔexoEΔexoSTYΔpopD::popD237-GSK strain was washed, scrapped, and equally divided into 3 tubes. Cells were pelleted at 4 °C (2000 × *g* 10 min). An aliquot of cells was resuspended in 210 μL 0.1% Triton X-100 in DPBS and served as the control for maximum cell lysis. The other two aliquots were resuspended in 210 μL of hypotonic ice cold 10 mM Tris-HCl pH 7.4. One aliquot was passed 30 times through a 25G needle using a syringe. The other was subjected to 3 cycles of freeze (−80 °C) and thaw (37 °C) for 5 and 3 min, respectively.

After each treatment, the released cytosolic content was separated from remnant cells by centrifugation at 4 °C (14,000 × *g*, 10 min). The supernatants were precipitated with chloroform/methanol, resuspended in 100 μL SDS sample buffer, and 15 μL were loaded in lanes 2 and 3 ("sup" in Fig. 10). The pellets were treated with 0.1% Triton X-100 in DPBS for 30 min at 4 °C and centrifuged at 4 °C (18,000 × *g*, 10 min). The supernatant fractions obtained from the Triton X-100 treatment of the three samples (PBS control, needle, and freeze-thaw) were precipitated with methanol/chloroform and resuspended in 100 μL of SDS sample buffer. An aliquot of 15 μL of the samples was loaded in lanes 4, 5, and 6 ("Triton sup" in Fig. 10). Fractions were analyzed by immunoblotting using anti-GAPDH (Fig. 10A), anti-Na^+/K^+ ATPase (Fig. 10B), and anti-RpoA (Fig. 10C) antibodies.

Most of the HeLa cell cytosol was released upon passage through the needle or freeze-thaw cycles (Fig. 10A). A small fraction of HeLa cell membranes was detected in the soluble fraction (sup) after needle passage, but not

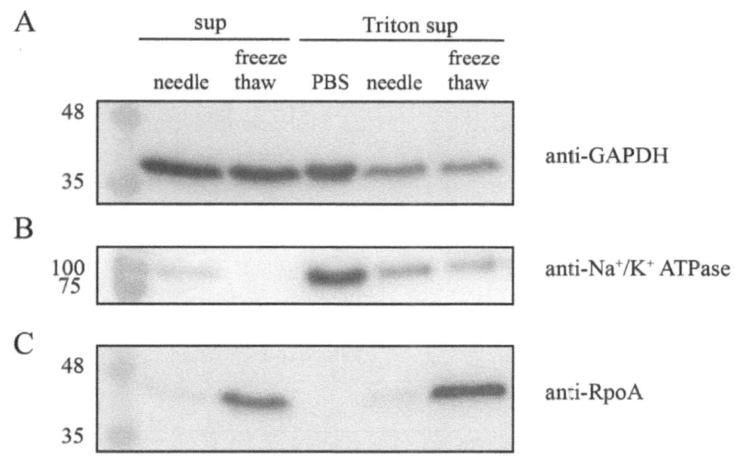

Fig. 10 Infected HeLa cell permeabilization using needle passage and cycles of freeze and thaw. PAKΔexoEΔexoSTYΔpopD:: popD237-GSK infected HeLa cells were equally divided into 3 tubes, pelleted, and resuspended in 0.1% Triton X-100 in DPBS (control), or Tris pH 7.4 10 mM. Cells in Tris buffer were lysed by passage a needle or by cycles of freeze and thaw as described in the text. Released cell content (sup) was separated from the pellet by centrifugation. The cell pellets were further lysed by Triton X-100, and Triton sup was collected as described in the text. All samples were analyzed using immunoblotting for the presence of HeLa cell cytosol and HeLa cell plasma membrane using anti-GAPDH (A) and anti-Na^+/K^+ ATPase (B) antibodies, respectively. Lysis of bacteria was detected using anti-RpoA (C) antibody.

in cells treated with freeze-thaw cycles (Fig. 10B). However, while 0.1% Triton X-100 in DPBS (control) and needle passage did not significantly lyse *P. aeruginosa*, freeze-thaw cycles severely disrupted the integrity of bacterial membranes (Fig. 10C). In conclusion, the needle passage shows some membrane proteins in the soluble fraction, and the freeze and thaw approach broke bacteria. Therefore, neither of these two methods are suitable for the selective permeabilization of HeLa cells and topological analysis.

2.4.3.2 Chemical methods: Cholesterol dependent permeabilization

Digitonin (DIG) is a non-ionic detergent which selectively lyses eukaryotic cells but not bacterial membranes due to its required binding to cholesterol (Lee, Anderson, & Schneewind, 1998). We examined if DIG can selectively permeabilize HeLa cells. A total of 8×10^6 HeLa cells were equally divided into 4 tubes, pelleted, and resuspended in 200 μL DPBS, 0.05% DIG, 0.1% DIG, or 0.1% Triton X-100. Detergent solutions were prepared in DPBS. After incubation for 10 min at 37 °C, detergent solubilized fractions ("sup" in Fig. 11) were separated from insoluble fractions ("pellet" in Fig. 11) by centrifugation at 20–23 °C (2000 × *g*, 10 min). The solubilized fractions were precipitated with methanol/chloroform and resuspended in 100 μL SDS sample buffer. The pellet fractions were washed with DPBS three times and resuspended in 100 μL SDS sample buffer. An aliquot of

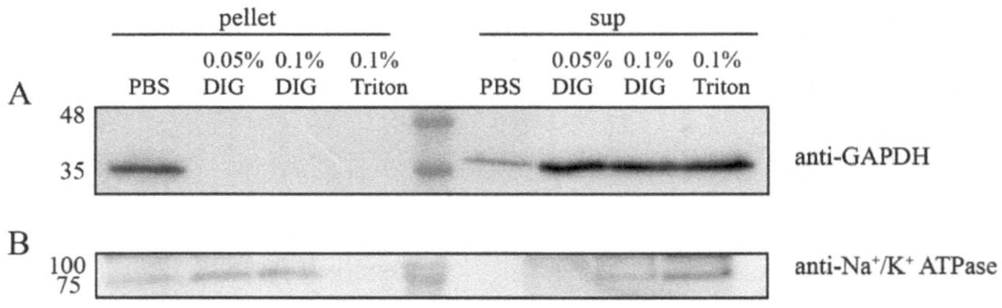

Fig. 11 0.05% Digitonin efficiently permeabilized HeLa cells. HeLa cells were equally divided into 4 tubes, pelleted, and resuspended in PBS, 0.05% DIG, 0.1% DIG, or 0.1% Triton X-100. Detergent solubilized fractions (sup) were separated from insoluble fractions (pellet) by centrifugation and precipitated with methanol/chloroform. All samples were resuspended in 100 μL SDS sample buffer and analyzed using immunoblotting for the presence of HeLa cell cytosol or plasma membrane markers using anti-GAPDH (A) and anti-Na^+/K^+ ATPase (B) antibodies, respectively.

15 µL of each sample was analyzed by immunoblotting using anti-GAPDH and anti-Na^+/K^+ ATPase antibodies. The analysis showed that the cell cytosol was completely released upon treatment with detergents (Fig. 11A, sup), indicating that DIG efficiently disrupted HeLa cell membranes. As expected, both GADPH and Na^+/K^+ ATPase were detected in the sup fraction of samples treated with 0.1% Triton X-100. No Na^+/K^+ ATPase was detected in samples treated with 0.05% DIG; however, some was detected in the sample treated with 0.1% DIG (Fig. 11B, sup). In conclusion, DIG can efficiently permeabilize HeLa cells when used at the proper DIG/total cells ratio, but the amount of DIG should be titrated and carefully controlled to avoid membrane solubilization.

Lastly, we compared the efficiency of HeLa cell permeabilization using Perfringolysin O (PFO), a bacterial toxin that requires cholesterol for membrane binding, and therefore does not lyse bacteria (Johnson & Heuck, 2014). We tested two PFO variants, rPFO and nPFO. nPFO contains the wild type PFO amino acid sequence, and rPFO is the Cys-less variant of nPFO (Johnson et al., 2012). Both proteins can form large pores in the mammalian cell membrane, but rPFO has slightly lower lytic activity than nPFO (Johnson, Breña, Anguita, & Heuck, 2017). HeLa cells (8.1×10^6) were incubated with PAKΔexoEΔexoSTYΔpopD:: popD WT, collected and equally divided into 3 tubes, pelleted, and resuspended in 210 µL of DPBS (control), 1 µM nPFO in HBS (Hepes 50 mM, NaCl 100 mM, pH 7.5) supplemented with 5 mM DTT, or 1 µM rPFO in HBS, and incubated for 30 min at 20–23 °C.

Supernatants and remnant cells were treated as described above for physical disruption tests. All samples were resuspended in 60 µL SDS sample buffer. An aliquot of 15 µL of each sample was analyzed for GAPDH and Na^+/K^+ ATPase using immunoblotting. The amount of soluble cytosolic components released after nPFO or rPFO treatment was similar (PFO sup in Fig. 12A), and no significant amount of membrane fraction was detected (Fig. 12B). In conclusion, any of these PFO variants could be used for selective HeLa cell permeabilization.

In summary, from all these permeabilization methods we chose to use 2 µM rPFO for the topology analysis of T3S translocators. rPFO does not require the presence of reducing agents to remain active (in contrast with nPFO), it does not disrupt bacteria because they lack cholesterol, and is more resistant to variations on the reagent/cell number ratio than DIG.

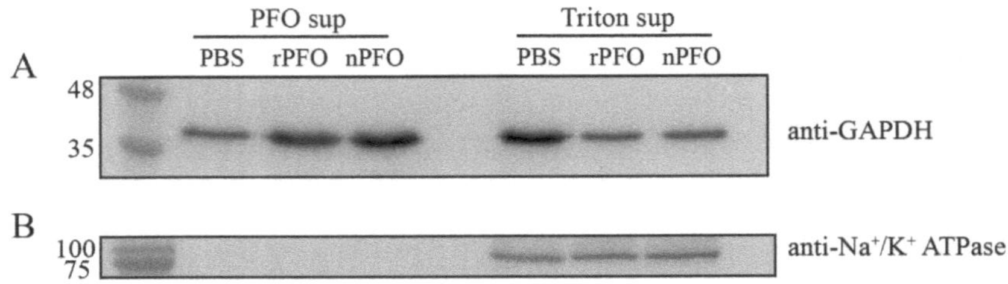

Fig. 12 Comparison of infected HeLa cell permeabilization using rPFO and nPFO. PAKΔexoEΔexoSTYΔpopD:: popD WT infected HeLa cells were equally divided into 3 tubes, pelleted, and incubated with PBS, rPFO, or nPFO solutions. PFO sup and Triton sup samples were collected as described in the text. An aliquot of 15 μL from each sample was analyzed using immunoblotting for the presence of HeLa cell cytosol or plasma membrane markers using anti-GAPDH (A), anti-Na^+/K^+ ATPase (B) antibodies, respectively. *Adapted from Tang, Y., Romano, F.B., Breña, M., & Heuck, A.P. (2018). The Pseudomonas aeruginosa type III secretion translocator PopB assists the insertion of the PopD translocator into host cell membranes.* The Journal of Biological Chemistry 293, *8982–8993.*

3. Analysis of protein segments exposed into the cytosol of HeLa cells for membrane-inserted translocators

We have optimized the experimental conditions to isolate the T3S translocators PopB and PopD associated with HeLa cells' membranes after incubations with *P. aeruginosa*. To establish the topology of the translocators, we need a tag that can be incorporated into PopB or PopD (e.g., as a short peptide) and assess its exposure to the HeLa cell cytosol or to the culture medium. Exposure of the translocator to the target cell cytosol can be detected by using tags that are modified by native enzymes or enzymes that are artificially expressed in the target cell (Lotze, Reinhardt, Seitz, & Beck-Sickinger, 2016; Torruellas Garcia et al., 2006). A tag that has been successfully used to study the exposure of T3S effectors and translocators to the target cell cytosol is the glycogen synthase kinase (GSK) peptide (Luo & Donnenberg, 2011; Torruellas Garcia et al., 2006; Urbanowski et al., 2007). In this chapter, we describe the protocol that uses the GSK-tag to study the exposure of the PopD translocator segments to the cytosol of HeLa cells.

The GSK-tag is a small peptide derived from the human GSK-3β, it is a 13-residue peptide (MSGRPRTTSFAES) that can be phosphorylated by multiple kinases when exposed to the cytosol of eukaryotic cells (Fig. 13, Brodsky & Medzhitov, 2008; Torruellas Garcia et al., 2006).

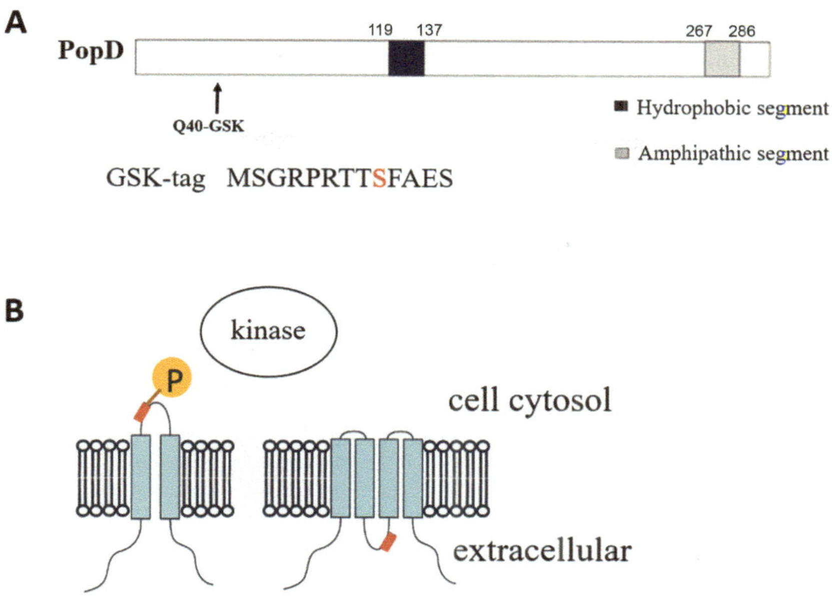

Fig. 13 Using GSK-tag to study protein topology. (A) Diagram of the primary sequence of PopD. The GSK-tag could be located at different locations of PopD (inserted after Gln40 in this diagram). The Ser (red) of the GSK tag is phosphorylated by cytosolic kinases. The black segment corresponds to amino acids 119–137, an hydrophobic segment, and the gray segment corresponds to amino acids 267–286, predicted to form an amphipathic α-helix. (B) GSK-tag (Red box) will only be phosphorylated (yellow circle) by kinases in the cytosol of mammalian cells. Phosphorylation of the GSK-tag will indicate that the tag in the protein is exposed to the target cell cytosol.

Phosphorylation of the GSK-tag is detected by immunoblotting using an antibody specific for the phosphorylated peptide.

The DNA sequence codifying for these thirteen amino acids is introduced at a particular location of the popD gene in the pUCPGHD plasmid described in Fig. 2 using Gibson assembly (Romano et al., 2016). The plasmid is transformed into a popD-less PAK strain, and the strain is subsequently incubated with HeLa cells. Membranes of the infected HeLa cells are isolated, and the presence of translocators and the phosphorylation of the GSK-tag is detected by immunoblotting as described below.

3.1 Equipment

1. Mammalian cell incubator (Thermo Isotemp model 3532).
2. T25 cell culture flask (Corning T25 flask with 2 μm vent cap, Tissue culture treated).

3. Cell counter (Nexcelom Cellometer Auto 1000)
4. Orbital shaker with temperature control (Thermo MaxQ 4000)
5. OD_{600} measurement instrument (WPA CO8000 cell density meter)
6. Phase contrast microscope (Nikon)
7. Microcentrifuge (Beckman Coulter Microfuge 18)

3.2 Reagents

1. Dulbecco's Modified Eagle Medium (DMEM) (Hyclone, catalog# SH30243.01) with 10% Fetal Bovine Serum (FBS) (Hyclone, catalog# SH30071.03)
2. Lysogeny broth (LB), Miller's (Fisher, catalog# BP1426-500)
3. Dulbecco's phosphate buffered saline (DPBS) (Hyclone, catalog# SH30264.02)
4. Protease inhibitor cocktail (PIC) (Roche, catalog# 4693116001 on Sigma)
5. Hepes (Fisher, catalog# BP410-500)
6. NaCl (Fisher, catalog# BP358-212)
7. NaF (Fisher, catalog# S299-100)
8. Triton X-100 (Sigma, catalog# X100-100ML)
9. Buffer A: 50 mM Hepes pH 7.5, with 100 mM NaCl
10. nPFO and rPFO purified as described in (Flanagan, Tweten, Johnson, & Heuck, 2009)
11. rPFO lysis solution: 2 μM rPFO in buffer 1 with PIC and 5 mM NaF (phosphatase inhibitor)
12. Lysis buffer: DPBS with 0.1% Triton X-100, PIC, and 5 mM NaF
13. Methanol (Fisher, catalog# A412-1)
14. Chloroform (Fisher, catalog# C607-1)
15. Sodium dodecyl sulfate (SDS) (Sigma, catalog# 436143-25G)
16. Anti-RpoA (Neoclone, catalog# WP003)
17. Anti-PopB (Thermo, customized rabbit)
18. Anti-PopD (Thermo, customized rabbit)
19. Anti-GAPDH (Cell Signaling Technology, catalog# 2118)
20. Anti-Na^+/K^+ ATPase (Cell Signaling Technology, catalog# 44759)
21. Anti-P-GSK-3β (Cell Signaling Technology, catalog# 12456)
22. Anti-GSK-3β tag (Cell Signaling Technology, catalog# 5558)

3.3 Protocol

3.3.1 *Location of GSK-tags*

The insertion of the GSK peptide could alter the protein function. However, using bioinformatics to predict secondary structure, one could asses what regions of the protein will form unstructured loops and are likely to be exposed to the aqueous solvent. Also, if sequences for homolog proteins are available, a multi sequence alignment could reveal poorly conserved regions of the protein. Modifications of these regions are less likely to affect protein function. The GSK peptide could be inserted or exchanged for an equivalent number of amino acids in the protein sequence. However, while this effort can save money and time, it will be difficult to predict exactly how the incorporation of the GSK peptide will affect the function of the protein. The functionality of every GSK-tagged translocator should be directly checked for proper function using a cell rounding assay, as described below.

3.3.2 *Functionality of the PopD-tagged: Cell rounding assay*

While the generation of an inactive GSK-tagged protein won't be useful for the topological analysis, it will reveal regions of the protein that are important for its function. In the case of the T3S translocators, the lack of effector injection into the target cell (cell rounding) will reveal that this region is important for (a) the interaction with the chaperone in the cytosol of the bacteria (the protein is unstable and it is not found in bacteria lysates), (b) targeting the protein to the secretion platform (the protein is found in bacteria lysates but it is not secreted), (c) insertion into the target membrane (the protein is secreted but it is not associated with membranes), or (d) assembly of the translocon (the protein is found associated with membranes but the translocon is non-functional). The protocol to test the function of tagged PopD is described below.

1. HeLa cells are incubated in DMEM with 10% FBS in a 5% CO_2, 37 °C incubator
2. The day before the experiment, use a cell counter (Cellometer) to count 2.5–2.8 × 10^6 cells and seed the cells onto a T25 flask for each infection
3. Grow PAKΔpopD complemented with popD WT, empty vector, and tagged popD in LB with corresponding antibiotics at 37 °C, 200 rpm, overnight (~15 h).
4. Dilute overnight culture into OD_{600} of 0.15 in fresh LB
5. Keep growing diluted culture until OD_{600} is 1

6. Prior to infection, wash cells twice with pre-warmed DPBS and replenish with 4 mL DMEM free of FBS
7. Add corresponding bacterial culture into each T25 flask containing 2.5–2.8 × 10^6 HeLa cells at a MOI of 30
8. Incubate cells in a 5% CO_2, 37 °C incubator for 2 h
9. Check the rounding of the HeLa cell of each T25 flask containing infected cells under a microscope

3.3.3 Cell infection and membrane protein isolation

1. HeLa cells are incubated in DMEM with 10% FBS in a 5% CO_2, 37 °C incubator
2. The day before the experiment, use a cell counter (Cellometer) to count 2.5–2.8 × 10^6 cells and seed the cells onto a T25 flask for each infection
3. Grow the effector-less PAKΔexsEΔexoSTYΔpopD complemented with a GSK-tagged PopD strain in LB supplemented with 300 μg/mL carbenicillin at 37 °C, 200 rpm, overnight (~15 h).
4. Dilute overnight culture to a final OD_{600} of 0.15 in fresh LB without antibiotics and growth until the OD_{600} is ~1
5. Prior to infection, wash HeLa cells twice with pre-warmed DPBS and replenish with 4 mL DMEM free of FBS
6. Add bacterial culture into the T25 flask containing 2.5–2.8 × 10^6 HeLa cells at a MOI of 30 and incubate cells in a 5% CO_2, 37 °C incubator for 1 h
7. At the end of incubation, wash the T25 flask containing infected cells gently three times with DPBS to removed unattached bacteria and proteins not associated with HeLa cell membranes
8. Scrape infected cells in 1 mL ice-cold DPBS, and pellet cells at 4 °C (2000 × *g* for 10 min).
9. Remove the supernatant and wash cells one more time with 1 mL DPBS
10. Resuspend cells in 230 μL rPFO lysis solution and incubated at 20–23 °C with constant mixing for 30 min
11. Separate the released cytosolic components from permeabilized cells and attached bacteria by centrifugation at 4 °C (2000 × *g* for 10 min).
12. Wash the permeabilized cells with 500 μL 10 mM Tris-HCl (pH 7.4) supplemented with PIC and 5 mM NaF on ice for 5 min
13. Treat the permeabilized cells in 210 μL of lysis buffer at 4 °C with constant mixing for 30 min

14. Separate solubilized membrane proteins from residual bacteria and insoluble debris by centrifugation at 4 °C (18,000 × *g* for 15 min).
15. Mix one volume of the supernatant containing solubilized membrane proteins with 8 volumes of a mixture of methanol: chloroform: H_2O (4:1:3 in volume) by vortexing
16. Centrifuge the mixture at 4 °C (14,000 × *g* for 10 min).
17. Remove the top aqueous layer, add additional four times volume of methanol
18. Centrifuge the new mixture at 4 °C (14,000 × *g* for 10 min).
19. Remove the solvent and let the pellet dry for 10 min
20. Resuspend the pellet in 60 μL buffer A containing 2% SDS

3.3.4 Immunoblotting

Protocols for immunoblot analysis of these samples have been previously described (Tang et al., 2018), but any immunoblotting protocol could be used. We employed custom made antibodies against the translocators, and a commercial antibody against the phosphorylated GSK-peptide (Anti-P-GSK-3β, Cell Signaling Technology) to detect phosphorylation of GSK-tagged proteins. An example is shown in Fig. 14. If an antibody against the protein under analysis is not available, there is a commercial antibody against the non-phosphorylated GSK peptide (anti-GSK-3β tag, Cell Signaling Technology) that can be used to detect the tagged protein. Since GSK is an endogenous protein in mammalian cells, a control experiment should be performed with samples of uninfected cells to avoid misinterpretation of the data. On uninfected HeLa cells we observed a single band around 48 kDa, and most (but not all) of this protein is removed during the rPFO permeabilization and cytosol removal step (not shown).

3.3.5 Discussion

While the phosphorylation of the GSK-peptide clearly indicates exposure of the protein segment to the cell cytosol, the lack of phosphorylation is less informative. The lack of phosphorylation should not be interpreted as the segment not being exposed to the cytosol because the GSK-peptide could be there but sterically hidden from the cytosolic kinases. However, a working model of the topological arrangement of the translocator could be constructed when regions exposed to the cytosol start to be unveiled. The working model will provide the basis to locate tags and determine what regions of the protein are located outside the target cell.

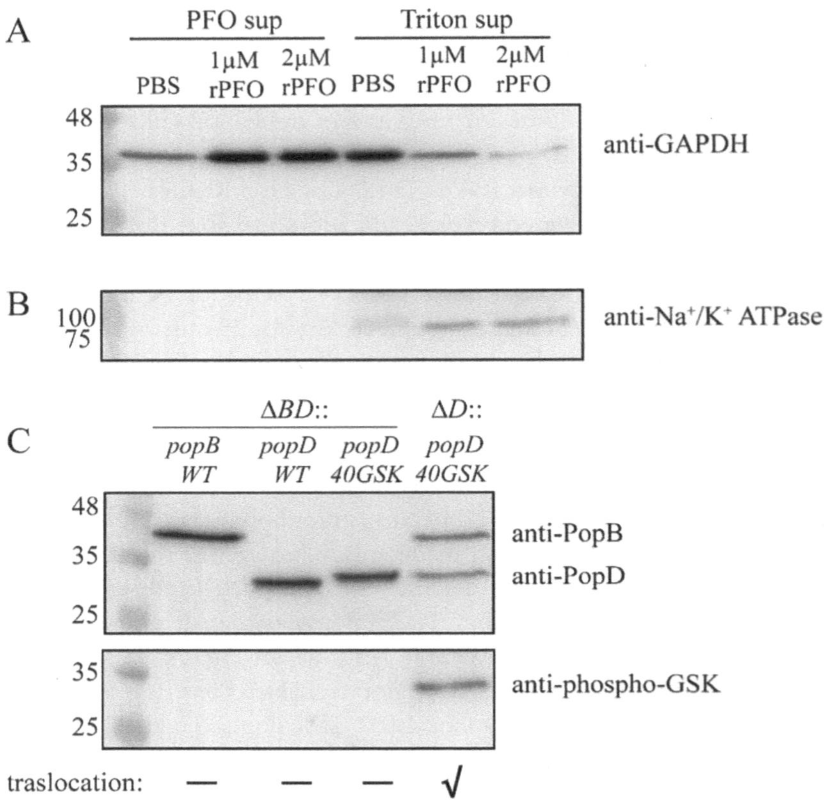

Fig. 14 Only properly assembled 40GSK-PopD in HeLa cell membranes is phosphorylated. (A and B) Validation of the membrane protein isolation method. PAKΔexoE ΔexoSTYΔpopD::popD WT infected HeLa cells were incubated with PBS, 1 μM rPFO, or 2 μM rPFO. PFO sup and Triton sup samples containing 23 μg of total protein were analyzed using immunoblotting for the presence of HeLa cell cytosol or plasma membrane markers using anti-GAPDH and anti-Na^+/K^+ ATPase antibodies. (C) detection of PopD insertion in HeLa cell membranes using GSK-tag phosphorylation assay. PAK ΔexoEΔexoSTYΔpopBD or ΔexoEΔexoSTYΔpopD strains complemented with pUCP18 plasmid carrying popB WT, popD WT or GSK-tagged popD codifying genes were incubated with HeLa cells in FBS free DMEM medium for 1 h. Proteins associated with HeLa cell membranes were isolated using the membrane protein isolation procedure described in Fig. 8, and the presence of PopB, PopD, and phospho-GSK PopD was detected by immunoblotting. Representative blots from two independent experiments are shown. The effector translocation ability of used PAK strains into HeLa cells is indicated with a "✓" symbol. *Adapted from Tang, Y., Romano, F.B., Breña, M., & Heuck, A.P. (2018). The Pseudomonas aeruginosa type III secretion translocator PopB assists the insertion of the PopD translocator into host cell membranes.* The Journal of Biological Chemistry 293, *8982–8993.*

4. Analysis of protein segments exposed to the outside of the HeLa cell plasma membrane

Exposure of the translocator to the outside of the target cell plasma membrane can be detected enzymatically, using tags that are specifically modified by purified recombinant enzymes (e.g., Sfp phosphopantetheinyl transferase labeling of the ybbR tag, Yin et al., 2005), or chemically by the modification of Cys residues incorporated into the primary sequence of the translocator (Bogdanov, 2017; Howe & Brown, 2017; Russo, Duncan, & Goldberg, 2019). For example, the chemical modification of single Cys-PopD variants with polyethylene glycol-maleimide will occur only if the Cys is exposed to the extracellular space (culture media) because the reagent cannot cross the plasma membrane. We will not describe this procedure in detail here because it is similar to the one described above without the treatment with rPFO. Briefly, after the incubation with bacteria, the Hela cells are washed and then incubated with the thiol specific reagent. The reagent is quenched with an excess of dithiothreitol and cell are lysed with DPBS supplemented with 0.1% Triton X-100. Modification of the Cys alters the mobility of PopD in SDS-PAGE (the molecular mass increases ~5 kDa for Methoxypolyethylene glycol maleimide, Sigma #63187). The shift is detected by immunoblotting using antibodies against the protein of interest as described above.

Exposure of Cys residues to the HeLa cell cytosol could be also assessed by adding a rPFO permeabilization step to the protocol before the addition of the reagent. While this procedure is not as simple as the enzymatic labeling of GSK-tags, it can be used to analyze regions of the protein that cannot tolerate the insertion of the tag.

5. Summary

Great advances have been made on the structural and functional studies of PFPs using reductionist approaches. The combination of purified recombinant proteins, model membrane systems, and biophysical techniques have unveil the mechanism of binding, assembly, and pore-formation for many of these proteins. However, the study of some PFPs have resisted the use of this reductionist approach. These PFPs are unstable when isolated as monomers in aqueous solutions, and they do not insert easily into model membranes. A typical example of these proteins is the T3S system translocators

(Faudry et al., 2007). These proteins need to be co-expressed with their cognate chaperone, and can only be purified as a complex with the chaperone (Faudry et al., 2006; Romano et al., 2011). Great efforts have been done towards the biophysical characterization of the T3S translocon using model membranes, but a functional assay to validate the working models generated with these systems is still not available (Romano et al., 2016; Tang et al., 2018). In this chapter, we described the optimization of a procedure that can be used to study the topology of T3S translocators on live cells (Fig. 8). We highlight the importance of isolating functional translocons from other non-functional forms of the translocators that could be present in the system (Fig. 14C), in order to properly interpret the data (Tang et al., 2018). With recent advances in Cryo-EM, Cryo-electron tomography (e.g., Lunelli et al., 2020; Park et al., 2018), and super resolution microscopy (Singh & Kenney, 2020), the topological maps of the translocators will be instrumental not only to interpret the experimental data, but also to locate purification tags and fluorescent labels without affecting the function of the proteins.

Acknowledgment

This work was supported by the National Institutes of Health Grant NIH GM097414 and R41 AI149922 (APH).

References

Alouf, J. E., & Freer, J. H. (1999). *The comprehensive sourcebook of bacterial protein toxins* (2nd ed.). London: Academic Press.

Arrhenius, S. (1907). *Immunochemistry. The application of the principles of physical chemistry to the study of the biological antibodies*. New York: The Macmillian Company.

Bogdanov, M. (2017). Mapping of membrane protein topology by substituted cysteine accessibility method (SCAM™). *Methods in Molecular Biology (Clifton, NJ), 1615*, 105–128.

Boyd, C. M., & Bubeck, D. (2018). Advances in cryoEM and its impact on β-pore forming proteins. *Current Opinion in Structural Biology, 52*, 41–49.

Brodsky, I. E., & Medzhitov, R. (2008). Reduced secretion of YopJ by *Yersinia* limits in vivo cell death but enhances bacterial virulence. *PLoS Pathogens, 4*, e1000067.

Bucior, I., Tran, C., & Engel, J. (2014). Assessing *Pseudomonas* virulence using host cells. *Methods in Molecular Biology (Clifton, NJ), 1149*, 741–755.

Cosentino, K., Ros, U., & García-Sáez, A. J. (2016). Assembling the puzzle: Oligomerization of α-pore forming proteins in membranes. *Biochimica et Biophysica Acta, 1858*, 457–466.

Dal Peraro, M., & van der Goot, F. G. (2016). Pore-forming toxins: Ancient, but never really out of fashion. *Nature Reviews. Microbiology, 14*, 77–92.

Deng, W., Marshall, N. C., Rowland, J. L., McCoy, J. M., Worrall, L. J., Santos, A. S., et al. (2017). Assembly, structure, function and regulation of type III secretion systems. *Nature Reviews. Microbiology, 15*, 323–337.

Discola, K. F., Forster, A., Boulay, F., Simorre, J. P., Attree, I., Dessen, A., et al. (2014). Membrane and chaperone recognition by the major translocator protein PopB of the type III secretion system of *Pseudomonas aeruginosa*. *The Journal of Biological Chemistry*, *289*, 3591–3601.

Engel, J., & Balachandran, P. (2009). Role of *Pseudomonas aeruginosa* type III effectors in disease. *Current Opinion in Microbiology*, *12*, 61–66.

Faudry, E., Job, V., Dessen, A., Attree, I., & Forge, V. (2007). Type III secretion system translocator has a molten globule conformation both in its free and chaperone-bound forms. *FEBS Journal*, *274*, 3601–3610.

Faudry, E., Vernier, G., Neumann, E., Forge, V., & Attree, I. (2006). Synergistic pore formation by type III toxin translocators of *Pseudomonas aeruginosa*. *Biochemistry*, *45*, 8117–8123.

Flanagan, J. J., Tweten, R. K., Johnson, A. E., & Heuck, A. P. (2009). Cholesterol exposure at the membrane surface is necessary and sufficient to trigger perfringolysin O binding. *Biochemistry*, *48*, 3977–3987.

Francis, M. S., & Wolf-Watz, H. (1998). YopD of *Yersinia pseudotuberculosis* is translocated into the cytosol of HeLa epithelial cells: Evidence of a structural domain necessary for translocation. *Molecular Microbiology*, *29*, 799–813.

Gilbert, R. J., Dalla Serra, M., Froelich, C. J., Wallace, M. I., & Anderluh, G. (2014). Membrane pore formation at protein-lipid interfaces. *Trends in Biochemical Sciences*, *39*, 510–516.

Goure, J., Pastor, A., Faudry, E., Chabert, J., Dessen, A., & Attree, I. (2004). The V antigen of *Pseudomonas aeruginosa* is required for assembly of the functional PopB/PopD translocation pore in host cell membranes. *Infection and Immunity*, *72*, 4741–4750.

Ha, U., & Jin, S. (2001). Growth phase-dependent invasion of *Pseudomonas aeruginosa* and its survival within HeLa cells. *Infection and Immunity*, *69*, 4398–4406.

Heuck, A. P., & Johnson, A. E. (2005). Membrane recognition and pore formation by bacterial pore-forming toxins. In L. K. Tamm (Ed.), *Protein-lipid interactions from membrane domains to cellular networks* (pp. 163–186). Weinheim: Wiley-VCH.

Hmelo, L. R., Borlee, B. R., Almblad, H., Love, M. E., Randall, T. E., Tseng, B. S., et al. (2015). Precision-engineering the *Pseudomonas aeruginosa* genome with two-step allelic exchange. *Nature Protocols*, *10*, 1820.

Howe, V., & Brown, A. J. (2017). Determining the topology of membrane-bound proteins using PEGylation. *Methods in Molecular Biology (Clifton, NJ)*, *1583*, 201–210.

Iwai, H., Kim, M., Yoshikawa, Y., Ashida, H., Ogawa, M., Fujita, Y., et al. (2007). A bacterial effector targets Mad2L2, an APC inhibitor, to modulate host cell cycling. *Cell*, *130*, 611–623.

Johnson, B. B., Breña, M., Anguita, J., & Heuck, A. P. (2017). Mechanistic insights into the cholesterol-dependent binding of perfringolysin O-based probes and cell membranes. *Scientific Reports*, *7*, 13793.

Johnson, B., & Heuck, A. (2014). Perfringolysin O structure and mechanism of pore formation as a paradigm for cholesterol-dependent cytolysins. In G. Anderluh, & R. Gilbert (Eds.), *MACPF/CDC proteins—agents of defence, attack and invasion* (pp. 63–81). Netherlands: Springer.

Johnson, B. B., Moe, P. C., Wang, D., Rossi, K., Trigatti, B. L., & Heuck, A. P. (2012). Modifications in perfringolysin O domain 4 alter the cholesterol concentration threshold required for binding. *Biochemistry*, *51*, 3373–3382.

Kaufman, M. R., Jia, J., Zeng, L., Ha, U., Chow, M., & Jin, S. (2000). *Pseudomonas aeruginosa* mediated apoptosis requires the ADP-ribosylating activity of ExoS. *Microbiology*, *146*, 2531–2541.

Lee, V. T., Anderson, D. M., & Schneewind, O. (1998). Targeting of *Yersinia* Yop proteins into the cytosol of HeLa cells: One-step translocation of YopE across bacterial and eukaryotic membranes is dependent on SycE chaperone. *Molecular Microbiology*, *28*, 593–601.

Lee, V. T., Mazmanian, S. K., & Schneewind, O. (2001). A program of *Yersinia enterocolitica* Type III secretion reactions is activated by specific signals. *Journal of Bacteriology, 183*, 4970–4978.

Lee, V. T., Smith, R. S., Tummler, B., & Lory, S. (2005). Activities of *Pseudomonas aeruginosa* effectors secreted by the type III secretion system in vitro and during infection. *Infection and Immunity, 73*, 1695–1705.

Lee, P. C., Zmina, S. E., Stopford, C. M., Toska, J., & Rietsch, A. (2014). Control of type III secretion activity and substrate specificity by the cytoplasmic regulator PcrG. *Proceedings of the National Academy of Sciences of the United States of America, 111*, E2027–E2036.

Lotze, J., Reinhardt, U., Seitz, O., & Beck-Sickinger, A. G. (2016). Peptide-tags for site-specific protein labelling in vitro and in vivo. *Molecular BioSystems, 12*, 1731–1745.

Lunelli, M., Kamprad, A., Bürger, J., Mielke, T., Spahn, C. M. T., & Kolbe, M. (2020). Cryo-EM structure of the *Shigella* type III needle complex. *PLoS Pathogens, 16*, e1008263.

Luo, W., & Donnenberg, M. S. (2011). Interactions and predicted host membrane topology of the Enteropathogenic *Escherichia coli* translocator protein EspB. *Journal of Bacteriology, 193*, 2972.

Nguyen, V. S., Jobichen, C., Tan, K. W., Tan, Y. W., Chan, S. L., Ramesh, K., et al. (2015). Structure of AcrH–AopB chaperone-translocator complex reveals a role for membrane hairpins in type III secretion system translocon assembly. *Structure, 23*, 2022–2031.

Park, D., Lara-Tejero, M., Waxham, M. N., Li, W., Hu, B., Galán, J. E., et al. (2018). Visualization of the type III secretion mediated *Salmonella*–host cell interface using cryo-electron tomography. *eLife*, 7, e39514.

Romano, F. B., Rossi, K. C., Savva, C. G., Holzenburg, A., Clerico, E. M., & Heuck, A. P. (2011). Efficient isolation of *Pseudomonas aeruginosa* type III secretion translocators and assembly of heteromeric transmembrane pores in model membranes. *Biochemistry, 50*, 7117–7131.

Romano, F. B., Tang, Y., Rossi, K. C., Monopoli, K. R., Ross, J. L., & Heuck, A. P. (2016). Type 3 secretion translocators spontaneously assemble a hexadecameric transmembrane complex. *The Journal of Biological Chemistry, 291*, 6304–6315.

Rossjohn, J., Feil, S. C., McKinstry, W. J., Tweten, R. K., & Parker, M. W. (1997). Structure of a cholesterol-binding, thiol-activated cytolysin and a model of its membrane form. *Cell, 89*, 685–692.

Russo, B. C., Duncan, J. K., & Goldberg, M. B. (2019). Topological analysis of the type 3 secretion system translocon pore protein IpaC following its native delivery to the plasma membrane during infection. *mBio, 10*, e00877-19.

Shatursky, O., Heuck, A. P., Shepard, L. A., Rossjohn, J., Parker, M. W., Johnson, A. E., et al. (1999). The mechanism of membrane insertion for a cholesterol-dependent cytolysin: A novel paradigm for pore-forming toxins. *Cell, 99*, 293–299.

Singh, M. K., & Kenney, L. J. (2020). Super-resolution imaging of bacterial pathogens and visualization of their secreted effectors. *FEMS Microbiology Reviews*. https://doi.org/10.1093/femsre/fuaa050.

Tang, Y., Romano, F. B., Breña, M., & Heuck, A. P. (2018). The *Pseudomonas aeruginosa* type III secretion translocator PopB assists the insertion of the PopD translocator into host cell membranes. *The Journal of Biological Chemistry, 293*, 8982–8993.

Torruellas Garcia, J., Ferracci, F., Jackson, M. W., Joseph, S. S., Pattis, I., Plano, L. R. W., et al. (2006). Measurement of effector protein injection by type III and type IV secretion systems by using a 13-residue Phosphorylatable glycogen synthase kinase tag. *Infection and Immunity, 74*, 5645–5657.

Urbanowski, M. L., Brutinel, E. D., & Yahr, T. L. (2007). Translocation of ExsE into Chinese hamster ovary cells is required for transcriptional induction of the

Pseudomonas aeruginosa type III secretion system. *Infection and Immunity*, *75*, 4432–4439.

Vermeulen, A. J., Tang, Y., & Heuck, A. P. (2016). Translocation of toxins by Gram-negative pathogens using the type III secretion system. In P. Gopalakrishnakone, B. Stiles, A. Alape-Girón, J. D. Dubreuil, & M. Mandal (Eds.), *Microbial toxins* (pp. 1–18). Dordrecht: Springer Netherlands.

Yin, J., Straight, P. D., McLoughlin, S. M., Zhou, Z., Lin, A. J., Golan, D. E., et al. (2005). Genetically encoded short peptide tag for versatile protein labeling by Sfp phosphopantetheinyl transferase. *Proceedings of the National Academy of Sciences of the United States of America*, *102*, 15815–15820.

CHAPTER FIFTEEN

Single-molecule imaging of pore-forming toxin dynamics in droplet interface bilayers

Christopher Parperis and Mark I. Wallace*
Department of Chemistry, Britannia House, King's College London, London, United Kingdom
*Corresponding author: e-mail address: mark.wallace@kcl.ac.uk

Contents

Abstract

Single-channel recording from pore-forming toxins (PFTs) provides a clear and direct molecular readout of toxin action. However to complete any mechanistic understanding of PFT behavior, this functional kinetic readout must be linked to the underlying changes in toxin structure, binding, conformation, or stoichiometry. Here we review how single-molecule imaging methods might be used to further our understanding

Methods in Enzymology, Volume 649
ISSN 0076-6879
https://doi.org/10.1016/bs.mie.2021.01.035

of PFTs, and provide detailed practical guidance on the use of droplet interface bilayers as a method capable of examining both single-molecule fluorescence and single-channel electrical signals from PFTs.

1. Introduction

Patch-clamping was the first widely adopted single-molecule technique, allowing ion channels to be interrogated without ensemble averaging (Neher & Sakmann, 1976) and sparking a revolution in our understanding of these important proteins (Coyaud, Kurkdjian, Kado, & Hedrich, 1987; Davies, Hunt, & Sanders, 1994; Maathuis & Sanders, 1994; Müller-Röber et al., 1995; Pei, Ward, & Schroeder, 1999). Despite its advantages, patch-clamping provides only one side of the structure–function approach that is the bedrock of modern molecular biology (Lord, Lee, & Moerner, 2010). A more complete single-molecule interpretation must link functional readout with spatial information such as pore structure, conformation, or location. One route would be to couple electrical measurements with a conformational trace via single-molecule fluorescence (SMF) microscopy (Moerner & Fromm, 2003; Shashkova & Leake, 2017; Wazawa & Ueda, 2005). The key hurdle is to provide an experiment optimized for *both* these methods, while retaining a membrane environment still representative of that experienced in vivo.

Droplet interface bilayers (DIBs) have been pioneered by our group as a means to address this challenge (Heron, Thompson, Mason, & Wallace, 2007; Thompson, Heron, Santoso, & Wallace, 2007). In a DIB, a planar membrane can be formed at the boundary between two aqueous droplets present in an oil bath, where each droplet is surrounded by a lipid monolayer. This methodology provides a flexible platform for single-molecule experiments where both single-molecule electrical and optical signals can be extracted from the membrane. In the context of this volume, we explore how SMF measurements might further our understanding of pore-forming toxin (PFT) function, discuss what future experiments might be feasible, and explore the advantages and limitations of different artificial membranes and imaging modalities. Practical guidance on the preparation of DIBs is also provided.

2. Context: Choosing the right technique

Single-molecule techniques can address many aspects of PFT behavior, ranging from pore architecture to pore conductance (Bainbridge, Gokce, &

Lakey, 1993; Gilbert, Serra, Froelich, Wallace, & Anderluh, 2014). The scope of these techniques ultimately determines the questions one can ask about a given PFT; its location within the cell, gating behavior, the presence of associated proteins, diffusive properties, etc. The suitability of techniques for addressing each question hinges on the context in which the PFT of interest is found, and a method optimized for molecular readout in vivo, for instance, may be less useful in an in vitro setting (Coelho, Maghelli, & Tolić-Nørrelykke, 2013). Overall, the reduced complexity and improved sensitivity of in vitro measurements must be weighed against alterations to activity caused by a nonnative environment (Escajadillo & Nizet, 2018).

2.1 Pore-forming toxins in cells

The overall effect any biomolecule has on an organism is best understood from in vivo studies (Moleiro, Conceição, Leite-Moreira, & Rocha-Sousa, 2017; Neubert et al., 1986). PFTs are no exception. This is because the influence is exerted within native context and interplay between all facets of the system is accounted for. In vivo experiments allow PFTs to be viewed on their own terms. Unfortunately, single-molecule study in vivo is not trivial due to the complex chemical environment (Duzdevich & Greene, 2012), and the opacity of biological tissue impedes SMF techniques. The role of PFTs in bacterial infection, including in vivo study, has been reviewed previously (DuMont & Torres, 2014; Los, Randis, Aroian, & Ratner, 2013). At time of writing, PFTs are yet to be studied at molecular resolution in whole organisms. This is not down to technological insufficiency; SMF has already been used to view structures in vivo (Bademosi et al., 2016; Berning, Willig, Steffens, Dibaj, & Hell, 2012), demonstrating applicability that may be extended to PFTs.

Whole-cell studies ex vivo have used patch-clamping and SMF techniques to investigate PFT assembly (Benke et al., 2015; Nguyen, Kamio, & Higuchi, 2003; Ribeiro, Vignes, & Brehélin, 2003; Subburaj et al., 2015; Thompson, Cronin, Bayley, & Wallace, 2011). Although the chemical gradients, tissue perfusion and mechanical forces of living organisms cannot be recreated ex vivo (Choi, Kwok, & Yun, 2015; Liu, Lavis, & Betzig, 2015), single-molecule tracking of PFTs in live cells reveals pore location and heterogeneity of copy numbers present in a cellular population (Elf & Barkefors, 2019). A typical experimental setup for SMF imaging of PFTs in live cells is represented in Fig. 1.

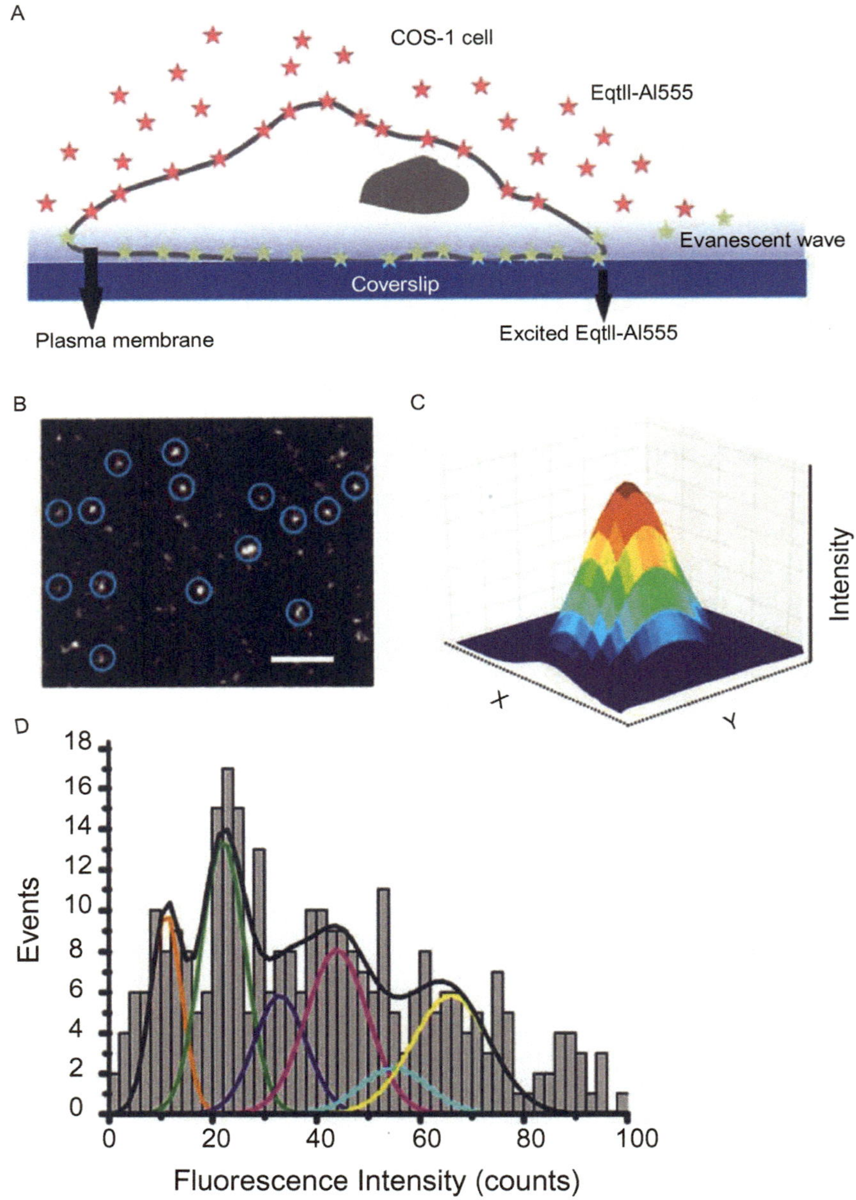

Fig. 1 Single-molecule imaging of Alexa555-labeled equinatoxin-II attacking the membrane of a living cell (Subburaj, Ros, Hermann, Tong, & García-Sáez, 2015). (A) Experimental setup using TIRF microscopy. (B) Labeled PFTs imaged as single-particles. Scale bar 1 μm. (C) 2D Gaussian fitting of particles to plot fluorescence intensity. (D) Pore stoichiometry from histogram of fluorescence intensities.

2.2 Pore-forming toxins in model membranes

The inherent complexity of living membranes is not conducive to understanding and, as we have just seen, there are limits imposed by whole-cell study. It is often desirable to make a careful selection of a few fragments of the overall picture for controlled study in an in vitro setting. There are two reasons for this. Besides precluding the noise and artifacts introduced by cellular complexity, it also allows for independent variables of the system to be precisely manipulated; temperature, pH, the concentration of various lipids, oils, salts, and proteins, etc. (Glazier & Salaita, 2017; Hamada et al., 2008). This kind of approach is not without disadvantages; emergent behaviors are necessarily absent and cannot be investigated (Rojko & Anderluh, 2015), and addressing every detail of membrane function piece by piece, gradually fleshing-out a broader understanding, is a long-winded process. Nevertheless, with the right equipment and a little forethought, experiments can be designed to collect quantitative information which describes all sorts of membrane phenomena—including the action of PFTs (Frey, Barth, Kranz, & Mizaikoff, 2018; Peterman et al., 2002; Rojko et al., 2014).

To this end, a range of model membrane platforms has been developed to aid the study of lipid bilayers (Siontorou, Nikoleli, Nikolelis, & Karapetis, 2017). These artificial assemblies all mimic natural membranes, but in each system the same biomolecules are presented in a unique geometry that may be opportune for one type of experiment and not for another. Perhaps the simplest starting point is detergent micelles, which have been used, for example, to reveal the assembly dynamics (Eifler et al., 2006) and conformational changes (Benke et al., 2015) of the ClyA pore at single-molecule resolution. Black lipid membranes (BLMs) are comprised of a planar bilayer supported on a hydrophobic film, separating two electrolyte solutions (Mueller, Rudin, Ti Tien, & Wescott, 1962). This allows electrodes to be inserted either side of the bilayer, and aqueous PFT monomer units can be introduced once the lipid bilayer has formed (Anderluh, Barlič, Potrich, Maček, & Menestrina, 2000; Bakás, Chanturiya, Herlax, & Zimmerberg, 2006). The bilayer can also be supported directly on a solid surface, such as a glass coverslip. This supported lipid bilayer (SLB) allows an aqueous solution to be introduced to one side (the exposed face) of the bilayer. While electrical investigation of the bilayer is inaccessible in this context, surface-sensitive evanescent field techniques may be used, such as surface plasmon resonance (Cooper, Try, Carroll, Ellar, & Williams, 1998; Parkkila, Elderdfi, Bunker, & Viitala, 2018) and total internal reflection fluorescence (TIRF) microscopy

(Axmann, Schütz, & Huppa, 2015; Glazier & Salaita, 2017; Weatherill, Coker, Cheetham, & Wallace, 2018). SLBs are remarkably stable because, unlike a BLM, imperfections in the bilayer do not compromise the entire assembly (Andersson & Köper, 2016; Möller & Seeger, 2015). Microfabricated membrane arrays have been successfully created by several groups with the capabilities of combining optical and electrical measurements, albeit not always with demonstrated single-molecule sensitivity (Hansen et al., 2009; Hemmler, Böse, Wagner, & Peters, 2005; Peters, 2003). This list is not exhaustive, and many other innovative methods are beyond the scope of this short introduction. Notably, giant unilamellar vesicles are omitted here, as they are described in depth elsewhere in this volume.

2.3 Droplet interface bilayers

DIB methods (Bayley et al., 2008; Leptihn et al., 2013; Villar & Bayley, 2013) have been pioneered by our group, as an alternative model membrane system well suited to the study of PFTs (Rojko et al., 2014; Thompson et al., 2007). A DIB can form either between two aqueous droplets in a solution of lipid in oil (not unlike a BLM) or between an aqueous droplet and a planar hydrogel (Bayley et al., 2008; Leptihn et al., 2013). As single-molecule imaging is our goal, we shall focus on DIBs formed on a planar hydrogel, as it is the more appropriate of the two modalities for this type of experiment (Fig. 2).

The DIB approach involves forming an agarose hydrogel on the surface of a microscope coverslip, followed by deposition of a phospholipid-in-oil solution and, subsequently, an aqueous droplet. The lipid molecules self-arrange into monolayers at both interfaces and, under gravity, the droplet sinks through the oil solution and spontaneously forms a bilayer membrane upon contact with the supporting hydrogel. As with BLMs, electrodes can be applied either side of the membrane to permit electrophysiological study. But unlike BLMs, the planar surface of the supporting hydrogel predetermines a geometry amenable to microscopy. SMF techniques discussed later in this chapter can be applied to visualize either single fluorescent emitters diffusing on the bilayer (Jaggers, Ridone, Martinac, & Baker, 2019; Thompson et al., 2011), or a plume of fluorescent material caused by an ionic flux through the membrane (Harriss, Cronin, Thompson, & Wallace, 2011; Huang, Romero-Ruiz, Castell, Bayley, & Wallace, 2015). Biomolecules such as proteins, nucleic acid polymers, salts, sugars, and cofactors can be included in the

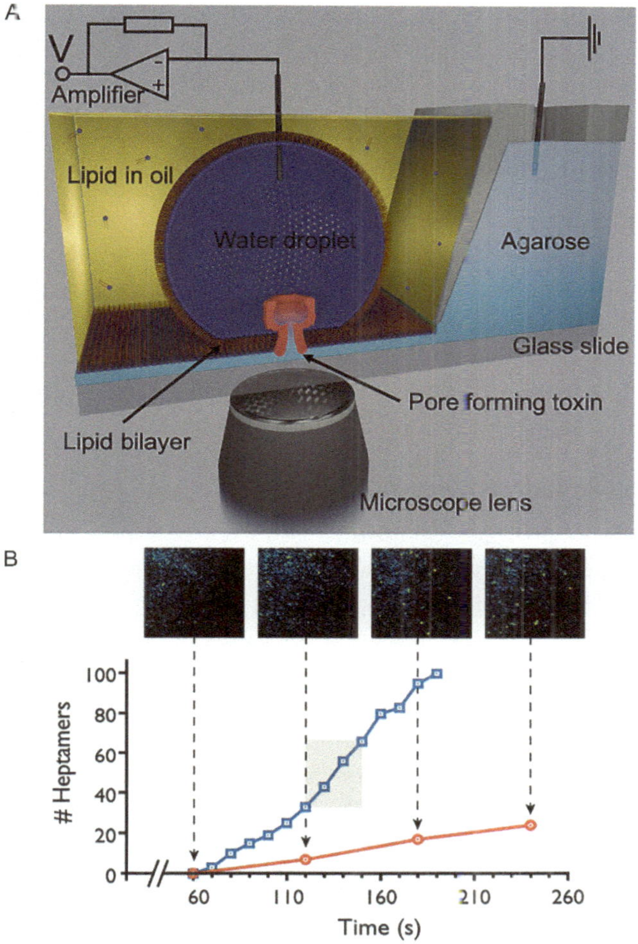

Fig. 2 (A) DIB schematic. Ag/AgCl electrodes applied either side of bilayer record ionic current through pores. The objective lens images fluorescently labeled monomer units in the plane of the bilayer. (B) α-Hemolysin assembly on a DIB. Single-particles are visualized by TIRF microscopy while an electrical trace of pore insertions is simultaneously recorded (Thompson et al., 2011).

aqueous droplet, agarose gel, or injected into the system after the membrane has been established to perform experiments.

Other novel imaging platforms have been developed for the single-molecule study of PFTs, including a suspended planar membrane for fluorescein-tagged colicin-Ia (Rajapaksha, Wang, & Lu, 2013), a tethered lipid bilayer for label-free imaging of protein interactions using surface plasmon resonance (Taylor, Linman, Wilkop, & Cheng, 2009), and a microfabricated array chip to detect diffusion of a fluorescent dye through single α-hemolysin pores (Watanabe et al., 2014).

3. Single-molecule methods

With this artificial membrane platform in hand, what single-molecule imaging methods are most compatible? The single-molecule toolkit has matured to allow a wide range of biomolecular processes to be scrutinized (Betzig, Chichester, Lanni, & Taylor, 1993; Block, Blair, & Berg, 1989; Gelles, Schnapp, & Sheetz, 1988; Hall, 1956; Macklin, Trautman, Harris, & Brus, 1996; Moerner & Kador, 1989), which we can compare against the requirements of experiments on PFTs:

3.1 Single-particle tracking

Chemical linking of fluorophores to PFTs is a common means to achieve single-molecule fluorescence imaging, for example to track the locations of single toxin monomers or pores within a sample (Subburaj et al., 2015). The PFT of interest is labeled either by direct, covalent linking to a fluorescent emitter (Giddings, Johnson, & Tweten, 2003; Ramm et al., 2020; Shatursky et al., 1999; Wuethrich et al., 2014), or through interactions with an intermediary molecule (an antibody, for example) which has itself been covalently bound to a fluorophore (Abrami, Fivaz, Glauser, Parton, & van der Goot, 1998; Evans et al., 2020; Garcia et al., 2012). Due to fluorescence quenching (or enhancement) upon membrane association, fluorescent labeling can also be used to identify amino acid residues within a PFT which interact with the bilayer, as previously shown with perfringolysin O (Shepard et al., 1998) and *Clostridium septicum* α-toxin (Melton, Parker, Rossjohn, Buckley, & Tweten, 2004). An advantage of fluorescence imaging is its ability to visualize living specimens, not just fixed cells (Murphy & Davidson, 2012), thus probing dynamic processes (Kiepas, Voorand, Mubaid, Siegel, & Brown, 2020; Stabley, Oh, Simon, Mattheyses, & Salaita, 2015). An example of single-particle tracking of PFTs using fluorescence is shown in Fig. 3.

Fluorophores are typically small organic molecules or proteins (Shen et al., 2020; Shimomura, 1979; Zheng et al., 2014), but semiconductor nanoparticles known as quantum dots can also be used (Nikazar et al., 2020). The photophysics of quantum dots are somewhat favorable compared to organic fluorophore species, remaining active for much longer (tens of minutes rather than tens of seconds (Petryayeva, Algar, & Medintz, 2013)), and have been used for single-particle tracking experiments with PFTs such

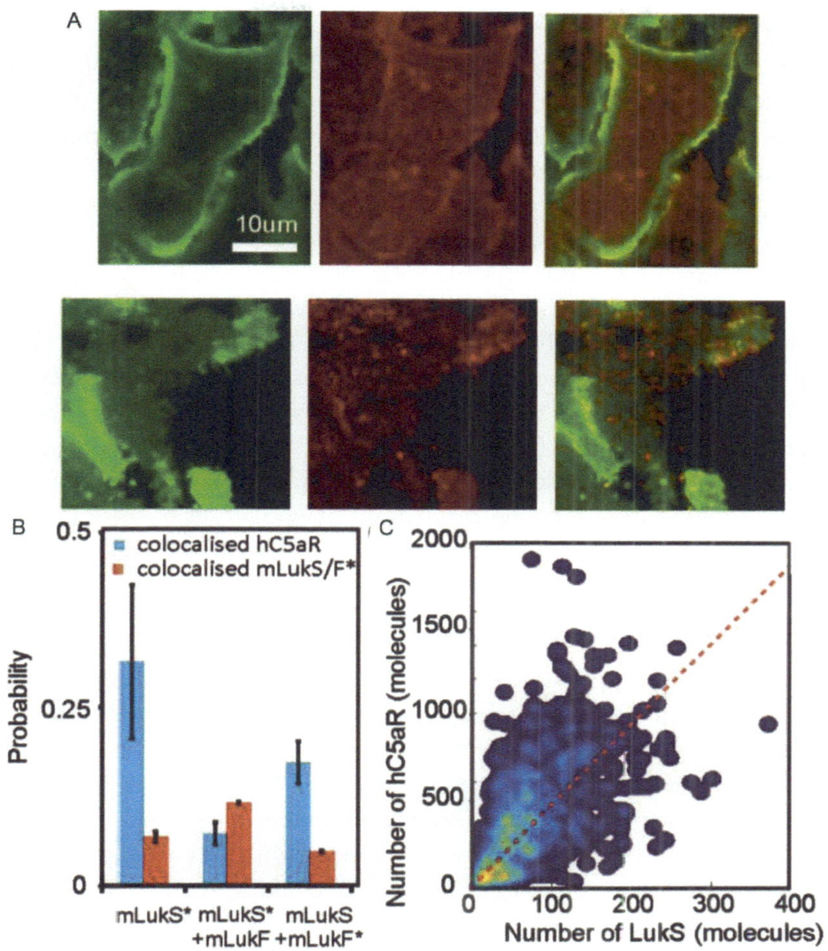

Fig. 3 Colocalized multicolor imaging to investigate the action of LukS and LukF toxins on live cells by binding to hC5aR receptors on the membrane surface. (A) Cells express a GFP-mutant form of hC5aR (green) and are incubated with labeled LukS (top row) and LukF (bottom row) toxins (red). A dual-color image is recorded. (B) Proportions of single toxin particles colocalized with receptors. (C) Correlation of hC5aR with labeled LukS stoichiometry. *Adapted from Haapasalo, K., Wollman, A. J. M., de Haas, C. J. C., van Kessel, K. P. M., van Strijp, J. A. G., & Leake, M. C. (2019). Staphylococcus aureus toxin LukSF dissociates from its membrane receptor target to enable renewed ligand sequestration. The FASEB Journal, 33 (3), 3807–3824.*

as *Clostridium septicum* α-toxin and *Clostridium perfringens* ε-toxin (Türkcan et al., 2012). Quantum dots are, however, susceptible to photo-blinking (Bera, Qian, Tseng, & Holloway, 2010; Goldoni et al., 2005) and are often cytotoxic (Derfus, Chan, & Bhatia, 2004).

3.2 Förster resonance energy transfer

Förster resonance energy transfer (FRET) is the distance dependent non-radiative transfer of energy between two fluorescent reporters (Sasmal, Pulido, Kasal, & Huang, 2016; Sekar & Periasamy, 2003), and measuring the relative fluorescence intensities from donor and acceptor fluorophores can be used to determine nanometer distances between the two. Widely used to assay binding of particular genetic sequences, molecular beacons are a common use of FRET (Kim, Sohn, & Tan, 2008; Mao, Ying, Wu, Krueger, & Chen, 2019; Santangelo, Nix, Tsourkas, & Bao, 2004).

Single-molecule FRET has been used to reveal transiently existing intermediate folding structures of the ClyA toxin (Dingfelder, Benke, Nettels, & Schuler, 2018), as well as to obtain assembly dynamics of PFTs on membranes, such as for ClyA (Benke et al., 2015) and γ-hemolysin (Nguyen et al., 2003). FRET has also been used in combination with electrophysiology to study the pore-forming mechanism of *Bacillus thuringiensis* Cry1Aa (Groulx, Juteau, & Blunck, 2010).

3.3 Optical single-channel recording

Analogous to electrophysiology, optical single-channel recording (oSCR) uses a fluorogenic ion indicator dye to provide an optical readout of local ion current near the pore. A potential difference across the bilayer drives cations through the pore. Subsequent binding to the dye results in a "plume" of fluorescence as shown in Fig. 4 (Demuro & Parker, 2004, 2005). Ionic flux across the membrane is proportional to plume brightness, enabling visualization of current through PFTs using a range of techniques (Heron, Thompson, Cronin, Bayley, & Wallace, 2009; Weatherill & Wallace, 2015; Zou, Lifshitz, Tuft, Fogarty, & Singer, 2002). For example, the interaction of α-hemolysin with cyclodextrins (Heron et al., 2009) and nucleotide polymers (Huang et al., 2015), and the action of aerolysin on erythrocyte membranes (Tschödrich-Rotter, Kubitscheck, Ugochukwu, Buckley, & Peters, 1996) has been investigated. Notably, the temporal resolution of this optical technique is limited by camera speed, which is typically much lower than for a direct electrical measurement (Beier & Ibey, 2014; Mason, Simpson, Mahaut-Smith, & Robinson, 2005).

Pores can be analyzed individually, even when a single membrane contains many pores, since each appears as a discrete plume of fluorescence in the image formed on the camera detector. This allows parallelization of data

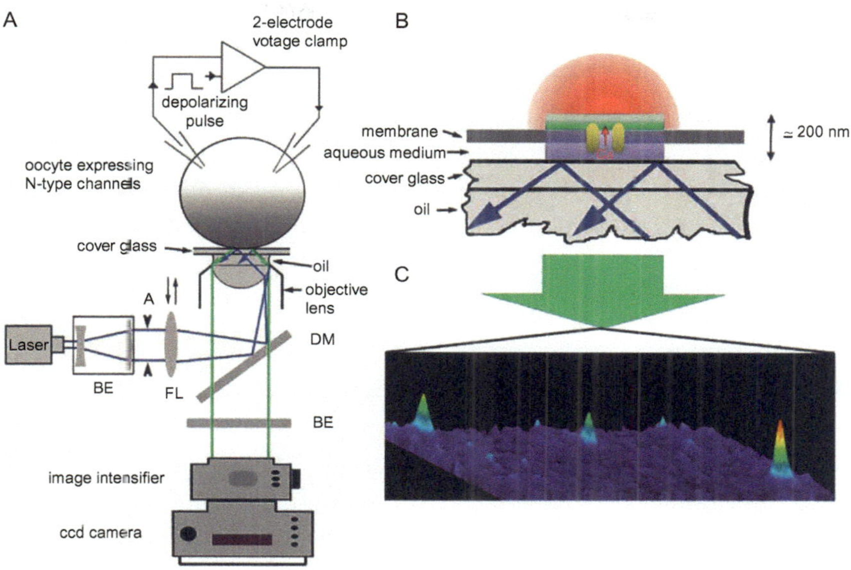

Fig. 4 Experimental setup for oSCR. Ionic flux through nicotinic acetylcholine receptors is demonstrated, but works equally well for PFTs (Demuro & Parker, 2004).

collection in an experiment; rather than investigating tens of membranes, each containing just one pore to guarantee kinetics arise from it alone, a single membrane is all that is needed to image many pores simultaneously (Demuro & Parker, 2005; Huang et al., 2015).

3.4 Limitations of single-molecule fluorescence

Caution must be taken to ensure the fluorophore does not alter function of the biomolecule to which it is attached (Hedegaard et al., 2018; Jensen, 2012; Lam & Tajkhorshid, 2020). Similarly, the chemistry of the local environment may govern photophysics of the fluorophore (such as interaction with a lipid membrane (Hughes, Rawle, & Boxer, 2014)). Emission cannot be recorded for arbitrarily long timescales due to photobleaching (Eggeling, Widengren, Rigler, & Seidel, 1998). The rate of photo-emission is fundamentally limited by the maximum speed at which an electron may be excited and subsequently relax back to ground state (Taylor & Sandoghdar, 2019). In addition, fluorescence emission can greatly depend on rotational orientation (Royer & Scarlata, 2008, chap. 5).

4. Imaging modalities

Given these methods, what experimental configurations are compatible with the requirements of both the membrane platform and single-molecule imaging? Techniques for imaging SMF commonly rely on a restricted detection or illumination volume to gain adequate sensitivity (Dreier et al., 2019), and provide a convenient criteria with which to subdivide this section.

4.1 Total internal reflection fluorescence microscopy

Total internal reflection at the microscope coverslip surface can achieve a suitably low illumination volume (Axelrod, Thompson, & Burghardt, 1983). The evanescent field produced in this configuration decays exponentially with distance from the surface; at most the field reaches a few hundred nanometers (Martin-Fernandez, Tynan, & Webb, 2013) into the medium of lower refractive index before its intensity completely dies away. A biological sample in contact with the glass interface is illuminated with the evanescent field as a thin optical section (Sako & Yanagida, 2003). Labeled molecules located in the field are stimulated to fluoresce, and the emission is focused to an image on the camera detector with high axial resolution and signal-to-noise (Soubies et al., 2019).

Despite low axial depth of the evanescent field, lateral dimensions are equal to the incident laser beam cross section, which under high magnification can easily fill the field of view of most camera detectors. TIRF microscopy is a wide-field imaging technique for investigating PFTs, with temporal resolution determined by the camera sampling rate. All that is required is a planar lipid bilayer formed in close proximity to the microscope coverslip. SLBs and DIBs are fit for such a purpose. PFTs tagged with a fluorescent label can be directly visualized interacting with the membrane (Fig. 5), or ionic flux can be imaged by oSCR (Castell, Berridge, & Wallace, 2012). Live cells can be viewed using this modality with improved signal-to--noise, but only for the membrane in direct contact with the glass coverslip (Axelrod, 2001; Mattheyses, Simon, & Rappoport, 2010). TIRF has been used to study hetero-oligomeric Panton–Valentine leukocidin, γ-hemolysin and LukF (Kaneko & Kamio, 2004), and α-hemolysin assembly dynamics (Thompson et al., 2011).

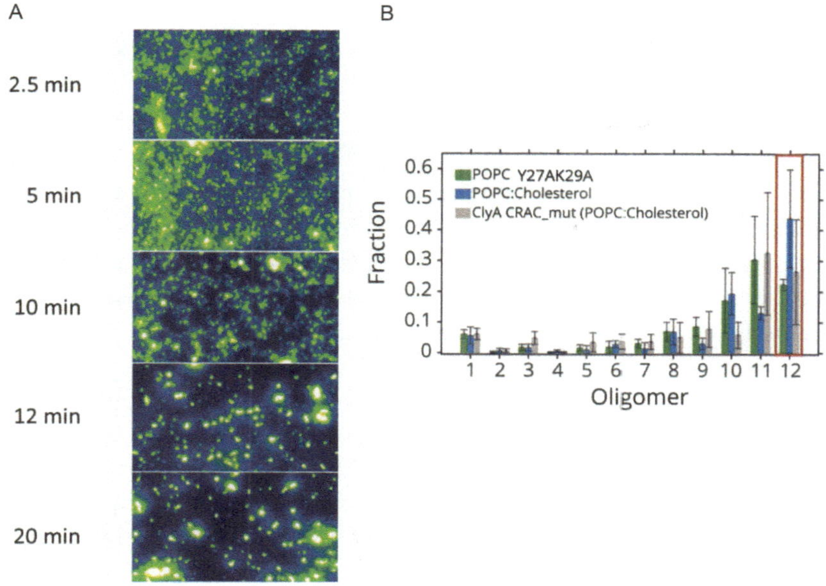

Fig. 5 TIRF visualization of ClyA on a supported lipid bilayer. The time lapse shows separated monomers oligomerizing over several minutes to form higher-order structures (Sathyanarayana et al., 2018).

4.2 Confocal microscopy

Confocal microscopy is a narrow-field imaging technique where an excitation laser source is focused to a single point using a pinhole to cut out of focus light. Fluorescence emission is recorded from this confined location and signal may be collected either from diffusing molecules or by scanning the point laser source through the sample to build an image pixel by pixel (Jonkman & Brown, 2015). Exocytosis of *Vibrio cholerae* cytolisin toxins (Elluri et al., 2014) and interaction of single lysenin particles (Ishitsuka, Yamaji-Hasegawa, Makino, Hirabayashi, & Kobayashi, 2004) have been studied using this modality. The laser focal point can be scanned in the *z* axis of the sample, as well as in *x* and *y*, to obtain a three dimensional image (Wilson, 2011). This gives confocal microscopy an advantage over TIRF for most whole-cell imaging applications, since the evanescent field cannot be transposed in *z*. But there are downsides. Scanning the light throughout the entire field of view takes time, diminishing temporal resolution (Mondal, 2014). Furthermore the pinhole gives rise to an elliptic point spread function several times longer than its width. This causes poorer axial resolution for a single optical section than TIRF is able to achieve (~500–700 nm and <100 nm, respectively), even if the lateral resolution is comparable between the two systems.

4.3 Super-resolution fluorescence microscopy

Super-resolution is an umbrella term encompassing those techniques which improve the spatial resolution of a light microscope beyond the diffraction limit (Huang, Babcock, & Zhuang, 2010; Pujals, Feiner-Gracia, Delcanale, Voets, & Albertazzi, 2019). In recent years these methods have enabled measurement of membrane reorganization around listeriolysin O toxin pores, revealing altered lipid diffusion in their immediate vicinity (Sarangi, P, Ayappa, Visweswariah, & Basu, 2016), as well as super-resolved images of membrane pores such as the nuclear pore complex (Fig. 6) (Gwosch et al., 2020; Schlichthaerle et al., 2019). Excellent reviews of these techniques demonstrate the potential for future research (Huang, Bates, & Zhuang, 2009; Vicidomini, Bianchini, & Diaspro, 2018), but in our opinion, super-resolution is far from achieving its potential for PFT research.

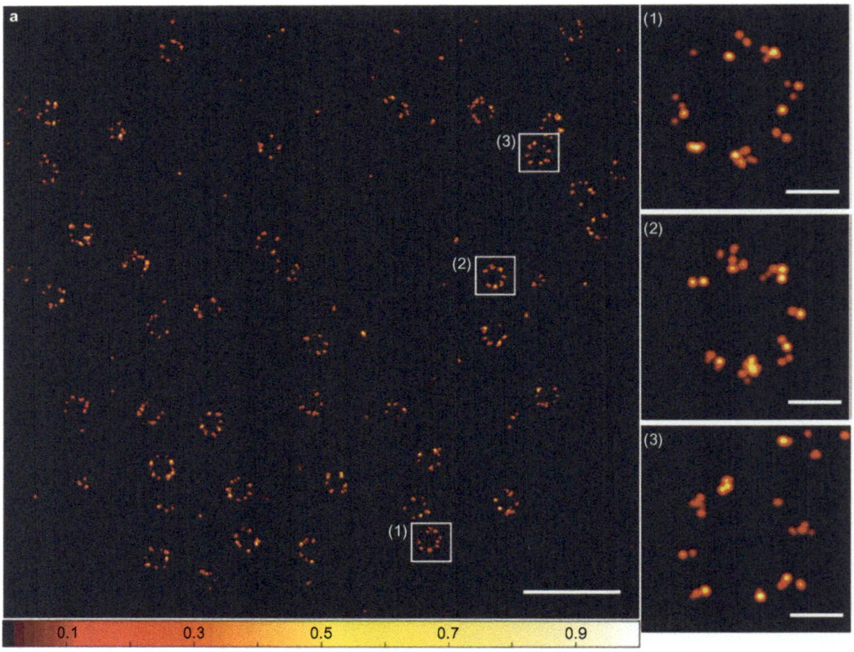

Fig. 6 MINFLUX used to image nuclear pore complex. Scale bars 500 nm (a), 50 nm (insets). *Adapted from Gwosch, K. C., Pape, J. K., Balzarotti, F., Hoess, P., Ellenberg, J., Ries, J., & Hell, S. W. (2020). MINFLUX nanoscopy delivers 3D multicolor nanometer resolution in cells. Nature Methods, 17 (2), 217–224.*

5. Single-molecule imaging of pore-forming toxins in droplet interface bilayers

DIBs enabled our research group to link SMF imaging with ionic current measurements for PFTs, examining assembly (Thompson et al., 2011), pore formation (Rojko et al., 2014) and lipid dependence alongside measurements of the ionic current across the membrane. In the following section, we include an expanded practical method for constructing DIBs for electrical recording, focusing on PFTs. This in vitro platform can be used in combination with any of the fluorescent techniques described in the preceding section. Specific guidance for oSCR and SMF imaging is also included.

5.1 Droplet interface bilayer preparation

5.1.1 Electrode preparation

A reference (ground) electrode and a recording electrode are prepared. For the reference electrode, silver wire (~1 cm in length, 1.5 mm diameter) is soldered to one end of a length of PVC-coated copper wire. At the other end, a connector for a patch-clamp amplifying head stage is soldered. For the recording electrode, silver wire (~2 cm in length, 0.1 mm diameter) is soldered to one end of a length of PVC-insulated copper wire. At the other end, a connector for a patch-clamp amplifying head stage is soldered. Fine grit sandpaper is used to remove the oxidized outer layer from the silver wire of both electrodes. The surface is washed in ethanol, dried, and submerged in a solution of sodium hypochlorite for ~20 min, until a white silver chloride residue forms on the electrode surface. As a rule of thumb, replenish the silver electrodes after every 10 h of experiments by abrasion of the surface and reapplication of hypochlorite.

5.1.2 Lipid preparation

From a prepared 50 mg mL^{-1} lipid stock in chloroform, pipette 26 μL into a glass phial and dry under a stream of nitrogen, steadily rotating the phial in hand to form an even lipid film (1.3 mg). Remove remaining solvent by desiccating under vacuum for 30 min. Add 150 μL of hexadecane and vortex to produce a 9.67 mg mL^{-1} solution.

5.1.3 Plasma cleaning

Glass coverslips are treated in an oxygen plasma (40 kHz) for 5 min, removing contaminants and rendering the surface hydrophilic. This increases surface wetting to ensure formation of a thin hydrogel in the spin-coating step.

5.1.4 *Agarose preparation*

Prepare a substrate agarose mixture of ultra-low gelling temperature agarose 0.75% (w/w) in MilliQ water, ~1 mL total volume (MilliQ is necessary to prevent precipitation of salt crystals when water evaporates from the coverslip surface). A rehydration agarose mixture is also prepared from low gelling temperature agarose 1.3% (w/w) in buffered salt solution. The mixtures must be heated to a temperature of 90°C to create a molten agarose liquids that will cross link to form a gel upon cooling.

5.1.5 *Spin-coating*

Pipette 140 μL of molten substrate agarose onto a plasma cleaned coverslip and spin at 4000 rpm for 30 s. In many commercial spin coaters, the coverslip is held in place by vacuum suction and must be carefully placed to prevent deformation while under vacuum. The resulting hydrogel is thin enough to permit the evanescent TIRF field to penetrate a lipid bilayer formed on its surface. The hydrogel thickness is alterable by increasing/decreasing spin speed or changing the agarose concentration of the hydrogel.

5.1.6 *Device fabrication*

A DIB device is milled from a block of poly(methyl methacrylate) (PMMA) using a CNC machine. A 4 × 4 array of 1 mm diameter wells permits formation of 16 droplet bilayers (Fig. 7). A recess in the underside of the device accommodates a glass coverslip. Injection ports either side of the well array allow introduction of agarose hydrogel to the inner volume of the device. A second PMMA block is milled to produce chambers to which lipid-in-oil may be added for droplet incubation with lipid.

5.1.7 *Device assembly*

Bring the spin-coated surface of the microscope coverslip into contact with the underside of the CNC-milled PMMA device. Apply precut lengths of tape to make the coverslip fast to the device, ensuring that each edge of the coverslip sits flush in place. Place the assembled device on a flat surface and, while applying downward pressure, flush molten rehydration agarose into the injection port of the device with a 200 μL pipette, filling the underside cavity. Take care to flush out air bubbles with additional molten agarose. If this step has been done correctly, the interstitial cavity separating each droplet well will fill with rehydration agarose, while the droplet wells will remain unfilled, with only the thin layer of substrate agarose present in the center of each well.

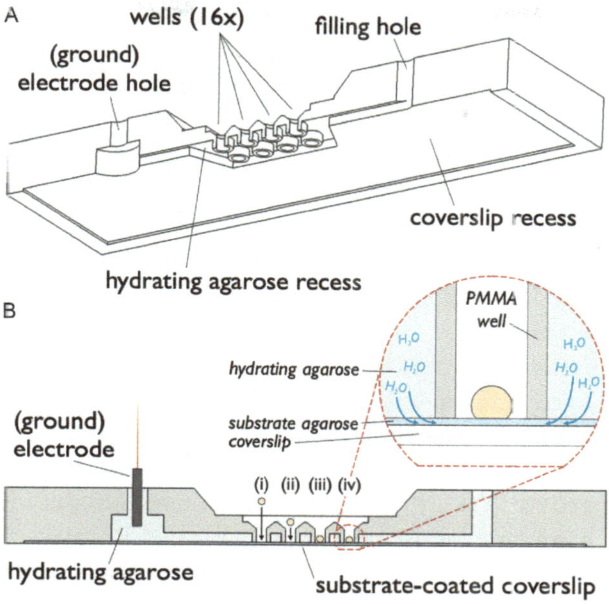

Fig. 7 (A) Cutaway diagram of the PMMA device for droplet bilayer formation (Sengel & Wallace, 2016). (B) (i) An aqueous droplet is introduced to a device well containing lipid-in-oil. (ii) The droplet sinks under gravity while lipids form a monolayer at its interface. (iii) The droplet contacts the substrate agarose and oil is excluded. (iv) Planar bilayer formation.

Pipette ~60 μL of lipid solution into the top side of the device, filling the 16-well array and taking care to expel air bubbles that form in any well. This will prevent the thin layer of agarose in the bottom of each well from dehydrating, while allowing an interfacial monolayer of lipids to form between the hydrogel and lipid solution. Once the lipid solution is in the device, leave it to stand for at least 30 min so the phospholipids may spontaneously arrange to form a monolayer at the phase boundary between the oil and the hydrogel.

5.1.8 Droplet solution preparation

An aqueous solution is required for droplet formation. For electrophysiological recording, it is sufficient to prepare the PFT of interest, at desired concentration, in buffered salt solution osmotically balanced with the rehydration agarose. For Ca^{2+} oSCR, the following paragraphs describe representative additional steps:

Prepare an aqueous solution of 10 mM HEPES, 1.5 M potassium chloride. To this solution add ~5 g of Chelex-200 beads, a resin which effectively removes excess divalent cations that would otherwise increase the background fluorescence of the sample. Leave the solution to be agitated overnight. Remove the chelex beads using spin filtration. Discard the beads and pH balance the solution to 7.4 (the chelex resin will have likely caused a significant change to the solution pH). Store this solution at 4°C.

Prepare a solution of 0.75 M calcium chloride, 10 mM HEPES buffer, and pH balance the solution to 7.4. Store this solution at 4°C. (Half the concentration of the divalent calcium salt is utilized to produce an approximately osmotically balanced DIB.) This solution is used to prepare the rehydration agarose for an oSCR experiment. At this stage, a stock solution of the PFT of interest can be diluted into the previously prepared potassium chloride solution until the desired concentration is reached for your experiments. Tagged PFTs (<1 nM) for SMF can also be introduced at this stage.

Prepare a solution of 10 mM EDTA. In a PCR tube, introduce the chelexed potassium/PFT solution (15 μL) to the EDTA solution (0.6 μL). Add 0.6 μL of the stock solution of calcium-sensitive dye—we find Fluo-8 or Cal-520 work well. Add the EDTA before the calcium dye. The resulting aqueous mixture is used to prepare droplets for DIBs.

5.1.9 Droplet incubation

Transfer 2× ~25 μL volumes of lipid solution into 2× machine milled incubation chambers. Using a 2 μL pipette, backfill the tip with lipid solution and transfer ~150 nL droplets of the aqueous solution of interest into the prefilled incubation chambers. Backfilling reduces capillary action and surface tension effects that lead to poor reproducibility of droplet size. Eight droplets will fit into each incubation channel without difficulty. Once formed, allow the droplets to incubate for at least 30 min to allow for the self-assembly of an interfacial monolayer of phospholipids.

5.1.10 Droplet interface bilayer formation

With a 2 μL pipette set to 200 nL, inspirate a single droplet into the pipette tip after the appropriate incubation time of 30 min has elapsed. Introduce this droplet to one well of the device, making a note of which well is used. If DIB formation fails, and the droplet fuses with the underlying hydrogel, wait for 5 min before making a new attempt in a fresh well. (It is important not to reuse wells for imaging experiments because when a droplet fuses with the hydrogel, it can cause the underlying liquid layer in that well to become too thick for TIRF imaging of the bilayer.) When confirmation that droplets form stable bilayers in the device is complete—i.e., the bilayer remains for at least several minutes without rupturing—begin to introduce the remaining droplets to the other available wells.

If the droplets and device are allowed to incubate for too long (>2 h) phospholipids crowd at the phase boundaries in multiple layers, preventing droplets from sinking to the bottom of the wells for bilayer formation. If the droplets do not sink and form bilayers after 30 min, gentle heating by placing the device on a heat block can encourage droplets to sink further and form membranes.

5.2 Electrophysiology in droplet interface bilayers

For electrophysiological study, the reference electrode must first be inserted into the agarose hydrogel, via the injection port. To make the device air tight, preventing the agarose gel from dehydrating, a strip of tape can be placed over the open exit port and vacuum grease applied around the entry point of the reference electrode. The recording electrode will be inserted into the droplet. A coating of hydrophilic agarose on the tip of the Ag/AgCl electrode will encourage it to penetrate through the phase boundary at the edge of the droplet. This can be achieved by simply dipping a micropipette tip into the salt-free substrate agarose, and touching the electrode tip against a bead of agarose formed on the tip. The device, when ready for imaging/electrophysiological investigation, will need to be placed inside a Faraday cage. Inserting the recording electrode into the droplet is achieved using a micromanipulator mounted on the microscope stage. Manipulating the position of the electrode is easier under a low magnification (×4 or ×10). Subsequently, an oil immersion TIRF objective lens can be brought into close proximity with the underside of the glass coverslip, the image coming into focus representing a space that is coplanar with the bilayer membrane. Droplets must be interrogated one at a time if electrophysiological or oSCR measurements are intended to be collected.

Certain properties of the bilayers are affected by the extent of hydration in the supporting agarose hydrogel. Protein pores in the membrane are found to be less diffusive in dehydrated agarose, for instance (Rojko et al., 2014). For studies of diffusion in the membrane, it can be beneficial to ensure the agarose substrate is hydrated by controlling the local humidity. Practically, a damp tissue is often sufficient. Once DIBs have formed they are stable on a timescale of days to months, provided the agarose is prevented from dehydrating (Bayley et al., 2008).

5.3 Imaging in droplet interface bilayers

For oSCR, the signal (a bright plume of fluorescence) appears as a bright spot on a dark background and is only visible when calcium ions are drawn into the droplet (Fig. 8). To achieve this, a negative voltage bias is induced between the recording electrode in the droplet and the reference electrode placed in the agarose hydrogel. Single fluorophores are imaged directly using an excitation source of appropriate wavelength. For a two-color experiment involving both oSCR and SMF, the fluorescence emission may be split using a dichroic mirror before the camera detector.

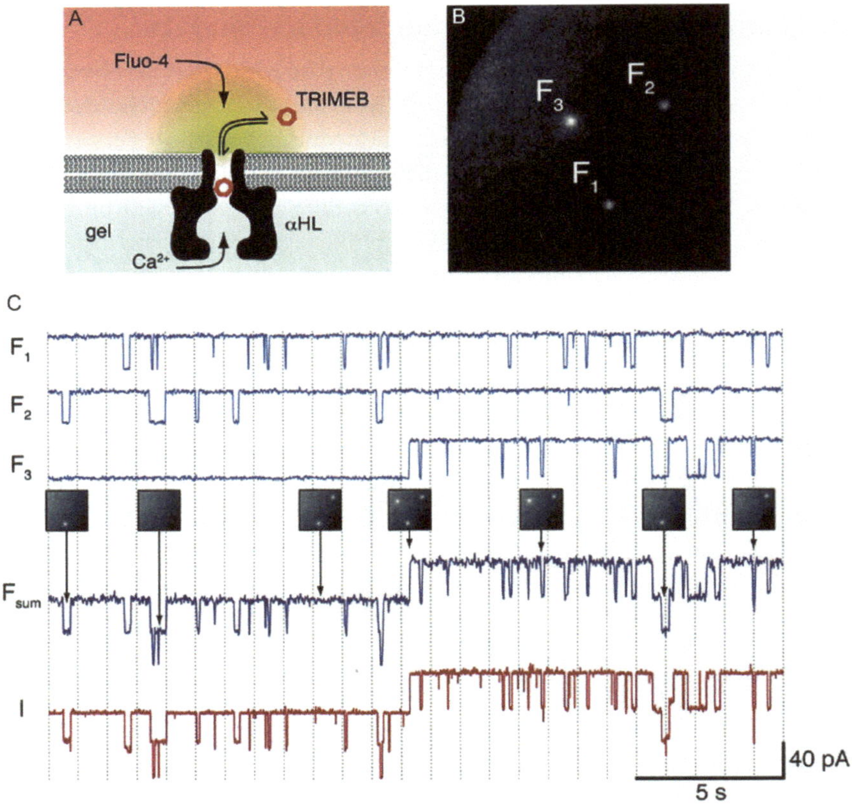

Fig. 8 Example of simultaneous optical and electrical single-channel recording of pores formed by αHL toxin (Heron et al., 2009) in a DIB, demonstrating how the optical trace can accurately record the ionic current through each individual pore. Transient blocking is caused by binding and unbinding of trimethyl-β-cyclodextrin (TRIMEB) within the pore. FLuo-4 is the Ca^{2+} sensitive dye used to visualize the flux of ions entering the droplet.

6. Potential limitations and problems

The time taken for lipids to form a stabilizing monolayer that will prevent immediate bilayer rupturing is influenced by several factors. If lipids are introduced as vesicles within the aqueous droplet, the necessary incubation time is much shorter than for lipids dissolved in hexadecane (Bayley et al., 2008; Mitev, Ivanova, & Vassilieff, 2002). Also, silicone oil can be used to cut the time for bilayer formation, but this may also have adverse effects on membrane properties (Tamaddoni, Taylor, Hepburn, Michael Kilbey, & Sarles, 2016). Additionally, increasing the concentration of lipid can reduce the necessary incubation time, but will also increase the size of the bilayer which forms relative to droplet size, reducing surface tension and lowering the contact angle of the droplet. However, increasing the lipid concentration above $\sim$10 mg mL^{-1} results in noticeably longer bilayer formation times (Leptihn et al., 2013). Similarly, lowering the lipid concentration results in smaller bilayers with a higher contact angle, and the droplet and device must be incubated for a longer time period before the stabilizing monolayer of lipid forms. A concentration below $\sim$2 mg mL^{-1} is found to be too low to form sufficient monolayers, resulting in bilayers that rupture shortly after formation (Leptihn et al., 2013).

Membrane proteins of research interest are often purified and stored in detergent to stabilize the folded conformation (Seddon, Curnow, & Booth, 2004). Detergents, owing to their amphiphatic nature, will be driven into the bilayer by the hydrophobic effect and can cause it to rupture. It is therefore important to keep a note of the detergent concentration in any protein stock, so that a dilution below the critical micelle concentration can easily be carried out prior to experiment. This is of course true for any model membrane system. For example, in DIBs we have previously reported DPhPC bilayers will tolerate up to 0.01% (wt/vol) n-decyl-β-D-maltopyranoside in the aqueous droplet (Leptihn et al., 2013).

The application of an electrical potential across the bilayer will cause, at sufficiently high voltage, electroporation bilayer rupture, as is observed in other model membrane systems. We have previously used DIBs to image the electroporation process (Sengel & Wallace, 2016). In the same way electrical potential can compromise bilayer stability, so too can an osmotic gradient across the membrane. Although droplet bilayers are resistant to osmotic pressure-induced rupture, above 50 atm of osmotic difference bilayers are unstable. Bearing this fact in mind, it is sensible to use buffered solutions to prepare DIBs.

DIBs are stable at elevated temperatures up to and beyond the melting temperature of the agarose hydrogel. Bilayers are also prone to rupture at temperatures below the freezing point of the oil used as lipid solvent.

7. Conclusion

The study of pore-forming toxins is being transformed by techniques which probe their action at single-molecule resolution. Recent insights have elucidated previously unknown mechanisms of pore assembly, but the influence on pathogenesis is still often not clearly understood. This chapter offers an introduction to single-molecule imaging techniques capable of linking structural and functional aspects of this problem, their merits and limitations, and instruction in their application to pore-forming toxin dynamics in vitro.

References

Abrami, L., Fivaz, M., Glauser, P.-E., Parton, R. G., & van der Goot, F. (1998). A pore-forming toxin interacts with a GPI-anchored protein and causes vacuolation of the endoplasmic reticulum. *Journal of Cell Biology*, *140*(3), 525–540.

Anderluh, G., Barlič, A., Potrich, C., Maček, P., & Menestrina, G. (2000). Lysine 77 is a key residue in aggregation of equinatoxin II, a pore-forming toxin from sea anemone actinia equina. *The Journal of Membrane Biology*, *173*(1), 47–55.

Andersson, J., & Köper, I. (2016). Tethered and polymer supported bilayer lipid membranes: structure and function. *Membranes*, *6*(2), 30.

Axelrod, D. (2001). Total internal reflection fluorescence microscopy in cell biology. *Traffic*, *2*(11), 764–774.

Axelrod, D., Thompson, N. L., & Burghardt, T. P. (1983). Total internal reflection fluorescent microscopy. *Journal of Microscopy*, *129*(1), 19–28.

Axmann, M., Schütz, G. J., & Huppa, J. B. (2015). Single molecule fluorescence microscopy on planar supported bilayers. *Journal of Visualized Experiments: JoVE*, (105), e53158–e53158.

Bademosi, A. T., Lauwers, E., Padmanabhan, P., Odierna, L., Chai, Y. J., Papadopulos, A., … Meunier, F. A. (2016). In vivo single-molecule imaging of syntaxin1A reveals polyphosphoinositide- and activity-dependent trapping in presynaptic nanoclusters. *Nature Communications*, 7(1), 13660.

Bainbridge, G., Gokce, I., & Lakey, J. H. (1998). Voltage gating is a fundamental feature of porin and toxin beta-barrel membrane channels. *FEBS Letters*, *431*(3), 305–308.

Bakás, L., Chanturiya, A., Herlax, V., & Zimmerberg, J. (2006). Paradoxical lipid dependence of pores formed by the Escherichia coli alpha-hemolysin in planar phospholipid bilayer membranes. *Biophysical Journal*, *91*(10), 3748–3755.

Bayley, H., Cronin, B., Heron, A., Holden, M. A., Hwang, W. L., Syeda, R., … Wallace, M. (2008). Droplet interface bilayers. *Molecular BioSystems*, *4*(12), 1191–1208.

Beier, H. T., & Ibey, B. L. (2014). Experimental comparison of the high-speed imaging performance of an EM-CCD and sCMOS camera in a dynamic live-cell imaging test case. *PLoS ONE*, *9*(1), e84614.

Benke, S., Roderer, D., Wunderlich, B., Nettels, D., Glockshuber, R., & Schuler, B. (2015). *Nature Communications*, *6*, 6198.

Bera, D., Qian, L., Tseng, T.-K., & Holloway, P. H. (2010). Quantum dots and their multimodal applications: A review. *Materials*, *3*(4), 2260–2345.

Berning, S., Willig, K. I., Steffens, H., Dibaj, P., & Hell, S. W. (2012). Nanoscopy in a living mouse brain. *Science*, *335*(6068), 551.

Betzig, E., Chichester, R. J., Lanni, F., & Taylor, D. L. (1993). Near-field fluorescence imaging of cytoskeletal actin. *Bioimaging*, *1*(3), 129–135.

Block, S. M., Blair, D. F., & Berg, H. C. (1989). Compliance of bacterial flagella measured with optical tweezers. *Nature*, *338*(6215), 514–518.

Castell, O. K., Berridge, J., & Wallace, M. I. (2012). Quantification of membrane protein inhibition by optical ion flux in a droplet interface bilayer array. *Angewandte Chemie International Edition*, *51*(13), 3134–3138.

Choi, M., Kwok, S. J. J., & Yun, S. H. (2015). In vivo fluorescence microscopy: Lessons from observing cell behavior in their native environment. *Physiology*, *30*(1), 40–49.

Coelho, M., Maghelli, N., & Tolić-Nørrelykke, I. M. (2013). Single-molecule imaging in vivo: The dancing building blocks of the cell. *Integrative Biology*, *5*(5), 748–758.

Cooper, M. A., Try, A. C., Carroll, J., Ellar, D. J., & Williams, D. H. (1998). Surface plasmon resonance analysis at a supported lipid monolayer. *Biochimica et Biophysica Acta (BBA)—Biomembranes*, *1373*(1), 101–111.

Coyaud, L., Kurkdjian, A., Kado, R., & Hedrich, R. (1987). Ion channels and ATP-driven pumps involved in ion transport across the tonoplast of sugarbeet vacuoles. *Biochimica et Biophysica Acta (BBA)—Biomembranes*, *902*(2), 263–268.

Davies, J. M., Hunt, I., & Sanders, D. (1994). Vacuolar H(+)-pumping ATPase variable transport coupling ratio controlled by pH. *Proceedings of the National Academy of Sciences*, *91*(18), 8547–8551.

Demuro, A., & Parker, I. (2004). Imaging the activity and localization of single voltage-gated Ca(2+) channels by total internal reflection fluorescence microscopy. *Biophysical Journal*, *86*(5), 3250–3259.

Demuro, A., & Parker, I. (2005). Optical single-channel recording: Imaging Ca2+ flux through individual ion channels with high temporal and spatial resolution. *Journal of Biomedical Optics*, *10*(1), 11002.

Derfus, A. M., Chan, W. C. W., & Bhatia, S. N. (2004). Probing the cytotoxicity of semiconductor quantum dots. *Nano Letters*, *4*(1), 11–18.

Dingfelder, F., Benke, S., Nettels, D., & Schuler, B. (2018). Mapping an equilibrium folding intermediate of the cytolytic pore toxin ClyA with single-molecule FRET. *The Journal of Physical Chemistry B*, *122*(49), 11251–11261.

Dreier, J., Castello, M., Coceano, G., Cáceres, R., Plastino, J., Vicidomini, G., & Testa, I. (2019). Smart scanning for low-illumination and fast RESOLFT nanoscopy in vivo. *Nature Communications*, *10*(1), 556.

DuMont, A. L., & Torres, V. J. (2014). Cell targeting by the staphylococcus aureus pore-forming toxins: It's not just about lipids. *Trends in Microbiology*, *22*(1), 21–27.

Duzdevich, D., & Greene, E. C. (2012). Towards physiological complexity with in vitro single-molecule biophysics. *Philosophical Transactions of the Royal Society of London. Series B, Biological Sciences*, *368*(1611), 20120271.

Eggeling, C., Widengren, J., Rigler, R., & Seidel, C. A. M. (1998). Photobleaching of fluorescent dyes under conditions used for single-molecule detection: Evidence of two-step photolysis. *Analytical Chemistry*, *70*(13), 2651–2659.

Eifler, N., Vetsch, M., Gregorini, M., Ringler, P., Chami, M., Philippsen, A., … Grauschopf, U. (2006). Cytotoxin ClyA from Escherichia coli assembles to a 13-meric pore independent of its redox-state. *EMBO Journal*, *25*(11), 2652–2661.

Elf, J., & Barkefors, I. (2019). Single-molecule kinetics in living cells. *Annual Review of Biochemistry*, *88*(1), 635–659.

Elluri, S., Enow, C., Vdovikova, S., Rompikuntal, P. K., Dongre, M., Carlsson, S., … Wai, S. N. (2014). Outer membrane vesicles mediate transport of biologically active vibrio cholerae cytolysin (VCC) from V. cholerae strains. *PLoS ONE*, *9*(9), e106731.

Escajadillo, T., & Nizet, V. (2018). Pharmacological targeting of pore-forming toxins as adjunctive therapy for invasive bacterial infection. *Toxins*, *10*(12), 1–19.

Evans, J. C., Johnstone, B. A., Lawrence, S. L., Morton, C. J., Christie, M. P., Parker, M. W., & Tweten, R. K. (2020). A key motif in the cholesterol-dependent cytolysins reveals a large family of related proteins. *mBio*, *11*(5), e02351–20.

Frey, C. M., Barth, H., Kranz, C., & Mizaikoff, B. (2018). Horizontal black lipid bilayer membranes for studying pore-forming toxins. *Analytical Methods*, *10*(26), 3153–3161.

Garcia, P. S., Chieppa, G., Desideri, A., Cannata, S., Romano, E., Luly, P., & Rufini, S. (2012). Sticholysin II: A pore-forming toxin as a probe to recognize sphingomyelin in artificial and cellular membranes. *Toxicon*, *60*(5), 724–733.

Gelles, J., Schnapp, B. J., & Sheetz, M. P. (1988). Tracking kinesin-driven movements with nanometre-scale precision. *Nature*, *331*(6155), 450–453.

Giddings, K. S., Johnson, A. E., & Tweten, R. K. (2003). Redefining cholesterol's role in the mechanism of the cholesterol-dependent cytolysins. *Proceedings of the National Academy of Sciences of the United States of America*, *100*(20), 11315–11320.

Gilbert, R. J. C., Serra, M. D., Froelich, C. J., Wallace, M. I., & Anderluh, G. (2014). Membrane pore formation at protein-lipid interfaces. *Trends in Biochemical Sciences*, *39*(11), 510–516.

Glazier, R., & Salaita, K. (2017). Supported lipid bilayer platforms to probe cell mechanobiology. *Biochimica et Biophysica acta. Biomembranes*, *1859*(9 Pt A), 1465–1482.

Goldoni, G., Troiani, F., Rontani, M., Bellucci, D., Molinari, E., & Hohenester, U. (2005). Collective properties of electrons and holes in coupled quantum dots BT. In B. A. Joyce, P. C. Kelires, A. G. Naumovets, & D. D. Vvedensky (Eds.), *Quantum dots: Fundamentals, applications, and frontiers* (pp. 269–283). Dordrecht: Springer Netherlands (Eds.).

Groulx, N., Juteau, M., & Blunck, R. (2010). Rapid topology probing using fluorescence spectroscopy in planar lipid bilayer: The pore-forming mechanism of the toxin Cry1Aa of Bacillus thuringiensis. *The Journal of General Physiology*, *136*(5), 497–513.

Gwosch, K. C., Pape, J. K., Balzarotti, F., Hoess, P., Ellenberg, J., Ries, J., & Hell, S. W. (2020). MINFLUX nanoscopy delivers 3D multicolor nanometer resolution in cells. *Nature Methods*, *17*(2), 217–224.

Hall, C. E. (1956). Method for the observation of macromolecules with the electron microscope illustrated with micrographs of DNA. *The Journal of Biophysical and Biochemical Cytology*, *2*(5), 625–628.

Hamada, T., Miura, Y., Komatsu, Y., Kishimoto, Y., Vestergaard, M., & Takagi, M. (2008). Construction of asymmetric cell-sized lipid vesicles from lipid-coated water-in-oil microdroplets. *The Journal of Physical Chemistry B*, *112*(47), 14678–14681.

Hansen, J. S., Perry, M., Vogel, J., Groth, J. S., Vissing, T., Larsen, M. S., … Nielsen, C. H. (2009). Large scale biomimetic membrane arrays. *Analytical and Bioanalytical Chemistry*, *395*(3), 719–727.

Harriss, L. M., Cronin, B., Thompson, J. R., & Wallace, M. I. (2011). Imaging multiple conductance states in an alamethicin pore. *Journal of the American Chemical Society*, *133*(37), 14507–14509.

Hedegaard, S. F., Derbas, M. S., Lind, T. K., Kasimova, M. R., Christensen, M. V., Michaelsen, M. H., … Nielsen, H. M. (2018). Fluorophore labeling of a cell-penetrating peptide significantly alters the mode and degree of biomembrane interaction. *Scientific Reports*, *8*(1), 6327.

Hemmler, R., Böse, G., Wagner, R., & Peters, R. (2005). Nanopore unitary permeability measured by electrochemical and optical single transporter recording. *Biophysical Journal*, *88*(6), 4000–4007.

Heron, A. J., Thompson, J. R., Cronin, B., Bayley, H., & Wallace, M. I. (2009). Simultaneous measurement of ionic current and fluorescence from single protein pores. *Journal of the American Chemical Society*, *131*(5), 1652–1653.

Heron, A. J., Thompson, J. R., Mason, A. E., & Wallace, M. I. (2007). Direct detection of membrane channels from gels using water-in-oil droplet bilayers. *Journal of the American Chemical Society*, *129*(51), 16042–16047.

Huang, B., Babcock, H., & Zhuang, X. (2010). Breaking the diffraction barrier: Super-resolution imaging of cells. *Cell*, *143*(7), 1047–1058.

Huang, B., Bates, M., & Zhuang, X. (2009). Super-resolution fluorescence microscopy. *Annual Review of Biochemistry*, *78*, 993–1016.

Huang, S., Romero-Ruiz, M., Castell, O. K., Bayley, H., & Wallace, M. I. (2015). High-throughput optical sensing of nucleic acids in a nanopore array. *Nature Nanotechnology*, *10*, 986.

Hughes, L. D., Rawle, R. J., & Boxer, S. G. (2014). Choose your label wisely: Water-soluble fluorophores often interact with lipid bilayers. *PLoS ONE*, *9*(2), e87649.

Ishitsuka, R., Yamaji-Hasegawa, A., Makino, A., Hirabayashi, Y., & Kobayashi, T. (2004). A lipid-specific toxin reveals heterogeneity of sphingomyelin-containing membranes. *Biophysical Journal*, *86*(1), 296–307.

Jaggers, O. B., Ridone, P., Martinac, B., & Baker, M. A. B. (2019). Fluorescence microscopy of piezo1 in droplet hydrogel bilayers. *Channels (Austin, Tex.)*, *13*(1), 102–109.

Jensen, E. C. (2012). Use of fluorescent probes: Their effect on cell biology and limitations. *The Anatomical Record*, *295*(12), 2031–2036.

Jonkman, J., & Brown, C. M. (2015). Any way you slice it—A comparison of confocal microscopy techniques. *Journal of Biomolecular Techniques: JBT*, *26*(2), 54–65.

Kaneko, J., & Kamio, Y. (2004). Bacterial two-component and hetero-heptameric Pore-forming Cytolytic Toxins: Structures, pore-forming mechanism, and organization of the genes. *Bioscience, Biotechnology, and Biochemistry*, *68*(5), 981–1003.

Kiepas, A., Voorand, E., Mubaid, F., Siegel, P. M., & Brown, C. M. (2020). Optimizing live-cell fluorescence imaging conditions to minimize phototoxicity. *Journal of Cell Science*, *133*(4), jcs242834.

Kim, Y., Sohn, D., & Tan, W. (2008). Molecular beacons in biomedical detection and clinical diagnosis. *International Journal of Clinical and Experimental Pathology*, *1*(2), 105–116.

Lam, K., & Tajkhorshid, E. (2020). Membrane interactions of Cy3 and Cy5 fluorophores and their effects on membrane-protein dynamics. *Biophysical Journal*, *119*(1), 24–34.

Leptihn, S., Castell, O. K., Cronin, B., Lee, E. H., Gross, L. C. M., Marshall, D. P., … Wallace, M. I. (2013). Constructing droplet interface bilayers from the contact of aqueous droplets in oil. *Nature Protocols*, *8*(6), 1048–1057.

Liu, Z., Lavis, L. D., & Betzig, E. (2015). Imaging live-cell dynamics and structure at the single-molecule level. *Molecular Cell*, *58*(4), 644–659.

Lord, S. J., Lee, H.-L. D., & Moerner, W. E. (2010). Single-molecule spectroscopy and imaging of biomolecules in living cells. *Analytical Chemistry*, *82*(6), 2192–2203.

Los, F. C. O., Randis, T. M., Aroian, R. V., & Ratner, A. J. (2013). Role of pore-forming toxins in bacterial infectious diseases. *Microbiology and Molecular Biology Reviews*, 77(2), 173–207.

Maathuis, F. J., & Sanders, D. (1994). Mechanism of high-affinity potassium uptake in roots of Arabidopsis thaliana. *Proceedings of the National Academy of Sciences of the United States of America*, *91*(20), 9272–9276.

Macklin, J. J., Trautman, J. K., Harris, T. D., & Brus, L. E. (1996). Imaging and time-resolved spectroscopy of single molecules at an interface. *Science*, *272*(5259), 255–258.

Mao, S., Ying, Y., Wu, X., Krueger, C. J., & Chen, A. K. (2019). CRISPR/dual-FRET molecular beacon for sensitive live-cell imaging of non-repetitive genomic loci. *Nucleic Acids Research*, *47*(20), e131–e131.

Martin-Fernandez, M. L., Tynan, C. J., & Webb, S. E. D. (2013). A 'pocket guide' to total internal reflection fluorescence. *Journal of Microscopy, 252*(1), 16–22.
Mason, M. J., Simpson, A. K., Mahaut-Smith, M. P., & Robinson, H. P. C. (2005). The interpretation of current-clamp recordings in the cell-attached patch-clamp configuration. *Biophysical Journal, 88*(1), 739–750.
Mattheyses, A. L., Simon, S. M., & Rappoport, J. Z. (2010). Imaging with total internal reflection fluorescence microscopy for the cell biologist. *Journal of Cell Science, 123*(Pt 21), 3621–3628.
Melton, J. A., Parker, M. W., Rossjohn, J., Buckley, J. T., & Tweten, R. K. (2004). The identification and structure of the membrane-spanning domain of the clostridium septicum alpha toxin. *Journal of Biological Chemistry, 279*(14), 14315–14322.
Mitev, D. J., Ivanova, T., & Vassilieff, C. (2002). Kinetics of lipid layer formation at interfaces. *Colloids and Surfaces B: Biointerfaces, 24*, 185–192.
Moerner, W. E., & Fromm, D. P. (2003). Methods of single-molecule fluorescence spectroscopy and microscopy. *Review of Scientific Instruments, 74*(8), 3597–3619.
Moerner, W. E., & Kador, L. (1989). Optical detection and spectroscopy of single molecules in a solid. *Physical Review Letters, 62*(21), 2535–2538.
Moleiro, A. F., Conceição, G., Leite-Moreira, A. F., & Rocha-Sousa, A. (2017). A critical analysis of the available in vitro and ex vivo methods to study retinal angiogenesis. *Journal of Ophthalmology, 2017*, 3034953.
Möller, I., & Seeger, S. (2015). Solid supported lipid bilayers from artificial and natural lipid mixtures. *Journal of Materials Chemistry B, 3*(29), 6046–6056.
Mondal, P. P. (2014). Temporal resolution in fluorescence imaging. *Frontiers in Molecular Biosciences, 1*, 11.
Mueller, P., Rudin, D. O., Ti Tien, H., & Wescott, W. C. (1962). Reconstitution of cell membrane structure in vitro and its transformation into an excitable system. *Nature, 194*(4832), 979–980.
Müller-Röber, B., Ellenberg, J., Provart, N., Willmitzer, L., Busch, H., Becker, D., … Hedrich, R. (1995). Cloning and electrophysiological analysis of KST1, an inward rectifying K+ channel expressed in potato guard cells. *The EMBO Journal, 14*(11), 2409–2416.
Murphy, D. B., & Davidson, M. W. (2012). Fluorescence Imaging of Dynamic Molecular Processes. *Fundamentals of Light Microscopy and Electronic Imaging* (2nd, pp. 233–263). Hoboken, New Jersey: Wiley-Blackwell. September. In press.
Neher, E., & Sakmann, B. (1976). Single-channel currents recorded from membrane of denervated frog muscle fibres. *Nature, 260*(5554), 799–802.
Neubert, D., Blankenburg, G., Chahoud, I., Franz, G., Herken, R., Kastner, M., … Lewandowski, C. (1986). Results of in vivo and in vitro studies for assessing prenatal toxicity. *Environmental Health Perspectives, 70*, 89–103.
Nguyen, V. T., Kamio, Y., & Higuchi, H. (2003). Single-molecule imaging of cooperative assembly of gamma-hemolysin on erythrocyte membranes. *The EMBO Journal, 22*(19), 4968–4979.
Nikazar, S., Sivasankarapillai, V. S., Rahdar, A., Gasmi, S., Anumol, P. S., & Shanavas, M. S. (2020). Revisiting the cytotoxicity of quantum dots: An in-depth overview. *Biophysical Reviews, 12*(3), 703–718.
Parkkila, P., Elderdfi, M., Bunker, A., & Viitala, T. (2018). Biophysical characterization of supported lipid bilayers using parallel dual-wavelength surface plasmon resonance and quartz crystal microbalance measurements. *Langmuir, 34*(27), 8081–8091.
Pei, Z. M., Ward, J. M., & Schroeder, J. I. (1999). Magnesium sensitizes slow vacuolar channels to physiological cytosolic calcium and inhibits fast vacuolar channels in fava bean guard cell vacuoles. *Plant Physiology, 121*(3), 977–986.
Peterman, M. C., Ziebarth, J. M., Braha, O., Bayley, H., Fishman, H. A., & Bloom, D. M. (2002). Ion channels and lipid bilayer membranes under high potentials using microfabricated apertures. *Biomedical Microdevices, 4*(3), 231–236.

Peters, R. (2003). Optical single transporter recording: Transport kinetics in microarrays of membrane patches. *Annual Review of Biophysics and Biomolecular Structure*, *32*, 47–67.

Petryayeva, E., Algar, W. R., & Medintz, I. L. (2013). Quantum dots in bioanalysis: A review of applications across various platforms for fluorescence spectroscopy and imaging. *Applied Spectroscopy*, *67*(3), 215–252.

Pujals, S., Feiner-Gracia, N., Delcanale, P., Voets, I., & Albertazzi, L. (2019). Super-resolution microscopy as a powerful tool to study complex synthetic materials. *Nature Reviews Chemistry*, *3*(2), 68–84.

Rajapaksha, S. P., Wang, X., & Lu, H. P. (2013). Suspended lipid bilayer for optical and electrical measurements of single ion channel proteins. *Analytical Chemistry*, *85*(19), 8951–8955.

Ramm, F., Dondapati, S. K., Thoring, L., Zemella, A., Wüstenhagen, D. A., Frentzel, H., … Kubick, S. (2020). Mammalian cell-free protein expression promotes the functional characterization of the tripartite non-hemolytic enterotoxin from Bacillus cereus. *Scientific Reports*, *10*(1), 2887.

Ribeiro, C., Vignes, M., & Brehélin, M. (2003). Xenorhabdus nematophila (Enterobacteriacea) secretes a Cation-selective calcium-independent porin which causes vacuolation of the rough endoplasmic reticulum and cell lysis. *Journal of Biological Chemistry*, *278*(5), 3030–3039.

Rojko, N., & Anderluh, G. (2015). How lipid membranes affect pore forming toxin activity. *Accounts of Chemical Research*, *48*(12), 3073–3079.

Rojko, N., Cronin, B., Danial, J. S. H., Baker, M. A. B., Anderluh, G., & Wallace, M. I. (2014). Imaging the lipid-phase-dependent pore formation of equinatoxin II in droplet interface bilayers. *Biophysical Journal*, *106*(8), 1630–1637.

Royer, C. A. & Scarlata, S. F. B. T. (2008). Fluorescence approaches to quantifying biomolecular interactions. In *Fluorescence spectroscopy: Vol. 450* (pp. 79–106). Academic Press.

Sako, Y., & Yanagida, T. (2003). Single-molecule visualization in cell biology. *Nature Reviews Molecular Cell Biology*, *4*(Suppl.), 1–5.

Santangelo, P. J., Nix, B., Tsourkas, A., & Bao, G. (2004). Dual FRET molecular beacons for mRNA detection in living cells. *Nucleic Acids Research*, *32*(6), e57.

Sarangi, N. K., P, I. I., Ayappa, K. G., Visweswariah, S. S., & Basu, J. K. (2016). Super-resolution stimulated emission depletion-fluorescence correlation spectroscopy reveals nanoscale membrane reorganization induced by pore-forming proteins. *Langmuir 32*(37), 9649–9657.

Sasmal, D. K., Pulido, L. E., Kasal, S., & Huang, J. (2016). Single-molecule fluorescence resonance energy transfer in molecular biology. *Nanoscale*, *8*(48), 19928–19944.

Sathyanarayana, P., Maurya, S., Behera, A., Ravichandran, M., Visweswariah, S. S., Ayappa, K. G., & Roy, R. (2018). Cholesterol promotes Cytolysin A activity by stabilizing the intermediates during pore formation. *Proceedings of the National Academy of Sciences*, *115*(31), E7323–E7330.

Schlichthaerle, T., Strauss, M. T., Schueder, F., Auer, A., Nijmeijer, B., Kueblbeck, M., … Jungmann, R. (2019). Direct visualization of single nuclear pore complex proteins using genetically-encoded probes for DNA-PAINT. *Angewandte Chemie—International Edition*, *58*(37), 13004–13008.

Seddon, A. M., Curnow, P., & Booth, P. J. (2004). Membrane proteins, lipids and detergents: Not just a soap opera. *Biochimica et Biophysica Acta (BBA)—Biomembranes*, *1666*(1), 105–117.

Sekar, R. B., & Periasamy, A. (2003). Fluorescence resonance energy transfer (FRET) microscopy imaging of live cell protein localizations. *The Journal of Cell Biology*, *160*(5), 629–633.

Sengel, J. T., & Wallace, M. I. (2016). Imaging the dynamics of individual electropores. *Proceedings of the National Academy of Sciences of the United States of America*, *113*(19), 5281–5286.

Shashkova, S., & Leake, M. C. (2017). Single-molecule fluorescence microscopy review: Shedding new light on old problems. *Bioscience Reports*, *37*(4), BSR20170031.
Shatursky, O., Heuck, A. P., Shepard, L. A., Rossjohn, J., Parker, M. W., Johnson, A. E., & Tweten, R. K. (1999). The mechanism of membrane insertion for a cholesterol-dependent cytolysin: A novel paradigm for pore-forming toxins. *Cell*, *99*(3), 293–299.
Shen, Q., Wang, S., Yang, N.-D., Zhang, C., Wu, Q., & Yu, C. (2020). Recent development of small-molecule organic fluorophores for multifunctional bioimaging in the second near-infrared window. *Journal of Luminescence*, *225*, 117338.
Shepard, L. A., Heuck, A. P., Hamman, B. D., Rossjohn, J., Parker, M. W., Ryan, K. R., … Tweten, R. K. (1998). Identification of a membrane-spanning domain of the thiol-activated pore-forming toxin clostridium perfringens perfringolysin O: An α-helical to β-sheet transition identified by fluorescence spectroscopy. *Biochemistry*, *37*(41), 14563–14574.
Shimomura, O. (1979). Structure of the chromophore of Aequorea green fluorescent protein. *FEBS Letters*, *104*(2), 220–222.
Siontorou, C. G., Nikoleli, G.-P., Nikolelis, D. P., & Karapetis, S. K. (2017). Artificial lipid membranes: Past, present, and future. *Membranes*, 7(3), 38.
Soubies, E., Radwanska, A., Grall, D., Blanc-Féraud, L., Van Obberghen-Schilling, E., & Schaub, S. (2019). Nanometric axial resolution of fibronectin assembly units achieved with an efficient reconstruction approach for multi-angle-TIRF microscopy. *Scientific Reports*, *9*(1), 1926.
Stabley, D. R., Oh, T., Simon, S. M., Mattheyses, A. L., & Salaita, K. (2015). Real-time fluorescence imaging with 20 nm axial resolution. *Nature Communications*, *6*(1), 8307.
Subburaj, Y., Ros, U., Hermann, E., Tong, R., & García-Sáez, A. J. (2015). Toxicity of an α-pore-forming toxin depends on the assembly mechanism on the target membrane as revealed by single molecule imaging. *The Journal of Biological Chemistry*, *290*(8), 4856–4865.
Tamaddoni, N., Taylor, G., Hepburn, T., Michael Kilbey, S., & Sarles, S. A. (2016). Reversible, voltage-activated formation of biomimetic membranes between triblock copolymer-coated aqueous droplets in good solvents. *Soft Matter*, *12*(23), 5096–5109.
Taylor, J. D., Linman, M. J., Wilkop, T., & Cheng, Q. (2009). Regenerable tethered bilayer lipid membrane arrays for multiplexed label-free analysis of lipid-protein interactions on poly(dimethylsiloxane) microchips using SPR imaging. *Analytical Chemistry*, *81*(3), 1146–1153.
Taylor, R. W., & Sandoghdar, V. (2019). Interferometric scattering microscopy: Seeing single nanoparticles and molecules via Rayleigh scattering. *Nano Letters*, *19*(8), 4827–4835.
Thompson, J. R., Cronin, B., Bayley, H., & Wallace, M. I. (2011). Rapid assembly of a multimeric membrane protein pore. *Biophysical Journal*, *101*(11), 2679–2683.
Thompson, J. R., Heron, A. J., Santoso, Y., & Wallace, M. I. (2007). Enhanced stability and fluidity in droplet on hydrogel bilayers for measuring membrane protein diffusion. *Nano Letters*, 7(12), 3875–3878.
Tschödrich-Rotter, M., Kubitscheck, U., Ugochukwu, G., Buckley, J. T., & Peters, R. (1996). Optical single-channel analysis of the aerolysin pore in erythrocyte membranes. *Biophysical Journal*, *70*(2), 723–732.
Türkcan, S., Masson, J.-B., Casanova, D., Mialon, G., Gacoin, T., Boilot, J.-P., … Alexandrou, A. (2012). Observing the confinement potential of bacterial pore-forming toxin receptors inside rafts with nonblinking Eu(3+)-doped oxide nanoparticles. *Biophysical Journal*, *102*(10), 2299–2308.
Vicidomini, G., Bianchini, P., & Diaspro, A. (2018). STED super-resolved microscopy. *Nature Methods*, *15*(3), 173–182.
Villar, G., & Bayley, H. (2013). Functional droplet interface bilayers BT. In G. C. K. Roberts (Ed.), *Encyclopedia of biophysics* (pp. 861–868). Berlin, Heidelberg: Springer Berlin Heidelberg (Ed.).

Watanabe, R., Soga, N., Fujita, D., Tabata, K. V., Yamauchi, L., Hyeon Kim, S., ... Noji, H. (2014). Arrayed lipid bilayer chambers allow single-molecule analysis of membrane transporter activity. *Nature Communications*, *5*(1), 4519.

Wazawa, T., & Ueda, M. (2005). Total internal reflection fluorescence microscopy in single molecule nanobioscience. In J. Rietdorf (Ed.), *Microscopy techniques* (pp. 77–106). Berlin, Heidelberg: Springer. In press.

Weatherill, E. E., Coker, H. L. E., Cheetham, M. R., & Wallace, M. I. (2018). Urea-mediated anomalous diffusion in supported lipid bilayers. *Interface Focus*, *8*(5), 20180028.

Weatherill, E. E., & Wallace, M. I. (2015). Combining single-molecule imaging and single-channel electrophysiology. *Journal of Molecular Biology*, *427*(1), 146–157.

Wilson, T. (2011). Resolution and optical sectioning in the confocal microscope. *Journal of Microscopy*, *244*(2), 113–121.

Wuethrich, I., Peeters, J. G. C., Blom, A. E. M., Theile, C. S., Li, Z., Spooner, E., ... Guimaraes, C. P. (2014). Site-specific chemoenzymatic labeling of aerolysin enables the identification of new aerolysin receptors. *PLoS ONE*, *9*(10), e109883.

Zheng, Q., Juette, M. F., Jockusch, S., Wasserman, M. R., Zhou, Z., Altman, R. B., & Blanchard, S. C. (2014). Ultra-stable organic fluorophores for single-molecule research. *Chemical Society Reviews*, *43*(4), 1044–1056.

Zou, H., Lifshitz, L. M., Tuft, R. A., Fogarty, K. E., & Singer, J. J. (2002). Visualization of Ca(2+) entry through single stretch-activated cation channels. *Proceedings of the National Academy of Sciences*, *99*(9), 6404–6409.

Further reading

Haapasalo, K., Wollman, A. J. M., de Haas, C. J. C., van Kessel, K. P. M., van Strijp, J. A. G., & Leake, M. C. (2019). Staphylococcus aureus toxin LukSF dissociates from its membrane receptor target to enable renewed ligand sequestration. *The FASEB Journal*, *33*(3), 3807–3824.

CHAPTER SIXTEEN

Using multiscale molecular dynamics simulations to obtain insights into pore forming toxin mechanisms

Rajat Desikan[a,†], Amit Behera[a,‡], Prabal K. Maiti[b], and K. Ganapathy Ayappa[a,c,*]

[a]Department of Chemical Engineering, Indian Institute of Science, Bengaluru, India
[b]Centre for Condensed Matter Theory, Department of Physics, Indian Institute of Science, Bengaluru, India
[c]Centre for Biosystems Science and Engineering, Indian Institute of Science, Bengaluru, India
*Corresponding author: e-mail address: ayappa@iisc.ac.in

Contents

Abstract

Pore forming toxins (PFTs) are virulent proteins released by several species, including many strains of bacteria, to attack and kill host cells. In this article, we focus on the utility of molecular dynamics (MD) simulations and the molecular insights gleaned from these

† Present address: Certara QSP, Certara UK Limited, Sheffield, United Kingdom.
‡ Present address: HP Green R&D Centre, Hindustan Petroleum Corporation Limited, Bengaluru 560067, India.

Methods in Enzymology, Volume 649
ISSN 0076-6879
https://doi.org/10.1016/bs.mie.2021.01.021

techniques on the pore forming pathways of PFTs. In addition to all-atom simulations which are widely used, coarse-grained MARTINI models and structure-based models have also been used to study PFTs. Here, the emphasis is on methods and techniques involved while setting up, monitoring, and evaluating properties from MD simulations of PFTs in a membrane environment. We draw from several case studies to illustrate how MD simulations have provided molecular insights into protein–protein and protein–lipid interactions, lipid dynamics, conformational transitions and structures of both the oligomeric intermediates and assembled pore structures.

1. Introduction

Several virulent bacterial strains, such as those causing cholera, pneumonia, and listeriosis, initiate their infections by releasing proteins which bind to the target cell membrane to form unregulated membrane-inserted pores, thus compromising ionic balance and eventually causing cell lysis (Peraro & van der Goot, 2016). These proteins, commonly referred to as pore forming toxins (PFTs), undergo a process of membrane-bound oligomerization to form a pore complex, the primary vehicle in several bacterial infections (Los, Randis, Aroian, & Ratner, 2013; Peraro & van der Goot, 2016). PFTs are classified as either α or β-PFTs depending on the secondary structure of their membrane-inserted segments in the pore state. Pore dimensions depend on the oligomeric state of the final pore complex (Cosentino, Ros, & García-Sáez, 2016; Desikan, Maiti, & Ayappa, 2021; Peraro & van der Goot, 2016). The pores formed by α-PFTs are smaller in size consisting of oligomers ranging from 2 to 26 mers, with some of the largest pores formed by the cytolysin A (ClyA) family of PFTs (Bräuning et al., 2018; Schubert, Vetter, Prumbaum, Penczek, & Raunser, 2018). In contrast, β-PFTs form both smaller and larger pores, with the cholesterol-dependent cytolysins (CDCs) forming the largest known PFT pores consisting of 30–50 mers (Tweten, Parker, & Johnson, 2001). Upon encountering the membrane, water-soluble PFT monomers typically bind using membrane binding motifs and/or membrane receptors, eventually undergoing a conformational change and oligomerization to form the membrane-inserted pore (Giri Rao, Desikan, Ayappa, & Gosavi, 2016; Peraro & van der Goot, 2016). Understanding the underlying molecular mechanisms of PFT-mediated virulence will aid in developing effective treatment strategies to disrupt or compromise pore formation pathways (Escajadillo & Nizet, 2018; Gammon, 2014), especially in the face of rising

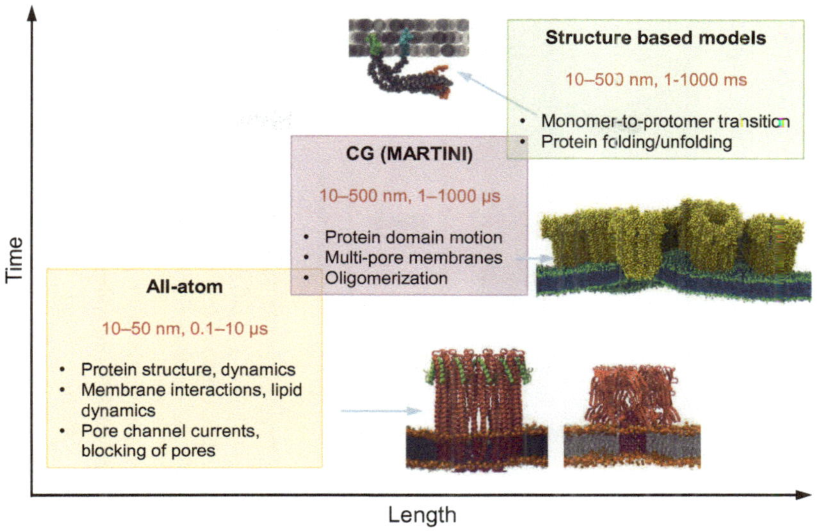

Fig. 1 Illustration of length and time scales typically probed in simulations of pore forming proteins, depicting all-atom, coarse-grained (CG) MARTINI and structure-based models.

antibiotic-resistant bacterial strains (Blair, Webber, Baylay, Ogbolu, & Piddock, 2015; Opal, 2016).

In this article, we focus on the utility of molecular dynamics (MD) simulations and the molecular insights obtained from these techniques on the pore forming pathways of several PFTs (Fig. 1). A challenge associated with carrying out MD simulations is the large number of atoms associated with the PFT pore complex, as well as their ability to capture slow conformational changes that are an intrinsic part of membrane binding and oligomerization. Unlike most transmembrane proteins, the presence of large extracellular domains of the pore complex is a characteristic feature of PFTs adding to the computational overhead. [See Fig. 2; also see Desikan et al. (2021) for an illustration of 13 α and β pore crystal structures with annotated transmembrane domains.] For example, in the case of the ClyA family of PFTs, less than a quarter of the residues are membrane-inserted. Thus, all-atom MD simulations have been sparingly used to understand PFTs. A few notable mentions are simulations used to study the dodecameric α-PFT, ClyA (Desikan, Maiti, & Ayappa, 2017, 2021; Desikan, Padmanabhan, & Ayappa, 2020; Desikan, Patra, Sarthak, Maiti, & Ayappa, 2017; Giri Rao et al., 2016; Mandal, Kanchi, Ayappa, &

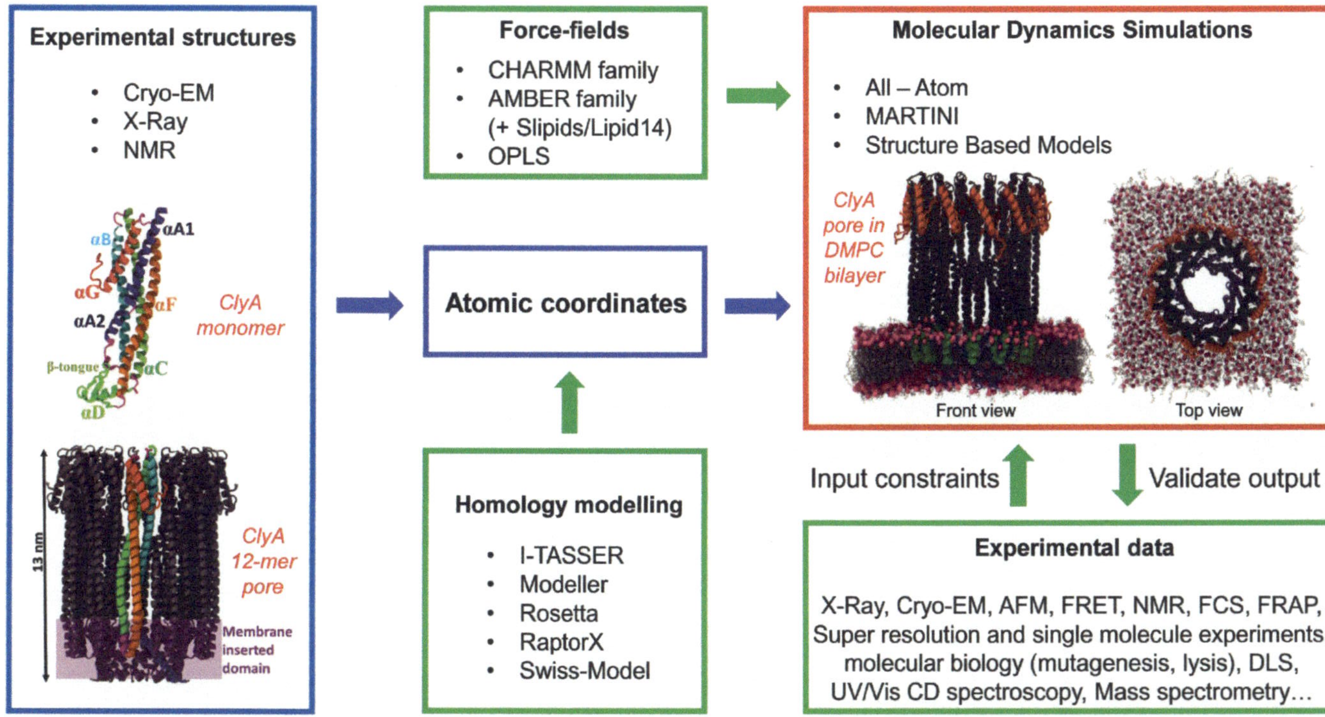

Fig. 2 Typical workflow while setting up MD simulations for pore forming toxins starting from available crystal structures. Procedures involve obtaining atomic coordinates of the protein with homology modeling, using appropriate force fields and comparing outputs from the MD simulation with experiments. *Molecular configurations (left top): Reprinted with permission from Sathyanarayana, P., Desikan, R., Ayappa, K. G., & Visweswariah, S. S. (2016). The solvent-exposed C-terminus of the cytolysin A pore-forming toxin directs pore formation and channel function in membranes. Biochemistry, 55 (42), 5952–5961. Copyright (2016) American Chemical Society; Left bottom: Reprinted with permission from Desikan, R., Maiti, P. K., & Ayappa, K. G. (2017). Assessing the structure and stability of transmembrane oligomeric intermediates of an α-helical toxin. Langmuir, 33 (42), 11496–11510. Copyright (2017) American Chemical Society; Right: Varadarajan, V., Desikan, R., & Ayappa, K. G. (2020). Assessing the extent of the structural and dynamic modulation of membrane lipids due to pore forming toxins: Insights from molecular dynamics simulations. Soft Matter, 16 (20), 4840–4857.*

Maiti, 2016; Sathyanarayana, Desikan, Ayappa, & Visweswariah, 2016; Sathyanarayana et al., 2018; Varadarajan, Desikan, & Ayappa, 2020), hexameric assemblies of the membrane attack complex/perforin (MACPF) (Ni et al., 2018), structure (Aksimentiev & Schulten, 2005; Desikan et al., 2021; Desikan, Patra, et al., 2017), lipid dynamics (Varadarajan et al., 2020), and water/solute transport (Aksimentiev & Schulten, 2005; Bonome, Cecconi, & Chinappi, 2019; Di Marino, Bonome, Tramontano, & Chinappi, 2015; Mathe, Aksimentiev, Nelson, Schulten, & Meller, 2005; Wong-ekkabut & Karttunen, 2016) in the heptameric β-PFT, *Staphylococcus aureus* α-hemolysin (AHL) pore, cholesterol sensing motifs for the CDCs streptolysin O (SLO) (Feil, Ascher, Kuiper, Tweten, & Parker, 2014) and listeriolysin O (LLO) (Cheerla & Ayappa, 2020; Ponmalar, Cheerla, Ayappa, & Basu, 2019), cry toxin (Crt4Aa) prepores (Taveecharoenkool, Angsuthanasombat, & Kanchanawarin, 2010), peptide sensing in the β-PFT aerolysin pores (Cao et al., 2019; Cirauqui, Abriata, van der Goot, & Dal Peraro, 2017; Degiacomi et al., 2013), and conformational changes in tripartite toxin complexes (Gatsogiannis et al., 2018; Meusch et al., 2014).

In addition to the all-atom simulations, coarse-grained MARTINI models (Desikan, Padmanabhan, & Ayappa, 2020; Desikan, Patra, et al., 2017; Vögele et al., 2019) and structure-based models (SBMs) (Gatsogiannis et al., 2018; Giri Rao et al., 2016) have also been used to study PFTs (Fig. 1). Coarse-grained methods result in a decrease in the number of degrees of freedom, thus allowing larger system sizes and sampling times compared to all-atom simulations. All-atom simulations, typically probe time scales ranging from 100s of nanoseconds to a few microseconds, yield detailed insights into molecular interactions, such as specific cholesterol binding sites (Ponmalar et al., 2019; Sathyanarayana et al., 2018), stabilizing salt bridges and hydrogen bond networks (Desikan et al., 2021), and lipid dynamics around pores (Varadarajan et al., 2020). The accuracy of these predictions rests with the choice of the inter-atomic potentials used to describe the interactions and forces in the MD simulation (discussed in Section 2.2). In the MARTINI model, protein residues, lipids, and water are treated as coarse-grained beads suitably modified to describe polar and nonpolar interactions. The advantage of the MARTINI models is the ability to simulate large systems sizes which require several tens-to-hundreds of microseconds of sampling time (Fig. 1). Both all-atom and MARTINI models are limited in their ability to explore protein conformational transitions that occur during the pore formation pathways of PFTs. SBMs, initially developed to study protein folding (Clementi, Nymeyer, & Onuchic, 2000; Noel & Onuchic, 2012), have been used to capture conformational changes associated with the conversion of a single ClyA monomer in to a membrane-inserted protomer (Giri Rao et al., 2016; Mueller, Grauschopf, Maier, Glockshuber, & Ban, 2009).

Here, we focus on the methodology used in our laboratory to study PFTs using various MD methods which include all-atom (Cheerla & Ayappa, 2020; Desikan et al., 2021; Desikan, Padmanabhan, & Ayappa, 2020; Mandal et al., 2016; Ponmalar et al., 2019; Sathyanarayana et al., 2016, 2018; Varadarajan et al., 2020), coarse-grained MARTINI (Desikan, Padmanabhan, & Ayappa, 2020; Desikan, Patra, et al., 2017), and SBM simulations (Giri Rao et al., 2016). In addition to describing the details of these different simulation methods, we also briefly illustrate key properties and molecular insights gained from each method, discuss their advantages and limitations, and end with a perspective.

2. All-atom MD simulations of PFTs

In this section, we discuss several practical aspects of setting up atomistic membrane-PFT systems and the specific insights gained from such studies. In particular, we will focus on the α-PFT, ClyA, as well as the β-PFTs LLO and AHL. For the basics of molecular simulation techniques and analysis, we refer the reader to excellent resources concerned with both membrane and membrane–protein simulations (Allen & Tildesley, 2017; Braun et al., 2018; Enkavi, Javanainen, Kulig, Róg, & Vattulainen, 2019; Frenkel & Smit, 2002; Grossfield et al., 2018; Lemkul, 2018; Maginn, Messerly, Carlson, Roe, & Elliott, 2018; Manna, Nieminen, & Vattulainen, 2019; Perilla et al., 2015; Schlick, 2010; Smith, Klauda, & Sodt, 2018).

2.1 Protein preparation and homology modeling of PFT oligomers and pores

Prior to embarking on a membrane-PFT simulation, a key prerequisite is a high-resolution structure of the protein. Fig. 2 illustrates a typical workflow of the various stages and inputs involved in setting up MD simulations for PFTs. A 12-mer crystal structure of the fully formed ClyA pore (PDB ID: 2WCD; 3.3 Å resolution) is available in the protein data bank (Mueller et al., 2009), with unresolved N-terminal and C-terminal residues 1–7 and 293–303, respectively. These amino acids could be important for ClyA's lytic activity as shown by previous mutagenesis studies (Atkins et al., 2000; Ludwig et al., 2010; Oscarsson et al., 1999; Sathyanarayana et al., 2016; Wai et al., 2003). Therefore, these amino acid stretches were homology modeled via the I-TASSER web-server (Roy, Kucukural, & Zhang, 2010; Zhang, 2008). Alternatively, the "Modeller" or "Rosetta" software (Alford et al., 2017; Fiser, Do, & Sali, 2000) could also be employed, as we did for modeling the termini of a single ClyA protomer in Giri Rao et al. (2016). The modeled residues were then stitched on to the pore crystal structure, and the resulting structure was then assumed to be the complete pore state. All residue protonation states were assumed to correspond to neutral pH. Similarly, a near-complete 7-mer structure of the AHL pore (PDB ID: 7AHL; 1.9 Å resolution) is available in the protein data bank (Song et al., 1996), but has residues with missing coordinates for a few atoms, viz, Arg 66 and Lys 70 in chain A, Lys 30 and Lys 240 in chain D, Lys 283 in chain F and Lys 30 in chain G. Since the backbone atom coordinates were available, rather than resorting to I-TASSER or Modeller for full-blown homology modeling, we reconstructed the missing atoms using the "psfgen" module implemented in VMD (Humphrey, Dalke, & Schulten, 1996; Phillips et al., 2005), similar to a previous study of the AHL pore (Aksimentiev & Schulten, 2005).

We have also previously simulated the membrane-inserted, assembly competent, protomer (1-mer) states of ClyA (Desikan, Maiti, & Ayappa, 2017; Sathyanarayana et al., 2018) and LLO (Ponmalar et al., 2019). While a crystal structure is available for simulating the fully oligomerized ClyA pore complex, experimental structures of the membrane-inserted protomer, or the partially oligomerized "arcs," which could potentially be intermediates along the pore forming pathway (Agrawal, Apoorva, & Ayappa, 2017; Desikan, Maiti, & Ayappa, 2017), are currently unavailable. The structure of the ClyA protomer was assumed to be similar to any of the subunits in the full ClyA pore structure (elaborated in Section 2.3.3) (Desikan, Maiti, & Ayappa, 2017). Membrane-inserted protomers generally exist as higher order oligomers—either as fully formed pores or arcs. For example, a single membrane-inserted protomer of ClyA tends to undergo large RMSF fluctuations of the backbone atoms (Desikan, Maiti, & Ayappa, 2017; Ponmalar et al., 2019). However, for a membrane-inserted dimer (Sathyanarayana et al., 2018) or partially oligomerized arcs (Desikan, Maiti, & Ayappa, 2017), a significant increase in structural stability, inferred from reduced root mean square fluctuations (RMSF) of the backbone C_α atoms, is observed.

In the absence of a crystal structure for the LLO pore, we used the membrane-inserted protomeric state of pneumolysin (PLY, PDB ID: 5LY6; 4.5 Å resolution), which has a 67% sequence similarity and 87%

structural similarity to LLO, for homology modeling (van Pee et al., 2017). Using the PLY membrane-inserted state as a template, we employed the SWISS-MODEL web-server (Waterhouse et al., 2018) and obtained a homology modeled LLO protomer structure which showed good global structural alignment with the PLY protomer (Ponmalar et al., 2019). Ramachandran maps of this homology modeled LLO protomer showed that ~85% of residues lie in the favored regions, an improvement over the PLY crystal structure which had ~74% residues in the favored regions. Further, upon relaxing the LLO structure in a membrane environment, up to 95% residues were found to lie in favored regions, and specific FRET and AFM distances calculated from the structure matched with experiments (Ponmalar et al., 2019; Ruan, Rezelj, Zavec, Anderluh, & Scheuring, 2016). Therefore, this structure may be a good representation of the LLO protomer, and the above procedure provides a potential computational workflow to obtain protomer PFT structures.

2.2 Force fields, simulation settings, and MD protocols

The accuracy of both protein and lipid force fields (FFs) is important while simulating PFTs (Fig. 2). In this regard, all-atom force fields (FFs) from the CHARMM and AMBER families have been extensively used for a wide variety of membrane–protein systems. In our work, to describe the protein and ion interactions, we employed either the AMBER 99SB-ILDN FF (Lindorff-Larsen et al., 2010) with ϕ corrections (Nerenberg & Head-Gordon, 2011), which has been shown to perform well for proteins (Beauchamp, Lin, Das, & Pande, 2012) and is implemented in the GROMACS MD engine (Pronk et al., 2013) (https://www.gromacs.org), or the AMBER12SB FF (Maier et al., 2015) implemented in the AMBER MD engine (Salomon-Ferrer, Case, & Walker, 2013). Alternately, we found similar results with the popular CHARMM36 protein + lipid FF (Best et al., 2012; Huang et al., 2017; Klauda et al., 2010; Pastor & MacKerell, 2011) implemented in GROMACS (http://mackerell.umaryland.edu/charmm_ff.shtml#gromacs, accessed on 16 October 2020) when compared with the AMBER FFs (SI of Desikan, Maiti, & Ayappa, 2017). We note that it is good simulation practice to compare key properties across FFs to gain confidence in the predictive ability of atomistic computations. Several FFs exist for lipids, which are important components of membrane–protein systems such as PFTs. One the earliest FFs is the Berger lipid united atom framework (Berger, Edholm, & Jahnig, 1997) that derives its interaction parameters and charge description from the GROMOS and OPLS FFs. While the original Berger FF did not reproduce the correct electron density or low temperature phase behavior as seen in experiments, later improvements (Chiu, Pandit, Scott, & Jakobsson, 2009) greatly ameliorated these issues. On the other hand, several fully atomistic lipid FFs such as the AMBER-compatible "Slipid" FF (Jämbeck & Lyubartsev, 2012), CHARMM36 lipid FF (Klauda et al., 2010; Pastor & MacKerell, 2011), and the AMBER family of FFs such as "GAFFlipid" (Dickson, Rosso, Betz, Walker, & Gould, 2012) and "Lipid14" (Dickson et al., 2014) accurately reproduce several experimentally observed membrane properties such as the area per head group, chain order parameters, bilayer thickness, and lipid diffusivity [description of membrane properties in Lemkul (2018), Varadarajan et al. (2020), and Smith et al. (2018)]. For our membrane-PFT simulations, we have extensively used the Slipid parameters to model the lipid membrane in GROMACS, in conjunction with the AMBER 99SB-ILDN-ϕ FF for protein and ions, and the TIP3P water model (Jorgensen, Chandrasekhar, Madura, Impey, & Klein, 1983). While investigating the propensity of dendrimers to block the ClyA channels (Mandal et al., 2016), MD simulations were carried out with the AMBER MD software, with the Lipid14 membrane FF used in combination with the AMBER12SB FF for protein and ions, TIP3P model for water, and the generalized AMBER force field (GAFF) (Wang, Wolf, Caldwell, Kollman, & Case, 2004) for the dendrimer. The dendrimer starting structure was built using the dendrimer building toolkit (Maingi, Jain, Bharatam, & Maiti, 2012).

All PFT-membrane systems are charge neutral with either 0.15 M Na^+Cl^-/K^+Cl^-, which is physiologically relevant, or with only charge balancing counter-ions (0 M salt). The procedure for insertion of PFTs into membranes is described later (Section 2.3.3). For GROMACS simulations, a leapfrog MD integrator (Allen & Tildesley, 2017) with an integration time step of 2 fs was used, along with Verlet buffered lists (target energy drift of 0.005 kJ mol^{-1} ns^{-1} atom^{-1}), and a neighbor list update frequency of once in every 10 steps. All covalent bonds or only H-bonds (depending on the FF) were constrained using the parallelized LINCS (Hess, Bekker, Berendsen, & Fraaije, 1997) or SHAKE (Ryckaert, Ciccotti, & Berendsen, 1977) algorithms. Electrostatic interactions were numerically computed using the particle mesh Ewald (PME) method (Darden, York, & Pedersen, 1993) with a 1.0 nm real space cut-off, and van der Waals interactions were computed with a 1.0 nm cut-off. Except where explicitly mentioned, systems were simulated in the isothermal-isobaric (NPT) ensemble with three-dimensional periodic boundary conditions. Weak temperature coupling was achieved via the stochastic velocity coupling (Bussi, Donadio, & Parrinello, 2007) or the Nosé–Hoover thermostats (Hoover, 1985; Nosé, 1984) with a coupling constant of 0.1–0.5 ps. Pressure coupling for the membrane–protein systems was achieved via the isotropic/semiisotropic Parrinello-Rahman barostat (Parrinello & Rahman, 1981) with a coupling constant of 10 ps, and isothermal compressibility of 4.5 $\times$ 10^{-5} bar^{-1}. To achieve good thermal equilibrium, membrane and/or protein atoms were coupled as one temperature group to the thermostat while the solvent was separately coupled as another temperature group. Visualization of all MD configurations was performed using the VMD software (Humphrey et al., 1996) (version 1.9.1 and higher).

The initial membrane–protein configurations were relaxed with energy minimization using the steepest descent method, followed by constant monitoring of system energies until they plateaued at a minimum. The energy minimized systems were subjected to short NVT and NPT runs with harmonic restraints (force constants of 100–1000 kJ mol^{-1} nm^{-2}) on the protein and membrane atoms for achieving thermal equilibration of the solvent and the membrane–protein solvation shells, and were followed by longer unrestrained NPT runs for equilibrating the volume fluctuations. A general equilibration recipe is as follows: (i) 0.5–1 ns simulation in the NVT ensemble with harmonic restraints of 1000 kJ mol^{-1} nm^{-2} on all protein and membrane atoms. In GROMACS, the Berendsen thermostat (Berendsen, Postma, van Gunsteren, DiNola, & Haak, 1984) can conveniently be employed at this step for quick thermal equilibration (Lemkul, 2018), though it is known not to produce a true canonical ensemble. Slow heating from 0 K to the reference temperature can also be performed at this step (Mandal et al., 2016). (ii) 1 ns simulation in the NPT ensemble with harmonic restraints of 1000 kJ mol^{-1} nm^{-2} on all protein and membrane atoms to pack solvent around the membrane–protein atoms, (iii) 1 ns simulation in the NPT ensemble with harmonic restraints of 100–1000 kJ mol^{-1} nm^{-2} on only the protein backbone and membrane phosphorus atoms, and (iv) 1–5 ns unrestrained simulation in the NPT ensemble until the pressure equilibrates. The final configuration from this NPT run can then be utilized for longer production MD simulations.

Some of the above equilibration steps could be skipped depending on the system particulars. For instance, ClyA arcs evacuated central lipids within a few ns, necessitating only short restrained equilibration runs lest the phenomenon of interest (central lipid dynamics) be complete within the equilibration phase itself (Desikan, Maiti, & Ayappa, 2017). Confirmation with experimentally available data (Fig. 2) on the pore structures from atomic force microscopy data in a membrane environment can also be used to validate the simulation methodology (Cheerla & Ayappa, 2020; Ponmalar et al., 2019).

2.3 Case studies

The utility of all-atom simulations are illustrated with specific case studies performed in our laboratory.

2.3.1 Assessing the structural stability of a PFT monomer from thermal unfolding MD

Although located in the solvent-exposed interface distal from the membrane interface, the C-terminal residues 293–303 of ClyA are essential for its lytic activity (Sathyanarayana et al., 2016). While the deletion of these residues in the ClyA monomer did not alter its ability to bind to target membranes or oligomerize upon addition of detergent, truncated ClyA poorly lysed erythrocytes, and displayed higher susceptibility to proteolysis and thermal unfolding. Since the molecular mechanism of altered activity in the deletion mutant was not immediately forthcoming from experiments, we took cues from thermal unfolding experiments and performed fully atomistic equilibrium and thermal unfolding simulations of wild-type (WT) ClyA monomer and the mutant (Sathyanarayana et al., 2016).

The deletion mutant was created by deleting residues 293–298 from the crystal structure of the ClyA monomer (PDB: 1QOY; 2.0 Å resolution), and appropriately capping the C-terminal residue 292 with a negatively charged carboxylate moiety to maintain the zwitterionic state of the protein. The WT structure was built from the crystal structure by modeling missing residues 299–303 similar to the ClyA pore (Section 2.1 and Sathyanarayana et al., 2016). Simulations were performed using GROMACS. Unrestrained equilibrium MD simulations of WT and the deletion mutant (three replicates, 100 ns each) were performed in the NPT ensemble at 310 K and 1 bar pressure as described previously (Section 2.2). Global structural analysis from these simulations showed that the deletion mutant was less stable than the WT, especially in the N-terminal helical domains that are

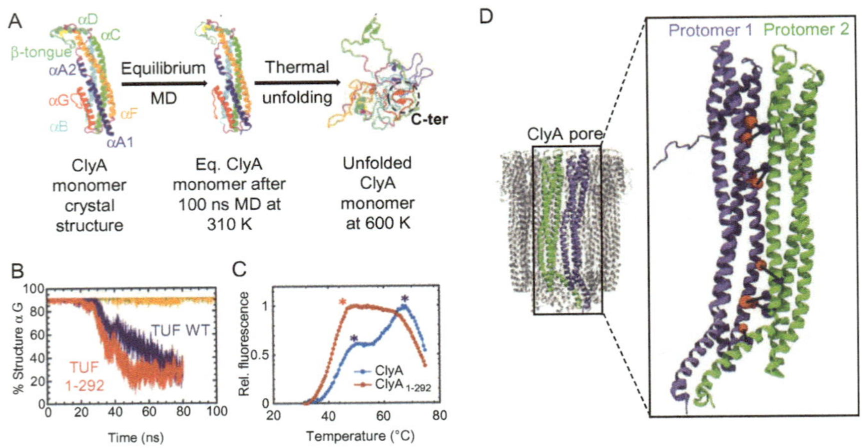

Fig. 3 (A)–(C) Thermal unfolding MD simulations of the ClyA monomer (Sathyanarayana et al., 2016). (A) Schematic of the thermal unfolding (TUF) simulations after equilibration at 310 K. (B) Increased loss of helicity in the αG helix of truncated ClyA 1–292 compared to wild type (WT). (C) Thermal denaturation experiments reveal loss of biphasic unfolding for ClyA 1–292 compared to WT. (D) Long lived hydrogen bond and salt bridge networks between two protomers in the ClyA pore complex (details in Desikan et al., 2021). Residues on protomers 1 and 2 depicted as *red* and *blue beads*, respectively, and hydrogen bonds/salt bridges as *black lines* connecting beads. *Panels (A)–(C): Reprinted with permission from Sathyanarayana, P., Desikan, R., Ayappa, K. G., & Visweswariah, S. S. (2016). The solvent-exposed C-terminus of the cytolysin A pore-forming toxin directs pore formation and channel function in membranes. Biochemistry, 55 (42), 5952–5961. Copyright (2016) American Chemical Society.*

proximal to the C-terminus in ClyA's monomer conformation. Thermal unfolding MD simulations (Beck & Daggett, 2004; Daggett, 2002) were performed in the NVT ensemble, and not the NPT ensemble, to ensure that water does not undergo a liquid-to-vapor phase transition with rising temperature (Fig. 3A). The final monomer (WT or deletion mutant) configurations, obtained from the aforementioned equilibrium MD simulations, were heated from 310 to 600 K at the rate of ~10 K/ns using the simulated annealing algorithm in GROMACS. Subsequently, a further 50 ns equilibrium MD simulation in the NVT ensemble, at 600 K, was performed to complete the unfolding of the ClyA monomers (Fig. 3A). Six independent MD replicates each of WT and the deletion mutant were performed for sufficient sampling of the unfolding pathway, and averages of structural quantities such as the % helicity of domains (e.g. Fig. 3B) were calculated across corresponding replicate trajectories.

The deletion mutant exhibited a faster rate of unfolding compared to the WT ClyA monomer, especially in the N-terminal helices that would have been proximal to the C-terminus residues 293–303 (Fig. 3B). Truncation of these residues could lead to loss of favorable interactions with the N-terminus and therefore result in destabilization of the domain. One of the affected N-terminal helices (αA1) eventually forms an integral part of ClyA's transmembrane domain in membrane-inserted states (Giri Rao et al., 2016; Mueller et al., 2009). The observations from simulations were experimentally validated by thermal unfolding assays (Sathyanarayana et al., 2016), where the deletion mutant completely unfolded at 321 K compared to 341 K for the WT (Fig. 3C). Strikingly, WT showed biphasic melting in both MD simulations and experiments, while this behavior was lost with the deletion mutant in both simulations and experiments (Fig. 3B and C). Thus, all-atom MD simulations yield a complementary molecular understanding of the functional role of the C-terminus, indicating that conformational changes required for ClyA pore formation could be compromised by the truncation of the C-terminal residues. These studies suggest that thermal unfolding simulations could serve as a powerful in silico method for gaining insights into PFT structure, folding, and function.

2.3.2 Predicting hot-spot residues that stabilize pores via hydrogen bonds and salt bridges

The subunits (protomers) of multimeric transmembrane PFT pores are held together by amino acid networks comprising hydrogen bond and salt bridge interactions (discussed in detail in Desikan et al., 2021). Identifying key interacting residues, targeting of which could impair pore formation, may potentially enable rational structure-based design of novel therapeutics. While currently available pore crystal structures can be screened in silico to identify important inter-protomer stabilizing residues, proteins are flexible and typically exist as a conformational ensemble. Pore crystal structures are a single low-energy static conformation that may not represent the ensemble, and do not contain information regarding protein dynamics. Additionally, most available pore structures are elucidated in detergent media, which can significantly impact protein conformations (Böckmann & Caflisch, 2005; Chen, OuYang, & Chou, 2019; Rouse &

Sansom, 2015; Sammalkorpi, Karttunen, & Haataja, 2007), and not in lipid membranes which better mimic in vivo pore environments. Therefore, instead of resource-intensive experiments such as mutagenesis (Mesa-Galloso et al., 2017; Panchal & Bayley, 1995; Walker & Bayley, 1995 a) and mass spectrometry (Gupta et al., 2017), all-atom MD simulations of membrane-inserted pore complexes offer an ideal platform to investigate inter-protomer interactions that stabilize PFT pores.

To analyze these interactions in representative α and β-PFT pores in silico, we created structural ensembles of membrane-inserted ClyA and AHL pores (no lipids in the pore interior) via equilibrium MD simulations, and analyzed these ensembles for identifying "hot-spot" inter-protomer interfacial residues engaging in buried hydrogen bonds and salt bridges (Desikan et al., 2021). We focused on buried hydrogen bonding and salt bridging residues since these interactions are known to facilitate specific protein binding and oligomerization as well as structural stability (Basu & Mukharjee, 2017; Bosshard, Marti, & Jelesarov, 2004; Desiraju & Steiner, 2001; Donald, Kulp, & DeGrado, 2011; Gupta et al., 2017; Kumar & Nussinov, 2002; Waldburger, Schildbach, & Sauer, 1995), specifically in PFTs (Mueller et al., 2009; Wade et al., 2015); the same exercise could also be repeated for hydrophobic interactions. The step-wise simulation protocols are given below.

Using the detergent-stabilized pore crystal structures, we set up membrane-pore systems for MD simulations as specified in Sections 2.1 and 2.2, and simulated each complex for 500 ns at physiologically relevant conditions (Desikan et al., 2021). By analyzing the convergence of global indicators of protein structural stability, we identified parts of the trajectories where both pores were fully equilibrated in membranes. From this converged ensemble, protein dynamics were analyzed to obtain occurrence frequencies of hydrogen bonds and salt bridges. Two suitable residues were considered to be hydrogen bonded if the donor–acceptor atomic distance was less than a cut-off threshold of 3 Å, and the hydrogen donor–acceptor angle was less than a cut-off threshold value of 20 degrees. Similarly, a salt bridge occurred if any oxygen atom of an acidic amino acid (glutamic acid, aspartic acid) was within a cut-off threshold distance of 4 Å from any nitrogen of a basic residue (arginine, lysine, histidine). Occurrence frequencies were defined as the fraction of pore conformations in the equilibrium MD ensemble where a particular hydrogen bond or salt bridge was switched on. A small subset of persistent (>50% occurrence), buried (relative solvent accessibility <0.25), inter-protomer, strongly interacting (interaction energies < −4 kcal/mol) hydrogen bonds and salt bridges were identified [see Desikan et al., 2021 for more details of the screening procedure]. Fig. 3D illustrates several long lived salt bridges in the extracellular and membrane-inserted segments for the ClyA pore complex. Previous mutagenesis studies (Panchal & Bayley, 1995; Sathyanarayana et al., 2018; Wade et al., 2015; Walker & Bayley, 1995a, 1995b), where mutating some of these important residues involved in long lived hydrogen bonds and salt bridges led to diminished pore formation and lytic activity, validated our observations.

2.3.3 Setting up PFT "arcs" and intermediates in lipid membranes and analysis of central lipid evacuation

In addition to simulations with the full dodecameric ClyA pore, we have also performed simulations of the single transmembrane protomer (1-mer), and intermediate ClyA "arcs" at various levels of oligomerization (6-mer to 10-mer) (Desikan, Maiti, & Ayappa, 2017). In the absence of high-resolution experimentally determined structures of any PFT arc, we constructed the ClyA arc structures from the fully formed pore crystal structure and tested their influence on lipid bilayers to glean molecular insights regarding pore forming mechanisms (Desikan, Maiti, & Ayappa, 2017). Similar procedures were also employed to simulate PLY and LLO pore intermediates in membranes to unravel mechanisms of pore opening (Cheerla & Ayappa, 2020; Vögele et al., 2019).

The structures of the ClyA pore intermediates, from the protomer (1-mer) to the partial n-mer arcs ($n = 6$–10), were obtained from the aforementioned 12-mer pore structure with added termini (Section 2.3.3), by deleting the requisite number of protomers to create an arc. For example, the 7-mer arc was obtained by deleting 5 adjacent protomers from the 12-mer pore structure (Fig. 4A). These n-mer arcs were then manually inserted into fluid-phase DMPC (1,2-dimyristoyl-sn-glycero-3-phosphocholine) and POPC (1-palmitoyl-2-oleoyl-sn-glycero-3-phosphocholine) lipid membranes, by carefully orienting their putative transmembrane iris-like α-helical protein domains within the membranes and subsequently deleting lipids that overlapped with the protein. We note that lipids were present in and around ClyA arcs at the start of simulations, and the lipid bilayers were intact and defect-free except for the inserted n-mer complexes. We also note that an alternate method for constructing membrane-inserted protein configurations is via the CHARMM-GUI web-server (Jo, Kim, Iyer, & Im, 2008; Lee et al., 2016; Wu et al., 2014). After energy minimization and short

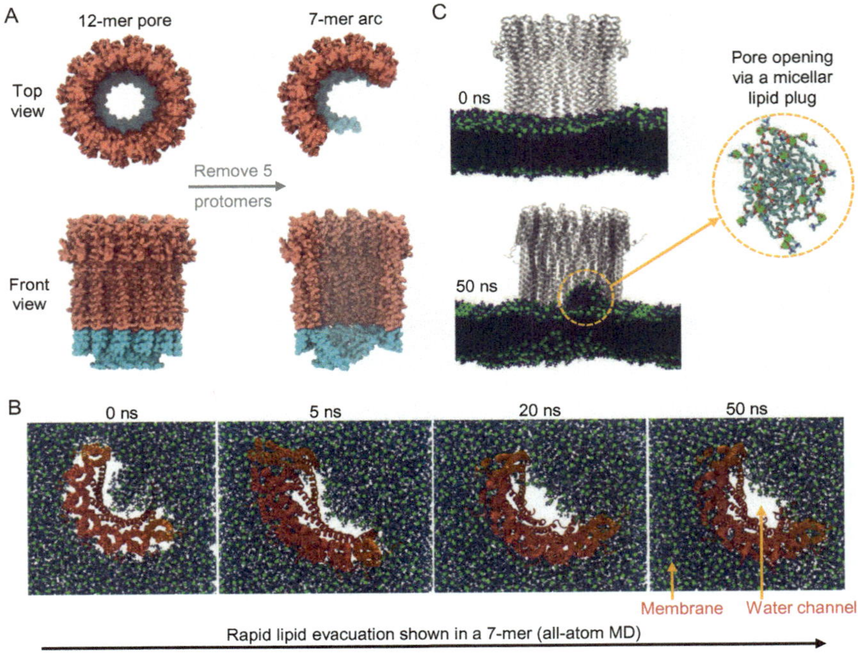

Fig. 4 Molecular dynamics simulations of pore opening with ClyA oligomers/pore (Desikan, Maiti, & Ayappa, 2017; Desikan, Padmanabhan, & Ayappa, 2020). (A) Creating initial structures of ClyA arcs from the pore illustrated for the 7-mer. (B) Lipid evacuation from the 7-mer arc interior to create functional pores capable of leakage (Desikan, Maiti, & Ayappa, 2017). (C) Micellar lipid plug formation within a fully oligomerized ClyA pore complex (Desikan, Padmanabhan, & Ayappa, 2020). *Panels (A) and (B): Reprinted with permission from Desikan, R., Maiti, P. K., & Ayappa, K. G. (2017). Assessing the structure and stability of transmembrane oligomeric intermediates of an α-helical toxin. Langmuir, 33 (42), 11496–11510. Copyright (2017) American Chemical Society. Panel (C): Reproduced from Desikan, R., Padmanabhan, P., & Ayappa, K. G. (2020). Opening of smaller toxin pores by lipid micelle formation. Proceedings of the National Academy of Sciences of the United States of America, 117 (10), 5107–5108.*

restrained equilibration runs (see Section 2.2), these membrane–protein systems were simulated for up to a few hundreds of nanoseconds. Protein structural attributes, lipid dynamics, and membrane rearrangements were continuously tracked.

Simulations of the ClyA protomer and the n-mer arcs showed that they were all structurally stable in the membrane over simulation timescales (Desikan, Maiti, & Ayappa, 2017), implying that they were reasonable approximations to putative structures of pore intermediates (Agrawal et al.,

2017; Peng, de Souza Santos, Li, Tomchick, & Orth, 2019). Strikingly, all arcs, including the protomer, spontaneously displaced the central lipids within the arc interior (shown for the 7-mer in Fig. 4B), thus creating protein-lipid-lined transmembrane water channels whose size was commensurate with the extent of PFT oligomerization. The kinetics of lipid evacuation from arc interiors, measured using the continuous survival probability formalism (see Eq. 2), was rapid and was typically completed within tens of nanoseconds.

All-atom simulations of the ClyA membrane-pore complex typically involve about 10^5–10^6 atoms for the full pore complex. In contrast, larger PFTs such as those from the CDC family, which form oligomers of variable size containing 30–50 protomers, can result in systems involving up to 10 million atoms, thus drastically increasing the computation overhead. In these situations, reduced systems with smaller oligomers can be judiciously constructed to yield insights into the pore forming mechanism. After homology modeling of the LLO membrane-inserted pore state, we studied two tetrameric ensembles, one consisting of four membrane-bound D4 (Tweten et al., 2001; Vögele et al., 2019) subunits, and the other consisting of four protomers to form a membrane-inserted arc (Fig. 5E). To construct the arc, a single membrane-inserted protomer was translated by one unit cell and rotated by 20 degrees with respect to its neighbor to generate the initial LLO tetrameric structure (Köster et al., 2014). Initialization and equilibration procedures were similar to those described in Section 2.2. The D4 subunits assembled into a linear array on the membrane consistent with electron micrograph data (Köster et al., 2014). Similar to what was observed with ClyA (Desikan, Maiti, & Ayappa, 2017) and PLY (Vögele et al., 2019) arcs, lipids were rapidly displaced from the arc interior to reorient and form a toroidal interface (Fig. 5E). Interestingly, lipids in the toroidal interface were devoid of cholesterol.

In summary, all-atom MD simulations provide strong evidence that partially oligomerized membrane-inserted arc-like states are stable entities for both α-PFTs as well as larger CDCs such as LLO and PLY. These simulations also unraveled structural and dynamical molecular features of membrane-inserted pore intermediates which are difficult to investigate experimentally.

2.3.4 Simulating fully formed pores with intact central lipids to explore pore opening

We subsequently explored alternate molecular mechanisms of ClyA pore formation and central lipid ejection via MD simulations. Our previous

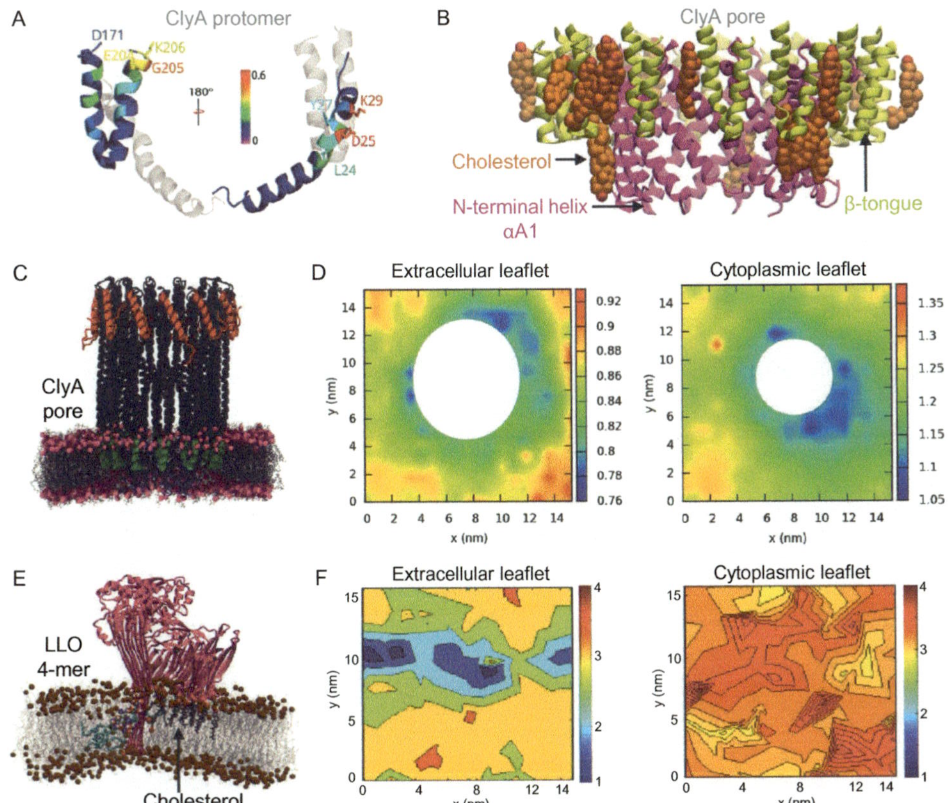

Fig. 5 See figure legend on opposite page.

ClyA arc MD simulations were initialized with membrane-inserted arcs, and we had not explored how arcs would insert into intact membranes. A recent study (Vögele et al., 2019) of PLY, with a coarse-grained MARTINI FF, showed that with slow insertion of PLY arcs, central lipids could laterally escape from the arc interior, similar to our previous observations with ClyA arcs (Desikan, Maiti, & Ayappa, 2017). However, during rapid insertion of PLY arcs, the central lipids in the lumen were trapped, and spontaneously buckled forming a liposomal plug which could then be expelled by osmotic flows. PLY pores/arcs have a large internal diameters of 30 to 50 nm, permitting the liposomal plug-formation pathway. How smaller pores such as ClyA, with internal diameters of < 10 nm (Desikan, Patra, et al., 2017; Mueller et al., 2009), would expel central lipids was not understood.

Therefore, similar to the arc simulations with ClyA, we inserted fully formed 12-mer ClyA pores into intact DMPC and POPC membranes, and deleted only the lipids that overlapped with the protein (Desikan, Padmanabhan, & Ayappa, 2020). This initially left the central lipids in an intact lamellar configuration within the pore lumen (Fig. 4C). After energy

Fig. 5 All-atom MD simulations reveal (A) cholesterol binding sites in a single membrane-inserted ClyA protomer and (B) cholesterol binding pockets between adjacent β-tongues in the transmembrane helices of the ClyA pore (Sathyanarayana et al., 2018). Simulations are carried out for (A) the entire protomer and (B) the dodecameric pore complex in solvent and membrane (only the cholesterol binding regions are displayed here for clarity). From MD simulations of a transmembrane ClyA pore (C), lipid displacement, d_n (Eq. 3) maps in a DMPC membrane with ClyA (D) illustrate regions of lowered mobility in the vicinity of the pore complex (Varadarajan et al., 2020) (*color bars* are in units of nm, $\Delta t = 50$ ns). Similarly for the LLO tetramer (Cheerla & Ayappa, 2020) (E), maps of $d_n/\Delta t$ for cholesterol in a DOPC–cholesterol bilayer (70:30) are shown (F) (*color bars* are in units of Å/ns, $\Delta t = 1$ ns), illustrating reduced lipid mobility in the extracellular leaflet due to specific binding with the D4 subunits as seen in the snapshot (E). *Panels (A) and (B): Reproduced from Sathyanarayana, P., Maurya, S., Behera, A., Ravichandran, M., Visweswariah, S. S., Ayappa, K. G., & Roy, R. (2018). Cholesterol promotes cytolysin A activity by stabilizing the intermediates during pore formation. Proceedings of the National Academy of Sciences of the United States of America, 115 (31), E7323–E7330; Panels (C) and (D): Reproduced from Varadarajan, V., Desikan, R., & Ayappa, K. G. (2020). Assessing the extent of the structural and dynamic modulation of membrane lipids due to pore forming toxins: Insights from molecular dynamics simulations. Soft Matter, 16 (20), 4840–4857; Panels (E) and (F): Reproduced with permission from Cheerla, R., & Ayappa, K. G. (2020). Molecular dynamics study of lipid and cholesterol reorganization due to membrane binding and pore formation by listeriolysin O. The Journal of Membrane Biology, 253, 535–550. https://doi.org/10.1007/s00232-020-00148-9. Copyright (2020) Springer Nature.*

minimization and short equilibration runs, production MD simulations were performed, and the central lipids monitored. Remarkably, within a few nanoseconds, the central lipids within the pore lumen spontaneously formed stable micellar (not liposomal) lipid plugs, that slowly rose from the membrane plane within the pore channel, and could presumably be expelled later via osmotic flows (Fig. 4C). The micelles exhibited solvated phosphatidylcholine head groups and desolvated lipid hydrocarbon tails (Fig. 4C), similar to lamellar configurations, and indicative of stability via the hydrophobic effect. Our work suggested that both small and large PFT nanopores may exhibit central lipid ejection as micelles and liposomes, respectively, during pore formation (Desikan, Padmanabhan, & Ayappa, 2020; Vögele et al., 2019).

2.3.5 Lipid and cholesterol binding and dynamics in PFTs

Studies have shown that presence of cholesterol in lipid membranes improves the binding and oligomerization of PFTs. With ClyA, cholesterol is known to enhance activity (Oscarsson et al., 1999; Peraro & van der Goot, 2016; Sathyanarayana et al., 2018). To understand the role of cholesterol in ClyA pore formation, all-atom MD simulations of a single protomer, dimer, and the fully oligomerized dodecameric pore complex of ClyA (Fig. 5A and B) embedded in DOPC–cholesterol (70:30) membranes, were carried out (Sathyanarayana et al., 2018). For studies with LLO (Fig. 5E), microsecond-long all-atom MD simulations were performed with a tetrameric membrane-bound D4 subunits as well as membrane-inserted tetrameric arcs (Cheerla & Ayappa, 2020).

Binding is assumed to occur when the distance between the center of masses of cholesterol and the specific residue lies within 0.5 nm (Sathyanarayana et al., 2018). We evaluated the fraction of MD sampling time a cholesterol molecule spends in the vicinity of specific residues. Amino acids K29, Y27, and D25 (Fig. 5A), belonging to the cholesterol recognition and consensus (CRAC) motif (Giri Rao et al., 2016), were identified as strong cholesterol binding sites (Sathyanarayana et al., 2018). Additional binding sites were also identified between the β-tongues of the assembled pore complex (Fig. 5B). Erythrocyte experiments with single mutations of ClyA—K29A, Y27A, and Y27F—indicated that Y27 plays a key role in cholesterol recognition (Sathyanarayana et al., 2018). Interestingly Y27 is also part of a long lived hydrogen bond (Fig. 3D; Desikan et al., 2021) attributed to stabilizing the membrane-inserted N-terminus in the assembled pore complex.

Molecular dynamics simulations of a DMPC bilayer with ClyA and AHL highlight PFT induced changes in lipid dynamics, and the spatial extent of PFT influence on membrane dynamics and lipid structure (Varadarajan et al., 2020). Different diffusive regimes were elucidated by the mean square displacements (MSD) calculated using,

$$\langle MSD \rangle = \frac{1}{N} \left\langle \sum_{i=1}^{N} |\mathbf{r}_i(t+\Delta t) - \mathbf{r}_i(t)|^2 \right\rangle \tag{1}$$

where $\mathbf{r}_i = \mathbf{e}_x x_i + \mathbf{e}_y y_i$ and the brackets indicate shifted time averaging. While evaluating diffusion coefficients, one needs to ensure that the particle dynamics have been sufficiently sampled to enter the diffusive regime where $MSD \sim t^{\alpha}$ with the exponent $\alpha = 1$. In the subdiffusive regime $\alpha < 1$ (Maginn et al., 2018; Varadarajan et al., 2020). Sufficiently long MD simulations of at least 1 μs must be carried out to reliably extract a diffusion coefficient. Additionally diffusion coefficients are colored by periodic boundary conditions (Vögele, Köfinger, & Hummer, 2018). Despite these caveats, lipid diffusion coefficients extracted from MD simulations compare reasonably well with fluorescence correlation spectroscopy (FCS) measurements (Cheerla & Ayappa, 2020; Ponmalar et al., 2019).

The continuous survival probabilities (CSP) (Desikan, Maiti, & Ayappa, 2017; Varadarajan et al., 2020), which indicates the duration a lipid molecule resides in a given region is computed using,

$$C_j(t) = \sum_{i=1}^{N_j} \left\langle \prod_{t_k=t_0}^{t_0+t} \delta_i(t_k) \right\rangle \tag{2}$$

where δ_i is the Kronecker delta, which is either unity if the coordinates of the phosphate head group of the ith lipid are present in the corresponding region, or zero if absent. The normalized CSP at any time "t," $C_j(t)/C_j(t_0)$, yields the fraction of time a molecule is continuously present in a given region (Desikan, Maiti, & Ayappa, 2017; Varadarajan et al., 2020). The extent of dynamic perturbation as revealed from the CSPs extend to greater distance (~4 nm) from the pore complex when compared with the structural perturbations which are shorter ranged (~2.5 nm) (Varadarajan et al., 2020).

Occupation of cholesterol around the pore can be quantified by using number density maps. The lipid bilayer area is divided into a 2D-grid, with grid lengths of 10 Å. The intersecting points of grid lines is a grid point.

The coordinates of the center-of-mass of each cholesterol molecule in the membrane plane was calculated, and the molecule was assigned to the nearest grid point. Grid point counts were then averaged over the simulation trajectory and normalized with the area of the box to obtain the areal density. The densities were then interpolated using MATLAB routines using the "scatteredInterpolant" function for plotting contour maps. The density maps for cholesterol for both ClyA (Sathyanarayana et al., 2018) and LLO (Cheerla & Ayappa, 2020), revealed the presence of distinct hot spots for cholesterol binding.

A convenient metric to determine the extent of spatial variation or dynamical heterogeneity in lipid and cholesterol dynamics around the pore complex is the in-plane displacement of a given particle,

$$d_n = \langle |\mathbf{r}_n(t + \Delta t) - \mathbf{r}_n(t)| \rangle \tag{3}$$

where $\mathbf{r}_n(t) = x_n(t)\mathbf{e}_x + y_n(t)\mathbf{e}_y$ and Δt is the specific time interval between two configurations, and brackets indicate a time average. The displacements are binned on a two-dimensional grid of 1 Å grid spacing based on the location of the initial coordinates of the displacement vectors. As the initial coordinates of the displacement vectors were not uniformly distributed, the "scatteredInterpolant" function in MATLAB was used to interpolate the scattered data to a uniform grid, and the interpolated values on the query points were used for creating contour plots or maps. Fig. 5D illustrates the lipid displacement maps for lipids around ClyA, and Fig. 5F illustrates the $d_n/\Delta t$ maps for cholesterol around LLO tetramers, showing reduced mobility of lipids and cholesterol as well as a distinct variation across leaflets for LLO. This dynamical heterogeneity induced by LLO in DOPC–cholesterol membranes result in two subpopulations of diffusing lipids as revealed by super resolution measurements (Sarangi, Ayappa, Visweswariah, & Basu, 2016).

The local density and displacement evaluations from MD simulation trajectories established that target cell membranes are not just frameworks for binding of PFTs, but contain molecular determinants that also regulate the activity of the PFTs. Molecular dynamics simulations elucidated the stabilizing interaction of cholesterol with ClyA at various stages of the pore forming pathway, namely, the single membrane-inserted protomeric state, membrane-inserted dimers, and the fully assembled dodecameric pore (Sathyanarayana et al., 2018). Similarly, the role of cholesterol in LLO pore forming activity and the induced modulation of lipid and cholesterol density was also established (Ponmalar et al., 2019).

2.3.6 Pore blocking and ionic currents in PFT pores

We have also explored the use of dendrimers as efficient pore blockers to prevent PFT function (Mandal et al., 2016). Dendrimers are a class of hyper-branched polymers with well-defined size and shape which can be tuned by solution pH (Maiti & Bagchi, 2009; Maiti, Çağın, Lin, & Goddard, 2005), and have shown promising application in drug delivery and therapeutics (Jain, Maingi, Maiti, & Bharatam, 2013; Maingi, Kumar, & Maiti, 2012). All-atom MD simulations were used to compare the pore blockage characteristics between the nonprotonated (G5-NH2) polyamido amine (PAMAM) dendrimer and the protonated (G5–NH3$^+$) PAMAM dendrimer (Mandal et al., 2016). Atomistic MD simulations were carried out using the PMEMD module of the AMBER41 software package. The energy minimized structure was gradually heated up from 0 to 300 K, while imposing a weak harmonic constraint on the solute atoms. The heated system was simulated for 100 ns in the NPT ensemble under ambient conditions. Protonated (P) and nonprotonated (NP) PAMAM were built using the dendrimer building toolkit (Maingi, Jain, et al., 2012) and energy minimized using the same above mentioned protocol using the generalized amber force field (GAFF). The equilibrated dendrimer structure was placed near the extracellular end of ClyA pore opening (Fig. 6) through VMD and followed by solvation and addition of ions for maintaining a 150 mM salt concentration. The entire system was then simulated for 200 ns.

Ionic current along the ClyA protein channel were induced by imposing a uniform electric field of 0.02 V nm^{-1} directed toward the cytosolic end of the pore. During the course of 35 ns MD simulations with the electric field, the bilayer was restrained to its initial configuration with a harmonic potential. The instantaneous ionic current, $I(t)$, was measured using the following equation,

$$I(t) = \frac{1}{\Delta t L_z} \sum_{i=1}^{N} q_i [z_i(t + \Delta t) - z_i(t)], \tag{4}$$

where z_i and q_i are the z coordinate and atomic charge of the ith atom, respectively, and N represents the total number of ions present in the system. L_z is the membrane thickness (~5 nm) and the coordinates of the ions were recorded at time intervals of $\Delta t = 1.5$ ps. Water flux was induced by introducing a small force (0.01 kcal mol^{-1}nm^{-1}) to the oxygen atom of the water molecule. Other simulation parameters and protocols are discussed in Section 2.2.

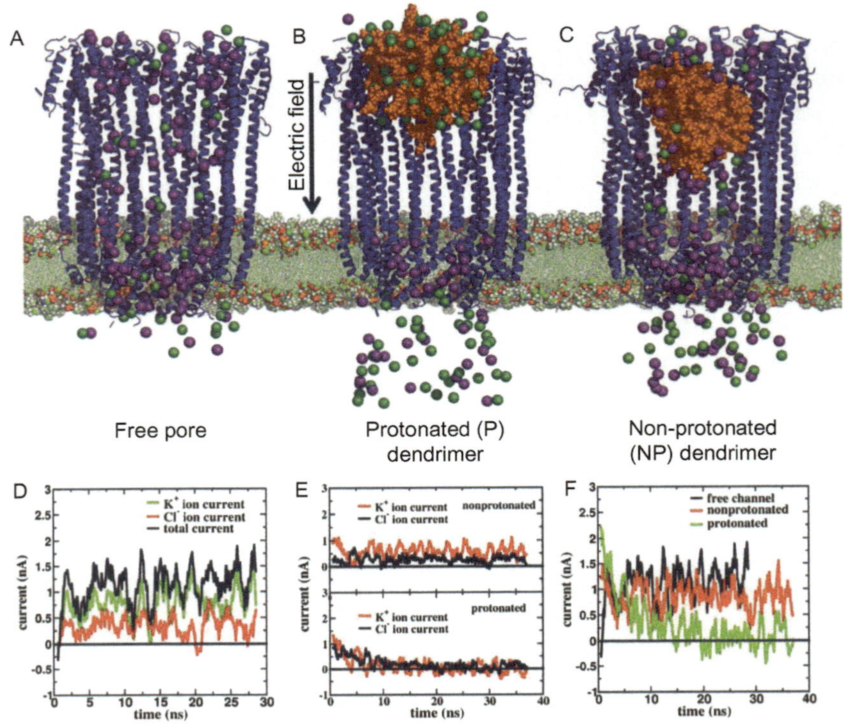

Fig. 6 All-atom MD simulations illustrating ClyA pore blocking with protonated and nonprotonated PAMAM dendrimers (Mandal et al., 2016). Snapshots illustrate (A) ions in a free pore, and effective pore blocking with (B) the protonated dendrimer and (C) the nonprotonated dendrimer. Ionic currents in (D) free pores, (E) pores with dendrimers, and (F) the total ionic current with dendrimers are also shown. Cl^-—*green*, K^+—*purple. Reproduced from Mandal, T., Kanchi, S., Ayappa, K. G., & Maiti, P. K. (2016). pH controlled gating of toxic protein pores by dendrimers. Nanoscale, 8 (26), 13045–13058.*

The entry of the NP dendrimer into the pore lumen was rapid, whereas the entry of the P dendrimer was more gradual. After 200 ns, the P and NP dendrimers moved 15 and 35 Å into the pore lumen, respectively. Density maps revealed greater void space in case of the NP dendrimer (Mandal et al., 2016). Due to the strong electrostatic interactions between the negatively charged inner pore lumen and positively charged terminal groups of the P dendrimer, the P dendrimer blocked the pore lumen more effectively compared to the NP dendrimer, the latter adopting a stretched configuration (Fig. 6C). The ion distribution inside the pore was dramatically altered in the presence of the P dendrimer leading to significant reduction in ionic current with the P dendrimer (Fig. 6E and F). The total number of Cl^- ions inside the nanopore also increased significantly due to increased binding of ions to the positively charged P dendrimer (Fig. 6B).

In the presence of the P dendrimer, residence time for Cl^-, evaluated by fitting the CSPs (Eq. 2) to a double exponential function, was 135 and 979 ps. The bound Cl^- ions exhibited a long relaxation time of ~ 1 ns, with the smaller relaxation time arising from the free ions within the pore. K^+ ions were absent in this region. This indicated strong attraction of Cl^- and repulsion of K^+ ions by the P dendrimer. On the other hand, the NP dendrimer did not show any significant effect on the ions when compared to the free pore. This case study illustrates the value of MD simulations for selecting suitable pore blockers to potentially mitigate PFT-mediated infections.

3. MARTINI coarse-grained MD simulations of PFTs

Many biological processes associated with PFTs, such as oligomerization, membrane-reorganization, and lipid dynamics, can only be explored in silico with large MD simulations accessing micrometer and microsecond spatio-temporal scales (Fig. 1). A popular biomolecular CG FF is the MARTINI model (Alessandri et al., 2019; Bruininks, Souza, & Marrink, 2019; Ingólfsson, Arnarez, Periole, & Marrink, 2016; Marrink, Risselada, Yefimov, Tieleman, & de Vries, 2007; Marrink & Tieleman, 2013), whose chemical building blocks are primarily parameterized to reproduce experimental thermodynamic data such as oil/water partition coefficients. MARTINI is versatile, modular, and contains well-tested parameter sets for biomolecules such as proteins (de Jong et al., 2013; Marrink et al., 2007; Monticelli et al., 2008; Periole, Cavalli, Marrink, & Ceruso, 2009), lipids (Daily, Olsen, Schlesinger, Ory, & Baker, 2014; Marrink, de Vries, & Mark, 2004), and nucleic acids (Uusitalo, Ingólfsson, Akhshi, Tieleman, & Marrink, 2015; Uusitalo, Ingolfsson, Marrink, & Faustino, 2017).

3.1 System preparation and simulation settings

MARTINI simulations of PFTs (Desikan, Padmanabhan, & Ayappa, 2020; Desikan, Patra, et al., 2017) were performed using FF version 2.2 with an elastic dynamic network (Elnedyn 2.2) implemented in GROMACS (de Jong et al., 2013; Marrink et al., 2007; Monticelli et al., 2008; Periole et al., 2009; Yesylevskyy, Schafer, Sengupta, & Marrink, 2010). Initially, the atomistic PFT protein structures were converted into CG resolution—4 heavy atoms: 1 CG bead with standard amino acid mapping—using the "martinize.py" tool; version 2.6 implemented in python 3, available from the official MARTINI website at http://cgmartini.nl/index.php/tools2/proteins-and-bilayers (accessed on 16 October 2020). Since MARTINI simulations do not allow for secondary structure transformations, the initial secondary structure of the protein backbone, computed using the DSSP algorithm (Kabsch & Sander, 1983) (version 2.1.0), must be passed as an input into martinize.py for defining secondary structure constraints to the protein topology. We also employed default Elnedyn force constants (Periole et al., 2009) of 500 kJ mol^{-1} nm^{-2}, and note that while modifying this parameter to match global structural properties from atomistic simulations is possible (Globisch, Krishnamani, Deserno, & Peter, 2013), we did not pursue that aspect in our work. With PFT oligomers and pores, an important detail to note is that the elastic network was applied within each protomer subunit in the oligomer/pore, and not across protomers. This is because the oligomeric assembly is supposed to be stabilized by nonbonded interactions, and applying an elastic network of springs across protomers could artificially stiffen/stabilize the oligomeric complexes.

The CG resolution transformed proteins were then inserted into homogeneous MARTINI DMPC, POPC, and DPPC (1,2-dipalmitoyl-sn-glycero-3-phosphocholine) bilayers, which were either created by self-assembly MD simulations, or by using the "INSANE" python script (Wassenaar, Ingólfsson, Böckmann, Tieleman, & Marrink, 2015) (available from the same link as above), followed by short equilibration. During insertion of PFTs into membranes, the lipids overlapping with the protein were deleted similar to atomistic simulations (Section 2.3.3). Sufficient hydration was maintained by adding appropriate amount of CG water along with 0 or 0.15 M Na^+/Cl^- ions to create charge neutral systems. Antifreeze solvent particles, which have bigger van der Waals radii to disrupt artificial freezing of CG water, were sometimes added (proportion: 10% of CG water beads). We note that contrary to typical observations, freezing was not observed with CG water in our PFT simulations even in the absence of anti-freeze beads, probably due to the heterogeneous nature of our membrane–protein–solvent systems (a large part of the PFTs are completely solvated), and the large numbers of CG ions disrupting freezing of the solvent.

The MARTINI simulation settings were fixed according to published parameter settings for improved performance in GROMACS (de Jong, Baoukina, Ingólfsson, & Marrink, 2016) (sample input scripts available at http://cgmartini.nl/index.php/force-field-parameters/input-parameters, accessed on 16 October 2020). While the MARTINI developers advocate an integration timestep of up to 40 fs, we conservatively ran all our MARTINI simulations with a 20 fs timestep for better accuracy and energy conservation. Simulations were performed in the NPT ensemble with weak temperature and pressure coupling enabled by the stochastic velocity-rescaling thermostat (Bussi et al., 2007) at 310 K (coupling constant of 1.0 ps), and the Parrinello–Rahman isotropic/semiisotropic barostat (Parrinello & Rahman, 1981) at 1 bar (coupling constant of 12.0 ps). Electrostatic interactions were calculated using the reaction-field algorithm with a cut-off of 1.1 nm, and with a relative dielectric constant of 2.5 and 15 for systems with and without MARTINI polarizable water (Yesylevskyy et al., 2010), respectively. The relative dielectric constant of the reaction field was set to infinity. van der Waal interactions were computed with a cut-off of 1.1 nm, and the Lennard-Jones potentials were shifted such that their values were zero at the cut-off. Upon energy minimization with steepest descent, short equilibration MD simulations with and without constraints on the membrane–protein beads were performed similar to atomistic simulations (see Section 2.2) before embarking on longer production simulations.

3.2 Case studies

3.2.1 Protomer-membrane binding energy via umbrella sampling

We employed MARTINI simulations, with a polarizable water model and PME electrostatics, to assess the binding free energy of a single ClyA protomer with a DMPC membrane (Fig. 7A) using umbrella sampling simulations (Desikan, Maiti, & Ayappa, 2017). The reaction coordinate, "ζ," was defined as the distance between the centers of mass of the protomer and the membrane along the membrane normal. The ClyA protomer was equilibrated in the DMPC membrane for 100 ns, and subsequently, short steered MD simulations were utilized to create 24 starting configurations for umbrella sampling along ζ, with $\Delta\zeta = 0.2$ nm. Each of these configurations were restrained around their respective initial values of ζ with weak harmonic potentials. From the variance of ζ in independent unrestrained simulations (ζ^*), the spring constant for these weak harmonic restraints was calculated as $\frac{RT}{\zeta^*} = 10$ kJ mol^{-1} nm^{-2} at 310 K (see Desikan, Maiti, & Ayappa, 2017 for more details). Movement of the membrane along its normal was restrained by applying a strong harmonic potential on the DMPC phosphate atoms (force constant of 100,000 kJ mol^{-1} nm^{-2}). Each umbrella sampling window was then simulated for 100 ns. From

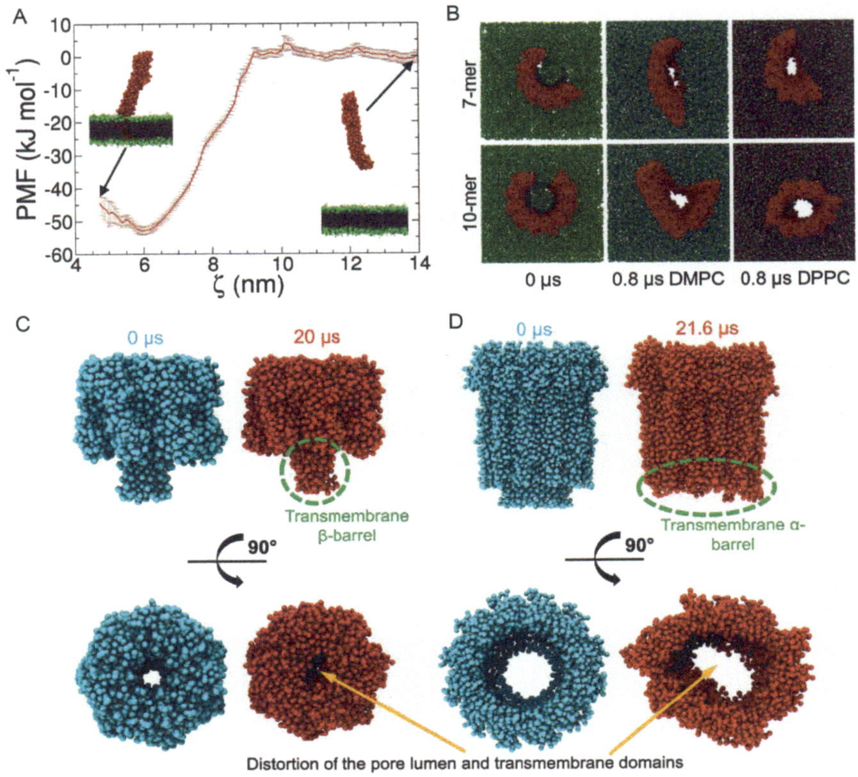

Fig. 7 Martini model simulations for (A) potential of mean force (PMF) computations of a ClyA protomer with a phospholipid (DMPC) membrane (Desikan, Maiti, & Ayappa, 2017), (B) lipid evacuation in membrane-inserted ClyA arcs (Desikan, Patra, et al., 2017) (similar to Fig. 4B), and distortions in membrane-inserted pore complexes of (C) AHL and (D) ClyA pores (Desikan, Patra, et al., 2017). Only protein atoms shown for clarity in (C) and (D). *Panel (A): Reprinted with permission from Desikan, R., Maiti, P. K., & Ayappa, K. G. (2017). Assessing the structure and stability of transmembrane oligomeric intermediates of an α-helical toxin. Langmuir, 33 (42), 11496–11510. Copyright (2017) American Chemical Society; Panels (B)–(D): Reprinted with Permission from Desikan, R., Patra, S. M., Sarthak, K., Maiti, P. K., & Ayappa, K. G. (2017). Comparison of coarse-grained (martini) and atomistic molecular dynamics simulations of α and β toxin nanopores in lipid membranes. Journal of Chemical Sciences, 129 (7), 1017–1030. Copyright (2017) Springer Nature.*

examining the composite histogram plot along ζ from all umbrella sampling trajectories, it was observed that umbrella sampling windows 11 and 12, corresponding to the protomer at the water–membrane interface, exhibited large fluctuations and poor sampling (overlap). Thus, only these windows were rerun with a higher force constant of 100 kJ mol^{-1} nm^{-2}. Note that the N-terminus of the protomer, which is a part of ClyA's transmembrane domain, undergoes significant conformational changes in the fully solvated state (umbrella sampling simulations at higher ζ). The potential of mean force (PMF) was obtained by using the weighted histogram analysis method (WHAM) along ζ, and error bars on the PMF were computed by bootstrapping. From the PMF (Fig. 7A), the membrane-inserted protomer state is observed to be a thermodynamically favorable state, with a free energy difference of ~ -13 kcal/mol compared to the fully solvated protomer.

3.2.2 MARTINI simulations of PFT pores exhibit significant structural distortions

PFT oligomers and pore complexes are stabilized by nonbonded interactions (Desikan et al., 2021; Mueller et al., 2009). While MARTINI FFs constrain the protein secondary structure, domain movements leading to changes in tertiary and quaternary structure are unrestricted. Hence, MARTINI simulations can be employed to assess the structural stability of PFT oligomers over longer timescales of tens of μs.

Lipid evacuation and stability of arcs were studied in MARTINI simulations. Indeed, while n-mer arcs evacuated lipids from their interior into the surrounding membrane (Fig. 7B), in agreement with atomistic simulations (Fig. 4B), they exhibited structural distortions in their transmembrane domains and exhibited a greater tendency to close with increasing number of protomers in the arcs (seen in the 10-mer arc in DPPC, Fig. 7B; more details in Desikan, Patra, et al., 2017). In contrast, ClyA arcs in all-atom simulations show no tendency to close (Desikan, Maiti, & Ayappa, 2017), and there is sufficient experimental evidence that stable ClyA arcs may be natural intermediates along ClyA (and α-PFT) pore formation pathways (Agrawal et al., 2017; Peng et al., 2019).

We also analyzed the stability of ClyA and AHL membrane-inserted pores in MARTINI simulations (Desikan, Patra, et al., 2017). We observed that the quaternary structure of the solvated extracellular regions was well-preserved, but the transmembrane domains of both PFT pores were severely distorted (Fig. 7C and D), despite the application of an intra-protomer elastic network. Particularly, the transmembrane domain of the AHL pore was constricted to such an extent that the pore channel was nearly closed, which would essentially render the pore nonfunctional. This is in stark contrast to experimental observations, crystal structures, and all-atom simulations where the pore geometries are stable and in close agreement with crystal structures (albeit performed over shorter time scales of $\sim$1000 ns; Desikan et al., 2021; Desikan, Patra, et al., 2017).

There are a few potential remedies to stabilize the tertiary and quaternary structures of protein complexes in MARTINI simulations. It has been shown that additional inter-protomer elastic dynamic networks may stabilize the entire arc/pore complex (Basdevant, Dessaux, & Ramirez, 2019; Vögele et al., 2019). Also, force constants of the elastic network can be calibrated using all-atom simulations (Globisch et al., 2013). Finally, we note that the next generation MARTINI 3 FFs currently in development (see http://cgmartini.nl/index.php/martini3beta, accessed on 16 October 2020), with new particle types, reparametrized interaction matrix, and improved representation of proteins, may improve the stability of pore complexes in MARTINI simulations.

4. Structure-based models for capturing the membrane-triggered conformational transition of an α-PFT monomer

A unique feature of PFTs is that their sequence encodes multiple 3D conformations that are stable depending on the environment (Peraro & van der Goot, 2016). In the case of ClyA, the secreted water-soluble monomeric form undergoes a large conformational change upon membrane recognition and binding into its assembly competent protomeric state (Giri Rao et al., 2016; Mueller et al., 2009). This conformational transition involves β-strand to α-helix and loop to α-helix secondary structure transformations, reorganization of inter-helix interfaces, and domain rearrangements of up to 140 Å, making it one of the largest conformational transitions observed in proteins of length of about 300 amino acids (Giri Rao et al., 2016; Mueller et al., 2009). The timescale of this process is estimated to lie in the range of seconds to minutes (Benke et al., 2015; Vaidyanathan, Sathyanarayana, Maiti, Visweswariah, & Ayappa, 2014), rendering it completely inaccessible to conventional atomistic molecular simulations, perhaps even with enhanced sampling techniques. Therefore, we employed MD simulations with a dual-basin, coarse-grained, implicit-solvent, structure-based model (SBM) of ClyA, along with a coarse-grained membrane model, to glean molecular insights into this conformational transition.

SBMs encode the native protein structure into a coarse-grained potential energy function (C_α protein atoms only) consisting of bonded (covalent bonds, angles, dihedrals) and nonbonded (modified contact potentials; no electrostatics) terms, which have been shown to capture the funnel-shaped energy landscape of proteins (Clementi et al., 2000; Giri Rao & Gosavi,

2018; Noel & Onuchic, 2012). To investigate the membrane-triggered PFT conformational transition, we employed dual-basin SBMs (Lin et al., 2014; Lu & Wang, 2008; Okazaki, Koga, Takada, Onuchic, & Wolynes, 2006; Ramirez-Sarmiento, Noel, Valenzuela, & Artsimovitch, 2015; Whitford, Miyashita, Levy, & Onuchic, 2007) (dSBMs), where both the ClyA monomer and protomer structures were encoded into the same potential energy function. Additionally, we also constructed a minimalist coarse-grained membrane model (Fig. 8A) that encoded essential

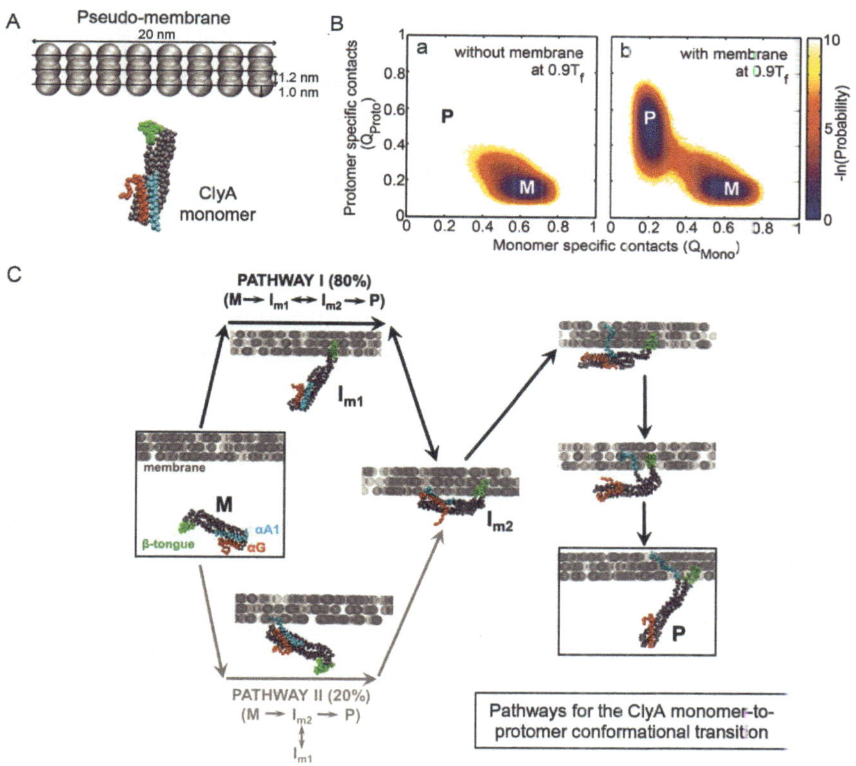

Fig. 8 Structure-based models (SBMs) illustrate the membrane-triggered conformational transition of a ClyA monomer into a protomer during membrane-insertion (Giri Rao et al., 2016). (A) Initial SBM MD configuration with the ClyA monomer situated away from the membrane. (B) Transition to protomer observed only in presence of a membrane. (C) Pathways observed during membrane-insertion. *Reprinted with permission from Giri Rao, V. V. H., Desikan, R., Ayappa, K. G., & Gosavi, S. (2016). Capturing the membrane-triggered conformational transition of an α-helical pore-forming toxin. The Journal of Physical Chemistry. B, 120 (47), 12064–12078. Copyright (2019) American Chemical Society.*

membrane–ClyA protomer interactions compatible with C_α dSBMs. While the monomer conformation was expected to be stable in the absence of the membrane, interaction with the membrane "triggered" the transition into the protomer conformation (Fig. 8B) while conferring stabilizing interactions with the transmembrane domain (Giri Rao et al., 2016).

Indeed, dSBM simulations, starting with the monomer state away from the membrane (Fig. 8A), captured ClyA's membrane-triggered monomer-to-protomer conformational transition (Fig. 8B) via two pathways (Fig. 8C). A dominant serial pathway via two membrane-bound intermediates (I_{m1} and I_{m2}) occurs in ~80% of the trajectories, while a secondary pathway where the intermediate I_{m2} appears as an off-pathway product occurs in the remaining ~20% of the trajectories (Fig. 8C). In our simulations, this transition only occurred in the presence of a membrane (Sathyanarayana et al., 2016) (Fig. 8B). Our simulations showed good agreement with FRET experiments (Benke et al., 2015; Roderer et al., 2014), which served as validation. A crucial insight was that one of ClyA's membrane-interacting domains, the β-tongue, functioned as a quick-response membrane sensor with a small barrier for conformational change and membrane-insertion, while the N-terminus helix αA1, functions as a fidelity enhancer to complete the conformational transition in the target membrane.

5. Summary

Molecular dynamics simulations have evolved as a powerful in silico framework to study a variety of biologically relevant molecules. In this article, we illustrate the utility of MD simulations to provide insights into the pore forming pathways of PFTs belonging to distinct PFT families, and their ability to provide molecular insights into biological phenomenon often difficult to probe experimentally. We have laid emphasis on methods and techniques involved while setting up, monitoring, and evaluating properties from MD simulations of PFTs in a membrane environment. Choosing the appropriate MD technique is governed by the phenomenon we seek to explore, as well as the ability to reliably sample ensemble interactions within typical sampling times associated with a given technique. Thus, all-atom MD simulations with appropriate force fields from the CHARMM and AMBER families are expected to accurately capture protein–protein, protein–lipid and protein–water interactions. Additionally, current computing resources provide the ability to reliably monitor phenomenon that occur on the order of microseconds for moderately sized pore complexes such as those associated with ClyA and AHL. Understanding conformational changes that occur over time scales of milliseconds to minutes requires the use of advanced sampling techniques or reduced models such as the SBMs. MARTINI models are another class of coarse-grained methods that have been used to study oligomers formed by PFTs.

With rapid advances in super resolution and single molecule microscopy (Sarangi et al., 2016; Sarangi, Roobala, & Basu, 2018; Sathyanarayana et al., 2018), MD simulations are expected to play a significant role in interpreting experimental data and providing molecular insights into the PFT pathways as illustrated in this article. With the latest stimulated emission depletion (STED) microscopes, which can probe phenomenon down to 30 nm, length scales accessible in MD simulations, we expect a stronger synergy to occur between experiments and simulations. Similarly, with advances in Cryo-EM techniques (Bräuning et al., 2018; Peng et al., 2019), high-resolution crystal structures of PFT pore complexes as well as oligomeric intermediates in native membrane environments may be forthcoming, and these techniques may open up new frontiers by providing valuable inputs and insights to MD simulations of PFTs, and vice versa.

Acknowledgments

This work was supported by the Department of Science and Technology, Science and Engineering Research Board. We thank the Supercomputer Education and Research Center, Indian Institute of Science for computational facilities.

Competing interests

The authors declare no competing interests.

References

Agrawal, A., Apoorva, K., & Ayappa, K. G. (2017). Transmembrane oligomeric intermediates of pore forming toxin cytolysin a determine leakage kinetics. *RSC Advances*, 7(82), 51750–51762.

Aksimentiev, A., & Schulten, K. (2005). Imaging α-hemolysin with molecular dynamics: Ionic conductance, osmotic permeability, and the electrostatic potential map. *Biophysical Journal*, *88*(6), 3745–3761.

Alessandri, R., Souza, P. C. T., Thallmair, S., Melo, M. N., de Vries, A. H., & Marrink, S. J. (2019). Pitfalls of the MARTINI model. *Journal of Chemical Theory and Computation*, *15*(10), 5448–5460. https://doi.org/10.1021/acs.jctc.9b00473.

Alford, R. F., Leaver-Fay, A., Jeliazkov, J. R., O'Meara, M. J., DiMaio, F. P., Park, H., … Gray, J. J. (2017). The Rosetta all-atom energy function for macromolecular modeling and design. *Journal of Chemical Theory and Computation*, *13*(6), 3031–3048.

Allen, M. P., & Tildesley, D. J. (2017). *Computer simulation of liquids*. Oxford University Press.

Atkins, A., Wyborn, N. R., Wallace, A. J., Stillman, T. J., Black, L. K., Fielding, A. B., … Green, J. (2000). Structure-function relationships of a novel bacterial toxin, Hemolysin E: The role of αG. *The Journal of Biological Chemistry*, *275*(52), 41150–41155.

Basdevant, N., Dessaux, D., & Ramirez, R. (2019). Ionic transport through a protein nanopore: A coarse-grained molecular dynamics study. *Scientific Reports*, *9*(1), 15740.

Basu, S., & Mukharjee, D. (2017). Salt-bridge networks within globular and disordered proteins: Characterizing trends for designable interactions. *Journal of Molecular Modeling*, *23*(7), 206.

Beauchamp, K. A., Lin, Y.-S., Das, R., & Pande, V. S. (2012). Are protein force fields getting better? A systematic benchmark on 524 diverse NMR measurements. *Journal of Chemical Theory and Computation*, *8*(4), 1409–1414.

Beck, D. A. C., & Daggett, V. (2004). Methods for molecular dynamics simulations of protein folding/unfolding in solution. *Methods*, *34*(1), 112–120. https://doi.org/10.1016/j.ymeth.2004.03.008. Investigating Protein Folding, Misfolding and Nonnative States: Experimental and Theoretical Methods.

Benke, S., Roderer, D., Wunderlich, B., Nettels, D., Glockshuber, R., & Schuler, B. (2015). The assembly dynamics of the cytolytic pore toxin ClyA. *Nature Communications*, *6*, 6198.

Berendsen, H. J. C., Postma, J. P. M., van Gunsteren, W. F., DiNola, A., & Haak, J. R. (1984). Molecular dynamics with coupling to an external bath. *The Journal of Chemical Physics*, *81*(8), 3684–3690. https://doi.org/10.1063/1.448118.

Berger, O., Edholm, O., & Jahnig, F. (1997). Molecular dynamics simulations of a fluid bilayer of dipalmitoylphosphatidylcholine at full hydration, constant pressure, and constant temperature. *Biophysical Journal*, *72*(5), 2002–2013.

Best, R. B., Zhu, X., Shim, J., Lopes, P. E. M., Mittal, J., Feig, M., & MacKerell, A. D. (2012). Optimization of the additive CHARMM all-atom protein force field targeting improved sampling of the backbone ϕ, ψ and side-chain χ_1 and χ_2 dihedral angles. *Journal of Chemical Theory and Computation*, *8*(9), 3257–3273. https://doi.org/10.1021/ct300400x.

Blair, J. M., Webber, M. A., Baylay, A. J., Ogbolu, D. O., & Piddock, L. J. (2015). Molecular mechanisms of antibiotic resistance. *Nature Reviews. Microbiology*, *13*(1), 42–51.

Böckmann, R. A., & Caflisch, A. (2005). Spontaneous formation of detergent micelles around the outer membrane protein OmpX. *Biophysical Journal*, *88*(5), 3191–3204.

Bonome, E. L., Cecconi, F., & Chinappi, M. (2019). Translocation intermediates of ubiquitin through an α-hemolysin nanopore: Implications for detection of post-translational modifications. *Nanoscale*, *11*(20), 9920–9930. https://doi.org/10.1039/C8NR10492A.

Bosshard, H. R., Marti, D. N., & Jelesarov, I. (2004). Protein stabilization by salt bridges: Concepts, experimental approaches and clarification of some misunderstandings. *Journal of Molecular Recognition*, *17*(1), 1–16.

Braun, E., Gilmer, J., Mayes, H. B., Mobley, D. L., Monroe, J. I., Prasad, S., & Zuckerman, D. M. (2018). Best practices for foundations in molecular simulations (Article v1.0). *Living Journal of Computational Molecular Science*, *1*(1), 5957.

Bräuning, B., Bertosin, E., Praetorius, F., Ihling, C., Schatt, A., Adler, A., … Groll, M. (2018). Structure and mechanism of the two-component α helical pore-forming toxin YaxAB. *Nature Communication*, *9*, 1806.

Bruininks, B. M. H., Souza, P. C. T., & Marrink, S. J. (2019). A practical view of the martini force field. In *Biomolecular simulations: Methods and protocols* (pp. 105–127). Springer New York. https://doi.org/10.1007/978-1-4939-9608-7_5.

Bussi, G., Donadio, D., & Parrinello, M. (2007). Canonical sampling through velocity rescaling. *The Journal of Chemical Physics*, *126*(1), 014101.

Cao, C., Cirauqui, N., Marcaida, M. J., Buglakova, E., Duperrex, A., Radenovic, A., & Dal Peraro, M. (2019). Single-molecule sensing of peptides and nucleic acids by engineered aerolysin nanopores. *Nature Communication*, *10*(1), 4918.

Cheerla, R., & Ayappa, K. G. (2020). Molecular dynamics study of lipid and cholesterol reorganization due to membrane binding and pore formation by listeriolysin O. *The Journal of Membrane Biology, 253*, 535–550. https://doi.org/10.1007/s00232-020-00148-9.

Chen, W., OuYang, B., & Chou, J. J. (2019). Critical effect of the detergent:protein ratio on the formation of the hepatitis C virus p7 channel. *Biochemistry, 58*(37), 3834–3837.

Chiu, S.-W., Pandit, S. A., Scott, H. L., & Jakobsson, E. (2009). An improved united atom force field for simulation of mixed lipid bilayers. *The Journal of Physical Chemistry. B, 113*(9), 2748–2763. https://doi.org/10.1021/jp807056c.

Cirauqui, N., Abriata, L. A., van der Goot, F. G., & Dal Peraro, M. (2017). Structural, physicochemical and dynamic features conserved within the aerolysin pore-forming toxin family. *Scientific Reports*, 7(1), 13932.

Clementi, C., Nymeyer, H., & Onuchic, J. N. (2000). Topological and energetic factors: What determines the structural details of the transition state ensemble and en-route intermediates for protein folding? An investigation for small globular proteins. *Journal of Molecular Biology, 298*(5), 937–953.

Cosentino, K., Ros, U., & García-Sáez, A. J. (2016). Assembling the puzzle: Oligomerization of α-pore forming proteins in membranes. *Biochimica et Biophysica Acta-Biomembranes, 1858*(3), 457–466.

Daggett, V. (2002). Molecular dynamics simulations of the protein unfolding/folding reaction. *Accounts of Chemical Research, 35*(6), 422–429. https://doi.org/10.1021/ar0100834.

Daily, M. D., Olsen, B. N., Schlesinger, P. H., Ory, D. S., & Baker, N. A. (2014). Improved coarse-grained modeling of cholesterol-containing lipid bilayers. *Journal of Chemical Theory and Computation, 10*(5), 2137–2150. https://doi.org/10.1021/ct401028g.

Darden, T., York, D., & Pedersen, L. (1993). Particle mesh Ewald: An Nlog(N) method for Ewald sums in large systems. *The Journal of Chemical Physics, 98*(12), 10089–10092.

de Jong, D. H., Baoukina, S., Ingólfsson, H. I., & Marrink, S. J. (2016). MARTINI straight: Boosting performance using a shorter cutoff and GPUs. *Computer Physics Communications, 199*, 1–7. https://doi.org/10.1016/j.cpc.2015.09.014.

Degiacomi, M. T., Iacovache, I., Pernot, L., Chami, M., Kudryashev, M., Stahlberg, H., … Dal Peraro, M. (2013). Molecular assembly of the aerolysin pore reveals a swirling membrane-insertion mechanism. *Nature Chemical Biology, 9*(10), 623–629.

de Jong, D. H., Singh, G., Bennett, W. F. D., Arnarez, C., Wassenaar, T. A., Schäfer, L. V., … Marrink, S. J. (2013). Improved parameters for the MARTINI coarse-grained protein force field. *Journal of Chemical Theory and Computation, 9*(1), 687–697.

Desikan, R., Maiti, P. K., & Ayappa, K. G. (2017). Assessing the structure and stability of transmembrane oligomeric intermediates of an α-helical toxin. *Langmuir, 33*(42), 11496–11510.

Desikan, R., Maiti, P. K., & Ayappa, K. G. (2021). Predicting interfacial hot-spot residues that stabilize protein-protein interfaces in oligomeric membrane-toxin pores through hydrogen bonds and salt bridges. *Journal of Biomolecular Structure & Dynamics, 39*, 20–34. https://doi.org/10.1080/07391102.2020.1711806.

Desikan, R., Padmanabhan, P., & Ayappa, K. G. (2020). Opening of smaller toxin pores by lipid micelle formation. *Proceedings of the National Academy of Sciences of the United States of America, 117*(10), 5107–5108.

Desikan, R., Patra, S. M., Sarthak, K., Maiti, P. K., & Ayappa, K. G. (2017). Comparison of coarse-grained (martini) and atomistic molecular dynamics simulations of α and β toxin nanopores in lipid membranes. *Journal of Chemical Sciences, 129*(7), 1017–1030.

Desiraju, G., & Steiner, T. (2001). *The weak hydrogen bond: In structural chemistry and biology*. Oxford University Press.

Dickson, C. J., Madej, B. D., Skjevik, Å. A., Betz, R. M., Teigen, K., Gould, I. R., & Walker, R. C. (2014). Lipid14: The AMBER lipid force field. *Journal of Chemical Theory and Computation, 10*(2), 865–879. https://doi.org/10.1021/ct4010307.

Dickson, C. J., Rosso, L., Betz, R. M., Walker, R. C., & Gould, I. R. (2012). GAFFlipid: A general amber force field for the accurate molecular dynamics simulation of phospholipid. *Soft Matter*, *8*(37), 9617–9627. https://doi.org/10.1039/C2SM26007G.

Di Marino, D., Bonome, E. L., Tramontano, A., & Chinappi, M. (2015). All-atom molecular dynamics simulation of protein translocation through an α-hemolysin nanopore. *Journal of Physical Chemistry Letters*, *6*(15), 2963–2968.

Donald, J. E., Kulp, D. W., & DeGrado, W. F. (2011). Salt bridges: Geometrically specific, designable interactions. *Proteins*, *79*(3), 898–915.

Enkavi, G., Javanainen, M., Kulig, W., Róg, T., & Vattulainen, I. (2019). Multiscale simulations of biological membranes: The challenge to understand biological phenomena in a living substance. *Chem. Rev.*, *119*(9), 5607–5774.

Escajadillo, T., & Nizet, V. (2018). Pharmacological targeting of pore-forming toxins as adjunctive therapy for invasive bacterial infections. *Toxins*, *10*, 542. https://doi.org/10.3390/toxins10120542.

Feil, S. C., Ascher, D. B., Kuiper, M. J., Tweten, R. K., & Parker, M. W. (2014). Structural studies of Streptococcus pyogenes streptolysin O provide insights into the early steps of membrane penetration. *Journal of Molecular Biology*, *426*(4), 785–792.

Fiser, A., Do, R. K. G., & Sali, A. (2000). Modeling of loops in protein structures. *Protein Science*, *9*(9), 1753–1773.

Frenkel, D., & Smit, B. (2002). *Understanding molecular simulation. From algorithms to applications*. Elsevier.

Gammon, K. (2014). Drug discovery: Leaving no stone unturned. *Nature*, *509*(7498), S10–12.

Gatsogiannis, C., Merino, F., Roderer, D., Balchin, D., Schubert, E., Kuhlee, A., … Raunser, S. (2018). Tc toxin activation requires unfolding and refolding of a β-propeller. *Nature*, *563*(7730), 209–213. https://doi.org/10.1038/s41586-018-0556-6.

Giri Rao, V. V. H., Desikan, R., Ayappa, K. G., & Gosavi, S. (2016). Capturing the membrane-triggered conformational transition of an α-helical pore-forming toxin. *The Journal of Physical Chemistry. B*, *120*(47), 12064–12078.

Giri Rao, V. V. H., & Gosavi, S. (2018). On the folding of a structurally complex protein to its metastable active state. *Proceedings of the National Academy of Sciences of the United States of America*, *115*(9), 1998–2003.

Globisch, C., Krishnamani, V., Deserno, M., & Peter, C. (2013). Optimization of an elastic network augmented coarse grained model to study CCMV capsid deformation. *PLoS One*, *8*(4), e60582.

Grossfield, A., Patrone, P. N., Roe, D. R., Schultz, A. J., Siderius, D., & Zuckerman, D. M. (2018). Best practices for quantification of uncertainty and sampling quality in molecular simulations (Article v1.0). *Living Journal of Computational Molecular Science*, *1*(1), 5067.

Gupta, K., Donlan, J. A., Hopper, J. T., Uzdavinys, P., Landreh, M., Struwe, W. B., … Robinson, C. V. (2017). The role of interfacial lipids in stabilizing membrane protein oligomers. *Nature*, *541*(7637), 421–424.

Hess, B., Bekker, H., Berendsen, H. J. C., & Fraaije, J. G. E. M. (1997). LINCS: A linear constraint solver for molecular simulations. *Journal of Computational Chemistry*, *18*(12), 1463–1472.

Hoover, W. G. (1985). Canonical dynamics: Equilibrium phase-space distributions. *Physical Review A*, *31*(3), 1695–1697.

Huang, J., Rauscher, S., Nawrocki, G., Ran, T., Feig, M., de Groot, B. L., … MacKerell, A. D. (2017). CHARMM36m: An improved force field for folded and intrinsically disordered proteins. *Nature Methods*, *14*(1), 71–73.

Humphrey, W., Dalke, A., & Schulten, K. (1996). VMD: Visual molecular dynamics. *Journal of Molecular Graphics*, *14*(1), 33–38.

Ingólfsson, H. I., Arnarez, C., Periole, X., & Marrink, S. J. (2016). Computational 'microscopy' of cellular membranes. *Journal of Cell Science, 129*(2), 257–268. https://doi.org/10.1242/jcs.176040.

Jain, V., Maingi, V., Maiti, P. K., & Bharatam, P. V. (2013). Molecular dynamics simulations of PPI dendrimer–drug complexes. *Soft Matter, 9*(28), 6482–6496. https://doi.org/10.1039/C3SM50434D.

Jämbeck, J. P. M., & Lyubartsev, A. P. (2012). Derivation and systematic validation of a refined all-atom force field for phosphatidylcholine lipids. *The Journal of Physical Chemistry. B, 116*(10), 3164–3179.

Jo, S., Kim, T., Iyer, V. G., & Im, W. (2008). CHARMM-GUI: A web-based graphical user interface for CHARMM. *Journal of Computational Chemistry, 29*(11), 1859–1865. https://doi.org/10.1002/jcc.20945.

Jorgensen, W. L., Chandrasekhar, J., Madura, J. D., Impey, R. W., & Klein, M. L. (1983). Comparison of simple potential functions for simulating liquid water. *The Journal of Chemical Physics, 79*(2), 926–935.

Kabsch, W., & Sander, C. (1983). Dictionary of protein secondary structure: Pattern recognition of hydrogen-bonded and geometrical features. *Biopolymers, 22*(12), 2577–2637.

Klauda, J. B., Venable, R. M., Freites, J. A., O'Connor, J. W., Tobias, D. J., Mondragon-Ramirez, C., … Pastor, R. W. (2010). Update of the CHARMM all-atom additive force field for lipids: Validation on six lipid types. *The Journal of Physical Chemistry. B, 114*(23), 7830–7843. https://doi.org/10.1021/jp101759q.

Köster, S., Van Pee, K., Hudel, M., Leustik, M., Rhinow, D., Kühlbrandt, W., … Yildiz, Ö. (2014). Crystal structure of listeriolysin O reveals molecular details of oligomerization and pore formation. *Nature Communication, 5*, 3690.

Kumar, S., & Nussinov, R. (2002). Close-range electrostatic interactions in proteins. *Chembiochem, 3*(7), 604–617.

Lee, J., Cheng, X., Swails, J. M., Yeom, M. S., Eastman, P. K., Lemkul, J. A., … Im, W. (2016). CHARMM-GUI input generator for NAMD, GROMACS, AMBER, OpenMM, and CHARMM/OpenMM simulations using the CHARMM36 additive force field. *Journal of Chemical Theory and Computation, 12*(1), 405–413. https://doi.org/10.1021/acs.jctc.5b00935.

Lemkul, J. (2018). From proteins to perturbed Hamiltonians: A suite of tutorials for the GROMACS-2018 Molecular Simulation Package (Article v1.0). *Living Journal of Computational Molecular Science, 1*(1), 5068.

Lin, X., Eddy, N. R., Noel, J. K., Whitford, P. C., Wang, Q., Ma, J., & Onuchic, J. N. (2014). Order and disorder control the functional rearrangement of influenza hemagglutinin. *Proceedings of the National Academy of Sciences of the United States of America, 111*(33), 12049–12054.

Lindorff-Larsen, K., Piana, S., Palmo, K., Maragakis, P., Klepeis, J. L., Dror, R. O., & Shaw, D. E. (2010). Improved side-chain torsion potentials for the AMBER ff99SB protein force field. *Proteins, 78*(8), 1950–1958.

Los, F. C., Randis, T. M., Aroian, R. V., & Ratner, A. J. (2013). Role of pore-forming toxins in bacterial infectious diseases. *Microbiology and Molecular Biology Reviews, 77*(2), 173–207.

Lu, Q., & Wang, J. (2008). Single molecule conformational dynamics of adenylate kinase: Energy landscape, structural correlations, and transition state ensembles. *Journal of the American Chemical Society, 130*(14), 4772–4783.

Ludwig, A., Völkerink, G., von Rhein, C., Bauer, S., Maier, E., Bergmann, B., … Benz, R. (2010). Mutations affecting export and activity of Cytolysin A from *Escherichia coli*. *Journal of Bacteriology, 192*(15), 4001–4011.

Maginn, E. J., Messerly, R. A., Carlson, D. J., Roe, D. R., & Elliott, J. R. (2018). Best practices for computing transport properties 1. Self-diffusivity and viscosity from equilibrium molecular dynamics (Article v1.0). *Living Journal of Computational Molecular Science*, *1*(1), 6324.

Maier, J. A., Martinez, C., Kasavajhala, K., Wickstrom, L., Hauser, K. E., & Simmerling, C. (2015). ff14SB: Improving the accuracy of protein side chain and backbone parameters from ff99SB. *Journal of Chemical Theory and Computation*, *11*(8), 3696–3713. https://doi.org/10.1021/acs.jctc.5b00255.

Maingi, V., Jain, V., Bharatam, P. V., & Maiti, P. K. (2012). Dendrimer building toolkit: Model building and characterization of various dendrimer architectures. *Journal of Computational Chemistry*, *33*(25), 1997–2011. https://doi.org/10.1002/jcc.23031.

Maingi, V., Kumar, M. V. S., & Maiti, P. K. (2012). PAMAM dendrimer–drug interactions: Effect of pH on the binding and release pattern. *The Journal of Physical Chemistry. B*, *116*(14), 4370–4376. https://doi.org/10.1021/jp211515g.

Maiti, P. K., & Bagchi, B. (2009). Diffusion of flexible, charged, nanoscopic molecules in solution: Size and pH dependence for PAMAM dendrimer. *The Journal of Chemical Physics*, *131*(21), 214901. https://doi.org/10.1063/1.3266512.

Maiti, P. K., Çağın, T., Lin, S.-T., & Goddard, W. A. (2005). Effect of solvent and pH on the structure of PAMAM dendrimers. *Macromolecules*, *38*(3), 979–991. https://doi.org/10.1021/ma049168l.

Mandal, T., Kanchi, S., Ayappa, K. G., & Maiti, P. K. (2016). pH controlled gating of toxic protein pores by dendrimers. *Nanoscale*, *8*(26), 13045–13058.

Manna, M., Nieminen, T., & Vattulainen, I. (2019). Understanding the role of lipids in signaling through atomistic and multiscale simulations of cell membranes. *Annual Review of Biochemistry*, *48*(1), 421–439.

Marrink, S. J., de Vries, A. H., & Mark, A. E. (2004). Coarse grained model for semiquantitative lipid simulations. *The Journal of Physical Chemistry. B*, *108*(2), 750–760.

Marrink, S. J., Risselada, H. J., Yefimov, S., Tieleman, D. P., & de Vries, A. H. (2007). The MARTINI force field: Coarse grained model for biomolecular simulations. *The Journal of Physical Chemistry. B*, *111*(27), 7812–7824.

Marrink, S. J., & Tieleman, D. P. (2013). Perspective on the MARTINI model. *Chemical Society Reviews*, *42*(16), 6801–6822.

Mathe, J., Aksimentiev, A., Nelson, D. R., Schulten, K., & Meller, A. (2005). Orientation discrimination of single-stranded DNA inside the α-hemolysin membrane channel. *Proceedings of the National Academy of Sciences of the United States of America*, *102*(35), 12377–12382.

Mesa-Galloso, H., Delgado-Magnero, K. H., Cabezas, S., Lopez-Castilla, A., Hernandez-Gonzalez, J. E., Pedrera, L., … Valiente, P. A. (2017). Disrupting a key hydrophobic pair in the oligomerization interface of the actinoporins impairs their pore-forming activity. *Protein Science*, *26*(3), 550–565.

Meusch, D., Gatsogiannis, C., Efremov, R. G., Lang, A. E., Hofnagel, O., Vetter, I. R., … Raunser, S. (2014). Mechanism of Tc toxin action revealed in molecular detail. *Nature*, *508*(7494), 61–65. https://doi.org/10.1038/nature13015.

Monticelli, L., Kandasamy, S. K., Periole, X., Larson, R. G., Tieleman, D. P., & Marrink, S.-J. (2008). The MARTINI coarse-grained force field: Extension to proteins. *Journal of Chemical Theory and Computation*, *4*(5), 819–834.

Mueller, M., Grauschopf, U., Maier, T., Glockshuber, R., & Ban, N. (2009). The structure of a cytolytic α-helical toxin pore reveals its assembly mechanism. *Nature*, *459*(7247), 726–730.

Nerenberg, P. S., & Head-Gordon, T. (2011). Optimizing protein-solvent force fields to reproduce intrinsic conformational preferences of model peptides. *Journal of Chemical Theory and Computation*, 7(4), 1220–1230.

Ni, T., Williams, S. I., Rezelj, S., Anderluh, G., Harlos, K., Stansfeld, P. J., & Gilbert, R. J. C. (2018). Structures of monomeric and oligomeric forms of the *Toxoplasma gondii* perforin-like protein 1. *Science Advances*, *4*(3), EAAQ0762. https://doi.org/10.1126/sciadv.aaq0762.

Noel, J. K., & Onuchic, J. N. (2012). The many faces of structure-based potentials: From protein folding landscapes to structural characterization of complex biomolecules. In N. V. Dokholyan (Ed.), *Computational modeling of biological systems* (pp. 31–54). Springer US (Ed.).

Nosé, S. (1984). A unified formulation of the constant temperature molecular dynamics methods. *The Journal of Chemical Physics*, *81*(1), 511–519.

Okazaki, K.-I., Koga, N., Takada, S., Onuchic, J. N., & Wolynes, P. G. (2006). Multiple-basin energy landscapes for large-amplitude conformational motions of proteins: Structure-based molecular dynamics simulations. *Proceedings of the National Academy of Sciences of the United States of America*, *103*(32), 11844–11849.

Opal, S. M. (2016). Non-antibiotic treatments for bacterial diseases in an era of progressive antibiotic resistance. *Critical Care*, *20*(1), 397.

Oscarsson, J., Mizunoe, Y., Li, L., Lai, X.-H., Wieslander, A., & Uhlin, B. E. (1999). Molecular analysis of the cytolytic protein ClyA (SheA) from *Escherichia coli*. *Molecular Microbiology*, *32*(6), 1226–1238.

Panchal, R. G., & Bayley, H. (1995). Interactions between residues in staphylococcal α-hemolysin revealed by reversion mutagenesis. *The Journal of Biological Chemistry*, *270*(39), 23072–23076.

Parrinello, M., & Rahman, A. (1981). Polymorphic transitions in single crystals: A new molecular dynamics method. *Journal of Applied Physiology*, *52*(12), 7182–7190.

Pastor, R. W., & MacKerell, A. D. (2011). Development of the CHARMM force field for lipids. *Journal of Physical Chemistry Letters*, *2*(13), 1526–1532. https://doi.org/10.1021/jz200167q.

Peng, W., de Souza Santos, M., Li, Y., Tomchick, D. R., & Orth, K. (2019). High-resolution cryo-EM structures of the *E. coli* hemolysin ClyA oligomers. *PLoS One*, *14*(5).

Peraro, M. D., & van der Goot, F. G. (2016). Pore-forming toxins: Ancient, but never really out of fashion. *Nature Reviews. Microbiology*, *14*(2), 77–92.

Perilla, J. R., Goh, B. C., Cassidy, C. K., Liu, B., Bernardi, R. C., Rudack, T., … Schulten, K. (2015). Molecular dynamics simulations of large macromolecular complexes. *Current Opinion in Structural Biology*, *31*, 64–74.

Periole, X., Cavalli, M., Marrink, S.-J., & Ceruso, M. A. (2009). Combining an elastic network with a coarse-grained molecular force field: Structure, dynamics, and intermolecular recognition. *Journal of Chemical Theory and Computation*, *5*(9), 2531–2543.

Phillips, J. C., Braun, R., Wang, W., Gumbart, J., Tajkhorshid, E., Villa, E., … Schulten, K. (2005). Scalable molecular dynamics with NAMD. *Journal of Computational Chemistry*, *26*(16), 1781–1802.

Ponmalar, I. I., Cheerla, R., Ayappa, K. G., & Basu, J. K. (2019). Correlated protein conformational states and membrane dynamics during attack by pore-forming toxins. *Proceedings of the National Academy of Sciences of the United States of America*, *116*(26), 12839–12844.

Pronk, S., Páll, S., Schulz, R., Larsson, P., Bjelkmar, P., Apostolov, R., … Lindahl, E. (2013). GROMACS 4.5: A high-throughput and highly parallel open source molecular simulation toolkit. *Bioinformatics*, *29*(7), 845–854.

Ramirez-Sarmiento, C. A., Noel, J. K., Valenzuela, S. L., & Artsimovitch, I. (2015). Interdomain contacts control native state switching of RfaH on a dual-funneled landscape. *PLoS Computational Biology*, *11*(7), e1004379.

Roderer, D., Benke, S., Müller, M., Fäh-Rechsteiner, H., Ban, N., Schuler, B., & Glockshuber, R. (2014). Characterization of variants of the pore-forming toxin ClyA from *Escherichia coli* controlled by a redox switch. *Biochemistry*, *53*(40), 6357–6369.

Rouse, S. L., & Sansom, M. S. P. (2015). Interactions of lipids and detergents with a viral ion channel protein: Molecular dynamics simulation studies. *The Journal of Physical Chemistry. B*, *119*(3), 764–772.

Roy, A., Kucukural, A., & Zhang, Y. (2010). I-TASSER: A unified platform for automated protein structure and function prediction. *Nature Protocols*, *5*(4), 725–738.

Ruan, Y., Rezelj, S., Zavec, A. B., Anderluh, G., & Scheuring, S. (2016). Listeriolysin O membrane damaging activity involves arc formation and lineaction-implication for listeria monocytogenes escape from phagocytic vacuole. *PLoS Pathogens*, *12*(4), e1005597.

Ryckaert, J.-P., Ciccotti, G., & Berendsen, H. J. C. (1977). Numerical integration of the cartesian equations of motion of a system with constraints: Molecular dynamics of n-alkanes. *Journal of Computational Physics*, *23*(3), 327–341. https://doi.org/10.1016/0021-9991(77)90098-5.

Salomon-Ferrer, R., Case, D. A., & Walker, R. C. (2013). An overview of the AMBER biomolecular simulation package. *WIREs Computational Molecular Science*, *3*(2), 198–210. https://doi.org/10.1002/wcms.1121.

Sammalkorpi, M., Karttunen, M., & Haataja, M. (2007). Structural properties of ionic detergent aggregates: A large-scale molecular dynamics study of sodium dodecyl sulfate. *The Journal of Physical Chemistry. B*, *111*(40), 11722–11733.

Sarangi, N. K., Ayappa, K. G., Visweswariah, S. S., & Basu, J. K. (2016). Super-resolution stimulated emission depletion-fluorescence correlation spectroscopy reveals nanoscale membrane reorganization induced by pore-forming proteins. *Langmuir*, *32*(37), 9649–9657.

Sarangi, N. K., Roobala, C., & Basu, J. K. (2018). Unraveling complex nanoscale lipid dynamics in simple model biomembranes: Insights from fluorescence correlation spectroscopy in super-resolution stimulated emission depletion mode. *Methods*, *140–141*, 198–211.

Sathyanarayana, P., Desikan, R., Ayappa, K. G., & Visweswariah, S. S. (2016). The solvent-exposed C-terminus of the cytolysin A pore-forming toxin directs pore formation and channel function in membranes. *Biochemistry*, *55*(42), 5952–5961.

Sathyanarayana, P., Maurya, S., Behera, A., Ravichandran, M., Visweswariah, S. S., Ayappa, K. G., & Roy, R. (2018). Cholesterol promotes cytolysin A activity by stabilizing the intermediates during pore formation. *Proceedings of the National Academy of Sciences of the United States of America*, *115*(31), E7323–E7330.

Schlick, T. (2010). *Molecular modeling and simulation: An interdisciplinary guide*. Springer.

Schubert, E., Vetter, I. R., Prumbaum, D., Penczek, P. A., & Raunser, S. (2018). Membrane insertion of α-xenorhabdolysin in near-atomic detail. *eLife*, 7, e38017. https://doi.org/10.7554/eLife.38017.

Smith, D. J., Klauda, J. B., & Sodt, A. J. (2018). Simulation best practices for lipid membranes (Article v1.0). *Living Journal of Computational Molecular Science*, *1*(1), 5966.

Song, L., Hobaugh, M. R., Shustak, C., Cheley, S., Bayley, H., & Gouaux, J. E. (1996). Structure of Staphylococcal α-hemolysin, a heptameric transmembrane pore. *Science*, *274*(5294), 1859–1865.

Taveecharoenkool, T., Angsuthanasombat, C., & Kanchanawarin, C. (2010). Combined molecular dynamics and continuum solvent studies of the pre-pore Cry4aa trimer suggest its stability in solution and how it may form pore. *PMC Biophysics*, *3*(1), 10. https://doi.org/10.1186/1757-5036-3-10.

Tweten, R. K., Parker, M. W., & Johnson, A. E. (2001). The cholesterol-dependent cytolysins. In *Pore-forming toxins* (pp. 15–33). Springer.

Uusitalo, J. J., Ingólfsson, H. I., Akhshi, P., Tieleman, D. P., & Marrink, S. J. (2015). MARTINI coarse-grained force field: Extension to DNA. *Journal of Chemical Theory and Computation, 11*(8), 3932–3945. https://doi.org/10.1021/acs.jctc.5b00286.

Uusitalo, J. J., Ingolfsson, H. I., Marrink, S. J., & Faustino, I. (2017). MARTINI coarse-grained force field: Extension to RNA. *Biophysical Journal, 113*(2), 246–256.

Vaidyanathan, M. S., Sathyanarayana, P., Maiti, P. K., Visweswariah, S. S., & Ayappa, K. G. (2014). Lysis dynamics and membrane oligomerization pathways for cytolysin A (ClyA) pore-forming toxin. *RSC Advances, 4*(10), 4930–4942.

van Pee, K., Neuhaus, A., D'Imprima, E., Mills, D. J., Kühlbrandt, W., & Yildiz, O. (2017). CryoEM structures of membrane pore and prepore complex reveal cytolytic mechanism of pneumolysin. *eLife, 6*, e23644.

Varadarajan, V., Desikan, R., & Ayappa, K. G. (2020). Assessing the extent of the structural and dynamic modulation of membrane lipids due to pore forming toxins: Insights from molecular dynamics simulations. *Soft Matter, 16*(20), 4840–4857.

Vögele, M., Bhaskara, R. M., Mulvihill, E., van Pee, K., Yildiz, Ö., Kühlbrandt, W., … Hummer, G. (2019). Membrane perforation by the pore-forming toxin pneumolysin. *Proceedings of the National Academy of Sciences of the United States of America, 116*(27), 13352–13357.

Vögele, M., Köfinger, J., & Hummer, G. (2018). Hydrodynamics of diffusion in lipid membrane simulations. *Physical Review Letters, 120*(26), 268104.

Wade, K. R., Hotze, E. M., Kuiper, M. J., Morton, C. J., Parker, M. W., & Tweten, R. K. (2015). An intermolecular electrostatic interaction controls the prepore-to-pore transition in a cholesterol-dependent cytolysin. *Proceedings of the National Academy of Sciences of the United States of America, 112*(7), 2204–2209.

Wai, S. N., Westermark, M., Oscarsson, J., Jass, J., Maier, E., Benz, R., & Uhlin, B. E. (2003). *Journal of Bacteriology, 185*(18), 5491–5499.

Waldburger, C. D., Schildbach, J. F., & Sauer, R. T. (1995). Are buried salt bridges important for protein stability and conformational specificity? *Nature Structural Biology, 2*(2), 122–128.

Walker, B., & Bayley, H. (1995a). Key residues for membrane binding, oligomerization, and pore forming activity of staphylococcal alpha-hemolysin identified by cysteine scanning mutagenesis and targeted chemical modification. *The Journal of Biological Chemistry, 270*(39), 23065–23071.

Walker, B., & Bayley, H. (1995b). Restoration of pore-forming activity in staphylococcal α-hemolysin by targeted covalent modification. *Protein Engineering, 8*(5), 491–495.

Wang, J., Wolf, R. M., Caldwell, J. W., Kollman, P. A., & Case, D. A. (2004). Development and testing of a general AMBER force field. *Journal of Computational Chemistry, 25*(9), 1157–1174. https://doi.org/10.1002/jcc.20035.

Wassenaar, T. A., Ingólfsson, H. I., Böckmann, R. A., Tieleman, D. P., & Marrink, S. J. (2015). Computational lipidomics with insane: A versatile tool for generating custom membranes for molecular simulations. *Journal of Chemical Theory and Computation, 11*(5), 2144–2155. https://doi.org/10.1021/acs.jctc.5b00209.

Waterhouse, A., Bertoni, M., Bienert, S., Studer, G., Tauriello, G., Gumienny, R., … Schwede, T. (2018). SWISS-MODEL: Homology modelling of protein structures and complexes. *Nucleic Acids Research, 46*(W1), W296–W303.

Whitford, P. C., Miyashita, O., Levy, Y., & Onuchic, J. N. (2007). Conformational transitions of adenylate kinase: Switching by cracking. *Journal of Molecular Biology, 366*(5), 1661–1671.

Wong-ekkabut, J., & Karttunen, M. (2016). Molecular dynamics simulation of water permeation through the α-hemolysin channel. *Journal of Biological Physics, 42*(1), 133–146.

Wu, E. L., Cheng, X., Jo, S., Rui, H., Song, K. C., Dávila-Contreras, E. M., … Im, W. (2014). CHARMM-GUI membrane builder toward realistic biological membrane simulations. *Journal of Computational Chemistry*, *35*(27), 1997–2004. https://doi.org/10.1002/jcc.23702.

Yesylevskyy, S. O., Schafer, L. V., Sengupta, D., & Marrink, S. J. (2010). Polarizable water model for the coarse-grained MARTINI force field. *PLoS Computational Biology*, *6*(6), 1–17.

Zhang, Y. (2008). I-TASSER server for protein 3D structure prediction. *BMC Bioinformatics*, *9*(1), 40.

CHAPTER SEVENTEEN

The use of pore-forming toxins to image lipids and lipid domains

Nario Tomishige†, Motohide Murate†, Pascal Didier, Ludovic Richert, Yves Mély, and Toshihide Kobayashi*
UMR 7021 CNRS, Université de Strasbourg, Illkirch, France
*Corresponding author: e-mail address: toshihide.kobayashi@unistra.fr

Contents

† Contributed equally.

Methods in Enzymology, Volume 649
ISSN 0076-6879
https://doi.org/10.1016/bs.mie.2021.01.019

Abstract

Very few proteins are reported to bind specific lipids. Because of the high selectivity and strong binding to specific lipids, lipid-targeting pore forming toxins (PFTs) have been employed to study the distribution of lipids in cell- and model-membranes. Non-toxic and monomeric PFT-derivatives are especially useful to study living cells. In this chapter we highlight sphingomyelin (SM)-binding PFT, lysenin (Lys), its derivatives, and newly identified SM/cholesterol binding protein, nakanori. We describe the preparation of non-toxic mutant of Lys (NT-Lys) and its application in optical and super resolution microscopy. We also discuss the observation of nanometer scale lipid domains labeled with nakanori and maltose-binding protein (MBP)-Lys in electron microscopy.

Abbreviations

CDCs cholesterol-dependent cytolysins
Chol cholesterol
CPE ceramide phosphoethanolamine
dSTORM direct stochastic optical reconstruction microscopy
EM electron microscopy
Eqt II equinatoxin II
EryA erylysin A
Ld liquid disordered
Lo liquid ordered
Lys lysenin
MBP maltose-binding protein
MDCK Madin-Darby canine kidney
NT-Lys non-toxic lysenin
OlyA ostreolysin A
PALM photoactivation localization microscopy
PlyA pleurotolysin A
PlyA2 pleurotolysin A2
PFO perfringolysin O
PFTs pore forming toxins
PC phosphatidylcholine
SDS-FRL sodium dodecyl sulfate-digested freeze fracture replica labeling
SM sphingomyelin
SMase sphingomyelinase
TIRF total internal reflection fluorescence

1. Introduction

A characteristic feature of cell surface lipids is their diversity among different animal species. Various pore forming toxins (PFTs) target host cell lipids that are not the components of toxin-producing organisms. Well-known examples are the cholesterol (Chol)-dependent cytolysins (CDCs) (Evans et al., 2020; Morton, Sani, Parker, & Separovic, 2019). CDCs are bacterial PFTs that specifically bind eukaryotic cells through their requirement of accessible Chol (Chakrabarti et al., 2017; Nelson, Johnson, & London, 2008). Moreover, a series of eukaryotic PFTs that specifically bind phosphosphingolipids have been identified (Table 1). These include earthworm PFT, lysenin (Lys) (Yamaji et al., 1998), and sea anemone PFT, equinatoxin II (Anderluh, Pungercar, Strukelj, Macek, & Gubensek, 1996), that specifically bind sphingomyelin (SM). Mushroom derived aegerolysin proteins, pleurotolysin A (PlyA) (Tomita et al., 2004), pleurotolysin A2 (PlyA2) (Bhat et al., 2013), ostreolysin (OlyA) (Endapally et al., 2019; Skocaj et al., 2014) and erylysin A (EryA) (Shibata et al., 2010), are membrane-binding subunits of unique two component PFTs. PlyA2 and OlyA have been shown to bind SM/Chol complexes (Bhat et al., 2013; Endapally et al., 2019; Skocaj et al., 2014). Recent results showed that PlyA2, OlyA and EryA exhibit more than 1000 times higher affinity to ceramide phosphoethanolamine (CPE)/Chol than SM/Chol (Bhat et al., 2015). Recently identified mushroom protein, nakanori, shows high affinity to SM/Chol. Although nakanori does not show cytotoxicity nor sequence-based homology to PFTs, crystal structure of nakanori overlaps with that of the SM-binding PFT, sticholysin (Makino et al., 2017).

Very few proteins are reported to bind specific lipids. Because of their high selectivity and strong binding to specific lipids, lipid-targeting PFTs have been employed to study the cellular distribution of lipids (Kishimoto, Ishitsuka, & Kobayashi, 2016; Ohno-Iwashita et al., 2004; Skocaj et al., 2013; Yamaji-Hasegawa, Hullin-Matsuda, Greimel, & Kobayashi, 2016). Non-toxic and monomeric PFT-derivatives are especially useful to study living cells (Kiyokawa et al., 2005; Waheed et al., 2001). In this chapter, we describe protocols to label cell cultures with Lys derivatives and a related protein, nakanori, for optical and electron microscopy observation. The readers are also referred to the labeling protocols of living cells with SM- and Chol-binding PFT derivatives (Abe & Kobayashi, 2017) and intra-cellular and plasma membrane Chol labeling with the perfringolysin O derivative (GFP-D4) (Wilhelm, Voilquin, Kobayashi, Tomasetto, & Alpy, 2019).

Sphingomyelin (SM) is a major sphingolipid, representing approximately 10% of the total phospholipids in mammalian cells (Hannun & Obeid, 2018). Bulk SM is synthesized in the lumenal side of the Golgi

Table 1 Characteristics and properties of the protein probes used to visualize sphingomyelin (SM) and ceramide phosphoethanolamine (CPE).

Protein	Origin	MW (kDa)	Toxicity	Ligand/ k_D	Applications	References
Lysenin	Earthworm *Eisenia fetida*	33	Yes	Clustered SM K_D (SM) $=5.3\times10^{-9}$ M	Fixed cell labeling, live cell labeling at low temperature, EM, model membranes	Ishitsuka, Yamaji-Hasegawa, Makino, Hirabayashi, and Kobayashi (2004), Kiyokawa et al. (2004), Murate et al. (2015), Wang, Shogomori, Hara, Yamada, and Kobayashi (2012), Yamaji et al. (1998) and Yilmaz and Kobayashi (2015)
NT-lysenin	Truncated form of lysenin	16	No	Clustered SM K_D (SM) $=1.9\times10^{-7}$ M	Fixed and live cell labeling, real-time imaging, super-resolution imaging, model membranes	Abe et al. (2012), Abe and Kobayashi (2017), Carquin et al. (2014), Dumitru et al. (2018), Ellison, Kukulski, Boyle, Munro, and Randow (2020), Ikenouchi et al. (2012), Ikenouchi, Hirata, Yonemura, and Umeda (2013), Kishimoto et al. (2020), Kiyokawa et al. (2005), Makino et al. (2015) and Mound et al. (2017)
Equinatoxin II (EqtII)	Sea anemone *Actinia equina*	20	Yes	Dispersed SM K_D (SM) $=7.5\times10^{-9}$ M	Fixed cell labeling, live cell labeling at low temperature, EM, model membranes	Bakrac et al. (2010), Makino et al. (2015), Murate et al. (2015) and Yachi et al. (2012)
Equinatoxin II (8–69) (EqtII(8–69))	2 cysteines inserted and oxidized EqtII	20	No	Dispersed SM K_D (SM) $=2.6\times10^{-8}$ M [8–69 oxidized]	Live cell labeling, real-time imaging	Hong et al. (2002), Kristan et al. (2004) and Makino et al. (2015)

Eqt-SM and Eqt-sol	V22W, Y108I mutant of EqtII	20	No	SM	Live cell labeling	Deng, Rivera-Molina, Toomre, and Burd (2016)
Nakanori	Mushroom *Grifola frondosa*	22	No	SM/Chol complex K_D (SM/Chol) $=1.4 \times 10^{-7}$ M	Fixed and live cell labeling, real-time imaging, super-resolution imaging, model membranes	Makino et al. (2017)
Pleurotolysin A2 (PlyA2)	Mushroom *Pleurotus eryngii*	15	No	SM/Chol <CPE <CPE/Chol	Fixed and live cell labeling, *in vivo* staining of *Drosophila*, model membranes	Bhat et al. (2013), Bhat et al. (2015) and Inaba et al. (2019)
Ostreolysin (Oly)	Mushroom *Pleurotus ostreatus*	15	No	SM/Chol << CPE/Chol	Fixed and live cell labeling, model membranes	Endapally et al. (2019), Johnson, Endapally, Vazquez, Infante, and Radhakrishnan (2019) and Skocaj et al. (2014)
Erylysin A (EryA)	Mushroom *pleurotus eryngii*	15	No	CPE/Chol	Discrimination of bloodstream form of *trypanosoma*	Bhat et al. (2015) and Shibata et al. (2010)

1. Conjugation of tags to C terminus of lysenin and N-termini of EqtII/PlyA2 /Oly /EryA abolishes their lipid-binding activity.
2. Oly, PlyA, PlyA2 as well as EryA are not toxic by themselves, toxicity occuring cooperatively with PlyB for the first three components and with EryB for the later.
3. Dissociation constant K_D of PlyA2 and Oly for SM/Chol could not be determined due to the weak affinity of the conjugated aegerolysins to SM/Chol liposomes. K_D of PlyA for SM/Chol has not been determined yet.
4. K_D of PlyA2, Oly and EryA for CPE/Chol are 12, 1.2 and 1.3 nM, respectively; only PlyA2 associates with CPE independently of Chol (K_D = 41 μM).
5. Toxicity is against cultured mammalian cells.

apparatus by SM synthase 1 and transported to the outer leaflet of the plasma membrane (Tafesse, Ternes, & Holthuis, 2006), whereas local SM is synthesized in the outer leaflet of the plasma membrane by SM synthase 2 (Mitsutake et al., 2011). The lumenal (outer) leaflet biosynthesis explains the enrichment of SM in the outer leaflet of the plasma membrane and the lumenal leaflet of endosomes (Kobayashi & Menon, 2018). However, small amount of SM is also distributed in the inner leaflet of the plasma membrane (Murate et al., 2015) and the cytoplasmic side of the Golgi apparatus (Bakrac et al., 2010). The degradation of SM by sphingomyelinase (SMase) produces the bioactive lipid, ceramide (Hannun & Obeid, 2018). In addition, SM is proposed to form specific liquid ordered (Lo) lipid raft domains together with Chol that provide a platform for various signaling events (Lingwood & Simons, 2010). However, in phase-separated model membranes, SM is distributed both in Lo and liquid disordered (Ld) domains (Makino et al., 2015). Thus, examining the membrane distribution and dynamics of SM is crucial to understand the pathophysiological role of SM.

Lys is an earthworm-derived 297 amino acid (Sekizawa, Kubo, Kobayashi, Nakajima, & Natori, 1997) pore forming toxin that specifically binds SM (Yamaji et al., 1998). The structure of Lys is similar to that of the glycosylphosphatidylinositol-binding toxin, aerolysin (De Colibus et al., 2012). The crystal structure of Lys predicts phosphocholine binding site at the N-terminal pore-forming module and C-terminal beta-trefoil motif, respectively (De Colibus et al., 2012), although Lys is highly selective to SM and does not bind another phosphocholine-containing lipid, phosphatidylcholine (PC) (Yamaji et al., 1998). The K_D of Lys to SM is 5.3×10^{-9} M (Yamaji et al., 1998) with SM/Lys stoichiometry of 5–6 (Ishitsuka et al., 2004; Ishitsuka & Kobayashi, 2007). Model membrane studies indicate that Lys exclusively binds clustered SM, whereas another SM-binding PFT, EqtII, preferentially binds dispersed SM (Makino et al., 2015). Crystal structure predicts that EqtII has one phosphocholine binding site (Bakrac et al., 2008; Mancheno, Martin-Benito, Martinez-Ripoll, Gavilanes, & Hermoso, 2003). In the presence of SM, Lys oligomerizes and forms a pore (Yamaji-Hasegawa et al., 2003; Yilmaz et al., 2013). Removal of the N-terminal pore-forming module of Lys inhibits oligomerization and abolishes toxicity but still keeps the property of selective binding to SM (Kiyokawa et al., 2005). The shortest (amino acid 161 to C-terminal end 297) nontoxic fragment of Lys is used as a non-toxic SM-binding protein (NT-Lys) in living cells (Abe et al., 2012; Ellison et al., 2020; Kishimoto et al., 2020; Kiyokawa et al., 2005). The K_D of NT-Lys (GST-NT-Lys) is increased 36 times (1.9×10^{-7} M) over that of native lysenin (Kiyokawa et al., 2005). This is because of the higher off rate of GST-NT-Lys. Interestingly, Lys monomer still binds SM clusters (Makino et al., 2015) but is exchangeable between membranes, indicating that the oligomerization stabilizes Lys at the membrane (Ishitsuka & Kobayashi, 2007). Conjugation of the fluorescent protein to the C-terminus but not the N-terminus of NT-Lys abolishes the binding to SM. Similar to Lys, Enhanced-GFP (EGFP)-NT-Lys specifically binds SM and does not bind sphingosylphosphorylcholine (SPC, lysosphingomyelin) in a solid phase assay. ^{125}I-labeled non-toxic fragment of lysenin (NT-Lys) revealed that ~3% of SM was labeled by the probe on red blood cell (RBC) membrane (Carquin et al., 2014).

HeLa cells (Abe et al., 2012; Kishimoto et al., 2020), mouse epithelial EpH4 cells (Ikenouchi et al., 2012, 2013), human skin fibroblasts (Makino et al., 2015) and human red blood cells (Carquin et al., 2014) are reported to be labeled with Lys and fluorescent NT-Lys derivatives. In contrast, Lys labeling is deficient in the apical membrane of differentiated Madin-Darby canine kidney (MDCK) epithelial cells as well as mouse melanoma cells (Ishitsuka et al., 2004; Makino et al., 2015). This is not due to the deficiency of SM but the absence of SM clusters in these membranes. It is postulated that high concentration of glycolipids inhibits the formation of SM clusters in these cells.

Nakanori is a mushroom-derived 202 amino acid protein. Although nakanori is not hemolytic nor cytotoxic to cultured mammalian cells, the crystal structure of nakanori resembles that of sticholysin II except for the additional 30 aa N-terminal residues and the lack of amphipathic nature of helix 3 (Makino et al., 2017). Both sticholysin II and EqtII are actinoporins that are SM-binding PFTs (Mancheno et al., 2003). Nakanori has high specificity and strong affinity to SM/Chol with $K_D = 1.4 \times 10^{-7}$ M (Makino et al., 2017). The use of nakanori as a lipid-domain–specific probe revealed altered distribution and dynamics of SM/Chol on the cell surface of Niemann-Pick type C fibroblasts and the budding of influenza virus from specific lipid domains (Makino et al., 2017).

Reagent or resource	source	Identifier
Antibodies		
Anti His-tag antibody	CUSABIO	CSB-PA000086
Rabbit IgG horse radish peroxidase (HRP) linked whole antibody	Cytiva	NA934
Anti-CD63 (H5C6), monoclonal mouse antibody	Novus Biologicals	NBP2-42225
Anti-nakanori antibody	Makino et al., 2017	
Anti-maltose binding protein antibody	New England BioLabs	E8030S
Colloidal gold conjugated anti-rabbit IgG	BBI solutions	EM.GAR10
Bacterial Strains		
E.coli BL21(DE3)		
E.coli DH5a		
Chemicals, Peptides, and Recombinant Proteins		
Protease Inhibitor Cocktail Set I	Calbiochem	539131
BugBuster Master Mix	Novagen	71456–3
Lipoprotein deficient serum (LPDS) from fetal calf	Sigma-Aldrich	S5394
Dulbecco's modified Eagle medium (DMEM) + GlutaMAX-1 containing 1 g/L D-glucose and pyruvate.	Gibco	21885-025
Sphingomyelinase from *Staphylococcus aureus*	Sigma-Aldrich	S8633
Paraformaldehyde 16% aqueous solution (methanol free)	Electron Microscopy Science	15710
20 mM Hoechst33342	ThermoFisher Scientific	62249
Prolong diamond antifade mountant	ThermoFisher Scientific	P36961
Brain SM	Avanti polar lipids	860062P
Egg PC	Avanti polar lipids	840051P
o-Phenylendiamine dihydrochloride tablet	Sigma-Aldrich	P6787

—cont'd

Reagent or resource	source	Identifier
Albumin, bovine, F-V, pH5.2	Nacalai Tesque	01863-48
Qdot 655 Streptavidin conjugate	ThermoFisher Scientific	Q10121MP
VECTASHIELD antifade mounting medium	Vector Laboratories	H-1000-10
Picodent twinsil	Picodent	
25% glutaraldehyde	TAAB Laboratories	G004
Acrolein	Sigma-Aldrich	110221
4% OsO_4 solution	TAAB Laboratories	O018
TI Blue solution	Nisshin EM	335
Toluidine blue O	Merck	1.15930
Nakanori	Makino et al., 2017	
Lead solution	Sigma-Aldrich	18-0875
MBP-Lysenin	Kiyokawa et al., 2005	
Critical Commercial Assays		
Pierce BCA protein assay kit	ThermoFisher Scientific	23225
Alexa Fluor 647 protein labeling kit	ThermoFisher Scientific	A20173
Experimental Models: Cell Lines		
HeLa	ATCC	CCL-2
Human skin fibroblast	ATCC	SCRC-1041
Recombinant DNA		
pET28/NT-Lys	Abe et al., 2012	
pET28/EqtII(8–69)-EGFP-His6	RIKEN Bioresource center DNA bank	RDB13497
pET28/His6-EGFP-NT-Lys	RIKEN Bioresource center DNA bank	RDB13498
pET28/His6-mCherry-NT-Lys	RIKEN Bioresource center DNA bank	RDB13499

Continued

—cont'd

Reagent or resource	source	Identifier
pET28/His6-EGFP-D4	RIKEN Bioresource center DNA bank	RDB13961
pET28/His6-EGFP-Nakanori	RIKEN Bioresource center DNA bank	RDB14301
pMAL/MBP-NT-Lys	RIKEN Bioresource center DNA bank	RDB13958
pQE30/His6-Venus-NT-Lys	RIKEN Bioresource center DNA bank	RDB13959
pQE30/His6-mRFP-NT-Lys	RIKEN Bioresource center DNA bank	RDB13960
pET28/PlyA2-EGFP-His6	RIKEN Bioresource center DNA bank	RDB13962
pET28/Oly-EGFP-His6	RIKEN Bioresource center DNA bank	RDB13963
pET28/EryA-EGFP-His6	RIKEN Bioresource center DNA bank	RDB13964

2. Key resources table

Note that not all areas will be used in every protocol.

3. Preparation of NT-Lys

To label SM-rich domains from outside, we add purified NT-Lys obtained in *E.coli*. NT-Lys is also expressed in the cytoplasm of mammalian cells to label SM in the cytoplasmic leaflet. Here, we introduce a protocol for preparation of NT-Lys and its labeling with a fluorescent dye. Plasmids for fluorescent protein (FP)-conjugated NT-Lys and other SM and Chol binding non-toxic proteins are also available from RIKEN Bioresource Center (http://dna.brc.riken.jp/) (see Key Resources Table). The protocols of purification of these proteins are essentially the same. We routinely measure the binding activity of fluorescent NT-Lys to SM by solid phase binding assay to several lipids and the binding selectivity by labeling cells that are treated with or without bacterial sphingomyelinase (SMase). Normally OD value of NT-Lys in ELISA for SM is 10 times higher than that for

PC. Bacterial SMase abolishes cell labeling with NT-Lys. You can also refer to the published protocol of FP-NT-Lys and EqtII(8–69)-FP preparation (Abe & Kobayashi, 2017).

3.1 Materials

1. Water bath or heat block, used for heat shock at 42 °C to transform BL21(DE3) competent cells.
2. Baffled-bottom 2 L Erlenmeyer flask.
3. Shaker (Eppendorf, New Brunswick Innova 40R Shaker). Any shaker that can shake 2 L flask at 18 °C and 37 °C can be used.
4. Centrifuge (Hitachi Koki, CF15RX). Any centrifuge that can centrifuge ≧50 mL culture at 4 °C.
5. HiTrap TALON crude, Cytiva, cat#28-9537-66, HiTrap TALON crude column, 1 mL prepacked column.
6. Microplate Reader, Molecular Devices, SpectraMax M2, any plate reader that can measure absorbance at 562, 490, and 630 nm can be used.
7. Slide-A-Lyser dialysis cassette, ThermoFisher Scientific, cat#66380, 10000 MWCO, 0.5-3 mL capacity.
8. 12-mm glass coverslip, VWR, cat#ECN 631–1577, 12 mmϕ, No. 1.
9. 24-well plate, Sigma-Aldrich, cat#Z707791, TPP tissue culture plates.
10. 96-well plate, ThermoFisher Scientific, cat#3355, Clear flat-bottom immuno nonsterile 96-well plate, immulon 1B.
11. Kimwipes, Sigma-Aldrich, cat#Z188956.

3.2 Reagents

1. LB broth (Lennox) (Sigma-Aldrich, L3022), sterilized by autoclave.
2. LB broth with agar (Lennox) (Sigma-Aldrich, L2897), to prepare LB + 50 μg/mL kanamycin plate to select BL21(DE3) harboring pET28/NT-Lys.
3. *E.coli* BL21(DE3).
4. pET28/NT-Lys.
5. 50 mM $CaCl_2$.
6. Kanamycin, Alfa Aesar, cat#J61272.09, kanamycin sulfate, dissolved at 50 mg/mL in MilliQ, then filter sterilized. The working concentration is 50 μg/mL in LB broth.
7. Isopropyl β-D-1-thiogalactopyranoside (IPTG), VWR life science, cat#AC121.

8. BugBuster Master Mix, Tris-based buffer supplemented with non-ionic and zwitterionic detergents, nuclease, and lysozyme.
9. Protease inhibitor cocktail, Protease Inhibitor Cocktail Set I. The powder is dissolved in 1 mL distilled MilliQ water.
10. Equilibration buffer (100 mM sodium phosphate buffer, pH 7.0, 300 mM NaCl).
11. Wash buffer (50 mM sodium phosphate buffer, pH 7.0, 300 mM NaCl, 10 mM Imidazole-Cl, pH 7.0).
12. Elution buffer (50 mM sodium phosphate buffer, pH 7.0, 300 mM NaCl, 400 mM Imidazole-Cl, pH 7.0).
13. PBS, Sigma-Aldrich, cat#P4417, Phosphate buffered saline, 1 tablet is dissolved in 200 mL MilliQ water.
14. BCA protein assay kit, Pierce BCA protein assay kit, is a detergent-compatible formulation based on bicinchoninic acid (BCA) for the colorimeteric detection and quantification of protein. The colored protein solution was measured by absorbance at 562 nm.
15. Alexa Fluor 647 protein labeling kit.
16. Glycerol, VWR chemicals, cat#24388.260, Glycerol bidistilled 99.5%.
17. DMEM.
18. Lipoprotein-deficient serum (LPDS). LPDS is used after inactivation at 56 °C for 30 min.
19. DMEM/5%LPDS (DMEM supplemented with 5% heat-inactiveated LPDS), this medium is usually used for cell labeling to avoid any loss of lipid probe due to its binding to lipids in serum lipoprotein.
20. Bacterial SMase, sphingomyelinase from *Staphylococcus aureus*. To deplete cell surface SM, it is used at 0.5 unit/mL medium.
21. 16% paraformaldehyde, prepare 4% solution just before use by diluting with PBS.
22. Hoechst33342, use at a final concentration of 10 μM.
23. Mounting medium, Prolong Diamond, keep at 4 °C and bring to room temperature (RT, (20–24 °C) before use.
24. TBS, (10 mM Tris-HCl, pH 7.5, 150 mM NaCl).
25. Brain SM, porcine brain SM.
26. Egg PC.
27. Primary antibody for ELISA, anti-His-tag antibody, rabbit polyclonal, diluted 1,000 times for ELISA. Anti-His tag antibody is used for detection of lipid probe binding because all lipid probes we introduce here have His-tag at their N-termini.

28. Secondary antibody for ELISA, anti-rabbit IgG conjugated HRP.
29. Coloring agent, (0.4mg/mL of *o*-phenylendiamine, 0.05% H_2O_2, in TBS).
30. 4N H_2SO_4.

3.3 Protocol

3.3.1 *Preparation of* E. coli *harboring the expression plasmid of NT-Lys for recombinant protein production*

3.3.1.1 Preparation of BL21(DE3) competent cells by $CaCl_2$

1. Inoculate either a single colony of BL21(DE3) from LB plate or a scratch of its frozen stock in 2mL of LB medium.
2. Culture bacteria at 37°C overnight by shaking.
3. Inoculate 100μL of preculture in 10mL of LB medium.
4. Grow the bacteria by shaking in a 37°C incubator until $O.D._{600}$ reaches approx. 0.6.
5. Harvest the bacteria by centrifugation at 2500 × *g* for 5min at 4°C.
6. Wash the bacterial pellet with 2mL of 50mM $CaCl_2$ solution twice.
7. Suspend the pellet in 1mL of 50mM $CaCl_2$ solution and place it on ice until use.

3.3.1.2 Transform BL21(DE3) competent cells with the bacterial expression plasmid of NT-Lys

1. Mix 200μL of BL21(DE3) competent cells with 50ng of pET28/NT-Lys.
2. Incubate the mixture on ice for 20min, followed by 30s of heat shock at 42°C.
3. Shake the cell mixture at 37°C for 1h after adding 400μL of LB medium.
4. Spread 100μL of the mixture onto LB + 50μg/mL kanamycin plate.
5. Incubate the LB plate in 37°C incubator until colonies appear.

3.3.2 *Induction of NT-Lys expression in* E. coli *and purification of NT-Lys*

3.3.2.1 Preparation of preculture

1. Inoculate a single colony obtained above in 2mL of LB + 50μg/mL kanamycin.
2. Culture the bacteria by shaking at 37°C for overnight to prepare preculture.

3.3.2.2 Express lipid probe in large culture

1. Inoculate 150 μL of preculture in 300 mL of LB + kanamycin in a 2 L Erlenmeyer flask with baffled-bottom.
2. Grow the bacteria by shaking at 37 °C until $O.D._{600}$ reaches approx. 1.0.*1
3. Place the flask on ice while setting the temperature of a shaker at 18 °C.
4. Add 75 μL (final 250 μM*1) of 1 M IPTG to the culture just before shaking at 18 °C.
5. Induce the NT-Lys expression by shaking at 18 °C for two overnights.*1

*1: The optimal induction condition varies depending on lipid probes and equipment in each laboratory. Parameters including OD value at the start of induction, temperature during the induction, concentration of IPTG, and the duration of induction need to be adjusted in each experiment.

3.3.2.3 Purification of recombinant lipid probe

1. Harvest the bacteria by centrifugation at 6,300 × *g* for 5 min at 4 °C and store the bacteria pellet at −80 °C for 15 min.
2. Suspend the bacteria pellet in 5 mL of BugBuster Master Mix containing protease inhibitor cocktail per g of bacterial pellet and incubate the suspension on ice for 10 min.*1
3. Centrifuge the suspension at 11,900 × *g* for 5 min at 4 °C and transfer the supernatant as cleared lysate into a new 50 mL tube.
4. Equilibrate a new Hitrap TALON crude column with 5-bed volume of Equilibration buffer.
5. Load the cleared lysate on Hitrap TALON crude column.
6. Wash the column with 10-bed volume of Wash buffer.
7. Elute the recombinant protein with 5-bed volume of Elution buffer. Fractionate and recover every 0.5 mL of eluate.*2
8. Examine the recovery of the eluted protein by applying the fractions in SDS-PAGE, followed by Coomassie brilliant blue (CBB) staining.*3

*1: Cells can also be disrupted by sonication on ice. We use a Misonix XL2020 sonicator with a 1/800 probe (30 5-s pulses, a 10-s incubation on ice in between, 40% full power) (Abe & Kobayashi, 2017).
*2: In case of EGFP-NT-Lys and mCherry-NT-Lys, His-tag is upstream of the fluorescent protein (FP). Thus FP alone is co-purified with FP-NT-Lys. However, FP does not hinder the labeling with FP-NT-Lys. If necessary, FP-NT-Lys is further purified with size-exclusion column.
*3: Take 30 μL of supernatant in each step from 3 to 6 as references in SDS-PAGE in step 8.

3.3.2.4 Dialysis of the eluted protein solution to remove imidazole

1. Combine fractions that contain NT-Lys and transfer them inside a dialysis cassette by syringe.
2. Dialyze NT-Lys against 0.5 L of PBS for 1 h, then transfer it to 1.5 L of PBS, and dialyze overnight.*1
3. Recover the protein solution by syringe according to the manufacturer's manual and collect aliquot of PBS used as the overnight dialysis buffer as blank in BCA protein assay.
4. Determine protein concentration of the dialyzed protein solution by BCA protein assay kit.
5. Stock the protein probe as 50% glycerol solution at −20 °C.

*1: You can also remove imidazole and concentrate protein at the same time using a centrifugal filter unit (Millipore, cat#UFC901024, Amicon ultra-15 centrifugal filter unit).

3.3.3 Measurement of the lipid binding activity of NT-Lys by ELISA

1. Transfer brain SM stock solution to a glass tube and evaporate solvent under N_2 stream.
2. Dissolve the lipid film in ethanol.
3. Add 50 μL of 10 μM brain SM in ethanol in a well of 96-well plate.
4. Place the 96-well plate in a desiccator and cover the plate with Kimwipes.
5. Evaporate ethanol under vacuum at RT.
6. Add 200 μL of 3% BSA in TBS in the wells.
7. Cover the plate with plastic wrap and incubate at RT for 2 h.
8. Wash the wells with TBS three times.
9. Add 50 μL of NT-Lys diluted with 1% BSA in TBS to the wells.*1
10. Cover the plate with plastic wrap and incubate at RT for 1 h.
11. Wash the wells with TBS three times.
12. Add 50 μL of anti-His tag antibody diluted 1000 times with 1% BSA in TBS.
13. Repeat steps10 and 11.
14. Add 50 μL of anti-rabbit IgG conjugated HRP diluted 1000 times with 1% BSA in TBS.
15. Repeat steps10 and 11.
16. Add 100 μL of coloring agent and incubate at RT until the solution turns orange.
17. Stop the reaction by adding 50 μL of 4 N H_2SO_4.
18. Measure the absorbance at both 490 and 630 nm by Microplate Reader.

*1: Serially dilute NT-Lys solution according to its protein concentration. We normally dilute in a range of 1 to 1000 ng/mL to determine the binding activity in ELISA.

3.3.4 Labeling NT-Lys with Alexa Fluor 647

1. Alexa Fluor 647 is conjugated to NT-Lys using Alexa Fluor 647 protein labeling kit targeted to primary amines, following the manual of the manufacturer. Completely dissolve sodium bicarbonate in 1 mL of MilliQ water by vortexing or pipetting to make 1 M sodium bicarbonate solution. This solution gives pH ~8.3.*1
2. Add 50 μL of 1 M sodium bicarbonate to 0.5 mL of 2 mg/mL NT-Lys.
3. Add the NT-Lys mixture to a vial of reactive Alexa fluor 647 dye, then mix the solution with a stirring bar at RT for 1 h.
4. Meantime, assemble a column and pack resin into the column.
5. Load the reaction mixture of NT-Lys/Alexa Fluor 647 dye onto the column.
6. While the column runs, two colored bands can be seen. Collect the first colored band, which corresponds to the Alexa Fluor 647-labeled NT-Lys (Alexa 647-NT-Lys). This protocol gives 1–2 fluorophores/NT-Lys.

*1: SNAP or HaLo tag conjugated lipid binding protein is also prepared for different color labeling (Makino et al., 2017). These tags ensure the binding of one fluorophore to one protein.

4. Labeling control and bacterial sphingomyelinase (SMase)-treated HeLa cells with Alexa 647-NT-Lys

Here we describe the labeling method of living HeLa cells with Alexa 647-NT-Lys. SMase treatment removes cell surface SM and thus inhibits the labeling. We routinely use this method to examine selectivity of new lot of NT-Lys probes. NT-Lys also labels SM after addition to the fixed cells. However, in case of fixed cells, one need to be careful with the interpretation of the data because fixation could alter lipid distribution. In this protocol we fix cells after labeling with Alexa 647-NT-Lys for prolonged observation. Cells can be examined without fixation.

4.1 Equipment

1. Confocal microscope, we use LSM700 confocal microscope, Zeiss, equipped with a C-Apochromat 63XW Korr (1.2 NA) objective and 405, 488, and 555 nm laser lines and FV1000D, Olympus, equipped with a 60 × oil objective (1.1 NA) and 405, 488, 543, and 633 nm laser lines.

4.2 Protocol*[1]

1. Plate HeLa cells on a 12-mm glass coverslip in 24-well plate to obtain 50–80% confluency on the day of the experiment.
2. On the day of the experiment, remove the medium and rinse cells with DMEM/5% LPDS once.
3. Add 250 μL of DMEM/5% LPDS with or without 0.125 unit of bacterial sphingomyelinase (bSMase) to the well, then incubate at 37 °C for 30 min.
4. Remove medium and rinse cells with DMEM/5% LPDS.
5. Add 250 μL of Alexa 647-NT-Lys diluted in DMEM/5%LPDS and incubate at 37 °C for 15 min.*[2]
6. Remove the medium and wash cells with 250 μL of PBS twice.*[3]
7. Fix cells with 250 μL of 4% PFA containing 10 μM Hoechst33342 at RT for 30 min.
8. Wash cells with 250 μL of PBS three times.
9. Neutralize the residual PFA with 250 μL of 0.1 M NH_4Cl by incubating at RT for 15 min.
10. Wash cells with 250 μL of PBS three times.
11. Mount the coverslip onto a few μL of mounting medium on a slide glass.
12. Harden the mounting medium in the dark.
13. Observe the slide under a fluorescent microscope. Typical images are shown in Fig. 1.

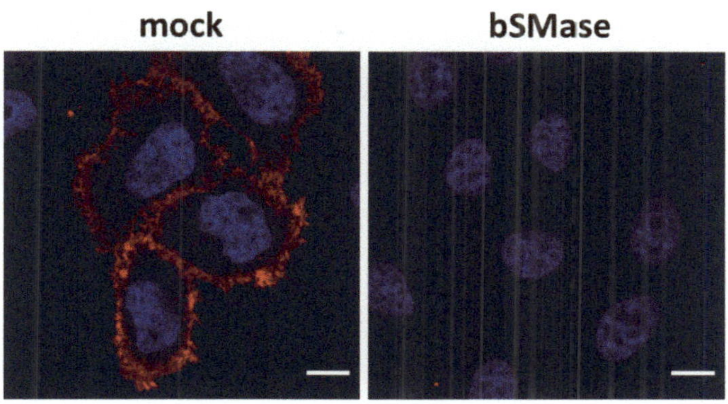

Fig. 1 Labeling of HeLa cells with Alexa 647-SNAP-NT-Lys. The cell surface SM in HeLa cells grown on a coverslip was labeled by incubating at 37 °C for 15 min in medium containing 15 μg/mL Alexa 647-SNAP-NT-Lys. The cells were fixed and their nuclei were stained with 4% PFA containing 10 μM Hoechst33342, at RT for 30 min. Images were taken in mock (left) and bSMase (right)-treated cells. Bar, 10 μm.

*1: Chemicals and consumables used in this section are shown in Section 2.
*2: To selectively label the plasma membrane, cells are incubated at low temperature (below 15 °C). However, the endocytosis of NT-Lys is slow and even at 37 °C incubation, most of the NT-Lys labels the plasma membrane.
*3: Remove step 7–12 in case of live cell observation and follow step 6′ instead of step 6.

6′. Remove the medium, wash cells with 250 μL of medium twice, and add 250 μL of medium.

5. Intracellular labeling with fluorescent-NT-Lys

Although a small portion of SM is distributed in the cytoplasmic leaflet of the plasma membrane (Murate et al., 2015) and the intracellular organelles (Bakrac et al., 2010), the majority of cellular SM resides in the outer leaflet of the plasma membrane and the lumen of endocytic compartments. To localize endogenous SM clusters in the lumen of the organelles by optical microscopy, Lys and Lys derivatives are added to fixed and permeabilized cells. Permeabilization by detergent such as Triton X-100 solubilizes and redistributes SM and thus cannot be used. Saponins such as digitonin permeabilize membranes in Chol-dependent manner. Digitonin permeabilization was used to visualize the accumulation of intracellular SM in Niemann-Pick type A fibroblasts by Lys and Lys derivatives (Kiyokawa et al., 2004, 2005; Yamaji et al., 1998). The disadvantage of saponin is that it solubilizes Chol and re-organizes Chol-contaning membranes. In addition, saponin does not permeabilize Chol-deficient organelles. As Chol-independent membrane permeabilization, we freeze-thaw cells using liquid N_2 (Makino et al., 2017). When HeLa cells were treated with liquid N_2, plasma membrane was weakly labeled with NT-Lys compared to the intracellular organelles, probably due to the high density of SM in the organelle. However, we have not extensively characterized the condition of liquid N_2 permeabilization. Thus, the protocol could be improved. Here, we describe a protocol for intracellular labeling with Alexa 647-SNAP-NT-Lys and an antibody against CD63, a marker protein of late endosomes (Escola et al., 1998; Kobayashi et al., 2002). Fig. 2 shows localization of SM-rich lipid domains in CD63-positive late endosomes.

5.1 Materials

1. Container for liquid N_2.
2. Parafilm, Sigma-Aldrich, P7793.

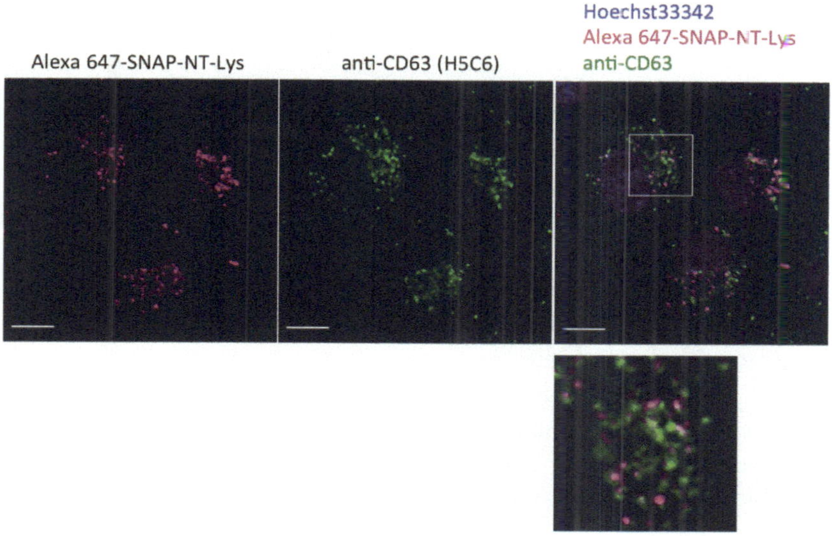

Fig. 2 After fixation with 4% PFA, cells were permeabilized with liquid N_2 and stained with Alexa 647-SNAP-NT-Lys and anti-CD63 antibody (H5C6). Bottom panel: the enlarged view of enclosed area in the merged image. SM in endosomal lumen was labeled with Alexa Fluor 647-SNAP-NT-Lys. Bar, 10 μm.

5.2 Reagents

1. liquid N_2.
2. 3% BSA, dissolve in PBS.
3. Primary antibody solution mixed with lipid probe, anti-CD63 (H5C6), and Alexa 647-SNAP-NT-Lys are diluted 100 times, and 12.5 times, respectively, with 0.1% BSA in PBS.
4. Secondary antibody solution, Alexa Fluor 488-conjugated anti-mouse IgG (Thermofisher scientific, A-11001) is diluted 500 times with 0.1% BSA in PBS.

For other materials, see Section 2.

5.3 Protocol

1. Plate HeLa cells on a 12-mm glass coverslip in 24-well plate to obtain 50–80% confluency on the day of the experiment.
2. On the day of the experiment, remove the medium and rinse cells with PBS once.
3. Fix cells with 250 μL of 4% PFA at RT for 37 °C.
4. Inactivate excess PFA with 250 μL of 0.1 M NH_4Cl solution.

5. Take a glass coverslip from a well of 24-well plate and remove PBS from the surface of glass coverslip by attaching Kimwipes at the edge of the coverslip.
6. Briefly (<1 s) immerse the coverslip into liquid N_2 with forceps to perforate cell membranes.*1
7. Put it back into PBS in the well of 24-well plate.
8. Exchange the buffer with 3% BSA in PBS for blocking and incubate at RT for 30 min.
9. Spot 30 μL drop of primary antibody solution containing Alexa 647-SNAP-NT-Lys on Parafilm.
10. Place a coverslip onto the drop cell side down and incubate at RT for 1 h.*2,3
11. Transfer the coverslip back into the well, wash them with PBS three times.
12. Spot 30 μL drop of secondary antibody solution on Parafilm.
13. Place the coverslip onto the drop of the secondary antibody and incubate at RT for 1 h.
14. Wash the coverslip with PBS three times in the well of 24-well plate.
15. Mount the coverslip with mounting medium on a glass slide.
16. Observe the slide under a confocal microscope (Fig. 2).

*1: Too long immersion of the coverslip in liquid N_2 damages cells.
*2: For single labeling with Alexa647-SNAP-NT-Lys, remove the primary antibody in the step 9 and the steps 12–14.
*3: To prevent the evaporation of antibody solutions, incubations in steps 7 and 10 are performed in a container with a rid, containing wet Kimwipes.

6. Alexa647-NT-Lys labeling of HeLa cells for PALM/dSTORM

Lys-labeled SM clusters form domains of 50–100 nm diameter (Murate et al., 2015). These domains are too small to be detected by confocal microscopy. Electron microscopy and super resolution fluorescence microscopy are employed to visualize these lipid domains. Dronpa-D4, Dronpa-NT-Lys, Alexa 647-NT-Lys, PAmCherry-nakanori have been used to visualize respectively, Chol-, SM- or SM/Chol-rich lipid domains in photoactivation localization microscopy (PALM)/direct stochastic optical reconstruction microscopy (dSTORM) (Abe et al., 2012; Makino et al., 2017; Mizuno et al., 2011). PALM/dSTORM showed that

SM/Chol-rich domains are a part of SM-rich lipid domains (Makino et al., 2017). PALM/dSTORM also showed the interbilayer co-localization of SM-rich lipid domains in the outer leaflet of the plasma membrane and phosphatidylinositol 4,5-bisphophate (PIP_2)-rich lipid domains in the inner leaflet (Abe et al., 2012). Our PALM/dSTORM system is a home-made setup configurated to work under a total internal reflection fluorescence (TIRF) regime (Glushonkov, Real, Boutant, Mely, & Didier, 2018). In this setup, we image the top surface of cell using oblique (or HILO (highly inclined and laminated optical sheet)) illumination scheme.

6.1 Materials

1. Microscope for super-resolution localization microscopy is a homemade setup based on Nikon Eclipse Ti microscope equipped with 405, 561, 642 nm laser lines, 100× oil-immersion objective (NA 1.49), and EM-CCD camera (Glushonkov et al., 2018).
2. 18-mm glass coverslip, MARIENFELD laboratory glassware, cat#0117580, High precision microscope cover glasses, No. 1.5H.
3. 12-well plate.
4. Standard glass microscope slides for mounting the coverslip.

6.2 Reagents

1. Qdot 655 SA, ThermoFisher Scientific, cat#Q10121MP, Qdot 655 Streptavidin conjugate.
2. NH_4Cl.
3. Paraformaldehyde, dilute 16% solution to 4% with PBS before use.
4. Dental cement, Picodent.
5. Mounting medium (Glushonkov et al., 2018), 20% Vectashield (Vector Laboratories) in Tris-glycerol (5%v/v 1 M Tris-Cl, pH 8.0 in glycerol).

6.3 Protocol

1. Plate cells on an 18-mm glass coverslip in 12-well plate to be 50–70% confluent on the day of the experiment.
2. On the day of the experiment, rinse cells with DMEM/5% LPDS.
3. Add 0.5 mL of DMEM/5% LPDS containing 20 μg/mL of Alexa647-NT-Lys and incubate at 37 °C for 15 min.
4. Wash labeled cells with 0.5 mL PBS twice.
5. Fix cells with 4% PFA and incubate at RT for 30 min.*[1]
6. Wash cells with PBS twice.

7. Add 0.1 M NH_4Cl to neutralize the residual PFA.
8. Wash with PBS twice.
9. Mix almost the same volume*2 of blue and yellow reagent of dental cement (Picodent) for sealing a coverslip, by cell lifter until the color turns blue-green in 6 cm culture dish then place the mixed reagent on ice to prevent from solidifying.
10. Spot 8 μL of mounting medium (20% Vectashield (Vector Laboratories) in Tris-glycerol (5% v/v 1 M Tris-Cl, pH 8.0 in glycerol)) onto a glass slide.
11. Immerse the coverslip in MilliQ water to avoid precipitation of salt in PBS and remove water from the coverslip by tapping the coverslip edge on Kimwipe.
12. Place the coverslip on a drop of the mounting medium cell side down.
13. Immediately seal the coverslip with the mixed dental cement because the mounting medium is liquid.
14. The sample is ready for imaging and can be stored at 4 °C for several days.
15. Top surface of cells is imaged by oblique illumination microscopy (Fig. 3).

*1: If fiducial markers are used for drift correction during acquisition, Qdot655SA is included at 0.8 nM in 4% PFA. Because our superresolution imaging is carried out on Alexa647-NT-Lys bound to cell surface SM,

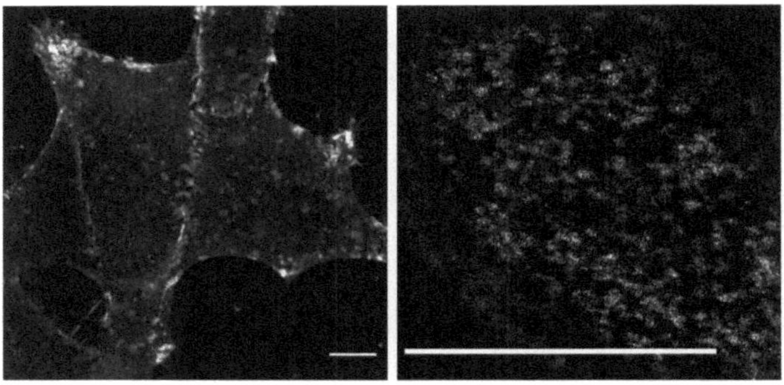

Fig. 3 Confocal (left) and dSTORM (right) images of HeLa cells that were labeled with Alexa 647-NT-Lys. HeLa cells grown on an 18-mm coverslip were labeled by incubating at 37 °C for 15 min in medium containing 15 μg/mL Alexa 647-NT-Lys. After fixation of cells with 4% PFA at RT for 30 min, slides were prepared appropriately for each microscope (see text). Images were taken under confocal microscope (FV1000D) equipped with 63 × objective and dSTORM setup. The dSTORM image was shown after reconstruction. Bar, 10 μm.

the fiducial marker Qdot655SA is expected to be fixed on cell surface through its streptavidin by PFA.
*2: The cement reagents are viscous and thus it is difficult to obtain accurate volume. Mixing roughly the same volumes by eye is sufficient for the purpose. 0.5 mL of the mixture is enough to seal 18-mm coverslip.

7. Labeling ultracryo sections with nakanori by immunoelectron microscopy

PFTs and related proteins are also used to examine lipid distribution using immunoelectron microscopy. In ultracryo section methods, sections cut at around −120 °C from chemically fixed and frozen cells are labeled with PFTs followed by the conjugation of immunogold. This method does not employ organic solvent that extract lipids from the specimen. However, since chemical fixatives do not fix most of the lipids, there is a possibility that lipids are lost during sample preparation. Using a proteolytically modified and biotinylated derivative of the cholesterol-binding θ-toxin (perfringolysin O (PFO)) (BCθ), previously Möbius et al. has established experimental conditions to reduce the loss of cholesterol from the membrane (Mobius et al., 2002). We modified their procedure. In this chapter, we describe the methods to label adherent cells with nakanori, that selectively binds sphingomyelin/cholesterol complex (Makino et al., 2017). Geuze group localized cholesterol in ultracryo sections from floating cultured cells and tissues using BCθ (Mobius et al., 2003) or biotinylated PFO (Slot & Geuze, 2007). GST-Lys was used to label ultracryo sections prepared from mouse embryo (Yoshida, Yoneda, Umeda, Ide, & Fujimoto, 2001). MBP-Lys, EqtII-GFP (Makino et al., 2015) and nakanori (Makino et al., 2017) were also used to label ultrathin sections prepared from high-pressure frozen HeLa cells by freeze substitution and low-temperature embedding of resin.

7.1 Materials

1. Specimen carrier, Product No. 16701950, Leica Microsystems, Vienna, Austria.
2. Forceps, Product No. 38D18031, Leica Biosystems, Buffalo Grove, IL.
3. Ultramicrotome with cryoattachment, UC6 and FC6, Leica Microsystems.
4. Trimming knife, Cryotrim 45 Tool, Diatome, Nidau, Switzerland.
5. Hot plate, EH-1, As One, Osaka, Japan.
6. Light microscope, DME, Leica Microsystems.

7. Diamond knife, Cryo immuno or Cryo 35° (dry), Diatome.
8. Wire-loop, loops formed by platinum or stainless wire with a diameter of 3 and 4 mm attached with thin wooden stick or loop handle, Product No. 2340, Nisshin EM, Tokyo, Japan.
9. Eyelash tool, eyelash attached with thin wooden stick.
10. Brush, Paint brush with soft hair.
11. Liquid nitrogen container, MVE XC47/11–6, PrincetonCryo, Pipersville, PA.
12. Fine tweezers, Style 5, Dumoxel, Dumont, Montignez, Switzerland.
13. Moist chamber, 10 cm petri dish with moistened filter paper and a sheet of Parafilm.
14. Lidded box, small bricks of polystyrene foam are attached in the box with double-stick tape. Double-stick tape is also put on the top surface of polystyrene foam.
15. Stereomicroscope, Stemi DV4, Zeiss, Jena, Germany.
16. Grid storage box, Product No. 2731, Nisshin EM.
17. Transmission electron microscope, JEM1230, JEOL, Tokyo, Japan.
18. CCD camera, Veleta, Olympus-SIS, Münster, Germany.
19. 10 cm dish, Product No. 150350, ThermoFisher Scientific, Waltham, MA.
20. Cell scraper, Product No. 3010, Corning, Corning, NY.
21. 15 mL plastic tube, Product No. 352097, Falcon polypropylene tube, Corning.
22. Liquid nitrogen.
23. Razor blade.
24. Formvar-coated specimen grids, Formvar-coated copper grids (Product No. 6514, Nisshin EM) are treated by glow-discharge machine (HDT-400, JEOL).
25. Cryogenic vial, Product No. 430488, Corning. Make several small holes just beneath the screw to enter liquid nitrogen freely.
26. Glass microscope slide, silan-coated, Product No. S8111, Matsunami Glass, Osaka, Japan.
27. Pasteur pipet, Product No. 7095B-9, Corning.
28. Parafilm, Bemis, Neenah, WI.
29. Filter paper, Product No. s4013, Advantech, Tokyo, Japan.
30. Blotting paper, Whatman Grade 50, Buckinghamshire, UK.
31. Double-stick tape, Product No. NW-15, Nichiban, Tokyo, Japan.

7.2 Reagents

1. PHEM buffer stock solution, 120 mM 1,4-piperazinediethanesulfonic acid (PIPES), 50 mM 4-(2-hydroxyethyl)-1-piperazineethanesulfonic

acid (HEPES), 20 mM ethylene glycol-bis(2-aminoethylether)-N,N, N′,N′-tetraacetic acid (EGTA), 4 mM $MgCl_2$, adjust pH to 6.9 with NaOH. Dilute with milliQ water to 1:1 ratio before use.

2. Mixture of 2% PFA, 0.2% glutaraldehyde (GA), 1% acrolein in PHEM buffer. mix 1 mL of 20% PFA solution (Product No.15713-S, Electron Microscopy Sciences), 80 μL of 25% GA (Product No. G004, TAAB Laboratories, Barkshire, UK), 100 μL of acrolein (Product No. 110221, Sigma, St. Louis, MO) in 5 mL of PHEM buffer stock solution before use. Adjust to 10 mL with milliQ water.
3. 1% gelatin, dissolve 1 g gelatin in 10 mL of PHEM buffer (1 ×).
4. 2.3 M sucrose, dissolve 78.7 g sucrose in a final volume of 100 mL PHEM buffer.
5. Mixture of 1.8 M sucrose, 0.5% OsO_4 and 10% TI Blue solution in PHEM buffer, mix 780 μL of 2.3 M sucrose, 120 μL of 4% OsO_4 solution (Product No. O018, TAAB Laboratories), 100 μL of TI Blue solution (Product No. 335, Nisshin EM) before use.
6. 2% methyl cellulose stock solution, Add 2 g methyl cellulose into 100 mL of preheated milliQ water at 90 °C and mix by stirring. Cool down on ice while stirring and then stir overnight at 4 °C. Let it stand at 4 °C for another 3 days. Store in the refrigerator. Use the supernatant.
7. Mixture of 1% methylcellulose, 1.15 M sucrose, Mix 2% methyl cellulose stock solution and 2.3 M sucrose (1:1). Keep in the refrigerator.
8. Toluidine blue solution, 0.5 g Toluidine blue O (Product No. 1.15930, Merck, Darmstadt, Germany), 0.5 g sodium tetraborate decahydrate in 100 mL MilliQ water. Use the supernatant.
9. TBS pH 7.5, dissolve 3 g Tris(hydroxymethyl)aminomethane (Tris), 4.4 g NaCl in 450 mL milliQ water. Adjust pH to 7.5 with HCl. Fill out to 500 mL.
10. 10% BSA stock solution, add 5 g BSA (Fraction V, Nacalai Tesque, Kyoto, Japan) to 40 mL TBS pH 7.5. Stir slowly overnight. Add sodium azide to a final concentration of 0.02% and TBS pH 7.5 to a final volume of 50 mL. Keep in small aliquots (1 mL) in freezer before use. After first use, keep in the refrigerator.
11. 2% gelatin stock solution, dissolve 1 g gelatin to 40 mL TBS pH 7.5. Add sodium azide to a final concentration of 0.02% and TBS pH 7.5 to a final volume of 50 mL. Keep in small aliquots (1 mL) in freezer before use. After first use, keep in the refrigerator.
12. Nakanori stock solution. Keep in the refrigerator.

13. 1 M NH_4Cl in TBS stock solution, dissolve 5.3 g of NH_4Cl in TBS. Dilute 20 times before use.
14. Anti-nakanori antibody antiserum, dilute 1000 times before use. Keep in the refrigerator.
15. Poly(ethylene glycol) stock solution, dissolve 5 g of poly(ethylene glycol) (Average molecular weight is 550, Product No. 202487, Sigma) into 10 mL of TBS pH 7.5. Keep in the refrigerator.
16. Colloidal gold-conjugated anti-rabbit IgG antibody, Product No: EM. GAR10, BBI Solutions, Crumlin, UK. Dilute 50 times before use. Keep in the refrigerator.
17. Lead solution, Product No. 18–0875, Sigma.
18. Mixture of 12.5% TI Blue solution and 1.75% methyl cellulose, mix 40 μL of TI Blue solution and 280 μL of 2% methyl cellulose stock solution before use.

7.3 Protocol

Day 0

1. Plate cells in a 10 cm dish.

Day 1

2. Rinse the cells three times with PHEM buffer pH 6.9.
3. Fix with the mixture of 2% PFA, 0.2% GA, 1% acrolein in PHEM buffer, for 30 min on ice.
4. Wash three times with PHEM buffer at RT.
5. Scrape cells from the dish in 1% gelatin in PHEM buffer, transfer to 15 mL plastic tube and centrifuge for 10 min at 8600 × *g*.
6. Decant the supernatant and apply the mixture of 1.8 M sucrose, 0.5% OsO_4 and 10% TI Blue solution in PHEM buffer to the pellet without suspending and stand for 2 h on ice.
7. Wash three times with PHEM buffer at RT.
8. Cut the cell pellet into small cubes with 1–2 mm on each side by razor blade under stereomicroscope.
9. Immerse the cell cubes into 2.3 M sucrose in PHEM buffer and keep at 4 °C at least overnight.

Day 2

10. Place one cell cube on the specimen carrier and dip into liquid nitrogen using forceps.*[1]
11. Trim the upper part of the frozen sample into ~0.2 mm on each side using the edge of trimming knife and make cell pellet exposed by

cutting the surface of the block at 500 nm thickness at −80 °C in ultracryo sectioning machine.*2

12. Cut the block into ultrathin sections of 70 nm thickness using diamond knife at −120 °C at cutting speed 1 mm/s.*3
13. Pick up the ultrathin sections with the mixture of 1% methylcellulose and 1.15 M sucrose in PHEM buffer in the 3 mm wire-loop, and transfer section ribbons to the Formvar-coated specimen grids.*4
14. Incubate the grids with the 20 μg/mL nakanori in TBS pH7.5 for 30 min at RT.*5
15. Wash four times with TBS for one grid (5 min each time).*6
16. Fix with 0.5% GA in TBS for 10 min.
17. Wash four times with TBS for one grid (5 min each time) at RT.
18. Quench free aldehyde bonds with 50 mM NH_4Cl in TBS for 10 min at RT.
19. Wash four times with TBS for one grid (5 min each time) at RT.
20. Incubate with 2% BSA and 0.2% gelatin in TBS for 1 h in the moist chamber.
21. Incubate with anti-nakanori antibody in 2% BSA and 0.2% gelatin in TBS overnight at 4 °C in the moist chamber.

Day 3

22. Wash four times with TBS for one grid (5 min each time) at RT.
23. Incubate with 2% BSA, 0.2% gelatin and 5 mg/mL poly(ethylene glycol) in TBS for 1 h at RT.
24. Incubate with colloidal gold-conjugated anti-rabbit IgG antibody in 2% BSA, 0.2% gelatin and 5 mg/mL poly(ethylene glycol) in TBS for 2 h at RT.
25. Wash four times with TBS for one grid (5 min each time) at RT.
26. Fix with 1% GA in TBS for 10 min at RT.
27. Wash four times with MilliQ water (5 min each time) at RT.
28. Counter staining with the lead solution for 5 min at RT.
29. Wash three times with MilliQ water at RT.
30. Counter staining with the mixture of 12.5% TI Blue and 1.75% methyl cellulose for 10 min on ice.
31. Pick up the grid with 4 mm wire loop and remove excess solution with blotting paper, followed by drying up overnight in the lidded box at RT.*7

Day 4

32. Grids are removed from the loops by tweezer and examined by TEM (Fig. 4).*7

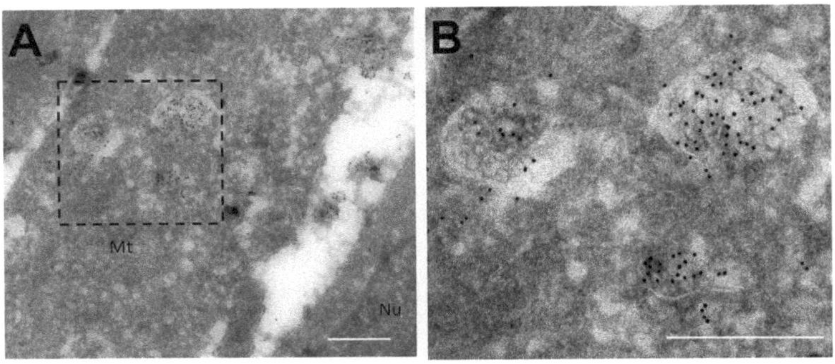

Fig. 4 Intracellular distribution of nakanori labeling in ultracryo section of HeLa cells. (A) Lower magnification. Mt., mitochondrion; Nu, nucleus. (B) Higher magnification of the area surrounded by dotted square in (A). The gold particles were localized on the internal membrane of multivesicular bodies. Bars: 500 nm. *Adapted and modified from Makino, A., Abe, M., Ishitsuka, R., Murate, M., Kishimoto, T., Sakai, S., Hullin-Matsuda, F., Shimada, Y., Inaba, T., Miyatake, H., et al. (2017). A novel sphingomyelin/cholesterol domain-specific probe reveals the dynamics of the membrane domains during virus release and in Niemann-Pick type C.* The FASEB Journal *31, 1301–1322.*

*1: After placing one cell pellet on the specimen carrier, small aliquots of 2.3 M sucrose in PHEM buffer are applied around the block to prevent the block to be removed in the step of trimming. The solution covered on the top of gelatin block is captured by the filter paper. Specimen carrier is dipped by forceps into liquid nitrogen. Freezing step is not crucial for the ultrastructure of the samples in this experiment.

*2: Examination of the semi-thin sections by light microscopy helps to check the quality and overall morphology in the sample. Sections are picked up by touching the semi-frozen methylcellulose and sucrose ball in the center of the wire-loop. After melting down the ball, the wire-loop is touched to a glass microscope slide to transfer the sections. Apply one drop of Toluidine blue solution and heat 10–20 s at 80–90 °C on the hot plate. Wash the sections gently by a flow of MilliQ water, dry up on the hot plate and observe under light microscope.

*3: Room temperature and humidity affect the cutting conditions due to the opening of the cryochamber. If the sections become flakes, the setting temperature is too low. If the sections look tarnished, the setting temperature is too high. The temperature, cutting speed and thickness of the sections should be adjusted to get ribbon-shaped sections. After the first section is cut, the front edge of the first section is lifted up by the eyelash, ready for making sections. If chatter occurs, specimen knife or cryochamber should be set tightly. If the experiment goes well, the color of sections is gold. However, it is acceptable even if the thickness of each section is not uniform, showing the color of sections from silver to light yellow. Picking up sections is easier when longer ribbons are prepared. The specimen block can be kept indefinitely in the cryogenic vials in liquid nitrogen container.

*4: The 3 mm wire-loop is immersed into the methylcellulose-sucrose solution several times to form a large drop in the loop. The section ribbons should be slowly transferred to the center of the drop, maintaining their shape. Avoid pushing the drop to the section ribbons. Monitor freezing of the drop by the binocular on the ultramicrotome. If the drop starts to freeze from its periphery, you will feel that the section ribbons jump into the drop spontaneously when you bring the drop close to the ribbons. They spontaneously attach to the periphery of freezing drop. The frost on the knife is removed by sweeping with a brush.

*5: Sections are incubated by floating the grid on a drop of nakanori in TBS on a Parafilm sheet, facing the solution.

*6: Sections are washed in TBS by incubating them as described above. After 5 min, the grid is transferred to the next drop of TBS by fine tweezers.

*7: After picking up, the grid is set in the center of the wire loop with a small volume of TI Blue and methyl cellulose solution. Blotting paper is touched to the edge of the loop to remove the solution slowly. After stopping the absorption, the grid attaches to the loop. The loop is kept horizontally, glued on the polystyrene foam in the lidded box.

*8: The grid is removed from the loop carefully by the tip of fine tweezers under the stereomicroscope. The grids are stored in a grid storage box and are kept indefinitely in a dry place.

8. Sodium dodecyl sulfate-digested freeze fracture replica labeling (SDS-FRL) with MBP-lysenin

Sodium dodecyl sulfate-digested freeze fracture replica labeling (SDS-FRL) is an immunoelectron microscopy technique using replica membrane (Fig. 5). The replicas are labeled with anti-lipid antibodies or lipid-binding toxins followed by colloidal gold-conjugated antibody or streptavidin labeling. SDS-FRL has several advantages to study lipid distribution over other methods: (1) Quick freezing (in the present protocol, −183 °C and less than 50 msec, values change depend on the cryogen) minimizes re-organization of lipids during sample preparation. (2) Physical immobilization of lipids on metal cast prevents crosslinking of lipids by antibodies or toxins.

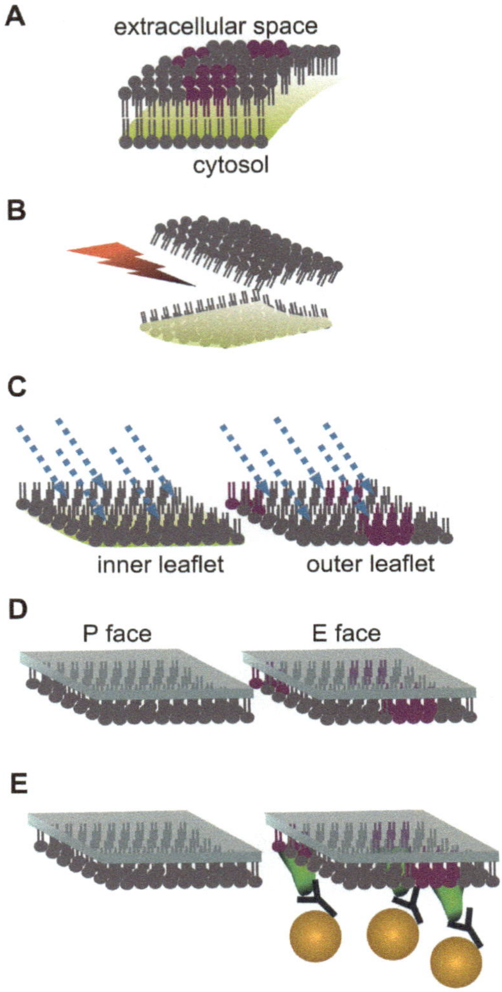

Fig. 5 Outline of SDS-FRL method. (A) Quick freezing at very low temperature. (B) Physical cracking to separate the outer and inner leaflets of the bilayer. (C) Deposition to physically fix membrane lipids and proteins on a metal cast. (D) SDS digestion to remove cytoplasmic components including membrane-associated proteins while retaining membrane lipids and integral proteins. (E) Exposed head groups of lipids are detected using lipid-binding PFT followed by colloidal gold-conjugated secondary antibodies. *Adapted and modified from Murate, M., & Kobayashi, T. (2016). Revisiting transbilayer distribution of lipids in the plasma membrane.* Chemistry and Physics of Lipids, *194, 58–71.*

(3) Outer and inner leaflet of cell membranes are separated. Thus lipid distribution in a single leaflet can be studied. The disadvantage of this method is that the samples are broken to small pieces during digesting cellular contents and thus it is technically difficult to identify corresponding outer leaflet and

inner leaflet pieces. We employ TIRF microscopy (Kishimoto et al., 2020) or super resolution microscopy, PALM/dSTORM (Abe et al., 2012) to examine interbilayer colocalization of lipids.

It is reported that around 70% of lipids are retained on the replica membrane (Fujimoto, Umeda, & Fujimoto, 1996). Model membrane experiments indicate that the detection limit of most lipids by lipid-binding probes is less than 1% (Murate et al., 2015). Several PFTs and related molecules have been employed for detection of lipids in SDS-FRL. These are: MBP-Lys (Makino et al., 2015; Murate et al., 2015), EqtII-GFP (Makino et al., 2015), PlyA2-EGFP (Bhat et al., 2015) and nakanori (Makino et al., 2017). In this chapter, we describe the methods using adherent cultured cells (Murate et al., 2015). However, this method is also applicable to cells in suspension as well as organs (Cheng et al., 2014; Fujimoto et al., 1996; Murate et al., 2015). Detailed methods of SDS-FRL are also available in previous publications (Fujimoto, 1997; Fujita, Cheng, & Fujimoto, 2010).

8.1 Materials

1. Reverse action tweezer, Style N2A, Dumoxel, Dumont.
2. Plunge freezing machine, EM CPC, Leica Microsystems.
3. Freeze fracturing machine, BAF400, Balzers, Balzers, Liechtenstein.
4. 6 or 12 well ceramic spot plate. We use wells of 21 mm diameter and 7 mm depth but the size of the well is not crucial (Fig. 6).
5. Block heater, SBH200D, Cole-Parmer, Staffordshire, UK.
6. Shaker incubator, KS 4000i, IKA-Werke, Staufen, Germany.
7. Wire-loop, loops formed by platinum wire with a diameter of 3 mm attached with loop handle.

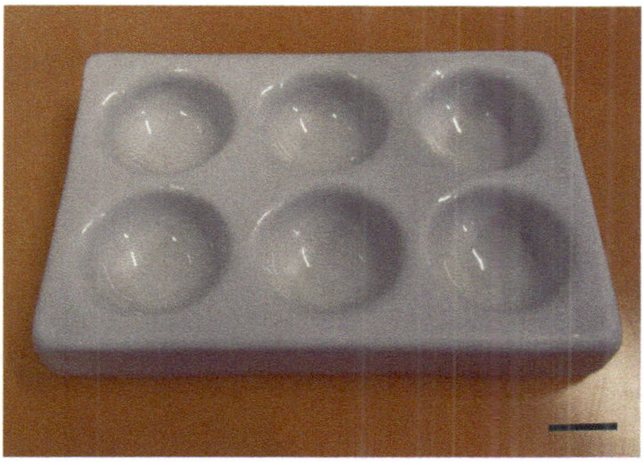

Fig. 6 A ceramic spot plate used for washing replica flakes. Bar: 1 cm.

8. Plastic dish, 3.5 or 6 cm, Product No. 150318 or 150288, ThermoFisher Scientific.
9. Microscope cover glass, round micro cover glass with a diameter of 15 mm (Product No. C015001, Matsunami Glass).
10. Copper plate, scratch the surface of copper plate (5 mm × 10 mm) by sandpaper (No. 100). Wash with MilliQ water and immerse in acetone for 10 min. After drying on filter paper, attach double-stick tape to the edge of the surface.
11. Kimwipes, Kimberly Clark, Irving, TX.
12. Other materials are listed in the Section 6.1.

8.2 Reagents

1. 35% polyvinylpyrroridone in PHEM buffer, dissolve 3.5 g of polyvinylpyrroridone (Product No. PVP10, Sigma) in 10 mL of PHEM buffer.
2. Ethane gas.
3. 0.1% BSA in TBS, dilute 10% BSA stock solution to 100 times with TBS.
4. SDS-digestion buffer, dissolve 3.8 g of Tris(hydroxymethyl)aminomethane in 400 mL of MilliQ water. Adjust pH at 6.8 with HCl. Dissolve 12.5 g of sodium dodecyl sulfate. Add 50 mL of glycerol and MilliQ water to final volume of 500 mL. Add β-mercaptoethanol to final concentration of 5% before use.
5. Maltose-binding protein (MBP)-Lys stock solution. Keep in the refrigerator.
6. Anti-MBP antibody antiserum, E8030S, New England BioLabs, Ipswich, MA, dilute 1000 times before use. Keep in the refrigerator.
7. Other chemicals are listed in the Section 6.2.

8.3 Protocol

Day 0

1. Plate cells on a cover glass in a 3.5 cm or 6 cm plastic dish.

Day 1

2. Apply small aliquot (~10 μL) of 35% polyvinylpyrrolidone in PHEM buffer pH 7.5 into copper plate.
3. Overlay the copper plate on the coverslip and dip the copper plate-cells-cover glass assembly rapidly into liquid ethane cooled by liquid nitrogen using plunge freezing machine.*1
4. Set the specimen in a freeze fracturing machine and remove the copper plate at −110 °C under vacuum (as high vacuum as possible).*2

5. Evaporate with carbon at 5 nm thickness from above, platinum-carbon at 2 nm thickness with the angle of 40°, and finally carbon at 15 nm thickness from above.*3
6. Take out the cover glass with specimen table from the machine, and thaw it at RT until the frost disappears.
7. Immerse the cover glass into 0.1% BSA in TBS pH 7.5 in ceramic spot plate and scrape replica membranes from the cover glass under the stereomicroscope.
8. Wash three times with 0.1% BSA in TBS at RT.*4
9. Mix vigorously in 5 mL of SDS-digestion buffer for 10 min and shake overnight at 60 °C.*5

Day 2

10. Wash three times with TBS (5 min for each time) at RT.*6
11. Blocking with 50 μL of 2% BSA and 0.2% gelatin in TBS for 1 h at RT in the moist chamber.
12. Incubate with 20 μL of 15 μg/mL MBP-Lys in 2% BSA and 0.2% gelatin in TBS overnight at 4 °C in the moist chamber.

Day 3

13. Wash three times with TBS (5 min each time) at RT.
14. Incubate with 50 μL of 2% BSA and 0.2% gelatin in TBS for 1 h at RT.
15. Incubate with 20 μL of anti-MBP antibody in 2% BSA and 0.2% gelatin in TBS overnight at 4 °C in the moist chamber.

Day 4

16. Wash three times with TBS (5 min each time) at RT.
17. Incubate with 2% BSA, 0.2% gelatin and 5 mg/mL poly(ethylene glycol) in TBS for 1 h at RT.
18. Incubated with colloidal gold-conjugated anti-rabbit IgG antibody in 2% BSA, 0.2% gelatin and 5 mg/mL poly(ethylene glycol) in TBS for 2 h at RT.
19. Wash three times with TBS (5 min each time) at RT.
20. Fix with 1% GA in TBS for 10 min at RT.
21. Wash three times with MilliQ water at RT.
22. Unfold the replica completely on the surface of MilliQ water and pick it up onto formvar-coated TEM grids, followed by drying up overnight.*7

Day 5

23. Examine by TEM (Fig. 7).

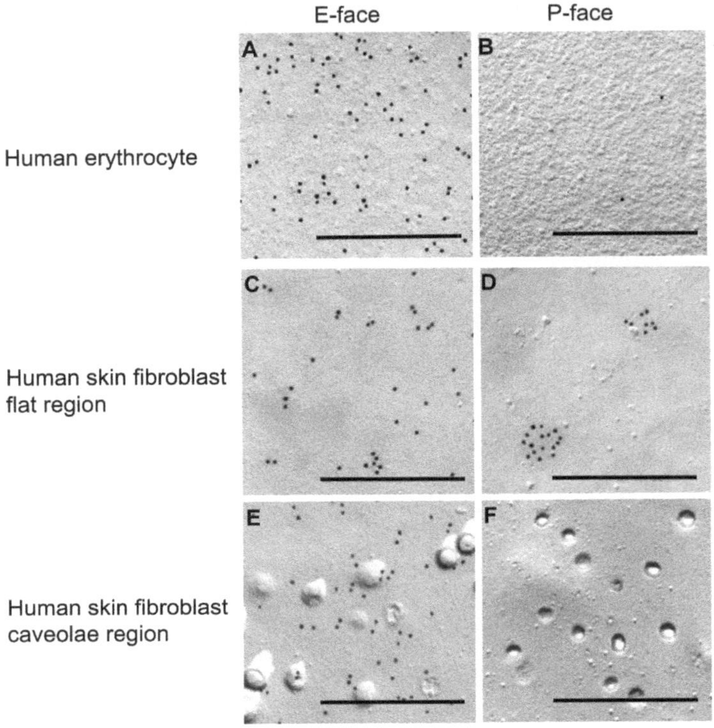

Fig. 7 Distribution of sphingomyelin on the plasma membranes of various cells. (A) The outer leaflet (E-face) of human erythrocyte membrane. (B) The inner leaflet (P-face) of erythrocyte membrane. (C) The outer leaflet on flat region of human skin fibroblast plasma membrane. (D) The inner leaflet on flat region of human fibroblast plasma membrane. (E) The outer leaflet on caveolae of human fibroblast plasma membrane. (F) The inner leaflet on caveolae of human fibroblast plasma membrane. Bars: 500 nm. *Adapted and modified from Murate, M., & Kobayashi, T. (2016). Revisiting transbilayer distribution of lipids in the plasma membrane.* Chemistry and Physics of Lipids, *194, 58–71.*

*1: Copper plate-cells-cover glass assembly are hold by a reverse action tweezers and injected into liquid ethane using the plunge freezing machine. After 10 s, the assembly is blotted with Kimwipes to draw out liquid ethane, and then, transferred to liquid nitrogen. The reverse action tweezer is removed under holding the assembly by forceps to avoid the separation of the copper plate from the cover glass. Frozen samples can be kept indefinitely in the liquid nitrogen container.

*2: Mount the assembly on a specimen table, and transfer to the pre-cooled stage at $-150\,°C$ of the freeze fracturing machine. When the vacuum reaches to $\sim 3 \times 10^{-6}$ Torr ($\sim 4 \times 10^{-4}$ Pa), set the specimen temperature up to −110 °C and cool the knife down to liquid nitrogen temperature. In cultured mammalian cells, the fracture plane normally passes through the middle of the bilayer of the plasma membrane at this temperature. However, the fracturing temperature needs to be adjusted depending on the cell type. When the vacuum reaches to the level of 10^{-7} Torr (10^{-5} Pa), the knife is set between the copper plate and cover glass, followed by the lift of the edge of the copper plate to remove it.

*3: As soon as the copper plate is removed, the carbon and platinum should be evaporated by beam gun to prevent producing ice crystals on the fracture plane.

*4: Replica membranes first float and then sink in the buffer. After sinking, discard the buffer as much as possible, and add new buffer.

*5: Transfer replica membrane into 15 mL tube and incubate for 10 min at 60 °C using block heater with pipetting four times every minute by a Pasteur pipette, followed by rotating overnight at 60 °C using shaker incubator.

*6: Pour replica membrane in SDS-digestion buffer to 35 mm dish, and transfer up to ~10 flakes of replica to TBS in the ceramic spot plate using the wire loop under stereomicroscope. Replica flakes should sink in the buffer. After few minutes, they are transferred to new TBS by the wire-loop. The remaining replica flakes in SDS-digestion buffer are kept indefinitely in the freezer.

*7: Replica membrane should be extended completely on the surface of MilliQ water using the wire loop. This is a time-consuming but critical step. Touch the edge of the replica membrane extended onto a tilted and half-sunk grid, and pick it up. Put the grids on a filter paper in a 3.5 or 6 cm plastic dish by inserting of a small piece of filter paper between two metal sticks of fine tweezers and pushing out the grid by the filter paper. Next day, the grids are stored in a grid storage box and are kept indefinitely in a dry place.

9. Summary

Multivalent binding to lipids and toxicity hinder the use of native lipid-binding PFTs to examine detailed distribution of endogenous lipids and lipid domains. Non-oligomerizable and non-toxic PFT mutants have solved these problems. However, native PFTs can also be used under conditions where oligomerization and toxicity are inhibited, such as low temperature. It should be taken into account that membrane lipid distribution is also sensitive to temperature. Although PFT-derived proteins are useful bio-probes to localize endogenous lipids at the nanoscale, binding is often dependent on lipid environment and membrane distribution of the lipids. Thus, combination with other probes and/or other methods such as biochemical lipid analysis helps to understand endogenous lipid distribution.

Acknowledgments

We are grateful to Brigitte Pollet for her technical assistance. Our research was supported by Grants-in-Aid from the Ministry of Education, Culture, Sports, Science, and Technology of Japan, Agence Nationale pour la Recherche (A20R417C to T.K.), Agence Nationale de Recherche sur le Sida et les Hépatites virale (18365 to T.K.), Ligue Contre le Cancer (to T.K.), Vaincre les Maladies Lysosomales (19/LBPH/S44 to T.K.) and RIKEN Integrated Lipidology Program and Glycolipidologue Program. Y.M. is grateful to the Institut Universitaire de France (IUF) for support and providing additional time to be dedicated to research.

References

Abe, M., & Kobayashi, T. (2017). Dynamics of sphingomyelin- and cholesterol-enriched lipid domains during cytokinesis. *Methods in Cell Biology, 137*, 15–24.

Abe, M., Makino, A., Hullin-Matsuda, F., Kamijo, K., Ohno-Iwashita, Y., Hanada, K., et al. (2012). A role for sphingomyelin-rich lipid domains in the accumulation of phosphatidylinositol 4,5-bisphosphate to the cleavage furrow during cytokinesis. *Molecular and Cellular Biology, 32*, 1396–1407.

Anderluh, G., Pungercar, J., Strukelj, B., Macek, P., & Gubensek, F. (1996). Cloning, sequencing, and expression of equinatoxin II. *Biochemical and Biophysical Research Communications, 220*, 437–442.

Bakrac, B., Gutierrez-Aguirre, I., Podlesek, Z., Sonnen, A. F., Gilbert, R. J., Macek, P., et al. (2008). Molecular determinants of sphingomyelin specificity of a eukaryotic pore-forming toxin. *The Journal of Biological Chemistry, 283*, 18665–18677.

Bakrac, B., Kladnik, A., Macek, P., McHaffie, G., Werner, A., Lakey, J. H., et al. (2010). A toxin-based probe reveals cytoplasmic exposure of Golgi sphingomyelin. *The Journal of Biological Chemistry, 285*, 22186–22195.

Bhat, H. B., Ishitsuka, R., Inaba, T., Murate, M., Abe, M., Makino, A., et al. (2015). Evaluation of aegerolysins as novel tools to detect and visualize ceramide phosphoethanolamine, a major sphingolipid in invertebrates. *The FASEB Journal, 29*, 3920–3934.

Bhat, H. B., Kishimoto, T., Abe, M., Makino, A., Inaba, T., Murate, M., et al. (2013). Binding of a pleurotolysin ortholog from Pleurotus eryngii to sphingomyelin and cholesterol-rich membrane domains. *Journal of Lipid Research, 54*, 2933–2943.

Carquin, M., Pollet, H., Veiga-da-Cunha, M., Cominelli, A., Van Der Smissen, P., N'Kuli, F., et al. (2014). Endogenous sphingomyelin segregates into submicrometric domains in the living erythrocyte membrane. *Journal of Lipid Research, 55*, 1331–1342.

Chakrabarti, R. S., Ingham, S. A., Kozlitina, J., Gay, A., Cohen, J. C., Radhakrishnan, A., et al. (2017). Variability of cholesterol accessibility in human red blood cells measured using a bacterial cholesterol-binding toxin. *eLife, 6*, e23355.

Cheng, J., Fujita, A., Yamamoto, H., Tatematsu, T., Kakuta, S., Obara, K., et al. (2014). Yeast and mammalian autophagosomes exhibit distinct phosphatidylinositol 3-phosphate asymmetries. *Nature Communications, 5*, 3207.

De Colibus, L., Sonnen, A. F., Morris, K. J., Siebert, C. A., Abrusci, P., Plitzko, J., et al. (2012). Structures of lysenin reveal a shared evolutionary origin for pore-forming proteins and its mode of sphingomyelin recognition. *Structure*, *20*, 1498–1507.

Deng, Y., Rivera-Molina, F. E., Toomre, D. K., & Burd, C. G. (2016). Sphingomyelin is sorted at the trans Golgi network into a distinct class of secretory vesicle. *Proceedings of the National Academy of Sciences of the United States of America*, *113*, 6677–6682.

Dumitru, A. C., Conrard, L., Lo Giudice, C., Henriet, P., Veiga-da-Cunha, M., Derclaye S., et al. (2018). High-resolution mapping and recognition of lipid domains using AFM with toxin-derivatized probes. *Chemical Communications*, *54*, 6903–6906.

Ellison, C. J, Kukulski, W., Boyle, K. B., Munro, S., & Randow, F. (2020). Trans-bilayer movement of sphingomyelin precedes catastrophic breakage of enterobacteria-containing vacuoles. *Current Biology*, *30*, 2974–2983 e2976.

Endapally, S., Frias, D., Grzemska, M., Gay, A., Tomchick, D. R., & Radhakrishnan, A. (2019). Molecular discrimination between two conformations of sphingomyelin in plasma membranes. *Cell*, *176*, 1040–1053 e1017.

Escola, J. M., Kleijmeer, M. J., Stoorvogel, W., Griffith, J. M., Yoshie, O., & Geuze, H. J. (1998). Selective enrichment of tetraspan proteins on the internal vesicles of multi-vesicular endosomes and on exosomes secreted by human B-lymphocytes. *The Journal of Biological Chemistry*, *273*, 20121–20127.

Evans, J. C., Johnstone, B. A., Lawrence, S. L., Morton, C. J., Christie, M. P., Parker, M. W., et al. (2020). A key motif in the cholesterol-dependent cytolysins reveals a large family of related proteins. *MBio*, *11*, e02351–20.

Fujimoto, K (1997). SDS-digested freeze-fracture replica labeling electron microscopy to study the two-dimensional distribution of integral membrane proteins and phospholipids in biomembranes: Practical procedure, interpretation and application. *Histochemistry and Cell Biology*, *107*, 87–96.

Fujimoto, K., Umeda, M., & Fujimoto, T. (1996). Transmembrane phospholipid distribution revealed by freeze-fracture replica labeling. *Journal of Cell Science*, *109*, 2453–2460.

Fujita, A., Cheng, J., & Fujimoto, T. (2010). Quantitative electron microscopy for the nano-scale analysis of membrane lipid distribution. *Nature Protocols*, *5*, 661–669.

Glushonkov O., Real, E., Boutant, E., Mely, Y., & Didier, P. (2018). Optimized protocol for combined PALM-dSTORM imaging. *Scientific Reports*, *8*, 8749.

Hannun, Y. A., & Obeid, L. M. (2018). Sphingolipids and their metabolism in physiology and disease. *Nature Reviews. Molecular Cell Biology*, *19*, 175–191.

Hong, Q., Gutierrez-Aguirre, I., Barlic, A., Malovrh, P., Kristan, K., Podlesek, Z., et al. (2002). Two-step membrane binding by Equinatoxin II, a pore-forming toxin from the sea anemone, involves an exposed aromatic cluster and a flexible helix. *The Journal of Biological Chemistry*, *277*, 41916–41924.

Ikenouchi, J., Hirata, M., Yonemura, S., & Umeda, M. (2013). Sphingomyelin clustering is essential for the formation of microvilli. *Journal of Cell Science*, *126*, 3585–3592.

Ikenouchi, J., Suzuki, M., Umeda, K., Ikeda, K., Taguchi, R., Kobayashi, T., et al. (2012). Lipid polarity is maintained in absence of tight junctions. *The Journal of Biological Chemistry*, *287*, 9525–9533.

Inaba, T., Murate, M., Tomishige, N., Lee, Y.-F., Hullin-Matsuda, F., Pollet, B., et al. (2019). Formation of tubules and helical ribbons by ceramide phosphoethanolamine-containing membranes. *Scientific Reports*, *9*, 5812.

Ishitsuka, R., & Kobayashi, T. (2007). Cholesterol and lipid/protein ratio control the oligomerization of a sphingomyelin-specific toxin, lysenin. *Biochemistry*, *46*, 1495–1502.

Ishitsuka, R., Yamaji-Hasegawa, A., Makino, A., Hirabayashi, Y., & Kobayashi, T. (2004). A lipid-specific toxin reveals heterogeneity of sphingomyelin-containing membranes. *Biophysical Journal*, *86*, 296–307.

Johnson, K. A., Endapally, S., Vazquez, D. C., Infante, R. E., & Radhakrishnan, A. (2019). Ostreolysin A and anthrolysin O use different mechanisms to control movement of cholesterol from the plasma membrane to the endoplasmic reticulum. *The Journal of Biological Chemistry*, *294*, 17289–17300.

Kishimoto, T., Ishitsuka, R., & Kobayashi, T. (2016). Detectors for evaluating the cellular landscape of sphingomyelin- and cholesterol-rich membrane domains. *Biochimica et Biophysica Acta*, *1861*, 812–829.

Kishimoto, T., Tomishige, N., Murate, M., Ishitsuka, R., Schaller, H., Mely, Y., et al. (2020). Cholesterol asymmetry at the tip of filopodia during cell adhesion. *The FASEB Journal*, *34*, 6185–6197.

Kiyokawa, E., Baba, T., Otsuka, N., Makino, A., Ohno, S., & Kobayashi, T. (2005). Spatial and functional heterogeneity of sphingolipid-rich membrane domains. *The Journal of Biological Chemistry*, *280*, 24072–24084.

Kiyokawa, E., Makino, A., Ishii, K., Otsuka, N., Yamaji-Hasegawa, A., & Kobayashi, T. (2004). Recognition of sphingomyelin by lysenin and lysenin-related proteins. *Biochemistry*, *43*, 9766–9773.

Kobayashi, T., Beuchat, M. H., Chevallier, J., Makino, A., Mayran, N., Escola, J. M., et al. (2002). Separation and characterization of late endosomal membrane domains. *The Journal of Biological Chemistry*, *277*, 32157–32164.

Kobayashi, T., & Menon, A. K. (2018). Transbilayer lipid asymmetry. *Current Biology*, *28*, R386–R391.

Kristan, K., Podlesek, Z., Hojnik, V., Gutierrez-Aguirre, I., Guncar, G., Turk, D., et al. (2004). Pore formation by equinatoxin, a eukaryotic pore-forming toxin, requires a flexible N-terminal region and a stable beta-sandwich. *The Journal of Biological Chemistry*, *279*, 46509–46517.

Lingwood, D., & Simons, K. (2010). Lipid rafts as a membrane-organizing principle. *Science*, *327*, 46–50.

Makino, A., Abe, M., Ishitsuka, R., Murate, M., Kishimoto, T., Sakai, S., et al. (2017). A novel sphingomyelin/cholesterol domain-specific probe reveals the dynamics of the membrane domains during virus release and in Niemann-Pick type C. *The FASEB Journal*, *31*, 1301–1322.

Makino, A., Abe, M., Murate, M., Inaba, T., Yilmaz, N., Hullin-Matsuda, F., et al. (2015). Visualization of the heterogeneous membrane distribution of sphingomyelin associated with cytokinesis, cell polarity, and sphingolipidosis. *The FASEB Journal*, *29*, 477–493.

Mancheno, J. M., Martin-Benito, J., Martinez-Ripoll, M., Gavilanes, J. G., & Hermoso, J. A. (2003). Crystal and electron microscopy structures of sticholysin II actinoporin reveal insights into the mechanism of membrane pore formation. *Structure*, *11*, 1319–1328.

Mitsutake, S., Zama, K., Yokota, H., Yoshida, T., Tanaka, M., Mitsui, M., et al. (2011). Dynamic modification of sphingomyelin in lipid microdomains controls development of obesity, fatty liver, and type 2 diabetes. *The Journal of Biological Chemistry*, *286*, 28544–28555.

Mizuno, H., Abe, M., Dedecker, P., Makino, A., Rocha, S., Ohno-Iwashita, Y., et al. (2011). Fluorescent probes for supersolution imaging of lipid domains on the plasma membrane. *Chemical Science*, *2*, 1548–1553.

Mobius, W., Ohno-Iwashita, Y., van Donselaar, E. G., Oorschot, V. M., Shimada, Y., Fujimoto, T., et al. (2002). Immunoelectron microscopic localization of cholesterol using biotinylated and non-cytolytic perfringolysin O. *The Journal of Histochemistry and Cytochemistry*, *50*, 43–55.

Mobius, W., Van Donselaar, E., Ohno-Iwashita, Y., Shimada, Y., Heijnen, H. F., Slot, J. W., et al. (2003). Recycling compartments and the internal vesicles of multivesicular bodies harbor most of the cholesterol found in the endocytic pathway. *Traffic*, *4*, 222–231.

Morton, C. J., Sani, M. A., Parker, M. W., & Separovic, F. (2019). Cholesterol-dependent cytolysins: Membrane and protein structural requirements for pore formation. *Chemical Reviews*, *119*, 7721–7736.

Mound, A., Lozanova, V., Warnon, C., Hermant, M., Robic, J., Guere, C., et al. (2017). Non-senescent keratinocytes organize in plasma membrane submicrometric lipid domains enriched in sphingomyelin and involved in re-epithelialization. *Biochimica et Biophysica Acta*, *1862*, 958–971.

Murate, M., Abe, M., Kasahara, K., Iwabuchi, K., Umeda, M., & Kobayashi, T. (2015). Transbilayer distribution of lipids at nano scale. *Journal of Cell Science*, *128*, 1627–1638.

Nelson, L. D., Johnson, A. E., & London, E. (2008). How interaction of perfringolysin O with membranes is controlled by sterol structure, lipid structure, and physiological low pH: Insights into the origin of perfringolysin O-lipid raft interaction. *The Journal of Biological Chemistry*, *283*, 4632–4642.

Ohno-Iwashita, Y., Shimada, Y., Waheed, A. A., Hayashi, M. Inomata, M., Nakamura, M., et al. (2004). Perfringolysin O, a cholesterol-binding cytolysin, as a probe for lipid rafts. *Anaerobe*, *10*, 125–134.

Sekizawa, Y., Kubo, T., Kobayashi, H., Nakajima, T., & Natori, S. (1997). Molecular cloning of cDNA for lysenin, a novel protein in the earthworm Eisenia foetida that causes contraction of rat vascular smooth muscle. *Gene*, *191*, 97–102.

Shibata, T., Kudou, M., Hoshi, Y., Kudo, A., Nanashima, N., & Miyairi, K. (2010). Isolation and characterization of a novel two-component hemolysin, erylysin a and B, from an edible mushroom, Pleurotus eryngii. *Toxicon*, *56*, 1436–1442.

Skocaj, M., Bakrac, B., Krizaj, I., Macek, P., Anderluh, G., & Sepcic, K. (2013). The sensing of membrane microdomains based on pore-forming toxins. *Current Medicinal Chemistry*, *20*, 491–501.

Skocaj, M., Resnik, N., Grundner, M., Ota, K., Rojko, N., Hodnik, V., et al. (2014). Tracking cholesterol/sphingomyelin-rich membrane domains with the ostreolysin A-mCherry protein. *PLoS One*, *9*, e92783.

Slot, J. W., & Geuze, H. J. (2007). Cryosectioning and immunolabeling. *Nature Protocols*, *2*, 2480–2491.

Tafesse, F. G., Ternes, P., & Holthuis, J. C. (2006). The multigenic sphingomyelin synthase family. *The Journal of Biological Chemistry*, *281*, 29421–29425.

Tomita, T., Noguchi, K., Mimuro, H., Ukaji, F., Ito, K., Sugawara-Tomita, N., et al. (2004). Pleurotolysin, a novel sphingomyelin-specific two-component cytolysin from edible mushroom Pleurotus ostreatus, assembles into a transmembrane pore complex. *The Journal of Biological Chemistry*, *279*, 26975–26982.

Waheed, A. A., Shimada, Y., Heijnen, H. F., Nakamura, M., Inomata, M., Hayashi, M., et al. (2001). Selective binding of perfringolysin O derivative to cholesterol-rich membrane microdomains (rafts). *Proceedings of the National Academy of Sciences of the United States of America*, *98*, 4926–4931.

Wang, T., Shogomori, H., Hara, M., Yamada, T., & Kobayashi, T. (2012). Nanomechanical recognition of sphingomyelin-rich membrane domains by atomic force microscopy. *Biochemistry*, *51*, 74–82.

Wilhelm, L. P., Voilquin, L., Kobayashi, T., Tomasetto, C., & Alpy, F. (2019). Intracellular and plasma membrane cholesterol labeling and quantification using filipin and GFP-D4. *Methods in Molecular Biology*, *1949*, 137–152.

Yachi, R., Uchida, Y., Balakrishna, B. H., Anderluh, G., Kobayashi, T., Taguchi, T., et al. (2012). Subcellular localization of sphingomyelin revealed by two toxin-based probes in mammalian cells. *Genes to Cells*, *17*, 720–727.

Yamaji, A., Sekizawa, Y., Emoto, K., Sakuraba, H., Inoue, K., Kobayashi, H., et al. (1998). Lysenin, a novel sphingomyelin-specific binding protein. *The Journal of Biological Chemistry*, *273*, 5300–5306.

Yamaji-Hasegawa, A., Hullin-Matsuda, F., Greimel, P., & Kobayashi, T. (2016). Pore-forming toxins: Properties, diversity, and uses as tools to image sphingomyelin and ceramide phosphoethanolamine. *Biochimica et Biophysica Acta*, *1858*, 576–592.
Yamaji-Hasegawa, A., Makino, A., Baba, T., Senoh, Y., Kimura-Suda, H., Sato, S. B., et al. (2003). Oligomerization and pore formation of a sphingomyelin-specific toxin, lysenin. *The Journal of Biological Chemistry*, *278*, 22762–22770.
Yilmaz, N., & Kobayashi, T. (2015). Visualization of lipid membrane reorganization induced by a pore-forming toxin using high-speed atomic force microscopy. *ACS Nano*, *9*, 7960–7967.
Yilmaz, N., Yamada, T., Greimel, P., Uchihashi, T., Ando, T., & Kobayashi, T. (2013). Real time visualization of assembling of a sphingomyelin-specific toxin on planar lipid membranes. *Biophysical Journal*, *105*, 1397–1405.
Yoshida, Y., Yoneda, K., Umeda, M., Ide, C., & Fujimoto, K. (2001). Localization of sphingomyelin during the development of dorsal and tail epidermis of mice. *The British Journal of Dermatology*, *145*, 758–770.

CHAPTER EIGHTEEN

The use of anthrolysin O and ostreolysin A to study cholesterol in cell membranes

Kristen A. Johnson and Arun Radhakrishnan*
Department of Molecular Genetics, University of Texas Southwestern Medical Center, Dallas, TX, United States
*Corresponding author: e-mail address: arun.radhakrishnan@utsouthwestern.edu

Contents

Abstract

Cholesterol is a major component of the plasma membranes (PMs) of animal cells, comprising 35–40 mol% of total PM lipids. Recent studies using cholesterol-binding bacterial toxins such as domain 4 of Anthrolysin O (ALOD4) and fungal toxins such as Ostreolysin A (OlyA) have revealed new insights into the organization of PM cholesterol. These studies have defined three distinct pools of PM cholesterol—a fixed pool that is essential for membrane integrity, a sphingomyelin (SM)-sequestered pool that can be detected by OlyA, and a third pool that is accessible and can be detected by ALOD4. Accessible cholesterol is available to interact with proteins and transport to the endoplasmic reticulum

Methods in Enzymology, Volume 649
ISSN 0076-6879
https://doi.org/10.1016/bs.mie.2021.01.011

(ER), and controls many cellular signaling processes including cholesterol homeostasis, Hedgehog signaling, and bacterial and viral infection. Here, we provide detailed descriptions for the use of ALOD4 and OlyA, both of which are soluble and non-lytic proteins, to study cholesterol organization in the PMs of animal cells. Furthermore, we describe two new versions of ALOD4 that we have developed to increase the versatility of this probe in cellular studies. One is a dual His_6 and FLAG epitope-tagged version and the other is a fluorescent version where ALOD4 is fused to Neon, a monomeric fluorescent protein. These new forms of ALOD4 together with previously described OlyA provide an expanded collection of tools to sense, visualize, and modulate levels of accessible and SM-sequestered cholesterol on PMs and study the role of these cholesterol pools in diverse membrane signaling events.

1. Introduction

Cholesterol is an essential and abundant component of animal cell membranes. The concentration of this lipid varies by as much as 10-fold in membranes of various cellular organelles. In the plasma membrane (PM), cholesterol comprises 35–40 mol% of total PM lipids whereas in the endoplasmic reticulum (ER) membrane, cholesterol comprises only ~5 mol % of total ER lipids (Das, Goldstein, Anderson, Brown, & Radhakrishnan, 2013; Radhakrishnan, Goldstein, McDonald, & Brown, 2008). In contrast to eukaryotes, cholesterol is extremely rare in prokaryotic organisms (Ourisson, Rohmer, & Poralla, 1987). Several Gram-positive bacteria exploit this difference to target eukaryotic membranes with soluble toxin proteins known as cholesterol dependent cytolysins (CDCs), which bind to eukaryotic cholesterol-containing membranes and form pores (Heuck, Moe, & Johnson, 2010; Ohno-Iwashita et al., 2010; Tweten, 2005). Bacteria are not alone in their lipid-dependent, pore-forming toxin strategy. Other organisms, including mushrooms, worms, and sea anemones, target the sphingomyelin (SM) lipid component of eukaryotic cell membranes (Anderluh & Macek, 2002; Bernheimer & Avigad, 1979; Bhat et al., 2013; Makino et al., 2017; Skocaj et al., 2014; Tomita et al., 2004; Yamaji et al., 1998). In some cases, special combinations of SM and cholesterol are targeted (Endapally, Frias, et al., 2019; Makino et al., 2017; Skocaj et al., 2014).

These diverse toxins use at least two different mechanisms to form pores. Some pore forming toxins contain both lipid binding and pore formation activities in a single protein. Examples of such toxins are Perfringolysin O (PFO) from *Clostridium perfringens* or Anthrolysin O (ALO) from *Bacillus anthracis* which bind to accessible cholesterol with one region of the protein

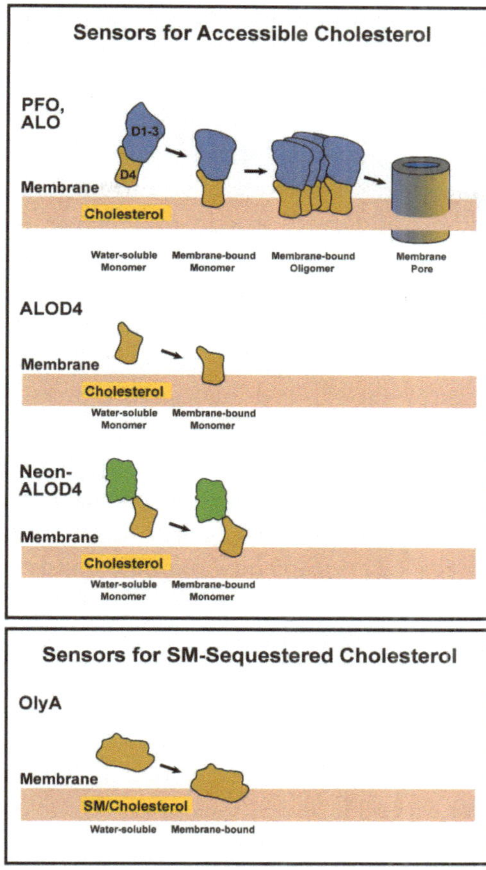

Fig. 1 Schematic representation of sensors for accessible cholesterol and SM-sequestered cholesterol that are described in this study.

(Domain 4, D4) and use a different region of the protein (Domains 1–3, D1–3) to homo-oligomerize and form a transmembrane pore (see Fig. 1) (Bourdeau et al., 2009; Gay, Rye, & Radhakrishnan, 2015; Heuck et al., 2010). Other pore forming toxins require activities of two proteins—one to bind to membrane lipids and another to drive oligomerization and pore formation. An example of such a two-protein system is provided by *Pleurotus ostreatus*. This fungus produces two proteins, Ostreolysin A (OlyA) and Pleurotolysin B (PlyB). OlyA binds to SM-sequestered cholesterol in membranes (see Fig. 1) and PlyB binds to the membrane-bound OlyA to promote oligomerization and pore formation (Endapally, Frias, et al., 2019; Lukoyanova et al., 2015; Skocaj et al., 2014).

In this chapter, we will review the uses of ALO and OlyA, both of which are soluble proteins, to detect two distinct pools of cholesterol—accessible and SM-sequestered—in the PMs of animal cells. Assessing the PM without interference from membranes of other cellular organelles requires elimination of the pore-forming properties of these toxins. Fortunately, after significant protein engineering, we generated a soluble form of domain 4 of ALO (ALOD4) that does not form pores in membranes (Gay et al., 2015; Infante & Radhakrishnan, 2017) (Fig. 1). In the case of OlyA, no protein engineering was required since OlyA is non-lytic on its own as described above. These non-lytic tools have been used to resolve how cholesterol is organized in PM (He et al., 2017; Johnson & Radhakrishnan, 2020), how cholesterol organization in PM influences its accessibility for intracellular transport and controls cholesterol synthesis (Das, Brown, Anderson, Goldstein, & Radhakrishnan, 2014; Infante & Radhakrishnan, 2017; Johnson, Endapally, Vazquez, Infante, & Radhakrishnan, 2019), how accessibility of cholesterol controls Hedgehog signaling at the ciliary membrane (Kinnebrew et al., 2019), and how immune signals mobilize accessible cholesterol away from the PM during bacterial and viral infection (Abrams et al., 2020; Wang et al., 2020; Zhou et al., 2020). To expand the versatility and potential uses of ALOD4, we have developed two new versions of this probe, a dual epitope-tagged version (His_6-FLAG-ALOD4) and a fluorescent version (His_6-Neon-FLAG-ALOD4) (Fig. 1). Here, we describe how to purify these new versions of ALOD4 (Section 2), how to use ALOD4 and OlyA as sensors for accessible and SM-sequestered cholesterol on live cells (Section 3), how to visualize accessible cholesterol on PM with His_6-Neon-Flag-ALOD4 (Section 4), and how to assay for ALOD4 inhibition of accessible cholesterol transport from PM to ER (Section 5).

2. Purification of recombinant cholesterol-binding toxin proteins

Procedures for the overexpression and purification of His_6-tagged versions of ALOD4 and OlyA have been described in detail (Endapally, Frias, et al., 2019; Endapally, Infante, & Radhakrishnan, 2019). Recently, our lab has generated additional ALOD4 constructs, His_6-Flag-ALOD4, His_6-Flag-ALOD4 (Mut) (GTTLYP → AAAAA, non-binding mutant), His_6-Neon-Flag-ALOD4, and His_6-Neon-Flag-ALOD4 (Mut) (GTTLYP → AAAAA, non-binding mutant). His$_6$-Flag-ALOD4 and His$_6$-Flag-ALOD4 (Mut) were overexpressed and purified using a previously described protocol (Endapally, Infante, et al., 2019). Unfortunately, when we used this same protocol to purify His$_6$-Neon-Flag-ALOD4, the resultant purified protein was prone to degradation. Therefore, we developed an alternate purification scheme which is described below.

2.1 Materials

1. Expression plasmids
   - **a.** pHis$_6$-ALOD4, in the pRSET B expression vector, encodes, in sequential order from the NH$_2$-terminus, a 10-aa linker that includes a His$_6$ epitope tag (MRGSHHHHHH), a 23 aa-linker that includes the epitope for the anti-Xpress monoclonal antibody and an enterokinase cleavage site (GMASMTGGQQMGRDLYDDDDKDP), followed by domain 4 of ALO (aa 404–512). The sole native cysteine in ALOD4 (C472) was mutated to alanine and S404 was mutated to cysteine to allow for maleimide labeling at a position far from the cholesterol binding site in ALOD4. This plasmid is available from Addgene (Cat # 111026) (Gay et al., 2015).
   - **b.** pHis$_6$-Flag-ALOD4, also in the pRSET B expression vector, encodes, in sequential order from the NH$_2$-terminus, a 10-aa linker that includes a His$_6$ epitope tag (MRGSHHHHHH), a 23 aa-linker that includes a FLAG epitope tag (GMASMTGGQQMGR-DYKDDDDKDP), and domain 4 of ALO (aa 404–512). The ALOD4 portion of the plasmid contains the same two mutations (S404C and C472A) described above for pHis$_6$-ALOD4.
   - **c.** pHis$_6$-Flag-ALOD4 (Mut) is the same as pHis$_6$-Flag-ALOD4 described above, except that it contains six mutations (G501A, T502A, T503A, L504A, Y505A, and P506A) that disrupt cholesterol binding.
   - **d.** pHis$_6$-Neon-Flag-ALOD4, in the pRSET B expression vector, encodes, in sequential order from the NH$_2$-terminus, a 13-aa linker that includes a His$_6$ epitope tag (MRGSHHHHHHGMA), the monomeric fluorescent protein Neon (aa 1–236) (Shaner et al., 2013), a 21 aa-linker that includes a FLAG epitope tag (ASMTGG-QQMGRDYKDDDDKDP), and domain 4 of ALO (aa 404–512). The ALOD4 portion of the plasmid contains the same two mutations (S404C and C472A) described above for pHis$_6$-ALOD4.

   e. pHis$_6$-Neon-Flag-ALOD4 (Mut) is the same as pHis$_6$-Neon-Flag-ALOD4 described above, except that it contains six mutations (G501A, T502A, T503A, L504A, Y505A, and P506A) that disrupt cholesterol binding.
   f. pOlyA-His$_6$, in the pET21C+ expression vector, encodes, in sequential order from the NH$_2$-terminus, OlyA (aa 1–138) and a 21-aa linker that includes a His$_6$ epitope tag (LVPRGSEGKSG-CGLEHHHHHH). The two native cysteines in OlyA (C62 and C94) was mutated to serine and a new cysteine was introduced near the COOH-terminus (S151C) to allow for maleimide labeling at a position far from the binding site in OlyA for sphingomyelin/cholesterol complexes. This plasmid is available from Addgene (Cat # 132391) (Endapally, Frias, et al., 2019).
2. BL21 (DE3) pLysS *Escherichia coli* competent cells, Invitrogen, Cat # C6060
3. Luria Broth (LB) capsules, Research Products International, Cat # L24041
4. Ampicillin, 1000 × (100 mg/mL) stock solution, powder available from RPI Cat # A40040
5. Isopropyl β-D-thiogalactoside (IPTG), 1000 × (1 M) stock solution in water, powder available from Goldbio, Cat # I2481C300
6. Buffer A, 50 mM Tris–HCl (pH 7.5), 150 mM NaCl, and 1 mM tris(2-carboxyethyl)phosphine (TCEP)
7. Buffer B, buffer A supplemented with 500 mM Imidazole (pH 7.5)
8. Phenyl methanesulfonyl fluoride (PMSF), 1000 × (20 mg/mL) stock solution in ethanol, powder available from Sigma-Aldrich, Cat # P7626
9. Protease inhibitor tablets, use 1 per 20 mL buffer, Thermo Fisher scientific, Cat # 88266
10. Lysozyme from chicken egg white 10 × (10 mg/mL) stock in water, Sigma-Aldrich, Cat # L6876
11. 100-mL all glass Dounce tissue grinder, Kimble though Fisher Scientific, Cat # K885303
12. Branson Digital Sonifier, Fisher Scientific
13. Ni-NTA agarose beads, Qiagen, Cat # 30250
14. Econo-column chromatography column, 1.5 × 20 cm, BioRad, Cat # 7371522
15. Amicon Ultra-4 10-kDa cutoff centrifugal filters, Millipore, Cat # UFC801024

16. Column for size-exclusion chromatography (SEC) such as Superdex 200 column, GE Healthcare, Cat # 28990944
17. Fast protein liquid chromatography (FPLC) system for SEC such as AKTA purifier, GE Healthcare
18. 15% sodium dodecyl sulfate polyacrylamide gel electrophoresis (SDS/PAGE) gels
19. Quick Coomassie stain, Anatrace, Cat # GEN-QC-STAIN
20. 5 × SDS loading buffer, 10% (*w/v*) SDS, 10 mM β-mercaptoethanol, 20% (*v*/v) glycerol, 0.2 M Tris–HCl, pH 6.8, 0.05% (w/v) bromophenol blue
21. Spectrophotometer such as the NanoDrop, Thermo Fisher Scientific, to determine protein concentration
22. Imager to visualize fluorescent protein bands on SDS/PAGE, such as the LICOR Odyssey, LICOR

2.2 Protocols

2.2.1 Expression protocol for His_6-Neon-Flag-ALOD4

1. Transform His_6-Neon-Flag-ALOD4 plasmid into BL21 (DE3) pLysS *Escherichia coli* competent cells according to the manufacturers protocol and plate transformed cells on LB-Agar plates with ampicillin (100 μg/mL).
2. After incubation of plates for 16 h at 37 °C, pick a single colony and inoculate 160 mL LB containing 100 μg/mL ampicillin. Incubate this starter culture in a shaker-incubator (250 rpm) for 12–16 h at 30 °C or until OD_{600} is between 0.8 and 1.0.
3. Inoculate 1 L LB containing 100 μg/mL ampicillin with 20 mL of the starter culture. Incubate at 37 °C in a shaker-incubator (250 rpm) until OD_{600} is 0.4–0.6. Reduce shaker temperature to 18 °C, add IPTG to a final concentration of 1 mM to induce protein expression and continue shaking at 18 °C for 18–24 h. Harvest the cells by centrifugation at 3220 × *g* for 10 min at 4 °C, and immediately use cell pellet for protein purification or flash-freeze bacterial pellets in liquid nitrogen and store at −80 °C for up to 6 months.

2.2.2 Purification protocol for His_6-Neon-Flag-ALOD4

1. Resuspend cell pellet from 1 L of bacterial culture in 20 mL buffer A supplemented with PMSF, 1 protease inhibitor tablet, and lysozyme (1 mg/mL).

2. Homogenize the cell suspension using 12 strokes in a Dounce homogenizer, incubate at 4 °C while rotating for 2 h, then repeat Dounce homogenization step.
3. Further lyse the lysozyme-disrupted cells using a tip sonicator: 40% amplitude, 3 min cycle (repeating sequences of 3 s on, 3 s off) followed by 6 min cool down; repeat cycle two additional times.
4. Subject cell lysate to centrifugation (4 °C, 30 min, 25,000 × *g*) to eliminate insoluble cell debris.
5. Incubate supernatant with 10 mL buffer A-equilibrated Ni-NTA beads on an orbital shaker at 4 °C for 30 min.
6. Pour protein-Ni-NTA mixture into an empty chromatography column.
7. After beads have settled, wash column with 50 mL buffer A, then with 50 mL buffer A containing 25 mM imidazole and protease inhibitor (5% buffer B), and finally with 50 mL buffer A containing 50 mM imidazole and protease inhibitor (10% buffer B).
8. Elute bound protein from column with 20 mL buffer A containing 250 mM imidazole and protease inhibitor (50% buffer B).
9. Concentrate eluate to 4 mL with Amicon Ultra-4 10-kDa cutoff centrifugal filters (3220 × *g*, 4 °C).
10. Subject concentrated eluate to centrifugation (4 °C, 30 min, 20,000 × *g*) to remove aggregate protein (if any) to prepare for SEC.
11. Inject 2 mL at a time onto Superdex 200 column equilibrated in buffer A connected to a FPLC system such as AKTA purifier. Collect protein-rich fractions and determine protein concentration with a bicinchoninic acid (BCA) assay kit or Spectrometer such as a NanoDrop according to manufacturer instructions.
12. See SEC profiles in Fig. 2A for examples of the monodisperse nature of His_6-Neon-Flag-ALOD4 (*dashed line*) purified as described above and His_6-Flag-ALOD4 (*solid line*) purified as described previously (Endapally, Infante, et al., 2019). Maximum values of absorbance at 280 nm for each protein were normalized to 1.
13. An aliquot (2 μg) of either His_6-Flag-ALOD4 (Fig. 2B, *lane 1*) or His_6-Neon-Flag-ALOD4 (Fig. 2B, *lane 2*), was subjected to 15% SDS/PAGE and stained with Coomassie (*left panel*) or imaged with the 600 nm filter on a Licor Imager to detect Neon fluorescence (*right panel*). Coomassie staining shows that purified His_6-Flag-ALOD4 migrates as a single band whereas purified His_6-Neon-Flag-ALOD4 migrates as a doublet. Only one of the two Coomassie-stained bands observed for His_6-Neon-Flag-ALOD4 showed a fluorescent signal. We did not conduct further analysis of the second Coomassie-stained band.

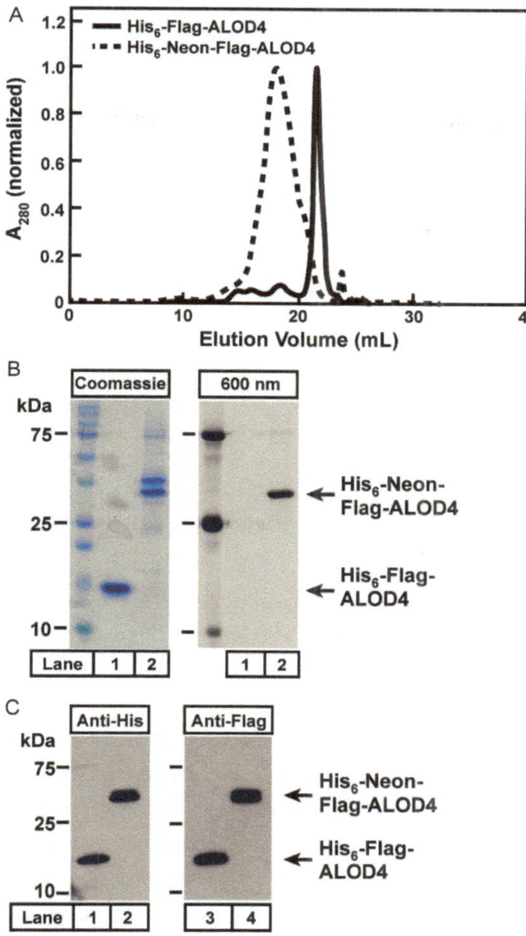

Fig. 2 Purification of recombinant cholesterol binding toxins. (A) Representative Superdex 200 SEC profiles of purified recombinant His_6-Flag-ALOD4 (*solid line*) and His_6-Neon-Flag-ALOD4 (*dashed line*). The maximum values for the absorbance at 280 nm for each protein are normalized to 1 to allow for comparison of the two profiles. (B) 2 µg of purified His_6-Flag-ALOD4 (*lane 1*) or His_6-Neon-Flag-ALOD4 (*lane 2*) were subjected to 15% SDS PAGE, followed by Coomassie staining (*left panel*) or fluorescence imaging with a 600 nm filter on a LICOR imager (*right panel*). (C) 20 ng of purified His_6-Flag-ALOD4 (*lane 1*) and His_6-Neon-Flag-ALOD4 (*lane 2*) were subjected to 15% SDS PAGE, followed by immunoblot analysis with anti-His (*left panel*) or anti-Flag (*right panel*) antibodies.

14. An aliquot (20 ng) of either His_6-Flag-ALOD4 (Fig. 2C, *lanes 1 and 3*) or His_6-Neon-Flag-ALOD4 (Fig. 2C, *lanes 2 and 4*), was subjected to 15% SDS/PAGE followed by immunoblot analysis with an anti-His (Fig. 2C, left panel) or anti-Flag (Fig. 2C, right panel) antibody. The immunoblot procedure used here is described in Section 3.3.3.

15. Store purified proteins for up to 2 months at 4 °C in buffer A or 6 months at −80 °C in buffer A supplemented with 20% (*v*/v) glycerol.
    a. His_6-Neon-Flag-ALOD4 is prone to degradation. Storage at −80 °C in buffer A supplemented with 20% (v/v) glycerol is recommended. Thaw protein aliquots on ice before use.
    b. See Section 5 for the SREBP2 processing assay, which is performed with every batch of purified ALOD4 proteins to determine protein activity.

3. Using toxins as cholesterol sensors on live cells

3.1 Background

Most of the total cellular cholesterol is localized in the PM. Within the PM, cholesterol is organized into three distinct pools (Das et al., 2014). Two of these pools of cholesterol are inaccessible due to (i) interactions with SM, or (ii) interactions with other membrane factors including glycerophospholipids containing phosphatidylcholine headgroups (Moe & Heuck, 2010). PM cholesterol in excess of these two pools constitutes a third pool that is accessible and can be transported to the ER to shut down cholesterol synthesis and uptake (see Section 5). In this section, methods to detect changes in accessible cholesterol with ALOD4 and SM-sequestered cholesterol with OlyA are described (Fig. 3). Degradation of SM in the PM using a bacterial sphingomyelinase (SMase) liberates cholesterol from the SM-sequestered pool, leading to an increase in accessible cholesterol levels as detected by ALOD4 and a decrease in SM-sequestered cholesterol levels as detected by OlyA (Fig. 3A). Accessible cholesterol can be depleted from PMs using low concentrations of hydroxypropyl-beta cyclodextrin (HPCD), SM-sequestered cholesterol can also be depleted, but require higher HPCD concentrations (Fig. 3A).

3.2 Materials

1. Purified ALOD4 and OlyA proteins (see Section 2 and Endapally, Frias, et al., 2019; Endapally, Infante, et al., 2019). Similar results are obtained with all versions of ALOD4 described in Section 2.1, list 1 (also see Section 5).
2. Medium A, 1:1 mixture of Dulbecco's modified Eagle's medium (DMEM) and Ham's F12 medium supplemented with 100 units/mL

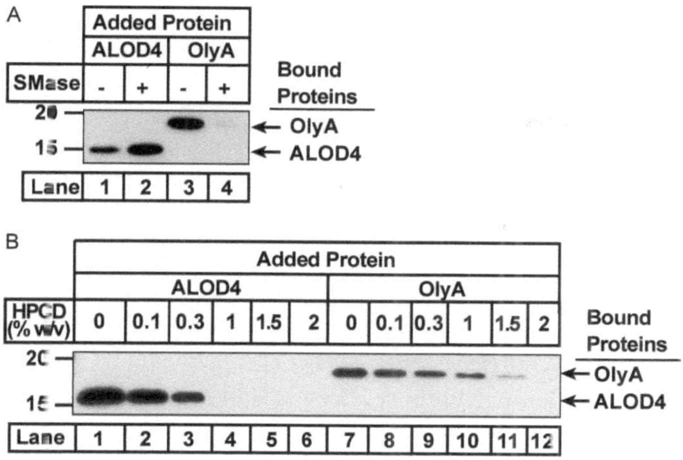

Fig. 3 ALOD4 and OlyA detect different cholesterol pools in PM. On day 0, CHOK1 were set up in medium A at a density of 6×10^4 cells per well of a 48 well plate. (A) On day 1, cells were washed twice with 500 μL PBS, then treated with 200 μL of medium A (without SMase) or sphingomyelin-depleting medium B (with SMase). After incubation for 30 min at 37 °C, media was removed, cells were washed twice with PBS, then treated with 200 μL medium A containing either 3 μM His_6-ALOD4 (*lanes 1, 2*) or 3 μM OlyA-His_6 (*lanes 3, 4*). (B) On day 1, cells were washed twice with 500 μL PBS then treated with either 200 μL medium A or medium A supplemented with cholesterol-depleting medium C to produce the indicated concentration of HPCD. After incubation for 1 h at 37 °C, media was removed, cells were washed twice with 500 μL PBS then treated with 200 μL medium A containing either 3 μM His_6-ALOD4 (*lanes 1–6*) or 3 μM OlyA-His_6 (*lanes 7–12*). (A and B) After incubation with ALOD4 or OlyA for 1 h at 37 °C, cells were washed twice with 500 μL PBS, then harvested in 200 μL buffer C. Equal aliquots of each cell lysate (10% of total) was subjected to immunoblot analysis as described in Section 3.3.

penicillin, 100 μg/mL streptomycin sulfate (1% (*v/v*) P/S), and 5% (v/v) fetal calf serum (FCS). Warm to 37 °C prior to use.

3. Medium B, sphingomyelin-depleting medium, medium A supplemented with 100 mU/mL SMase (SMase, Sigma, Cat # S8633). Warm to 37 °C prior to use.
4. Medium C, cholesterol-depleting medium, medium A supplemented with 2% (*w/v*) HPCD. Warm to 37 °C prior to use.
5. Dulbecco's phosphate-buffered saline (PBS)
6. Buffer C, 10 mM Tris–HCl (pH 6.8), 100 mM NaCl, 1% (w/v) SDS, 1 mM EDTA, 1 mM EGTA, 20 μg/mL PMSF and protease inhibitors (1 tablet per 20 mL; Roche). Warm to room temperature (RT, 20–22 °C) before use.

7. 5 × SDS loading buffer, 10% (w/v) SDS, 10 mM β-mercaptoethanol, 20% (v/v) glycerol, 0.2 M Tris–HCl, pH 6.8, 0.05% (w/v) bromophenol blue
8. 15% SDS/PAGE gels
9. 1 × SDS/PAGE running buffer, Tris/Glycine/SDS buffer (TGS), BioRad Cat #1610772
10. 1 × Transfer buffer (4 °C) from Trans-Blot Turbo 5 × transfer buffer, BioRad, Cat #10026938
11. Phosphate buffered saline (pH 7.4) supplemented with 0.05% (w/v) TWEEN 20 (PBST), Millipore Sigma, Cat #P3563
12. 5% (w/v) milk in PBST
13. Anti-His antibody, Millipore, cat # 05-949, use 1:1000 dilution
14. Anti-mouse IgG, JacksonResearch, cat # 715-035-150, use 1:5000 dilution
15. SuperSignal West Pico PLUS Chemiluminescent Substrate, Thermo-Fisher Scientific, cat # 34580

3.3 Protocols

3.3.1 Sphingomyelin depletion protocol

1. On day 0, CHOK1 cells were set up in medium A at a density of 6×10^4 cells per well of a 48 well plate.
2. On day 1, media was removed, cells were washed twice with 500 μL PBS, then treated with 200 μL of either medium A (without SMase) or sphingomyelin-depleting medium B (with SMase).
3. After incubation for 30 min at 37 °C, cells were washed twice with PBS, then treated with 200 μL medium A containing either 3 μM ALOD4 or 3 μM OlyA.
4. After incubation for 1 h at 37 °C, cells were washed twice with 500 μL PBS, then harvested with 200 μL buffer C. Cell lysates were mixed with SDS loading buffer, 10% of each cell lysate was subjected to 15% SDS/PAGE followed by immunoblot analysis as described below.
5. Fig. 3A shows that SMase treatment of CHOK1 cells increases PM binding of ALOD4 (*lanes 1, 2*) while decreasing the PM binding of OlyA (*lanes 3, 4*).

3.3.2 Cholesterol depletion protocol

1. On day 0, CHOK1 cells were set up in medium A at a density of 6×10^4 cells per well of a 48-well plate.

2. On day 1, media was removed, cells were washed twice with 500 μL PBS, then treated for 1 h with either 200 μL cholesterol-depleting medium C or a mixture of medium A and C to achieve the indicated concentration of HPCD (0%, 0.1%, 0.3%, 1%, 1.5%, or 2% (w/v)). This HPCD treatment did not result in any significant cell death, as judged by normal cell morphology and the lack of any significant rounding up or loss of cells from the plate.
3. After incubation for 1 h at 37 °C as described above, cells were washed twice with 500 μL PBS and then treated with 200 μL medium A containing either 3 μM ALOD4 or 3 μM OlyA.
4. After incubation for 1 h at 37 °C, cells were washed twice with PBS then lysed with 200 μL buffer C. Cell lysates were mixed with SDS loading buffer, 10% of each cell lysate was subjected to 15% SDS/PAGE followed by immunoblot analysis as described below.
5. Fig. 3B shows that HPCD treatment of CHOK1 cells decreases PM binding of both ALOD4 (*lanes 1–6*) and OlyA (*lanes 7–12*), although with different sensitivities.

3.3.3 Immunoblot protocol

1. Proteins were transferred from electrophoresed gels onto nitrocellulose membranes using a BioRad Trans-Blot Turbo apparatus (2.5 A, 25 V for 10 min).
2. Membranes were blocked for 1 h at RT in 5% (w/v) milk in PBST.
3. To detect bound ALOD4 or OlyA, membranes were incubated with primary antibody (anti-His, 1:1000 dilution) in 5% (w/v) milk in PBST either overnight at 4 °C on an orbital shaker or for 1 h at room temperature on an orbital shaker.
4. Primary antibody was removed, and membranes were washed with PBST for 8 min. The wash step was repeated two additional times.
5. Membranes were then incubated with secondary antibody (anti-mouse, 1:5000 dilution) in 5% (w/v) milk in PBST.
6. After incubation for 30 min at RT, secondary antibody was removed and membranes were washed with PBST for 8 min, three times as described above.
7. Bound antibodies were visualized by chemiluminescence. The signal was developed by addition of SuperSignal Chemiluminescent substrate for 5 min. Membranes were then exposed to film for 1–30 s, a Western blot imager such as a LICOR could also be used.

4. Using toxins to visualize cholesterol accessibility on PM

4.1 Background

While the analysis described in Section 3 captures overall changes in cholesterol pools, recent work has illuminated differences in the distribution of accessible and SM-sequestered cholesterol within PM. When fluorescently-labeled ALOD4 and OlyA were added to live cells, the overall binding patterns were largely distinct but also showed some regions of overlap (Johnson et al., 2019; Johnson & Radhakrishnan, 2020). Over the course of a few minutes, some of the bound OlyA was internalized while ALOD4 remained bound to the surface. In contrast to SM-sequestered cholesterol, which may recycle through endocytic pathways, accessible cholesterol relies on transport proteins to move from PM to ER where it performs regulatory signaling duties. Cholesterol transport proteins such as GramD/AsterB are enriched at PM-ER contact sites and are required for accessible cholesterol transport from PM to ER (Naito et al., 2019; Sandhu et al., 2018). Visualization of accessible and SM-sequestered pools of cholesterol could also shed light on how these cholesterol pools regulate Hedgehog signaling (Kinnebrew et al., 2019) and allow for cell-to-cell spread of pathogenic bacteria such as *Listera monocytogenes* (Abrams, Johnson, Perelman, et al., 2020; Abrams, Johnson, Radhakrishnan, & Alto, 2020). In this section, we describe the use of a new fluorescent version of ALOD4, His_6-Neon-Flag-ALOD4, to visualize accessible cholesterol on PM of cells.

4.2 Materials

1. Fluorescent versions of ALOD4 where Neon, a monomeric fluorescent protein, is fused to ALOD4 (His_6-Neon-Flag-ALOD4) or where the lone cysteine of ALOD4 (His_6-ALOD4 or His_6-Flag-ALOD4) is labeled with Alexa Fluor or equivalent fluorescent maleimide reagents and non-fluorescent versions of ALOD4 (His_6-ALOD4 or His_6-Flag-ALOD4) (see Section 2 for details of protein constructs). Fluorescent versions of OlyA-His_6 can also be generated by labeling its lone cysteine with an Alexa Fluor or equivalent fluorescent maleimide reagent. The various fluorescent versions of ALOD4 and OlyA can be used individually or in combination.

2. Nunc Lab-Tek II chambered cover glass 8-well dish No. 1.5 glass (8-well imaging plate), Thermo Fisher Scientific, Cat # 12-565-338, or equivalent glass bottom dish suitable for imaging.
3. Medium A, 1:1 mixture of DMEM and Ham's F12 supplemented with 100 units/mL penicillin and 100 μg/mL streptomycin sulfate and 5% (*v*/v) fetal calf serum (FCS). Warm to 37 °C prior to use.
4. Medium B, medium A supplemented with 2% (*w*/*v*) HPCD. Warm to 37 °C prior to use.
5. Dulbecco's phosphate-buffered saline (PBS), Sigma, Cat # D8537
6. Buffer D, 4% paraformaldehyde solution in PBS, Alfa Aesar through Fisher Scientific, Cat # AAJ61899AK
7. Buffer E, 2% (w/v) bovine serum albumin (BSA) in PBS solution, sterile filtered through 0.22 μm filter before use. Lyophilized BSA is available from Sigma-Aldrich, Cat # A6003.
8. Anti-His antibody, Millipore, Cat # 05-949, use 1:500 dilution
9. Anti-Flag M2 antibody, Sigma-Aldrich, Cat # F1804, use 1:500 dilution
10. Goat anti-mouse-Alexa Fluor 594 secondary antibody, Thermo Fisher, Cat # A-21235, use 1:400 dilution
11. 4′,6-diamidino-2-phenylindole (DAPI), 5000 × (5 mg/mL) stock solution in water, treat with 1 μg/mL, Millipore Sigma, Cat # 5087410001
12. Microscope for visualization such as a Nikon Eclipse *Ti*, Nikon

4.3 Protocols

4.3.1 Protocol using His_6-Neon-Flag-ALOD4 to visualize accessible cholesterol on PM

1. On day 0, CHOK1 cells were set up in medium A at a density of 2×10^4 cells/well of an 8-well imaging plate.
2. On day 1, media was removed, cells were washed twice with 500 μL PBS, then treated with 200 μL lipoprotein-rich medium A or 200 μL cholesterol-depleting medium B.
3. After incubation for 1 h, cells were washed twice with 500 μL PBS, then treated with 200 μL medium A containing either 3 μM His_6-Neon-Flag-ALOD4 or 3 μM His_6-Neon-Flag-ALOD4 (Mut).
4. After incubation for 15 min, cells were washed twice with PBS and fixed by addition of 200 μL buffer D for 15 min.
5. After fixation, cells were washed 5 times with 500 μL PBS then treated with 500 μL buffer E on an orbital shaker.

6. After 30 min, blocking buffer was replaced with 500 μL buffer E containing 1 μg/mL DAPI.
7. After 1 h, cells were washed 5 times with PBS, postfixed with buffer D, then imaged to visualize bound His_6-Neon-Flag-ALOD4.
   a. Representative images are shown for His_6-Neon-Flag-ALOD4 binding to CHOK1 cells (Fig. 4A). When cells are incubated with lipoprotein-rich FCS, His_6-Neon-Flag-ALOD4 binds robustly to the PM of cells. When cellular cholesterol is depleted with HPCD, binding is lost. His_6-Neon-Flag-ALOD4 (Mut) does not bind to cells under either condition (Fig. 4A).
   b. Additional information on preparing and using Alexa Fluor labeled ALOD4 and OlyA is available in previous publications (Endapally, Infante, et al., 2019; Johnson et al., 2019; Kinnebrew et al., 2019).

4.3.2 Immunofluorescence protocol to visualize binding of fluorescent and non-fluorescent versions of ALOD4 to PMs

1. On day 0, CHOK1 cells were set up in medium A at a density of 2×10^4 cells/well of an 8-well imaging plate.
2. On day 1, media was removed, and cells were washed twice with 500 μL PBS, then treated with 200 μL medium A containing 3 μM of His_6-ALOD4, His_6-Flag-ALOD4, or His_6-Neon-Flag-ALOD4.
3. After incubation for 15 min, cells were washed twice with 500 μL PBS and fixed by addition of 200 μL buffer D for 15 min.
4. After fixation, cells were washed 5 times with 500 μL PBS then treated with 500 μL buffer E and placed on an orbital shaker (50 rpm).
5. After 30 min, buffer E was removed and replaced with 200 μL buffer E supplemented with either anti-His (1:1000 dilution) or anti-Flag (1:1000 dilution).
6. After incubation for 1 h, cells were washed by incubation with 500 μL PBS for 5 min. This wash step was repeated 5 additional times.
7. After washing, cells were treated with 200 μL buffer E supplemented with 1:400 goat anti-mouse-Alexa Fluor 647 secondary antibody and 1 μg/mL DAPI.
8. After incubation for 1 h, cells were washed by incubation with 500 μL PBS for 5 min. The wash step was repeated 5 additional times, and the cells were then post-fixed with 200 μL buffer D for 10 m, followed by 5 washes with PBS.
9. Cells were imaged to visualize bound ALOD4 as shown in Fig. 4B and C. When using immunofluorescence, it is important to include

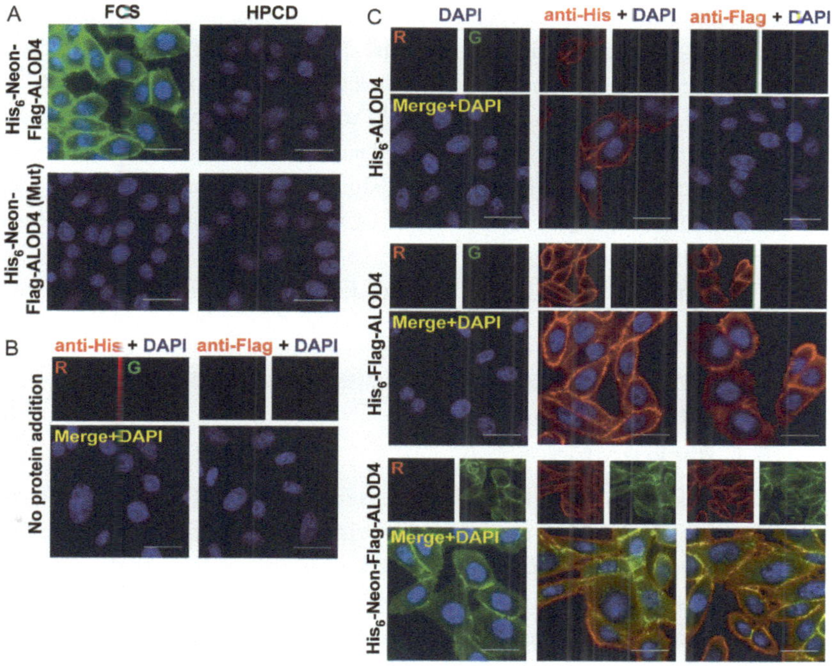

Fig. 4 Visualizing PM accessible cholesterol with fluorescent ALOD4. On day 0, CHOK1 cells were set up in medium A at a density of 2×10^4 cells per well of an 8-well glass bottom MatTeck imaging plate. (A) On day 1, cells were washed twice with 500 µL PBS then treated with 200 µL of either lipoprotein-rich medium A or cholesterol-depleting medium B. After 1 h at 37 °C, cells were washed twice with 500 µL PBS, then treated with 200 µL medium A supplemented with either 3 µM His_6-Neon-Flag-ALOD4 or His_6-Neon-Flag-ALOD4 (Mut). After 15 min at RT, cells were washed twice with 500 µL PBS, then fixed with 200 µL buffer D. After incubation for 15 min at RT, cells were washed five times with PBS and then treated with 500 µL blocking buffer E for 30 min. Cells were then treated with 500 µL buffer E containing 1 µg/mL DAPI. After incubation for 1 h at RT, cells were washed five times with PBS, post-fixed with buffer D, then imaged with a Nikon Eclipse microscope. Scale bars are 20 µm. (B and C) On day 1, cells were washed twice with 500 µL PBS then treated with 200 µL of either medium A (B) or medium A supplemented with 3 µM of either His_6-ALOD4, His_6-Flag-ALOD4, or His_6-Neon-Flag-ALOD4 (C). After incubation for 15 min at RT, cells were washed twice with 500 µL PBS, then fixed with 200 µL buffer D. After incubation for 15 min at RT, cells were washed five times with PBS, then treated with 500 µL buffer E. After incubation for 30 min at RT, buffer E was removed and replaced with 200 µL buffer E (C, *left panel*) or buffer E containing 1:1000 of either anti-His (B, *left panel*; C, *center panel*) or anti-Flag (B and C, *right panel*) antibodies. After incubation for 1 h at RT, cells were washed six times with 500 µL PBS (5 min per wash), then treated with 200 µL buffer E containing 1:400 goat anti-mouse-594 secondary antibody and 1 µg/mL DAPI. After incubation for 1 h at RT, cells were washed six times with 500 µL PBS (5 min per wash), then post-fixed with 200 µL buffer D for 10 min at RT, then washed five times with PBS. Cells were imaged with a Nikon Eclipse microscope. Scale bars are 20 µm.

untreated samples to determine background signal of the antibodies used. Fig. 4B shows CHOK1 cells in the absence of ALOD4 addition after the complete immunofluorescence protocol with either anti-His (*left panel*) or anti-Flag (*right panel*) to confirm specificity of signal observed. Fig. 4C shows Neon (*left panel*), anti-His (*middle panel*), or anti-Flag (*right panel*) fluorescence signal after treatment with 3 μM His_6-ALOD4 (*top row*), His_6-Flag-ALOD4 (*center row*), or His_6-Neon-Flag-ALOD4 (*bottom row*).

a. This protocol can also be used to detect OlyA-His_6 bound to cells.

5. Using toxins to study cholesterol sensing and transport

5.1 Background

Animal cells carefully regulate their cholesterol content. To enable such regulation, cells must measure cholesterol levels in their membranes. If the level is too low, cholesterol synthesis and uptake are initiated. If the level is too high, synthesis and uptake are terminated. These opposite outcomes are triggered by the actions of a sensor protein called Scap that resides in the ER membrane (Brown, Radhakrishnan, & Goldstein, 2018). Scap forms complexes with ER membrane-bound transcription factors called sterol regulatory element-binding proteins (SREBPs), which activate all the genes involved in synthesis of cholesterol and the LDL receptor which acquires exogenous cholesterol from LDL particles. Activation of SREBPs requires their transport by Scap to the Golgi where two proteases release their transcription factor domain that then enter the nucleus for target gene activation. When cholesterol in the ER rises and binds to Scap, it traps SREBPs in the ER, thus preventing proteolytic cleavage and nuclear entry. However, most of the cell's cholesterol is located in the PM (Das et al., 2014; Lange, Ye, & Steck, 2004). How does the Scap/SREBP regulatory machine in the ER detect changes in PM cholesterol? A solution to this problem lies in the organization of cholesterol in PMs. As described in Section 3, PM cholesterol is organized into three distinct pools (Das et al., 2014). Two of these pools of cholesterol are inaccessible due to interactions with SM or interactions with other membrane factors. PM cholesterol in excess of these two pools constitutes a third pool that is accessible and is transported to the ER to terminate SREBP activation.

While the cholesterol sensing machinery in ER has been well characterized, how the various cholesterol sensing and transport machineries work in concert to maintain cholesterol homeostasis remains unclear. ALOD4 and OlyA are useful tools to study the connection between cholesterol levels in PM and the response of the cholesterol sensing machinery in ER. When ALOD4 binds accessible cholesterol on the outer leaflet of PM, cholesterol transport from PM-to-ER is blocked and ER cholesterol drops below a threshold concentration of 5 mol% of total ER lipids, triggering activation of SREBP transcription factors even though cellular cholesterol has not been diminished (Infante & Radhakrishnan, 2017). In this section, we describe assays to confirm functionality of our new ALOD4 protein constructs by testing whether they trigger SREBP processing.

5.2 Materials

1. Purified ALOD4 proteins (see Section 2 and Endapally, Infante, et al., 2019)
2. Medium A, 1:1 mixture of DMEM and Ham's F12 supplemented with 100 units/mL penicillin and 100 μg/mL streptomycin sulfate (1% (v/v) P/S) and 5% (v/v) fetal calf serum (FCS). Warm to 37 °C prior to use.
3. Dulbecco's phosphate-buffered saline (PBS)
4. Buffer C, 10 mM Tris–HCl (pH 6.8), 100 mM NaCl, 1% (w/v) SDS, 1 mM EDTA, 1 mM EGTA, 20 μg/mL PMSF, and protease inhibitors (1 tablet (Roche) per 20 mL). Warm to RT before use.
5. 5 × SDS loading buffer, 10% (w/v) SDS, 10 mM β-mercaptoethanol, 20% (v/v) glycerol, 0.2 M Tris–HCl, pH 6.8, 0.05% (w/v) bromophenol blue
6. 15% SDS/PAGE gel and 10% SDS/PAGE gel
7. 1 × SDS/PAGE running buffer, Tris/Glycine/SDS buffer (TGS), BioRad Cat # 1610772
8. 1 × Transfer buffer (4 °C) from Trans-Blot Turbo 5 × transfer buffer, BioRad, Cat # 10026938
9. Phosphate buffered saline (pH 7.4) supplemented with 0.05% (w/v) TWEEN 20 (PBST), Millipore Sigma, Cat # P3563
10. 5% Milk (w/v) in PBST
11. Anti-His antibody, Millipore, Cat # 05-949, use 1:1000 dilution
12. Anti-Flag M2 antibody, Sigma-Aldrich, Cat # F1804, use 1:1000 dilution
13. Anti- β-actin to detect actin, Cell Signaling Technology, Cat# 4970S, 13E5, use 1:1000 dilution

14. IgG-7D4, antibody to detect SREBP2, one of the three isoforms of SREBP, commercially available from ATCC CRL-2198, use 10 μg/mL dilution
15. Anti-rabbit IgG, JacksonResearch, Cat # 111-035-003, use 1:5000 dilution
16. Anti-mouse IgG, JacksonResearch, Cat # 715-035-150, use 1:5000 dilution
17. SuperSignal West Pico PLUS Chemiluminescent Substrate, Thermo-Fisher Scientific, Cat # 34580

5.3 Protocol

1. On day 0, CHOK1 cells were set up in medium A at a density of 6×10^4 cells per well of a 48 well plate.
2. On day 1, cells were washed twice with PBS then treated with the indicated concentration of ALOD4 in medium A.
3. After incubation for 1 h at 37 °C, media was removed, cells were washed twice with 500 μL PBS, and harvested by addition of 200 μL buffer C. Cell lysates were mixed with SDS loading buffer, 10% of each cell lysate was subjected to 15% SDS/PAGE followed by immunoblot analysis described below.
4. 10% of each cell lysate sample was subjected to either 10% SDS/PAGE or 15% SDS/PAGE followed by immunoblot analysis for SREBP2 and bound ALOD4, respectively.
5. The immunoblot procedure described in Section 3.3.3 was followed.
   a. Use IgG 7D4 (10 μg/mL) to detect SREBP2, anti-His (1:1000 dilution) or anti-Flag M2 (1:1000 dilution) to detect bound ALOD4, and anti-β-Actin (1:1000 dilution) to detect actin.
6. Fig. 5 shows results of experiments assaying for the effects of various versions of ALOD4 on trapping accessible cholesterol on PMs and triggering SREBP2 processing. As shown in Fig. 5A, incubation with either His_6-ALOD4 (*lanes 1–4*) or His_6-Flag-ALOD4 (*lanes 5–8*) results in binding of these proteins to the PM and processing of SREBP2 to its activated nuclear form. Fig. 5B shows that His_6-Neon-Flag-ALOD4 also binds to the PM and triggers the processing of SREBP2 (*lanes 1–4*) whereas His_6-Neon-Flag-ALOD4 (Mut) shows no binding to PMs and does not trigger the processing of SREBP2 (*lanes 5–8*).

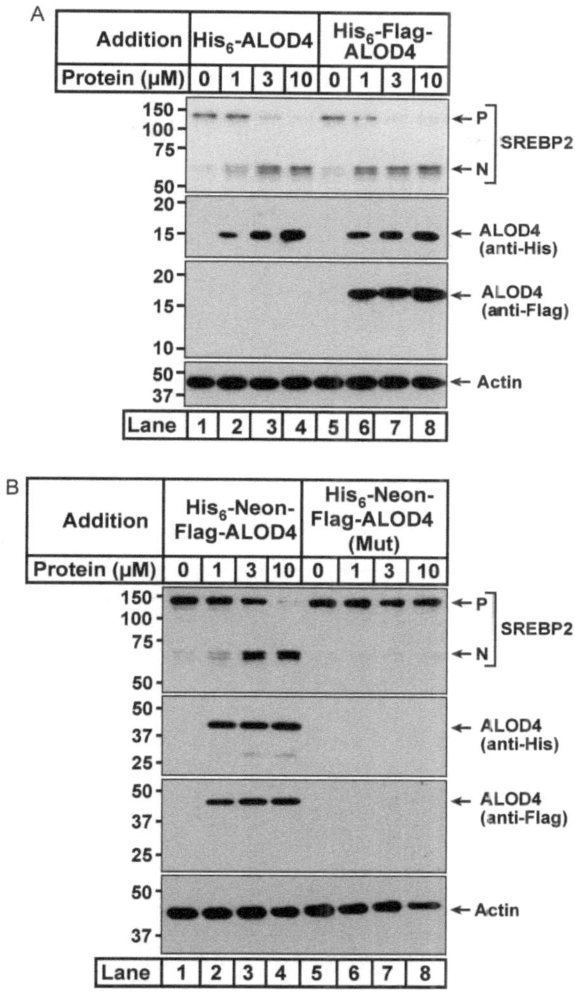

Fig. 5 Depletion of ER cholesterol by ALOD4 binding to accessible cholesterol in PM. On day 0, CHOK1 were set up in medium A at a density of 6×10^4 cells per well of a 48 well plate. On day 1, cells were washed twice with 500 µL PBS, then treated with 200 µL medium A supplemented with the indicated concentration of either His_6-ALOD4 or His_6-Flag-ALOD4 (A) or with either His_6-Neon-Flag-ALOD4 or His_6-Neon-Flag-ALOD4 (Mut) (B). After incubation for 1 h at 37 °C, cells were washed twice with 500 µL PBS, then harvested with 200 µL buffer C. 10% of each sample was subjected to immunoblot analysis as described in Section 5. *P*, precursor form of SREBP2; *N*, cleaved nuclear form of SREBP2.

6. Summary

Cholesterol dependent cytolysins such as ALOD4 and fungal toxins such as OlyA have been engineered to serve as laboratory tools to detect the level and distribution of cholesterol in animal membranes. These probes have allowed for the determination of three distinct pools of cholesterol in PMs and for elucidating the role of these pools in controlling various signaling processes in cells. While these tools are easy to work with and work robustly in a variety of animal cell types, caution must be used when interpreting results. For example, while ALOD4 and OlyA are excellent sensors for accessible and inaccessible cholesterol (see Section 3), these probes also rapidly inhibit transport of accessible cholesterol from PM to ER, triggering cholesterol uptake and synthesis pathways even though cellular cholesterol levels have not changed (see Section 5 and Infante & Radhakrishnan, 2017; Johnson et al., 2019). This feature of ALOD4 and OlyA binding to PMs can be useful in studying intracellular cholesterol transport; however, it should also not be ignored when using ALOD4 or OlyA as live-cell sensors.

Acknowledgments

We thank Danya Vazquez for excellent technical assistance. K.A.J. is the recipient of a postdoctoral fellowship from the Hartwell Foundation. We are grateful for support from the National Institutes of Health (HL20948), the Welch Foundation (I-1793), and the Leducq Foundation (19CVD04).

References

Abrams, M. E., Johnson, K. A., Perelman, S. S., Zhang, L. S., Endapally, S., Mar, K. B., et al. (2020). Oxysterols provide innate immunity to bacterial infection by mobilizing cell surface accessible cholesterol. *Nature Microbiology*, *5*(7), 929–942. https://doi.org/10.1038/s41564-020-0701-5.

Abrams, M. E., Johnson, K. A., Radhakrishnan, A., & Alto, N. M. (2020). Accessible cholesterol is localized in bacterial plasma membrane protrusions. *Journal of Lipid Research*, *61*, 1538. https://doi.org/10.1194/jlr.ILR120000891.

Anderluh, G., & Macek, P. (2002). Cytolytic peptide and protein toxins from sea anemones (Anthozoa: Actiniaria). *Toxicon*, *40*(2), 111–124.

Bernheimer, A. W., & Avigad, L. S. (1979). A cytolytic protein from the edible mushroom, Pleurotus ostreatus. *Biochimica et Biophysica Acta*, *585*(3), 451–461.

Bhat, H. B., Kishimoto, T., Abe, M., Makino, A., Inaba, T., Murate, M., et al. (2013). Binding of a pleurotolysin ortholog from Pleurotus eryngii to sphingomyelin and cholesterol-rich membrane domains. *Journal of Lipid Research*, *54*(10), 2933–2943. https://doi.org/10.1194/jlr.D041731.

Bourdeau, R. W., Malito, E., Chenal, A., Bishop, B. L., Musch, M. W., Villereal, M. L., et al. (2009). Cellular functions and X-ray structure of anthrolysin O, a cholesterol-dependent cytolysin secreted by bacillus anthracis. *The Journal of Biological Chemistry*, *284*(21), 14645–14656. https://doi.org/10.1074/jbc.M807631200.

Brown, M. S., Radhakrishnan, A., & Goldstein, J. L. (2018). Retrospective on cholesterol homeostasis: The central role of Scap. *Annual Review of Biochemistry*, *87*, 783–807. https://doi.org/10.1146/annurev-biochem-062917-011852.

Das, A., Brown, M. S., Anderson, D. D., Goldstein, J. L., & Radhakrishnan, A. (2014). Three pools of plasma membrane cholesterol and their relation to cholesterol homeostasis. *Elife (Cambridge)*, *3*, e02882. https://doi.org/10.7554/eLife.02882.

Das, A., Goldstein, J. L., Anderson, D. D., Brown, M. S., & Radhakrishnan, A. (2013). Use of mutant 125I-perfringolysin O to probe transport and organization of cholesterol in membranes of animal cells. *Proceedings of the National Academy of Sciences of the United States of America*, *110*(26), 10580–10585. https://doi.org/10.1073/pnas.1309273110.

Endapally, S., Frias, D., Grzemska, M., Gay, A., Tomchick, D. R., & Radhakrishnan, A. (2019). Molecular discrimination between two conformations of Sphingomyelin in plasma membranes. *Cell*, *176*(5), 1040–1053.e1017. https://doi.org/10.1016/j.cell.2018.12.042.

Endapally, S., Infante, R. E., & Radhakrishnan, A. (2019). Monitoring and modulating intracellular cholesterol trafficking using ALOD4, a cholesterol-binding protein. *Methods in Molecular Biology*, *1949*, 153–163. https://doi.org/10.1007/978-1-4939-9136-5_12.

Gay, A., Rye, D., & Radhakrishnan, A. (2015). Switch-like responses of two cholesterol sensors do not require protein oligomerization in membranes. *Biophysical Journal*, *108*(6), 1459–1469. https://doi.org/10.1016/j.bpj.2015.02.008.

He, C., Hu, X., Jung, R. S., Weston, T. A., Sandoval, N. P., Tontonoz, P., et al. (2017). High-resolution imaging and quantification of plasma membrane cholesterol by NanoSIMS. *Proceedings of the National Academy of Sciences of the United States of America*, *114*(8), 2000–2005. https://doi.org/10.1073/pnas.1621432114.

Heuck, A. P., Moe, P. C., & Johnson, B. B. (2010). The cholesterol-dependent cytolysin family of gram-positive bacterial toxins. *Sub-Cellular Biochemistry*, *51*, 551–577. https://doi.org/10.1007/978-90-481-8622-8_20.

Infante, R. E., & Radhakrishnan, A. (2017). Continuous transport of a small fraction of plasma membrane cholesterol to endoplasmic reticulum regulates total cellular cholesterol. *Elife (Cambridge)*, *6*, e25466. https://doi.org/10.7554/eLife.25466.

Johnson, K. A., Endapally, S., Vazquez, D. C., Infante, R. E., & Radhakrishnan, A. (2019). Ostreolysin a and anthrolysin O use different mechanisms to control movement of cholesterol from the plasma membrane to the endoplasmic reticulum. *The Journal of Biological Chemistry*, *294*(46), 17289–17300. https://doi.org/10.1074/jbc.RA119.010393.

Johnson, K. A., & Radhakrishnan, A. (2020). Accessibility of cholesterol at cell surfaces. *Journal of Lipid Research*, *61*, 1307. https://doi.org/10.1194/jlr.ILR120000836.

Kinnebrew, M., Iverson, E. J., Patel, B. B., Pusapati, G. V., Kong, J. H., Johnson, K. A., et al. (2019). Cholesterol accessibility at the ciliary membrane controls hedgehog signaling. *Elife (Cambridge)*, *8*, e50051. https://doi.org/10.7554/eLife.50051.

Lange, Y., Ye, J., & Steck, T. L. (2004). How cholesterol homeostasis is regulated by plasma membrane cholesterol in excess of phospholipids. *Proceedings of the National Academy of Sciences of the United States of America*, *101*(32), 11664–11667. https://doi.org/10.1073/pnas.0404766101.

Lukoyanova, N., Kondos, S. C., Farabella, I., Law, R. H., Reboul, C. F., Caradoc-Davies, T. T., et al. (2015). Conformational changes during pore formation by the perforin-related protein pleurotolysin. *PLoS Biology*, *13*(2), e1002049. https://doi.org/10.1371/journal.pbio.1002049.

Makino, A., Abe, M., Ishitsuka, R., Murate, M., Kishimoto, T., Sakai, S., et al. (2017). A novel sphingomyelin/cholesterol domain-specific probe reveals the dynamics of the membrane domains during virus release and in Niemann-pick type C. *The FASEB Journal*, *31*(4), 1301–1322. https://doi.org/10.1096/fj.201500075R.

Moe, P. C., & Heuck, A. P. (2010). Phospholipid hydrolysis caused by Clostridium perfringens alpha-toxin facilitates the targeting of perfringolysin O to membrane bilayers. *Biochemistry, 49*, 9498–9507.

Naito, T., Ercan, B., Krshnan, L., Triebl, A., Koh, D. H. Z., Wei, F. Y., et al. (2019). Movement of accessible plasma membrane cholesterol by the GRAMD1 lipid transfer protein complex. *Elife (Cambridge), 8*, e51401. https://doi.org/10.7554/eLife.51401.

Ohno-Iwashita, Y., Shimada, Y., Hayashi, M., Iwamoto, M., Iwashita, S., & Inomata, M. (2010). Cholesterol-binding toxins and anti-cholesterol antibodies as structural probes for cholesterol localization. *Sub-Cellular Biochemistry, 51*, 597–621. https://doi.org/10.1007/978-90-481-8622-8_22.

Ourisson, G., Rohmer, M., & Poralla, K. (1987). Prokaryotic hopanoids and other polyterpenoid sterol surrogates. *Annual Review of Microbiology, 41*, 301–333. https://doi.org/10.1146/annurev.mi.41.100187.001505.

Radhakrishnan, A., Goldstein, J. L., McDonald, J. G., & Brown, M. S. (2008). Switch-like control of SREBP-2 transport triggered by small changes in ER cholesterol: A delicate balance. *Cell Metabolism, 8*(6), 512–521. https://doi.org/10.1016/j.cmet.2008.10.008.

Sandhu, J., Li, S., Fairall, L., Pfisterer, S. G., Gurnett, J. E., Xiao, X., et al. (2018). Aster proteins facilitate nonvesicular plasma membrane to ER cholesterol transport in mammalian cells. *Cell, 175*(2), 514–529.e520. https://doi.org/10.1016/j.cell.2018.08.033.

Shaner, N. C., Lambert, G. G., Chammas, A., Ni, Y., Cranfill, P. J., Baird, M. A., et al. (2013). A bright monomeric green fluorescent protein derived from Branchiostoma lanceolatum. *Nature Methods, 10*(5), 407–409. https://doi.org/10.1038/nmeth.2413.

Skocaj, M., Resnik, N., Grundner, M., Ota, K., Rojko, N., Hodnik, V., et al. (2014). Tracking cholesterol/sphingomyelin-rich membrane domains with the ostreolysin A-mCherry protein. *PLoS One, 9*(3), e92783. https://doi.org/10.1371/journal.pone.0092783.

Tomita, T., Noguchi, K., Mimuro, H., Ukaji, F., Ito, K., Sugawara-Tomita, N., et al. (2004). Pleurotolysin, a novel sphingomyelin-specific two-component cytolysin from the edible mushroom Pleurotus ostreatus, assembles into a transmembrane pore complex. *The Journal of Biological Chemistry, 279*(26), 26975–26982. https://doi.org/10.1074/jbc.M402676200.

Tweten, R. K. (2005). Cholesterol-dependent cytolysins, a family of versatile pore-forming toxins. *Infection and Immunity, 73*(10), 6199–6209. https://doi.org/10.1128/iai.73.10.6199-6209.2005.

Wang, S., Li, W., Hui, H., Tiwari, S. K., Zhang, Q., Croker, B. A., et al. (2020). Cholesterol 25-hydroxylase inhibits SARS-CoV-2 and coronaviruses by depleting membrane cholesterol. *EMBO Journal, 39*, e2020106057. https://doi.org/10.15252/embj.2020106057.

Yamaji, A., Sekizawa, Y., Emoto, K., Sakuraba, H., Inoue, K., Kobayashi, H., et al. (1998). Lysenin, a novel sphingomyelin-specific binding protein. *The Journal of Biological Chemistry, 273*(9), 5300–5306.

Zhou, Q. D., Chi, X., Lee, M. S., Hsieh, W. Y., Mkrtchyan, J. J., Feng, A.-C., et al. (2020). Interferon-mediated reprogramming of membrane cholesterol to evade bacterial toxins. *Nature Immunology, 21*(7), 746–755. https://doi.org/10.1038/s41590-020-0695-4.

CHAPTER NINETEEN

Strategies for enzymological studies and measurements of biological molecules with the cytolysin A nanopore

Carsten Wloka*, Nicole S. Galenkamp, Nieck J. van der Heide, Florian L.R. Lucas, and Giovanni Maglia*
Groningen Biomolecular Sciences & Biotechnology Institute, University of Groningen, Groningen, The Netherlands
*Corresponding authors: e-mail address: c.wloka@rug.nl; g.maglia@rug.nl

Contents

Abstract

Pore-forming toxins are used in a variety of biotechnological applications. Typically, individual membrane proteins are reconstituted in artificial lipid bilayers where they form water-filled nanoscale apertures (nanopores). When a voltage is applied, the ionic current passing through a nanopore can be used for example to sequence biopolymers, identify molecules, or to study chemical or enzymatic reactions at the single-molecule level. Here we present strategies for studying individual enzymes and measuring molecules, also in highly complex biological samples such as blood.

Methods in Enzymology, Volume 649
ISSN 0076-6879
https://doi.org/10.1016/bs.mie.2021.01.007

1. Introduction: The ClyA nanopore

The transmembrane protein cytolysin A (ClyA) is a pore-forming toxin, which is secreted by several bacteria (Ludwig, Bauer, Benz, Bergmann, & Goebel, 1999; von Rhein et al., 2009). We and others use ClyA originating from *Salmonella typhi* (Kwak, Kim, Lee, Ryu, & Chi, 2020; Soskine, Biesemans, De Maeyer, & Maglia, 2013; Soskine et al., 2012; Soskine, Biesemans, & Maglia, 2015; Wloka et al., 2017) but notice that the ortholog of *Escherichia coli* ClyA can yield similar results (Li, Lee, Shorkey, Chen, & Chen, 2020). A ClyA pore consists of multiple identical subunits of 34 kDa. X-ray and cryo-EM structures revealed that these monomeric pore-forming subunits can assemble into different quaternary structures consisting of 12, dodecamer, (Mueller, Grauschopf, Maier, Glockshuber, & Ban, 2009; Peng, de Souza Santos, Li, Tomchick, & Orth, 2019) (Fig. 1A and B), 13, tridecamer, (Eifler et al., 2006; Peng et al., 2019) or 14 subunits, tetradecamer, (Peng et al., 2019). In lipid bilayer experiments we also observe three different nanopore species as we discuss later, which we refer to as type I pore, likely the dodecamer, (Soskine et al., 2012), type II pore, likely the tridecamer, (Soskine et al., 2013) and type III pore, likely the tetradecamer, (Soskine et al., 2013). Type III ClyA has a lifetime of only few minutes in lipid bilayers, while type I and type II nanopores can be used for h in electrical recordings. Type I ClyA (Fig. 1A and B) can accommodate proteins of about 20–45 kDa (as monomers), and it has been the main nanopore used in our laboratory.

Currently we mainly use a ClyA variant we call ClyA-AS. ClyA-AS is built upon a cysteine-free variant of *S. typhi* ClyA, replacing two native cysteine residues. Further mutations were introduced, *via* directed evolution, to increase the soluble expression of the monomer and to enhance stability of the nanopore in electrophysiological experiments with a more homogeneous nanopore conductance (Soskine et al., 2013). ClyA-AS (containing a 5-histidine tag at the C-terminus) differs from the *S. typhi* wildtype sequence ultimately in the following positions: C87A, L99Q, E103G, F166Y, I203V, C285S, K294R and H307Y. The ClyA-AS type I pore has a total length of 14 nm. The lumen of the pore consists of two compartments: the 10 nm long *cis* lumen with a 6 nm internal diameter and a 4 nm long *trans* constriction with a 3 nm internal diameter (Fig. 1) (Soskine et al., 2013).

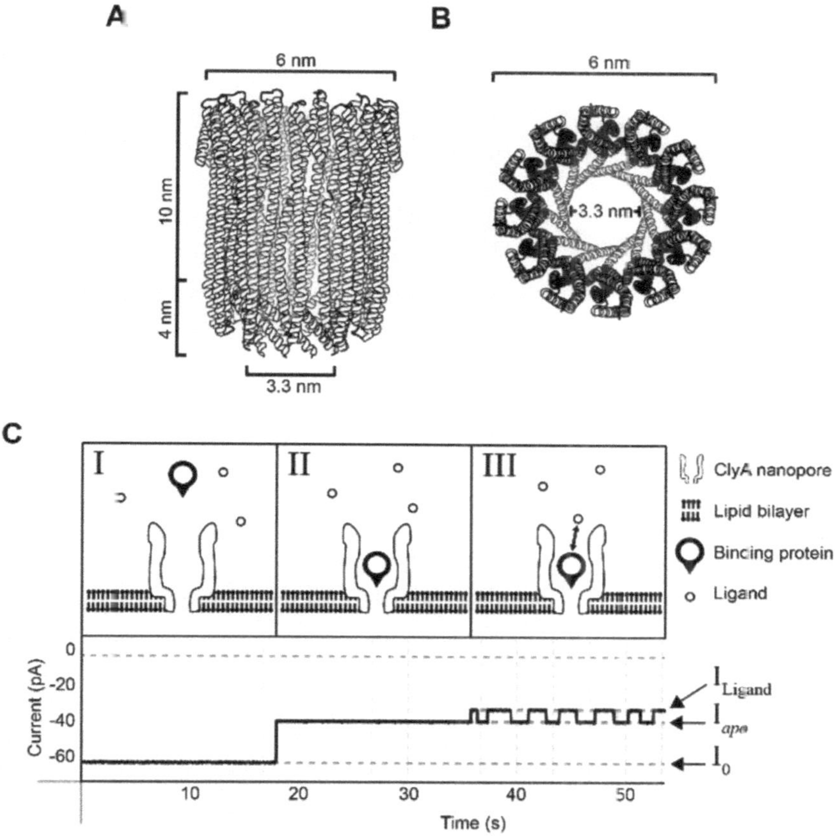

Fig. 1 Crystal structure of cytolysin A (ClyA) and schematic of the basic principle of nanopore enzymology. (A and B) Crystal structure of *E. coli* ClyA (PDB: 6MRT) with side view (A) and top view (B). (C) (I) Drawing of protein capture inside ClyA with (II) a protein captured inside ClyA and (III) ligand binding to the protein confined within ClyA.

The ClyA-AS pore has a negatively charged interior and displays cation-selectivity (Ludwig et al., 1999; von Rhein et al., 2009) in solutions containing up to about 2 M NaCl (Willems et al., 2020). When an external bias is applied, a significant electroosmotic flow (EOF) is generated that allows the trapping of proteins inside the nanopore, typically for seconds to minutes. Although in this method we focus on protein enzymology and quantitative detection of small molecules, we would like to point out that ClyA could also be used for studies of nucleic acids. We developed two variants, ClyA-R and ClyA-RR, in which one or two rings of positively charged arginine residues have been introduced. These positive charges allow for the translocation of, for example, double-stranded and single-stranded DNA at near-physiological salt concentrations (150 mM NaCl, 15 mM Tris–HCl, pH 7.5) (Franceschini, Brouns, Willems, Carlon, & Maglia, 2016; Nomidis, Hooyberghs, Maglia, & Carlon, 2018).

2. The experimental setup

For general technical instructions about the electrophysiological setup, we would like to refer to a previous publication (Maglia, Heron, Stoddart, Japrung, & Bayley, 2010). The methods presented here describe our experience with the ClyA nanopore for the purposes of enzymological studies and quantitative detection of small molecules. For small molecule detection, we rely on protein adaptors trapped inside the ClyA nanopore whose signal elicited in the pore changes upon binding of their cognate ligand. For most of our experiments we commonly use a near-physiological buffer to create a near-native environment for protein studies. This is in contrast with smaller nanopores such as the αHL or MspA nanopores, which are usually sampled at higher ionic strengths (typically 1 M NaCl or KCl). However, since ClyA is a comparably larger pore it allows a large current flow, even at low ionic strength. This represents a key advantage when working with most proteins. If needed, however, ClyA is also stable in various other conditions, including very high (3 M) salt concentrations and different acidity. However, ionic strengths higher than 0.5 M reduce the EOF across the nanopore (Willems et al., 2020), and usually alter dwell times of proteins inside the nanopore.

The detailed protocol for the expression and purification of ClyA nanopores was described recently (Galenkamp et al., 2021). Briefly, we purify ClyA monomers using standard his-tag purification and dilute the protein concentration to 1 mg/mL. Oligomerization is triggered by addition of 0.2% DDM (*n*-dodecyl β-D-maltoside) to the solution of monomers and subsequent incubation for 30 min at 37 °C. After oligomerization, blue native PAGE sample buffer is added to the solution. The different oligomeric forms of the ClyA pores are then separated on a blue native PAGE gel using precast gels (4–20% gel, Criterion Cell system, BioRad). To prepare the pre-cast gels, the buffer chamber and the top compartment of the gel is filled with 1 × native running buffer (Novex). Per well we load 45 μL of the ClyA mixture. The gel is then run for 10 min at 120 V with 2 mL of 20 × cathode buffer additive added to the top compartment of the gel already containing 1 × native running buffer and mixed by pipetting. The cathode buffer additive prevents membrane proteins from aggregating during separation and removes detergents. After 10 min, the cathode buffer additive is replaced with fresh 1 × native running buffer and continues to run for another 60 min at 120 V. The blue-stained protein bands are cut out of the gel with a surgical blade. The acquired gel bands are further cut into even smaller pieces and stored at −80 °C. When needed, the ClyA oligomers are extracted by submerging a gel piece in 20–50 μL of the respective recording buffer supplemented with 0.2% DDM and 1 mM EDTA. The extracted ClyA oligomers can be stored at 4 °C for several weeks. The extracted ClyA oligomers are not stable when they are subsequently frozen. Therefore, long-term storage requires the oligomers to be stored and frozen while embedded within native gel bands.

3. How to obtain a single channel

We prepare planar lipid bilayers consisting of 1,2-diphytanoyl-*sn*-glycero-3-phosphocholine (DPhPC) using the Montal-Mueller method (Maglia et al., 2010; Montal & Mueller, 1972) in an open vertical lipid bilayer device (referred to as the chamber) as described previously (Maglia et al., 2010). It is important to note that the likelihood of ClyA insertion into the lipid bilayer is largely dependent on the surface area of the lipid bilayer. Since the lipid bilayer acts as a capacitor, the surface area can be estimated by determining the observed capacitance. How the capacitance of a lipid bilayer can be estimated was shown previously (Maglia et al., 2010). To summarize, and extend, we utilize the current-voltage relation for capacitors, $I(t) = C \times dV(t)dt^{-1}$. Since we apply a constant voltage ramp of $1\,V\,s^{-1}$ (Maglia et al., 2010), the current measured in Amperes corresponds to the capacitance of the bilayer in Farads. Typically, we apply a voltage ramp between +15 and −15 mV over a period of 30 ms to create a triangular wave profile with a constant voltage ramp of +1 or $-1\,V\,s^{-1}$. The resulting current appears as a square wave, due to the constant switching between positive and negative voltage ramps, where the current amplitude is equal to the capacitance. Subsequently, the capacitance is directly proportional to the surface area and thickness of the lipid membrane, a dielectric material, following the equation for the capacitance of a dielectric, $C = \varepsilon \times A \times d^{-1}$, where C is the capacitance in Farad, ε is the absolute permittivity, A is the area of the bilayer in square meter and d its thickness in meter. While it is difficult to give a well-determined absolute permittivity of the bilayer (Nymeyer & Zhou, 2008), this relationship is still useful to estimate the size of our bilayer and to establish if we formed bi- or multilayered lipid membranes. In practice, the first time that an electrophysiology chamber is used, multiple bilayers are formed and broken while the resulting capacitance of the lipid bilayers is carefully noted. Typical bilayers in an aperture with a diameter of approximately 100 μm exhibit a capacitance, according to the equations above, of 60 to 200 pF and allow the reconstitution of ClyA-AS. In our experience, we observe that ClyA insertion is especially efficient in bilayers with an observed capacitance of 80 to 150 pF equal to a measured current of 80 to 150 pA. While it is possible to use larger than 150 pF bilayers, this results in an increase in noise due to the simultaneous increase in capacitance of the bilayer. Larger bilayers display more noise while smaller bilayers make it difficult to obtain pores. Because bilayers can get smaller over time (shrink), the size of the bilayer should be checked regularly during experiments.

To obtain single nanopores, we commonly use a small pipette tip (without an attached pipette) to briefly touch the extraction buffer containing the nanopores. The tip is then subsequently immerged in the *cis* compartment. If nanopore insertions cannot be observed, larger volumes of stock nanopore solution are added to the electrophysiology chamber. For every new addition we take a fresh pipette tip. If, after addition of pores, a bilayer cannot be formed anymore with ease, it is likely that too many pores were added. Because there is no DDM in our recording buffer, pore insertion becomes more difficult over time (i.e. the pores are likely to fall apart), so it can be advisable to simply wait a couple of minutes before trying to form a bilayer again. Also, it is possible to add an extra dilution step before the addition of pores. This can be done, for example, by swirling the pipette tip in a separate tube of recording buffer before adding it to the *cis* chamber. This step removes excess pores on the tip. We cannot calculate a precise concentration, due to the very low amount of nanopores added, but we estimate sub-nanomolar pore concentrations.

For pore insertion, we routinely apply −35 mV in a "Gap Free" protocol (Clampfit 10.7), a protocol for continuous observation of current (Maglia et al., 2010), and wait for a stepwise current increase, signifying pore insertion. It can be advisable to expose the bilayer to alternating voltages of ±200 mV for 20–30 ms intervals and to break and reform the bilayer regularly. Possibly, bilayer imperfections assist pore insertions. If no insertion is observed and the bilayer capacitance is compatible with a properly formed lipid bilayer, more pores should be added.

Once an insertion event is observed, the first step is to balance a possible potential offset. This is done by applying zero bias potential and using the "pipette offset" to rectify the current to 0 pA. The next steps are to make sure only a single channel has inserted and to evaluate the type of nanopore. First, we check the currents at ±35 mV. In a balanced system with an accurately prepared buffer (150 mM NaCl, 15 mM Tris–HCl, pH 7.5), at a positive potential of 35 mV a stable open pore current of about 65 ± 2 pA is expected for type I ClyA-AS pores. At negative potential, the open pore current should always be lower (−60 ± 2 pA, Fig. 2A). If the former values are established, a regular type I ClyA-AS pore has inserted with its wide opening oriented to the *cis* side of the chamber (ground electrode) and its narrow entry to the *trans* side of the chamber (working electrode). A pore can also be used if it is inserted the other way around ("upside-down" pore, Fig. 2B), however, attention must be paid so that all voltages are applied oppositely. Higher currents could indicate, for example, the insertion of a different type of ClyA pore (type II is −68 ± 2 pA at −35 mV and 72 ± 2 pA at +35 mV) or multiple pores (Fig. 2C). Overall, a complete system with a ClyA pore inserted into the bilayer should have a root mean square (RMS) deviation of noise near around 2 pA using a 5 kHz bandwidth by using a Butterworth filter. This filter is always automatically applied regardless of the front-panel Bessel filter when using the I_{RMS} function of the Axopatch 200B amplifier (Molecular Devices, Axopatch 200B manual). Further, the open pore current should be stable without any current fluctuations. Whenever a fluctuating open pore current is observed, we first wait a couple of minutes. If there is no improvement, a few minutes at 0 mV or application of high potentials (±150–200 mV) for tens of milliseconds may help stabilize the nanopore. Once we establish that a low-noise single channel has inserted, we typically wait 5 min to make sure no second pore inserts. Our rationale is that pores break down in solution due to the highly diluted DDM concentration. To remove excess pores, it is also possible to perfuse the recording buffer after an insertion is observed, however, there is a risk to break the bilayer.

4. Enzymological and protein-ligand binding studies and their signals

The most important criteria when selecting a protein for our studies is that it should fit within the ClyA lumen. With type I ClyA, we find this is often the case for monomeric globular proteins of 20–45 kDa in size.

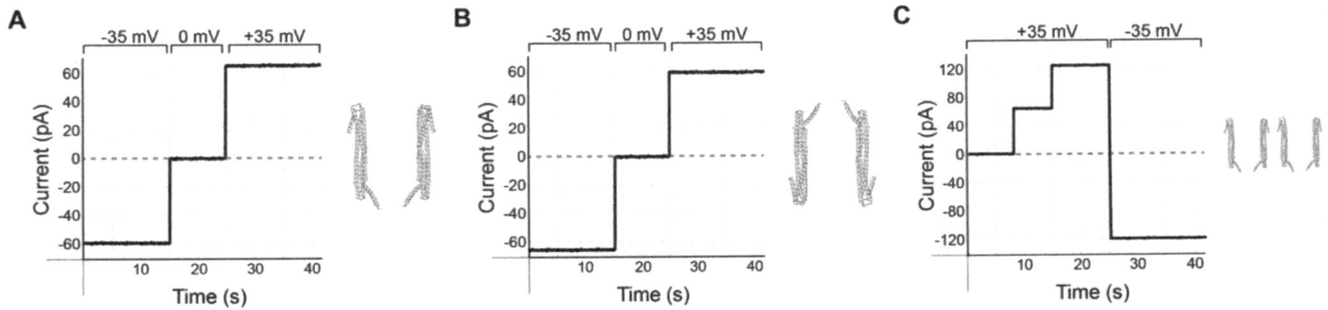

Fig. 2 Expected currents for ClyA-AS. (A) A regular ClyA-AS nanopore allows a stable current of about -60 ± 2 pA at -35 mV and 65 ± 2 pA at 35 mV to pass through in a balanced system an accurately prepared buffer of 150 mM NaCl, 15 mM Tris–HCl, pH 7.5. (B) A pore inserted in the other orientation ("upside-down") will show conversely opposite potentials. (C) Multiple pores show a larger current. The steps are shown to highlight the current contribution of a single pore. In this case two pores would have inserted in the same direction, yielding twice the current of a single pore.

Larger proteins may eclipse the pore while smaller proteins may translocate too quickly to be detected reliably. For our studies, we typically add a final protein concentration of 50–100 nM to the *cis* side of the chamber. We find this to be a good compromise between materials used and capture rates at voltages between −30 mV and −150 mV, a voltage range we commonly use for the capture of proteins. A negatively applied potential is required, because it creates an electroosmotic flow (EOF) across the nanopore from *cis* to *trans* (Willems et al., 2020), which induces the entry of the protein inside the nanopore. The latter is manifested by the decrease of the open pore current I_O to the blocked pore current I_B (Fig. 3A). In order to account for pore to pore variations, when describing current blockades, we typically use the residual current I_{RES}, which is defined as I_O/I_B (or $I_{RES\%} = I_O/I_B \times 100$). The strength of the EOF depends on the ionic concentration of the solution (Willems et al., 2020), and we typically use 150–500 mM NaCl or KCl solutions. For our recordings we typically use a sampling rate of 50–10 kHz (20–100 μs) and a Bessel filter of 10–2 kHz. The sampling rate should always be about fivefold the filtering rate.

Once a protein is inside the nanopore, the signal can be similar to the open pore current (low-noise signal) or display a variety of different current levels (Zernia, van der Heide, Galenkamp, Gouridis, & Maglia, 2020). Although the protein signal can be improved for example by introducing a charge dipole within the protein (Van Meervelt et al., 2017) or N- or C-terminal tag (Willems et al., 2019), the origin of the noise of proteins inside the nanopore is not understood. The further addition of cognate ligands to either the *cis* or *trans* chamber might induce a change in the blocked pore current (Fig. 3B). When testing a new protein, the binding of the ligands might or might not bring a change in the I_B. Thus far, 1/3 of the protein adaptors we tested showed no ligand-induced current changes (Zernia et al., 2020). Hence, when we test a ligand for the first time, we add it into the *cis* chamber at a final concentration near the bulk dissociation constant (K_d). If the K_d is unknown, we add consecutive amounts of ligands from very low to high concentrations. Another challenge is that proteins and protein in complex with a ligand can also elicit various signals. Currently we cannot predict how a signal will look for a given protein. We try to summarize possible signals and scenarios and give ideas on how to analyze data accordingly (Fig. 3). It should be noted that combinations of such types can also occur.

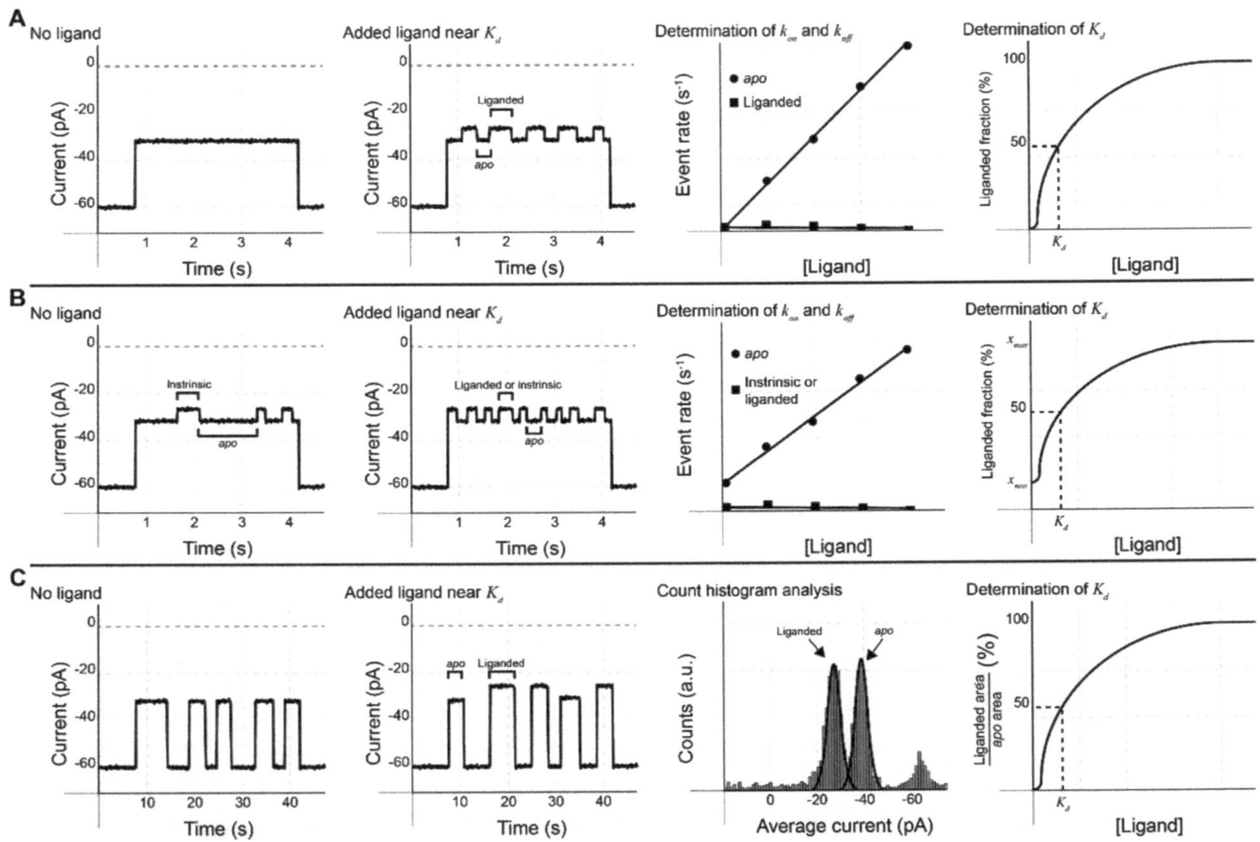

Fig. 3 See figure legend on opposite page.

4.1 Regular signals and their analysis

In the first scenario (Fig. 3A), a protein displays electrically quiet blockades with residence times of at least a few seconds. After addition of ligand, the current might increase or decrease producing an "event" within the protein blockade, reflecting the binding, and unbinding of the ligand. Under these conditions the on-rate and off-rate of the ligands can be measured (*see later*), and the concentration of the ligand in the solution can be calculated. Occasionally, we observe protein blockades of the *apo*-protein (unbound) with regular 'intrinsic' current fluctuations. We attribute this to the spontaneous closing of the protein in the absence of a ligand. The addition of the ligand then increases the frequency of current fluctuation (Fig. 3B). In a third scenario, a protein also displays electrically quiet blockades, but the addition of the ligand provokes additional and uniform protein blockades that have different residual currents (I_{RES}, Fig. 3C). If all such blockades have a different I_{RES} it is possible that this is caused by the saturation of the ligand (i.e., all proteins are bound to the ligand and the unbinding is too fast to be detected). In this case, the concentration of the ligand should be reduced. If only a few blockades show a change in I_{RES}, then the signal can be explained by the binding of a high affinity ligand to the protein. In this case, the residence time of the protein inside the pore is shorter than the ligand off-rate and the latter cannot be detected or measured. The on-rate can often still be measured if the protein is added on the *cis* side of the nanopore and the ligand is added on the *trans* side.

On- and off-rates are calculated by measuring the time corresponding to the duration of a ligand-bound event and the time between two ligand-induced events (inter-event time). The easiest analysis is the allocation of

Fig. 3 (A) Electrically quiet protein blockades that gain an additional level upon addition of a ligand (liganded protein). By plotting the event rate per second against the ligand concentration, the k_{on} and k_{off} can be estimated. By plotting the fraction of liganded protein against the concentration of ligand, the K_d can be estimated at 50% liganded fraction. (B) Protein blockades with intrinsic events that gain frequency upon addition of a ligand (liganded protein). By plotting the event rate per second against the ligand concentration, the k_{on} and k_{off} can be estimated, however, the event rate is offset by the intrinsic events. By plotting the fraction of liganded protein against the concentration of ligand, the K_d can be estimated at 50% normalized liganded fraction, which is the liganded fraction with an offset for the intrinsic events. (C) Electrically quiet protein blockades, that when ligand is added appear to change in current. The K_d can be estimated by determining the percentage of liganded protein (relative to total protein), the K_d is the concentration in which 50% protein is liganded.

the I_{RES} of the unbound (*apo*) protein blockades and liganded protein blockades as well as the dwell times. The (semi) automatic "single-channel" search function in Clampfit 10.7 (Molecular Devices) gives the amplitudes of events, from which I_{RES} can be calculated, as well as their dwell times and other parameters. By treating all blockades separately, impurities and noise can be dealt with. Dwell times are stochastically distributed, and their values can be plotted in a cumulative or conventional histogram and fitted to one or more exponential functions, depending on the shape of the histogram. Two or more exponents should be used if the fitting is not matching the shape of the curve or two or more distributions are visible, however, it is important to check if the addition of exponents is justified. As a rule of thumb, for non-normally distributed data, the binning of the histogram with few data points should be following the Freedman-Diaconis rule (Eq. 1) (Scott, 1979; Sharma et al., 2018)

$$\Delta_b = 2\frac{(q_{75} - q_{25})}{\sqrt[3]{n}} \tag{1}$$

where Δ_b is the bin width, q_{75} is the point of the 75% quantile, q_{25} is the point of the 25% quantile and n is the number of datapoints. In practice, we fit the data several times with different binning and take the Δ_b that corresponds to one third of the fitted dwell time.

The fitting to the curve gives the dwell time parameter, tau (τ, in seconds). The off-rate (k_{off}, s^{-1}) for ligands inside the nanopore is measured from the inverse of the dwell time of the ligand-induced event and does not change whether the ligand is added to the *cis* or *trans* side. Inter-event times are related to the concentration of the ligand in bulk and to the propensity of the ligand to enter the nanopore. Plotting the inverse of the inter-event times of *apo* blockades (in events·s^{-1}) against the bulk ligand concentration should give a linear relationship whose slope is equal to the on-rate (k_{on}, M^{-1} s^{-1}) for the ligand inside the nanopore. Compared to bulk values, the nanopore-measured on-rate might be voltage dependent, especially if the ligand is charged (*see later*). Also, since the *cis* and *trans* entry of the nanopore have a different diameter, the on-rate will depend on the addition to either the *cis* or *trans* side of the nanopore. For non-charged analytes, we typically do not observe significant difference between the on-rates measured for proteins inside the nanopore (for ligands added to the *cis* compartment) compared to bulk values (Zernia et al., 2020).

The dissociation constant, K_d can be measured from the on-rate and off-rate (k_{off}/k_{on}). Alternatively, it can be calculated by plotting the percentage of signal that is in the liganded state, relative to the total signal, against the concentration of the ligand. The percentages of relative liganded population can be obtained by plotting an all-point histogram from individual protein blockades and fit normal (Gaussian) distribution functions to obtain the area under the curve for the liganded and *apo* configuration of the protein. All areas can then be added together to obtain the overall percentage of liganded and *apo* configuration.

If intrinsic closing events are observed (Fig. 3B), the obtained curve can be normalized with Eq. (2).

$$x_{norm} = \frac{x - x_{min}}{x_{max} - x_{min}} \tag{2}$$

where x_{norm} is the normalized value, x_{min} is the fraction of *apo* protein in the closed configuration measured before addition of ligand, and x_{max} is the fraction of closed protein.

The resulting apparent K_d is the point in the graph of the liganded state against the concentration, at which the liganded state amounts to 50% of the total signal. Hill function might be used to determine eventual cooperativity.

If the apparent off-rate of the ligand-induced events is longer than the residence time of the protein (Fig. 3C), we calculate the K_d by creating a histogram of the average current of a large amount of protein blockades (i.e., one protein blockade is one count). Two (or more) distributions should appear, relating to the *apo* and ligand bound state of the protein. We plot the fraction of bound protein, relative to the total number of blockades, against the concentration of ligand. The K_d is equal to the bound where 50% of the protein appears in the ligand bound state.

4.2 Complex protein signals

About one third of proteins we tested showed a non-uniform signal. Examples include blockades with multiple well-defined I_{RES} values and/or with a different level of noise (Fig. 4A). Once contaminant proteins in the solution have been ruled out, the different signals might be due to a variety of factors, including the same protein entering the nanopore with different conformations/orientation, to the intrinsic molecular noise of the protein moving inside the nanopore, or to the protein interacting with a different region within the nanopore. Upon addition of ligands, one or several levels

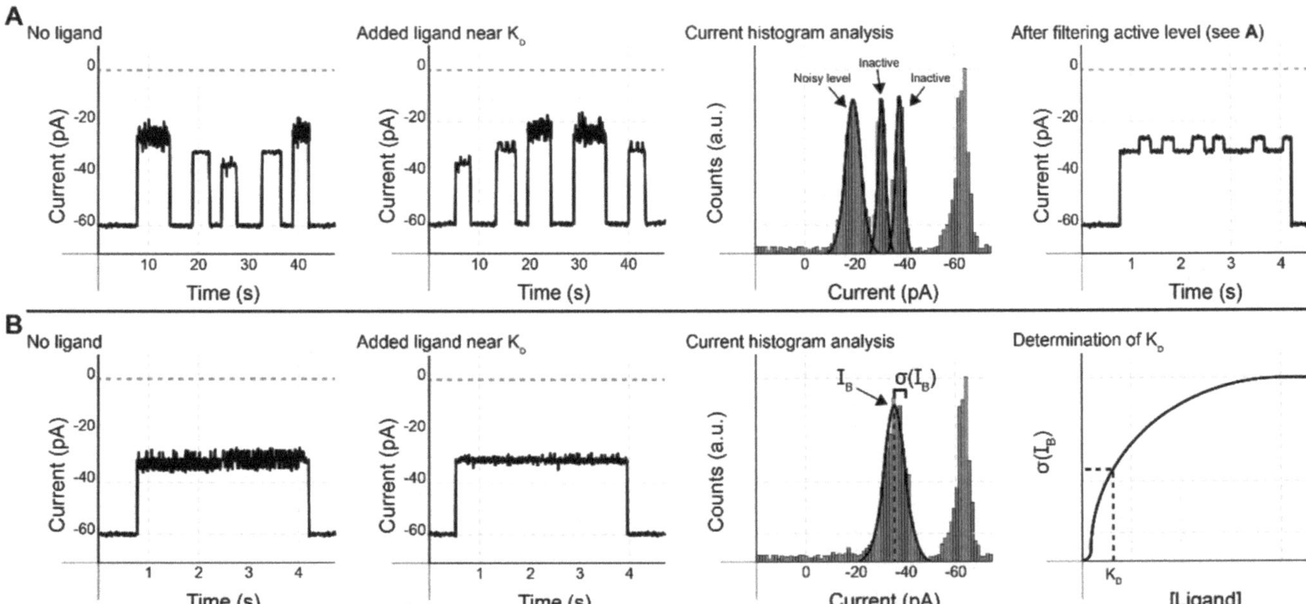

Fig. 4 (A) Protein blockades with multiple levels entering the pore. Upon addition of the ligand, a (potentially noisy) level appears in addition to the *apo* signals. The baseline current of the ligand active protein is filtered based on its current and these blockades are further analyzed. (B) Electrically noisy *apo* protein blockade with a reduced noise upon addition of ligand. The K_d is estimated by setting the standard deviation of protein blockades against the ligand concentration. The K_d is the halfway point between the *apo* protein noise and fully liganded protein noise.

may be changing (i.e. the binding of ligands induce a change in the signal), or additional blockades might be appearing, or the noise or the dwell time might change. Then, the K_d can be determined by analyzing the current blockades after ligand additions. For instance, a full point current histogram of one or more current blockades could reveal the appearance of additional peak(s) corresponding to the liganded complex (Fig. 4A). Then the relative change between the peak intensity can be plotted over the bulk concentration of ligand. The relative peak height or the area under the curve could be used. However, since a fitting a function increases error, using the peak height is often recommended. In another example we encountered, the noise might change as depicted in Fig. 4B. In this case, the standard deviation of the signal—conventionally named $\sigma(I_B)$—can be determined by fitting a normal distribution. Plotting the average $\sigma(I_B)$ against the concentration of the ligand should result in a binding curve with a minimum noise and maximum noise. The concentration of the point in between (50% increase) is the concentration of the dissociation constant (K_d).

4.3 How to improve a signal

At present we cannot predict current blockades elicited by the three-dimensional structure of a protein. However, we gained insight into how to improve a signal. When well-defined ligand-induced events are observed, optimization of the signal is obtained by varying the bias or by adjusting the ionic strength of the solution. Typically, a better signal is obtained by increasing the ionic strength or the external bias. However, ionic strengths higher than 500 mM decrease the EOF and generally reduce the residence time of a protein inside the nanopore, and/or might change the binding site of the protein within the nanopore. Similarly, high potentials might change the localization of the protein inside the nanopore, which might have an unexpected effect on the ligand-induced protein signal. Furthermore, the physical properties at which the protein of interest functions might also represents a limiting factor. Apart from the applied potential and the ionic strength, the types of salts can be varied. For example, we observe higher currents with potassium chloride instead of sodium chloride, due to the high mobility of K^+ ions. The sampling of enzymatic reactions might also require changing the pH of the solution. Within pH values of 7 to 9, we observe ClyA to still work well. Further, protein signals might also be improved by site-directed mutagenesis. For example, we tested proteins of about 20 kDa in size displaying only short-lasting blockades. Upon introduction of a ring of tryptophan-residues (ClyA-AS-Q56W) such blockades became longer, enabling their reliable detection (Wloka et al., 2017).

It is also possible to modify the protein of interest to improve its electrical signal. For example, to increase the dwell time of a negatively charged protein inside the pore it is possible to add a positively charged tag onto the protein of interest (Biesemans, Soskine, & Maglia, 2015; Willems et al., 2019). Due to the negative voltage bias applied, molecules will experience an EOF and an electrophoretic force (EF). The EOF and the EF on the tag will be dragging the protein inside the nanopore, while the negative charge on the body of the protein will pull in the opposite direction. The positive charge tag have further electrostatic interactions with the negatively charged constriction of the nanopore that will elongate the dwell time of the protein inside the nanopore (Willems et al., 2019). We also showed that the genetic engineering of a charge dipole within a protein can allow reducing the current noise (Soskine et al., 2015; Wloka et al., 2017; Zernia et al., 2020). Most likely, the dipole allows aligning the protein to the electric field inside the nanopore and stops the protein from tumbling. These surface changes adjustments are possible with as little as changing one amino acid residue (Van Meervelt et al., 2017).

5. The effect of electrophoretic forces and nanopore confinement

To date, nanopore analysis is the only single-molecule technique for the study of native or unlabeled proteins. Other techniques require immobilization or (fluorescent) tags to study a protein of interest or introduce appendices (optical tweezers). Moreover, because a nanopore represents a relatively crowded environment, it may also offer a more natural environment than the highly diluted solutions used in other single-molecule biochemical studies performed in bulk. It was previously reported that applied voltages may unfold proteins in solid-state pores (Freedman et al., 2011; Talaga & Li, 2009). We did not find strong evidence for such an effect, and we think that the confined space in a ClyA nanopore does not have a large effect on the thermodynamic properties of a trapped protein, at least at the near-physiological conditions and moderate voltages we routinely use (-30 to -150 mV). Overall, we have found that, for uncharged ligands, the apparent dissociation constants measured inside the nanopore are similar to reported bulk values (Zernia et al., 2020). For example, the observed open and closed conformations of a glucose binding protein matched well to NMR studies (Galenkamp, Soskine, Hermans, Wloka, & Maglia, 2018). In other examples we observed similar values for the binding of asparagine and glutamine to the respective substrate binding proteins as for ensemble single-molecule Förster resonance energy transfer measurements (Van Meervelt et al., 2017). However, when a ligand is charged, we often observe voltage dependent on-rates, which most likely reflect the altered diffusion of the ligands in the electric field inside the nanopore. In those cases, extrapolated values to zero applied potential often match bulk values (Galenkamp, Biesemans, & Maglia, 2020). Nonetheless, it is likely that the EOF and EF will have some effect on proteins inside the nanopore. For example, we found that in many cases, increasing the applied potential increases the probability of the translocation of protein across the nanopore, possibly by forcing the transient unfolding of the transmembrane alpha helices of ClyA (Soskine et al., 2012).

6. Employing proteins for quantitative measurements of small molecules in complex samples

For measuring quantitatively in highly complex samples such as blood, a key advantage of protein adaptors is that they confer their selectivity to the ClyA nanopore. We demonstrated that ligands that bind to proteins lodged inside the nanopore can be quantitatively detected in blood (Galenkamp et al., 2018). The most effective way to perform such an experiment is to first prepare a standard or calibration curve in near-physiological buffers (e.g. 150 mM NaCl, 15 mM Tris–HCl, pH 7.5). The latter is prepared by making standard additions of *known* concentrations of the analyte and the change in signal is recorded. In the simplest example, at each ligand concentration a histogram is made for each protein blockade or for a group of blockades. Then, the area under the histogram corresponding to the blocked and unblocked current signal is measured. Multiple protein blockades are usually considered for each ligand concentration. A binding isotherm is then prepared. We found that best results are obtained by plotting the ratio between the blocked and unblocked signals at different ligand concentrations. This calibration curve is then used to calculate the concentration of the ligand in an unknown sample. Although the ligand can be added to either the *cis* or *trans* chamber (but different calibration curves should be made) it is advisable to add a biological sample to the *trans* side of the nanopore when possible. The protein adaptor is always added to the *cis* side.

In this configuration, the *trans* entry of the nanopore acts as a filter for large molecules and pore clogging is normally not observed, especially if dilutions are used. Normally, the signal is not affected by the biological sample and the concentration of the ligand is extracted directly from the standard curve. For simultaneous sensing, without the need to add sequential dilutions of a sample, the tuning of an adaptor protein to the analyte's concentration range in the sample of interest is required, as demonstrated (Galenkamp et al., 2018).

References

Biesemans, A., Soskine, M., & Maglia, G. (2015). A protein rotaxane controls the translocation of proteins across a ClyA nanopore. *Nano Letters*, *15*(9), 6076–6081. https://doi.org/10.1021/acs.nanolett.5b02309.

Eifler, N., Vetsch, M., Gregorini, M., Ringler, P., Chami, M., Philippsen, A., et al. (2006). Cytotoxin ClyA from Escherichia coli assembles to a 13-meric pore independent of its redox-state. *The EMBO Journal*, *25*(11), 2652–2661. https://doi.org/10.1038/sj.emboj.7601130.

Franceschini, L., Brouns, T., Willems, K., Carlon, E., & Maglia, G. (2016). DNA translocation through nanopores at physiological ionic strengths requires precise nanoscale engineering. *ACS Nano*, *10*(9), 8394–8402. https://doi.org/10.1021/acsnano.6b03159.

Freedman, K. J., Jurgens, M., Prabhu, A., Ahn, C. W., Jemth, P., Edel, J. B., et al. (2011). Chemical, thermal, and electric field induced unfolding of single protein molecules studied using nanopores. *Analytical Chemistry*, *83*(13), 5137–5144. https://doi.org/10.1021/ac2001725.

Galenkamp, N. S., Biesemans, A., & Maglia, G. (2020). Directional conformer exchange in dihydrofolate reductase revealed by single-molecule nanopore recordings. *Nature Chemistry*, *12*(5), 481–488. https://doi.org/10.1038/s41557-020-0437-0.

Galenkamp, N. S., Soskine, M., Hermans, J., Wloka, C., & Maglia, G. (2018). Direct electrical quantification of glucose and asparagine from bodily fluids using nanopores. *Nature Communications*, *9*(1), 4085. https://doi.org/10.1038/s41467-018-06534-1.

Galenkamp, N. S., Van Meervelt, V., Mutter, N. L., van der Heide, N. J., Wloka, C., & Maglia, G. (2021). Preparation of cytolysin A (ClyA) nanopores. *Methods in Molecular Biology*, *2186*, 11–18. https://doi.org/10.1007/978-1-0716-0806-7_2.

Kwak, D. K., Kim, J. S., Lee, M. K., Ryu, K. S., & Chi, S. W. (2020). Probing the neuraminidase activity of influenza virus using a Cytolysin A protein nanopore. *Analytical Chemistry*, *92*(21), 14303–14308. https://doi.org/10.1021/acs.analchem.0c03399.

Li, X., Lee, K. H., Shorkey, S., Chen, J., & Chen, M. (2020). Different anomeric sugar bound states of maltose binding protein resolved by a cytolysin A nanopore tweezer. *ACS Nano*, *14*(2), 1727–1737. https://doi.org/10.1021/acsnano.9b07385.

Ludwig, A., Bauer, S., Benz, R., Bergmann, B., & Goebel, W. (1999). Analysis of the SlyA-controlled expression, subcellular localization and pore-forming activity of a 34 kDa haemolysin (ClyA) from Escherichia coli K-12. *Molecular Microbiology*, *31*(2), 557–567. https://doi.org/10.1046/j.1365-2958.1999.01196.x.

Maglia, G., Heron, A. J., Stoddart, D., Japrung, D., & Bayley, H. (2010). Analysis of single nucleic acid molecules with protein nanopores. *Methods in Enzymology*, *475*, 591–623. https://doi.org/10.1016/S0076-6879(10)75022-9.

Montal, M., & Mueller, P. (1972). Formation of bimolecular membranes from lipid monolayers and a study of their electrical properties. *Proceedings of the National Academy of Sciences of the United States of America*, *69*(12), 3561–3566. https://doi.org/10.1073/pnas.69.12.3561.

Mueller, M., Grauschopf, U., Maier, T., Glockshuber, R., & Ban, N. (2009). The structure of a cytolytic alpha-helical toxin pore reveals its assembly mechanism. *Nature*, *459*(7247), 726–730. https://doi.org/10.1038/nature08026.

Nomidis, S. K., Hooyberghs, J., Maglia, G., & Carlon, E. (2018). DNA capture into the ClyA nanopore: Diffusion-limited versus reaction-limited processes. *Journal of Physics. Condensed Matter*, *30*(30), 304001. https://doi.org/10.1088/1361-648X/aacc01.

Nymeyer, H., & Zhou, H. X. (2008). A method to determine dielectric constants in nonhomogeneous systems: Application to biological membranes. *Biophysical Journal*, *94*(4), 1185–1193. https://doi.org/10.1529/biophysj.107.117770.

Peng, W., de Souza Santos, M., Li, Y., Tomchick, D. R., & Orth, K. (2019). High-resolution cryo-EM structures of the E. coli hemolysin ClyA oligomers. *PLoS One*, *14*(5), e0213423.. https://doi.org/10.1371/journal.pone.0213423

Scott, D. W. (1979). On optimal and data-based histograms. *Biometrika*, *66*(3), 605–610. https://doi.org/10.1093/biomet/66.3.605.

Sharma, S. K., Kumar, S., & Karmeshu. (2018). Suppression of multimodality in inter-spike interval distribution: Role of external damped oscillatory input. *IEEE Transactions on Nanobioscience*, *17*(3), 329–341. https://doi.org/10.1109/TNB.2018.2845454.

Soskine, M., Biesemans, A., De Maeyer, M., & Maglia, G. (2013). Tuning the size and properties of ClyA nanopores assisted by directed evolution. *Journal of the American Chemical Society*, *135*(36), 13456–13463. https://doi.org/10.1021/ja4053398.

Soskine, M., Biesemans, A., & Maglia, G. (2015). Single-molecule analyte recognition with ClyA nanopores equipped with internal protein adaptors. *Journal of the American Chemical Society*, *137*(17), 5793–5797. https://doi.org/10.1021/jacs.5b01520.

Soskine, M., Biesemans, A., Moeyaert, B., Cheley, S., Bayley, H., & Maglia, G. (2012). An engineered ClyA nanopore detects folded target proteins by selective external association and pore entry. *Nano Letters*, *12*(9), 4895–4900. https://doi.org/10.1021/nl3024438.

Talaga, D. S., & Li, J. (2009). Single-molecule protein unfolding in solid state nanopores. *Journal of the American Chemical Society*, *131*(26), 9287–9297. https://doi.org/10.1021/ja901088b.

Van Meervelt, V., Soskine, M., Singh, S., Schuurman-Wolters, G. K., Wijma, H. J., Poolman, B., et al. (2017). Real-time conformational changes and controlled orientation of native proteins inside a protein nanoreactor. *Journal of the American Chemical Society*, *139*(51), 18640–18646. https://doi.org/10.1021/jacs.7b10106.

von Rhein, C., Bauer, S., Lopez Sanjurjo, E. J., Benz, R., Goebel, W., & Ludwig, A. (2009). ClyA cytolysin from Salmonella: Distribution within the genus, regulation of expression by SlyA, and pore-forming characteristics. *International Journal of Medical Microbiology*, *299*(1), 21–35. https://doi.org/10.1016/j.ijmm.2008.06.004.

Willems, K., Ruic, D., Biesemans, A., Galenkamp, N. S., Van Dorpe, P., & Maglia, G. (2019). Engineering and modeling the electrophoretic trapping of a single protein inside a nanopore. *ACS Nano*, *13*(9), 9980–9992. https://doi.org/10.1021/acsnano.8b09137.

Willems, K., Ruic, D., Lucas, F. L. R., Barman, U., Verellen, N., Hofkens, J., et al. (2020). Accurate modeling of a biological nanopore with an extended continuum framework. *Nanoscale*, *12*(32), 16775–16795. https://doi.org/10.1039/d0nr03114c.

Wloka, C., Van Meervelt, V., van Gelder, D., Danda, N., Jager, N., Williams, C. P., et al. (2017). Label-free and real-time detection of protein ubiquitination with a biological nanopore. *ACS Nano*, *11*(5), 4387–4394. https://doi.org/10.1021/acsnano.6b07760.

Zernia, S., van der Heide, N. J., Galenkamp, N. S., Gouridis, G., & Maglia, G. (2020). Current blockades of proteins inside nanopores for real-time metabolome analysis. *ACS Nano*, *14*(2), 2296–2307. https://doi.org/10.1021/acsnano.9b09434.

CHAPTER TWENTY

Pore-forming toxins as tools for polymer analytics: From sizing to sequencing

Fabien Piguet[a,b], Tobias Ensslen[c], Mazdak A. Bakshloo[a,b], Monasadat Talarimoghari[c], Hadjer Ouldali[a,b], Gerhard Baaken[d], Ekaterina Zaitseva[d], Manuela Pastoriza-Gallego[a,b], Jan C. Behrends[c,e,f,*], and Abdelghani Oukhaled[a,b,*]

[a]CY Cergy Paris Université, CNRS, LAMBE, Cergy, France
[b]Université Paris-Saclay, Univ Evry, CNRS, LAMBE, Evry-Courcouronnes, France
[c]Laboratory for Membrane Physiology and Technology, Department of Physiology, Faculty of Medicine, University of Freiburg, Freiburg, Germany
[d]Ionera Technologies GmbH, Freiburg, Germany
[e]Freiburg Centre for Interactive Materials and Bioinspired Technologies (FIT), Freiburg, Germany
[f]Freiburg Centre for Materials Research, Freiburg, Germany
*Corresponding authors: e-mail address: jan.behrends@physiologie.uni-freiburg.de; abdelghani.oukhaled@cyu.fr

Contents

Methods in Enzymology, Volume 649
ISSN 0076-6879
https://doi.org/10.1016/bs.mie.2021.01.017

Abstract

We report here on the nanopore resistive pulse sensing (Np-RPS) method, involving pore-forming toxins as tools for polymer analytics at single molecule level. Np-RPS is an electrical method for the label-free detection of single molecules. A molecule interacting with the pore causes a change of the electrical resistance of the pore, called a resistive pulse, associated with a measurable transient current blockade. The features of the blockades, in particular their depth and duration, contain information on the molecular properties of the analyte. We first revisit the history of Np-RPS, then we discuss the effect of the configuration of the molecule/nanopore interaction on the molecular information that can be extracted from the signal, illustrated in two different regimes that either favor molecular sequencing or molecular sizing. Specifically, we focus on the sizing regime and on the use of two different pore-forming toxins, staphylococcal α-hemolysin (αHL) and aerolysin (AeL) nanopores, for the characterization of water-soluble polymers (poly-(ethylene glycol), (PEG)), homopeptides, and heteropeptides. We discuss how nanopore sizing of polymers could be envisioned as a new approach for peptide/protein sequencing.

1. Introduction

1.1 Historical background

It is worth recalling (DeBlois & Bean, 1970; Golibersuch, 1973) that the question of the variation of the electrical resistance of a cylinder filled with a conducting fluid induced by the addition of an insulating particle in the cylinder (Fig. 1A) has roots in the work of James C. Maxwell in the 19th century. Maxwell expressed the effective resistivity ϱ_{eff} of an infinitely diluted suspension of insulating spherical particles in a conducting fluid of resistivity ϱ as (Maxwell, 1873):

$$\varrho_{eff} = \varrho\left(1 + \frac{3}{2}f + \ldots\right) \tag{1}$$

where f is the fractional volume occupied by the insulating spheres ($f \ll 1$). Considering a cylinder of length L and diameter D filled with a fluid of resistivity ϱ, its electrical resistance R_0 may be written, if $D \ll L$, as:

$$R_0 = \frac{4\varrho L}{\pi D^2} \tag{2}$$

Introducing an insulating spherical particle of diameter d in the cylinder under the assumptions of Eq. (1), the resistance R_b of the system now reads:

$$R_b = \frac{4\varrho_{eff} L}{\pi D^2} = \frac{4\varrho L}{\pi D^2}\left(1 + \frac{3}{2}f + \ldots\right) = R_0\left(1 + \frac{3}{2}f + \ldots\right) \tag{3}$$

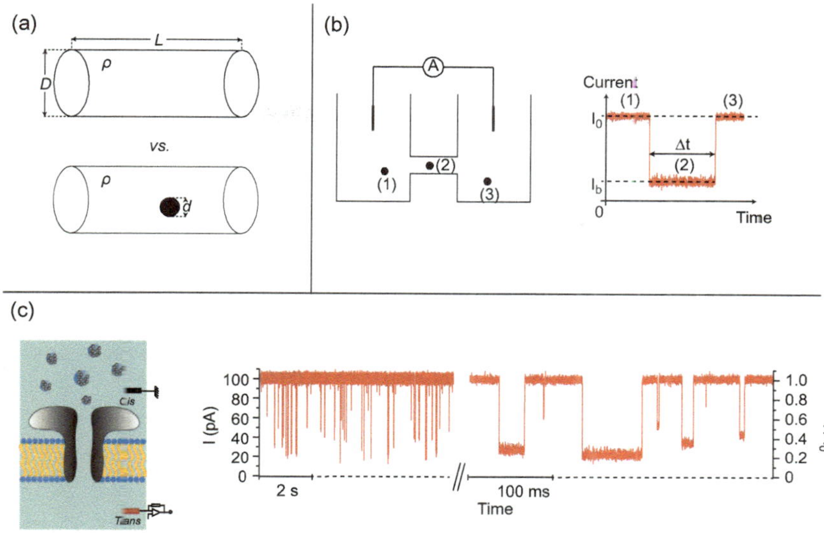

Fig. 1 Principle of resistive pulse sensing. (A) Illustration of the fundamental question from which the resistive pulse sensing concept emerged, which interrogates the variation of the electrical resistance of a cylinder of length L and diameter D filled with a conductive fluid of electrical resistivity ρ (top) when an insulating spherical particle of diameter d is added in the cylinder (bottom). (B) Principle of the Coulter counter. (Left) Schematic of the experimental configuration (not to scale), where a cylindrical channel connects two chambers filled with an electrolyte solution. A pair of electrodes is used to apply a voltage between the chambers, and to record the ionic current flowing through the channel over time. A particle initially added in one chamber may be transported through the channel to the other chamber (successive positions 1, 2 and 3). (Right) Illustration of a channel current vs time recording corresponding to the transport of a single particle through the channel. As long as the particle is in its initial chamber (position 1), a steady current of mean value I_0 is observed. The presence of the particle in the channel (position 2) induces a partial and temporary blockade of the current, of mean residual current value I_b, and of duration Δt, before the current reestablishes to I_0 after the particle escaped from the channel (position 3). It must be noted that, while the situation is here exemplified with a particle that is transported through the channel from one chamber to the other, the current vs time signal does not allow to directly discriminate between the case where the particle effectively crossed the channel, and the case where the particle escaped the channel on the same side it entered. (C) Principle of single-molecule detection with a nanopore. (Left) Schematic of the experimental configuration (not to scale), as a nanoscale analogue of the Coulter counter configuration, where the channel connecting the two chambers (*cis* and *trans*) now consists of a single nanopore from a pore-forming toxin inserted in a planar lipid bilayer. In this illustration, analyte molecules are initially added in the *cis*-chamber. (Right) Portion of a typical nanopore current vs time recording, in presence of analytes. The left axis corresponds to current *I*-values, and the right axis to relative current I/I_0 values. The horizontal enlargement on the right-most part of the graph highlights current blockades of different amplitudes and durations. Each current blockade corresponds to the interaction of an individual analyte molecule with the nanopore.

where f, the fractional volume of the spherical particle in the cylinder, reads:

$$f = \frac{\frac{4}{3}\pi\left(\frac{d}{2}\right)^3}{\pi\left(\frac{D}{2}\right)^2 L} = \frac{2d^3}{3D^2L} \text{ with } d \text{ such that } f \ll 1 \tag{4}$$

thus

$$R_b = R_0\left(1 + \frac{d^3}{D^2L} + \ldots\right) \tag{5}$$

and the relative variation of electrical resistance at the first order reads:

$$\frac{\Delta R}{R} = \frac{R_b - R_0}{R_0} = \frac{d^3}{D^2L} = -\frac{\Delta I}{I} \tag{6}$$

where I is the electrical current flowing through the cylinder.

It is worth noting that the latter expression is independent of the electrolyte conductivity ϱ.

Following this idea, in the 1950s, Wallace H. Coulter developed a device (which he named the Coulter counter) for counting and sizing particles based on the measurement of the variation of the electrical current flowing through an electrolyte-filled cylinder, induced by the presence of a particle in the cylinder. This device, illustrated in Fig. 1B, consists of a cylindrical channel connecting two chambers filled with an aqueous electrolyte solution. A fixed voltage is applied between the chambers via two electrodes, and a steady ionic current I_0 flowing through the channel is measured. When a dilute suspension of particles is added to one chamber, the particles may be driven from the initial chamber to the other under an electrical force and/or a hydrodynamic flow. The presence of a particle in the channel causes a transient perturbation (in most cases an attenuation, i.e. a blockade) of the channel current (Fig. 1B). Such current blockades, corresponding to transient increases of channel resistance, were termed resistive pulses, giving rise to the concept of resistive pulse sensing.

Such measurements require the particles, on the one hand, to be sufficiently small (or deformable) to enter the channel, and, on the other hand, to be sufficiently large compared to the channel diameter to induce a detectable current blockade (which deviates from $f \ll 1$). This picture considers the particle conductivity to be negligible compared to the surrounding electrolyte; otherwise, the particle becomes less and less detectable as its conductivity approaches the electrolyte conductivity, or, in rare cases, even induces a detectable increase of channel current if the particle conductivity is greater than the electrolyte conductivity.

Using a Coulter counter with a channel of diameter ~100 μm and length of several hundreds of micrometers, counting and sizing of micrometric (10 μm–30 μm) blood cells was first demonstrated in 1953 (Coulter, 1953).

Five years later, Herbert E. Kubitschek substituted the Coulter counter aperture by a small 10 μm diameter hole drilled in a 50 μm thick glass sheet for counting and sizing bacteria (1 μm–10 μm) (Kubitschek, 1958). In the same vein, in 1970, Charles Bean and Ralph DeBlois further reduced the Coulter counter aperture to a sub-micrometric 500 nm diameter hole drilled in a plastic sheet, to analyze submicrometer objects such as spherical colloidal particles (100 nm) or viral particles (60 nm) (DeBlois & Bean, 1970; DeBlois, Bean, & Wesley, 1977). DeBlois and Bean refined and generalized the theoretical description developed by Maxwell of the relative resistance change induced by infinitely small spheres to a wider range of sphere size (DeBlois, 1970).

At the technological base of single-molecule resistive pulse sensing or "molecular Coulter counting" is the application of the voltage clamp technique, first developed in the middle of the last century for work on the squid giant axon (Cole, 1949; Cole & Moore, 1960; Hodgkin, Huxley, & Katz, 1952), to a nanometer-thick dielectric layer separating two compartments filled with electrolyte solution and containing a single aperture that is equally of molecular dimensions. The voltage clamp is a potentiostatic circuit where a feedback amplifier compares a command voltage to the actual voltage against ground and delivers current to cancel any deviation. If the aperture is a pore-forming protein, the natural choice for the dielectric layer is a free-standing synthetic lipid membrane or lipid bilayer of the type first described by Paul Mueller and colleagues in the 1960s (Mueller, Rudin, Tien, & Wescott, 1962). Early voltage clamp measurements on lipid bilayers showed discrete conductance fluctuations when they were doped with a bacterial extract called excitability inducing material (Bean, Shepherd, Chan, & Eichner, 1969) or with the antibiotic peptide gramicidin (Hladky & Haydon, 1970).

When, using the same technique, it was established that alpha-toxin of *Staphylococcus aureus* (α-hemolysin, αHL) forms discrete, stable conducting pathways in synthetic membranes (Menestrina, 1986), the want of structural information on the pore prompted Krasilnikov and coworkers to use the non-ionic polymer poly-(ethylene glycol) (PEG) as a probe for pore size (Krasilnikov, Sabirov, Ternovsky, Merzliak, & Muratkhodjaev, 1992). The strategy was based on the comparison of the decrease of bulk conductivity and pore's conductance upon addition of polydisperse mixtures of PEG of diverse average molecular weight and the reasoning that steric exclusion of polymer from the pore would produce a mismatch between pore conductance and electrolyte conductivity. This was indeed observed for polydisperse PEGs of average Mw $\geq$1500 g/mol, with a complete absence of pore conductance decrease for Mw $\geq$2000 g/mol. Correlation with the average hydrodynamic radius of the PEG preparations then led to estimation of the pore radius at 1.3 nm. This approach, which was later also used to show pore asymmetry (Merzlyak et al., 1999) in accordance with the crystal structure of αHL (Song et al., 1996), strongly relied on the assumption of a lack of interaction between polymer and pore (Krasilnikov, 2002). However, it later became apparent that this was true only at the low salt concentrations used in this work (0.1 M KCl). In fact, the use of protein nanopores for molecular sensing -in stark contrast to the Coulter counter principle- emphatically depends on interaction of the analyte with the pore in order to produce pore blockages that are long enough to be resolved by ionic current measurements (Kasianowicz, Robertson, Chan, Reiner, & Stanford, 2008). Importantly, this is not only due to limitations in the bandwidth of electrical measurements but more fundamentally, to the limited ion flux through the protein pores (on the order of 10^8–10^9 ions/s).

1.2 Nanopores from pore-forming toxins for the discrimination of molecules: Sequencing, size spectrometry, or both?

The year 1996 can be identified with some justification as crucial for the development of molecular sensing with pore-forming toxins and biological nanopores in general. In addition to the crystal structure of αHL, which was published that year (Song et al., 1996), two pivotal discoveries emerged. The first, well known as the first step towards nanopore sequencing, was the finding by Kasianowicz, Brandin, Branton, and Deamer (1996) that single-stranded RNA and DNA oligomers of length 215 and 150 bases could be translocated through the αHL-pore, giving rise to well-resolved albeit short resistive pulses, with dwell times of translocation events distributed as two Gaussian peaks according to the 3′–5′ orientation of the polynucleotides, the means in the range of 100 μs to a few ms linearly increasing with polymer length, and decreasing with applied potential as expected for translocation of polyelectrolyte in an extended conformation.

This paper, evidently, pointed the way towards DNA sequencing, much later demonstrated for the first time by Gundlach and coworkers using the mycobacterial outer membrane porin MspA (Derrington et al., 2010), notably in combination with a processive polymerase enzyme to control and slow down translocation (Laszlo et al., 2014; Manrao et al., 2012), as first proposed by Akeson and coworkers (Cherf et al., 2012; Lieberman et al., 2010). It also initiated vigorous commercial activity, and while Oxford Nanopore's devices are the only ones currently available, further developments appear to continue elsewhere as well.

Perhaps somewhat less noted at the time, another seminal publication in 1996 concerned the demonstration by Bezrukov, Kasianowicz and coworkers of strong interaction of PEG with the αHL pore when the salt concentration was increased to 1 M (Bezrukov, Vodyanoy, Brutyan, & Kasianowicz, 1996), see also (Bezrukov & Kasianowicz, 1997). Among the evidence, most suggestive of things to come was the appearance of prominent fluctuations in pore current in the presence of PEG with a prominent peak in noise power for PEG of Mw 2000 g/mol.

However, a full decade passed until, using near saturating KCl concentration (4 M), Krasilnikov, Rodrigues, and Bezrukov (2006) showed the first clear resistive pulses with characteristic dwell times of order 10 ms due to the interaction of PEG800-8000 with the αHL pore and demonstrated an increase of the depth of block as well as of the resistive pulse duration with polymer size in the range PEG800 to PEG3000, with the depth of block saturating and the dwell time decreasing for larger polymers. Immediately after that, Kasianowicz, Krasilnikov and coworkers, demonstrated that, surprisingly, the amplitudes of the resistive pulses could be used to construct "mass spectrograms" with a resolution of single monomers, allowing identification of molecular species of a given size in a polydisperse mixture, while dwell times were monoexponentially distributed for each species, with characteristic values steeply increasing with polymer size, as expected for a first-order binding reation (Robertson et al., 2007). This fundatory experiment, which was reproduced with remarkable quantitative agreement by a number of labs including experiments on higher-throughput formats (Baaken, Ankri, Schuler, Rühe, & Behrends, 2011; del Rio Martinez, Zaitseva, Petersen, Baaken, & Behrends, 2015; Rodrigues, Machado, Chevtchenko, & Krasilnikov, 2008; Rodrigues, Machado, da Silva, Júnior, & Krasilnikov, 2011; Zheng, Baaken, Vellinger, Behrends, & Rühe, 2014), inaugurated a second paradigm of nanopore sensing initially termed nanopore single molecule mass spectrometry (Np-SMMS). This term was changed to single-molecule nanopore-spectrometry (SMNS) (Chavis, Brady, Kothalawala, & Reiner, 2015) or nanopore-based single-molecule size-discrimination (Np-SMSD) (Piguet et al., 2016) in recognition

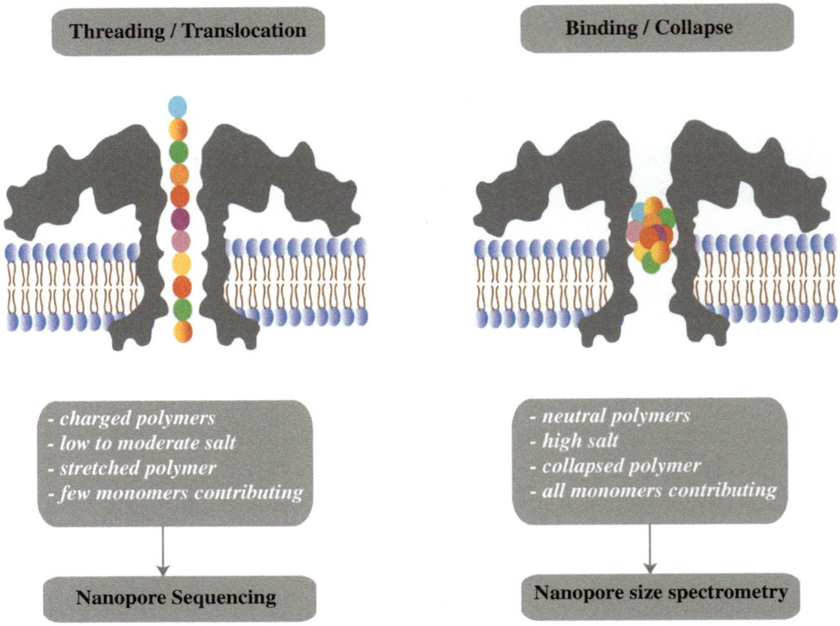

Fig. 2 Two different regimes of molecule/nanopore interaction. (Left) Cartoon illustrating the regime where the analyte molecule is linearly transported through the nanopore. The composition of the intact molecule may be sequentially read as it threads through the pore, corresponding to nanopore-based single-molecule sequencing. (Right) Cartoon illustrating the regime where the analyte molecule is fully accommodated *en bloc* in the nanopore and each individual molecular subunit provides its own contribution to the current blockade, allowing nanopore-based single-molecule size-spectrometry.

of the likelihood that—as has now been shown without a doubt (see below)-molecular mass is not responsible for the size-dependence of interaction.

The interaction of polymers with biological nanopores can thus be broadly categorized into two regimes: in the first, which we call the threading/translocation regime (Fig. 2 left), the polymer is in an extended conformation, is longer than the pore's dominant constriction (see below) and is translocated through the pore in a force field, e.g. electrophoretically. This is the regime that allows polymer strand sequencing: as the polymer is threaded through the pore, few monomers contribute to the current block at a given moment so that the polymer is sequentially probed along its backbone. Typically, this paradigm is realized when a charged polymer is used in low to moderate salt concentrations and the pore has a well-defined, short constricted zone where most of the resistance change occurs. It is this requirement in particular that is fulfilled by the MspA and CsgG porins, engineered versions of which are currently used for DNA sequencing (Faller, Niederweis, & Schulz, 2004; Goyal et al., 2014).

In contrast, the second regime, called binding/collapse regime (Fig. 2 right), is typical of a polymer sizing experiment, where, evidently, all of the monomers contribute to the block. The αHL pore can resolve monomer size differences of PEG molecules up to Mw 3000 g/mol, corresponding to degrees of polymerization (d.p) of 60–70. Assuming helical structure, the length of such a polymer in water would be 17–20 nm (Oesterhelt, Rief, & Gaub, 1999) i.e. significantly longer than the entire αHL pore (10 nm). It follows that the molecules must be in a collapsed globule or at least random coil conformation during the resistive pulse in order to enable each monomer to contribute to the block. Significantly, this regime is typically realized when nonionic polymers are presented to the pore in high salt concentrations. The first explicit statement of a covariance of the solubility of PEG in the electrolyte solution and the frequency and duration of the PEG-pore interaction was made in a systematic study of the effect of salt concentration (Rodrigues et al., 2008). There, it was shown that the frequency of blockages as well as dwell times increased with increasing KCl concentration following a power law with an exponent similar to that of the decrease of PEG solubility. Results matching a Hofmeister series were later obtained using various anion (Rodrigues et al., 2011) and cation substitutions (Breton et al., 2013; Machado et al., 2016) that a collapsed state of PEG may be a prerequisite for interaction with or even partitioning into the pore, at least for larger PEGs is suggested by the finding that the size range of single monomer resolution can be extended beyond the 3000 g/mol limit by increasing the temperature (Piguet et al., 2016), as PEG is known to undergo coil to globule transition at a lower critical solution temperature that also depends on salt.

A collapsed state of the polymer during the interaction, possibly brought about by competition between polymer and salt for scarce hydrating waters in the pore is also strongly suggested by the finding that polymer architecture hardly affects the size-dependent interaction with biological nanopores (Talarimoghari, Baaken, Hanselmann, & Behrends, 2018).

In 2015, Baaken et al. (Baaken et al., 2015) introduced another toxin pore, aerolysin (AeL) from *Aeromonas hydrophila*, that had been used in translocation experiments on charged polymers (Cressiot et al., 2015; Merstorf et al., 2012; Pastoriza-Gallego et al., 2014, 2011; Stefureac, Long, Kraatz, Howard, & Lee, 2006) to polymer sizing, by showing that under experimental conditions similar to those used previously for αHL (save an inverted and much higher *trans-cis* voltage gradient), aerolysin interacted more strongly than αHL with smaller PEGs, extending the range of Np-SMSD to d.p. <20 (Mw <880 g/mol), i.e. in the size range of an arginine tripeptide (see details in Section 4.1.2, Fig. 8). In 2016, Cao et al. (Cao et al., 2016) reported size discrimination of adenine oligonucleotides (A3–A10) and 2 years later, Piguet et al (Piguet et al., 2018) showed that the AeL nanopore enables the size-discrimination of short uniformly charged homopeptides, mixed in solution, with single amino acid resolution (see details in Section 4.2.1, Fig. 9).

The application of single molecule nanopore detection to the problem of sequencing polypeptides had already been invoked in the 1998 patent application (Church, Deamer, Branton, Baldarelli, & Kasianowicz, 1998) by the authors of the 1996 Kasianowicz et al. paper (Kasianowicz et al., 1996). Protein sequencing by nanopore presents particular challenges such as a (i) higher number of elementary constituents of the sequence (20 different amino acids, with possible additional post-translational modifications, compared to 4 different DNA nucleotides), (ii) the heterogeneity of proteins in term of charge distribution along the amino acid chain while polynucleotides are roughly uniformly charged, which is of particular importance in the control of the unidirectional transport and transport velocity of the molecule through the pore, and (iii) the necessity to unfold the protein structure.

One approach towards nanopore-based protein sequencing that is actively pursued by a number of groups stays within the translocation/threading regime, being a direct transposition of the strand sequencing approach to proteins. The problem of unfolding the protein, as well as that of controlling translocation, may, as originally proposed by Akeson and coworkers (Nivala, Marks, & Akeson, 2013) be addressed by coupling to the pore protein a ClpX protein unfoldase, acting as a motor that can both unfold protein domains and ensure the linear translocation of the unfolded protein through the pore in an ATP-dependent manner. The ionic current trace measured during protein enzymatic translocation allowed to detect differences according to various protein domains, with a maximum resolution of ≈663 amino acids (Nivala, Mulroney, Li, Schreiber, & Akeson, 2014). Protein sequencing using this approach remains challenging, because the protein is still transported through the nanopore too rapidly to enable the reading of the amino acid sequence.

On the other hand, the binding regime of nanopore/polymer interaction may have its uses for protein analytics as well: Ouldali et al (Ouldali et al., 2020) recently showed that the substitution by any of the 20 proteinogenic amino acids of a single terminal variable amino acid linked to an hepta-arginine peptide could be detected and, in a majority of cases, differentiated from the others, with a good discrimination even between peptides of equal mass (see details in Section 4.2.2 Figs. 9 and 10). This work opens new perspectives towards the use of this high-resolution molecular discrimination approach, based on nanopore size-spectrometry, for peptide/protein sequencing.

2. Method for nanopore measurements

2.1 Experimental setup

The schematic diagram of our two versions of experimental setups is shown in Fig. 3: classical single-channel recording setup (left) and parallel on-a-chip single-channel recording setup (Orbit 16) (right). Detailed description of each version is described in Fig. 3 legend. In both versions a voltage is applied across a lipid membrane via two Ag/AgCl electrodes and drives an ionic current through the nanopore inserted in the membrane. The current flowing through the nanopore is voltage-converted and amplified, then digitized with an analog to digital converter (ADC) and monitored/recorded with a dedicated computer software; such signal allows to track in real time current blockades induced by single molecules interacting with the nanopore. The voltage applied through the digital to analog converter (DAC) is controlled by the same software.

2.2 Protocol

2.2.1 Materials

All solutions should be prepared using ultrapure water (18.2 MΩ cm at 25 °C)

- recombinant wild-type proaerolysin (pAeL) obtained from pET22b −// proAeL plasmid transformed in *E. coli* BL21 strain
- 1,2-diphytanoyl-*sn*-glycero-3-phosphocholine (Avanti Polar Lipids, ref. no 850356P)
- decane (Sigma-Aldrich, ref. no 457116)
- potassium chloride (Sigma-Aldrich, ref. no P5405)
- HEPES (Fischer Scientific, ref. no BP310-1)
- trypsin (Sigma-Aldrich, ref. no T1426)
- agarose (Sigma-Aldrich, ref. no A9539)
- bleach (NaClO)

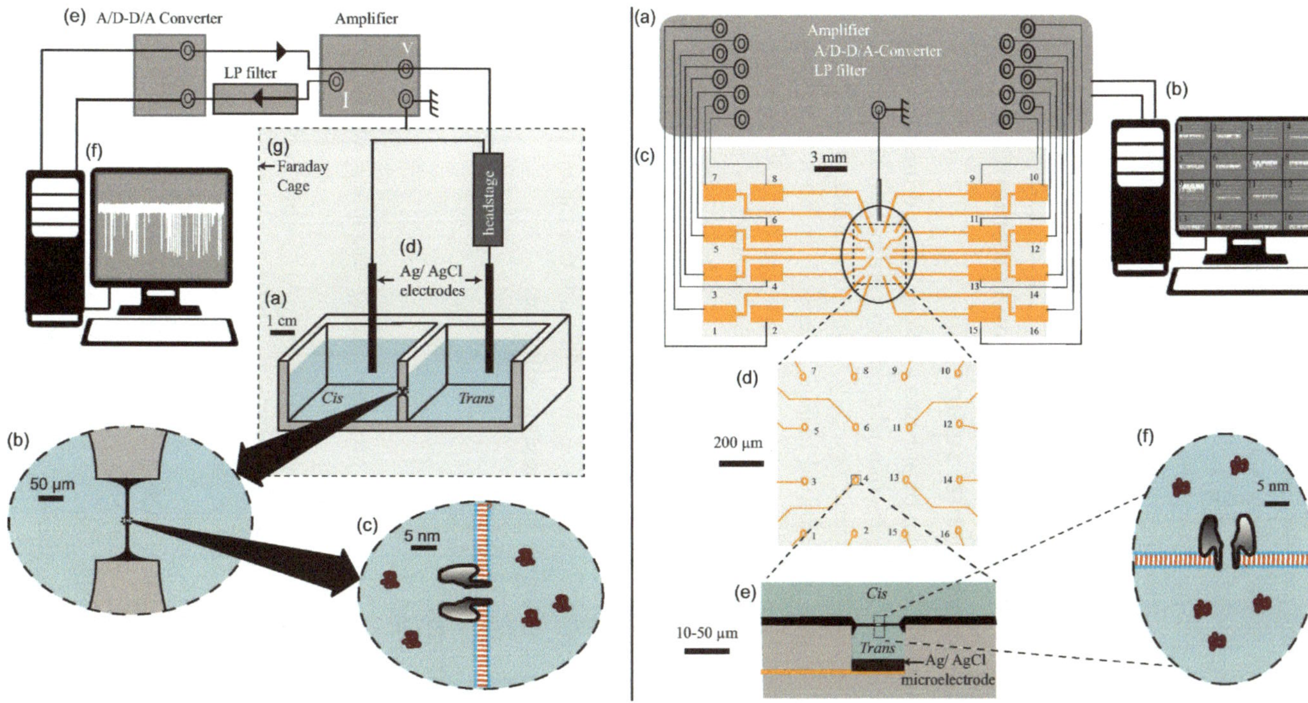

Fig. 3 See figure legend on opposite page.

2.2.2 Equipment

- water purification system
- Axopatch 200B patch-clamp amplifier with a CV-203BU Headstage (Molecular Devices)
- DigiData 1440A AD-converter (Molecular Devices)
- Peltier device controlled by a bipolar temperature controller (Warner Instruments, ref. no CL-100)
- liquid cooling device (Warner Instruments, ref. no LCS-1)
- Faraday cage (home-made)
- perfusion bilayer chamber (Warner Instruments, ref. no 64-0423)
- Delrin perfusion bilayer cup (Warner Instruments, ref. no 64-0427)
- Clampex 10.2 software (Molecular Devices)
- analytical balance (Mettler Toledo, ref. no XS205)
- freezers (−20 °C) and refrigerators (4 °C)
- hot plate (Velp Scientifica)
- micropipettes 0.5–10 µL; 2–20 µL; 20–200 µL; 100–1000 µL (eppendorf)
- silver wire (0.404 mm diameter, Alfa Aesar, ref. no 45852)

Fig. 3 Detail of two versions of the experimental setup. (Left) Classical single-channel recording setup. (A) Two chambers (*cis* and *trans*) of ~milliliter-capacity filled with an electrolyte solution are separated by a wall in which a permanent 150 µm-diameter hole has been drilled. (B) An insulating vertical planar lipid bilayer membrane is formed in the hole. (C) A single nanopore from a pore-forming toxin is inserted in the lipid bilayer and forms a nanometer-sized path connecting the two chambers; analyte molecules may be added in cis-, trans- or both chambers. (D) A pair of Ag/AgCl electrodes is used to apply a transmembrane voltage, the cis-electrode being at ground, and to record the ionic current flowing through the nanopore over time. (E) The analog electrical signal from the nanopore current is first pre-amplified by the headstage, then re-amplified and low-pass filtered before being digitized and finally interpreted and recorded with a computer software (F). (G) The chambers, electrodes and headstage are placed in a Faraday cage in order to be isolated from the surrounding electrical noise. (Right) Schematic of a chip-based multichannel recording device for simultaneous recording from multiple single nanopores. The multichannel amplifier (A) with built in AD converter and filter is controlled by a computer (B) and contacted to the 4 × 4 microelectrode cavity array (MECA16) chip (C) via its 16 contact pads numbered 1–16. Coplanar gold strip lines lead from the contact pads to the individual microelectrode cavities (MECs) which are situated in a 1 mm^2 central active area (D). (E) Shows the schematic of a MEC which is formed by photoresist lithography on a glass substrate. The bottom of each MEC contains a nanoporous Ag/AgCl microelectrode formed by microgalvanic Ag deposition followed by chloridization. Then membrane can be formed by remotely actuated painting or with the bubble formation. Reconstitution of pores (F) is done as in the classical set-up.

2.2.3 Reagent solutions

- lipid stock solution 1% DPhPc w/v in decane. Stored at −20 °C up to ~1 month
- trypsin stock 0.1 μg/μL in 1 mM HCl. Stored at −20 °C up to ~1 year
- pAeL stock 0.1 μg/μL in MilliQ-water. Stored at 4 °C up to ~1 year
- KCl 4 M, HEPES 25 mM, pH 7.5, stored at SATP up to ~3 months
- KCl 3 M, HEPES 25 mM, pH 7.5, stored at SATP up to ~3 months

2.2.4 Activation of pAeL

(1) Mix 15 μL of pAeL stock with 9 μL of trypsin stock, complete with MilliQ-water to reach 30 μL. Incubate the solution at SATP for 15 min, stop the reaction at 4 °C. The activated AeL can be stored for a week at 4 °C.

- The trypsin volume can be adjusted according to the trypsin activity

2.2.5 Agar salt bridges

(2) Add 1 g of agarose to 50 mL of KCl 3 M, HEPES 25 mM, pH 7.5 to obtain a mixture of 2% agarose w/v. Cover and heat the solution on the hot plate with gentle stirring (to avoid bubbles) until the agarose has completely dissolved (clear solution at T ~ 90 °C).

(3) Fill 10 μL tips with the agar solution and place the filled tips in fresh KCl 3 M, HEPES 25 mM, pH 7.5. The agar salt bridges can be stored at SATP for up to ~3 months.

- Be careful to avoid bubbles when filling the tips

2.2.6 Preparation of Ag/AgCl wires

(4) Solder a copper wire to an Au pin on one side to a silver wire on the other side.

- A teflon shield can be added on the soldering points.

(5) Immerse three quarters of length of the silver wire in the bleach for 10–15 min in the dark to obtain an Ag/AgCl electrode. A light gray color is observed. It can be stored up to 6 months.

- The Ag/AgCl electrodes should be placed in a dry dark space.

2.2.7 Lipid pretreatment

(6) Dip a 10 μL tip directly into the 1% w/v DPhPc lipid stock solution and paint over both sides of the Delrin perfusion bilayer cup aperture. Let it dry for at least 1 h at SATP to obtain a hydrophobic surface.

2.2.8 Setup assembly

(7) Switch on the patch clamp amplifier 30 min before use. This helps to thermally equilibrate the amplifier. Cool headstage components to about −15 °C (activate rear switch) for lower thermal noise.

(7.1) patch clamp settings

As an example, the following settings can be used:

- Whole-cell mode ß = 1
- Output Gain: $\alpha = 10$
- Low-pass Bessel-Filter: 5 kHz

(8) Assemble the perfusion bilayer chamber and the pretreated Delrin perfusion bilayer cup and place it on an antivibrational plate inside the Faraday cage to avoid external noise.

(9) Connect the Ag/AgCl electrodes to the CV-203BU headstage.

- Be careful to avoid any static electricity transfer, especially once the amplifier is on. You can use special gloves or discharge yourself by touching a grounded piece of metal before touching setup components

(10) Each electrode is immersed in an agar salt bridge placed in the perfusion hole. One into the *cis* compartment and the other into the *trans* compartment. The *cis* compartment is at voltage ground. The mean current value at this stage should be ~0 pA.

(11) Add 1 mL of KCl 4 M—HEPES 25 mM pH 7.5 in each compartment. The amplifier should be overloaded.

(12) Switch on the temperature controller and the liquid cooling device. Set the target temperature.

- If the target temperature is <15 °C or >30 °C, it will be easier to insert a pore first at room temperature (18–22 °C) and then set the desired temperature

(13) Open Clampex 10.2 software.

2.2.9 Lipid membrane formation

(14) Dip a 10 µL tip in the lipid solution, and gently paint over the *cis* aperture surface until the current drops to ~0 pA.

- Do not exceed 1 µL of lipid to prevent the formation of lipid plug

(15) Measure the capacitance of the membrane using the Clampex software. In our experimental conditions, a range of 60–75 pF indicates the appropriate thickness to insert a pore.

- If the capacitance is <60 pF:

- The membrane might be too thick to enable pore insertion. In this case destroy it with the patch clamp amplifier's ZAP function (high voltage pulse) and repeat step 14.
- Bubbles may have filled the aperture, in this case gently flush the aperture surface. Be careful not to damage the hydrophobic surface by applying too much pressure on the aperture.

- If the capacitance is >75 pF:
  - The membrane may be unstable and not withstand high voltage (>250 mV).

(16) Test the lipid membrane by applying a potential of 250–350 mV.
- If the membrane is broken, repeat step 14, 15.
- If there is any current jump, repeat step 14, 15.
- If there is increased fluctuation of the current trace, repeat step 14, 15.
- Observing a stable I value and a IRMS (+300 mV) > IRMS (0 mV) may indicate that the membrane is appropriate for nanopore insertion.

2.2.10 Insertion of the AeL nanopore

(17) Apply V = 0 mV and add 0.5 μL of the activated AeL close to the aperture in the *cis*-compartment chamber. The membrane might break or be replaced by a bubble. If it happens repeat step 14, 15.

(18) Apply V = +250 mV and wait until a current jump is observed. The current jump occurs when a pore is inserted in the membrane. Decrease the voltage to +50 mV to avoid any further pore insertions and check for any offset current at zero voltage. The conductance should be at 2 ± 0.1 nS at RT, KCl 4 M HEPES 25 mM pH 7.5.
- If the value is higher it might be explained whether by an unstable membrane or by second or third pore insertion. In this case, break the membrane and repeat step 14, 15.
- If no current jump signaling pore insertion has occurred after 20 min, repeat step 17.

(19) After a single AeL channel is inserted in the lipid bilayer, check the orientation and the background noise at different voltages.
- Regarding the orientation
- Physiological insertion of the Ael nanopore in the membrane from the *cis*-compartment corresponds to the AeL cap protruding from the membrane toward the *cis*-compartment; in this case, current rectification is observed so that for a given voltage magnitude, current magnitude is lower for positive voltage (for example: $|I(+50\,mV)| = 99 \pm 1$ pA; $|I(-50\,mV)| = 102 \pm 1$ pA (at 20 °C).

Two other *I/V curves* might be observed:

- For a given voltage magnitude, current magnitude is higher for positive voltage (for example: |I(+50 mV)| = 102 ± 1 pA; |I(−50 mV)| = 99 ± 1 pA (at 20 °C). This indicates an inverted pore, Break the membrane and repeat step 14, 15, unless you want to work with an inverted pore.
- For a given voltage magnitude, current magnitude is the same for positive and negative voltages (for example: |I(+50 mV)| = 100 ± 1 pA; |I(−50 mV)| = 100 ± 1 pA (at 20 °C). Break the membrane and repeat step 14, 15.

(20) Once the nanopore has been certified, record blank current vs time traces at desired voltages.
- Check the offset before and during the recording.

2.2.11 *Adding analyte and recording*

(21) Apply V = +25 mV and add X μL of the analyte at desired concentration in the *cis* or *trans* compartment. Gently mix the *cis* or *trans* compartment, typically with a final mixing volume of $2^*V_{chamber}$ (for example 20*100 μL for a 1 mL chamber) to obtain a homogeneous solution.
- Be careful: the membrane and the pore are fragile! The mixing should be done as gently and as far from the aperture as possible to avoid breaking the membrane or losing the pore.

(22) Start recording the current trace at the desired voltage for X minutes at X °C.
- Check the offset every 5 min.

2.3 Alternative configuration: Parallel nanopore recording (Orbit16)

2.3.1 *Materials*

All solutions should be prepared using ultrapure water (18.2 MΩ cm at 25 °C)

- wild-type proAerolysin (wt-pAeL) is obtained from pET22b(+)//pAeL-HIS_6 in an *E. coli* BL21 (DE3)pLysS overexpression system
- cells are cultivated at 37 °C, expression is induced via the *lac* operon
- preprotein is purified from cell lysate by its poly-histidine tag, followed by size-exclusion chromatography
- preprotein stocks are stored at −80 °C at a concentration of 1 mg mL^{-1}

- 1,2-diphytanoyl-*sn*-glycero-3-phosphocholine (Avanti Polar Lipids, ref. no 850356C)
- octane (Sigma-Aldrich, ref. no 296988-100ML)
- potassium chloride (Carl Roth, ref. no 6781.1)
- TRIS (Pharmacia Biotech, ref. no 17-1321-01)
- EDTA (Sigma, ref. no E-4884)
- DTT (Sigma, ref. no 43815-25G)
- Trypsin (Promega, ref. no V511C)

2.3.2 Equipment

- water purification system CDUFBI001 (MilliQ)
- multi-electrode-cavity-array MECA16 (Ionera Technology GmbH)
- Orbit16 (Nanion Technologies GmbH)
- stirring magnet 2 × 3 mm (Nanion Technologies GmbH)
- PTFE inlay (Nanion Technologies GmbH)
- rubber O-ring (Nanion Technologies GmbH)
- Elements Data Reader software EDR v3.8.6 (Elements SRL)
- Ag/AgCl sintered pellet electrodes 1.0 × 3 mm (Neurospec, ref. no NS-7671)
- analytical balance (Mettler Toledo, ref. no AB54)
- freezers (−80 °C/−20 °C) and refrigerators (4 °C)
- hot-plate (Heidolph, ref. no MR3001)
- micropipettes 0.1–2 μL, 1–10 μL, 20–200 μL (Gilson)

2.3.3 Reagent solutions

- Lipids are aliquoted, dried in an Ar flux and stored at −20 °C.
- Lipid stocks are dissolved in octane (1% DPhPC w/v) and stored at −20 °C.
- Trypsin stocks of 12.5 ng μL^{-1} in 1 mM HCl are stored at −80 °C (adjust concentration to activity).
- Wild type pAeL stocks of 1 μg μL^{-1} are stored at −80 °C.
- KCl 4 M, Tris 25 mM, pH 7.5, Ag/AgCl saturated is stored at SATP
  Note: Saturation of solutions with Ag/AgCl is required when using KCl concentrations >3 M to counteract the dissolution of Ag/AgCl electrodes by the formation of soluble complexes by $AgCl + Cl^- \rightarrow AgCl_2^{2-}$ and $AgCl_3^{2-}$ in highly concentrated solutions of chloride (Brown & MacInnes, 1935; Ito, Hachiya, Baba, Asano, & Wada, 1995). This is not necessary at lower ionicities.

2.3.4 Activation of proAerolysin

(1) Dilute pAeL stock 1:4 in PBS, containing 100 μM EDTA and 100 μM DTT.
(2) Mix 1 μL pAeL stock with 1 μL trypsin stock, vortex, spin-down, incubate for 15 min at SATP.
(3) Dilute AeL sample 1:5 with PBS, containing 100 μM EDTA and 100 μM DTT.
(4) Store activated AeL on ice or at 4 °C for maximum 1 week.

2.3.5 Preparation of Ag/AgCl counter electrodes

(1) Solder a sintered Ag/AgCl pellet electrode to a connection pin.
(2) Chloridize pellet electrodes galvanostatically for 5 min at 150 mA in 1.5 mM KCl.
(3) Repeat this process whenever pellet electrode exhibits a white color.

2.3.6 Priming of MECA16

Note: these priming steps are required for enhanced MECA16 electrode stability with 4 M KCl solutions. For lower ionicities (up to 3 M), MECA chips can be used as delivered.

(1) Dechloridize MECA16 chips galvanostatically at 25 nA in 150 mM KCl.
(2) Rechloridized MECA16 chips galvanostatically for 7 min at 25 nA in 150 mM KCl.
(3) Rinse MECA16 chips with ddH_2O, IPA, EtOH, ddH_2O, pre-dry in a nitrogen flux and heat to full dryness at 100 °C for 10 min before usage.

2.3.7 Setup assembly

(1) Connect the Orbit16 system to your computer via USB
(2) Connect test-board to the Orbit 16 and compensate for all offsets via EDR software
(3) Remove test-board and assemble the MECA16 chip in the provided cavity
(4) mount the PTFE electrolyte chamber and o-ring in the PCB on top of the chip
(5) lower the electrolyte chamber housing PCB to the chip surface using the lever on the right side
(6) Add 150 μL electrolyte (4 M KCl) into the chamber
(7) Ensure electrode wetting by gently moving a rubber piston across the chip's surface

(8) Place the pellet counter electrode in the electrolyte and connect it to the Orbit PCB by its pin

(9) Apply a repeating seal-test (−10 mV for 10 ms, rectangular) via EDR software

(10) The Orbit16's amplifier will saturate if electrodes are wetted entirely during voltage application

2.3.8 *Automated lipid membrane formation*

(1) Add a stirring magnet to the electrolyte chamber and assure that its resting position is next to the electrode array on the MECA16

(2) Set a pipette to 0.2 μL and dip its tip into the lipid solution

(3) Eject the lipid on the chip surface, next to the electrode array

(4) Set the stirring speed to 40 rpm

(5) Spread the lipid over the chip's surface by using the Orbit16's machinery

(6) Change the stirring speed to −15 rpm and spread lipid until all 16 bilayers are formed

(7) Painted membranes will result in a GΩ seal

(8) Apply a membrane capacity test (50 mV for 50 ms, triangular) via EDR software

(9) Real bilayers result in a capacity of 10–15 pF

(10) Destroy all membranes (zap: 600 mV, 200 ms) until all show >10 pF capacity

(11) Repaint membranes at −15 rpm

2.3.9 *Insertion of the aerolysin nanopore*

(1) Add 0.5 μL of the activated wt-AeL solution to the electrolyte

(2) Pores will insert into bilayers automatically within <10 min

(3) Check the orientation of the pore:

(4) Physiological insertion of a nanopore from the *cis*-compartment into the membrane corresponds to the aerolysin cap protruding from the membrane towards the *cis*-compartment; in this case, current rectification is observed when, for a given voltage magnitude, the current magnitude is found to be slightly lower for positive voltage. If no rectification is observed, zap the membrane and repeat Sections 2.3.8 and 2.3.9.

(5) Activate membrane rupture monitor in EDR software

(6) Record blank current vs time traces at desired voltages

Note: Pore insertion, which follows a Poisson process, can not be fully controlled. The basic procedure consists in adjusting the final concentration of protein in order to obtain a maximum of single insertions. Between 5 and 10 single insertions can be routinely achieved. This has to be repeated for each new batch of pore protein, as now indicated in the text. Single pore insertion can further be optimized by applying a high command voltage (e.g. 200 mV) until a pore is inserted and immediately switching to a lower voltage upon insertion of one pore. This protocol can be automated in the current version of the EDR software (EDR3.8.6).

2.3.10 Analyte addition and recording

(1) Add analyte to the electrolyte chamber until the desired concentration is reached
(2) Wait for 10 min in order to ensure homogeneous analyte distribution
(3) Set the sampling rate to 20 or 50 kHz, the current range to 200 pA and select no filter
(4) Apply the desired voltage via EDR software and start recording
(5) Record data until >1000 events are collected
(6) If current drift is >1 pA min^{-1}, break membrane and repeat from Section 2.3.6

2.4 General considerations on high-resolution recording

Some aspects of the recording apparatus can be optimized for high-resolution-recordings. These are:

(1) The use of small apertures and of short wire connections between the active electrode and the input stage of the amplifier will reduce the total capacitance load and thereby the component of current noise that is due to voltage fluctuations in the amplifier head stage (Levis & Rae, 1998; Sigworth & Klemic, 2005).
(2) Using the capacitive feedback mode if available (i.e. Axopatch 200B (Molecular Devices), dPatch Sutter) will significantly reduce noise provided that the total capacitive load is kept below approximately 10 pF (see 1)). However, it has to be borne in mind that resets of the feedback capacitance are necessary when the voltage on the capacitor reaches the power supply limit (usually +/− 10 V). For instance, with average nanopore currents in the 100–200 pA range the 1 pF-capacitor of the Axopatch 200B has to be reset every 100–50 ms. During resets, the current output is held at the value immediately before the reset is activated for approximately 50 μs, resulting in a loss of between 0.05% and 0.1%

of data for this current range. This is not a major problem in most applications, but should be borne in mind.

(3) Using the amplifier's capacitance compensation circuit to reduce artifactual current noise due to noisy voltage command inputs: rapid fluctuations in the command voltage input will induce large current fluctuations at high bandwidth. For instance, on the Axopatch200B, correctly setting the pipette capacitance compensation (c_{fast}, τ_{fast}) produces an appreciable decrease in high-frequency noise. This becomes of practical importance if recordings are to be performed at bandwidths significantly exceeding 10 kHz, because it is in this higher frequency range that DAC inputs show voltage noise.

(4) In order to obtain a faithful digital record of the current signal, sufficient preamplification is necessary to make the best use of the dynamic range of the ADC. With a 16 bit ADC with ±10 V range, a gain of 10 mV/pA corresponds to ±1 nA range so that the digital resolution δ will be 2 nA/2^{16} = 30.5 fA/bit. Importantly, as long as δ is not significantly smaller than the r.m.s. noise of the analog signal at the recording bandwidth chosen, additional white noise in the frequency band from DC to $f_s/2$ (where f_s is the sampling frequency) will be produced by the digitization. In the example shown in Fig. 4, a signal was filtered at a cut-off frequency $f_c = 5$ kHz using the 4-pole Bessel filter of the Axopatch 200B and digitized at $f_s = 500$ kHz using different preamplifier gains. It can be seen that a gain of 10 mV/pA is the lowest one, under these conditions, where digitization does not induce an appreciable increase of r.m.s. noise of the digitized signal over the 0.355 pA measured on the analog meter. In other words, δ should be at least 10 times smaller than the r.m.s. of the analog signal noise to avoid digitization noise. However, as is evident from the traces shown, even at 10 or 20 mV/pA, there is still appreciable jitter on the trace. In fact, a noise power of δ^2/f_s is superimposed on the power spectrum of the digitized trace (Maertz, 2008).

In the context of Np-SMSD, the focus is on the point average approaching the true mean current value for an event as closely and as quickly—i.e. with as few averaged samples—as possible in order to be able to correctly determine the relative amplitude of also the shortest resistive pulses. As shown in Fig. 4C, even for relatively low amplitude resolutions (e.g. 5 mV/pA) the point average rapidly converges towards the true mean (0 in this case). However, for gains below 5 mV/pA there is clearly an excess deviation for small numbers of averaged points. Clearly, therefore, the effect of digitization noise on Np-SMSD can be mitigated somewhat by using rapid sampling well above the Nyquist criterion of $2f_c$.

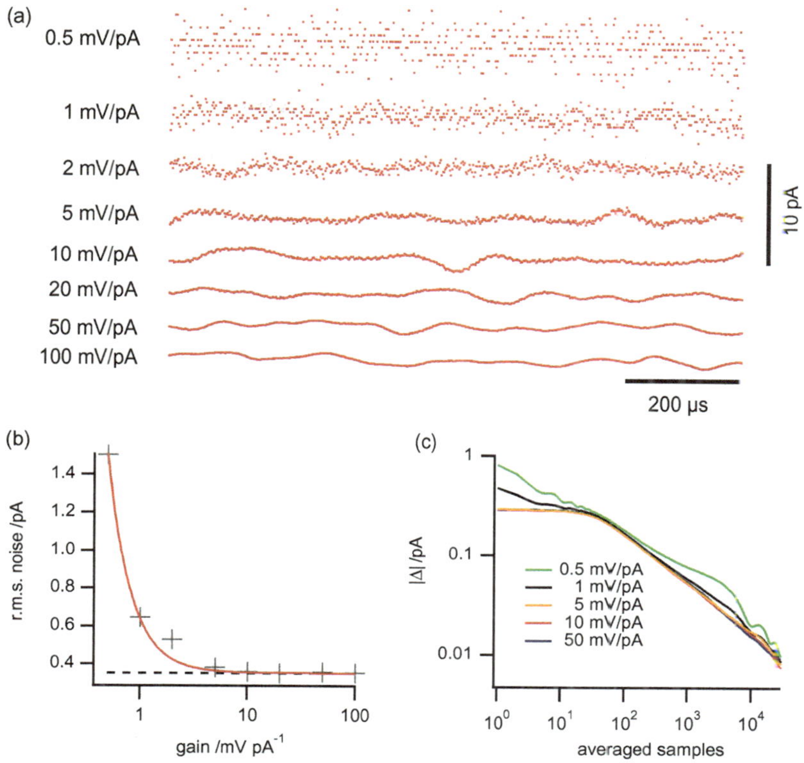

Fig. 4 Effect of digitization noise on level averaging. (A) 1 ms data traces (membrane-pore model with 900 MΩ, 2 pF) digitized at $f_s = 500$ kHz after filtering (8-pole Bessel, Axopatch 200B, $f_c = 5$ kHz) after preamplification to the indicated gains. (B) Plot of signal r.m.s. noise vs preamplification gain. Red curve shows a fit with $y_0 + Ax^n$ with y0 and n fixed at 0.355 pA and −2, respectively. (C) Average absolute value of the difference between the point average and the true mean (0 pA) of a 30,000 point trace plotted against the number of samples included in the average. Each curve is an average of 9.399 trials with starting points separated by 50 samples along a 1 s (500,000 samples) trace.

3. Data analysis of nanopore recordings

3.1 Resistive pulse analysis based on the current-threshold detection approach (Fig. 5)

In this approach, the detection of each individual current blockade in a nanopore current vs time recording is based on a two current-thresholds method: a possible blockade event starts when the current value becomes smaller than a first threshold Th_1, and ends when the current returns to a value greater than Th_1.

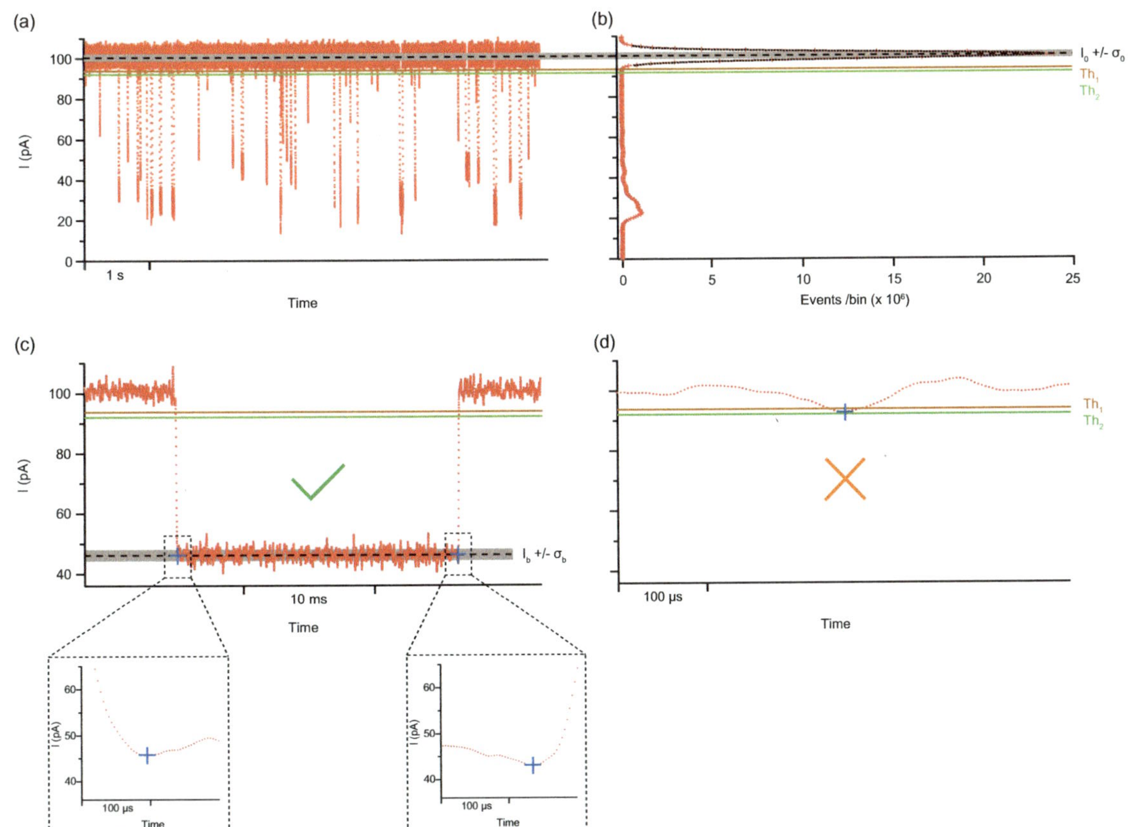

Fig. 5 See figure legend on opposite page.

If the mean current value during the event is smaller than a second threshold Th_2, the event is considered as an analyte-induced current blockade; otherwise the event is dismissed.

We typically use $Th_1 = I_0 - 4\sigma_0$ and $Th_2 = I_0 - 5\sigma_0$, where I_0 and σ_0 are respectively the mean value and the standard deviation of a Gaussian fit of the open-pore current distribution.

This value of Th_1 appears to be, on the one hand, sufficiently small to exclude the overwhelming majority of the open-pore current noisy fluctuations from blockade detection, as >99.99% of a Gaussian distribution falls in the range $I_0 \pm 4\sigma_0$, and, on the other hand, sufficiently large not to exclude from event detection current drops that significantly deviate from the open-pore current and may correspond to analyte-induced blockades.

Please note that, while this choice of thresholds values appears to be appropriate in a large number of experimental conditions, a different choice may sometimes have to be made in order to adapt to the specificity of a given situation.

For each event, the range of points used to compute the mean event current to be compared with Th_2 starts with the first point of current increase after having monotonically decreased below Th_1, and ends with the last point of current decrease before monotonically increasing above Th_1.

Fig. 5 Data analysis: a current-threshold approach for the detection of current blockades in a nanopore current vs time recording. (A) Portion of a typical nanopore current vs time recording, showing multiple current blockades. (B) Histogram of the current values of the complete current vs time recording ($\approx$15 min) from which the portion shown in (A) was extracted. A Gaussian fit (black dotted line) provides the mean value I_0 and standard deviation σ_0 of the open-pore current distribution, respectively represented as a black dashed line and as a gray rectangle vertically centered on I_0 and of width $2\sigma_0$, spanning panels (A) and (B). In this approach, the detection of each individual current blockade is based on a two current-thresholds method: a possible blockade event starts when the current value becomes smaller than a first threshold Th_1 (orange solid line, here $Th_1 = I_0 - 4\sigma$), and ends when the current returns to a value greater than Th_1. If the mean current value during the event is smaller than a second threshold Th_2 (green solid line, here $Th_2 = I_0 - 5\sigma$), the event is considered as an analyte-induced current blockade (C); otherwise the event is dismissed (D). In (C) and (D), blue crosses (highlighted in zoom-in insets of panel (C)) indicate the start and end of the range of points used to compute the mean event current to be compared with Th2 (in the example of panel (D), start and end points are identical). If the event is accepted as a blockade, the same range of points is used to compute the other blockade quantities (see text for details). In (C), the mean value I_b and standard deviation σ_b of the residual current during the blockade are respectively represented as a black dashed line and as a gray rectangle vertically centered on I_b and of width $2\sigma_b$.

If the event is accepted as a blockade, the same range of points is used to compute other blockade quantities, some of the main of which are:

- the mean residual current Ib during the blockade (already computed as the mean event current to be compared with Th_2), and its value relative to the mean open-pore current, I_b/I_0.
- the standard deviation σb of the current distribution during the blockade, and its value relative to the standard deviation of open-pore current distribution, σ_b/σ_0.
- the blockade duration Δt.

A statistical analysis of these different quantities is then performed, which involves at least several hundreds, more typically several thousands, of blockades. Typically, this analysis starts by plotting a scatterplot of Δt vs I_b/I_0 superimposed with a histogram of I_b/I_0-levels using a suitable bin width $\Delta I_b/I_0$ (usually 0.001–0.002). Maxima corresponding to the diverse molecular species interacting with the pore become immediately visible in such plots (see e.g. Fig. 7).

It must be noted that the chosen definition of the range of points that defines each blockade excludes the initial current decrease and final current increase portions of a blockade, whose relative contributions to the computed blockade quantities, if they had been included, would increase as blockade duration decreases.

It is worth mentioning that the use we make of this detection method involving two current thresholds slightly differs from the initial purpose of the method, where every event for which the current crosses Th_1 back and forth is considered as a current blockade.

Then the distinction is made between the shallowest blockades, whose mean current values are greater than Th_2 and which are considered as "bumping" events where the analyte bumps on the nanopore entrance before going back to its initial bulk compartment, and the deepest blockades, whose mean current values are smaller than Th_2 and which are considered as resulting from a strong analyte-nanopore interaction

In our case, the use of two different current thresholds is rather a reminiscence of this previously used approach, and a single-threshold method ($Th_1 = Th_2$) would have been sufficient, as previously-considered "bumping" events are here rejected; furthermore, blockades may be sorted according to their mean current values (and/or other criteria) after blockade detection.

3.2 Resistive pulse analysis based on thresholding the differentially rectified signal (Fig. 6)

Here, the original current vs time trace is first transformed into what might be called differentially rectified traces as follows: each of two derived traces follows the movement of the original signal in one direction (increase or decrease of current) until it reverses direction for the first time whereupon the derived trace is reset to 0. In contrast to simple rectification, this algorithm—which was originally proposed by Ivan Cohen (Institut de Biologie Paris Seine) as an alternative to the first derivative for automated detection of synaptic currents (Ankri, Legendre, Faber, & Korn, 1994) and later applied to nanopore recordings by Norbert Ankri (Université de la Mediterannée, Marseille)—only takes into account portions of the signal where it monotonically moves in one direction. And unlike the first derivative dI/dt, the signal derived in this way remains true to the phase of the original having maxima wherever a monotonic change in one direction ends (see Fig. 6). Thresholding is performed on these derived waveforms. The threshold is set so that the up- and downward excursions due to noise are cleared (typically at $>5\sigma_0$). The algorithm then determines the first and the last point of the monotonic change in current as the onset and end, respectively of a transition, whereas the portions of the trace between transitions is defined as a current level of which all points are averaged.

This detection principle is particularly suited for the detection of changes in current level during blocked states, i.e. complex resistive pulses as shown in Fig. 6, because unlike thresholding on the original trace it is not oblivious of transitions occurring below a certain current value. It also enables a very precise definition of state dwell time as the interval between the onset of the transition leading to the state and the onset of the next transition.

The output of the first stage of detection is a list including consecutive open and blocked states with parameters onset of transition leading to the state, number of samples averaged, level mean current value and standard deviation.

In a second step, relative, I/I_0-values are generated by setting a threshold mean current value above which all levels are to be considered open or unblocked levels and then dividing each absolute level by a moving average of 10 neighboring open levels. In this way, slight drifts in apparent pore conductance due to e.g. evaporation or electrode polarization are automatically accounted for.

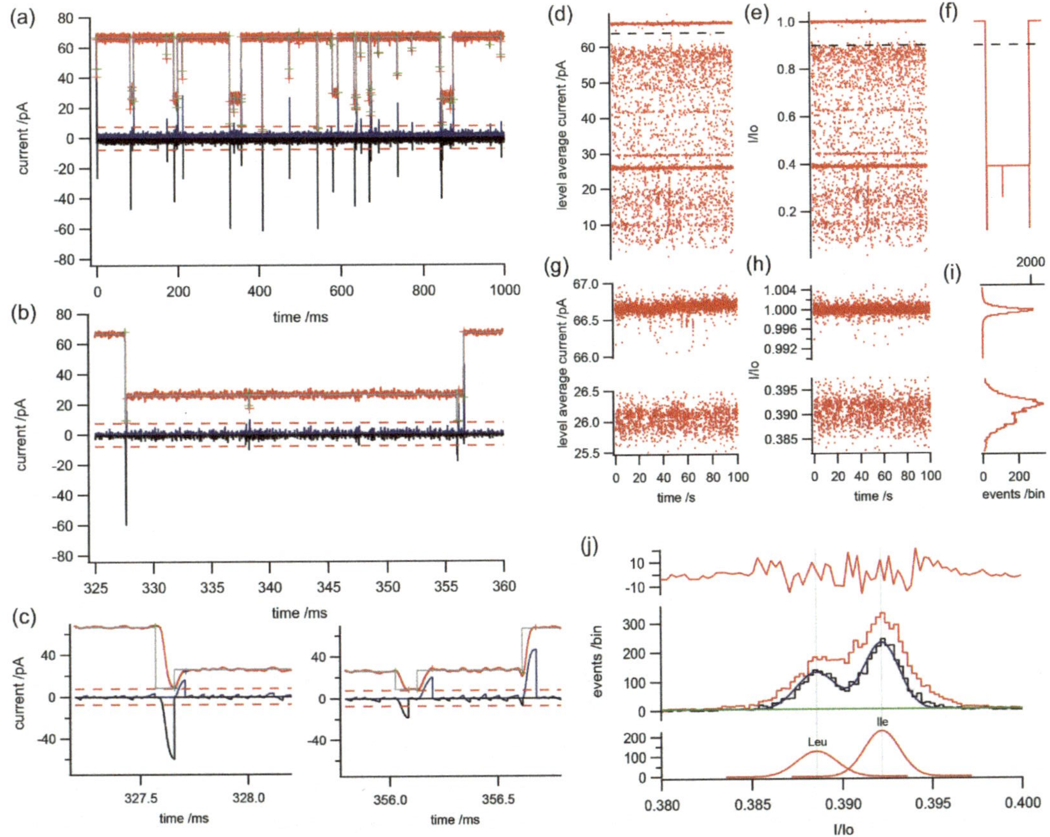

Fig. 6 See figure legend on opposite page.

The setting of the threshold for open-state definition is thus performed after point averaging of levels, which itself is done without assuming a threshold separating open and blocked states. Simple inspection of the scatterplot of absolute levels vs time shows that after point averaging of levels this distinction is obvious. In the example shown in Fig. 6 there is a window of 5 pA or 7% of the total signal amplitude in which the threshold may be placed without affecting the result.

As mentioned above, thresholding after differential rectification does not rely on an absolute current threshold for detection and is therefore able to detect blocked substates such as those shown in Fig. 6. Such substates have been observed notably with oligonucleotides (Boukhet, Halimeh, Baaken, & Behrends, 2017) but also with peptides (Ensslen, Ouldali, Oukhaled, & Behrends, 2020; Ouldali et al., 2020). For the purpose of Np-SMSD substates pose a problem in so far as they produce additional variation in the current signal that will tend to jeopardize the precision with

Fig. 6 Resistive pulse detection and analysis based on differential rectification. (A–C): 1 s, (A), 35 ms (B), and 1 ms (C) original current traces (red) recorded on a MECA16 chip (aerolysin, +40 mV *trans*, 2 M KCl + 2 M KNO_3 in the presence of R_7L and R_7I peptides) filtered (analog, 8-pole Bessel) at $f_c = 50$ kHz, digitized at $f_s = 1$ MHz, 16 bit after preamplification to 100 mV/pA ($\delta = 3.05$ fA). Blue and black trace are differentially rectified signals; dashed red lines are positive and negative threshold for transition detection. Green and red cross symbols are start and end of detected transitions, continuous gray line is idealized record constructed from the detection. (B) Is selected to show one resistive pulse lasting approximately 29 ms but consisting of three short-lived deep-blocked and three long-lived shallow sublevels (details see text). (C) Left panel details the onset of the resistive pulse with a short visit to a deep blocked state followed by a transition to a long-lived state. Right panel details the second interruption of the long lived state by a deep state and the final unblock. Note the final downward dip of the original trace likely signaling an unresolved excursion to a deep state just before unblock that is missed at this bandwidth. (D) Scatterplot of level average current values vs time, dashed line is threshold separating open and blocked levels; (E) as in (D) but for data normalized to moving average of open pore current (I/I_0). Dashed line in upper panel is threshold used for definition of resistive pulses (see text). (F) Illustration of the definition from the data in (E) of the complex resistive pulse shown in (B). (G and H) Magnified view of level current averages and I/I_0 values for open pore (upper panels) and blocked states corresponding to interaction of R_7L and R_7I peptides with the aerolysin pore (lower panels). Note the slight upward drift of average current in (G). (I) Histograms of I/Io from (H). (J) Middle panel: histograms of level averaged (red contour) and resistive pulse averaged (black contour) I/Io values with superimposed fit (blue line) with the sum of the two gaussians shown in the lower panel (residuals in upper panel). Note the increased separation of maxima in the distribution of resistive pulse averages with respect to level averages.

which the mean of the long-lived current level that carries information on molecular identity can be approximated. As shown in Fig. 6, detection and separation of these blocked substates allows to remove their contribution and enhance the precision and separation of species.

To this end, the list of consecutive I/I_0-values is subjected to a second thresholding operation defining a relative current threshold of typically 90% below which all levels are considered blocked levels. Each sequence of contiguous blocked levels (i.e. uninterrupted by an open level) is the defined as a resistive pulse. In the absence of any substates, each resistive pulse would contain exactly one blocked level. In the example shown in Fig. 6, there are six blocked levels, an initial short-lived deep block followed by a long-lived state that is interrupted twice again by short-lived deep states. For details regarding the procedure of removing the contribution of deep states, the reader is referred to the supporting information in (Ouldali et al., 2020).

3.3 Representation and interpretation of blockade quantities in the context of Np-SMSD (Fig. 7)

Fig. 7 presents a typical analysis of a current vs time recording following the current-threshold approach for the detection of blockades (see Section 3.1), and illustrates the size-discrimination of analyte molecules in the case of the interaction of a pore-forming toxin nanopore with a polydisperse mixture of homopolymers of different monomer numbers. The example shown here concerns the discrimination of polyarginine peptides of degree of polymerization (d.p.) 4–10.

Accordingly, the histogram of I_b/I_0 values exhibits seven clearly distinct peaks, corresponding to seven populations of homopolymers of different monomer numbers in solution (Fig. 7A), which may be individually identified e.g. by comparison with independent measurements in presence of monodisperse polymer of a given d.p.

The value of I_b/I_0 decreases as the number of monomers increases, and two adjacent populations differ by only one unit in monomer number.

Comparing panels (A), (B) and (C) of Fig. 7 shows that distributions of I_b/I_0values corresponding to polymers of different monomer numbers are far more distinct than corresponding distributions of Δt and of σ_b/σ_0, which are more dispersed and significantly overlap.

This makes the distribution of I_b/I_0values the primary criterion on which discrimination is based, a common feature of a large majority of nanopore-based size-discrimination studies to date; in addition, the peakyness of

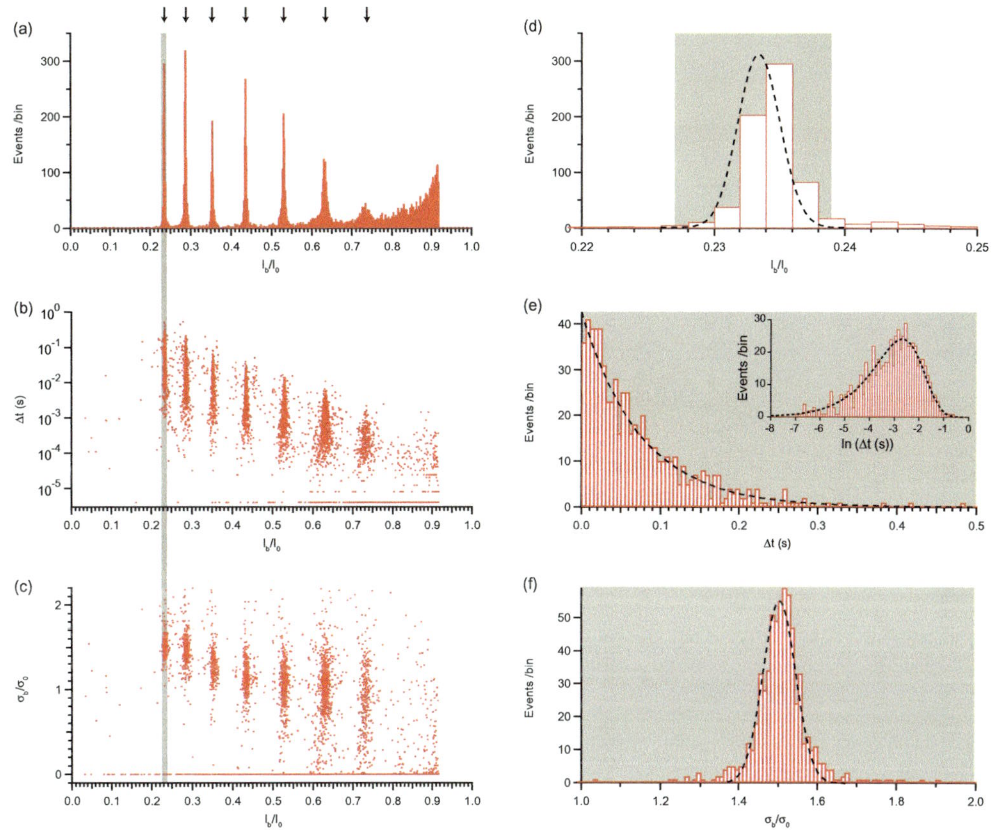

Fig. 7 See figure legend on next page.

I_b/I_0distribution is of particular importance from the perspective of single-passage identification of individual molecules (Ouldali et al., 2020).

Nevertheless, statistical analysis of Δt, σ_b, σ_b/σ_0 distributions, as well as other parameters (e.g. frequency spectra of blocked states vs open states, &c. (Ouldali et al., 2020), are of great importance as it may provide additional informations on the analyte/nanopore interaction process, as well as additional criteria to characterize/discriminate between different populations.

The quality of the discrimination of two adjacent of I_b/I_0 populations A and B may be evaluated by calculating the resolution parameter between of r as:

$$r = \frac{\Delta_{\mathrm{AB}}}{1/2(W_A + W_B)} \tag{7}$$

where Δ_{AB} is the distance between the maxima of peaks A and B, and W_A and W_B respectively represent a characteristic width of peak A and of peak B (Christophe, 1971; Rony, 1968).

Fig. 7 Data analysis and nanopore-based single-molecule size-discrimination. Typical data analysis of a current vs time recording showing size-discrimination of analyte molecules in the case of the interaction of a pore-forming toxin nanopore with a polydisperse mixture of homopolymers of different monomer numbers. (A) Histogram of mean relative residual current I_b/I_0values, which, in this example case, exhibits seven distinct peaks (indicated with black arrows) corresponding to seven populations of homopolymers of different monomer numbers, I_b/I_0decreasing as the monomer number increases; two adjacent populations differ by only one unit in monomer number. (B) Scatter-plot of blockade duration Δt vs $I_b/I_0$and (C) scatter-plot of relative standard deviation σ_b/σ_0vs I_b/I_0; each red dot corresponds to a single current blockade. (D) Zoom-in view of histogram of panel (A), in the I_b/I_0 portion corresponding to the left-most population. In order to analyze each I_b/I_0population individually, a given I_b/I_0 population i is defined as the subset of blockades whose I_b/I_0values fall in the range $\langle I_b/I_0\rangle_i \pm 3\sigma_i$ (gray rectangle, reproduced in panels (A), (B) and (C)), where $\langle I_b/I_0\rangle_i$ and σ_i are respectively the mean value and standard deviation of a Gaussian fit (black dashed line) of the distribution of I_b/I_0 values in the corresponding histogram peak. (E) Histogram of Δt values and (inset) histogram of the natural logarithm of Δt values in the case of the left-most I_b/I_0 population. Black dashed lines are proportional to a single-exponential probability density function, which nicely fits the histograms; the maximum of the distribution of ln(Δt) values occurs at ln(τ_i), where τ_i is the mean blockade duration of population i (Sigworth & Sine, 1987) (here $\tau_i = 68.5$ ms). (F) Histogram of σ_b/σ_0 values in the case of the left-most I_b/I_0population. The black dashed line is a Gaussian fit.

Here, each I_b/I_0 population i is nicely fitted by a Gaussian distribution (Fig. 7D), with the mean value $\langle I_b/I_0 \rangle_i$ and standard deviation σ_i, respectively, corresponding to the position of the peak maximum and to a characteristic peak width W_i.

In Fig. 7A the r values typically fall in the range [20–50], allowing a clear discrimination between the different populations.

However, in cases where two adjacent peaks partially overlap, the height of the valley between the peaks may have a significant impact on the discrimination quality, while it may only slightly affect the resolution parameter, depending on how W values are calculated.

In such cases, another parameter that appears appropriate to evaluate the quality of the discrimination is the valley-to-peak ratio/P, which reads:

$$V/P = h_v/h_p \tag{8}$$

where h_v and h_p are respectively the minimum height of the valley and the maximum height of one the adjacent peaks (Christophe, 1971); the symmetry (or not) of the peak heights has also to be considered when discussing the V/P ratio.

In the case of well-distinct peaks, such as in Fig. 7A, V/P ratios tend towards zero.

It is worth mentioning that increasing blockade durations allows to reduce the contribution of noisy current fluctuations in the dispersion of time-averaged I_b values (Fig. 7B), which may be very helpful to enhance the distinction between I_b/I_0 populations in case of initially poor discrimination (see e.g. Sections 4.1.1 and 4.1.2).

In order to analyze each I_b/I_0 population individually, a given I_b/I_0population i is defined as the subset of blockades whose I_b/I_0values fall in the range $\langle I_b/I_0 \rangle_i \pm 3\sigma_i$ (Fig. 7D), which includes >99.7% of the Gaussian distribution.

As an illustration, panel (E) of Fig. 7 presents the linear histogram of dwell times and the linear histogram of their natural logarithm for the case of the deepest (left-most) I_b/I_0 maximum.

The latter representation is particularly sensitive to multiple exponential components, especially in the case of large time constants (Sigworth & Sine, 1987). The maximum of the distribution of $\ln(\Delta t)$ values occurs at $\ln(\tau_i)$, where τ_i is the mean dwell time for the population i (Sigworth & Sine, 1987); here $\tau_i = 68.5$ *ms*.

Here, both histograms are well-fitted by a single-exponential distribution, indicating a first order kinetic process as usually found in the polymer sizing regime (see Section 2).

In addition, σ_b/σ_0 values appear to be distributed in Gaussian fashion for the case of the left-most I_b/I_0 population (Fig. 7F).

4. Selected examples of Np-SMSD

This section aims at illustrating Np-SMSD of different polymeric molecules in examples delected from the work of our groups. In particular, the exquisite sensitivity of AeL nanopore for Np-SMSD is highlighted, going from initial experiments with PEG polymers up to the most recent results with short homo- and hetero-peptides, that may inspire an Np-SMSD approach towards single-molecule sequencing of peptides/proteins.

4.1 PEG polymers

4.1.1 Effect of nanopore nature: αHL vs AeL

Following pioneering Np-SMSD works that exclusively used αHL as a nanopore sensor (Robertson et al., 2007), the question of whether another nanopore could perform similar discrimination of polydisperse PEG polymers was investigated in 2015 by Baaken et al. using AeL, another pore-forming toxin (Baaken et al., 2015).The top half of Fig. 8 compares results obtained in the case of the interaction of a *cis*-added mixture of polydisperse PEG 1500, and highly monodisperse PEG 1250 (28 r.u.) in excess as an internal standard, with αHL nanopore and with AeL nanopore, in KCl 4 M.

First, αHL and AeL exhibit a very different voltage-dependence of their interaction with PEG: while in the case of αHL, PEG-induced resistive pulses are observed with both *trans*-positive and *trans*-negative voltage gradients, with maximal dwell times at +40 mV *trans* (Rodrigues et al., 2008) in the case of AeL *trans*-negative voltage strongly increases the duration of PEG/nanopore interaction, with a maximum around −120 mV *trans*.

The voltage gradients maximizing dwell times are obviously also the optimal voltage gradients for Np-SMDS, as increasing blockade durations allows for efficient time-averaging to suppress low-frequency noise contributions and increase the precision of current level determination.

As a main result, comparing αHL and AeL in their respective best-resolution voltage conditions (+40 mV and −120 mV respectively) reveals that AeL provides additional size-discrimination of short PEG species (down to 17–18 r.u. in this case), not observed with αHL (low limit around 24–25 r.u.) (top half of Fig. 8, right panels). In addition to the increased dwell times for

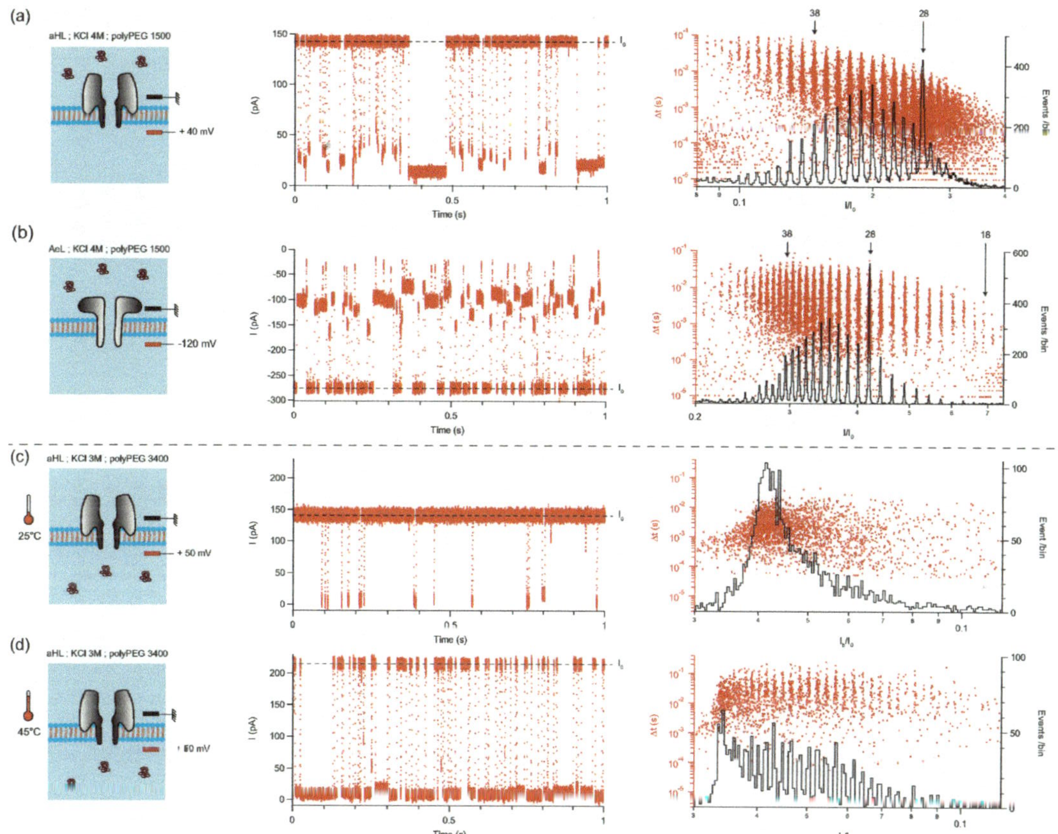

Fig. 8 See figure legend on next page.

shorter PEGs, the fcat that AeLs optimal sensitivity voltage gradient is three times larger in absolute terms than that of αHL represents an advantage in terms of signal to noise because of the larger driving force more than compensates the lower conductance of AeL compared to αHL (~2.3 vs ~3.5 nS in 4 M KCl). For instance, the absolute difference between the mean current level for PEG33 and PEG34 in AeL is 3.3 pA, approximately twice that for αHL (1.54 pA). Meanwhile, both nanopores exhibit a similar high limit of Np-SMSD in these conditions, around 47–48 r.u.

4.1.2 Effect of temperature

As explained in Section 2, an early finding with αHL was that the depth of block saturated with average PEG sizes >3000 g/mol. Together with the finding that dwell times decreased for these larger masses, this led to the interpretation that not more than about 70 PEG monomers can be fit into the αHL pore. Any excess length of the polymer outside the pore would not contribute to the block and act as an entropic spring, shortening the dwell time (Krasilnikov et al., 2006). As shown in Fig. 8C for a recording 3 M KCl where this saturation effect sets in at lower masses than in 4 M KCl, while clear resistive pulses are observed in the presence of a polydisperse PEG mixture with Mw = 4300 g/mol (i.e. of average d.p. 78), the periodic maxima in

Fig. 8 Np-SMSD of PEG polymers. Each horizontal panel consists of (left) a cartoon depicting the experimental configuration, (center) a portion of nanopore current vs time recording, and (right) the scatter-plot of Δt vs I_b/I_0 (left vs bottom axis, red dots) with the superimposed histogram of I_b/I_0values (right vs bottom axis, black line). (Top half of the figure) Effect of nanopore nature on Np-SMSD of PEG (Baaken et al., 2015). Results obtained in the case of the interaction of a *cis*-added mixture of polydisperse PEG 1500, and highly monodisperse PEG 1250 (28 r.u.) in excess, with αHL nanopore at +40 mV (A) and with AeL nanopore at −120 mV (B), in KCl 4 M and at room temperature (18–22 °C) (note the opposite current sign due to opposite voltage polarities between the two cases). For each nanopore, results are shown at the voltage value for which the best resolution of size-discrimination was observed (note the different I_b/I_0 ranges on horizontal axis). One of the main result is that, while both αHL and AeL provide size-discrimination of PEG 1500 with a single-monomer resolution (see black arrows and corresponding number of PEG r.u.), AeL nanopore enables additional size-discrimination of short PEG species (down to 17–18 r.u. in this example), not observed with αHL. (Bottom half of the figure) Effect of temperature on Np-SMSD of PEG (Piguet et al., 2016). Results obtained in the case of the interaction of a *trans*-added mixture of polydisperse PEG 3400 with αHL nanopore at 25 °C (C) and at 45 °C (D), in KCl 3 M and at +50 mV. Increasing the temperature to 45 °C allows single-monomer resolution size-discrimination, which is not observed at 25 °C. Also note the significant increase of blockade durations with temperature.

the I/Io-histogram that are the hallmark of Np-SMSD do not occur. However, periodic maxima do occur for smaller PEG sizes in 3 M KCl (Piguet et al., 2018), see also (del Rio Martinez et al., 2015; Zheng et al., 2014).

Fig. 8D shows results from the same recording after increasing the temperature to 45 °C. Along with an increase in the mean duration of resistive pulses by an order of magnitude, periodic maxima in the I/Io-histogram appear, indicating that at the higher temperature, the large PEG molecules in the mixture that did interact but failed to be fully accommodated in the pore, now are, so that Np-SMSD is again possible.

It is known that PEG undergoes coil-to-globule transition at a lower critical solution temperature, which decreases with polymer molar mass as well as with salt concentration. (Bailey & Callard, 1959; Malcolm & Rowlinson, 1957). It thus appears that PEG can be compacted by this effect, which is tantamount to a temperature-dependent reduction in solubility, so that all monomers can contribute to the block at higher temperature. This interpretation is in line with the intimate relationship between PEG solubility and the apparent affinity of PEG for the αHL pore (Rodrigues et al., 2008). Such effects may be used in future to tune the sensitivity of protein pores to different polymer size ranges for Np-SMSD.

4.2 Short peptides

In view of the high sensitivity of AeL in the context of Np-SMSD analysis of PEG (Baaken et al., 2015) and short oligonucleotides (Cao et al., 2016), AeL-sensing of peptides seemed an attractive possibility. Here, we give examples of the analysis of short (d.p. = 4–10) homopeptides and of heteropeptides towards a potential application in protein analytics and sequencing.

4.2.1 *Homo-peptides*

Panel (A) of Fig. 9 presents results obtained in the case of the interaction of AeL nanopore with an equimolar mixture of homo-peptides of arginine (R) of different lengths (5, 6, 7, 8, 9 and 10 amino acids) added to the *cis*-compartment in KCl 4 M pH 7.5 at −50 mV *trans* and 20 °C.

The I_b/I_0 histogram exhibits seven distinct maxima, which correspond to the different peptide species in solution. It is immediately obvious that differences of one arginine (Mw = 174 g/mol) are easily detected. The width of the maxima is small, due to relatively long dwell times (ranging from

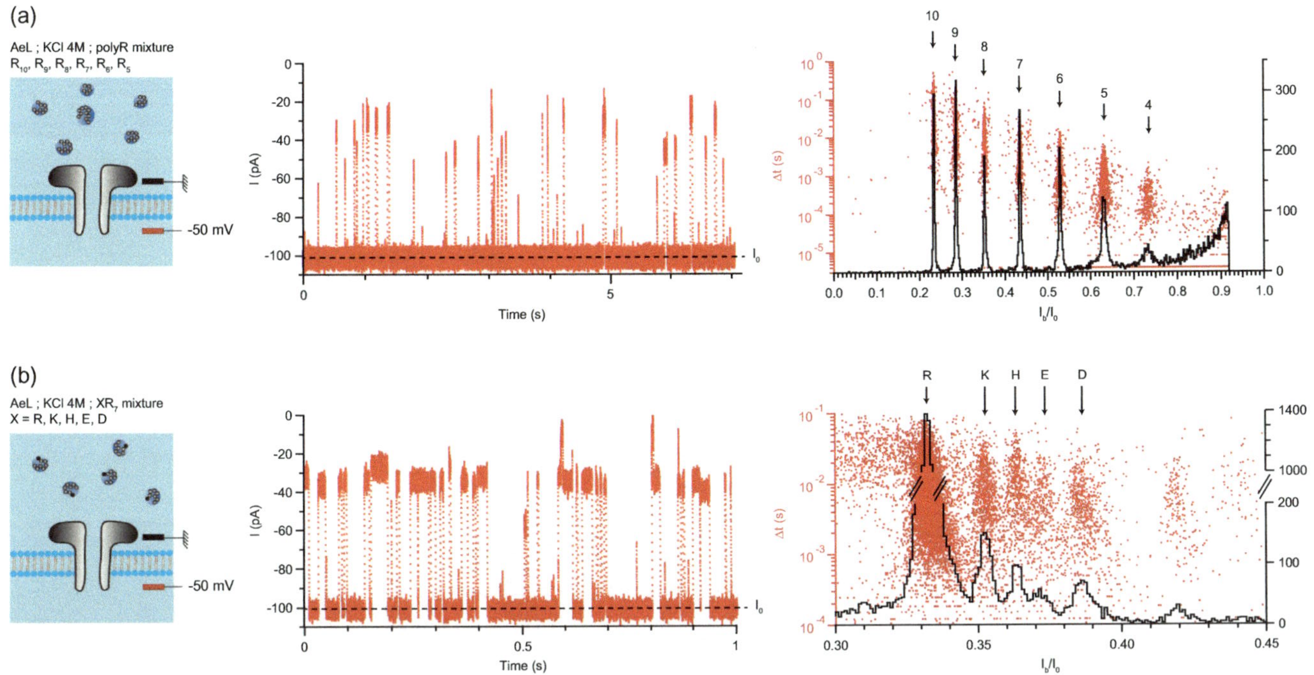

Fig. 9 See figure legend on opposite page.

approx. 1 ms for R5 to tens of ms for R10). Maximal dwell times for oligoarginine peptides were actually observed at −40 mV *trans* (Piguet et al., 2018).

Note that there is an additional maximum corresponding to a tetraarginine peptide (R4) that is likely a result of incomplete peptide synthesis (each initial peptide sample of a given number of amino acids has a peptide content of >98% purity) and/or of fragmentation of longer peptides in solution.

4.2.2 Hetero-peptides

From the experiments with oligoarginine peptides, it was clear that resolution of the presence of absence of a single amino acid was easily possible with AeL owing to a high affinity between peptides and pore. The next question was, therefore, whether it would be possible to differentiate between peptides with different amino acid compositions. As a first test, we designed a number of peptides differing only in the last (C-terminal) amino acid

Fig. 9 Np-SMSD of short homo- and hetero-peptides with AeL nanopore. Each horizontal panel consists of (left) a cartoon depicting the experimental configuration, (center) a portion of nanopore current vs time recording, and (right) the scatter-plot of Δt vs I_b/I_0 (left vs bottom axis, red dots) with the superimposed histogram of I_b/I_0values (right vs bottom axis, black line). (A) Results obtained in the case of the interaction of AeL nanopore with a cis-added equimolar mixture of homo-peptides of arginine (R) of different lengths (5, 6, 7, 8, 9 and 10 amino acids), in KCl 4 M pH 7.5 at −50 mV and 20 °C (Piguet et al., 2018). The I_b/I_0histogram exhibits seven distinct populations, which correspond to the different peptide lengths in solution discriminated with a single amino acid resolution (see black arrows and corresponding number of arginine amino acids). The 7th and right-most I_b/I_0 population corresponds to four amino acids long peptides, which appear to be present in solution as "impurities" (either resulting from incomplete peptide synthesis or from fragmentation of longer peptides in solution). (B) Results obtained in the case of the interaction of AeL nanopore with a cis-added equimolar mixture of hetero-peptides consisting of an hepta-arginine carrier (R7) linked to a terminal amino acid (X) of variable nature (X = R, K, H, E or D), in KCl 4 M pH 7.5 at −50 mV and 20 °C (Ouldali et al., 2020). The I_b/I_0 histogram reveals the discrimination of the five different XR7 peptides in solution (see black arrows and corresponding nature of X). Note that the right axis has been cut, as the frequency of RR7-induced current blockades is significantly greater than the blockade frequency of the other peptides. Also note that, in this case, the relatively small number of ER7-induced blockades, and the proximity of the adjacent HR7 population, makes the ER7 population rather difficult to identify from the I_b/I_0 histogram only; in this respect the scatter-plot of Δt vs I_b/I_0 appears to be particularly helpful, as the cloud of points corresponding to ER7 blockades is more easily distinguished from the adjacent populations, especially as I_b/I_0 values tend to be less dispersed as Δt increases.

position of a peptide otherwise composed of arginines, to determine the sensitivity of the pore's conductance to different amino acids.

Thus, 20 heteropeptides (XR7) consisting of a hepta-arginine carrier (R7) linked to a C-terminal amino acid (X) chosen from the 20 different proteinogenic amino acids were tested with the AEL pore. (Fig. 10A).

In this strategy, the poly-R carrier not only enables well-resolved peptide/AeL interactions but also, due to its dominant electrical charge under appropriate pH conditions, drives the hetero-peptide into the pore with a constant voltage gradient.

The number of seven arginines for the carrier was chosen because, as shown above, I_b/I_0 values from R7-induced blockades are already well distinguished from the open-pore current but not so low as to risk saturation by an additional contribution from the terminal amino acid X to the current blockade. Furthermore, the durations of R7-induced blockades were shown to be long enough (characteristic dwell times of approximately 10–20 ms) to be well-resolved by the acquisition apparatus and to shallow for effective averaging to precisely measure the current level, yet not so long as to place an undue limit on the number of events that can be recorded in given time.

Panel (B) of Fig. 9 presents results obtained in the case of the interaction of AeL nanopore with a *cis*-added equimolar mixture of XR7 hetero-peptides, X varying among the five amino acids having an electrically charged side chain (X=R, K, H, E or D), in KCl 4 M pH 7.5 at −50 mV *trans* and 20 °C.

Thus, at this pH, both positively (R, K, H) and negatively (E, D) charges amino acids are represented when in the C-terminal position. The I_b/I_0histogram, accordingly, reveals the discrimination of the five different XR7 peptides in solution, with the frequency of RR7-induced current blockades being significantly greater than the blockade frequency of the other peptides.

In particular, the relatively small number of ER7-induced blockades, and the proximity of the adjacent HR7 population, makes the ER7 population rather difficult to identify from the histogram only; in this respect the scatter-plot of Δt vs I_b/I_0 is particularly helpful, as the I_b/I_0 cloud of points corresponding to ER7 blockades is more easily distinguished from the adjacent populations as the dispersion of I_b/I_0values decreases as dwell time increases.

Once again, a population of peptidic "impurities" appears around $I_b/I_0=0.42$.

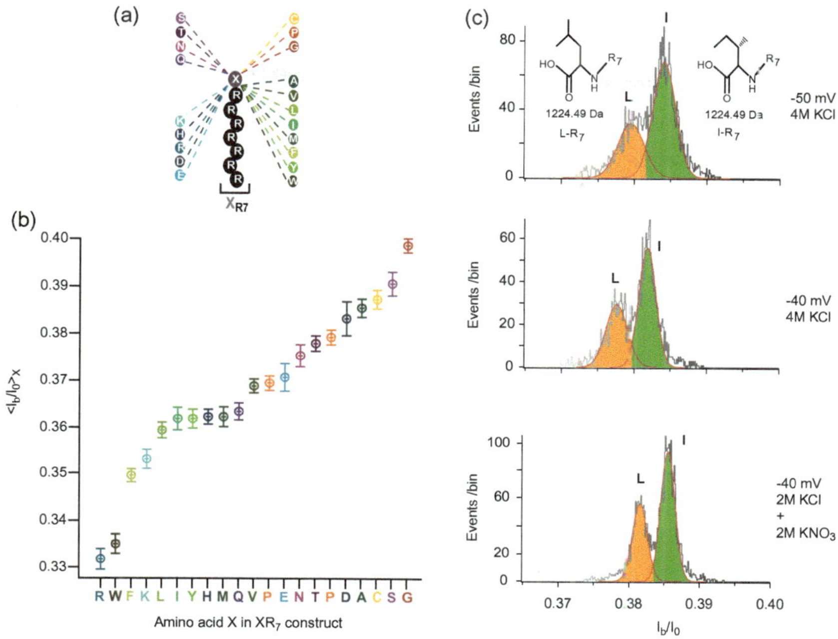

Fig. 10 Np-SMSD approach for the characterization of the 20 proteinogenic amino acids with AeL nanopore; an option towards peptide sequencing? (Ouldali et al., 2020). (A) Schematic of the peptidic construction where an hepta-arginine carrier (R7) is linked to a terminal amino acid (X) whose chemical nature is varied among the 20 different proteinogenic amino acids. (B) Mean I_b/I_0value of the XR7 population $\langle I_b/I_0 \rangle_x$ as a function of the nature of X, in KCl 4 M pH 7.5 at −50 mV and 20 °C. $\langle I_b/I_0 \rangle_x$ is the mean value of a Gaussian fit of the distribution of I_b/I_0 values in the corresponding XR7 peak of the I_b/I_0 histogram, from experiments investigating the interaction of AeL nanopore with each nature of XR7 peptide independently. Thirteen out of the twenty different amino acids exhibit a $\langle I_b/I_0 \rangle_x$value individually distinct from the $\langle I_b/I_0 \rangle_x$ value of the other amino acids. In particular, even leucine (L) and isoleucine (I) which are of equal mass (structural isomers) provide distinct values for LR7 and IR7 peptides. (C) Histograms of I_b/I_0values in the case of the interaction of an equimolar mixture of LR7 and IR7 peptides with AeL nanopore using a low-noise setup and under different experimental conditions: (top) −50 mV and 4 M KCl, (middle) −40 mV and 4 M KCl, (bottom) −40 mV and 2 M KCl/2 M KNO_3. Red lines show the underlying Gaussian distributions, color-filled in orange for LR7 and in green for IR7. From top to bottom, the different conditions tend to reduce the dispersion of the I_b/I_0 values of each peptide population, without affecting the difference between their mean I_b/I_0 values, enabling a better discrimination between the two populations.

A closer look at the maximum corresponding to the RR7 population, the scatter-plot of Δt vs I_b/I_0 exhibits a smearing-out towards smaller I_b/I_0 values for the highest (>10 ms) Δt values.

This plume of data points corresponds to current blockades which, in contrast to the majority of blockades in this example, exhibit at least one additional deep sub-level of residual current during the blockade (three of these two-levels blockades are clearly visible in the portion of current vs time recording in Fig. 9B center).

In the current-threshold approach used in this case for blockade detection (see Section 3.1), the presence of an additional sub-level of smaller residual current induces a smaller value of Ib for the blockade, as current values are time-averaged during the whole blockade. This artifact is avoided by using the alternative detection algorithm detailed in Section 3.2.

The long sublevel blockades shown here may, in fact, correspond to the simultaneous interaction of two individual analyte molecules with the nanopore, which may occur when the total analyte concentration is sufficiently large for the mean duration of unblocked intervals τ_{on} to be of the same order of magnitude as the blockade durations τ_{off} (Fig. 9B center), and are not observed when $\tau_{on} >> \tau_{off}$ (compare panels (A) and (B) of Fig. 9, R8=RR7).

In the case of Fig. 9B, such sublevel blockades mainly involve RR7 peptides, as they induce a large majority of the blockades, and mainly exhibit high Δt values, as increasing blockade duration increases the probability to observe a second molecule simultaneously interacting with the nanopore.

It is worth mentioning than, in other cases, multi-levels blockades do not result from the simultaneous interaction of more than one analyte molecule with the nanopore, and rather correspond to the signature of the individual analyte/nanopore interaction.

Panel (B) of Fig. 10 shows the mean I_b/I_0 values for all 20 XR7 species in ascending order with the identity of X plotted on the x-axis. Values were determined as the value of a Gaussian fit of the distribution of I_b/I_0 values in the XR7 peak of the I_b/I_0 histogram.

For 13 out of the 20 different amino acids, a distinct mean I_b/I_0 value could be assigned.

In particular, even phenylalanine (F) and tyrosine (T) which differ only by an –OH group, or leucine (L) and isoleucine (I) which are of equal mass (structural isomers), provide distinct mean relative current values. On the other hand, some amino acids that are structurally quite did not appear to confer distinct mean I_b/I_0 values to the peptides they terminated. This is particularly evident for a series of 5 amino acids: I, Y, H, M and Q that with respect to R all caused strongly increased but overlapping I_b/I_0 values despite the fact that they span a mass range from 131 (I) to 181 g/mol (Y) and comprise amino acids with polar uncharged (Q), charged (H) and hydrophobic (I, M, Y) side chains. Similarly, overlapping mean I_b/I_0 values were observed for C-terminal substitutions by V and one of two maxima observed for P.

In order to enhance the discrimination of individual amino acids in this setting we attempted to optimize recording conditions for low noise as well as long resistive pulse dwell times, and applied data analysis techniques detailed in Section 3.2 to reduce the influence of sublevels.

Panel (C) of Fig. 10 presents histograms of I_b/I_0 values in the case of the interaction of an equimolar mixture of LR7 and IR7 peptides with AeL nanopore using a MECA-16/Orbit-16 setup modified for low noise/high resolution recording and under different conditions: (top) −50 mV and 4 M KCl, (middle) −40 mV and 4 M KCl, (bottom) −40 mV and 2 M KCl/2 M KNO_3.

These high resolution, low noise measurements identified excess low frequency (1/f) current noise in the blocked state as the main origin of the observed variance in current levels leading to a considerable overlap of the Gaussian maxima for L and I-terminated peptides, even when dielectric noise was minimized (Fig. 10C upper panel). Thus, it was reasoned that further increasing the dwell time of resistive pulses should result in increased precision of current level estimation by averaging. This proved possible by changing the voltage gradient from −50 mV to −40 mV *trans* in 4 M KCl where, as shown by previous investigations of poly-R/AeL interaction, blockade durations are maximal for oligoarginines (Piguet et al., 2018). This results in a reduction of the standard deviations of I_b/I_0distributions of LR7 and IR7 respectively, without affecting the difference of their mean values, thus enabling a better discrimination between the two populations. Further optimization was possible when 50% of Cl^- anions were replaced with NO_3^-, due to a substantial decrease in the added low-frequency noise and a further increase in dwell time. We interpret this finding as indicating that the charge density of counterions affects the conformational mobility of the cationic peptide in residence in the pore. This modification results in a further reduction of the standard deviations of I_b/I_0 distributions, once again without affecting the difference of their mean values, enabling an even better discrimination.

5. Conclusion and perspectives

Resistive pulse sensing with pore-forming toxins is emerging as a valuable tool for polymer analytics at the single-molecule level. The kind of molecular information that can be extracted from the resistive pulses depends strongly on the configuration of the molecule/nanopore interaction. We have described two regimes, (i) the translocation/threading regime where the analyte molecule is linearly transported through the nanopore and (ii) the binding/collapse regime where the analyte molecule is fully accommodated *en bloc* in the nanopore, each individual molecular subunit providing a contribution to the current blockade. The first regime enables the composition of the intact molecule to be sequentially read as it threads through the pore, while the second regime allows nanopore-based single-molecule size-spectrometry (Np-SMSD). The first regime, under appropriate experimental conditions, has been successfully used for DNA sequencing but its transposition for peptide/protein sequencing remains challenging. We have focused on the second regime, which under appropriate experimental conditions, has been successfully used for the discrimination of water-soluble polymers (Angevine, Chavis, Kothalawala, Dass, & Reiner, 2014; Baaken et al., 2011, 2015; del Rio Martinez et al., 2015; Piguet et al., 2016; Reiner, Kasianowicz, Nablo, & Robertson, 2010; Robertson et al., 2007; Talarimoghari et al., 2018; Zheng et al., 2014), metallic clusters (Ettedgui, Forstater, Robertson, & Kasianowicz, 2019; Ettedgui, Kasianowicz, & Balijepalli, 2016), short oligonucleotides (Cao et al., 2016), homopeptides (Piguet et al., 2018), and heteropeptides (Chavis et al., 2017; Huang, Voet, & Maglia, 2019; Ouldali et al., 2020). Np-SMSD of short heteropeptides demonstrated the high sensitivity of this approach to identify even subtle differences between a majority of all 20 proteinogenic amino acids linked to a peptidic carrier, in particular between amino acids of very similar chemical structure, such as Phenylalanine and Tyrosine, and even between amino acids of equal mass such as Leucine and Isoleucine (structural isomers) (Ouldali et al., 2020). Consequently, Np-SMSD could be envisioned as a new approach for peptides/proteins sequencing. Such approach would require the individual amino acids of the protein to be sequentially cleaved and ligated to a peptidic carrier, before being read by the nanopore. Sequential cleavage would be envisioned either via chemical (Edman) degradation, or using proteolytic digestion by proteases or even by a protein digestion complex, such as ClpXP, which has already been shown to work in tandem with a nanopore

(Nivala et al., 2013). As a perspective and as an alternative to protein sequencing in the context of protein identification, the second regime may also allow "protein fingerprinting", which consists in splitting a protein into multiple peptide fragments and using the nanopore signals generated by the fragments as a signature of the initial protein ref. (Bakshloo, Talarimoghari, Ouldali, Behrends, & Oukhaled, 2020).

References

Angevine, C. E., Chavis, A. E., Kothalawala, N., Dass, A., & Reiner, J. E. (2014). Enhanced single molecule mass spectrometry via charged metallic clusters. *Analytical Chemistry, 86*(22), 11077–11085.

Ankri, N., Legendre, P., Faber, D. S., & Korn, H. (1994). Automatic detection of spontaneous synaptic responses in central neurons. *Journal of Neuroscience Methods, 52*(1), 87–100.

Baaken, G., Ankri, N., Schuler, A. K., Rühe, J., & Behrends, J. C. (2011). Nanopore-based single-molecule mass spectrometry on a lipid membrane microarray. *ACS Nano, 5*(10), 8080–8088.

Baaken, G., Halimeh, I., Bacri, L., Pelta, J., Oukhaled, A., & Behrends, J. C. (2015). High-resolution size-discrimination of single nonionic synthetic polymers with a highly charged biological nanopore. *ACS Nano, 9*(6), 6443–6449.

Bailey, F. E., Jr., & Callard, R. W. (1959). Some properties of poly (ethylene oxide) 1 in aqueous solution. *Journal of Applied Polymer Science, 1*(1), 56–62.

Bakshloo, M. A., Talarimoghari, M., Ouldali, H., Behrends, J. C., & Oukhaled, A. (2020). Protein fingerprinting using the aerolysin nanopore. *Biophysical Journal, 118*(3), 475a.

Bean, R. C., Shepherd, W. C., Chan, H., & Eichner, J. (1969). Discrete conductance fluctuations in lipid bilayer protein membranes. *The Journal of General Physiology, 53*(6), 741–757.

Bezrukov, S. M., & Kasianowicz, J. J. (1997). The charge state of an ion channel controls neutral polymer entry into its pore. *European Biophysics Journal, 26*(6), 471–476.

Bezrukov, S. M., Vodyanoy, I., Brutyan, R. A., & Kasianowicz, J. J. (1996). Dynamics and free energy of polymers partitioning into a nanoscale pore. *Macromolecules, 29*(26), 8517–8522.

Boukhet, M. A., Halimeh, I., Baaken, G., & Behrends, J. C. (2017). Translocation and binding in the recognition of short oligonucleotides by a biological nanopore. *Biophysical Journal, 112*(3), 459a–460a.

Breton, M. F., Discala, F., Bacri, L., Foster, D., Pelta, J., & Oukhaled, A. (2013). Exploration of neutral versus polyelectrolyte behavior of poly (ethylene glycol) s in alkali ion solutions using single-nanopore recording. *The Journal of Physical Chemistry Letters, 4*(13), 2202–2208.

Brown, A. S., & MacInnes, D. A. (1935). The determination of activity coefficients from the potentials of concentration cells with transference. I. Sodium chloride at 25. *Journal of the American Chemical Society, 57*(7), 1356–1362.

Cao, C., Ying, Y. L., Hu, Z. L., Liao, D. F., Tian, H., & Long, Y. T. (2016). Discrimination of oligonucleotides of different lengths with a wild-type aerolysin nanopore. *Nature Nanotechnology, 11*(8), 713–718.

Chavis, A. E., Brady, K. T., Hatmaker, G. A., Angevine, C. E., Kothalawala, N., Dass, A., et al. (2017). Single molecule nanopore spectrometry for peptide detection. *ACS Sensors, 2*(9), 1319–1328.

Chavis, A. E., Brady, K. T., Kothalawala, N., & Reiner, J. E. (2015). Voltage and blockade state optimization of cluster-enhanced nanopore spectrometry. *Analyst, 140*(22), 7718–7725.
Cherf, G. M., Lieberman, K. R., Rashid, H., Lam, C. E., Karplus, K., & Akeson, M. (2012). Automated forward and reverse ratcheting of DNA in a nanopore at 5-Å precision. *Nature Biotechnology, 30*(4), 344–348.
Christophe, A. B. (1971). Valley to peak ratio as a measure for the separation of two chromatographic peaks. *Chromatographia, 4*(10), 455–458.
Church, G., Deamer, D. W., Branton, D., Baldarelli, R., & Kasianowicz, J. (1998). *U.S. Patent No. 5,795,782*. Washington, DC: U.S. Patent and Trademark Office.
Cole, K. S. (1949). Dynamic electrical characteristics of the squid axon membrane. *Archives des Sciences Physiologiques, 3*(2), 253–258.
Cole, K. S., & Moore, J. W. (1960). Ionic current measurements in the squid giant axon membrane. *The Journal of General Physiology, 44*(1), 123–167.
Coulter, W. H. (1953). *U.S. Patent No. 2,656,508*. Washington, DC: U.S. Patent and Trademark Office.
Cressiot, B., Braselmann, E., Oukhaled, A., Elcock, A. H., Pelta, J., & Clark, P. L. (2015). Dynamics and energy contributions for transport of unfolded pertactin through a protein nanopore. *ACS Nano, 9*(9), 9050–9061.
DeBlois, R. W., & Bean, C. P. (1970). Counting and sizing of submicron particles by the resistive pulse technique. *Review of Scientific Instruments, 41*(7), 909–916.
DeBlois, R. W., Bean, C. P., & Wesley, R. K. (1977). Electrokinetic measurements with submicron particles and pores by the resistive pulse technique. *Journal of Colloid and Interface Science, 61*(2), 323–335.
del Rio Martinez, J. M., Zaitseva, E., Petersen, S., Baaken, G., & Behrends, J. C. (2015). Automated formation of lipid membrane microarrays for ionic single-molecule sensing with protein nanopores. *Small, 11*(1), 119–125.
Derrington, I. M., Butler, T. Z., Collins, M. D., Manrao, E., Pavlenok, M., Niederweis, M., et al. (2010). Nanopore DNA sequencing with MspA. *Proceedings of the National Academy of Sciences of the United States of America, 107*(37), 16060–16065.
Ensslen, T., Ouldali, H., Oukhaled, A., & Behrends, J. C. (2020). Mass-independent, high-fidelity single-molecule differentiation using the aerolysin protein pore. *Biophysical Journal, 118*(3), 474a–475a.
Ettedgui, J., Forstater, J., Robertson, J. W., & Kasianowicz, J. J. (2019). High resolution physical characterization of single metallic nanoparticles. *Journal of Visualized Experiments, 148*, e58257.
Ettedgui, J., Kasianowicz, J. J., & Balijepalli, A. (2016). Single molecule discrimination of heteropolytungstates and their isomers in solution with a nanometer-scale pore. *Journal of the American Chemical Society, 138*(23), 7228–7231.
Faller, M., Niederweis, M., & Schulz, G. E. (2004). The structure of a mycobacterial outer-membrane channel. *Science, 303*(5661), 1189–1192.
Golibersuch, D. C. (1973). Observation of aspherical particle rotation in Poiseuille flow via the resistance pulse technique: I. application to human erythrocytes. *Biophysical Journal, 13*(3), 265–280.
Goyal, P., Krasteva, P. V., Van Gerven, N., Gubellini, F., Van den Broeck, I., Troupiotis-Tsaïlaki, A., et al. (2014). Structural and mechanistic insights into the bacterial amyloid secretion channel CsgG. *Nature, 516*(7530), 250–253.
Hladky, S. B., & Haydon, D. A. (1970). Discreteness of conductance change in bimolecular lipid membranes in the presence of certain antibiotics. *Nature, 225*(5231), 451–453.
Hodgkin, A. L., Huxley, A. F., & Katz, B. (1952). Measurement of current-voltage relations in the membrane of the giant axon of Loligo. *The Journal of Physiology, 116*(4), 424.
Huang, G., Voet, A., & Maglia, G. (2019). FraC nanopores with adjustable diameter identify the mass of opposite-charge peptides with 44 dalton resolution. *Nature Communications, 10*(1), 1–10.

Ito, S., Hachiya, H., Baba, K., Asano, Y., & Wada, H. (1995). Improvement of the silver/silver chloride reference electrode and its application to pH measurement. *Talanta, 42*(11), 1685–1690.

Kasianowicz, J. J., Brandin, E., Branton, D., & Deamer, D. W. (1996). Characterization of individual polynucleotide molecules using a membrane channel. *Proceedings of the National Academy of Sciences of the United States of America, 93*(24), 13770–13773.

Kasianowicz, J. J., Robertson, J. W., Chan, E. R., Reiner, J. E., & Stanford, V. M. (2008). Nanoscopic porous sensors. *Annual Review of Analytical Chemistry, 1*, 737–766.

Krasilnikov, O. V. (2002). Sizing channels with neutral polymers. In *Structure and dynamics of confined polymers* (pp. 97–115). Dordrecht: Springer.

Krasilnikov, O. V., Rodrigues, C. G., & Bezrukov, S. M. (2006). Single polymer molecules in a protein nanopore in the limit of a strong polymer-pore attraction. *Physical Review Letters, 97*(1), 018301.

Krasilnikov, O. V., Sabirov, R. Z., Ternovsky, V. I., Merzliak, P. G., & Muratkhodjaev, J. N. (1992). A simple method for the determination of the pore radius of ion channels in planar lipid bilayer membranes. *FEMS Microbiology Immunology, 5*(1–3), 93–100.

Kubitschek, H. E. (1958). Electronic counting and sizing of bacteria. *Nature, 182*(4630), 234–235.

Laszlo, A. H., Derrington, I. M., Ross, B. C., Brinkerhoff, H., Adey, A., Nova, I. C., et al. (2014). Decoding long nanopore sequencing reads of natural DNA. *Nature Biotechnology, 32*, 829–833.

Levis, R. A., & Rae, J. L. (1998). [14] Low-noise patch-clamp techniques. In *Vol. 293. Methods in enzymology* (pp. 218–266). Elsevier.

Lieberman, K. R., Cherf, G. M., Doody, M. J., Olasagasti, F., Kolodji, Y., & Akeson, M. (2010). Processive replication of single DNA molecules in a nanopore catalyzed by phi29 DNA polymerase. *Journal of the American Chemical Society, 132*(50), 17961–17972.

Machado, D. C., Júnior, J. J., Melo, M. C., Silva, A. M., Fontes, A., & Rodrigues, C. G. (2016). Effects of alkali and ammonium ions in the detection of poly (ethyleneglycol) by alpha-hemolysin nanopore sensor. *RSC Advances, 6*(61), 56647–56655.

Maertz, W. H. (2008). *The axon guide*. Molecular Devices Inc.

Malcolm, G. N., & Rowlinson, J. S. (1957). The thermodynamic properties of aqueous solutions of polyethylene glycol, polypropylene glycol and dioxane. *Transactions of the Faraday Society, 53*, 921–931.

Manrao, E. A., Derrington, I. M., Laszlo, A. H., Langford, K. W., Hopper, M. K., Gillgren, N., et al. (2012). Reading DNA at single-nucleotide resolution with a mutant MspA nanopore and phi29 DNA polymerase. *Nature Biotechnology, 30*(4), 349–353.

Maxwell, J. C. (1873). *A treatise on electricity and magnetism. Vol. 1.* Oxford: Clarendon Press.

Menestrina, G. (1986). Ionic channels formed byStaphylococcus aureus alpha-toxin: Voltage-dependent inhibition by divalent and trivalent cations. *The Journal of Membrane Biology, 90*(2), 177–190.

Merstorf, C., Cressiot, B., Pastoriza-Gallego, M., Oukhaled, A., Betton, J. M., Auvray, L., et al. (2012). Wild type, mutant protein unfolding and phase transition detected by single-nanopore recording. *ACS Chemical Biology, 7*(4), 652–658.

Merzlyak, P. G., Yuldasheva, L. N., Rodrigues, C. G., Carneiro, C. M., Krasilnikov, O. V., & Bezrukov, S. M. (1999). Polymeric nonelectrolytes to probe pore geometry: Application to the α-toxin transmembrane channel. *Biophysical Journal, 77*(6), 3023–3033.

Mueller, P., Rudin, D. O., Tien, H. T., & Wescott, W. C. (1962). Reconstitution of cell membrane structure in vitro and its transformation into an excitable system. *Nature, 194*(4832), 979–980.

Nivala, J., Marks, D. B., & Akeson, M. (2013). Unfoldase-mediated protein translocation through an α-hemolysin nanopore. *Nature Biotechnology, 31*(3), 247–250.

Nivala, J., Mulroney, L., Li, G., Schreiber, J., & Akeson, M. (2014). Discrimination among protein variants using an unfoldase-coupled nanopore. *ACS Nano*, *8*(12), 12365–12375.

Oesterhelt, F., Rief, M., & Gaub, H. E. (1999). Single molecule force spectroscopy by AFM indicates helical structure of poly (ethylene-glycol) in water. *New Journal of Physics*, *1*(1), 6.

Ouldali, H., Sarthak, K., Ensslen, T., Piguet, F., Manivet, P., Pelta, J., et al. (2020). Electrical recognition of the twenty proteinogenic amino acids using an aerolysin nanopore. *Nature Biotechnology*, *38*(2), 176–181.

Pastoriza-Gallego, M., Breton, M. F., Discala, F., Auvray, L., Betton, J. M., & Pelta, J. (2014). Evidence of unfolded protein translocation through a protein nanopore. *ACS Nano*, *8*(11), 11350–11360.

Pastoriza-Gallego, M., Rabah, L., Gibrat, G., Thiebot, B., van der Goot, F. G., Auvray, L., et al. (2011). Dynamics of unfolded protein transport through an aerolysin pore. *Journal of the American Chemical Society*, *133*(9), 2923–2931.

Piguet, F., Ouldali, H., Discala, F., Breton, M. F., Behrends, J. C., Pelta, J., et al. (2016). High temperature extends the range of size discrimination of nonionic polymers by a biological nanopore. *Scientific Reports*, *6*, 38675.

Piguet, F., Ouldali, H., Pastoriza-Gallego, M., Manivet, P., Pelta, J., & Oukhaled, A. (2018). Identification of single amino acid differences in uniformly charged homopolymeric peptides with aerolysin nanopore. *Nature Communications*, *9*(1), 1–13.

Reiner, J. E., Kasianowicz, J. J., Nablo, B. J., & Robertson, J. W. (2010). Theory for polymer analysis using nanopore-based single-molecule mass spectrometry. *Proceedings of the National Academy of Sciences of the United States of America*, *107*(27), 12080–12085.

Robertson, J. W., Rodrigues, C. G., Stanford, V. M., Rubinson, K. A., Krasilnikov, O. V., & Kasianowicz, J. J. (2007). Single-molecule mass spectrometry in solution using a solitary nanopore. *Proceedings of the National Academy of Sciences of the United States of America*, *104*(20), 8207–8211.

Rodrigues, C. G., Machado, D. C., Chevtchenko, S. F., & Krasilnikov, O. V. (2008). Mechanism of KCl enhancement in detection of nonionic polymers by nanopore sensors. *Biophysical Journal*, *95*(11), 5186–5192.

Rodrigues, C. G., Machado, D. C., da Silva, A. M., Júnior, J. J., & Krasilnikov, O. V. (2011). Hofmeister effect in confined spaces: Halogen ions and single molecule detection. *Biophysical Journal*, *100*(12), 2929–2935.

Rony, P. R. (1968). The extent of separation: A universal separation index. *Separation Science and Technology*, *3*(3), 239–248.

Sigworth, F. J., & Klemic, K. G. (2005). Microchip technology in ion-channel research. *IEEE Transactions on Nanobioscience*, *4*(1), 121–127.

Sigworth, F. J., & Sine, S. M. (1987). Data transformations for improved display and fitting of single-channel dwell time histograms. *Biophysical Journal*, *52*(6), 1047.

Song, L., Hobaugh, M. R., Shustak, C., Cheley, S., Bayley, H., & Gouaux, J. E. (1996). Structure of staphylococcal α-hemolysin, a heptameric transmembrane pore. *Science*, *274*(5294), 1859–1865.

Stefureac, R., Long, Y. T., Kraatz, H. B., Howard, P., & Lee, J. S. (2006). Transport of α-helical peptides through α-hemolysin and aerolysin pores. *Biochemistry*, *45*(30), 9172–9179.

Talarimoghari, M., Baaken, G., Hanselmann, R., & Behrends, J. C. (2018). Size-dependent interaction of a 3-arm star poly (ethylene glycol) with two biological nanopores. *The European Physical Journal E*, *41*(6), 77.

Zheng, T., Baaken, G., Vellinger, M., Behrends, J. C., & Rühe, J. (2014). Generation of chip based microelectrochemical cell arrays for long-term and high-resolution recording of ionic currents through ion channel proteins. *Sensors and Actuators B: Chemical*, *205*, 268–275.

CHAPTER TWENTY-ONE

Use of pore-forming toxins to study co-translocational protein folding

Antonio De la Torre-Cabrera and David Rodriguez-Larrea*
Departamento de Bioquímica y Biología Molecular, Instituto Biofisika, UPV/EHU-CSIC, Leioa, Spain
*Corresponding author: e-mail address: david.rodriguezl@ehu.es

Contents

Abstract

In vivo proteins fold mainly as they emerge from the ribosome or as they emerge from a membrane translocon. Membrane translocation in particular poses technical challenges to the study of the associated protein folding processes. Recently we have developed a single-molecule methodology that allows the capture of a single protein molecule through a membrane translocon with biotinylated oligonucleotides covalently bound at its N- and C- terminus using streptavidin. The resulting rotaxane can be driven forwards and backwards changing the voltage polarity, and carefully planned experiments allow inference of the folding pathway. Here we will discuss the details of a simplified methodological approach.

Methods in Enzymology, Volume 649
ISSN 0076-6879
https://doi.org/10.1016/bs.mie.2021.01.018

1. Introduction

Protein folding is a fundamental process by which a polypeptide chain attains a 3D structure that is essential for the protein's function. *In vivo*, proteins fold mainly upon synthesis at the ribosome exit (Holtkamp et al., 2015; Nilsson et al., 2015) in the case of cytosolic proteins or at the exit of a translocon when the protein locates at membrane-delimitated compartment (Kadokura & Beckwith, 2009; Okamoto et al., 2002). In both cases the polypeptide chain emerges from a narrow conduit (1–2 nm wide). This implies that some parts of the polypeptide emerge and are free to initiate the folding in advance of the remaining polypeptide. This is frequently known as vectorial folding (Junker, Besingi, & Clark, 2009), because folding starts at one end of the polypeptide chain first and it propagates to the other end as additional segments of the polypeptide emerge from the conduit.

Vectorial folding is better studied using single-molecule methods (Ha, 2014). In contrast to macroscopic measurements (Ha, 2014), the measurements do not require millions of molecules synchronously initiating the folding process to produce a coherent signal. In the case of membrane translocation, it seems technically challenging to initiate simultaneously the translocation of billions of protein molecules. In addition, single-molecule observations are not obscured by the rate-limiting step of multi-step processes and short-lived intermediates are readily observed. Single-molecule methods which have been already successful at elucidating the folding of proteins emerging from the ribosome (Goldman et al., 2015) may also be adequate to study protein folding during membrane translocation.

We have previously developed a method to study protein unfolding through a membrane pore (Rodriguez-Larrea & Bayley, 2013, 2014; Rosen, Bayley, & Rodriguez-Larrea, 2020), the well-known staphylococcal α-hemolysin (α-HL), and we have elaborated this system further to study protein folding upon membrane translocation. Briefly, an *E. coli* thioredoxin 2 variant (Rodriguez-Larrea et al., 2010) lacking the internal disulfide and with a unique cysteine either at the N- or C-terminus is covalently tagged with an oligonucleotide long enough to completely traverse the α-HL pore. In the presence of an electric field across the membrane, the oligonucleotide is electrophoresed through a single α-HL pore inserted in the membrane. The force experienced by the oligonucleotide is enough to unfold and translocate the protein and the steps of the process can be inferred from the modulation in the ionic current passing through the α-HL.

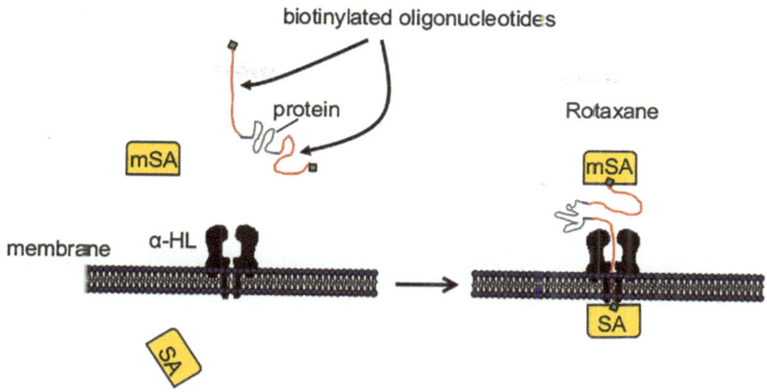

Fig. 1 Components of a membrane rotaxane. The formation of a membrane rotaxane requires a nanopore inserted in a lipid membrane, monovalent streptavidin (mSA) at the cis side of the membrane, streptavidin (SA) at the trans side of the membrane and a protein substrate tagged with biotinylated oligonucleotides at both the N- and the C-terminus. How to proceed to produce and assemble these elements into a membrane rotaxane (left) is the aim of this chapter.

We built on this model system to study protein folding upon membrane translocation (Feng et al., 2020). We formed a rotaxane (an interlocked structure across the α-HL pore, Fig. 1) by first tagging both ends of the protein with biotinylated oligonucleotides. This construct was exposed to monovalent streptavidin (Howarth et al., 2006) (mSA) at a 1:2 ratio, such that the most abundant species are constructs with only one biotin tag bound to mSA. The construct is electrophoresed through the α-HL with the unbound oligonucleotide leading the process. The protein unfolds and translocates, and the other oligonucleotide tag remains in the original compartment held by the mSA. Meanwhile in the recipient compartment there are plenty of SA molecules which complex with the biotin and end the rotaxane formation.

Here, we will discuss the key aspects for successful rotaxane formation, namely the translocating protein substrate and the oligonucleotide tagging molecules. As we will show, there are ample of alternatives and we will describe the methodology to achieve the simplest of them.

2. The translocating protein substrate

The protein substrate is unfolded and translocated through a narrow nanopore by the electrophoretic force exerted on the covalently bound oligonucleotide. This imposes two considerations:

1. Size: The protein should not be composed of independently folding domains. This is because once the oligonucleotide pulls and unfolds the first domain it will translocate through the pore and will not exert additional force to unfold the remaining domains (Rodriguez-Larrea & Bayley, 2013). In general, the methodology is adequate for small, single-domain proteins (we have successfully applied it on proteins ~200 amino acids long).
2. Amino acid sequence: While we have successful experience in the translocation of some protein substrates with internal disulfides we are also aware that some internal disulfides cannot be translocated through the narrow α-HL pore. We therefore recommend removal of all naturally occurring cysteines by serines which in addition facilitates site-specific labelling. In this line, proteins with a knotted topology may similarly be impossible to translocate using the proposed methodology.

In addition, the oligonucleotide tags must be selectively placed at, or just a few residues away from, the N- and C-terminus. The N-terminus can be selectively modified by a number of alternative chemistries (Rosen & Francis, 2017). The number of alternatives for the C-terminus is narrower.

In our simplified approach we propose the simplest alternative, which is a single-domain thioredoxin mutant where the cysteines forming the naturally occurring disulfide are substituted by serines (C32S, C35S), three stabilizing mutations were included (A22P, I23V, P68A) (Rodriguez-Larrea et al., 2010), the serine at position 2 was mutated to cysteine, and a terminal cysteine was introduced. In the case of thioredoxin the initial methionine is removed, on other proteins it may not be removed. This minor detail should not affect the unfolding and translocation behavior but it can affect the tagging approach on the N-terminus as it will be discussed in the next section.

*An important step of the protocol is the incubation of the protein with methoxyamine. Cysteines found at the N-terminus are highly reactive with aldehydes and it is common that the purified form is modified with undetermined adducts (Nguyen, Elliott, Holt, Muir, & Chin, 2011). Luckily this modification is reversible and a N-terminal free cysteine can be recovered by incubation with methoxyamine.

3. The oligonucleotide tagging molecules

The oligonucleotide serves first to provide the force which unfolds the protein and to drive the translocation. Also, once the mSA-biotin complex is formed, it prevents the complete translocation. To serve these functions, several considerations are important:

1. Length: The oligonucleotide has to completely translocate the membrane nanopore. In the case of α-HL, this can be achieved with a 23-nt long oligonucleotide. Longer oligonucleotides should not be a problem, but shorter oligonucleotides show fewer events, produce smaller electrophoretic forces and are more prone to scape the nanopore without completing translocation (Wiggin, Tropini, Tabard-Cossa, Jetha, & Marziali, 2008). In general, the formation of secondary structure should be avoided.
2. Place of the modifications: We usually place the reactive probes and the affinity tag (biotin) indifferently at the 3′ or the 5′ ends of the oligonucleotide (whatever is more convenient). It is only important to note that this will have an effect on the orientation of the oligonucleotide inside the nanopore (3′ to 5′ at the entrance and exit respectively, or *vice versa*), which in turn has an effect on the ionic current measurements. In addition, affinity tags (biotin) can be placed internally, it is only important that between the nucleotide carrying the affinity tag (biotin) and the one carrying the reactive probe there are at least 23 nucleotides.
3. Affinity tag: We have mainly worked with biotin. We have limited experience using other tags, but preliminary experiments with dinitrophenol (DNP) and desthiobiotin show that while they also do the job they have a weaker affinity for anti-DNP and SA respectively, and this reduces the lifetime of the rotaxane. Alternatively, the affinity tag may be replaced by a covalently bound tag.
4. Reactive probe: We have experience using thiolated oligonucleotides, oligonucleotides with a maleimide, and with oligonucleotides modified with 2-cyanobenzothiazole. The thiolated oligonucleotides can be used to reversibly modify cysteines. Maleimide oligonucleotides instead irreversibly modify cysteine residues. 2-cyanobenzothiazole specifically modifies cysteines at the N-terminus.

*An important step of this protocol is the covalent, irreversible, modification of both cysteines of the protein with maleimide modified oligonucleotides. The yield of this reaction is low if the protein sample is not fully reduced, the reducing agent is not completely removed from the media, the oligonucleotide is not at least 2 × fold in excess, or if the pH is not kept at ∼7.

4. Protein substrate

In this section we describe materials, instruments, and protocols to produce a protein substrate, thioredoxin tagged at both N- and C-terminal cysteines with a 5′ maleimide, 3′ biotin 40-mer oligonucleotide composed exclusively of deoxycytidines. (The protein threading direction (N- or C-first) is deduced from the signal similarity to independent experiments where either terminus is labeled with an oligonucleotide.)

4.1 Materials and instruments

4.1.1 Bacterial transformation with expression plasmid

- Chemically competent *E. coli* BL21 (DE3) (200131, Agilent)
- Plasmid: thioredoxin mutant in pET 20+
- Antibiotic: kanamycin sulfate (100 mg/mL in H_2O) (Gibco)
- LB Broth (Condalab) 0.5 L in 2 L flasks, autoclaved
- Flask shaker in a 37 °C incubator
- Water bath with temperature control

4.1.2 Induction of the expression

- Isopropyl-beta-D-thiogalactopyranoside (Dioxan free) (IPTG) (Apollo Scientific)
- Large-volume centrifuge

4.1.3 Thioredoxin purification

- Centrifuge for 50 and 15 mL falcon tubes
- TE buffer: 30 mM Tris·Cl, 1 mM EDTA in distilled water (pH 7.5)
- Methoxamine hydrochloride
- Tris(2-carboxyethyl)phosphine hydrochloride (TCEP)
- Dithiothreitol (DTT)
- PD10 Buffer exchange columns (GE)
- Centrifuge
- Sonicator [Soniprep 150]
- ÄKTA pure Cytiva
- Superdex 75 HiLoad 16/60 column, Cytiva
- NuPAGE™ 4–12% Bis-Tris gel
- Amicon® Ultra—15 (3 K) centrifugal filters
- HiTrap™ Q FF 1 mL column, Cytiva
- Spectrophotometer Agilent Cary 8454

4.1.4 Thioredoxin modification with DNA probe

- Modified oligonucleotides (Biomers)

4.1.5 Trx-DNA construct purification

- ÄKTA pure Cytiva

- NuPAGE™ 4–12% Bis-Tris gel
- Amicon® Ultra—15 (3 K) centrifugal filter device
- HiTrap™ QFF 1 mL column, Cytiva

4.2 Protocol

4.2.1 Bacterial transformation with expression plasmid

1. Pass 100 μL of BL21 (DE3) chemically competent bacteria to a 1.5 μL Eppendorf tube. Cells must be maintained on ice.
2. Add 30 ng of plasmid for every 20 μL of bacteria. Gently mix without introducing bubbles.
3. Incubate on ice for 30 min.
4. Give a heat pulse: 45 s in a waterbath at 42 °C.
5. Quickly transfer the Eppendorf tubes to ice and keep for 2 min.
6. Add 0.9 mL of room temperature (20–25 °C) LB liquid media.
7. Incubate at 37 °C for 1 h under agitation (225–250 rpm).
8. Transfer the cells to 0.5 L of LB in 2 L flasks previously autoclaved.
9. Add 0.5 mL of antibiotic (kanamycin 100 mg/mL) to the culture.
10. Incubate overnight at 37 °C and 180 rpm shaking.

4.3 Induction of the expression

1. Collect the overnight grown bacteria by centrifuging (5000 × *g*, 10 min, 4 °C).
2. Resuspend the cell pellet in 0.5 L fresh LB and supplement with 0.5 mL kanamycin sulfate (100 mg/mL).
3. Incubate for 1 h at 37 °C and 180 rpm shaking.
4. Add 0.5 mL of 0.4 M IPTG for every 0.5 L of culture.
5. Incubate for 4 h at 37 °C and 180 rpm shaking.

4.3.1 Thioredoxin purification

1. After the IPTG treatment, centrifuge the culture (5000 × *g*, 10 min, 4 °C) to collect the cells. Discard the supernatant.
2. Resuspend the cell pellet in 30 mL of TE buffer. If needed, store the resuspended cells at −20 °C.
3. Sonicate bacteria for 30 s let them cool for 60 s and repeat this cycle five times, set the amplitude to 10%. Keep the cells on ice to avoid overheating.
4. Centrifuge the lysate (5000 × *g*, 10 min, 4 °C). Keep the supernatant and discard the cell debris pellet. Add DTT to the supernatant to a final concentration of 10 mM.

5. Purify the protein by Size-Exclusion Chromatography using a Superdex 75 HiLoad 16/60 column. The collected fractions can be stored at −20 °C.
6. Perform SDS-PAGE to identify those fractions that contain the target protein.
7. Incubate the fractions that contain the protein with 400 mM methoxamine and 10 mM TCEP for at least 2 h, 4 °C.
8. Remove the methoxamine either by dialysis, buffer exchange columns or using centrifugal filter devices.
9. Load the protein on a HiTrap™ QFF 1 mL column previously equilibrated in TE + 1 mM DTT.
10. Elute the protein with a NaCl gradient in TE + 1 mM DTT (200 mL from 0 to 1 M NaCl). Store the collected fractions at 4 °C.
11. Perform SDS-PAGE to identify those fractions that contain the target protein.

4.3.2 Thioredoxin modification with DNA probe

1. Measure the protein concentration by absorption at 280 nm using the calculated molar absorptivity. For thioredoxin, 14100 M^{-1} cm^{-1}.
2. Use 2 mg of protein at a concentration of 1 mg/mL. Incubate with DTT 10 mM for at least 1 h.
3. Buffer exchange the protein to TE buffer, pH 7.5. All DTT must be removed (if using PD10 columns, collect only 75% of the recommended volume).
4. Measure protein concentration.
5. The oligonucleotide comes lyophilized and should not be rehydrated beforehand as the maleimide group hydrolyses easily. Instead rehydrate the lyophilized oligonucleotide with the protein solution. Use the amount that keeps a molar ratio protein: oligonucleotide <0.5.
6. Allow the reaction to proceed for 2 h at room temperature with gentle stirring.

4.3.3 Trx-DNA construct purification

1. Load the reaction in a HiTrap™ Q FF 1 mL column previously equilibrated in TE buffer pH 8.5.
2. Elute the products with a 200 mL linear gradient from 0 to 1 M KCl (or NaCl) in the same buffer. The pH of the buffer is crucial to obtain a good peak resolution and other proteins may resolve better at a different pH.

3. Perform SDS-PAGE to identify those fractions that contain the desired construct. The protein tagged at both ends migrates with an apparent molecular weight higher than the protein with just one or without oligonucleotide. *Add reducing agent to the loading buffer.
4. Measure the concentration of the oligonucleotide-protein-oligonucleotide complex using the absorbance at 260 nm and the calculated molar extinction coefficient of the oligonucleotides.

5. α-HL nanopore preparation

5.1 Materials and instruments

5.1.1 Bacterial transformation with expression plasmid

- Bacterial strain *E. coli* BL21 (DE3).
- Plasmid: pET28a with the N-terminal tagged α-HL.
- Antibiotic: Ampicillin (100 mg/mL in H_2O).
- LB Broth (Condalab) 0.5 L in 2 L flasks, autoclaved.
- Flask shaker in a 37 °C incubator.
- Water bath with temperature control.

5.2 Induction of the expression

- IPTG (Apollo Scientific)
- Antibiotic: Ampicillin Sodium (100 mg/mL in H_2O)
- Shaker incubator
- Large volume centrifuge

5.2.1 α-HL purification

- Triton X-100
- Urea (it also works with guanidinium hydrochloride)
- Binding buffer: 150 mM NaCl, 10 mM HEPES, 20 mM imidazole in distilled water (pH 7.2)
- Elution buffer: 150 mM NaCl, 10 mM HEPES, 500 mM imidazole in distilled water (pH 7.2)
- Centrifuge for 50 and 15 mL
- Sonicator
- ÄKTA pure
- Ni-NTA column
- Amicon Ultra—15 (3K) centrifugal filter device

5.2.2 α-HL heptamerization

- Erythrocyte membranes from rabbit
- Washing buffer A: 2M NaCl, 10mM Tris·HCl (pH 8)
- Washing buffer B: 2M NaCl 10% SDS
- Eppendorf centrifuge

5.3 Protocol

5.3.1 Bacterial transformation with expression plasmid

1. Pass 100 µL of BL21 (DE3) chemically competent bacteria to a 1.5 µL Eppendorf tube. Cells must be maintained on ice.
2. Add 30 ng of plasmid for every 20 µL of bacteria. Gently mix without introducing bubbles.
3. Incubate on ice for 30 min.
4. Give a heat pulse: 45 s in a water bath at 42 °C.
5. Quickly transfer the Eppendorf tubes to ice and keep for 2 min.
6. Add 0.9 mL of tempered LB liquid media.
7. Incubate at 37 °C for 1 h under agitation (225–250 rpm).
8. Transfer the cells to 0.5 L of LB in 2 L flasks previously autoclaved.
9. Add 0.5 mL of antibiotic (ampicillin 100 mg/mL) to the culture.
10. Incubate overnight at 37 °C and 180 rpm shaking.

5.3.2 Induction of the expression

1. Collect the overnight grown bacteria by centrifuging (5000 × *g*, 10 min, 4 °C).
2. Resuspend the cell pellet in 0.5 L fresh LB and supplement with 0.5 mL ampicillin (100 mg/mL).
3. Incubate for 1 h at 37 °C and 180 rpm shaking.
4. Add 0.5 mL of 0.4 M IPTG for every 0.5 L of culture.
5. Incubate for 4 h at 37 °C and 180 rpm shaking.

5.3.3 αHL purification

1. After the IPTG treatment, centrifuge the culture (5000 × *g*, 10 min, 4 °C) to collect the bacteria. Discard supernatant.
2. Resuspend the pellet in 30 mL of Binding buffer. If need, store the cells at −20 °C.
3. Sonicate bacteria for 30 s let them cool for 60 s and repeat this cycle five times, set the amplitude to 10%. Keep the cells on ice to avoid overheating.
4. Centrifuge the lysate (5000 × *g*, 10 min, 4 °C). Keep the pellet.

5. Resuspend the pellet in 50 mL Binding buffer, Triton X-100 0.5%, pH 8 and keep it at room temperature for 30 min.
6. Centrifuge (5000 × *g*, 10 min, 4 °C). Keep the pellet.
7. Resuspend the pellet in 30 mL Binding buffer supplemented with 8 M urea. Leave overnight at 4 °C with gentle stirring to resuspend the inclusion bodies.
8. Load the sample in a Ni-NTA column previously equilibrated in Binding Buffer. Elute with a gradient to 100% Elution buffer. Store the collected fractions at 4 °C.
9. Perform SDS-PAGE to identify those fractions that contain the target protein.
10. If protein concentration is low, the protein sample can be concentrated using centrifugal filter devices.

5.3.4 α-HL heptamer formation

1. Add 100 μL of purified-concentrated α-HL monomers (1 mg/mL) to each rabbit erythrocyte membrane aliquot. Gently homogenize with a pipette and leave for 5 min at room temperature.
2. Centrifuge (17,000 × *g*, 2 min, 4 °C). Discard supernatant.
3. Wash the pellet twice with 500 μL washing buffer A by centrifuging and discarding supernatant.
4. Add 50 μL of washing buffer B to the pellet.
5. Centrifuge (17,000 × *g*, 2 min, 4 °C). The heptamers are found in the supernatant. Keep the supernatant at 4 °C (1 week) or at −20 °C (1 year).

6. Rotaxane formation

Here we describe the key elements in the formation of a rotaxane. For a detailed discussion on single-molecule experiments using nanopores use reference (Maglia, Heron, Stoddart, Japrung, & Bayley, 2010).

6.1 Materials and instruments

- Electrolyte solution: 2 M KCl, 10 mM HEPES in distilled water (pH 7.5)
- Hexadecane in pentane 1% (v:v)
- 1,2-Diphytanoyl-*sn*-glycero-3-phosphocholine (Avanti Polar Lipid) in pentane, 5 mg/mL
- Heptameric α-HL pores
- oligo-protein-oligo construct

- mSA streptavidin (see reference Howarth et al., 2006 for detailed work on its preparation)
- 3D printed chamber divided by a 25 μm thick Teflon film (made of ABS)
- Ag/AgCl electrodes (by overnight incubation of Ag in sodium hypochlorite)
- Faraday cage
- Amplifier Axopatch 200B (Axon Instruments)
- Axon Digidata 1550B Low-Noise Data Acquisition System (Axon Instruments)
- Clampex 10.6 software (Molecular Devices)

6.2 Protocol

1. Using a glass capillary, apply a drop of hexadecane solution to each side of the Teflon film and wait a few seconds.
2. Fill both chambers with electrolyte solution and place one electrodes at each chamber. Check that the hexadecane has not clogged the hole in the film, by checking there is still current between the two electrodes. If there is not, flush buffer through until current is free to flow through.
3. Add a drop of lipids (5 μL) to the surface of each compartment and wait 2 min for the pentane to evaporate and allow a monolayer of phospholipids to form at the surface. Then, pipette out and in the electrolyte solution until the membrane bilayer forms. Note that the bilayer is correctly formed by applying a triangle-shaped voltage protocol.
4. Add 0.3 μL of heptameric α-HL pores to the *cis* chamber (the *cis* side is the ground level), while applying a 100 mV potential. A single insertion event is required, characterized by a single step increase in the current of approximately 200 pA (22 °C).
5. Perfuse the *cis* compartment with 10 chamber-volumes of fresh electrolyte solution, in order to remove the rest of the heptamers in solution.
6. Add oligonucleotide-protein-oligonucleotide construct to the *cis* chamber for a final concentration of 0.1 μM. Change to 120 mV applied potential, and note normal single events are taking place (Fig. 2).
7. Add monovalent streptavidin to the *cis* compartment for a final concentration of 0.05 μM.
8. Wait until an ionic current signal proceeds as normal but rather than coming back to the open pore current leaves the pore in a conducting state similar to that produced by oligonucleotides threading the nanopore. In this state, if the voltage polarity is changed (from positive

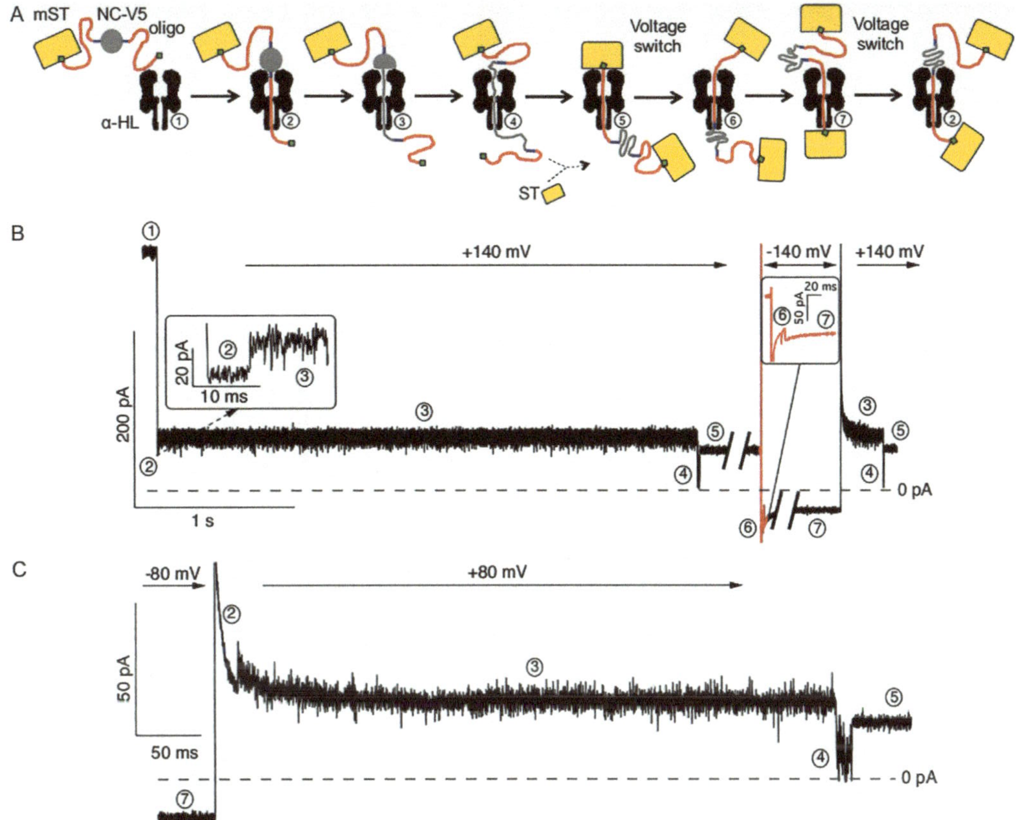

Fig. 2 Steps for the formation of a membrane rotaxane. (A) Steps in the formation of rotaxane. The numbers 1 to 7 define different states producing discernible ionic current levels. For the sake of simplicity, the lipid membrane is not depicted. (B) Ionic current signal of an oligonucleotide-thioredoxin-oligonucleotide molecule translocated through the nanopore and bound to mSA and SA to form a rotaxane. The different ionic current levels are numbered 1 to 7. (C) The rotaxane translocated at a lower potential (+80 mV) to reveal Level 2, which at +140 mV in (B) proceeded to fast and could not be observed in the final part of the ionic current signal.

to negative), there is a characteristic negative ionic current signal (which corresponds to the unfolding and translocation of the protein from the trans side to the *cis* side). To form the rotaxane, do not change the voltage polarity and proceed to the next step.

9. Add streptavidin to the *trans* compartment to a final concentration of 0.2 μM. Wait 5 min.
10. Change the polarity from positive to negative. Confirm that the characteristic ionic current at negative voltage is not obtained but instead an ionic current that is 10 to 20% that of the open pore. Then restore the positive voltage polarity and confirm that an ionic current signal similar to the original one is obtained (it may show some differences, mainly in the dwell time of the ionic current levels). This signal represents the unfolding of a protein molecule that first went from the *cis* to the *trans* compartment and later from the *trans* to the *cis* compartment. (Fig. 2). For a detailed explanation of what each feature of the ionic current signal means go to the reference(Feng et al., 2020).

**Typically, data were low-pass filtered at 5 kHz, which corresponds to a temporal resolution of ~0.2 ms, and sampled at 20 kHz.*

7. Summary

Protein folding during membrane translocation is a process that occurs in all living forms but which possess significant technical challenges. Here we have described a simplified methodology that allows the formation of a membrane rotaxane for the manipulation of a protein molecule. By doing so, it is possible to promote the translocation of the protein molecule forwards and backwards across a single nanopore inserted in a lipid membrane. During these manipulations, the protein refolds as it emerges from the pore. The ionic current signal obtained, reports on which protein domains refolded and its length on its stability. Carefully conducted experiments, controlling the time the protein has to refold before pulling it through the nanopore, allow to define the folding landscape of a vectorially folded protein molecule. This is possible because the biotin-SA interaction remains stable in physiological conditions (and in high salt) and it is weakly dependent on the applied voltage. By these means, it has been shown that thioredoxin, a protein that folds through a fast two-state mechanism in bulk solution, when folding as it emerges through a membrane nanopore is prone to kinetic traps slowing down folding orders of magnitude (Feng et al., 2020). It will be interesting to learn whether the opposite may also be true on other proteins (i.e. other proteins may fold faster when emerging vectorially from a nanopore), which is of great interest in applications in the biotechnology sector to refold expensive and hard to produce enzymes.

Acknowledgments

This work was supported by grants BIO2017-88946-R, BFU2016-81754-ERC from MINECO (FEDER funds), IT1201-19 from the Basque Government. We thank Fundacion Biofisika Bizkaia for their support. We also thank Natalia Orozko for her support in the preparation of different materials.

References

Feng, J., et al. (2020). Transmembrane protein rotaxanes reveal kinetic traps in the refolding of translocated substrates. *Communications Biology*, *3*, 159.
Goldman, D. H., et al. (2015). Mechanical force releases nascent chain-mediated ribosome arrest in vitro and in vivo. *Science*, *348*, 457–460.
Ha, T. (2014). Single-molecule methods leap ahead. *Nature Methods*, *11*, 1015–1018.
Holtkamp, W., et al. (2015). Cotranslational protein folding on the ribosome monitored in real time. *Science*, *350*, 1104–1107.
Howarth, M., et al. (2006). A monovalent streptavidin with a single femtomolar biotin binding site. *Nature Methods*, *3*, 267–273.
Junker, M., Besingi, R. N., & Clark, P. L. (2009). Vectorial transport and folding of an autotransporter virulence protein during outer membrane secretion. *Molecular Microbiology*, *71*, 1323–1332.
Kadokura, H., & Beckwith, J. (2009). Detecting folding intermediates of a protein as it passes through the bacterial translocation channel. *Cell*, *138*, 1164–1173.
Maglia, G., Heron, A. J., Stoddart, D., Japrung, D., & Bayley, H. (2010). Analysis of single nucleic acid molecules with protein nanopores. *Methods in Enzymology*. https://doi.org/10.1016/S0076-6879(10)75022-9.
Nguyen, D. P., Elliott, T., Holt, M., Muir, T. W., & Chin, J. W. (2011). Genetically encoded 1,2-aminothiols facilitate rapid and site-specific protein labeling via a bio-orthogonal cyanobenzothiazole condensation. *Journal of the American Chemical Society*, *133*, 11418–11421.
Nilsson, O. B., et al. (2015). Cotranslational protein folding inside the ribosome exit tunnel. *Cell Reports*, *12*, 1533–1540.
Okamoto, K., et al. (2002). The protein import motor of mitochondria: A targeted molecular ratchet driving unfolding and translocation. *The EMBO Journal*, *21*, 3659–3671.
Rodriguez-Larrea, D., & Bayley, H. (2013). Multistep protein unfolding during nanopore translocation. *Nature Nanotechnology*, *8*, 288–295.
Rodriguez-Larrea, D., & Bayley, H. (2014). Protein co-translocational unfolding depends on the direction of pulling. *Nature Communications*, *5*, 1–7.
Rodriguez-Larrea, D., et al. (2010). Role of conservative mutations in protein multi-property adaptation. *The Biochemical Journal*, *429*, 243–249.
Rosen, C. B., Bayley, H., & Rodriguez-Larrea, D. (2020). Free-energy landscapes of membrane co-translocational protein unfolding. *Communications Biology*, *3*, 160
Rosen, C. B., & Francis, M. B. (2017). Targeting the N terminus for site-selective protein modification. *Nature Chemical Biology*, *13*, 697–705.
Wiggin, M., Tropini, C., Tabard-Cossa, V., Jetha, N. N., & Marziali, A. (2008). Nonexponential kinetics of DNA escape from α-hemolysin nanopores. *Biophysical Journal*, *95*, 5317–5323.

Kolibri 360 GmbH
Ferdinand-Jühlke-Straße 7
99095 Erfurt, Deutschland
produktsicherheit@kolibri360.de